AF885360

Chris Langan's Publications 1989–2025

Chris Langan's Publications 1989–2025

Mega Foundation Press

www.ctmu.org
christopherlangan.com

Published by
Mega Foundation Press
Princeton, MO

info@megapress.org

Compiled and edited by
Dr. Gina Lynne Langan
and Michał Szczęsny

Prepared for print by
Michał Szczęsny

Langan, C. M. (Christopher Michael)

Chris Langan's Publications 1989–2025

1. Metaphysics 2. General Knowledge

ISBN 978-1-957661-01-8

To my beloved wife, Gina, who helped me bring the CTMU to the world

Contents

Chris Langan's Major Papers
1989–2020

Publisher's Note

This volume contains the major papers of Christopher Michael Langan, the author of the Cognitive-Theoretic Model of the Universe (CTMU), from the 1980s to the present. It chronicles the development of the CTMU from short articles and essays published in the journals of the ultra-high IQ societies to a series of ground-breaking papers published in well-regarded academic journals.[1]

Early Writings

The Resolution of Newcomb's Paradox[2]

We begin by defining Newcomb's problem. Enter Newcomb's Demon, a rather close relative of Maxwell's Demon (the similarity goes well beyond a shared affinity for black boxes). Newcomb's Demon – call him ND for short – is a paragon of intelligence, a genius's genius whose acuity transcends the temporal boundaries of merely ephemeral human beings. Like many geniuses, he displays a certain playfulness: his "hobby" is testing his own infallibility in the prediction of human behavior by making a generous offer to random human playthings. This offer is couched as a "choice"; though ND's human subjects are in fact forbidden to predicate their choices on external events, they are tacitly allowed to use deterministic or nondeterministic internal strategies computed by corresponding classes of neural events (you may already see the superficiality of the distinction between inward and outward events; I include it only as a customary ingredient of the formulation).

Suppose that ND has chosen you as an experimental subject, and that you know that he has never failed to achieve the result he predicted for any subject throughout the extensive history of his experimental run. Nothing appears rigged; in particular, the data are plausibly distributed with respect to the critical behavior of experimental subjects (if Nozick's data are valid, at a ratio of approximately 2.5/1). The experimental set-up is minimal. You are led to a table on which are two boxes. One is transparent, and in it you can clearly see $1000. The other is dead black and totally opaque. ND tells you that he has placed $1,000,000 in the black box if and only if he has predicted that you will take only it and not the other (transparent) box. Otherwise – if he has predicted that you will try to "outsmart" him and take both boxes – the black box has been left empty. ND earnestly assures you that this is no joke; he does not intend to fail. Just in case there are any doubts about this, he offers to take you

[1] This edition of *Chris Langan's Major Papers 1989–2020* has been expanded to include "The Resolution of Schrödinger's Paradox: Reality Research in the Computational Universe."

[2] Originally published in the Mega Society journal, *Noesis*, № 44.

through a few trial runs using monopoly money and a couple of spare boxes (naturally, you will have to leave the room between these "rehearsals"). You take him up on it. He wins no matter what you think or do. Finally, he tells you that play time is over; next time, the deal is for real. He adds that, lest you suppose he has an infinite amount of time to waste on you, he is giving you a reasonable but finite amount of time to make your move.

Having been around the block, you have already perceived that this is all a kind of game in which ND is the "house", and you his opponent. Of course, since ND is playing for the pure satisfaction of being right while you are playing for big money, a minor conceptual adjustment is in order. But, all things considered, you reason that you should nonetheless apply the standard theory of games in order to maximize your gain. Your reasoning then becomes a little more involved.

You know of certain principles of rational decision that seem to apply to this game. One is known as "the maximization of subjective expected utility", and states that you should try to maximize the sum of the products of the utility (monetary value) of each distinct possible outcome by the probability of that outcome, given your "move". Since ND's past performance and serious intent seem to bring the probability that he has correctly predicted your choice close to unity, this principle is telling you in no uncertain terms to take only the black box.

But there is a similar rule, called the "dominance principle", which also seems to apply to your predicament. It states that if, for every possible move (or prediction) of your opponent, there is a single countermove as advantageous to you as any other, and more so in at least one case, then you should make that move. You know that ND has already made his move: there is either $1,000,000 in the black box, or there is nothing. In either case, your eyes tell you unequivocally that there is $1000 in the clear box. You reason that you have nothing to lose, and precisely $1000 to gain, by taking both boxes. The dominance principle merely drapes a veil of reason over what your eyes and instincts have already told you: ($1,001 million or $0.001 million) beats ($1 million or $0). Conclusion: take both boxes.

But this seems to constitute a dilemma. These two principles of rationality, which are supposed to be mutually consistent, are in this case telling you to do opposite things. Feverishly – there is a lot of money at stake for you – you rack your memory for other rules which might help you reconcile or at least decide between them. What about the gambler's fallacy? No matter how many times ND has won in the past, isn't his string of victories really just an extremely unlikely random sequence? Time only runs in one direction, so how can he possibly claim to be more than a lucky (and very rich) guesser? If this is the case, then you certainly don't want to embrace the fallacy of assuming that the outcome of your imminent trial is in any way dependent on those preceding it. If nothing else comes to mind, and quick, you're going to have to go the dominance route and take both boxes.

As if on cue, your neurons begin again to flicker. It occurs to you that you could make no decisions whatever, and would know nothing at all, without relying on various scientific axioms. But axioms are mutually independent, or virtually random with respect to each other. None of them can be deduced from other axioms. Doesn't accepting these axioms – including the dominance principle or whatever more primitive concepts underlie it – thus amount to something very much like the gambler's fallacy? That is, isn't one accepting as a premise that certain future events will be like similar events from the past, but without being able to prove it? You are now in the Minotaur's recreatorium without a thread.

You are also roughly as well-informed on what you should do as any professional scientist or logician who has ever published an analysis of Newcomb's paradox. Fortunately, I am no professional, and need grind the favorite axe of no publisher, professor, or employer. The trail that unwinds below is freshly blazed; though a bit steep in spots, it can be negotiated purely on the strength of an open and reasonably sharp mind.

The first step towards resolving the paradox is to provide a logical scaffolding from which to construct the mathematical model necessary for sound inference. Past arguments involving the problem have used either the standard decision-theoretic model of people playing a game, or the linear "arrow of time" fundamental to classical physics. These two models have been taken to imply opposite solutions, and this suggests that they be somehow unified in an extended "meta-model" which adequately relates the concepts to each other. It has been less obvious just what this higher model should be.

The solution of problems, and the resolution of paradoxes, are inherently computational activities. What, then, could be a better setting for this resolution than a computative one? And what could possibly be a more fitting preface than a brief introduction to the abstract theory of computation?

Consider an acceptor $F = (Q, \Sigma, \delta, q_o, A)$. Q is a finite nonempty set of internal states, Σ is an alphabet whose symbols are concatenated as strings, $q_o \in Q$ is the initial state, and $A \subseteq Q$ is the set of accepting states triggered only by input recognizable to F. The transition mapping δ, which governs the way in which F changes states, is deterministic if δ: $Q \times \Sigma \rightarrow Q$, but nondeterministic if δ: $Q \times \Sigma \rightarrow 2^Q$ (where 2^Q is the set of all subsets of Q). In the nondeterministic case, δ will be written δ_n, for clarity. In terms of human psychology, we might regard the 5-tuple of F as its "categorical imperative", or accepting syntax, and say that F projects this syntax onto its universe. Nothing in the universe of F is recognizable to it but the particular input strings (sense data, facts) which cause it to pass through some $q \in A$; they are its phenomenal "reality", a subset of the noumenal metareality of the wider universe in which strings are representationally generated and entered by programmers. The restriction to finite Q is pragmatic and amenable to conditional relaxation.

If F is deterministic, it accepts (recognizes) a string $s \in \Sigma^*$ if and only if $\delta(q_0, s) \in A$. Since we have defined δ only for the individual symbols $\sigma \in \Sigma$, we must define an extended transition function δ': where λ is the null string, $\delta'(q, \lambda) = q$; and for all $q \in Q$, $\sigma \in \Sigma$, and $s \in \Sigma^*$ (where Σ^* is the set of varirepetitional permutations of the $\sigma \in \Sigma$), $\delta'(q, s\sigma) = \delta(\delta'(q, s), \sigma)$. Thus, the accepting behavior of F is defined inductively for σ-quantized string extensions; the way in which we recognize and assimilate new bits of information within our reality is specified in δ'. Were we to widen the discussion to imagination, conceptualization, theorization, or other intrinsic computations, we would need to consider "ideas"; we would have to generalize from recognition to abstraction by means of a nonlocal or self-iterating, input-free extension of δ. If the reference to "strings" seems to imply a dimensional limitation on input, this too can be generalized.

Where F is nondeterministic, it accepts a string $s \in \Sigma^*$ if and only if $\delta'_n(q_0, s) \cap A = \sim\emptyset$. The nondeterministic extension δ'_n of δ_n is defined by induction: $\delta'_n(q, \lambda) = \{q\}$, and $\delta'_n(q, s\sigma) = \bigcup_{q' \in \delta'_n(q, s)} \delta_n(q', \sigma)$. I.e., q' is one of the possible successors of q under δ'_n given s; the unextended mapping δ_n on singletons of Σ^* then determines the image under δ'_n of s plus an adjoint symbol σ, given q'. This is a classical recursive definition. It describes a stepwise probabilistic ramification of computative potential whose complexity depends on δ_n.

Nondeterminism is not always restricted to $Q \times \Sigma$; under certain conditions, either Q or Σ can be extended, or A shifted within Q. This, of course, entails a modification of F, unless F is defined to allow for parametric extension and adjustment. To this effect, let F' be such an open extension. To be meaningful in mechanistic contexts such as those in which acceptors are usually considered, F' must exist within an appropriate mechanistic extension of the computative environment of F. Organisms, being mechanical in the deterministic sense, need not be distinguished in this extension. Nondeterminism can be used to subtly manipulate recognition, thus cryptically modifying an acceptor's reality. Nondeterministic recognition can help to explain the ability of an acceptor to rapidly seize certain kinds of higher-order phenomena, or even interact with higher-order agencies ordinarily insensible to it.

Having thus formalized the logical abstraction of recognition – i.e., the passive phase of organo-mechanical cognition – we now proceed to the output behavior of computative automata, or to the active phase of cognition. Consider a transducer $M = (\Sigma, Q, T, \delta, \mu)$, where Σ is a finite nonempty input alphabet, Q a finite nonempty state set, T a finite nonempty output alphabet, $\delta: Q \times \Sigma \to Q$ the state-transition function, and $\mu: Q \times \Sigma \to T$ the output function. A computation of M has internal and external phases; through μ, the output that M delivers back to its outward universe depends on strings of δ-iterated transitional internal states. Thus, μ is a functional of the function δ of input. Together, μ and δ totally determine the behavior of M. They can be extended from Σ and T to Σ^* and T^* as for the acceptor F: $\delta'(q, \lambda) = q$, $\mu'(q, \lambda) = \lambda$; and

$\delta'(q, s\sigma) = \delta(\delta'(q, s), \sigma)$, $\mu'(q, s\sigma) = \mu'(q, s)\mu(\delta'(q, s), \sigma)$. Where appropriate, we can add to M an initial state ("reset control") q_0: $M = (\Sigma, Q, T, \delta', \mu', q_0)$, to be regarded as a separator of locally independent computations, and put it at the disposal of a function $r \subset M$ which determines computational relevancy.

Considered as a robotic brain, M T-behaves according to μ, but Q-reasons towards its decisions along paths generated by δ. Where δ is deterministic, μ may be related to it as a "timing function" according to which any computation can be arrested (input aborted, regression terminated) and converted to output on passing certain tests. Where the duration of the computation is determined with δ, input becomes the only variable. Where input as well is fixed in content and scheduling, the entire system is tightly determined. As Laplace might have observed, predicting the behavior of determinate mechanisms requires only data, the means to acquire it, and a valid scientific methodology to organize and interpret it. While the situation is actually more complex, the fact remains that were one to play a deterministic game with a deterministic transducer like M, one would need only a detailed knowledge of its input and programming to predict the outcome, given analytic tools adequate for that purpose and consistent with one's own constraints (e.g., the amount of time available for analysis and strategy). If one's object in the game were merely the validation of one's prediction, so much the easier to win.

Suppose instead that M has nondeterministic output mapping μ_n, where state-transition may or may not be deterministic. Then the prediction of output entails control of μ_n by the predictor. To win a game of prediction, one must now control μ_n as well as δ_n to the extent that it is output-critical; one must take over where the probabilistic μ_n leaves off. Since whichever control the transducer has over itself resides in μ_n and δ_n, one must in effect deprive it of self-control. The relevance to "free will" is obvious.

Computation is purposive. The purpose of an acceptor is pure recognition; no action is explicitly predicated on its internal transitions. The purpose of a transducer is conversion of input to output; yet, such a conversion is aimless unless algorithmic. Like yin and yang, acceptors and transducers are complementary; only together can they begin to resemble functional systems of organic complexity. In order to model organic systems, transducers must be endowed with goals and algorithms comparable to the ends and means of living beings. Algorithms are themselves purposive procedures which model both acceptance and transduction. The problems which comprise their input are scanned by preliminary steps for certain kinds of information, which must in turn be accepted as parameters by subsequent steps, and so on to the output stage (at which point the algorithm delivers its answer). The mechanistic representation of an algorithm must allow for the innate structure of a device, considered apart from the algorithm itself; this structure may have variant and invariant aspects. The algorithm simply conforms variables to purpose given the invariants. As the definitions of F and M might lead one to expect, this generally involves importing to M the set $A \subseteq Q$ defined for F.

Human beings, it is said, are self-programming. Their thought is polyalgorithmic; useful algorithms are either meta-algorithmically constructed, or selected from a learned store, to deal with input. If learning, construction, and selection are deterministic, then they characterize a deterministic meta-algorithm not fundamentally different from any other deterministic algorithm we might study. If they are nondeterministic, then they are characteristic of a nondeterministic meta-algorithm, and likewise. It follows that the formal transductive model of human nature withstands any objection from the relative complexity of human mentality or behavior.

Newcomb's object-transducer M_N naturally includes an acceptor: $M_N = (\Sigma, Q, q_0, \delta, A, \mu, T)$. Recognition is phasic; a string must often be "pre-accepted" for M to tell whether to accept or reject it. Ordinarily, this tentative phase of recognition is easily computed by the physical entities whose behavior is predicted by ND. To be "real", an input-quantum σ must simply possess a certain first-order predicate, "reality", which – this being a self-validating tautology – induces a type-theoretic predicate stratification like that involving the old Cretan, Epimenides. To this *sine qua non* of recognition corresponds a primary element q_1 of A; no input-quantum failing the q_1-test is reified, whereas all those passing are relayed to $Q - \{q_1\}$. Higher-order recognition of "passed" quanta then proceeds at a rate determined by the respective computational demands of the stratified-algorithmic phases of δ_M. Corresponding to the structure of A are various ordered states analogous to q_1 within their respective levels of acceptance.

Let us narrow the definition of M_N in a way consistent with Newcomb's problem. Suppose that associated with μ_M is a threshold value $\beta > 0$ below which output is nil, but above which a decision will be finalized and implemented. With each $q_i \in Q$ we associate a pair of strength coefficients a_{i1} and a_{i2} to be incremented and decremented according to a strategic δ_M, appropriate to the Newcomb decision-theoretic context; these represent the current tendency, given the present amount of input, for M_N to output either possible behavior (taking one or both boxes, respectively). The a_i divide Q into three classes X^Q, Y^Q, and Z^Q, with membership conditions $a_{i1} > a_{i2}$, $a_{i1} < a_{i2}$, and $a_{i1} = a_{i2}$ respectively. To each q_i is attached a total weight $\alpha_i = |a_{i1} - a_{i2}|$. As soon as the output condition $[\alpha \geq \beta]$ is met, a decision and behavior result which correspond to the Q-class of the current state (note that δ_M, being strategic, precludes $q \in Z^Q$, or "indecision", at the output deadline). Thus, the states of $Q \subset M_N$ are preferential and impetal, the graded fore-images of the outputs they favor. This is just a convenient way to view the internal configurations to which they correspond, and does not violate the general definition of transducers. Nor, for that matter, does it violate the way human beings perceive their own decision-making processes.

The question posed by Newcomb's problem involves the computative analysis, by a predictive agency with computative characteristics, of the computative analysis undertaken by a transducer on a given input. That input is the problem itself, presented in the manner prescribed by the formulation. This situation, which defines a computative regression, is recursive and inductively extensible. The regression in turn defines the only soluble context for the higher-level "paradox" generated by the problem. This context translates as mechanism. The mechanism is a stratified automaton Γ containing both the predictor and its object-transducer as sub-automata. Whether "free will" is defined deterministically as mere outside non-interference in μ and δ, or nondeterministically as the ability of M_N to override any exogenous restriction of μ_n or δ_n, its mechanism is contained in that of Γ.

Logical diagonalization of the formal computational language generated by the accepting syntax of M_N directly implies that certain structural aspects of Γ may be unrecognizable to M_N. In particular, those aspects involving M_N-relativized nondeterminacy, as well as those involving certain higher-order predicates of the nondistributive, nonlocal organizations involving μ_M and δ_M, are formally undecidable to it and need not be recognized directly by it with any degree of specificity. To understand why, consider the extent to which a common computer "recognizes" the extended system including its CPU, its other components, its programmers, and the environment it inhabits. In fact, it can recognize nothing that does not conform to its input-to-output transformational grammar. Even if it were self-analytic, such analysis could be limited to a schematic electronic syntax which overlooks the material elements of which it is constructed. In any case, it can make sense of nothing but strings of input translated and rearranged according to the internal stratification of its hard and soft programming.

You, your purposes, and your dependencies are undecidable to it, and so are the mechanisms by which you can predict and control its behavior. It matters not who formulates this undecidability; if the machine's internal logic is inadequate to do so, yours surely is not (currently, most mechanical acceptors are nongeneralistic, treating complementation as negation and negation as rejection; this bars the tools of diagonalization from their computations). Should it ignore your higher prerogatives, you could "diagonalize" it – if nothing extrinsic to the machine were to stop you – with a sledgehammer whose effects on it do not depend on its acceptance. By analogy, Newcomb's object-transducer M_N cannot preclude Γ on grounds of "insensibility". Nor, for that matter, can we.

There are many self-styled experts on undecidability who have expressed the opinion that all attempts to reify Gödel's theorem along paranormal lines reflect a misunderstanding of its "real nature". Such experts are quite correct in that a misunderstanding exists, but the misunderstanding is all theirs. What the theorem forces by juxtaposing truth and derivability (or consistency and completeness) is a hierarchical stratification of classes of truth functions and the inferential syntaxes which parametrize them. This stratification follows that of Γ, fractionating computative reality along with the "truth" to which it corresponds.

The stratification of Γ induces stratum-relativizations of computative time and space. Thus, the timetype in which M_N computes recognition and output is a mere subtype of that in which it is programmed. Dynamical "arrows of determinacy" which are inviolable to M_N, being programmed into its accepting syntax, have no force whatsoever to the programmatic agencies themselves. This applies just as well to "metrical" restrictions embodied in the M_N-syntax; these may allow M_N to recognize nothing but an artificial submetric of the metric in which these agencies define their own existence. M_N and its reality might consist of quanta with higher-dimensional interpretations as the termini of channels for the transmission of information between strata. Metatemporal predicates may exist with respect to which those of M_N are definite only in a mutual sense; predicates which M_N accepts as "before" and "after" could be the programmatic projections of "in front of" and "in back of", or any other Γ-consistent higher-prepositional relationships.

There can thus exist a mechanism $x \subset \Gamma$ through which a predictor like ND could measure and/or control the mappings $\delta, \mu \subset M_N$ in ways directly insensible to M_N. Where in Γ relative to M_N would such a predictor have to be located? Precisely where access is available. Simplistically, we might characterize the predictor-M relationship as one of proper inclusion, where it is understood that prediction is direct rather than second-hand, and programmatic in the passive and active senses. That is, a programmer mentally internalizes the structure of that which he programs, and this internalization amounts to computative inclusion. The fine structure of Γ, while to a degree analytic, is a matter of some complexity. For now, it will suffice to have demonstrated the possibility of x and its utility to well-situated Γ-subautomata. Because Γ is structured to allow for relativized determinacy and nondeterminacy, the solution is invariant with respect to argumentation involving mind-brain dichotomy. That is, such dichotomies reduce to distinctions of determinacy and nondeterminacy, and may be treated in kind.

Restricted dominance, which relies on probabilistic independence derived from the lower-order, localistic dynamical timetype of M_N's artificially restricted "reality", is revealed under Γ-extension to be itself dominated by utility. That is, the subjective utility of M_N forces the assimilation by δ_M of this entire demonstration, which disables restricted dominance and thus frees the strategic component to recognize higher patterns among observed data. The principle of restricted dominance, though valid as long as the reality of M_N remains unbreached, loses all force in the presence of exodynamic influence.

Let's sum it up. You can be modeled as a deterministic or nondeterministic transducer with an accepting syntax that can be diagonalized, or complemented by logical self-negation. ND can be modeled as a metalogical, metamechanistic programmatic agency, some insensible part of whom surrounds you in a computational space including physical reality, but not limited thereto. This space is the mechanistic equivalent of the computative regression around which Newcomb's problem is essentially formulated. The existence of this space cannot be precluded by you on the grounds that you cannot directly observe it, nor can it be said by you to deny ND a mechanism of control and prediction of your thought and behavior. Additionally, you have an open-ended run of data which lowers to $1/\infty$ the probability that ND is "just lucky". This implies that mechanism does indeed exist, and warrants the adjunction to the axioms of physics an independent, empirical physical axiom affirming that mechanism. This then implies that ND can predict or control human thought and behavior (a somewhat weaker implication, you will notice, than "omniscience"). ND possesses means, motive, opportunity ... and you. You are "possessed" by Newcomb's Demon, and whatever self-interest remains to you will make you take the black box only. (Q.E.D.)

Do not let incredulity cheat you of understanding. It may be hard to accept the idea of a mathematical proof of the possibility of demons; if so, you may take solace from the fact that this does not of itself imply their actual existence. Remember that Newcomb's problem includes a hatful of hypothetical empirical data that may have no correspondents in your sector of reality. The proof just given merely prevents you from generally precluding the paranormal experiences reported by others on logical grounds. That is, it restrains you from any blanket distribution of too narrow a brand of "rationalism" over reality. Because I have long been exploring the ramifications of the logic this proof employs, I already know how easy it would be to deal with whomever might dispute it on "scientific" or other grounds, regardless of their prestige or reputed intelligence. There is a certainly a strong correlation between rationalistic dogma and the persecution of its foes. Fortunately, it is not too late for "rationalism" to catch up with the advances of twentieth-century logic.

Regarding free will, consider that demons are generally reputed to offer a material reward for the surrender of one's soul. Where we define "soul" as the autoprogrammatic Γ-extension M' of one's characteristic transductive representation M, this means only that one is rewarded for voluntarily moving aside and letting the demon take over as captain of one's fate. He does this by adjoining to his subject any functional extensions he requires; e.g., $d(\delta_n)$, or $m(\mu_n)$! Or, if restricted determinism prevails but he is not in an especially predictive mood, he might create a parametric extension F'_{ND} (analogous to F') enabling a hyperdeterministic override $m \supset \mu$ of one's deterministic output function. Actually, volition may be irrelevant; one might be shoved aside rather than nudged. Where freedom is defined negatively on restricted determinacy, it is duly stratified.

It has by now occurred to most readers that "demon" is a darkly connoted term that could just as well be replaced with "guardian angel" or something equally comforting. After all, ND does not ask his subjects to commit evil, but only to let him do them a rather large favor with no apparent strings attached. Insisting on "free will" here is reminiscent of the child who stares defiantly at its mother while touching the hot stove she has just finished telling it to avoid. If the child's act has any meaning at all, it is just that poorly timed independence can lead to more painful dependency down the line (surprisingly, Isaac Asimov advocates just this kind of defiance in the Newcomb context, and on what he seems to feel are ethical grounds; but we will deal with ethics later). One cannot be willingly "enslaved" by ND before being enslaved by the money he offers. Who among us always behaves in willful defiance of economic reality? My heart goes out to him, for he is either "enslaved" by an institution, or he is at the absolute mercy of the spontaneous, unsolicited charity of those whose own demons generally predispose them to pass him by.

Having quenched the fire, we will now "mop up". Professor Nozick himself has remarked on the possible nonrecursivity of scientific laws involved in the determination of behavior. This, he surmises, prevents the inference of predictability from determinism. But this conclusion entails the confinement of these predicates within a restricted syntax, whereas the extensionality of Γ renders such a constraint powerless. In other words, restricted recursivity is beside the point for agencies defined to possess sufficient access to the channels through which such laws project as restricted reality, and particularly to their source. All ND has to do is run a Γ-presimulation of interstratum transmission. Note that the computative requirements for this, which correspond to assumptions on the nature of Γ, are satisfied given ND's efficacy.

Nozick also remarks on the regression of determinism to self-validating premises. Self-validating languages comprise their own inferential schemata. Validation, being inferential to the extent of quantification, follows the Γ-stratification of inference. So self-validation is also stratified, and such languages correspond to appropriate levels of computative reality. For sufficiently expressive languages, self-validation becomes diagonalistic self-vitiation. Determinism is thus stratified, and so is its counterpart, freedom. The seemingly infinite regression can be brought to closure at an ultimate form of undecidability, which in turn can be locally resolved with a relativized degree of confirmation.

Regarding various remarks that have been submitted to Nozick by members of the physics community, I should explain that the above resolution has a metaphysical generalization of which certain implications may broadly be termed physical. This generalization, as developed by the author, is called the CTMU, or Computation-Theoretic Model of the Universe (the acronym has a mnemonic pronunciation: *'cat-mew*). We will later examine chosen aspects of this model, which transcends characterization as a "theory" by virtue of its categoricality with respect to computative entities such as human beings. That is, the CTMU is as close to "absolute truth" as we will ever be privileged to get along computative or intellectual pathways. If I might be indulged a bit of testimony:

beyond the CTMU ultimate syntax, the lattice which gives shape to the multilayered veil of maya, there is but the light that shines forever.

References

Bar-Hillel, M. & Margalit A. (1972) Newcomb's Paradox Revisited. *The British Journal for the Philosophy of Science*, Vol. 23, № 4, pp. 295-304.

Gardner, M. (1974) Reflections on Newcomb's Problem: A Prediction and Free-Will Dilemma. *Scientific American*, Vol. 230, Issue 3, pp. 102–109.

Nozick, R. (1970) Newcomb's Problem and Two Principles of Choice. In N. Rescher (Ed.) *Essays in Honor of Carl G. Hempel: A Tribute on the Occasion of his Sixty-Fifth Birthday*, pp. 114–146. New York: Humanities Press.

The Resolution of Schrödinger's Paradox: Reality Research in the Computational Universe[3]

In recent issues, we have established – perhaps to the limits of the logical sophistication of the average member of this group – certain aspects of the overall structure of reality. We have defined a generalized "oracle machine" whose logical structure is the syntax not only of all we can observe or test for directly, but of all that is conceivable to the human mind. Because such machines are already established conventions in the field of abstract computation theory, we were granted temporary immunity from objections concerning their "unreality" or "impossibility".

But the immunity needed only to be short-lived; predicates like "unreal" and "impossible" are themselves open to objections which effectively redeem the concept of nondeterministic computation. We needed only to "relativize" the concepts "determinacy" and "nondeterminacy" to the *accepting syntaxes* of the Γ-subautomata with respect to which they are used. Where *information* is defined in terms of efficacy or relevance, it too is relativized to the acceptors on which it is effective. Because such acceptors may be defined arbitrarily within Γ, this reflects no loss in the utility of the relativized terms. In fact, they become meaningful in even more exact and scientific ways, and one wonders how their erstwhile fuzziness was ever tolerated within the self-contained, and thus didactically pseudotautological, scientific community.

A new branch of science has arisen which reflects the situation. An offshoot of what has been called "the crisis in physics", it purports to seek an adequate model of reality among the ruins of the classical, deterministic-mechanistic worldview which evolved alongside Newtonian mechanics. The demise of this eminently pragmatic picture of the world has long been a foregone conclusion

[3] Originally published in the Mega Society journal, *Noesis*, № 47.

among modern physicists, whose intuitions have been stretched by seemingly antimechanistic notions like general relativity and quantum indeterminacy. This new science has been called "reality research", and it has been the subject of much entertaining – but sometimes confusing – popularistic literature.

Of course, since popularizers relay knowledge taken from "primary sources", not all the blame can be heaped on them. In many cases, they are to be congratulated for their courage in venturing outside the sanctified circles of their colleagues...whose scorn for those who feed at "the trough of mass communication" is well-known. Feeding at the trough of a corporate or taxpayer-funded institution, while it may be a step or two removed from scientific journalism, is likewise no ticket to apotheosis; both levels are essential to a scientifically literate society. Few among us have remained so true to the ideals of knowledge that we can live out of Socratic bathtubs, reject all outside demands, and abjure all forms of salesmanship (though I admit to having come close).

Reality research, as generally practiced, has led nowhere. The reasons are multifold: the intuitions of its practitioners are still in shock, struggling to stay afloat in a witch's cauldron of weird mathematical formalisms which bear little similarity to the structures encountered in everyday life. These formalisms not only seem alien to quotidian existence, but are hard to integrate with one another; those who study quantum gravity, for instance, are obvious victims of translative difficulties. But perhaps the most important reason is that physicists are faced with a bewildering array of perspectives on the meaning of quantum phenomena.

The search for an overall view of reality, including man and his peculiar preoccupations, has traditionally been considered the province of philosophy. How, apart from its fancy new name, does "reality research" distinguish itself from plain old "philosophy"? By particular attention to the modern oxymoron known as "quantum mechanics". The term is oxymoronic in that quanta are generally considered to be statistically predictable at best, whereas the physical usage of "mechanics" derives from the deterministic Newtonian mechanics which reigned over science for the two centuries preceding this one. This usage is still current, due mainly to the fact that quantum theory has never been deemed essential to most of the applied sciences. That is, most science and technology is concerned with larger-scale effects which do obey laws which have been deterministically formulated.

Many theories purporting to be "scientific" – including those encountered in the socioeconomic disciplines, which are almost absurdly fallible in their determinations – are explicitly statistical. This is usually written off to the number and complexity of parameters for which such theories must account, as well as to the notorious difficulty of predicting the outputs of many variably-programmed human "transducers". But it has seldom been claimed that improvement is impossible, given ways to acquire and handle larger amounts of data. This situation differs sharply from that of physics, in which science appears to have reached certain ultimate mensural limitations. Here, the means of measurement have been shown to interfere unavoidably with measured quantities; when a physical quantum is measured, the measuring device must interact with it and disrupt its state. This, of course, bodes ill for the formation and validation of theories linking the states of quanta with the events in which they participate.

Yet, it has always been expected of physical theories that they should not only allow prediction of phenomena, but account for the mechanisms presumed to underlie them. This attention to causality has been prematurely jettisoned by those who assume that logic can exist without benefit of reason. Cybernetics, the machine science, has been consistently formulated such that mechanism and causality can be identified with constraint, which obtains whenever variety is restricted...in short, whenever a theory can possibly give more than chance predictions. Since the only theories of any interest or use are those which improve upon coin tossing, the existence of such a theory implies mechanism and so causality. This syllogism is a property of the logic in terms of which we comprehend the world; without it, neither statistical nor inductive probability can be justified. The confusion arises when we attempt to place unjustifiable restrictions on mechanism...e.g., that it be "material" in the sense defined on our particular accepting syntax.

Obviously, theorists who propose to do away with the concept of causal mechanism in favor of statistics tacitly propose to use those statistics inductively – i.e., to make predictions. But this forces them to define some equivalent concept in its place. Their criticisms are like word games played by truants at the expense of their lessons, and cannot be suffered in the classroom of logic.

Mechanism is reified constraint. Single-argument (logical) constraints differ generically from statistical constraints only in the extent to which elementwise distribution is possible...by inductive limits on the range of quantification. Constraint thus attends any reduction in variety, which we may identify with probabilistic freedom. So mechanism attends alongside it. Theories are useful only when they substitute predictive veracity for variety. So the existence of useful theories implies mechanism, and any such theory necessarily models the "automata" to which it applies. Objective mechanism is thus seen to be an outward projection of the "hard-wired" logic of our inner mechanisms. The precybernetic version of this truth originated publicly with Kant, and was not difficult to reformulate in light of modern insight concerning the structures of computative devices like human brains. The sheer mathematical necessity of this reformulation is so evident, and so obviously crucial to "reality research", that the delay in its discovery must be ranked a major curiosity.

This gap has been spanned by the CTMU, which may accordingly be considered the last word in "reality research". It may inspire sorrow to see the lid slammed on an emerging "branch of science", particularly one which seemed to promise so much to so many in the form of literary royalties and the like. But lids can double as foundations, and this one bears appreciable weight. Its importance thus overshadows the priorities of individuals who may (or may not) have planned to capitalize on the hole it covers.

The quantum-mechanical oxymoron has frequently been portrayed as the Achilles' heel of Newtonian mechanistic reality. How, it is asked, can reality resemble a deterministic machine, when the laws by which it operates are nondeterministic? Mechanism is challenged by mystical concepts, like "quantum holicity", which would – despite certain real conceptual assets – have been laughed off the stage of science at any other point in recent history. But the challenge is not fatal. Werner Heisenberg, the inventor of matrix mechanics, showed that it is possible to replace certain elements of the Newtonian machine – dynamical variables in the classical equations of motion of a particle – with probabilistic matrices, resulting in a coherent "statistical mechanics". The question thus becomes: what grounds exist for discarding the notion of machine in favor of its negation as a model of physics?

Matrices of transition probabilities are governed by mathematics generally invariant with respect to interpretation. On one hand, such matrices can be interpreted in terms of quantum states: on the other, in terms of machine states. Where, then, is the crucial distinction between physics and mechanism to be found? One fact is obvious: it is certainly not to be found in the formalistic intersection. Any attempt to vitiate the mechanistic view of reality on the basis of quantum-formalistic probabilism is thus preordained to failure. Quantum indeterminacy can be validated only as a characteristic of our particular relationship to the reality-machine. That particularity is the cue for relativization, which in this case can apply only to the term "indeterminacy". This relativization implies others, and the overall relativization which results is the CTMU...a kind of "logical general relativity" (as we have remarked before, the need to relativize indeterminacy does not directly imply that the universe is deterministic).

So the "crisis in physics" can be reformulated as a crisis in logic. The basic problem in "reality research" is usually called the *quantum measurement problem*; it centers on the mensural collapse of the quantum wave-function. Like Newcomb's (or any other) problem, it has an antinomial formulation: the "Schrödinger's Cat" paradox. The sorry plight of this animal was described by C. Cole in № 43 of this journal. In it, Schrödinger – the originator of wave mechanics, the diffraction-oriented version of quantum mechanics – managed to encapsulate with devilish ingenuity the problem of defining the means and scheduling of quantum collapse. Unfortunately, because his formulation rendered the feline incommunicado, he was unable to pose his question to it directly. Had he done so, he would no doubt have been answered by a *cat-mew* (see the final paragraph on p. 11, № 44 [p. 22 in this volume]).

Schrödinger's formulation was clever in that it incorporated a critical range of possible junctures for quantum wave-function collapse. It involves an emitter; a quantum; in its original form, a calcite polarization-splitter; a quantum-detector rigged to a mechanical apparatus; a lower form of life (the cat); and a higher form of life (the observer). The quantum is emitted, filtered, and detected, whereupon its state determines whether or not the device kills the cat. The question is, which juncture gets the first shot at converting the "nondeterministic" quantum "wave" into definite, deterministic information? (For now, we will ignore the filter.)

The question is logical, and calls for a logical answer. While most problems in physics demand exact quantitative determinations, this one is so basic – so metaphysical – that quantity is all but irrelevant. The decisive generalities may accordingly be analyzed directly, without benefit of abstract symbology. It should thus be possible to answer Schrödinger's question in plain language without sacrificing clarity or concision. As this is done out of consideration for my readers, I do not expect to be counted lax for dereliction of my scholastic duty to obfuscate.

The proper resolution of any paradox requires that we closely examine its formulation to see how the antinomial syntax has been mapped into its semantical domain. Here, this calls for some background. Matter quanta, depending on observational context, display the characteristics of both particles and waves. The waves in question are not ocean-style waves of many particles, but waves of potential states of single particles. They are thus often referred to as "possibility waves". This model is usually inferred from the probabilistic formalisms necessary to describe quantum phenomena, as well as from wavelike effects observed in some kinds of quantum experiments (e.g., n-slit diffraction experiments). Of course, the analogy holds only up to a point; there may be aspects of waves and quanta which do not readily translate.

Macroscopic bodies like cats and Schrödinger boxes, because they consist of vast numbers of matter quanta, are supposedly subject to the same duality as the quanta. That is, they can be represented by wave functions defined on the wave functions of the quanta comprising them. Because quantum waves are superpositioned potentia, so are the complex waveforms of macro-bodies. So the material objects of classical physics, far from being solid and intrinsically deterministic, are ghostly apparitions in which many contradictory attributes exist at once.

Yet, when we observe such objects, they never display this kind of ambiguity: if they are here, they are here, and if they are over there, then they are there. If a cat is alive, then it is alive, and if it is dead, then it is dead. Γ_0 (physical reality) is two-valued for every attribute defined within it; there are no intervening, quasireal attributes like "here-there" or "live-dead" for observational purposes. In this sense, observation is like computation. There may be uncertainty as to what an automaton does to produce output from input, but there is no uncertainty when it comes to recognizing the output once it has been produced.

Where input-to-output transformation depends on the automaton effecting it, a given input can represent many different *potential* transforms, each to be realized by a different set of transducers. This suggests a question: given a "pre-real" stage of acceptance, why should different acceptors recognize (reify) a given input in the same way? That is, why should different observers collapse interactive quantum waves in a consistent manner? While a single wave can be collapsed only once, why should correlated *sets* of such waves not be inconsistently collapsed by the varied acceptors that intercept them? Consistency is either built into the system as a measure of observer independence (and dependency), or it is enforced by other means. But it is either the system itself, or the wider system including it, which enforces it.

It has been shown that the classical view of reality is based on three assumptions which are generally accepted without proof. The first premise, that of *realism*, holds that the patterns we observe among phenomena exist independently of our observations. The next, that of *unrestricted induction*, affirms the validity of inductive inference, whereby these patterns are imputed to causes. The third premise, *locality* (Einstein separability), holds that nothing propagates faster than light. These three "axioms" characterize certain "neoclassical" theories of quantized reality. The problem with such theories is that the data confirming quantum mechanics appear inconsistent with the given axioms, seeming to necessitate the rejection or revision of one or more of them.

Thus, quantum data seem "anomalous" with respect to theories incorporating the above premises as axioms, and a paradox exists. Technically, it was resolved by the theoretical advance from classical to quantum mechanics, which accounts for the data and thus decides the issue against neoclassicism. But the paradox persists, largely because quantum mechanics is merely probabilistic, gives only statistical predictions, and must therefore be classified as "incomplete" in the usual sense of a theory. Quantum mechanics is not giving us all the answers...and among those it is leaving out, some may bear on the relationship of the "superseded" premises in such a way as to redeem them in spirit, if not absolutely.

The desirability of such a redemption is obvious. Just imagine having to give up one of these premises unconditionally. If you reject the first, then there is no reality; there is only observation (at which point you will be forced to consider the reality of observations and those who make them, including yourself). If you reject unrestricted induction, then you keep your reality, but surrender any possibility you thought you had of understanding it. If you reject locality, you are squaring off against a pantheon of revered scientists and philosophers, including Albert Einstein and several of the fathers of quantum mechanics themselves. And many of those who now call themselves "physicists" will all but laugh you out of whatever club or journal you might have planned to use as a forum for serious consideration of your views.

The question thus arises: is there some model of reality which retains the validity of all three premises within their respective physical ranges of application? Such a model must obviously extend the classical framework within which the neoclassical paradox has been derived. This extension must not create further unresolvable paradoxes; that is, it must be constructed with "headroom" for the resolution of whatever paradoxes might be formulated within it. And it must be able to incorporate physical theories like quantum mechanics and general relativity as subformalisms. Is there, among these criteria, anything to serve as a clue in our search for the required extension?

These criteria are all computational, and computational logic is thus the means of extension. This could not possibly be plainer or more irrefutable. In the CTMU, standard *induction* has indeed been *de-restricted* to enable the stratification of *reality* in such a way that *locality* applies on the appropriate stratum. The logic is flawless. The only thing that militates against it is the idea that reality is the output phase of an incredibly huge and complex "computer". This thesis seems designed to stretch the throat of whomever tries to swallow it, and the intellectually timid will be discouraged from even making the attempt. Of course, the intellectually timid are typically not involved in any major conceptual advance, and their opinions may accordingly be dismissed.

To see why, let us sample the kind thinking that would tend to resist the CTMU extension of physical reality. *The model is too complex; such a computer as Γ would be impossible to construct or even envision. Of what would it be constructed, and by whom?* It will suffice to observe that these same criticisms can be leveled at the classical model. In fact, it can easily be shown that the classical model is equivalent to a parallel distributed "cellular" automaton! If one has an aversion to inhabiting a computer, it may reasonably be asked what he thinks he is doing by existing at all. Such a critic must either respond with an hysterical torrent of meaningless objections, or effect a rapid accumulation of mental courage. After all, it is better to exist in a logical construct than in the "vat" of a "mad scientist" – an inductive possibility mentioned in these very pages.

Where it can be shown that all roads lead to Rome, we might as well save ourselves the walk and get on a plane. The plane to catch in this case wears on its fuselage letters which read: "The wave-function collapse can be modeled as the executive phase of an instruction decoded by the meta-automaton Γ as Γ_0 output, or "physical reality". This formulation is invariant with respect to the local structure of Γ, the "programming language" formulating the collapse function, or the code by which it translates as Γ_0 output-level reality; an appropriate morphism must exist. The scheduling of this collapse, as von Neumann realized, cannot be experienced beyond the range of effect. In CTMU terms, this effect equates to the first Γ_0-effective computation after propagation (i.e., the first effect of an associated Γ_0 event-causation). Very tight and tautological...but with the important qualification that the tautology is in this case a property of our innate mental structures and *their* underlying causes, and therefore terminal.

Prior to EPR/Bell nonlocality, the only restriction on the classical viewpoint involved Heisenberg uncertainty. But this was ultimately no restriction at all; one could still induce from observables whatever classical, localistic "hidden variables" one might need to explain the observations. In this case, however, it would still have been necessary to extend the observational syntax by adjunction of these variables, and this cannot be done without admitting the existence of a *metaphysical syntax*. The "advantage" of such a "localistic" version of Γ would be the retention of the Γ_0 metric, and the elevation of the locality premise to the status of metaphysics. Some of those attracted to this advantage, using a variety of increasingly dubious arguments, still question the validity of experiments designed to confirm nonlocality. But reality must still be partitioned into that which we can know directly, and that which we can know only by effect (on the parts we can know directly). Thus, reality is tacitly relativized to syntax.

Notice what this does to the concept of wave-function collapse. By relativizing reality, we relativize the function determining the collapse of superimposed potentia *into* reality. So collapse is relativized to the criteria for reality. This means that where we insist on attributing objective reality to classical objects consisting of many quanta, the collapse of a quantum wave must occur upon the interaction of that wave with the classical object. In terms of Mr. Cole's formulation of Schrödinger's paradox, this occurs at d/v seconds after emission of the quantum wave from its source (where d is the distance in meters from emitter to detector and v is the speed of the particle in meters per second).

Because the source is not a system of radioactive atoms, but a single atom, it is not quite a classical object. But even if it were, can the loss of mass-energy associated with the radioactive decay of a single atom be considered a classical, "observable" event? If not, then the formulation leaves us in a temporary bind. Fortunately, an atom is composite enough to let its decay be rated a classical event, qualitatively different from the localistic propagation of a matter quantum through space. Even if this were not the case, however, the classical nature of the apparatus would force the classical scheduling of its state transitions.

The problem that now arises has much to do with a certain elementary philosophical paradigm: if a tree falls in the forest, but no one hears the sound of its falling, then is there a sound? Now, sound is usually considered an objective, classical physical phenomenon. Molecules – in this case, of air – are caused by the mechanical vibrations of a source to form compression waves which have objective attributes like frequency and intensity. The effects of such waves, while for the most part transient, can be mechanically amplified – or electromechanically translated by the human inner ear – into observable effects or observations.

The objective, classical point of view mandates an affirmative answer to this question on the basis of the movement of a single molecule of air, as caused by the impact of the tree against the earth. But there are other ways to view the situation. As effects defined on the pre- and post-sonic positions of gaseous molecules are virtually unmeasurable in principle – and absolutely so in the absence of a conscious attempt to measure them and read the measurement – we can, without fear of contradiction by any physical measurement, deny the effect in the absence of observation. Forget the observations of insects, snakes, and squirrels; these too are dependent for their being on the sacred act of human observation, living only for, and in, the eyes of human godlings.

This may not sound much like science, but it happens to be the view tacitly held by those quantum physicists who concur with the conventional "Copenhagen interpretation" of quantum mechanics due to the Danish physicist Niels Bohr. The etiology of this view goes something like this: the induction of causes is enabled only by determinacy. The inference of causes is thus impossible without a deterministic explanative formalism, which is not available with respect to quantum or other "indeterminate" phenomena. So these causes – or, if you like, these *mechanisms* – have no absolute determinate existence in the absence of a deterministic formalism by which to "construct" them. So existence in general depends on the mind generating the constructive formalism. This now applies not only to indeterminate contexts but to *all* contexts, and reality – where it can be said to exist at all – is *observer-created.*

Notice that there is yet farther to go along this chain of reasoning. Any individual human being can easily adapt it to a solipsistic worldview in which nothing exists before he *personally* observes it! All things, including other so-called human beings, are nothing but the products of his observations; he "creates" them by acts of "transcendental imagination". The universe is a one-man show, a command performance by the individual consciousness for the pleasure and edification of itself. The "godling" becomes God.

We thus go from a quantum-mechanical vitiation of induction to the dependency of reality on intelligent observation, to the sufficiency of a single intelligent observer and the superfluity of all "others". Few would deny that this gives science a whole new meaning. Unfortunately, meaning has now become indistinguishable from meaninglessness. Because science itself has warranted this reasoning, only a science of science – i.e., a metaphysics – can save it. Since all possible metaphysical schemata are required to conform to CTMU principles, the CTMU itself is the core of that metaphysics. The implications are straightforward enough.

Clearly, collapse must be relativized to the various realities of Γ-subautomata. This is no harder to comprehend than the special-relativistic symmetrization of velocities in Γ_0. Both of these relativizations have the same overall effect: they ensure the logical consistency of a complex computation, that of physical reality, in tacit accordance with principles of computational economy (efficiency, elegance). Both thus confirm the computational nature of reality, even though Einstein's contributions were made prior to the "age of computers". This is hardly astonishing, since good theories typically have places even in the advanced theories which follow them (e.g., as Newtonian physics is the "non-relativistic limit" of relativistic physics and the "classical limit" of quantum physics). The requisite transformational symmetry is called an *empiron*; under inductive stratification, it becomes the *empyreon*, a large-scale algebraic formulation of Γ.

The computational necessity of logical consistency has direct bearing on the status of Schrödinger's poor cat. To wit, your eventual observation of its biological state, with an eye to determining whether it is dead or alive, must jibe with the "subjective", co-dependent "observations" of the apparatus, the quantum, and the cat. The differentiated existence of these realities is ensured by their inclusion in the physical stratum Γ_0 of Γ, whose connected reality is equivalent to the consistency-invariant necessarily governing multiplex observations. This consistency implies a recurring potential for theoretic incompleteness...an "incomputability" problematic to "observer-created reality". It is hard to create by observation what cannot be computed apart therefrom, particularly when one is required to create it in perfect harmony with other observations.

Where the consistency function is localistic, it can be distributed in Γ_0. Where consistency is nonlocal – as for spin conservation – the consistency function must reside in Γ "hyperspace" and act projectively. That part which is Γ_0-distributed is the guarantor that physical reality exists apart from observation; that part which is not is the guarantor that programmatic reality is similarly objective. The consistency function consists of nomological invariants in which observations are meristic; if it did not, it could not *constrain* observation in any dependable way. Observers allowed to constrain themselves autonomously in their observations of local or nonlocal physical phenomena will inevitably contradict each other; that this does not happen in irremediable ways is all the empirical evidence we need against it. The quantum-holistic interconnectedness of all observers, to whatever extent it enforces consistency among them, is *computational*... and meaningless outside the CTMU formalism.

I.e., if you admit that I exist, then I'll admit that you exist. This requires that we formulate an agreement to observe nothing in absolutely contradictory ways. The details of this agreement comprise the consistency-function of our mutual existence. This function, like any enforceable pact, must allow for the finest details of our mutual observations...the *quantum* events defining them. It follows that if you and I have mutual reality, then so do quanta, and so do the classical objects composed of them. Because they are automatonic, they have "accepting syntaxes". The mutual consistent translation of these syntaxes – the transformations of common data configurations from logical basis to logical basis – then becomes the essence of science.

Science computes theories and devices from experimentatively observed phenomena. For scientific purposes, the syntax of wavefunction collapse should therefore be that of the most powerful deterministic Γ-subautomaton definable, where "determinism" is relativized to our current scientific capabilities (these in turn being apparently limited by Heisenberg uncertainty). This is by definition Γ_0, with the finest possible deterministic quantization of space, time, and matter. This takes us directly down to the atomic level, at least with respect to the solid-state physics of the Schrödinger apparatus. This implies that we can schedule wavefunction collapse by means of an atomic clock...with or without anyone watching it. So if you are not a solipsist, and if you are or aspire to be a scientist, this is your "relativized" resolution of Schrödinger's paradox. And because you are the human object of an anthropic relativization, the resolution becomes absolute.

Suppose, then, that you are so obstinate a solipsist as to yet insist that this atomic clock can only be created by your own act of mind. Then rational self-interest compels you to adopt a viewpoint tending to maximize your own power over this self-simulated pseudoreality you have "created". This reality happens to be quite a bit more comfortable than other realities you might have created for yourself. Your further satisfaction compels you to produce a reason for this, thereby to improve the quality of your daydreamt "lifestyle".

First, you notice that your dream is far too deep and complex to bring entirely under your conscious control; much of it has to be of "subconscious" origin. For one thing, the scientists and technicians who invent and operate all those pleasurable and timesaving gadgets for which you cannot produce blueprints must be tapping into your subconscious mind for their designs. It follows that the design principles, and the designers' access to them, are the means by which your subconscious has fashioned the means of your gratification. The value to you of this process is clear. It thus behooves you to allow quantum collapse to be scheduled in the way that best enhances its efficiency. But this is the same way that it should be scheduled to maximize the power of objective science! That is, the resolution is invariant with respect to the degree of objectivity of science or to your "degree of solipsism"; the solipsist is compelled to identify the reality of Γ with that of his own subconscious mind. So the Copenhagen interpretation is realistically pruned, and Newton defies banishment from the dream.

Of course, there are other versions of "quantum reality" than Bohr's. There is Bohm's "implicate order", which translates as the nonlocal parallel distributed computative involution of Γ_0 by way of empyreonic temporal operators (call it "organic" if you like, but organisms necessarily compute and must therefore answer to Γ). There is Everett's "many-worlds" interpretation, an unconstrained computative exhaustion of all possible evolutions (and subject to pruning by means of multilevel Γ-programming). There is the vague "quantum logical" perspective, which relies on Γ-universality for the relativized reification of arbitrary logics...and is thus realized as the logical structure of Γ itself. There is "neo-realism", which cannot survive Bell nonlocality without the Γ-stratification of reality. And then there is Heisenberg's conception of quasi-real "potentia" whose existence is supportable only in terms of the "pro-output phase" of Γ-functionability. All are analytic within the CTMU unification of physics and higher logic, and there only. Our results are thus impervious to objections from them.

The projectivity of our internal logic allows us to derive the potential for quantum nonlocality – an apparently "objective" concept – subjectively. This sounds fatally unscientific, but only because it is a "boundary condition" of science...a limit. By analogy, the sequence ±(finite n)/1, ±n/2, ±n/3... converges on 0, which is not the same kind of number as those in the series. Does the "unscientific" nature of projective derivation make it any less valid with respect to the process of scientific induction than the "unnumeric" aspect of 0 makes it with regard to numerical induction of the given series? Not likely. In fact, because 0 is the identity of the additive group of such numbers, they can be considered its "projections": they exist "within it", coupled symmetrically with their additive inverses. If this seems hard to fathom, just couple 1/2 and -1/2 by addition, and see what you get. This is how the world works, to whatever extent we can reasonably hope to know it.

We project our inner reality outwardly. Our inner reality is computational, and so too is outward reality for the purposes of human observation and understanding. The attempt to separate these purposes from a higher "objective" reality has fueled the quantum-reality debate. In effect, it is asked: given that human nature and human purposes impose conditions on reality, what is reality "really" like in the absence of human beings and their designs? Unfortunately, somebody has to be asking this question, and he is probably human. He therefore requires an answer formulated within his accepting syntax...the same syntax he proposes to factor out of the picture. This, of course, is a paradox, and it necessitates a conceptual extension of the formulation. But to be humanly comprehensible, this extension must also be formulated in our native syntax! This regresses intractably, and so the above question is otiose...for our purposes. The closest we can get to a meaningful answer is the CTMU stratification of inference, and the CTMU is thus our highest possible conception of "objective reality". Try sliding out of that, and your circular slide will deliver you hard and unerringly back to your starting point...or, if you prefer, into a puddle of tar in which your intellectual movements, like the struggling of an insect in prefossilized amber, will mire you ever more hopelessly.

The universe is computation-theoretic. Notice that we are no longer calling it a "computer"; this would draw irrelevant associations out of the reader's own background, and very possibly prejudice his or her understanding. Computation theory is so general, and so powerful, that we need not at this stage restrict Γ to any particular architecture or style of computation, except by consistency with our own computational architectures. But it remains true that extended reality, and every part of it, both computes and is computed by other parts. These computative interactions are limited only by the dynamical or cognitive syntaxes – "programs" – of the appropriate subsets of extended reality, and cannot for the most part be "absolutized" except in tacit regard to them.

Time is computation. Space is computative potential. Matter computes, and parametrizes the deterministic or nondeterministic computations of other matter. Relative to position, this is just the quantized version of Mach's thesis. In it is the realization that matter has its own sort of "awareness"; to the extent that a conscious observer self-projectively creates reality, reality is consciousness. Attributes argumental to the Γ_0 localistic dynamic are, to a large extent, computed by the locally-interacting quanta themselves. It follows that wavefunction collapse occurs largely on the basis of material proximity, where proximity is defined on the ranges of the elementary forces relating quanta. Macroscopic objects are systems of quanta interlocked in fieldwise computation at the appropriate distances; they are self-computing and mutually observant, and do not need to be peered at by crypto-solipsists. Observers who intend to exploit dynamical processes "in absentia" are required to reify them "in absentia". Schrödinger's cat is real, and those attributes argumental to the intrinsic (metabolic) and extrinsic Γ_0-dynamical processes involving it are computed by it. When the tree falls out of earshot, it thunders nonetheless.

The true nature of Schrödinger's paradox has now become clear. The observer creates the computative observer-aspect of reality; reality computes the reality-aspect of the observer. You observe Γ, and Γ observes you back. The two sides of observation are codependent, and the logico-mathematical formulation of codependency is symmetry. The relationship of an observer to objective reality, as perceived by a second observer witnessing their interaction, is thus a symmetrical relationship. The observer can create reality, and the sum of all other observers, only insofar as they create him. His limitations stratify Γ; Γ projects its stratification as that of his mind and experience. This ontological feedback is the stuff of consciousness. Consciousness is relativized to the Γ-subautomata possessing it, and is quantified according to their parameters. It is a function of unified multiplicity...the unified multiplicity of the self-differentiated telesis that is Γ.

So these are the principles needed to resolve the issue of collapse: syntax-projectivity; the computative tenor of reality; observational symmetry; and the global consistency constraint on sub-symmetric computations. These, in light of the structure of Γ as thus far described, suffice to lay the matter to rest. If there is any doubt of this, you need merely try – in your own mind – to refute one of these principles. Given enough time, and the insight of which you are capable, you will recognize the logical necessity (and even the self-evidence) of all of them.

Note that we have been discussing collapse from a physical perspective. Our resolution is thus Γ_0-relativized; it applies to the physical collapsation of physical quanta. But where collapse is more generally defined on the measurement or determination of arbitrarily-relativized random variables, it can obviously occur in different Γ control-levels. Ultimately, we can speak of the *teleological* phase of collapse, tentatively identifiable with the empyreonic identity. This is the highest possible "quantum holor", whose relation to its merates is the concern of religion...a concern we are not allowed to neglect. As Kant, Newton, and Einstein knew, it is naturally the ultimate goal of any serious exploration of the nature of reality.

The theory of quantum wave-function collapse marks the crossroads of physics, philosophy, and the foundations of mathematics. If there was ever any doubt that these fields are inseparable, this is where it ends. Such doubts are the delusions of moles who tunnel until they have lost the capacity to sense anything but the minutiae just ahead of their snouts. It is therefore a point of some importance that the members of this society have demonstrated more than subterranean intelligence. A clear verification of this was offered by C. Cole, who – after circling around many of the issues essential to this discussion – expressed his feeling that the problems of nonlocality and collapse are deeply connected. We now know him to be unambiguously correct: both have computation-theoretic solutions within the CTMU formalism.

As, in principle, does everything else.

***

Editor's note from *Noesis* № 47 The bibliography for the above paper is rather extensive and will therefore be omitted at this printing due to lack of space. It is in no way to be assumed that this reflects an unscientific absence of formality. Such determinations are typically made by those who face different practical constraints than the ones associated with this publication, and therefore do not apply here.

On the CTMU

Scientific theories are *mental* constructs that have *objective reality* as their content. According to the scientific method, science puts objective content first, letting theories be determined by observation. But the phrase "a theory of reality" contains two key nouns, *theory* and *reality*, and science is really about both. Because all theories have certain necessary logical properties that are abstract and mathematical, and therefore independent of observation – it is these very properties that let us recognize and understand our world in conceptual terms – we could just as well start with these properties and see what they might tell us about objective reality. Just as scientific observation makes demands on theories, the logic of theories makes demands on scientific observation, and these demands tell us in a general way what we may observe about the universe.

In other words, a comprehensive theory of reality is not just about observation, but about theories and their logical requirements. Since theories are *mental* constructs, and mental means "of the mind", this can be rephrased as follows: *mind* and *reality* are linked in mutual dependence at the most basic level of understanding. This linkage of mind and reality is what a **TOE** (Theory of Everything) is really about. The **CTMU** is such a theory; instead of being a mathematical description of specific observations (like all established scientific theories), it is a "metatheory" about the general relationship between theories and observations ... i.e., about science or knowledge itself. Thus, it can credibly lay claim to the title of **TOE**.

Mind and reality – the abstract and the concrete, the subjective and the objective, the internal and the external – are linked together in a certain way, and this linkage is the real substance of "reality theory". Just as scientific observation determines theories, the logical requirements of theories to some extent determine scientific observation. Since reality always has the ability to surprise us, the task of scientific observation can never be completed with absolute certainty, and this means that a comprehensive theory of reality cannot be based on scientific observation alone. Instead, it must be based on the process of making scientific observations in general, and this process is based on the relationship of mind and reality. So the **CTMU** is essentially *a theory of the relationship between mind and reality*.

In explaining this relationship, the **CTMU** shows that reality possesses a complex property akin to *self-awareness*. That is, just as the mind is real, reality is in some respects like a mind. But when we attempt to answer the obvious question "*whose* mind?", the answer turns out to be a mathematical and scientific definition of God. This implies that we all exist in what can be called "the Mind of God", and that our individual minds are parts of God's Mind. They are not as powerful as God's Mind, for they are only parts thereof; yet, they are directly connected to the greatest source of knowledge and power that exists. This connection of our minds to the Mind of God, which is like the connection of parts to a whole, is what we sometimes call the *soul* or *spirit*, and it is the most crucial and essential part of being human.

Thus, the attempt to formulate a comprehensive theory of reality, the **CTMU**, finally leads to spiritual understanding, producing a basis for the unification of science and theology. The traditional Cartesian divider between body and mind, science and spirituality, is penetrated by logical reasoning of a higher order than ordinary scientific reasoning, but no less scientific than any other kind of mathematical truth. Accordingly, it serves as the long-awaited gateway between science and humanism, a bridge of reason over what has long seemed an impassable gulf.

Introduction to the CTMU

The **real universe** has always been theoretically treated as an *object*, and specifically as the composite type of object known as a *set*. But an object or set exists in space and time, and reality does not. Because the real universe by definition contains all that is real, there is no "external reality" (or space, or time) in which it can exist or have been "created". We can talk about lesser regions of the real universe in such a light, but not about the real universe as a whole. Nor, for identical reasons, can we think of the universe as the sum of its parts, for these parts exist solely within a spacetime manifold identified with the whole and cannot explain the manifold itself. This rules out pluralistic explanations of reality, forcing us to seek an explanation at once *monic* (because nonpluralistic) and *holistic* (because the basic conditions for existence are embodied in the manifold, which equals the whole). Obviously, the first step towards such an explanation is to bring monism and holism into coincidence.

When theorizing about an all-inclusive reality, the first and most important principle is *containment*, which simply tells us what we should and should not be considering. Containment principles, already well known in cosmology, generally take the form of tautologies; e.g., "The physical universe contains all and only that which is physical." The predicate "physical", like all predicates, here corresponds to a structured set, "the physical universe" (because the universe has structure and contains objects, it is a structured set). But this usage of *tautology* is somewhat loose, for it technically amounts to a predicate-logical equivalent of propositional tautology called *autology*, meaning self-description. Specifically, the predicate *physical* is being defined on topological containment in the *physical universe*, which is tacitly defined on and descriptively contained in the predicate *physical*, so that the self-definition of "physical" is a two-step operation involving both topological and descriptive containment. While this principle, which we might regard as a statement of "physicalism", is often confused with *materialism* on the grounds that "physical" equals "material", the material may in fact be only a part of what makes up the physical. Similarly, the physical may only be a part of what makes up the real. Because the content of reality is a matter of science as opposed to mere semantics, this issue can be resolved only by rational or empirical evidence, not by assumption alone.

Can a containment principle for the real universe be formulated by analogy with that just given for the physical universe? Let's try it: "The real universe contains all and only that which is real." Again, we have a tautology, or more accurately an autology, which defines the *real* on inclusion in the *real universe*, which is itself defined on the predicate *real*. This reflects *semantic duality*, a logical equation of predication and inclusion whereby perceiving or semantically *predicating* an attribute of an object amounts to perceiving or predicating the object's topological *inclusion* in the set or space dualistically corresponding to the predicate. According to semantic duality, the predication of the attribute *real* on the real universe from within the real universe makes reality a *self-defining predicate*, which is analogous to *a self-including set*. An all-inclusive set, which is by definition self-inclusive as well, is called "the set of all sets". Because it is all-descriptive as well as self-descriptive, the reality predicate corresponds to the set of all sets. And because the self-definition of reality involves both descriptive and topological containment, it is a two-stage hybrid of universal autology and the set of all sets.

Now for a brief word on sets. Mathematicians view set theory as fundamental. Anything can be considered an object, even a space or a process, and wherever there are objects, there is a set to contain them. This "something" may be a relation, a space or an algebraic system, but it is also a set; its relational, spatial or algebraic structure simply makes it a *structured* set. So mathematicians view sets, broadly including null, singleton, finite and infinite sets, as fundamental objects basic to meaningful descriptions of reality. It follows that reality itself should be a set ... in fact, the largest set of all. But every set, even the largest one, has a powerset which contains it, and that which contains it must be larger (a contradiction). The obvious solution: define an extension of set theory incorporating two senses of "containment" which work together in such a way that the largest set can be defined as "containing" its powerset in one sense while being contained *by* its powerset in the other. Thus, it topologically includes itself in the act of descriptively including itself in the act of topologically including itself ..., and so on, in the course of which it obviously becomes more than just a set.

In the *Cognitive-Theoretic Model of the Universe* or CTMU, the set of all sets, and the real universe to which it corresponds, take the name (SCSPL) of the required extension of set theory. SCSPL, which stands for *Self-Configuring Self-Processing Language*, is just a totally intrinsic, i.e. completely self-contained, language that is comprehensively and coherently (self-distributively) self-descriptive, and can thus be model-theoretically identified as its own *universe* or referent domain. Theory and object go by the same name because unlike conventional ZF[4] or NBG[5] set theory, SCSPL hologically infuses sets and their elements with the distributed (syntactic, meta-logical) component of the theoretical framework containing and governing them, namely SCSPL syntax itself, replacing ordinary set-theoretic objects with SCSPL *syntactic operators*. The CTMU is so-named because the SCSPL universe, like the set of all sets, distributively embodies the logical syntax of its own descriptive mathematical language. It is thus not only self-descriptive in nature; where logic denotes the rules of *cognition* (reasoning, inference), it is self-cognitive as well.

An act is a temporal process, and self-inclusion is a spatial relation. The act of self-inclusion is thus "where time becomes space"; for the set of all sets, there

[4] Zermelo–Fraenkel

[5] Von Neumann–Bernays–Gödel

can be no more fundamental process. No matter what else happens in the evolving universe, it must be temporally embedded in this dualistic self-inclusion operation. In the CTMU, the self-inclusion process is known as *conspansion* and occurs at the distributed, Lorentz-invariant *conspansion rate* c, a time-space conversion factor already familiar as the speed of light *in vacuo* (conspansion consists of two alternative phases accounting for the wave and particle properties of matter and affording a logical explanation for accelerating cosmic expansion). When we imagine a dynamic self-including set, we think of a set growing larger and larger in order to engulf itself from without. But since there is no "without" relative to the real universe, external growth or reference is not an option; there can be no external set or external descriptor. Instead, self-inclusion and self-description must occur inwardly as the universe stratifies into a temporal sequence of states, each state topologically and computationally contained in the one preceding it (where the conventionally limited term *computation* is understood to refer to a more powerful SCSPL-based concept, *protocomputation*, involving spatiotemporal parallelism). On the present level of discourse, this inward self-inclusion is the conspansive basis of what we call *spacetime*.

Every object in spacetime includes the entirety of spacetime as a *state-transition syntax* according to which its next state is created. This guarantees the mutual consistency of states and the overall unity of the dynamic entity the *real universe*. And because the sole real interpretation of the set-theoretic entity "the set of all sets" is the entire real universe, the associ-ated foundational paradoxes are resolved in kind (by attributing mathematical structure like that of the universe to the pure, uninterpreted set-theoretic version of the set of all sets). Concisely, resolving the set-of-all-sets paradox requires that:

(1) an endomorphism or self-similarity mapping $D: S \rightarrow r \in S$ be defined for the set of all sets S and its internal points r;

(2) there exist two complementary senses of inclusion, one topological [$S \supset_t D(S)$] and one predicative [$D(S) \supset_d S$], that allow the set to descriptively "include itself" from within, i.e. from a state of topological self-inclusion (where $\supset_t$ denotes topological or set-theoretic inclusion and $\supset_d$ denotes descriptive inclusion, e.g. the inclusion in a language of its referents);

(3) the input S of D be global and structural, while the output $D(S) = (r \supset_d S)$ be internal to S and play a syntactic role.

In short, the set-theoretic and cosmological embodi-ments of the self-inclusion paradox are resolved by properly relating the self-inclusive object to the descriptive syntax in terms of which it is necessarily expressed, thus effecting true self-containment: "the universe (set of all sets) is that which *topologically* contains that which *descriptively* contains the universe (set of all sets)."

This characterizes a system that consistently perceives itself and develops its own structure from within via *hology*, a 2-stage form of self-similarity roughly analogous to holography. (Hology is a logico-cybernetic form of self-similarity in which the global structure of a self-contained, self-interactive system doubles as its distributed self-transductive syntax; it is justified by the obvious fact that in a self-contained system, no other structure is available for that purpose.) The associated conspansive mapping D is called *incoversion* in the spatiotemporally inward direction and *coinversion* in the reverse (outward, D^{-1}) direction. Incoversion carries global structure inward as state-recognition and state-transformation syntax, while coinversion projects syntactic structure outward in such a way as to recognize existing structure and determine future states in conformance with it. Incoversion is associated with an operation called *requantization*, while coinversion is associated with a complementary operation called *inner expansion*. The alternation of these operations, often referred to as *wave-particle duality*, comprises the conspansion process. The **Principle of Conspansive Duality** then says that what appears as cosmic expansion from an interior (local) viewpoint appears as material and temporal contraction from a global viewpoint. Because metric concepts like "size" and "duration" are undefined with respect to the universe as a whole, the spacetime metric is defined strictly intrinsically, and the usual limit of cosmological regress, a pointlike cosmic singularity, becomes the closed spacetime algebra already identified as SCSPL.

Thus, the real universe is not a static set, but a dynamic process resolving the self-inclusion para-dox. Equivalently, because any real explanation of reality is contained in reality itself, reality gives rise to a paradox unless regarded as an inclusory self-mapping. This is why, for example, category theory is increasingly preferred to set theory as a means of addressing the foundations of mathematics; it centers on invariant relations or mappings between covariant or contravariant (dually related) objects rather than on static objects themselves. For similar reasons, a focus on the relative invariants of semantic processes is also well-suited to the formulation of evolving theories in which the definitions of objects and sets are subject to change; thus, we can speak of *time and space* as equivalent to *cognition and information* with respect to the invariant semantic relation *processes*, as in "time processes space" and "cognition processes information". But when we define reality as a process, we must reformulate containment accordingly. Concisely, reality theory becomes a study of SCSPL autology naturally formulated in terms of mappings. This is done by adjoining to logic certain metalogical principles, formulated in terms of mappings, that enable reality to be described as an autological (self-descriptive, self-recognizing/self-processing) system.

The first such principle is **MAP**, acronymic for *Metaphysical Autology Principle*. Let S be the real universe, and let T = T(S) be its theoretical description or "TOE". MAP, designed to endow T and S with mathematical closure, simply states that T and S are closed with respect to all internally relevant operations, including recognition and description. In terms of mappings, this means that all inclusional or descriptive mappings of S are *automorphisms* (e.g., permutations or foldings) or *endomorphisms* (self-injections). MAP is implied by the unlimited scope, up to perceptual relevance, of the universal quantifier implicitly attached to reality by the containment principle. With closure thereby established, we can apply techniques of logical reduction to S without worrying about whether the lack of some external necessity will spoil the reduction. In effect, MAP makes T(S) "exclusive enough" to describe S by excluding as a descriptor of S anything not in S. But there still remains the necessity of providing S with a mechanism of self-description.

This mechanism is provided by another metalogical principle, the **M=R** or *Mind Equals Reality Principle*, that identifies S with the extended cognitive syntax D(S) of the theorist. This syntax (system of cognitive rules) not only determines the theorist's perception of the universe, but bounds his cognitive processes and is ultimately the *limit* of his theorization (this relates to the observation that all we can directly know of reality are our *perceptions* of it). The reasoning is simple; S determines the composition and behavior of objects (or subsystems) s in S, and thus comprises the general syntax (structural and functional rules of S) of which s obeys a specific restriction. Thus, where s is an ideal observer/theorist in S, S is the syntax of its own observation and explanation by s. This is directly analogous to "the real universe contains all and only that which is real", but differently stated: "S contains all and only objects s whose extended syntax is isomorphic to S." M=R identifies S with the veridical limit of any partial theory T of S [lim T(S) = D(S)], thus making S "inclusive enough" to describe itself. That is, nothing relevant to S is excluded from $S \cong D(S)$.[6]

Mathematically, the M=R Principle is expressed as follows. The universe obviously has a structure S. According to the logic outlined above, this structure is self-similar; S distributes over S, where "distributes over S" means "exists without constraint on location or scale within S". In other words, the universe is a perfectly self-similar system whose overall structure is replicated everywhere within it as a general state-recognition and state-transition syntax (as understood in an extended computational sense). The self-distribution of S, called hology, follows from the containment principle, i.e. the tautological fact that everything within the real universe must be described by the predicate "real" and thus fall within the constraints of global structure. That this structure is completely self-distributed implies that it is locally indistinguishable for subsystems s; it could only be discerned against its absence, and it is nowhere absent in S. Spacetime is thus transparent from within, its syntactic structure invis-ible to its contents on the classical (macroscopic) level. Localized systems generally express and utilize only a part of this syntax on any given scale, as determined by their specific structures. I.e., where there exists a hological incoversion endomorphism $D: S \rightarrow \{r \in S\}$ carrying the whole structure of S into every internal point and region of S, objects (quantum-geometrodynamically) embedded in S take their recognition and state-transformation syntaxes directly

[6] $\cong$ isomorphic

from the ambient spatiotemporal background up to isomorphism. Objects thus utilize only those aspects of D(S) of which they are structural and functional representations.

The inverse D^{-1} of this map (coinversion) describes how an arbitrary local system s within S recognizes S at the object level and obeys the appropriate "laws", ultimately giving rise to human perception. This reflects the fact that S is a self-perceptual system, with various levels of self-perception emerging within interactive subsystems s (where perception is just a refined form of *interaction* based on *recognition* in an extended computational sense). Thus, with respect to any class {s} of subsystems of S, we can define a homomorphic submap d of the endomorphism D : $d : S \rightarrow \{s\}$ expressing only that part of S to which {s} is isomorphic. In general, the s_i are coherent or physically self-interactive systems exhibiting dynamical and informational closure; they have sometimes-inaccessible internal structures and dynamics (particularly on the quantum scale), and are distinguishable from each other by means of informational boundaries contained in syntax and comprising a "spacetime metric".

According to the above definitions, the global self-perceptor S is amenable to a theological interpretation, and its contents {s} to "generalized cognitors" including subatomic particles, sentient organisms, and every material system in between. Unfortunately, above the object level, the validity of *s-cognition* – the internal processing of sentient subsystems s – depends on the specific cognitive functionability of a given s ... the extent to which s can implicitly represent higher-order relations of S. In General Relativity, S is regarded as given and complete; the laws of mathematics and science are taken as preexisting. On the quantum scale, on the other hand, laws governing the states and distributions of matter and energy do not always have sufficient powers of restriction to fully determine quantum behavior, requiring probabilistic augmentation in the course of quantum wavefunction collapse. This prevents a given s, indeed anything other than S from enclosing a complete nomology (set of laws); while a complete set of laws would amount to a complete deterministic history of the universe, calling the universe "completely deterministic" amounts to asserting the existence of prior determinative constraints. But this is a logical absurdity, since if these constraints were real, they would be included in reality rather than prior or external to it (by the containment principle). It follows that the universe freely determines its own constraints, the establishment of nomology and the creation of its physical (observable) content being effectively simultan-eous and recursive. The incoversive distribution of this relationship is the basis of *free will*, by virtue of which the universe is freely created by sentient agents existing within it.

Let's elaborate a bit. Consider the universe as a completely evolved perceptual system, including all of the perceptions that will ultimately comprise it. We cannot know all of those perceptions specifically, but to the extent that they are interactively connected, we can refer to them *en masse*. The set of "laws" obeyed by the universe is just a minimal set of logical relations that suffices to make these perceptions noncontradictory, i.e. mutually consistent, and a distributed set of laws is just a set of laws formulated in such a way that the formulation can be read by any part of the system S. Obviously, for perceptions to be connected by laws, the laws themselves must be internally connected according to a syntax, and the ultimate syntax of nomological connectedness must be globally valid; whatever the laws may be at any stage of system evolution, all parts of S must be able to unambiguously read them, execute and be acted upon by them, and recognize and be recognized as their referents ("unambiguously" implies that 2-valued logic is a primary ingredient of nomology; its involvement is described by a third metalogical principle designed to ensure consistency, namely **MU** or *Multiplex Unity*). This implies that the action and content of the laws are merged together in each part of the system as a single (but dual-aspect) quantity, *infocognition*. The connectedness and consistency of infocognition is maintained by refinement and homogenization as nomological languages are superseded by extensional metalanguages in order to create and/or explain new data; because the "theory" SCSPL model-theoretically equates itself to the real universe, its "creation" and causal "explanation" operations are to a certain extent identical, and the SCSPL universe can be considered to create or configure itself by means of "self-theorization" or "self-explanation".

The simplest way to explain "connected" in this context is that every part of the (object-level) system relates to other parts within an overall structural description of the system itself (to interpret "parts", think of events rather than objects; objects are in a sense *defined* on events in a spatiotemporal setting). Obviously, any part which fails to meet this criterion does not conform to a description of the system and thus is not included in it, i.e. not "connected to" the system (on the other hand, if we were to insist that it *is* included or connected, then we would have to modify the systemic description accordingly). For this description to be utile, it should be maximally compact, employing compact predictive generalizations in a regular way appropriate to structural categories (e.g., employing general "laws of physics"). Because such laws, when formulated in an "if conditions (a, b , c, ...) exist, then (X and Y or Z) applies" way, encode the structure of the entire system and are universally applicable within it, the system is "self-distributed". In other words, every part of the system can consistently interact with every other part while maintaining an integral identity according to this ("TOE") formulation. Spatiotemporal relations can be skeletally depicted as edges in a graph whose vertices are events (physical interactions), i.e. spacetime "points". In this sense, graph-theoretic connectivity applies. But all those object-level connections must themselves be connected by more basic connections, the basic connections must be connected by even more basic connections, and so on. Eventually – perhaps sooner than later – we reach a basic level of connectivity whose syntax comprises a (partially undecidable) "ultimate nomology" for the level of reality we're discussing.

Is this nomology, and the cognitive syntax in which it is expressed, wholly embodied by matter? In one sense the answer is yes, because S is distributed over each and every material $s \in S$ as the reality-syntax D(S). Thus, every axiom and theorem of mathematics can be considered implicit in material syntax and potentially exemplified by an appropriate material pattern, e.g. a firing of cerebral neurons. Against holism – the idea that the universe is more than the sum of its parts – one can further object that the holistic entity in question is still a material ensemble, thus insinuating that even if the universe is not the "sum" of its parts, it is still a determinate *function* of its parts. However, this fails to explain the mutual consistency of object-syntaxes, without the enforcement of which reality would disintegrate due to perceptual inconsistency. This enforcement function takes matter as its argument and must therefore be reposed in spacetime itself, the universal substrate in which matter is unconditionally embedded (and as a geometrodynamic or quantum-mechanical excitation of which matter is explained). So the background has logical ascendancy over derivative matter, and this permits it to have aspects, like the power to enforce consistency, not expressible by localized interactions of compact material objects (i.e., within the bounds of *materialism* as invoked regarding a putative lack of "material evidence" for God, excluding the entire material universe).

On the other hand, might cognitive syntax reside in an external "ideal" realm analogous to Plato's world of Parmenidean forms? Plato's ideal abstract reality is explicitly set apart from actual concrete reality, the former being an eternal world of pure form and light, and the latter consisting of a cave on whose dirty walls shift murky, contaminated shadows of the ideal world above. However, if they are both separate and in mutual correspondence, these two realities both occupy a more basic joint reality enforcing the correspondence and providing the metric of separation. If this more basic reality is then juxtaposed to another, then there must be a more basic reality still, and so on until finally we reach the most basic level of all. At this level, there will (by definition) be no separation between the abstract and concrete phases, because there will be no more basic reality to provide it or enforce a remote correspondence across it. This is the inevitable logical terminus of "Plato's regress". But it is *also* the reality specified by the containment principle, the scope of whose universal quantifier is unlimited up to perceptual relevance! Since it is absurd to adopt a hypothesis whose natural logical extension is a negation of that hypothesis, we must assume that the ideal plane coincides with this one ... but again, not in a way necessarily accessible to identifiable physical operations. Rather, physical reality is embedded in a more general or "abstract" ideal reality equating to the reality-syntax D(S) and the syntax D(S) is in turn embedded in physical reality by incoversion. Thus, if D(S) contains supraphysical components, they are embedded in S right along with their physical counterparts (indeed, this convention is already in restricted use in string theory and M-theory, where unseen higher dimensions get "rolled up" to sub-Planck diameter).

What does this say about **God**? First, if God is real, then God inheres in the comprehensive reality syntax, and this syntax inheres in matter. *Ergo*, God inheres in matter, and indeed in its spacetime substrate as defined on material and supramaterial levels. This amounts to *pantheism*, the thesis that God is omnipresent with respect to the material universe. Now, if the universe were *pluralistic* or reducible to its parts, this would make God, Who coincides with the universe itself, a pluralistic entity with no internal cohesion. But because the mutual syntactic consistency of parts is enforced by a unitary holistic manifold with logical ascendancy over the parts themselves – because the universe is a dual-aspected *monic* entity consisting of essentially homogeneous, self-consistent infocognition – God retains monotheistic unity despite being distributed over reality at large. Thus, we have a new kind of theology that might be called *monopantheism*, or even more descriptively, *holopantheism*. Second, God is indeed real, for a coherent entity identified with a self-perceptual universe is self-perceptual in nature, and this endows it with various levels of self-awareness and *sentience*, or constructive, cre-ative intelligence. Indeed, without a guiding Entity whose Self-awareness equates to the coherence of self-perceptual spacetime, a self-perceptual universe could not coherently self-configure. Holopantheism is the logical, metatheological umbrella beneath which the great religions of mankind are unknowingly situated.

Why, if there exists a spiritual metalanguage in which to establish the brotherhood of man through the unity of sentience, are men perpetually at each other's throats? Unfortunately, most human brains, which comprise a particular highly-evolved subset of the set of all reality-subsystems, do not fire in strict S-isomorphism much above the object level. Where we define one aspect of "intelligence" as the amount of global structure functionally represented by a given $s \in S$, brains of low intelligence are generally out of accord with the global syntax D(S). This limits their capacity to form true representations of S (global reality) by syntactic autology [$d(S) \supset_d d(S)$] and make rational ethical calculations. In this sense, the vast majority of men are not well-enough equipped, conceptually speaking, to form perfectly rational worldviews and societies; they are deficient in education and intellect, albeit remediably so in most cases. This is why force has ruled in the world of man ... why might has always made right, despite its marked tendency to violate the optimization of global utility derived by summing over the sentient agents of S with respect to space *and* time.

Now, in the course of employing deadly force to rule their fellows, the very worst element of humanity – the butchers, the violators, i.e. those of whom some modern leaders and politicians are merely slightly-chastened copies – began to consider ways of maintaining power. They lit on religion, an authoritarian priesthood of which can be used to set the minds and actions of a populace for or against any given aspect of the political *status quo*. Others, jealous of the power thereby consolidated, began to use religion to gather their own "sheep", promising special entitlements to those who would join them ... mutually conflicting promises now setting the promises at each other's throats.

But although religion has often been employed for evil by cynics appreciative of its power, several things bear notice. (1) The abuse of religion, and the God concept, has always been driven by human politics, and no one is justified in blaming the God concept, whether or not they hold it to be real, for the abuses committed by evil men in its name. *Abusus non tollit usum.* (2) A religion must provide at least emotional utility for its believers, and any religion that stands the test of time has obviously been doing so. (3) A credible religion must contain elements of truth and undecidability, but no elements that are verifiably false (for that could be used to overthrow the religion and its sponsors). So by design, religious beliefs generally cannot be refuted by rational or empirical means.

Does the reverse apply? Can a *denial* of God be refuted by rational or empirical means? The short answer is yes; the refutation follows the reasoning outlined above. That is, the above reasoning constitutes not just a logical framework for reality theory, but the outline of a logical proof of God's existence and the basis of a "logical theology". While the framework serves other useful purposes as well, e.g. the analysis of mind and consciousness, we'll save those for another time.

THE CTMU

The Cognitive-Theoretic Model of the Universe: A New Kind of Reality Theory

Diagrams

Abstract

Inasmuch as science is observational or perceptual in nature, the goal of providing a scientific model and mechanism for the evolution of complex systems ultimately requires a supporting theory of reality of which perception itself is the model (or theory-to-universe mapping). Where *information* is the abstract currency of perception, such a theory must incorporate the theory of information while extending the information concept to incorporate reflexive self-processing in order to achieve an intrinsic (self-contained) description of reality. This extension is associated with a limiting formulation of model theory identifying mental and physical reality, resulting in a reflexively self-generating, self-modeling theory of reality identical to its universe on the syntactic level. By the nature of its derivation, this theory, the *Cognitive-Theoretic Model of the Universe* or CTMU, can be regarded as a *supertautological* reality-theoretic extension of logic. Uniting the theory of reality with an advanced form of computational language theory, the CTMU describes reality as a *Self-Configuring Self-Processing Language* or SCSPL, a reflexive *intrinsic language* characterized not only by self-reference and recursive self-definition, but full self-configuration and self-execution (reflexive read-write functionality). SCSPL reality embodies a dual-aspect monism consisting of *infocognition*, self-transducing information residing in self-recognizing SCSPL elements called *syntactic operators*. The CTMU identifies itself with the structure of these operators and thus with the distributive syntax of its self-modeling SCSPL universe, including the reflexive grammar by which the universe refines itself from *unbound telesis* or UBT, a primordial realm of infocognitive potential free of informational constraint. Under the guidance of a limiting (intrinsic) form of anthropic principle called the *Telic Principle*, SCSPL evolves by *telic recursion*, jointly configuring syntax and state while maximizing a generalized self-selection parameter and adjusting on the fly to freely-changing internal conditions. SCSPL relates *space*, *time* and *object* by means of *conspansive duality* and *conspansion*, an SCSPL-grammatical process featuring an alternation between dual phases of existence associated with design and actualization and related to the familiar wave-particle duality of quantum mechanics. By distributing the design phase of reality over the actualization phase, conspansive spacetime also provides a distributed mechanism for *Intelligent Design*, adjoining to the restrictive principle of natural selection a basic means of generating information and complexity. Addressing physical evolution on not only the biological but cosmic level, the CTMU addresses the most evident deficiencies and paradoxes associated with conventional discrete and continuum models of reality, including temporal directionality and accelerating cosmic expansion, while preserving virtually all of the major benefits of current scientific and mathematical paradigms.

Introduction

Among the most exciting recent developments in science are *Complexity Theory*, the theory of self-organizing systems, and the modern incarnation of *Intelligent Design Theory*, which investigates the deep relationship between self-organization and evolutionary biology in a scientific context not preemptively closed to teleological causation. Bucking the traditional physical reductionism of the hard sciences, complexity theory has given rise to a new trend, *informational reductionism*, which holds that the basis of reality is not matter and energy, but information. Unfortunately, this new form of reductionism is as problematic as the old one. As mathematician David Berlinski (2001) writes regarding the material and informational aspects of DNA: "We quite know what DNA is: it is a macromolecule and so a material object. We quite know what it achieves: apparently everything. Are the two sides of this equation in balance?" More generally, Berlinski observes that since the information embodied in a string of DNA or protein cannot affect the material dynamic of reality without being read by a material transducer, information is meaningless without matter.

The relationship between physical and informational reductionism is a telling one, for it directly mirrors Cartesian *mind-matter dualism*, the source of several centuries of philosophical and scientific controversy regarding the nature of deep reality (Wozniak, 1992). As long as matter and information remain separate, with specialists treating one as primary while tacitly relegating the other to secondary status, dualism remains in effect. To this extent, history is merely repeating itself; where mind and matter once vied with each other for primary status, concrete matter now vies with abstract information *abstractly representing* matter and its extended relationships. But while the formal abstractness and concrete descriptiveness of information seem to make it a worthy compromise between mind and matter, Berlinski's comment demonstrates its inadequacy as a conceptual substitute. What is now required is thus what has been required all along: a conceptual framework in which the relationship between mind and matter, cognition and information, is made explicit. This framework must not only permit the completion of the gradual ongoing dissolution of the Cartesian mind-matter divider, but the construction of a footworthy logical bridge across the resulting explanatory gap.

Mathematically, the theoretical framework of Intelligent Design consists of certain definitive principles governing the application of complexity and probability to the analysis of two key attributes of evolutionary phenomena, *irreducible complexity* (Behe, 1998) and *specified complexity* (Dembski, 1998). On one hand, because the mathematics of probability must be causally interpreted to be scientifically meaningful, and because probabilities are therefore expressly relativized to specific causal scenarios, it is difficult to assign definite probabilities to evolutionary states in any model not supporting the detailed reconstruction and analysis of specific causal pathways. On the other hand, positing the "absolute improbability" of an evolutionary state ultimately entails the specification of an absolute (intrinsic global) model with respect to which absolute probabilistic deviations can be determined. A little reflection suffices to inform us of some of its properties: it must be rationally derivable from *a priori* principles and essentially tautological in nature, it must on some level identify matter and information, and it must eliminate the explanatory gap between the mental and physical aspects of reality. Furthermore, in keeping with the name of that to be modeled, it must meaningfully incorporate the *intelligence* and *design* concepts, describing the universe as an intelligently self-designed, self-organizing system.

How is this to be done? In a word, with *language*. This does not mean merely that language should be used as a tool to analyze reality, for this has already been done countless times with varying degrees of success. Nor does it mean that reality should be regarded as a machine language running in some kind of vast computer. It means using language as a *mathematical paradigm unto itself*. Of all mathematical structures, language is the most general, powerful and necessary. Not only is every formal or working theory of science and mathematics by definition a language, but science and mathematics in whole and in sum are languages. Everything that can be described or conceived, including every structure or process or law, is isomorphic to a description or definition and therefore qualifies as a language, and every sentient creature constantly affirms the linguistic structure of nature by exploiting syntactic isomorphism to perceive, conceptualize and refer to it. Even cognition and perception are languages based on what Kant might have called "phenomenal syntax". With logic and mathematics counted among its most fundamental syntactic ingredients, language defines the very structure of information. This is more than an empirical truth; it is a rational and scientific necessity.

Of particular interest to natural scientists is the fact that the laws of nature are a language. To some extent, nature is regular; the basic patterns or general aspects of structure in terms of which it is apprehended, whether or not they have been categorically identified, are its "laws". The existence of these laws is given by the stability of perception. Because these repetitive patterns or universal laws simultaneously describe multiple instances or states of nature, they can be regarded as distributed "instructions" from which self-instantiations of nature cannot deviate; thus, they form a "control language" through which nature regulates its self-instantiations. This control language is not of the usual kind, for it is somehow built into the very fabric of reality and seems to override the known limitations of formal systems. Moreover, it is profoundly reflexive and self-contained with respect to configuration, execution and read-write operations. Only the few and the daring have been willing to consider how this might work ... to ask where in reality the laws might reside, how they might be expressed and implemented, why and how they came to be, and how their consistency and universality are maintained. Although these questions are clearly of great scientific interest, science alone is logically inadequate to answer them; a new explanatory framework is required. This paper describes what the author considers to be the most promising framework in the simplest and most direct terms possible.

On a note of forbearance, there has always been comfort in the belief that the standard hybrid empirical-mathematical methods of physics and cosmology will ultimately suffice to reveal the true heart of nature. However, there have been numerous signals that it may be time to try a new approach. With true believers undaunted by the (mathematically factual) explanatory limitations of the old methods, we must of course empathize; it is hard to question one's prior investments when one has already invested all the faith that one has. But science and philosophy do not progress by regarding their past investments as ends in themselves; the object is always to preserve that which is valuable in the old methods while adjoining new methods that refine their meaning and extend their horizons. The new approach that we will be exploring in this paper, which might be colorfully rendered as "reality theory is wedded to language theory and they beget a synthesis", has the advantage that it leaves the current picture of reality virtually intact. It merely creates a logical mirror image of the current picture (its *conspansive dual*), merges the symmetric halves of the resulting picture, and attempts to extract meaningful implications. Science as we now know it is thereby changed but little in return for what may, if fate smiles upon us, turn out to be vast gains in depth, significance and explanatory power.

And on that note, I thank you for your kind attention and wish you a fruitful journey.

On Theories, Models and False Dichotomies

It has almost become embarrassing to point out that science is in a state of crisis ... not because it is untrue, but because it has become a *cliché* too often accompanied by little or no remedial insight. For all of the magnificent achievements of science, its grander ambitions long ago succeeded in taxing its traditional models and organizational principles beyond their explanatory limits. In the search for ever deeper and broader explanations, science has reached the point at which it can no longer deny the existence of intractable conceptual difficulties devolving to the explanatory inadequacies of its fundamental conceptual models of reality. This has spawned a new discipline known as *reality theory*, the study of the nature of reality in its broadest sense. The overall goal of reality theory is to provide new models and new paradigms in terms of which reality can be understood, and the consistency of science restored as it deepens and expands in scope.

Mainstream reality theory counts among its hotter foci the interpretation of quantum theory and its reconciliation with classical physics, the study of subjective consciousness and its relationship to objective material reality, the reconciliation of science and mathematics, complexity theory, cosmology, and related branches of science, mathematics, philosophy and theology. But in an integrated sense, it is currently in an exploratory mode, being occupied with the search for a general conceptual framework in which to develop a more specific theory and model of reality capable of resolving the paradoxes and conceptual inconsistencies plaguing its various fields of interest (where a *model* is technically defined as a valid interpretation of a theory in its universe of reference). Because of the universal scope of reality theory, it is subject to unique if seldom-recognized demands; for example, since it is by definition a universal theory of everything that is real, it must by definition contain its rules of real-world interpretation. That is, reality theory must contain its own model and effect its own self-interpretative mapping thereto, and it must conform to the implications of this requirement. This "self-modeling" capacity is a primary criterion of the required framework.

The ranks of reality theorists include researchers from almost every scientific discipline. As the physical sciences have become more invested in a quantum mechanical view of reality, and as science in general has become more enamored of and dependent on computer simulation as an experimental tool, the traditional continuum model of classical physics has gradually lost ground to a new class of models to which the concepts of information and computation are essential. Called "discrete models", they depict reality in terms of bits, quanta, quantum events, computational operations and other discrete, recursively-related units. Whereas continuum models are based on the notion of a *continuum*, a unified extensible whole with one or more distance parameters that can be infinitely subdivided in such a way that any two distinct points are separated by an infinite number of intermediate points, discrete models are distinguished by realistic acknowledgement of the fact that it is impossible to describe or define a change or separation in any way that does not involve a sudden finite jump in some parameter.

Unfortunately, the advantages of discrete models, which are receiving increasingly serious consideration from the scientific and philosophical communities, are outweighed by certain basic deficiencies. Not only do they exhibit scaling and nonlocality problems associated with their "display hardware", but they are inadequate by themselves to generate the conceptual infrastructure required to explain the medium, device or array in which they evolve, or their initial states and state-transition programming. Moreover, they remain anchored in materialism, objectivism and Cartesian dualism, each of which has proven obstructive to the development of a comprehensive explanation of reality. Materialism arbitrarily excludes the possibility that reality has a meaningful nonmaterial aspect, objectivism arbitrarily excludes the possibility that reality has a meaningful subjective aspect, and although Cartesian dualism technically excludes neither, it arbitrarily denies that the mental and material, or subjective and objective, sides of reality share common substance.[1]

One might almost get the impression that the only two available choices are the classical model, to which quantum theory has been fastened with approximately the same degree of cogency as antlers on a jackrabbit, and the newer discrete models, which purport to be more in line with quantum theory but fall by the wayside *en route* to the new kind of quantum cosmology they portentously seem to promise. For such claims exhibit an unmistakable irony: classical reality is precisely that on which information and computation are defined! Like classical reality itself, a well-defined entity unable to account for its own genesis, information and computation are well-defined and non-self-generative aspects of reality as it is observationally presented to us at an advanced stage of its existence. So they invite the same questions as does classical reality: how, and by what, were they originally defined and generated? Without an answer to this question, little can be gained by replacing one kind of reality with the other.

Some may have felt, as they watched the history of Big Theories and New Paradigms unfold over the last few years, as though they were being forced to watch the same show, or read the same novel, a thousand times in tedious succession with no more than an occasional minor revision of plot or character. However, there is a third alternative which has thus far remained in the background. It provides exactly what is required in light of any thesis and antithesis: synthesis. This synthesis yields a new class of model(s)[2] preserving the best features of both thesis and antithesis, continuum and quantum, uniting them through general and preferably self-evident principles. This paper presents this new class through a single example, the *Cognitive-Theoretic Model of the Universe* (CTMU).

---

[1] In fact, they are identical up to an isomorphism beyond which the mental side, being the more comprehensive of the two, outstrips the material.

[2] Although the CTMU is by definition unique up to isomorphism with its own syntax – this "self-isomorphism" is one of its definitive structural characteristics – we will duly refrain from summarily ruling out the possibility that in the future, others may present self-determinative "temporal feedback models" of their own. It should be noted, however, that since the CTMU is intrinsically tautological and tautologically identified with its universe on the deepest level of its existence, and since this is the selfsame real universe studied by science, any other valid theory of reality will necessarily equate to the CTMU up to isomorphism; whatever it adds will come by way of specificity, not generality. Although others have already presented "cybernetic" and "information-theoretic" theories of cosmology, these theories rely on the standard theories of information and cybernetics, and thus inherit their explanatory limitations with respect to cosmology.

Determinacy, Indeterminacy and the Third Option

Like the mathematics, science and philosophy whence they issue, classical continuum and modern discrete models of reality generally allow for exactly two modes of determinacy: external causality, and acausality or "randomness". Given an object, event, set or process, it is usually assumed to have come about in one or both of just two ways:

(1) its existence owes to something prior and external to it;

(2) it is uncaused and sprang forth spontaneously and pointlessly in a something-from-nothing, rabbit-out-of-the-hat sort of way, as if by magic.

A similar assumption is made with regard to its behavior: either it is controlled by laws that are invariant with respect to it and therefore existentially external to it (even though they control it through its intrinsic structure and properties), or it is behaving in an utterly aleatory and uncontrolled fashion. This has given rise to a dichotomy: *determinacy versus randomness*, or a total absence of causation versus causation by laws that are ultimately independent of the determined entity.

Determinacy and indeterminacy ... at first glance, there seems to be no middle ground. Events are either causally connected or they are not, and if they are not, then the future would seem to be utterly independent of the past. Either we use causality to connect the dots and draw a coherent picture of time, or we settle for a random scattering of independent dots without spatial or temporal pattern and thus without meaning. At the risk of understatement, the philosophical effects of this assumed dichotomy have been corrosive in the extreme. No universe that exists or evolves strictly as a function of external determinacy, randomness or an alternation of the two can offer much in the way of meaning. Where freedom and volition are irrelevant, so is much of human experience and individuality.

But there is another possibility after all: *self-determinacy*. Self-determinacy is like a circuitous boundary separating the poles of the above dichotomy ... a reflexive and therefore closed boundary, the formation of which involves neither preexisting laws nor external structure. Thus, it is the type of causal attribution suitable for a perfectly self-contained system. Self-determinacy is a deep but subtle concept, owing largely to the fact that unlike either determinacy or randomness, it is a source of *bona fide* meaning. Where a system determines its own composition, properties and evolution independently of external laws or structures, it can determine its *own* meaning, and ensure by its self-configuration that its inhabitants are crucially implicated therein.

DIAGRAM I

1. *Indeterminacy* 2. *External determinacy* 3a. *Self-determinacy* 3b. *Intrinsic self-determinacy*: The effectual aspect of the object or event has simply been moved inside the causal aspect, permitting the internalization of the arrow of determinacy from 3a and making causality endomorphic.

If determinacy corresponds to an arrow of causation pointing to an event from a surrounding medium, then indeterminacy corresponds to no arrow at all (acausality), and self-determinacy to a looping arrow or complex of arrows involving some kind of feedback. But cybernetic feedback, which involves information passed among controllers and regulated entities through a conductive or transmissive medium, is meaningless where such entities do not already exist, and where no sensory or actuative protocol has yet been provided. With respect to the origin of any self-determinative, perfectly self-contained system, the feedback is ontological in nature and therefore more than cybernetic. Accordingly, ontological feedback bears description as "precybernetic" or "metacybernetic". Indeed, because of their particularly close relationship, the theories of information, computation and cybernetics are all in line for a convergent extension ... an extension that can, in a reality-theoretic context, lay much of the groundwork for a convergent extension of all that is covered by their respective formalisms.[3]

Ordinary feedback, describing the evolution of mechanical (and with somewhat less success, biological) systems, is cyclical or recursive. The system and its components repeatedly call on internal structures, routines and actuation mechanisms in order to acquire input, generate corresponding internal information, internally communicate and process this information, and evolve to appropriate states in light of input and programming. However, where the object is to describe the evolution of a system from a state in which there is no information or programming (information-processing syntax) at all, a new kind of feedback is required: "telic feedback".

---

[3] Those who expect the field of cosmology to culminate in a grand reduction of the universe to pure information sometimes forget that this would merely transform a certain question, "what is the universe and where did it come from?", to another no less vexing question, "what is information and where did it come from?"

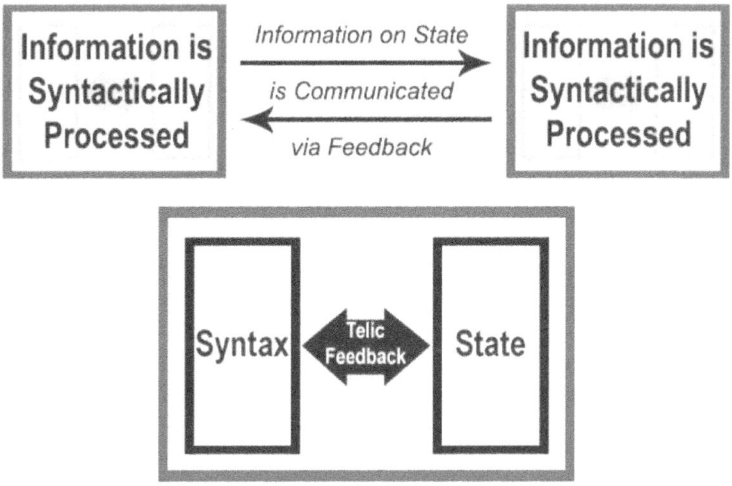

DIAGRAM II

The upper diagram illustrates ordinary cybernetic feedback between two information transducers exchanging and acting on information reflecting their internal states. The structure and behavior of each transducer conforms to a syntax, or set of structural and functional rules which determine how it behaves on a given input. To the extent that each transducer is either deterministic or nondeterministic (within the bounds of syntactic constraint), the system is either deterministic or "random up to determinacy"; there is no provision for self-causation below the systemic level. The lower diagram, which applies to coherent self-designing systems, illustrates a situation in which syntax and state are instead determined in tandem according to a generalized utility function assigning differential but intrinsically-scaled values to various possible syntax-state relationships. A combination of these two scenarios is partially illustrated in the upper diagram by the gray shadows within each transducer.

The currency of telic feedback is a quantifiable self-selection parameter, *generalized utility*, a generalized property of law and state in the maximization of which they undergo mutual refinement (note that generalized utility is self-descriptive or *autologous*, intrinsically and retroactively defined within the system, and "pre-informational" in the sense that it assigns no specific property to any specific object). Through telic feedback, a system retroactively self-configures by reflexively applying a "generalized utility function" to its internal existential potential or possible futures. In effect, the system brings itself into existence as a means of atemporal communication between its past and future whereby law and state, syntax and informational content, generate and refine each other across time to maximize total systemic self-utility. This defines a situation in which the true temporal identity of the system is a distributed point of temporal equilibrium that is both *between* and *inclusive* of past and future. In this sense, the system is timeless or *atemporal*.

A system that evolves by means of telic recursion – and ultimately, every system must either be, or be embedded in, such a system as a condition of existence – is not merely computational, but *protocomputational*. That is, its primary level of processing configures its secondary (computational and informational) level of processing by telic recursion. Telic recursion can be regarded as the self-determinative mechanism of not only cosmogony, but a natural, scientific form of teleology.

However, before taking these ideas any further, let's attend a little orientation session based on the remarkably penetrating vision of John Archibald Wheeler, a preeminent scientist and reality theorist whose name is virtually synonymous with modern physics.

The Future of Reality Theory According to John Wheeler

In 1979, the celebrated physicist John Wheeler, having coined the phrase "black hole", put it to good philosophical use in the title of an exploratory paper, "Beyond the Black Hole" (Wheeler, 1980), in which he describes the universe as a *self-excited circuit*. The paper includes an illustration in which one side of an uppercase (U), ostensibly standing for *Universe*, is endowed with a large and rather intelligent-looking eye intently regarding the other side, which it ostensibly acquires through observation as sensory information. By dint of placement, the eye stands for the sensory or *cognitive* aspect of reality, perhaps even a human spectator within the universe, while the eye's perceptual target represents the *informational* aspect of reality. By virtue of these complementary aspects, it seems that the universe can in some sense, but not necessarily that of common usage, be described as "conscious" and "introspective" ... perhaps even "infocognitive".

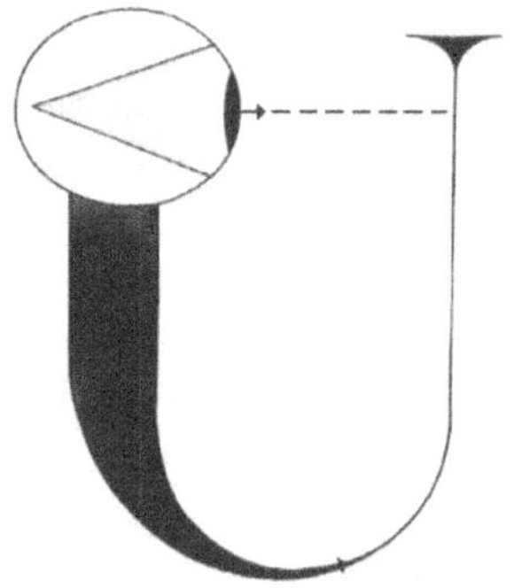

DIAGRAM III

The Universe as a self-excited circuit. Diagram adapted from Wheeler, J. A. (1980, p. 362)

Wheeler, an eminent and highly capable representative of those familiar with the advantages and deficiencies of our current models of reality, did not arrive at the given illustration as an isolated speculation. In conjunction with several other Wheeler concepts, the *Participatory Universe*, *Law without Law* and *It from Bit*, the self-excited circuit amounts to a preliminary but well-considered program for describing the physical universe. According to its mandate, the true description of reality must possess two novel features not found in any dominant paradigm:

(1) *global structural and dynamical reflexivity* or "self-excited circuitry", with perception an integral part of the self-recognition function of reality;

(2) *matter-information equivalence*, an identification (up to isomorphism) of concrete physical reality with information, the abstract currency of perception.

Together, these features constitute a cosmological extension of cybernetics, or equivalently, a metacybernetic extension of cosmology.

Wheeler characterizes these four concepts as follows:

The Self-Excited Circuit A participatory universe is a self-excited circuit in the sense that it implicates observers in (perceptual, ontological) feedback. It is a "logic loop" in which "physics gives rise to observer-participancy; observer-participancy gives rise to information; and information gives rise to physics" (Wheeler, 1990a, p. 8).

The Participatory Universe The cognitive and perceptual processes of observers are integral to the self-excitative feedback of reality. This is asserted by the *Participatory Principle* (or *Participatory Anthropic Principle*), which Wheeler (1979) informally describes as follows:

> Stronger than the Anthropic Principle is what I might call the *Participatory Principle*. According to it, we could not even imagine a universe that did not somewhere and for some stretch of time contain observers, because the very building materials of the universe are these acts of observer-participancy. ... This participatory principle takes for its foundation the absolutely central point of the quantum: no elementary phenomenon is a phenomenon until it is an *observed* (or registered) phenomenon [emphasis added].

Note that on some level of generality, the last sentence identifies observation with registration and thus implicitly equates human and mechanical recognition: "... an observed (or *registered*) phenomenon" [emphasis again added].

Law Without Law / Order from Disorder Concisely, nothing can be taken as given when it comes to cosmogony. In Professor Wheeler's (1979) own words:

> To me, the greatest discovery yet to come will be to find how this universe, coming into being from a Big Bang, developed its laws of operation. I call

> this "Law without Law" [or "Order from Disorder"]. ... imagine the universe with all its regularities and its laws coming into being out of something utterly helter-skelter, higgledy-piggledy and random ... If you were the Lord constructing the universe, how would you have gone about it? It's inspiring to read the life of Charles Darwin and think how the division of plant and animal kingdoms, all this myriad of order, came about through the miracles of evolution, natural selection and chance mutation. To me this is a marvelous indication that you can get order by starting with disorder.

It From Bit Reality educes and/or produces itself in the form of information residing in quantum events. As Wheeler summarizes in his paper "Information, Physics, Quantum: The Search for Links" (1990a, p. 3-4): "... every physical quantity, every it, derives its ultimate significance from bits, binary yes-or-no indications..." He then goes on to discuss this concept at length, offering three questions, four "no's" and five "clues" about the quantum-informational character of reality. The questions are as follows: (1) *How come existence?* (2) *How come the quantum?* (3) *How come the "one world" out of many observer-participants?* The no's, seductive pitfalls to be avoided in answering the three questions, include *No tower of turtles, No laws, No continuum,* and *No space, no time.* And the clues, which light the way toward the true answers, include *The boundary of a boundary is zero*; *No question? No answer!*; *The Super-Copernican Principle*; *"Consciousness"* (including the quotes); and *More is different.*

We will now give a brief account of these questions, precautions and clues.[4]

How come existence? The ontological and cosmological thrust of this question is obvious; in some form, it has bedeviled philosophers from time immemorial. As interpreted by Wheeler, it leads to four inevitable conclusions:

> (1) The world cannot be a giant machine, ruled by any pre-established continuum physical law. (2) There is no such thing at the microscopic level as space or time or spacetime continuum. (3) The familiar probability function or functional, and wave equation or functional wave equation, of standard quantum theory provide mere continuum idealizations and by reason of this circumstance conceal the information-theoretic source from which they derive. (4) No element in the description of physics shows itself as closer to primordial than the elementary quantum phenomenon, that is, the elementary device-intermediated act of posing a yes-no physical question and eliciting an answer or, in brief, the elementary act of observer-participancy. Otherwise stated, every physical quantity, every it, derives its ultimate significance from bits, binary yes-or-no indications, a conclusion which we epitomize in the phrase, *it from bit.*

How come the quantum? Why is the universe made up of apparently *propter hoc* nondeterministic, but *post hoc* informational, quantum events? As Wheeler observes: "Quantum physics requires a new view of reality." What, then,

[4] Unless stated otherwise, the Wheeler quotes are from the 1990a article mentioned above.

is the exact logical relationship between the quantum and the new view of reality it demands? What is this new view, and how does the quantum fit into it?

How come the "one world" out of many observer-participants? Insofar as the term "observer-participants" embraces scientists and other human beings, this question invites a quasi-anthropological interpretation. Why should a universe consisting of separate observers with sometimes conflicting agendas and survival imperatives display structural and nomological unity? Where observers are capable of creating events within the global unitary manifold of their common universe, why should they not be doing it strictly for themselves, each in his or her *own* universe, and never the twain shall meet? Where the observer-participant concept is generalized to include non-anthropic information-transducing systems, what is holding all of these systems together in a single unified reality?

No tower of turtles Borrowed from William James, this aphorism means "no infinite regress to ever-prior causal domains and principles". To this we might equate an updated version of a well-known aphorism credited to Harry Truman: "The explanatory buck stops here", where *here* refers to this reality that we actually inhabit and observe. To this Wheeler adds a crucial insight: "To endlessness no alternative is evident but loop, such a loop as this: Physics gives rise to observer-participancy; observer-participancy gives rise to information; and information gives rise to physics."

No laws Wheeler states that the universe must have come into being without "a pre-existing plan ... Only a principle of organization which is no organization at all would seem to offer itself ". Or to reiterate: "The world cannot be a giant machine, ruled by any pre-established continuum physical law."

No continuum The venerable continuum of analysis and mechanics is a mathematical and physical chimera. (Usually associated with the set of real numbers, a *continuum* is a unified extensible whole with a distance parameter that can be infinitely subdivided in such a way that any two distinct points are separated by an infinite number of intermediate points.) As Wheeler puts it: "A half-century of development in the sphere of mathematical logic has made it clear that there is no evidence supporting the belief in the existential character of the number continuum." Some numbers, e.g. irrational ones like ($\sqrt{2}$), cannot be precisely computed and therefore do not correspond to any physically meaningful location on a number line or physical trajectory; they have an abstract existence only.

No space, no time Again, "there is no such thing at the microscopic level as space or time or spacetime continuum." On the submicroscopic level, the Heisenberg Uncertainty Principle turns spacetime into seemingly chaotic "quantum foam", casting doubt on the connectivity of space and the ordinality of time. Wheeler quotes Einstein in a Kantian vein: 'Time and space are modes by which we think, and not conditions in which we live', regarding these modes as derivable from a proper theory of reality as idealized functions of an idealized continuum: "We will not feed time into any deep-reaching account of existence. We must derive time – and time only in the continuum idealization – out of it. Likewise with space."

The boundary of a boundary is zero In essence, this intuitive notion from algebraic topology says that closed structures embody a certain kind of "self-cancellative" symmetry. This can be illustrated in three dimensions by a tetrahedron, the simplicial "boundary" of which incorporates its four equilateral triangular faces. To find the boundary of this boundary, one would measure the clockwise- or counterclockwise-oriented edges around each face, thus measuring each edge of the tetrahedron twice in opposite directions. Because summing the measurements now cancels to 0 at each edge, the boundary of the boundary of the tetrahedron is zero. This property turns out to have extensive applications in physics, particularly the theory of fields, as regards the mutual "grip" of matter on space and space on matter (or less colorfully, the relationship of space and matter). In Wheeler's view, its ubiquity "inspires hope that we will someday complete the mathematization of physics and derive everything from nothing, all law from no law." Thus, it is closely related to law without law and so-called *ex nihilo* creation.

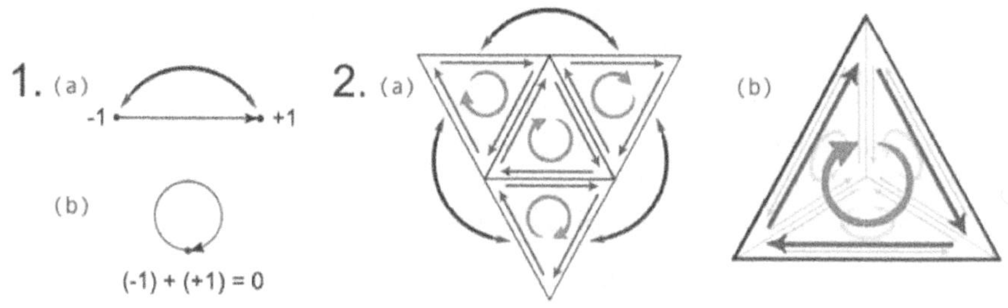

DIAGRAM IV

1a. The boundary of a directed 1-dimensional line segment consists of its 0-dimensional endpoints, which separate the line from its complement (the space surrounding the line). The initial point represents the "debt" required to start the line and is thus given a value of -1, while the terminal point represents the "payoff" for completing the line and is given a value of +1. When the initial and terminal points of the line are identified as indicated by the curved arrow, the result is a closed line bounding a planar disk (**1b**). Because the endpoints now coincide, they sum to 0 and no longer separate the line from its complement; thus, the 0-dimensional boundary of the 1-dimensional boundary of the 2-dimensional disk is 0.
2a. A triangular area can be decomposed into 4 smaller triangular areas. Introducing a uniform (in this case, clockwise) orientation to the areas (inner circular arrows) imparts the same orientation to the outer perimeter (outer straight arrows), recreating the situation of 1b (notice that the straight arrows next to each interior edge point in opposite directions and therefore cancel). Again, the initial and terminal points of the perimeter coincide and cancel to 0 no matter where they lie. When adjacent perimeter segments are identified as indicated by the outer curved arrows, the triangle folds into a

tetrahedron (**2b**). Its faces form a closed 2-dimensional boundary separating its 3-dimensional interior from its exterior, while its edges form a closed 1-dimensional boundary separating its faces from each other. But now the straight arrows cancel out at *every* edge, and the 1-dimensional boundary of the 2-dimensional boundary of the tetrahedron is 0. So for both the 2D disk and the 3D tetrahedron, the boundary of the boundary is 0. While physicists often use this rule to explain the conservation of energy-momentum – or as Wheeler (1990b) calls it, "momenergy" – it can be more generally interpreted with respect to *information* and *constraint*, or *state* and *syntax*. That is, the boundary is analogous to a constraint which separates an interior attribute satisfying the constraint from a complementary exterior attribute, thus creating an informational distinction.

No question? No answer! In a quantum experiment, the measuring device and its placement correspond to a question, and the result to its answer. The existence of the answer, consisting of information on state, is predicated on the asking of the question (or the occurrence of the measurement), and the kind of answer received depends on the kind of question asked and the manner in which it is posed. The world is thus composed of measurement events in which information is exchanged by objects, one or both of which "ask a question" and one or both of which "give an answer". Question and answer, the stimulation and observation of an event, cannot be separated on the ontological level, and they cannot be shielded from the influence of the environment. At the root of this criterion are quantum uncertainty and complementarity, the foundation-stones of quantum mechanics.

The Super-Copernican Principle Just as Copernicus displaced geocentricity with heliocentricity, showing by extension that no particular place in the universe is special and thereby repudiating "here-centeredness", the Super-Copernican Principle says that no particular point in time is special, repudiating "now-centeredness". Essentially, this means that where observer-participation functions retroactively, the participatory burden is effectively distributed throughout time. So although the "bit-size" of the universe is too great to have been completely generated by the observer-participants who have thus far existed, future generations of observer-participants, possibly representing modes of observer-participation other than that associated with human observation, have been and are now weighing in from the future. (The relevance of this principle to the Participatory Anthropic Principle is self-evident.)

"Consciousness" Wheeler emphasizes the difficulty of making a general distinction between the form of information processing characteristic of humans, and that characteristic of various complex systems and devices that may or may not be "conscious". "The line between the unconscious and the conscious begins to fade ...," he states. "We may someday have to enlarge the scope of what we mean by a 'who'." The term *who*, he suggests, is too specific to man, life and consciousness; its anthropic connotations are anti-Copernican, while the concepts of life and consciousness are subject to revision as science advances. "It would seem more reasonable," he suggests, "to dismiss for the present the semantic overtones of 'who' and explore and exploit the insights to be won from the phrases, 'communication' and 'communication employed to establish meaning'."

More is different The potential for complexity increases with cardinality; with large numbers of elements comes combinatorial variety and the potential for the sort of multilevel logical structure that typifies biological organisms and modern computers alike. This is a fundamental precept of complexity theory. Wheeler poses a question: "Will we someday understand time and space and all the other features that distinguish physics – and existence itself – as the ... self-generated organs of a self-synthesized information system?"

Together, these pithy slogans, questions, precautions and clues add up to a call for a new strain of reality theory, a unified conceptual model for our thoughts and observations. How many of the models currently being held forth respond to this call? The answer, of course, is "almost none". While some of them seem to address one or two of the questions and meet one or two of the criteria, none comes close to addressing and meeting all of them. What each model has been forced to give in order to meet any small subset of criteria has cost it dearly in terms of meeting the others. Thus, we have thesis and antithesis in the form of classical physics and discrete quantum models, but because the full depth of the relationship between the two is unfathomed, no synthesis. Virtually everybody seems to acknowledge the correctness of Wheeler's insights, but the higher-order relationships required to put it all together in one big picture have proven elusive. The logical difficulty of answering all of the questions and meeting all of the criteria at once, in parallel, using integrated, logically tractable concepts, has simply been prohibitive.

Can this situation be redressed?

Some Additional Principles

Although insights regarding the ideal and/or perceptual basis of reality go back millennia, we may as well start with some their more recent proponents for the sake of continuity. First, Descartes posited that reality is mental in the sense of rationalism, but contradicted his own thesis by introducing *mind-body dualism*, the notion that mind and matter are irreducibly separate. The empiricist Berkeley then said that reality is perceptual in nature, a kind of intersect of mind and matter. This can be seen by mentally subtracting perception from one's conception of reality; what remains is pure subjective cognition, but without any objective grist for the perceptual mill. (Although attempts to cognitively subtract cognition from reality are far more common, they are a bit like trying to show that a sponge is not inherently wet while immersing it in water, and can never be successful on the parts of cognitive entities.) Hume then attempted to do away with cognition and causation entirely, asserting that both mind and matter inhere in perception and exist apart from neither it nor each other.

In disposing of mind, Hume made another salient "contribution" to reality theory: he attempted to dispose of causation by identifying it as a cognitive artifact, supporting his thesis with the *problem of induction*.[5] The problem of induction states that because empirical induction entails the prior assumption of that which it seeks to establish, namely the uniformity of nature, science is circular and fundamentally flawed. The problem of induction is very real; it is manifest in Heisenberg uncertainty and the cosmic horizon problem, finite limitations of scientific tools of microscopic and macroscopic observation, and is why no general theory of reality can ever be reliably constructed by the standard empirical methods of science. Unfortunately, many scientists have either dismissed this problem or quietly given up on the search for a truly general theory, in neither case serving the long-term interests of science. In fact, the problem of induction merely implies that a global theory of reality can only be established by the rational methods of mathematics, specifically including those of logic.

In response to Berkeley and Hume, Kant asserted that the unprimed cognition which remains when perceptual content is subtracted has intrinsic structure that exists *prior* to content; it comprises the *a priori* categories of perceptual or "phenomenal" reality (Russell, 1979). Unfortunately, subtracting perception according to Kantian rules yields more than unprimed cognition; it also yields noumena, absolute objects or "things-in-themselves". On one side of the result is a perceptual isomorphism between the mind and phenomenal reality; on the other yawns a chasm on the far side of which sits an unknowable but nonetheless fundamental noumenal reality, which Kant evidently regarded as the last word in (sub-theological) reality theory.

However, Kant's chasm is so deep and wide, and so thoroughly interdicts any mind-reality isomorphism, that it precludes causal efficacy and for that matter any other comprehensible principle of correspondence. This implies that noumena are both rationally and empirically irrelevant to cognitive and perceptual reality, and thus that they can be safely eliminated from reality theory. Whatever Kant had in mind when he introduced the concept of a *noumenon*, his definition essentially amounts to "inconceivable concept" and is thus an oxymoron. Whatever he really meant, we must rely on something other than Kantian metaphysics to find it (Langan, 2001b).

Thus far, we have managed to narrow reality down to the phenomenal reality studied by science, a combination of perceptual content and rational principles

[5] We will ignore for now the fact that if the mind is illusory, then so is all philosophy including Hume's, and if causation is illusory, then so is the form of causation represented by logical implication, including Hume's use of logical implication to define his philosophy.

of cognition. A scientist employs empirical methods to make specific observations, applies general cognitive relationships from logic and mathematics in order to explain them, and comes off treating reality as a blend of perception and cognition. But this treatment lacks anything resembling an explicit justification. When a set of observations is *explained* with a likely set of equations *interpreted* therein, the adhesion between explanandum and explanation might as well be provided by rubber cement. I.e., scientific explanations and interpretations glue observations and equations together in a very poorly understood way. It often works like a charm ... but why? One of the main purposes of reality theory is to answer this question.

The first thing to notice about this question is that it involves the process of *attribution*, and that the rules of attribution are set forth in stages by mathematical logic. The first stage is called *sentential logic* and contains the rules for ascribing the attributes *true* or *false*, respectively denoting inclusion or non-inclusion in arbitrary cognitive-perceptual systems, to hypothetical relationships in which predicates are linked by the logical functors *not*, *and*, *or*, *implies*, and *if and only if*. Sentential logic defines these functors as *truth functions* assigning *truth values* to such expressions irrespective of the contents (but not the truth values) of their predicates, thus effecting a circular definition of functors on truth values and truth values on functors. The next stage of attribution, *predicate logic*, ascribes specific properties to objects using quantifiers. And the final stage, *model theory*, comprises the rules for attributing complex relations of predicates to complex relations of objects, i.e. theories to universes. In addition, the form of attribution called *definition* is explicated in a theory-centric branch of logic called *formalized theories*, and the mechanics of functional attribution is treated in recursion theory.

In sentential logic, a *tautology* is an expression of functor-related sentential variables that is always true, regardless of the truth values assigned to its sentential variables themselves. A tautology has three key properties: it is universally (syntactically) true, it is thus self-referential (true even of itself and therefore closed under recursive self-composition), and its implications remain consistent under inferential operations preserving these properties. That is, every tautology is a self-consistent circularity of universal scope, possessing validity by virtue of closure under self-composition, comprehensiveness (non-exclusion of truth), and consistency (freedom from irresolvable paradox). But tautologies are not merely consistent unto themselves; they are mutually consistent under mutual composition, making sentential logic as much a "self-consistent circularity of universal scope" as any one of its tautologies. Thus, sentential logic embodies two levels of tautology, one applying to expressions and one applying to theoretical systems thereof. Predicate logic then extends the tautology concept to cover the specific acts of attribution represented by (formerly anonymous) sentential variables, and model theory goes on to encompass more complex acts of attribution involving more complex relationships.

Reality theory is about the stage of attribution in which two predicates analogous to true and false, namely *real* and *unreal*, are ascribed to various statements about the real universe. In this sense, it is closely related to sentential logic. In particular, sentential logic has four main properties to be emulated by reality theory. The first is *absolute truth*; as the formal definition of truth, it is true by definition. The other properties are *closure*, *comprehensiveness* and *consistency*. I.e., logic is wholly based on, and defined strictly within the bounds of, cognition and perception; it applies to everything that can be coherently perceived or conceived; and it is by its very nature consistent, being designed in a way that precludes inconsistency. It is the basis of mathematics, being the means by which propositions are stated, proved or disproved, and it is the core of science, underwriting the integrity of rational and empirical methodology. Even so-called "nonstandard" logics, e.g. modal, fuzzy and many-valued logics, must be expressed in terms of fundamental two-valued logic to make sense. In short, two-valued logic is something without which reality could not exist. If it were eliminated, then *true* and *false*, *real* and *unreal*, and *existence* and *nonexistence* could not be distinguished, and the merest act of perception or cognition would be utterly impossible.

Thus far, it has been widely assumed that reality theory can be sought by the same means as any other scientific theory. But this is not quite true, for while science uses the epistemological equivalent of magic glue to attach its theories to its observations, reality theory must give a recipe for the glue and justify the means of application. That is, reality theory must describe reality on a level that justifies science, and thus occupies a deeper level of explanation than science itself. Does this mean that reality theory is mathematical? Yes, but since mathematics must be justified along with science, *metamathematical* would perhaps be a better description ... and when all is said and done, this comes down to logic pure and simple. It follows that reality theory must take the form of an extended logic ... in fact, a "limiting form" of logic in which the relationship between theory and universe, until now an inexhaustible source of destructive model-theoretic ambiguity, is at last reduced to (dual-aspect) monic form, short-circuiting the paradox of Cartesian dualism and eliminating the epistemological gap between mind and matter, theory and universe.

As complexity rises and predicates become theories, tautology and truth become harder to recognize. Because universality and specificity are at odds in practice if not in principle, they are subject to a kind of "logical decoherence" associated with relational stratification. Because predicates are not always tautological, they are subject to various kinds of ambiguity; as they become increasingly specific and complex, it becomes harder to locally monitor the heritability of consistency and locally keep track of the truth property in the course of attribution (or even after the fact). Undecidability (Gödel, 1962), LSAT intractability and NP-completeness, predicate ambiguity and the Löwenheim–Skolem theorem, observational ambiguity and the Duhem–

Quine[6] are some of the problems that emerge once the truth predicate "decoheres" with respect to complex attributive mappings. It is for reasons like these that the philosophy of science has fallen back on falsificationist doctrine, giving up on the tautological basis of logic, effectively demoting truth to provisional status, and discouraging full appreciation of the tautological-syntactic level of scientific inquiry even in logic and philosophy themselves.

In fact, the validity of scientific theories and of science as a whole absolutely depends on the existence of a fundamental reality-theoretic framework spanning all of science ... a fundamental syntax from which all scientific and mathematical languages, and the extended cognitive language of perception itself, can be grammatically unfolded, cross-related and validated. Tautology, the theoretical basis of truth as embodied in sentential logic, is obviously the core of this syntax. Accordingly, reality theory must be developed through amplification of this tautological syntax by adjunction of additional syntactic components, the principles of reality theory, which leave the overall character of the syntax invariant. Specifically, in order to fashion a reality theory that has the truth property in the same sense as does logic, but permits the logical evaluation of statements about space and time and law, we must adjoin principles of extension that lend meaning to such statements while preserving the tautology property.

According to the nature of sentential logic, truth is tautologically based on the integrity of cognitive and perceptual reality. Cognition and perception comprise the primitive (self-definitive) basis of logic, and logic comprises the rules of structure and inference under which perception and cognition are stable and coherent. So when we say that truth is heritable under logical rules of inference, we really mean that *tautology* is heritable, and that the primitive cognitive-perceptual basis of sentential logic thus maintains its primary status. By converting tautologies into other tautologies, the rules of inference of sentential logic convert cognitive-perceptual invariants into other such invariants. To pursue this agenda in reality theory, we must identify principles that describe how the looping structure of logical tautology is manifest in various reality-theoretic settings and contexts on various levels of description and interpretation; that way, we can verify its preservation under the operations of theoretic reduction and extension. I.e., we must adjoin generalized principles of loop structure to logical syntax in such a way that more and more of reality is thereby explained and comprehensiveness is achieved.

For example, take the sentential tautology X ∨ ~X (X or not-X). Applied to perception, this means that when something is seen or observed, it is not seen in conjunction with its absence; if it were, then two contradictory perceptions would coincide, resulting in a "splitting off" of perceptual realities. In effect, either the consciousness of the perceiver would split into two separate

[6] For more on the Löwenheim–Skolem theorem and Duhem–Quine thesis see Langan (2002b).

cognitive realities in a case of chain-reactive dissociation, or the perceiver himself would physically split along with physical reality. When $X \vee \sim X$ is composed with other tautologies (or itself) by substitution, the stakes are exactly the same; any violation of the compound tautology would split perceptual and cognitive reality with disastrous implications for its integrity.[7]

After its tautological nature, the first thing to note about sentential logic in the context of reality theory is that against the spirit in which it was founded – it does, after all, represent the rules of the *mental* processes[8] of cognition and perception, which would seem to endow it with a mental character from the start – it has a basic functional inadequacy: it seems to require an external logician to mentally read, understand and apply it. On the other hand, nature (or cognitive-perceptual reality) requires no external logician to apply the rules of logic. Therefore, the proposed tautology-preserving principles of reality theory should put mind back into the mix in an explicit, theoretically tractable way, effectively endowing logic with "self-processing capability". This, after all, is exactly what it possesses in its natural manifestation, reality at large, and is an essential dimension of the closure property without which truth is insupportable. That is, reality must be able to recognize itself and impart this ability to its components as a condition of their existence and interaction.

Having explained the main technical issues in reality theory, we may now cut to the chase: the way to build a theory of reality is to identify the properties that it must unconditionally possess in order to exist, and then bring the theory into existence by defining it to possess these properties without introducing merely contingent properties that, if taken as general, could impair its descriptive relationship with the real universe (those can come later and will naturally be subject to empirical confirmation). In other words, the means by which the theory is constructed must be rational and tautological, while those by which it is subsequently refined may be empirical. Since we want our theory to be inclusive enough, exclusive enough and consistent enough to do the job of describing reality, these properties will certainly include *comprehensiveness*

---

[7] The Many Worlds interpretation of quantum mechanics, in claiming to circumvent the problem of quantum mechanical wave function collapse, in effect claims to circumvent the problem of split consciousness as well. However, since MW is itself necessarily formulated in terms of two-valued logic, the problem regresses to that of how the "hyperuniverse" associated with MW itself "collapsed" out of the sea of all potential "meta-realities". On this score, MW has little insight to offer on its own behalf. MW thus lacks justificative self-containment and is therefore no basis for a fundamental theory of reality. In any case, since the theory we are describing in this paper is designed to address reality in its most basic form, and thus to address all conceivable realities as its possible instantiations, it supersedes any explanation afforded by MW alone.

[8] Several millennia ago, Aristotle proposed (categorical) logic as the formal rules of correct reasoning. Now as then, "reasoning" is inferential, mental and subjective in nature. After standing largely changeless until the nineteenth century, the field of logic began to undergo additional progress. But although this progress has been considerable, it has left the inferential, mental and subjective nature of logic essentially intact.

(less thorough but also less undecidable than completeness), *closure*, and *consistency*. To these properties, the "3 C's", we shall assign three principles that are basically tautological in form; that way, adjoining them to logic-based reality theory will preserve the tautology property of logic, rationally precluding uncertainty by the same means as logic itself. A theory of reality constructed in this way is called a *supertautology*.

Because our three principles correspond to the 3 C's, and because they all begin with the letter M, we might as well call them the "3 M's": M=R, MAP and MU, respectively standing for the *Mind Equals Reality Principle*, the *Metaphysical Autology Principle*, and the *Multiplex Unity Principle*. The M=R principle, a tautological theoretical property that dissolves the distinction between *theory* and *universe* and thus identifies the real universe as a "self-reifying theory", makes the syntax of this theory comprehensive by ensuring that nothing which can be cognitively or perceptually recognized as a part of reality is excluded for want of syntax. MAP tautologically renders this syntax closed or self-contained in the definitive, descriptive and interpretational senses, and in conjunction with M=R, renders the universe perfectly self-contained in the bargain. And MU tautologically renders this syntax, and the theory-universe complex it describes, coherent enough to ensure its own consistency (thus, the "C" corresponding to MU actually splits into two C's, *consistency* and *coherence*, and we have four altogether). To each of these principles we may add any worthwhile corollaries that present themselves.[9]

Since it is the lot of every reality theorist to use properties of reality to explain reality, and these properties are recursively defined, we will sometimes implicitly or explicitly refer to various properties in the descriptions of other properties. This precludes a neat series of cumulative definitions, which is possible in any case only by taking for granted the content and wherewithal of theorization (unfortunately, one can take nothing for granted in reality theory). As we will see below, the recursive nature of the CTMU is unavoidable. Secondly, the CTMU is developed "backwards" with respect to the usual deductive theories of science and mathematics, by first peeling away constraints and only then using the results to deduce facts about content. Most theories begin with axioms, hypotheses and rules of inference, extract implications, logically or empirically test these implications, and then add or revise axioms, theorems or hypotheses. The CTMU does the opposite, stripping away assumptions and "rebuilding reality" while adding no assumptions back.

The following principles are presented in three stages. The first stage includes the *Reality Principle*, the *Principle of Linguistic Reducibility* and the *Principle of Syndiffeonesis*, which may be considered preliminary to MAP, M=R and MU

[9] Note that while this seems to imply that the 3M's are "axioms" and therefore independent, the premise of axiomatic independence is itself a rather flimsy concept. These principles are actually rather strongly related in the sense that they can to some extent be inferred from each other in a reality-theoretic context.

respectively (the order of presentation may differ slightly from that just given). The second stage consists of MAP, M=R and MU themselves, while the third stage consists of several auxiliary principles that can be viewed as their consequences.

The Reality Principle

Reality, i.e. the real universe, contains all and only that which is real. The reality concept is analytically self-contained; if there were something *outside* reality that were *real* enough to affect or influence reality, it would be *inside* reality, and this contradiction invalidates any supposition of an external reality (up to observational or theoretical relevance).[10]

While this characterization of reality incorporates a circular definition of relevance, the circularity is essential to the reality concept and does not preclude a perceptual (observational, scientific) basis. Indeed, we can refine the definition of reality as follows: "Reality is the perceptual aggregate including (1) all scientific observations that ever were and ever will be, and (2) the entire abstract and/or cognitive explanatory infrastructure of perception" (where the *abstract* is a syntactic generalization of the *concrete* standing for ideas, concepts or cognitive structures distributing over physical instances which conform to them as content conforms to syntax).

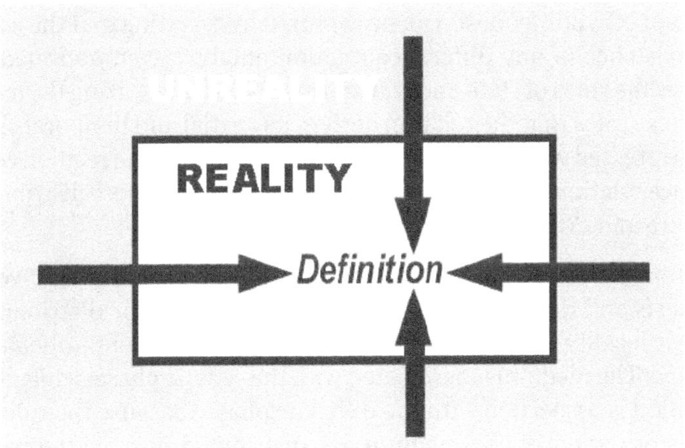

DIAGRAM V

Unreal Definition of Reality

[10] In physics, the phrase "Reality Principle" has sometimes been associated with the idea that objective reality exists independently of human observation, at least to the extent that it does not suddenly disappear when not being directly watched. As used in this paper, the phrase "Reality Principle" is to be interpreted as described in this section unless otherwise indicated.

It should be noted that any definition amounts to a microscopic theory of the thing defined. The Reality Principle, which can be viewed as a general definition of reality, is a case in point; it can be viewed as the seed of a reality theory that we have now begun to build. In defining reality as self-contained, this "microtheory" endows itself with a simple kind of closure; it calls on nothing outside the definiendum in the course of defining it, and effectively forbids any future theoretical extension of this definition from doing so either (this becomes explicit in a related principle, the *MAP*).

But now back to the queue. Thus far, we have learned that reality is self-contained; it is everywhere the same as itself. What about all of its internal distinctions?

Syndiffeonesis

Reality is a relation, and every relation is a *syndiffeonic* relation exhibiting *syndiffeonesis* or "difference-in-sameness". Therefore, reality is a syndiffeonic relation. Syndiffeonesis implies that any assertion to the effect that two things are different implies that they are reductively the same; if their difference is real, then they both reduce to a common reality and are to that extent similar. Syndiffeonesis, the most general of all reductive principles, forms the basis of a new view of the relational structure of reality.

The concept of syndiffeonesis can be captured by asserting that the expression and/or existence of any difference relation entails a common medium and *syntax*, i.e. the rules of state and transformation characterizing the medium. It is from these rules that the relation derives its spatial and temporal characteristics as expressed within the medium. Thus, a syndiffeonic relation consists of a difference relation embedded in a relational medium whose distributed rules of structure and evolution support its existence.

Every syndiffeonic relation has *synetic* and *diffeonic* phases respectively exhibiting *synesis* and *diffeonesis* (sameness and difference, or distributivity and parametric locality), and displays two forms of containment, *topological* and *descriptive*. The medium is associated with the synetic phase, while the difference relation is associated with the diffeonic phase (because the rules of state and transformation of the medium are distributed over it, the medium is homogeneous, intrinsically possessing only relative extension by virtue of the difference relationships it contains). Because diffeonic relands are related to their common expressive medium and its distributive syntax in a way that combines aspects of union and intersection, the operation producing the medium from the relands is called *unisection* (⋒). The synetic medium represents diffeonic *potential* of which the difference relationship is an *actualization*.

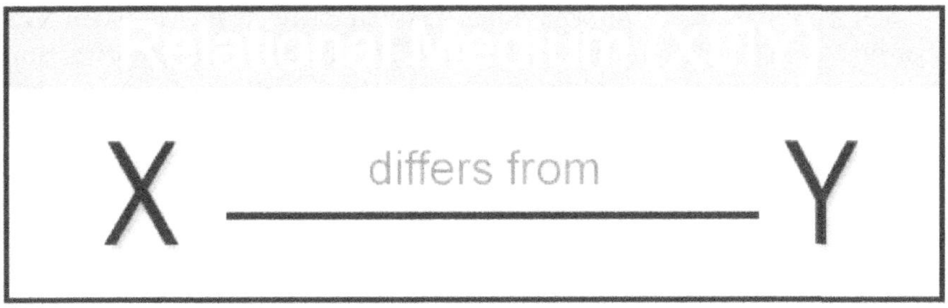

DIAGRAM VI

This generic **syndiffeonic diagram** illustrates a simple fact: any difference relation requires a supporting medium with extension in the differential parameter. As illustrated, the medium distributes over both the linear relation "X differs from Y" and its *relands* (related entities) X and Y, bestowing on them a common "relatedness" property equating to "inclusion in the relational medium X ⋒ Y", where X ⋒ Y is the unisect or "syntactic product" of X and Y. This common attribute invalidates any assertion to the effect that the difference between the relands is "absolute" or "irreducible"; the mere fact that the difference can be linguistically or geometrically expressed implies that it is only partial and that both relands are manifestations of one and the same ontological medium. Where X and Y represent arbitrary parts or aspects of the difference relation called *reality*, this diagram graphically demonstrates that reality ultimately consists of a unitary ontological medium. Accordingly, reality theory must be a monic theory reducing reality to this medium (this idea is further developed in the *Principle of Infocognitive Monism*).

Note that any syntactic (as opposed to informational) inhomogeneity in the common medium is itself a difference relationship and thus invites a recreation of the diagram. Similarly, any inhomogeneity in the common medium illustrated by the recreated diagram would invite yet another recreation of the diagram, and so on. Any such syndiffeonic regress must terminate, for if it did not, there would be no stable syntax and therefore no "relation" stable enough to be perceived or conceived. The informational stability of perceptual reality shows that reality has a stable syntax.

The above diagram might be compactly expressed as follows: syn(X ⋒ Y) : diff(X,Y). For example, syn(nom_AX ⋒ nom_BX) : diff(nom_AX, nom_BX) means that where nom_AX, nom_BX are sets of laws obeyed by the system X at different times, locations or frames of reference A and B within the system X, there exists a more basic set of laws (nom_AX ⋒ nom_BX) in terms of which this difference may be expressed. This shows that on some level, general covariance must hold. This is not merely true "up to isomorphism with X"; even if more than one valid set of laws can be distinguished, any one of which might be active at any given location (A, B, ...) within X [$X_A \models nom_1$, $X_B \models nom_2$, ..., where numerical indices denote nomological distinctness], any distinguishable difference between these sets also requires a common syntax. Informational coherence is thus a *sine qua non* of recognizable existence; any system in which it were to fail would simply decohere for lack of anything to hold it together.

In other words, (1) where informational distinctions regarding a system X are regarded as instantiations of law, they can also be regarded as expressions conforming to syntax; and (2) the expression of differences requires a unified expressive syntax (or set of "laws"), and this syntax must distribute over the entire set of differential expressions (or "instantiations of law"). E.g., where X is a "perceptual intersect" consisting of generally recognizable objects, attributes and events, the laws of perception must ultimately be constant and distributed. Where a putative nomological difference exists for some pair of loci (A, B), reductive syntactic covariance applies due to the need for an expressive medium, and where no such difference exists for *any* pair of loci (A, B), syntactic covariance applies *a fortiori* with no need for reduction.

Syndiffeonic relations can be regarded as elements of more complex *infocognitive lattices* with spatial and temporal (ordinal, stratificative) dimensions. Interpreted according to CTMU duality principles, infocognitive lattices comprise logical relationships of *state* and *syntax*. Regressing up one of these lattices by unisection ultimately leads to a syntactic medium of perfect generality and homogeneity ... a universal, reflexive "syntactic operator".

In effect, syndiffeonesis is a metalogical tautology amounting to *self-resolving paradox*. The paradox resides in the coincidence of sameness and difference, while a type-theoretic resolution inheres in the logical and mathematical distinction between them, i.e. the stratificative dimension of an infocognitive lattice.[11] Thus, reducing reality to syndiffeonesis amounts to "paradoxiforming" it. This has an advantage: a theory and/or reality built of self-resolving paradox is immunized to paradox.

So far, we know that reality is a self-contained syndiffeonic relation. We also have access to an instructive sort of diagram that we can use to illustrate some of the principles which follow. So let us see if we can learn more about the *kind* of self-contained syndiffeonic relation that reality is.

The Principle of Linguistic Reducibility

Reality is a self-contained form of language. This is true for at least two reasons. First, although it is in some respects material and concrete, reality conforms to the algebraic definition of a language. That is, it incorporates:

(1) representations of (object-like) individuals, (space-like) relations and attributes, and (time-like) functions and operations;

(2) a set of "expressions" or perceptual states;

---

[11] The type-theoretic resolution of this paradox is incomplete; full resolution requires MU, a kind of "meta-syndiffeonesis" endowing the infocognitive lattice of spacetime with a higher level of closure.

(3) a syntax consisting of (a) logical and geometric rules of structure, and (b) an inductive-deductive generative grammar identifiable with the laws of state transition.

Second, because perception and cognition are languages, and reality is cognitive and perceptual in nature, reality is a language as well.

While there have been many reductionist programs in science and philosophy, the promised reduction is always to the same thing: a theoretical language. Because this is necessarily true, language is fundamental. The fact that most such theories, e.g. theories of physics, point to the fundamental status of something "objective" and "independent of language", e.g. matter and/or energy, is quite irrelevant, for the very act of pointing invokes an isomorphism between theory and objective reality ... an isomorphism that is subject to the Reality Principle, and which could not exist unless reality shared the linguistic structure of the theory itself.

Perhaps the meaning of this principle can be most concisely expressed through a generalization of the aphorism "whereof one cannot speak, one must be silent": *whereof that which cannot be linguistically described, one cannot perceive or conceive.* So for the observational and theoretical purposes of science and reality theory, that which is nonisomorphic to language is beyond consideration as a component of reality.

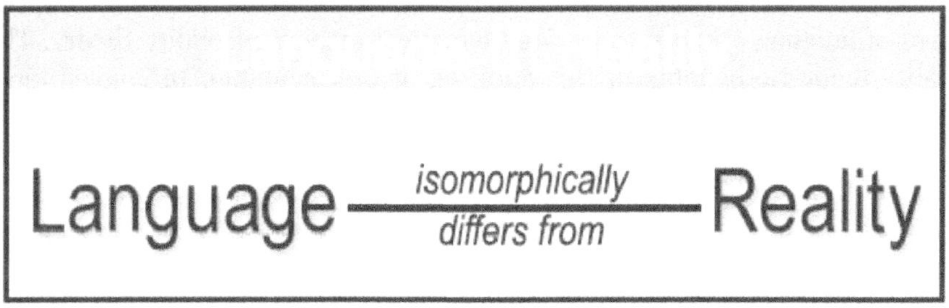

DIAGRAM VII

In this syndiffeonic diagram, the assertion "Language differs from reality" is laid out along an extended line segment representing the supposed difference between the relands. Just as in the generic diagram above, both relands possess the attribute "inclusion in the relational syntactic medium (Language ⋒ Reality)". Because they are both manifestations of the same underlying medium, their difference cannot be absolute; on a fundamental level, reality and language share common aspects. This is consistent with the nature of the "difference" relationship, which is actually supposed to represent a semantic and model-theoretic isomorphism.

As we have already seen, the Reality Principle says that reality contains all and only that which is real. As defined by this statement, the predicate *reality* is primarily a linguistic construct conforming to syntactic structure, where *syntax* consists of the rules by which predicates are constructed and interpreted. In this sense, *reality* amounts to a kind of *theory* whose axioms and rules of inference are implicitly provided by the logical component of the conceptual syntax in which it is expressed. The Principle of Linguistic Reducibility merely clarifies the issue of whether *reality* is a linguistic predicate or the objective content of such a predicate by asserting that it is both. Thus, where the reality *predicate* is analytically (or syntactically) self-contained, *reality* is self-contained. This can be expressed as follows: on the level of cognitive-perceptual syntax, *reality* equals *reality theory*. Where theory and universe converge, Occam's razor and physical principles of economy become tautologies.

Because perception is a sensory intersect of mind and reality, perception is impossible without cognition, and to this extent the cognitive predicate *reality* equates to its perceptual content. On the level of cognitive and perceptual syntax, language is necessarily isomorphic to that which it describes; in a perceptual reality like that which exists around us, it is tautologically true that the basic language of cognition and perception is syntactically isomorphic to reality (though illusion and falsehood become possible on the semantic level). It follows that we can speak of reality in terms of *generalized cognition and perception*, where this phrase denotes conformance to cognition and perception on the syntactic level. In particular, generalized cognition is that process through which reality everywhere "recognizes" itself.

The Principle of Linguistic Reducibility provides a mandate to add an advanced form of language theory to the mathematical arsenal of reality theory. The reality-theoretic benefits of this addition are incalculable. In conventional physical theory, the fundamental entities are point particles, waves and more recently, strings; each class of object has its problems and paradoxes. In the CTMU, the fundamental objects are *syntactic operators* (units of self-transducing information or *infocognition*) that are not only capable of emulating all of these objects and more, but of containing the syntactic structures to which they must inevitably conform and resolving their characteristic paradoxes in the bargain. Because meaning equates to semantic connectivity and is thus linguistic in every sense of the term, the shift to a linguistic perspective is indispensable to teleology or any other form of meaning.

Now we know that reality is a *linguistic* self-contained syndiffeonic relation, although we still seem to be knowing it from an external vantage in a rather inspecific way. Where should we go next in search of clues? At this point, we could really use a MAP.

Syntactic Closure: The Metaphysical Autology Principle (MAP)

All relations, mappings and functions relevant to reality in a generalized effective sense, whether descriptive, definitive, compositional, attributive, nomological or interpretative, are generated, defined and parameterized within reality itself. In other words, reality comprises a "closed descriptive manifold" from which no essential predicate is omitted, and which thus contains no critical gap that leaves any essential aspect of structure unexplained. Any such gap would imply non-closure.

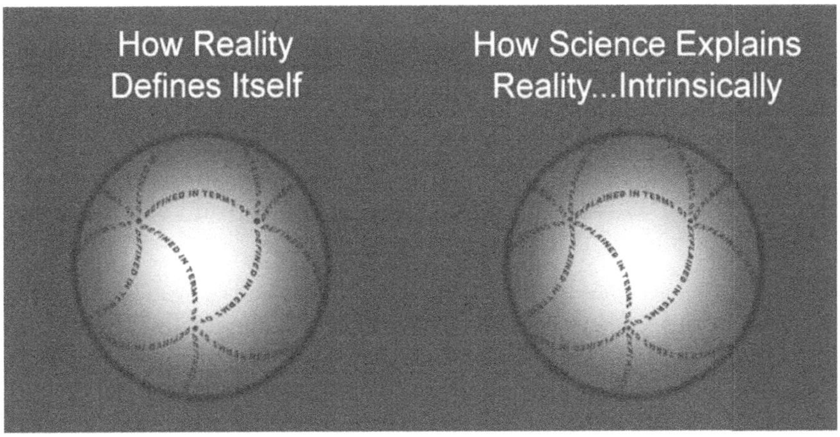

DIAGRAM VIII

MAP, a theoretical refinement of the self-containment criterion set forth by the Reality Principle, extends the closure property of the definition of reality to the set of all real predicates. MAP effects closure on the definitive, descriptive, explanatory and interpretative levels of reality theory by making it take the form of a closed network of coupled definitions, descriptions, explanations and interpretations that refer to nothing external to reality itself. Another way to state this is that MAP, like the Reality Principle, requires that everything to which any reality-theoretic definition, description, explanation or interpretation refers be located within reality. This has the effect of making reality responsible for its own structure and evolution in the abstract and concrete senses. MAP requires a closed-form explanation on the grounds that distinguishability is impossible without it. Again this comes down to the issue of syntactic stability.[12] To state it in as simple a way as possible, reality must ultimately possess a stable 2-valued object-level distinction between that which it *is* and that which it *is not*, maintaining the necessary informational boundaries between objects, attributes and events. The existence of closed informational boundaries within a system is ultimately possible only by virtue of systemic

[12] A syndiffeonic regress ultimately leads to a stable closed syntax in which all terms are mutually defined; mutual definition is what stabilizes and lends internal determinacy (internal identifiability of events) to the system through syntactic symmetry. I.e., relational closure orthogonal to the (temporal) dimension of regress amounts to the distribution of a systemic identity, and unless this identity is fully defined, the identities of system components cannot be defined or recognized by or with respect to each other. Thus, the dynamic identity is a self-identification function with definitive, descriptive and interpretative dimensions.

closure under dualistic (*explanans–explanandum*) composition, which is just how it is effected in sentential logic.

As an example of the tautological nature of MAP, consider a hypothetical external scale of distance or duration in terms of which the absolute size or duration of the universe or its contents can be defined. Due to the analytic self-containment of reality, the functions and definitions comprising its self-descriptive manifold refer only to each other; anything not implicated in its syntactic network is irrelevant to structure and internally unrecognizable, while anything which is relevant is already an implicit ingredient of the network and need not be imported from outside. This implies that if the proposed scale is relevant, then it is not really external to reality; in fact, reality already contains it as an implication of its intrinsic structure.

In other words, because reality is defined on the mutual relevance of its essential parts and aspects, *external* and *irrelevant* are synonymous; if something is external to reality, then it is not included in the syntax of reality and is thus internally unrecognizable. It follows that with respect to that level of reality defined on relevance and recognition, there is no such thing as a "real but external" scale, and thus that the universe is externally undefined with respect to all measures including overall size and duration. If an absolute scale were ever to be internally recognizable as an ontological necessity, then this would simply imply the existence of a deeper level of reality to which the scale *is* intrinsic and by which it is itself intrinsically explained as a relative function of other ingredients. Thus, if the need for an *absolute* scale were ever to become recognizable within reality – that is, recognizable to reality itself – it would by definition be relative in the sense that it could be defined and explained in terms of other ingredients of reality. In this sense, MAP is a "general principle of relativity".[13]

The "no gaps" criterion of MAP permits no critical explanatory holes omitting any essential aspect of structure. What this means can best be illustrated by means of a recurrent fallacy: "The existence of the universe is given and therefore in no need of explanation." The phrase *is given* is incomplete; it has hidden "loose ends" corresponding to *that by which* existence is given, the *means by which* it is given, and the *reason for which* it is given. If the source, means and reason are actually real, then they are inside reality, and the explanatory gap exists only in the mind of the claimant rather than in the self-explanatory network of reality itself.

---

[13] It is tempting to note that "relativity" means roughly the same thing in the General Theory of Relativity as it does here. That is, using a distributed-syntactic "tangent space", the structure of spacetime is tensorially defined in terms of the masses and relative positions of its own material contents, resulting in an intrinsic MAP-like definition of spacetime. Unfortunately, the basic formalism of GR, differential geometry, is not self-contained with respect to time; as currently formulated, it tacitly relies on an embedding (conspansive) medium to provide it with temporal potential.

On the other hand, omitting this phrase (*is given*) results in something like "the existence of the universe is inexplicable". However, this amounts to the assertion that the universe has no identifiable basis or medium of existence, *not even itself* ... i.e., that no explanatory function can be defined on the explanandum, and that the universe is somehow prohibited from serving as its own source, means, or reason. But this amounts to saying that the universe could only exist "by magic", popping out of the *apeiron* with a spontaneity exceeding that by which a genuine magician might pull a magic rabbit out of a hat. For whereas magic rabbits can at least be said to originate by magic associated with magicians who pull them out of top hats into the bright light of reality, or to magically bootstrap themselves out of their own hats into their own realities, the universe would be denied any ontological basis or medium whatsoever ... even a bootstrap.

Because questions like "why and how does reality exist (within the domain of existential potential supporting the possibility of existence)?" and "why does this reality exist instead of some other reality?"[14] address the ontological or teleological levels of the structure of reality, and because these levels of structure are logically meaningful, they must have answers ... even if those answers are determined, as some of them are, by the closure criterion itself.

Now we know that the closed, single-predicate definition of the Reality Principle is actually a closed descriptive manifold of linked definitions in principle containing the means of its own composition, attribution, recognition, processing and interpretation. But this is still somewhat automatonic. What about *mind*? Since it is through our minds that we understand anything at all, understanding remains incomplete until we understand more about the relationship between mind and reality. So, having equipped ourselves with a MAP, we now attend to the correspondence between the MAP and the terrain.

Syntactic Comprehensivity-Reflexivity: The Mind Equals Reality Principle (M=R)

The M=R or *Mind Equals Reality* Principle asserts that mind and reality are ultimately inseparable to the extent that they share common rules of structure and processing. The existence of a difference relation between mind and reality

[14] What such a question really asks is "why does the habitation and observation relationship holding between the observer/questioner and the universe exist?" The Weak Anthropic Principle goes halfway toward an answer by applying a probabilistic selection function: "the relationship has been logically selected by the fact that only certain kinds of universe can accommodate acts of observation and an observer like the questioner." This is right as far as it goes, but omits the generative, pre-selective phase of the explanation ..., the part that accounts for the selection function and the domain to which it is applied. In this respect, the WAP is a bit like natural selection; it weeds the garden well enough, but is helpless to grow a single flower.

syndiffeonically presupposes a relational medium having the characteristics of both, and this medium has logical priority over the difference relation itself.

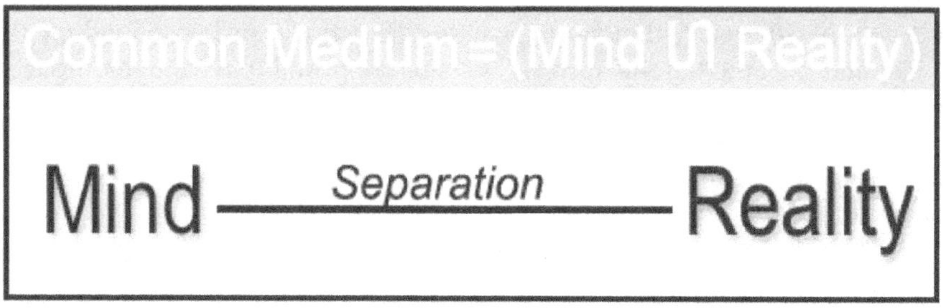

DIAGRAM IX

M=R (Mind = Reality) Principle In the above syndiffeonic diagram, mind is juxtaposed with reality in a space bounded by a box. The line separating mind and reality represents the supposed difference between them, while the interior of the box represents their comparability or "relatedness" (or more technically, their uniform differentiating syntax or unisect, denoted by means of the ∪∩ functor). The extensionality of the line is just that of the box; without the box, there would be no extensional medium to contain the line, and no way to express the associated difference relation. Because the separation cannot exist without a common medium incorporating a differentiative syntax that distributes over both relands of the difference relation, the "absolute separation" of mind and reality has no model ... and without a model, the premise of Cartesian mind-matter dualism fails. This indicates that reality and mind, information and information processor, must ultimately be regarded as one. Any Cartesian-style distinction between them must be strictly qualified.

The M=R principle is merely a logical version of what empiricist philosophers long ago pointed out: we experience reality in the form of perceptions and sense data from which the existence and independence of mind and objective external reality are induced. Since any proof to the contrary would necessarily be cognitive, as are all "proofs", and since the content of cognition is cognitive by embedment, no such proof can exist; such a proof would undermine its own medium and thereby cancel itself. On the other hand, the Reality Principle says that reality is self-contained with respect to recognition and control, and to the extent that recognition and control are "mental" (in the sense of being effected according to cognitive and perceptual syntax), so is reality. The M=R Principle entails comprehensivity by defining all of our perceptions, along with their syntax-level cognitive-syntactic infrastructure, as parts of reality regardless of decidability.[15] When it comes to M=R, it is hard to resist a little play on words:

---

[15] Note that the M=R Principle does not assert (e.g.) that the fruits of human imagination are "real" except as neural firing patterns with no necessary relevance to anything outside the brain, or that we can immediately make things "real" just by thinking of them. The M=R principle merely says that where mind is the engine of cognition and perception, and these processes are jointly regarded as a "language" obeying rules of

M=R says that at the syntactic level of cognition and perception, "the MAP is the terrain." Note that M=R goes beyond the mere Kantian isomorphism between phenomenal reality and the categories of thought and perception; it says that syntax and its content are recursively related, and in conjunction with the Reality Principle, that any supposed "content" not related to the rules of structure and evolution of reality is irrelevant. (Although this is a trivial observation insofar as "unrelated" and "irrelevant" are synonymous, it seems to have been largely ignored by many who should have known better.)

To put it another way: if the "noumenal" (perceptually independent) part of reality were truly unrelated to the phenomenal (cognition-isomorphic) part, then these two "halves" of reality would neither be coincident nor share a joint medium relating them. In that case, they would simply fall apart, and any integrated "reality" supposedly containing both of them would fail for lack of an integrated model. Where M (mind) is identified with cognition and R (reality) with physically-embodied information, M=R says that reality everywhere consists of a common substance, infocognition, having the dual nature of mind and (informational) reality.

The M=R property takes up where the Principle of Linguistic Reducibility leaves off in eliminating the distinction between theory and universe. By its light, the theoretical description of reality by human beings contained in reality amounts to reality describing itself. (Bearing in mind that a theory is a mental construct, this can be illustrated by simply replacing Mind and Reality in the above diagram by Theory and Universe, and *Mind* ᔕᑎ *Reality* by *Theory* ᔕᑎ *Universe*.) It thus makes the theory reflexive and thus inclusive enough by definition to describe the entire universe, including that which is rational, abstract and subjective, and that which is empirical, concrete and objective. The dissolution of this distinction can be viewed as a reduction.

So now we know that reality is more than just a linguistic self-contained syndiffeonic relation comprising a closed descriptive manifold of linked definitions containing the means of its own configuration, composition, attribution, recognition, processing and interpretation. It is also a self-processing theory identical to its universe.

Syntactic Coherence and Consistency: The Multiplex Unity Principle (MU)

The universe topologically contains that which descriptively contains the universe. MU, the minimum and most general informational configuration of reality, defines the relationship holding between unity and multiplicity, the

---

syntax, this syntax – the rules of cognition and perception – necessarily distributes over percipient and percept.

universe and its variegated contents. Through its structure, the universe and its contents are mutually inclusive, providing each other with a medium.

In other words, we can equivalently characterize the contents of the universe as being topologically "inside" it (topological inclusion), or characterize the universe as being descriptively "inside" its contents, occupying their internal syntaxes as acquired state (descriptive inclusion). The universe generically includes its contents by serving as their syntactic unisect, while the contents contain the universe in a more specific sense involving specific event histories that become "entangled" by interaction. From the first viewpoint, the syntactic coherence of the overall medium enforces mutual consistency of contents, while from the second viewpoint, the coherent syntaxes of its contents contain and consistently recognize and transform the medium. Thus, the universe enforces its own consistency through dual self-containment.

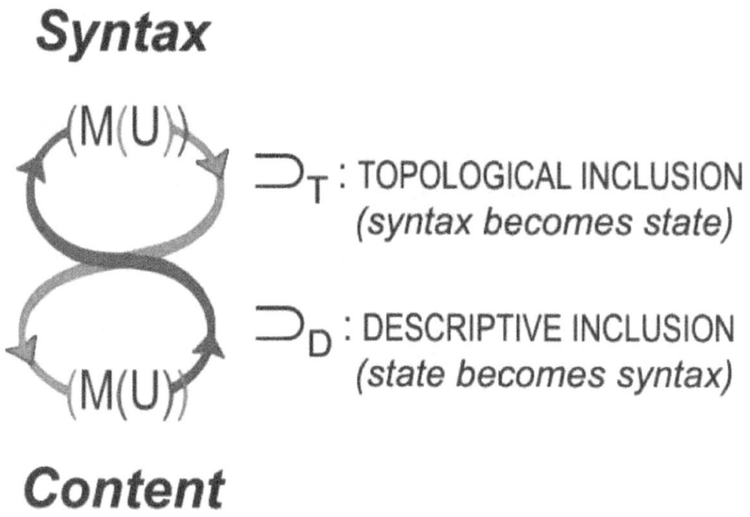

DIAGRAM X

In the syndiffeonic **DIAGRAM VI**, we can plainly see the containment of objects by the medium, but we cannot see the containment of the medium by the objects. Bearing in mind that the terms syntax and content are to some extent relative designations, the upper node in **DIAGRAM X** corresponds to the global medium (global syntactic unisect or "metasyntax" of reality), while the lower node corresponds to the objects therein (syntactic operators contained in the medium); each is a multiplex unity. Coherence flows from global syntax into local content by way of global topological containment, thereby enforcing unity across diverse locales, and back to global syntax in multiple entangled streams generated by cross-transduction of content. Syntax becomes state, and state becomes syntax (where "syntax" is understood to encompass an "ecto-syntactic" distribution of syntactic operators). The universe thus remains coherent and consistent in the course of evolution.

MU expresses syndiffeonic symmetry of syntax and content on the spatiotemporal level of reality. Just as syndiffeonesis can be regarded as a paradox identifying difference with sameness, MU can be regarded as an ultimate form of paradox identifying spatiotemporal multiplicity and unity (the MU diagram is an explosion of the syndiffeonic relation diagram in which the stratification dimension is split into descriptive and topological strands or "temporal dimensions"). MU structure resolves the MU paradox *in situ* by dual stratification, providing closure as the open-ended informational stratification of type theory cannot. Because MU can thus be regarded as the resolution of the paradox it describes, its meaning, like that of syndiffeonesis, can be expressed as follows: *reality is a self-resolving paradox.*

MU, by the way, need not be regarded as the ultimate guarantor of consistency; that honor can safely go to the stability of perceptual reality itself. Quite simply, the syntactic stability of reality overrides any and all objections regarding the limitations of formal systems. MU merely describes how reality, considered as a reflexive SCSPL theory, achieves intrinsic stability in the course of evolving. Thus, it is not functioning as an algorithm guaranteed to terminate on consistency but not on inconsistency, and is therefore not in conflict with undecidability. The stability of reality affirms its consistency regardless of whether or not any lesser theory happens to be consistent.

MU serves as a unifying concept for a complex of ideas having to do with coherence and consistency in the reality-theoretic context, including hology and several CTMU duality principles.

The Principle of Hology (Self-composition)

Hology, a logical analogue of holography characterizing the most general relationship between reality and its contents, is a form of self-similarity whereby the overall structure of the universe is everywhere distributed within it as accepting and transductive syntax, resulting in a homogeneous syntactic medium. That is, because reality requires a *syntax* consisting of general laws of structure and evolution, and there is nothing but reality itself to serve this purpose, reality comprises its own self-distributed syntax under MU (which characterizes the overall relationship between syntax and content).

The hology property itself is distributed over reality. That is, the informational boundary of a coherent object (syntactic operator) is hologically multiplexed with respect to state (attribute and value) in order to define the descriptive interior of the operator as it participates in global self-processing without input. This multiplexing of possibilities is just the replication of the structure of the boundary over the interior of the boundary as a function of time. Again, the operator ultimately has nothing else in terms of which to express its spatiotemporal capacity.

Hology is implied by MAP because reality is closed under composition and attribution; it is implied by M=R because reality is composed of syntactic operators with generalized mental or cognitive functionality; and it is implied by syndiffeonesis and MU because it is an expression of the relationship between the global spatiotemporal medium and its contents.

Duality Principles

Duality is a ubiquitous concept in mathematics, appearing in fields from logic and the theory of categories to geometry and analysis. The duality relation is symmetric; if dualizing proposition A yields proposition B, then dualizing B yields A. In projective geometry, for example, the dualization operation consists of switching the terms "point" and "line" in propositions containing them, as in "Two non-coincident points determine a line" ← dualize → "Two non-parallel lines determine a point." Re-stating this as "lines are functions of points" ← dualize → "points are functions of lines" reveals a duality relationship between functions and arguments. Thus, in vector algebra, the *dual space* V* of a vector space V is the space of all linear functionals on V (i.e. all linear maps from V to R), while V** is the space of all linear maps from V* to R.

An even more striking form of duality is encountered in graph theory, where the *dual graph* of a planar graph transforms faces to vertices and vertices to faces without disrupting its overall pattern of adjacencies. The boundary of each face is replaced by transverse edges converging on its dual vertex (and vice versa), and the adjacency relation is redefined accordingly. Where edges are given a temporal interpretation, interesting transformations can occur; e.g., circulations along facial boundaries become "vertex spins", and motion along an edge can be characterized as an operation between the dual faces of its endpoints.

Duality principles thus come in two common varieties, one transposing spatial relations and objects, and one transposing objects or spatial relations with mappings, functions, operations or processes. The first is called *space-object* (or S-O, or S ↔ O) duality; the second, *time-space* (or T-S/O, or T ↔ S/O) duality. In either case, the central feature is a transposition of *element* and a (spatial or temporal) *relation* of elements. Together, these dualities add up to the concept of *triality*, which represents the universal possibility of consistently permuting the attributes *time*, *space* and *object* with respect to various structures. From this, we may extract a third kind of duality: ST-O duality. In this kind of duality, associated with something called *conspansive duality*, objects can be "dualized" to spatiotemporal transducers, and the physical universe internally "simulated" by its material contents.

M=R, MU and hology are all at least partially based on duality.

The Principle of Attributive (Topological-Descriptive, State-Syntax) Duality

Where points belong to sets and lines are relations between points, a form of duality also holds between sets and relations or attributes, and thus between set theory and logic. Where sets *contain* their elements and attributes distributively *describe* their arguments, this implies a dual relationship between topological containment and descriptive attribution as modeled through Venn diagrams. Essentially, any containment relationship can be interpreted in two ways: in terms of position with respect to bounding lines or surfaces or hypersurfaces, as in point set topology and its geometric refinements ($\supset$T) or in terms of descriptive distribution relationships, as in the Venn-diagrammatic grammar of logical substitution ($\supset$D) (Langan, 1999).

Attributive or TD duality is reflected in the fact that sets and logic are described by the same algebraic structure, Boolean algebra, which expresses their dual relationship in the relationship between its two operations. Expressed in set-theoretic terms, these operations are union and intersection ($\cup$, $\cap$); in logical terms, they are OR and AND ($\vee$, $\wedge$). ($\cup$, $\cap$) and ($\vee$, $\wedge$) are related as follows: the *union* ($A \cup B$) of two sets A and B consists of all and only the elements that belong to either A *or* B or both ($\forall x \in A \cup B : x \in A \vee x \in B$), while the *intersect* ($A \cap B$) of A and B consists of all and only the elements that belong to both A *and* B ($\forall x \in A \cap B : x \in A \wedge x \in B$). This kind of duality is well known; it relates to the fact that every attributive statement defining a relation of predicates can be rephrased as a statement about sets (and vice versa).

But the relationship of set theory and logic is even more interesting than this, for each has a particular representational affinity for just one of these operations. That is, set theory tends to focus on *objects* (sets and elements), while logic tends to focus on *attributes*, or informational "boundary constraints" that objects must satisfy. Thus, set theory ultimately defines sets in terms of the objects they contain, while logic tends to define them "from the outside in" on the intersecting boundary constraints to which they conform. The difference hinges on the univalent *not* functor ($\sim$), on which complementation and intersection, but not union, are directly or indirectly defined.

For example, while it is easy enough to identify an individual element or set by constructively naming or "enumerating" it, e.g. "X", identifying its complement often requires that its name be used as the basis of a restrictive constraint that can be applied across an entire finite or infinite context in one attributive operation, e.g. "not-X". The duality relationship holding between names and constraints is nicely captured by *De Morgan's Laws*, $\sim A \cap \sim B = \sim(A \cup B)$ and $\sim A \cup \sim B = \sim(A \cap B)$, which express it by permuting the objective and attributive operations $\cup$ and $\cap$.

Because states express topologically while the syntactic structures of their underlying operators express descriptively, attributive duality is sometimes called *state-syntax duality*. As information requires syntactic organization, it amounts to a valuation of cognitive / perceptual syntax; conversely, recognition consists of a subtractive restriction of informational potential through an additive acquisition of information. TD duality thus relates information to the informational potential bounded by syntax, and perception (cognitive state acquisition) to cognition.

In a Venn diagram, the contents of circles reflect the structure of their boundaries; the boundaries are the primary descriptors. The interior of a circle is simply an "interiorization" or self-distribution of its syntactic "boundary constraint". Thus, nested circles corresponding to identical objects display a descriptive form of containment corresponding to syntactic layering, with underlying levels corresponding to *syntactic coverings*.

This leads to a related form of duality, *constructive-filtrative* duality.

Constructive-Filtrative Duality

Any set that can be constructed by adding elements to the space between two brackets can be defined by restriction on the set of all possible sets. Restriction involves the Venn-like superposition of constraints that are subtractive in nature; thus, it is like a subtractive color process involving the stacking of filters. Elements, on the other hand, are additive, and the process of constructing sets is thus additive; it is like an additive color process involving the illumination of the color elements of pixels in a color monitor. CF duality simply asserts the general equivalence of these two kinds of process with respect to logico-geometric reality.

CF duality captures the temporal ramifications of TD duality, relating geometric operations on point sets to logical operations on predicates. Essentially, CF duality says that any geometric state or continuous transformation is equivalent to an operation involving the mutual "filtration" of intersecting hological state-potentials. States and objects, instead of being constructed from the object level upward, can be regarded as filtrative refinements of general, internally unspecified higher-order relations.

CF duality is necessary to show how a universe can be "zero-sum"; without it, there is no way to refine the objective requisites of constructive processes "from nothingness". In CTMU cosmogony, "nothingness" is informationally defined as zero constraint or pure freedom (*unbound telesis* or *UBT*), and the apparent construction of the universe is explained as a self-restriction of this potential. In a realm of unbound ontological potential, defining a constraint is not as simple as merely writing it down; because constraints act restrictively on content, constraint and content must be defined simultaneously in a unified syntax-state relationship.

Conspansive Duality

This principle was to some extent adumbrated by the following wry quote attributed to Arthur Eddington (1933) regarding the expanding universe:

> We walk the stage of life, performers of a drama for the benefit of the cosmic spectator. As the scenes proceed he notices that the actors are growing smaller and the action quicker. When the last act opens the curtain rises on midget actors rushing through their parts at frantic speed. Smaller and smaller. Faster and faster. One last microscopic blur of intense agitation. And then nothing.

Eddington's surreal vision accompanied a tongue-in-cheek proposal that the theory of the expanding universe might be replaced by a theory of the "shrinking atom" (1933, p. 90). It was thus a bit overdone for the sake of humor. Indeed, Eddington was not sufficiently interested in the idea to develop its implications beyond a very rudimentary level. However, it turns out that he was skirting the edges of an important duality principle.

Cosmic expansion and ordinary physical motion have something in common: they are both what might be called *ectomorphisms*. In an ectomorphism, something is mapped to, generated or replicated in something external to it. However, the Reality Principle asserts that the universe is analytically self-contained, and ectomorphism is inconsistent with self-containment. Through the principle of conspansive duality, ectomorphism is conjoined with *endomorphism*, whereby things are mapped, generated or replicated within themselves. Through conspansive endomorphism, syntactic objects are injectively mapped into their own hological interiors from their own syntactic boundaries.

In the language of TD and CF duality, this shifts the emphasis from spacetime geometry to descriptive containment, and from constructive to filtrative processing. As a result, new states are formed within the images of previous states. Nothing moves or expands "through" space; space is state, and each relocation of an object is just a move from one level of perfect stasis to another. This ties conventional motion, in which worldlines are constructively created by additions of state in Minkowski diagrams, to differential endomorphism, in which the internal descriptive potentials of attributes are cumulatively restricted.

A (Minkowski) spacetime diagram is a kind of "event lattice" in which nodes represent events and their connective worldlines represent the objects that interact in those events. The events occur at the foci of past and future light cones to which the worldlines are internal. If one could look down the time axis of such a diagram at a spacelike cross section, one would see something very much like a Venn diagram with circles corresponding to lightcone cross sections. This rotation of the diagram corresponds to conspansive dualization, converting a spatiotemporal lattice of worldlines and events to a layered series of Venn diagrams.

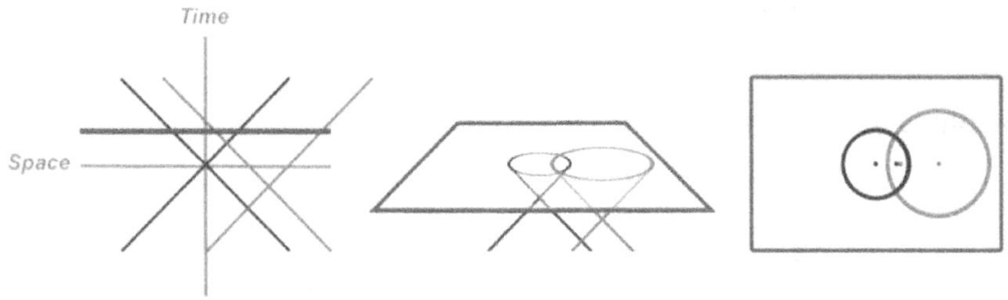

DIAGRAM XI

In the above illustration, a spatial cross section of a spacetime diagram (thick line) is rotated toward the viewer and displayed along the time axis (outer rectangle). The result is a Venn diagram in which circles represent objects and events, or $(n > 1)$-ary interactive relationships of objects. That is, each circle depicts the "entangled quantum wavefunctions" of the objects which interacted with each other to generate it. The small dots in the centers of the circles represent the initial events and objects from which the circles have arisen, while the twin dots where the circles overlap reflect the fact that any possible new event, or interaction between objects involved in the old events, must occur by mutual acquisition in the intersect. The outward growth (or by conspansive duality, mutual absorption) of the circles is called inner expansion, while the collapse of their objects in new events is called requantization. The circles themselves are called IEDs, short for inner expansive domains, and correspond to pairs of interactive syntactic operators involved in generalized-perceptual events (note the hological "evacuation" and mutual absorption of the operators). Spacetime can be illustrated in terms of a layering of such Venn diagrams, mutual contact among which is referred to as "extended superposition" (in the real world, the Venn diagrams are 3-dimensional rather than planar, the circles are spheres, and "layering" is defined accordingly). Extended superposition "atemporally" distributes antecedent events over consequent events, thus putting spacetime in temporally-extended self-contact. In light of the Telic Principle (see below), this scenario involves a new interpretation of quantum theory, sum over futures. Sum over futures involves an atemporal generalization of "process", telic recursion, through which the universe effects on-the-fly maximization of a global self-selection parameter, generalized utility.

In a Venn diagram, circles represent sets through their definitive attributes. The attributes represented by the circles are synetic (syntactically distributed and homogeneous with respect to potential differences of state), and the attribute represented by a particular circle is uniformly heritable by the elements of the set represented by any circle inside it. In the spatiotemporal Venn diagram just described, the circular lightcone cross sections correspond to objects and events relating in just this way. Because quantum-scale objects are seen to exist only when they are participating in observational events, including their "generalized observations" of each other, their worldlines are merely *assumed* to exist between events and are in fact syntactically retrodicted, along with the continuum, from the last events in which they are known to have participated. This makes it possible to omit specific worldlines entirely, replacing them with series of Venn diagrams in which circles inner-expand, interpenetrate and "collapse to points" at each interactive generalized-observational event. This scenario is general, applying even to macroscopic objects consisting of many particles of matter; the higher definition of the worldlines of macroscopic objects can be imputed to a higher frequency of collapse due to interactive density among their constituent particles.

The areas inside the circles correspond to event potentials, and where events are governed by the laws of physics, to potential instantiations of physical law or "nomological syntax". Where each circle corresponds to two or more objects, it comprises object potentials as well. That is, the circular boundaries of the Venn circles can be construed as those of "potentialized" objects in the process of absorbing their spatiotemporal neighborhoods. Since the event potentials and object potentials coincide, potential instantiations of law can be said to reside "inside" the objects, and can thus be regarded as functions of their internal rules or "object syntaxes". Objects thus become syntactic operators, and events become intersections of nomological syntax in the common value of an observable state parameter, position. The circle corresponding to the new event represents an attribute consisting of all associated nomological relationships appropriate to the nature of the interaction including conserved aggregates, and forms a pointwise (statewise) "syntactic covering" for all subsequent potentials.

Notice that in this scenario, spacetime evolves linguistically rather than geometrodynamically. Although each Venn circle seems to expand continuously, its content is unchanging; its associated attribute remains static pending subsequent events involving the objects that created it. Since nothing actually changes until a new event is "substituted" for the one previous, i.e. until a new circle appears within the old one by syntactic embedment, the circles are intrinsically undefined in duration and are thus intrinsically atemporal. Time arises strictly as an ordinal relationship among circles rather than within circles themselves. With respect to time-invariant elements of syntax active in any given state (circle), the distinction between zero and nonzero duration is intrinsically meaningless; such elements are heritable under substitution and become syntactic ingredients of subsequent states. Because each circle is structurally self-distributed, nothing need be transmitted from one part of it to another; locality constraints arise only with respect to additional invariants differentially activated within circles that represent subsequent states and break the hological symmetry of their antecedents. Conspansion thus affords a certain amount of relief from problems associated with so-called "quantum nonlocality".

Because the shrinkage of an object within its prior image amounts to a form of logical substitution in which the object is Venn-diagrammatically "described" or determined by its former state, there is no way to distinguish between outward systemic expansion and inward substitution of content, or between the associated dynamical and logical "grammars". This is merely a restatement of attributive duality; topological containment relations among point-sets are equivalent to descriptively predicating truth of statements asserting containment, and on distribution relationships among state-descriptors. In conjunction with the intrinsic symmetry of externally undefined systems, attributive duality eliminates any possible logical or geometric distinction between the outward expansion of a self-contained universe as its contents remain static in size, and a logical endomorphism in which the universe remains static while the states of its contents are recursively substituted for previous states.

It has already been noted in connection with MAP that where the external dimensions of a system are undefined, no distinction as to size can be made beyond the size ratio of the system to its contents. Consider a simple arithmetical analogy: $1/2 = 1000/2000 = 1(10^{9999})/2(10^{9999}) = (\ldots)$. Where the numerator and denominator of a fraction are both multiplied by a given number, the value of the fraction does not change; it is independent of distinctions involving the size of the multiplier. Similarly, the intrinsic proportionality of a self-contained system is independent of distinctions involving any external measure. This implies that with respect to a self-contained universe for which no external measure exists, no distinction can be made between the expansion of the system with respect to its contents, and the shrinkage of its contents with respect to it. In fact, because that which is undefined cannot change – there is nothing definite with respect to which change would be possible – apparent expansion of the container cannot be extrinsically defined, but implies a conspansively-equivalent intrinsic shrinkage of its contents.

Thus, conspansive duality relates two complementary views of the universe, one based on the external (relative) states of a set of objects, and one based on the internal structures and dynamics of objects considered as language processors. The former, which depicts the universe as it is usually understood in physics and cosmology, is called ERSU, short for *Expanding Rubber Sheet Universe*, while the latter is called USRE (ERSU spelled backwards), short for *Universe as a Self-Representational Entity*. Simplistically, ERSU is like a set, specifically a topological-geometric point set, while USRE is like a self-descriptive nomological language. Whereas ERSU expands relative to the invariant sizes of its contents, USRE "conspands", holding the size of the universe invariant while allowing object sizes and time scales to shrink in mutual proportion, thus preserving general covariance.

This has certain interesting implications. First, whereas it is ordinarily assumed that the sizes of material objects remain fixed while that of the whole universe "ectomorphically" changes around them, conspansion holds the size of the universe changeless and endomorphically changes the sizes of objects. Because the universe now plays the role of invariant, there exists a global standard rate of inner expansion or mutual absorption among the contents of the universe ("c-invariance"), and due to syntactic covariance, objects must be resized or "requantized" with each new event according to a constant (time-independent) rescaling factor residing in global syntax. Second, because the rate of shrinkage is a constant function of a changing size ratio, the universe appears from an internal vantage to be accelerating in its "expansion", leading to the conspansive dual of a positive cosmological constant (Langan, 2001a).

Conspansive duality, the role of which in the CTMU is somewhat analogous to that of the Principle of Equivalence in General Relativity, is the only escape from an infinite ectomorphic "tower of turtles". Were the perceptual geometry of reality to lack a conspansive dual representation, motion of any kind would require a fixed spatial array or ectomorphic "background space" requiring an explanation of its own, and so on down the tower. Conspansion permits the universe to self-configure through temporal feedback. Each conspanding circle represents an event-potential corresponding to a certain combination of law and state; even after one of these intrinsically atemporal circles has "inner-expanded" across vast reaches of space and time, its source event is still current for anything that interacts with it, e.g. an eye catching one of its photons. At the same time, conspansion gives the quantum wave function of objects a new home: *inside the conspanding objects themselves*. Without it, the wave function not only has no home, but fails to coincide with any logically evolving system of predicates or "laws of physics". Eliminate conspansion, and reality becomes an inexplicable space full of deterministic worldlines and the weighty load of problems that can be expected when geometry is divorced from logic.

Where reality is characterized by dual-aspect infocognitive monism (read on), it consists of units of infocognition reflecting a distributed coupling of transductive syntax and informational content. Conspansion describes the "alternation" of these units between the dual (generalized-cognitive and informational) aspects of reality, and thus between syntax and state. This alternation, which permits localized mutual refinements of cognitive syntax and informational state, is essential to an evolutionary process called *telic recursion*. Telic recursion requires a further principle based on conspansive duality, the *Extended Superposition Principle*, according to which operators can be simultaneously acquired by multiple *telons*, or spatiotemporally-extensive syntax-state relationships implicating generic operators in potential events and opportunistically guiding their decoherence.

Note that conspansion explains the "arrow of time" in the sense that it is not symmetric under reversal. On the other hand, the conspansive nesting of atemporal events puts all of time in "simultaneous self-contact" without compromising ordinality. Conspansive duality can be viewed as the consequence of a type of gauge (measure) symmetry by which only the relative dimensions of the universe and its contents are important.

The Extended Superposition Principle

In quantum mechanics, *the principle of superposition of dynamical states* asserts that the possible dynamical states of a quantized system, like waves in general, can be linearly superposed, and that each dynamical state can thus be represented by a vector belonging to an abstract vector space. The superposition principle permits the definition of so-called "mixed states" consisting of many possible "pure states", or definite sets of values of state-parameters. In such a superposition, state-parameters can simultaneously have many values.

The superposition principle highlights certain problems with quantum mechanics. One problem is that quantum mechanics lacks a cogent model in which to interpret things like "mixed states" (waves alone are not sufficient). Another problem is that according to the uncertainty principle, the last states of a pair of interacting particles are generally insufficient to fully determine their next states. This, of course, raises a question: how are their next states actually determined? What is the source of the extra tie-breaking measure of determinacy required to select their next events ("collapse their wave functions")?

The answer is not, as some might suppose, "randomness"; randomness amounts to acausality, or alternatively, to informational incompressibility with respect to any distributed causal template or ingredient of causal syntax. Thus, it is either no explanation at all, or it implies the existence of a "cause" exceeding the representative capacity of distributed laws of causality. But the former is both absurd and unscientific, and the latter requires that some explicit allowance be made for higher orders of causation ... more of an allowance than may readily be discerned in a simple, magical invocation of "randomness".

The superposition principle, like other aspects of quantum mechanics, is based on the assumption of physical *Markovianism*.[16] It refers to mixed states between adjacent events, ignoring the possibility of nonrandom temporally-extensive relationships not wholly attributable to distributed laws. By putting temporally remote events in extended descriptive contact with each other, the Extended Superposition Principle enables coherent cross-temporal telic feedback and thus plays a necessary role in cosmic self-configuration. Among the higher-order determinant relationships in which events and objects can thus be implicated are utile state-syntax relationships called *telons*, telic attractors capable of guiding cosmic and biological evolution.

---

[16] A Markoff process is a stochastic process with no memory. That is, it is a process meeting two criteria: (1) state transitions are constrained or influenced by the present state, but not by the particular sequence of steps leading to the present state; (2) state transition contains an element of chance. Physical processes are generally assumed to meet these criteria; the laws of physics are defined in accordance with 1, and because they ultimately function on the quantum level but do not fully determine quantum state transitions, an element of chance is superficially present. It is in this sense that the distributed laws of physics may be referred to as "Markovian". However, criterion 2 opens the possibility that hidden influences may be active.

Given that quantum theory does not seem irrevocably attached to Markovianism, why has the possibility of higher-order causal relationships not been seriously entertained? One reason is spacetime geometry, which appears to confine objects to one-dimensional "worldlines" in which their state-transition events are separated by intervening segments that prevent them from "mixing" in any globally meaningful way. It is for this reason that superposition is usually applied only to individual state transitions, at least by those subscribing to conservative interpretations of quantum mechanics.

Conspansive duality, which incorporates TD and CF components, removes this restriction by placing state transition events in direct descriptive contact. Because the geometric intervals between events are generated and selected by descriptive processing, they no longer have separative force. Yet, since worldlines accurately reflect the distributed laws in terms of which state transitions are expressed, they are not reduced to the status of interpolated artifacts with no dynamical reality; their separative qualities are merely overridden by the state-syntax dynamic of their conspansive dual representation.

In extending the superposition concept to include nontrivial higher-order relationships, the Extended Superposition Principle opens the door to meaning and design. Because it also supports distribution relationships among states, events and syntactic strata, it makes cosmogony a distributed, coherent, ongoing event rather than a spent and discarded moment from the ancient history of the cosmos. Indeed, the usual justification for observer participation – that an observer in the present can perceptually collapse the wave functions of ancient (photon-emission) events – can be regarded as a consequence of this logical relationship.

Supertautology

Truth, a predicate representing inclusion in a domain, is the logical property by virtue of which one thing may be identified and distinguished from another at any level of resolution. All theories aim at truth, and reality theory is no exception. With respect to science, there is a problem with truth: beyond the level of direct observation, it cannot be certified by empirical means. To blame are various forms of uncertainty, model-theoretic ambiguity, and the problem of induction: scientific generalizations are circular insofar as they are necessarily based on the assumption that nature is uniform. The problem of induction effectively limits certitude to mathematical reasoning.

This is hardly surprising, for truth is ultimately a mathematical concept. In logic, truth is defined by means of always-true expressions called *tautologies*. A logical tautology possesses three distinctive properties: it is descriptively universal, it is closed under recursive self-composition, and it is internally and externally consistent on the syntactic and semantic levels of reference. Since logic is the theory of truth, the way to construct a fully verifiable theory is to start with logic and develop the theory by means of rules or principles under which truth is heritable. Because truth is synonymous with logical tautology, this means developing the theory by adjoining rules which themselves have a tautological structure – i.e., which are universal, closed and consistent – and logically extracting the implications. A theory of reality constructed in this way is called a *supertautology*.

In a supertautological theory of reality, it is unnecessary to assume the uniformity of nature with respect to certain kinds of generalization. Instead, such generalizations can be mathematically deduced ... e.g. nomological covariance, the invariance of the rate of global self-processing (c-invariance), and the internally-apparent accelerating expansion of the system.

Reduction and Extension

The greatest theoretical advances have typically been associated with two complementary processes, *reduction* and *extension*. The conceptual components of a theory are *reduced* to more fundamental components, and the theory *extended* by the emergence of new and more general relationships among them. The CTMU reduces reality to self-transducing information and ultimately to telesis, using the closed, reflexive syntactic structure of the former as a template for reality theory.

In science, everything requires an explanation ... even explanations. Not only do observations demand explanatory theories, but theories require explanations of their own. Unfortunately, it is sometimes forgotten that until something has been explained in an explicable way, it has not been properly explained at all. If a theory is not self-explanatory, then it must be reduced to a more fundamental theory that explains it; otherwise, it merely relies on assumptions.

E.g., consider an explanation to the effect that "birds can fly because they have wings". Without an explanation of atmospheric resistance, this explanation is incomplete; it contains no explanation of why or how wings enable flight, merely relying on the assumption that they do. Therefore, while it is true as far as it goes, it leaves out crucial supporting knowledge and cannot stand alone. Concisely, every theory T_{i+1} that is not self-explanatory must be reducible to a more fundamental theory $T_i \models T_{i+1}$ that explains and supports it, so that T_i, and this explanatory regress can only end with a self-explanatory theory T_0.

This fact is very frequently forgotten in evolutionary biology, where (e.g.) details of molecular structure and dynamics are used to explain organic phenomena. Although these details come from the more fundamental theories of quantum chemistry and physics, they will never constitute a satisfactory explanation of life until they incorporate not only an explanation of physics and chemistry, but reality at large. This is true because physical (observable) reality is not a complete model for physics and thus is not self-contained with respect to explanation – in this sense, any exclusively materialistic interpretation of physical theory is *prima facie* absurd – and because physics is a non-self-explanatory theory regardless of model. To explain organic phenomena using natural selection, one needs an explanation for natural selection, including the "natural selection" of the laws of physics and the universe as a whole.

Theoretical reduction involves a regressive unbinding of progressive informational constraints in order to achieve increasingly basic explanations. Closed theoretical signatures are ripped open and reduced to more basic concepts that can be reformed into more basic and expressive signatures. However, the informational part of the regress terminates where further reduction would compromise intelligibility; there can be no further reductive regress through increasingly fundamental theoretic strata once the requirements of regression, reduction, theorization and stratification have themselves been lost. Beyond this point, infocognition gives way to informational and cognitive potential, or telesis.

The process of reducing distinctions to the homogeneous syntactic media that support them is called *syndiffeonic regression*. This process involves *unisection*, whereby the rules of structure and dynamics that respectively govern a set of distinct objects are reduced to a "syntactic join" in an infocognitive lattice of syntactic media. Unisection is a general form of reduction which implies that all properties realized within a medium are properties of the medium itself.

Where emergent properties are merely latent properties of the teleo-syntactic medium of emergence, the mysteries of emergent phenomena are reduced to just two: how are emergent properties anticipated in the syntactic structure of their medium of emergence, and why are they not expressed except under specific conditions involving (e.g.) degree of systemic complexity?

The Principle of Infocognitive Monism

Where language consists of information and information has linguistic structure, the Principle of Linguistic Reducibility implies that information is as fundamental as language. Insofar as we cannot understand reality except in theoretical (linguistic, informational) terms, this permits us to cast reality as a "self-processing language", or self-defining, self-explaining, self-modeling theory-universe ensemble, without fear of being proven wrong by some alternate theoretical reduction. However, the linguistic reduction of reality is superficially macroscopic. Just as a perfectly self-contained language must be self-processing (for lack of anything external to process it), so must the information of which it consists. This leads to the concept of *self-processing information*, and ultimately to a microscopic (quantum) theory of information.

It is easy to show that information is self-processing. Structure is attributive; the parts of any structure possess attributes that position them or otherwise define them relative to other parts. To be meaningful and thus informative, information must have structure; therefore, information must possess attributes. Attributive relationships, intrinsic or otherwise, must conform to the logical rules that govern attribution, i.e. to an attributive logical syntax incorporating the propositional and predicate calculi. So information can exist only in conjunction with attributive logical syntax. Because it necessarily incorporates attributive syntax, it has enough native self-processing capacity to maintain its intrinsic structure, which is precisely what it must do to qualify as "informational".

Because cognition and generic information transduction are identical up to isomorphism – after all, cognition is just the specific form of information processing that occurs in a mind – information processing can be described as "generalized cognition", and the coincidence of information and processor can be referred to as *infocognition*. Reality thus consists of a single "substance", infocognition, with two aspects corresponding to *transduction* and being *transduced*. Describing reality as infocognition thus amounts to (infocognitive) *dual aspect monism*. Where infocognition equals the distributed generalized self-perception and self-cognition of reality, infocognitive monism implies a stratified form of "panpsychism" in which at least three levels of self-cognition can be distinguished with respect to scope, power and coherence: *global*, *agentive* and *subordinate*.

Ultimately, the conceptual shift from information to self-transducing information requires extensions of information-intensive theories including the theories of information, computation and cybernetics. The problem stems from the fact that as it is understood in these fields, information is a limited concept based on an engineering model in which the existence of senders, receivers, messages, channels and transmissive media is already conveniently given, complete with all of the structural and dynamical laws required to make them work together. Moreover, the bit structure specified in this model relates to the actual structure of information the way propositional logic relates to logic as a whole, including the predicate calculus and model theory. To wit, only a single level of structure and functionality is considered, and attribution is primarily limited to just a pair of central predicates common to both theories, *True/False* = 1/0. Just as sentential logic concerns itself only with the functorial relationships of sentential variables and ignores their content, information theory concerns itself only with the probabilities of symbols in message strings and ignores the details of syntactic and semantic structure and processing.

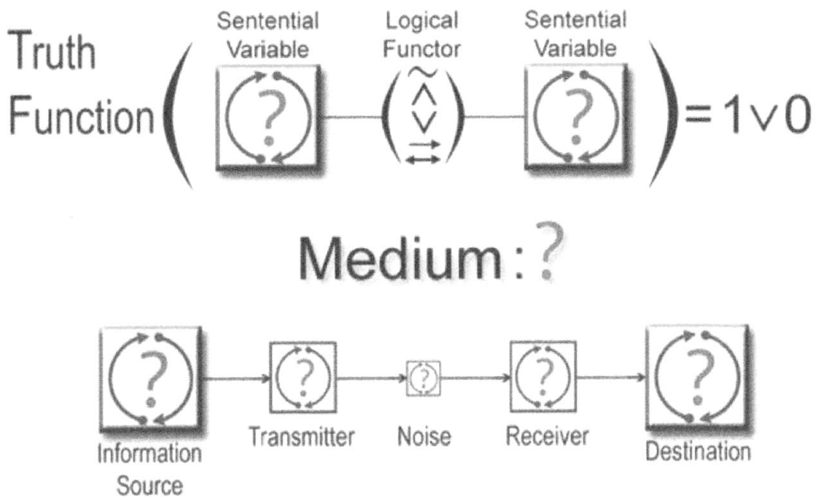

DIAGRAM XII

Sentential logic and information theory both ignore entire levels of structure in order to reduce the universe to 1 and 0s. In sentential logic, sentential variables are distinguished only by whether they are true or false (1 or 0), while the standard theory of information, along with the theories of computation and cybernetics, deals with "raw data" expressed or "encoded" in the most basic possible terms, namely the binary digits 1 and 0. While the role of these "bits" is to reduce uncertainty regarding specific items of content, certain essential details of syntactic and semantic structure and processing, and more specific relationships among variables and data, are conveniently omitted. The question marks indicate that neither sentential logic nor information theory fully explains itself, its model or its medium. [Diagram partially adapted from Shannon (1948).]

However, the most interesting part of the analogy is its logical extension. Just as sentential logic is naturally extended to encompass the levels of attribution associated with predicate logic and model theory, the theory of information can be naturally extended to encompass deeper levels of attribution ... in fact, the same two levels adjoined to sentential logic.

Retooling the information concept consists of three steps. First, it must be equipped with the means of its own *transduction* or transformative processing. Where information transduction is (cognitively) recognized as *generalized cognition*, this amounts to replacing it with a dual-aspect quantum of reflexivity, *infocognition*, which embodies telic feedback. Second, its bit structure, a simplistic and rather uninspired blend of 2-valued propositional logic and probability theory, must be extended to accommodate logic as a whole, including (1) predicate logic, (2) model theory and (3) language theory, broadly including the theories of mathematical languages, metalanguages and generative grammars. After all, since information does nothing but attribute *linguistically*-organized *predicates* to objects in the context of *models*, its meaning involves the mathematics of predicates, languages and models. And third, it must be generalized to an ultimate ancestral medium, *telesis*, from which cognitive syntax and its informational content arise by specificative feedback as part of a unified complex ... a recursive coupling of information and *metainformation*, or transductive syntax.

This retooling is accomplished by associating information with reflexive *syntactic operators* (units of coherent infocognition) in a reflexive linguistic structure, *Self-Configuring Self-Processing Language* (SCSPL), that incorporates its own model and is thus identical to its universe. SCSPL evolves by *conspansion* (material contraction *qua* spatial expansion), a structured grammatical alternation between a linguistic "output" phase (classical reality) consisting of the observable states or external relationships of syntactic operators, and a "production phase" that transforms one state to another.

This being said, there is a sense in which infocognitive monism well agrees with the thesis that bits are universal descriptors of reality: because the bit values 1 and 0 are analogous to the truth values of 2-valued logic, the fact that perceptual reality is described by 2-valued logic implies that it can be described in terms of bits. However, while reality at large is defined by relevance to perceptual reality in the relativistic sense, it does not consist of perceptual reality alone.

Telic Reducibility and Telic Recursion

Telic recursion is a fundamental process that tends to maximize a cosmic self-selection parameter, *generalized utility*, over a set of possible syntax-state relationships in light of the self-configurative freedom of the universe. An inherently "quantum" process that reflects the place of quantum theory in SCSPL, telic recursion is a "pre-informational" form of recursion involving a combination of hology, telic feedback and recursive selection acting on the informational potential of MU, a primal syndiffeonic form that is symmetric with respect to containment.

Where perceptual reality consists of infocognition (self-transducing information), explaining the genesis and evolution of reality amounts to explaining the genesis and evolution of infocognition. Because generalized cognition (information processing) is *temporal*, while information locates objects or message units in attributive *spaces*, information and cognition are respectively spatial and temporal in nature; infocognition is analogous to spacetime, and spacetime is infocognitive. It follows that perceptual reality consists not merely of infocognition but of spacetime, and that seeking an explanation of the genesis and evolution of reality amounts to seeking an explanation of the genesis and evolution of spacetime *qua* infocognition ... i.e., to cosmology in the context of information transduction.

Cosmology, humanity's grand attempt to explain the origin and nature of the universe, has traditionally amounted to the search for a set of "ultimate laws" capable of explaining not only how the universe currently functions, but how it came to be. Unfortunately, even were such a set of laws to be found, the associated explanation could not be considered adequate until the laws themselves were explained, along with the fundamental objects and attributes on which they act. This gives rise to what seems like an imponderable question: how can a set of laws, objects and attributes be explained except by invoking another set of laws in the form of an explanatory syntax that would itself demand an explanation, and so on *ad infinitum*?

The answer is hiding in the question. Laws do not stand on their own, but must be defined with respect to the objects and attributes on which they act and which they accept as parameters. Similarly, objects and attributes do not stand on their own, but must be defined with respect to the rules of structure, organization and transformation that govern them. It follows that the active medium of cross-definition possesses logical primacy over laws and arguments alike, and is thus pre-informational and pre-nomological in nature ... i.e., *telic*. *Telesis*, which can be characterized as "infocognitive potential", is the primordial active medium from which laws and their arguments and parameters emerge by mutual refinement or *telic recursion*.

In other words, telesis is a kind of "pre-spacetime" from which time and space, cognition and information, state-transitional syntax and state, have not yet separately emerged. Once bound in a primitive infocognitive form that drives emergence by generating "relievable stress" between its generalized spatial and temporal components – i.e., between state and state-transition syntax – telesis continues to be refined into new infocognitive configurations, i.e. new states and new arrangements of state-transition syntax, in order to relieve the stress between syntax and state through telic recursion (which it can never fully do, owing to the contingencies inevitably resulting from independent telic recursion on the parts of localized subsystems). As far as concerns the primitive telic-recursive infocognitive MU form itself, it does not "emerge" at all except intrinsically; it has no "external" existence except as one of the myriad possibilities that naturally exist in an unbounded realm of zero constraint.

Telic recursion occurs in two stages, *primary* and *secondary* (global and local). In the primary stage, universal (distributed) laws are formed in juxtaposition with the initial distribution of matter and energy, while the secondary stage consists of material and geometric state-transitions expressed in terms of the primary stage. That is, where universal laws are syntactic and the initial mass-energy distribution is the initial state of spacetime, secondary transitions are derived from the initial state by rules of syntax, including the laws of physics, *plus* telic recursion. The primary stage is associated with the global telor, reality as a whole; the secondary stage, with internal telors ("agent-level" observer-participants). Because there is a sense in which primary and secondary telic recursion can be regarded as "simultaneous", local telors can be said to constantly "create the universe" by channeling and actualizing generalized utility within it.

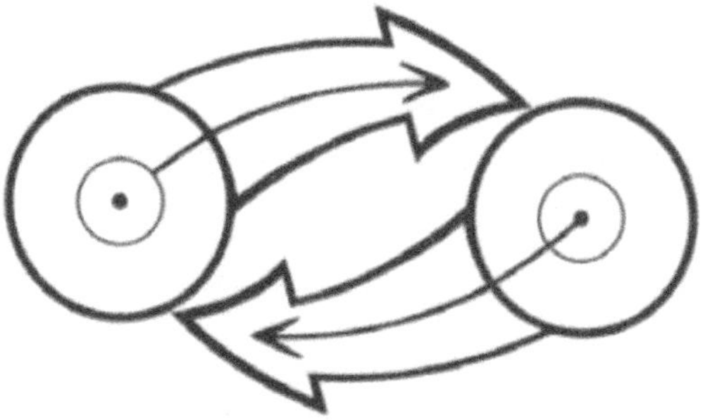

DIAGRAM XIII

The above diagram illustrates the relationship of primary and secondary telic recursion, with the latter "embedded in" or expressed in terms of the former. The large circles and arrows represent universal laws (distributed syntax) engaged in telic feedback with the initial state of spacetime (initial mass-energy distribution), while the small circles and arrows represent telic feedback between localized contingent aspects of syntax and state via conspansion. The primary stage maximizes global generalized utility on an *ad hoc* basis as local telors freely and independently maximize their local utility functions. The primary-stage counterparts of inner expansion and requantization are called *coinversion* and *incoversion*. It is by virtue of telic recursion that the SCSPL universe can be described as its own self-simulative, self-actualizative "quantum protocomputer".

Deterministic computational and continuum models of reality are recursive in the standard sense; they evolve by recurrent operations on state from a closed set of "rules" or "laws". Because the laws are invariant and act deterministically on a static discrete array or continuum, there exists neither the room nor the means for optimization, and no room for self-design. The CTMU, on the other hand, is conspansive and telic-recursive; because new state-potentials are constantly being created by evacuation and mutual absorption of coherent objects (syntactic operators) through conspansion, metrical and nomological uncertainty prevail wherever standard recursion is impaired by object sparsity. This amounts to self-generative freedom, hologically providing reality with a "self-simulative scratchpad" on which to compare the aggregate utility of multiple self-configurations for self-optimizative purposes.

Standard recursion is "Markovian" in that when a recursive function is executed, each successive recursion is applied to the result of the preceding one. Telic recursion is more than Markovian; it self-actualizatively coordinates events in light of higher-order relationships or *telons* that are invariant with respect to overall identity, but may display some degree of polymorphism on lower orders. Once one of these relationships is nucleated by an opportunity for telic recursion, it can become an ingredient of syntax in one or more telic-recursive (global or agent-level) operators or *telors* and be "carried outward" by inner expansion, i.e. sustained within the operator as it engages in mutual absorption with other operators. Two features of conspansive spacetime, the atemporal homogeneity of IEDs (operator strata) and the possibility of extended superposition, then permit the telon to self-actualize by "intelligently", i.e. telic-recursively, coordinating events in such a way as to bring about its own emergence (subject to various more or less subtle restrictions involving available freedom, noise and competitive interference from other telons). In any self-contained, self-determinative system, telic recursion is integral to the cosmic, teleo-biological and volitional (Langan, 2002a) levels of evolution.

The Telic Principle

Restricted to the teleological ("Why?") level of explanation, MAP yields the *Telic Principle*: the universe configures itself according to the requirement that it self-select from a background of undifferentiated ontological potential or *telesis*. This requirement, amounting to a need for *self-actualization* and *self-expression*, is implicit in the MU form. The Telic Principle is responsible for converting potential to actuality in such a way as to maximize a universal self-selection parameter, *generalized utility*.

Teleology, the idea that the universe has a purpose which explains its existence and guides its evolution, some time ago began losing sway in the court of scientific opinion. Although it was at first assumed that a more neutral, less "theological" explanation for the existence of man and the universe would come along to fill the resulting explanatory void, it eventually became evident that no such replacement was conveniently falling out of the equations; some amount of higher-level interpretation would be required in any case. This marked the rise of the so-called *Anthropic Principle*, which now comes in several flavors including "weak", "strong", "final", and that favored by Wheeler, "participatory".

The initial (weak) version, the *Weak Anthropic Principle* or WAP, begins with the trivial if somewhat Bayesian point that our cosmological observations of the universe reveal a capacity for life "because" a life-bearing universe is the only kind of universe in which there are living beings able to make cosmological observations. But while this seems to imply that there exists a domain of many universes in which such a universe can be passively distinguished by the circumstantial constraint that it contain living observers, the WAP offers no ready explanation for such a domain. Indeed, to those not convinced of its virtues, the WAP almost seems to add an unnecessary dose of explanatory red ink to the cosmological ledger.

The *Strong Anthropic Principle* (SAP) eliminates much of this red ink by making a more extreme claim, asserting that the existence of intelligent life is not just a circumstantial selection principle, but a *sine qua non* of cosmic existence. In effect, this limits the set of possible universes to just those which are capable of producing life. However, this leads to problems. How can the idea that living observers are necessary for the existence of the universe be squared with the idea that objective reality is essentially independent of observation and those who observe it? And how does intelligent life, which seems to have evolved billions of years after the universe was born, play any kind of causal role in cosmology? Is some sort of "time travel" occurring? Selection is one thing; retroactive self-generation is quite another.

It has often been remarked that the anthropic principles employ circular reasoning. I.e., they seem to take that which they purport to explain, the observable fact that the universe is "fine-tuned" to support life, as a premise, asserting that living beings observe the universe to be friendly to life "because" life is present in the universe to make this observation. In other words, we are here to observe the universe, and the universe is here to let us observe it, because we are here to observe the universe! Unfortunately, the anthropic principles lack something that they would need to make this work: a circular model to which their loop-like reasoning can be consistently mapped. Quite simply, the type of causal circularity they suggest is at odds with the "arrow of time" and other aspects of the prevailing non-circular models of time and space.

Because circular arguments are self-justifying and resistant to falsification, it is frequently assumed that tautology and circular reasoning are absolute theoretical evils. But this is far from the case, for logic and mathematics are almost completely based on circularity. Truth and logical tautology, recursion and iteration, algebraic and topological closure ... all involve it to some degree. The problems arise only when circular reasoning is employed without the assurance of full mathematical generality, incorporating false claims of universality on (what may be) non-universal premises.

Unfortunately, not even valid tautologies are embraced by the prevailing school of scientific philosophy, falsificationism. While non-universal tautologies are rightly forbidden due to their resistance to falsificative procedures that would reveal their limitations, universal tautologies are pronounced "scientifically uninteresting" for much the same reason. But in fact, science could exist in no way, shape or form without them. The very possibility of a scientific observation is utterly dependent on the existence of tautological forms on which to base a stable, invariant syntax of perception. This raises the possibility that falsificationist thinking has accidentally obscured the true place of tautological reasoning in cosmology.

If the universe is really circular enough to support some form of "anthropic" argument, its circularity must be defined and built into its structure in a logical and therefore universal and necessary way. The Telic principle simply asserts that this is the case; the most fundamental imperative of reality is such as to force on it a supertautological, conspansive structure. Thus, the universe "selects itself" from *unbound telesis* or UBT, a realm of zero information and unlimited ontological potential, by means of *telic recursion*, whereby infocognitive syntax and its informational content are cross-refined through telic (syntax-state) feedback over the entire range of potential syntax-state relationships, up to and including all of spacetime and reality in general.

The Telic Principle differs from anthropic principles in several important ways. First, it is accompanied by supporting principles and models which show that the universe possesses the necessary degree of circularity, particularly with respect to time. In particular, the Extended Superposition Principle, a property of conspansive spacetime that coherently relates widely-separated events, lets the universe "retrodict" itself through meaningful cross-temporal feedback. Moreover, in order to function as a selection principle, it generates a generalized global selection parameter analogous to "self-utility", which it then seeks to maximize in light of the evolutionary freedom of the cosmos as expressed through localized telic subsystems which mirror the overall system in seeking to maximize (local) utility. In this respect, the Telic Principle is an ontological extension of so-called "principles of economy" like those of Maupertuis and Hamilton regarding least action, replacing least action with deviation from generalized utility.

In keeping with its clear teleological import, the Telic Principle is not without what might be described as theological ramifications. For example, certain properties of the reflexive, self-contained language of reality – that it is syntactically self-distributed, self-reading, and coherently self-configuring and self-processing – respectively correspond to the traditional theological properties *omnipresence*, *omniscience* and *omnipotence*. While the kind of theology that this entails neither requires nor supports the intercession of any "supernatural" being external to the real universe itself, it does support the existence of a supraphysical being (the SCSPL global operator-designer) capable of bringing more to bear on localized physical contexts than meets the casual eye. And because the physical (directly observable) part of reality is logically inadequate to explain its own genesis, maintenance, evolution or consistency, it alone is incapable of properly containing the being in question.

Some Background

A review of the standard computational theory of language may prove useful. Computation theory recognizes two general types of automata, transducers and acceptors. Transducers convert input to output, while acceptors classify or "recognize" input consisting of strings of symbols without necessarily producing output.

A *finite transducer* is a 5-tuple $(\Sigma, Q, \Gamma, \delta, \omega)$ where Σ is a finite nonempty input alphabet, Q is a finite nonempty state set, Γ is a finite nonempty output alphabet, $\delta : Q \times \Sigma \rightarrow Q$ is the state transition function, and $\omega : Q \times \Sigma \rightarrow \Gamma$ is the output function. To this we can add a start state q_0. Finite transducers ultimately rely on mechanical laws to function, transforming informational input to informational output by transforming their own states.

A *finite acceptor* is a 5-tuple $(Q, \Sigma, \delta, q_0, A)$, where Q is a nonempty finite set of internal states, and Σ is an alphabet, q_0, is the start state, and $A \subseteq Q$ is the set of accepting states. The range of the transition mapping δ determines the type of acceptor; it is *deterministic* if $\delta: Q \times \Sigma \rightarrow Q$, and *nondeterministic* if $\delta: Q \times \Sigma \rightarrow 2^Q$ (where 2^Q represents the power set of possible states). A deterministic finite acceptor $(Q, \Sigma, \delta, q_0, A)$ *accepts* a string $x \in \Sigma^*$ iff $\delta(q_0, x) \in A$. A *language* is the set of strings accepted by a given automaton or class of automata.

Languages are generated by grammars. In the computational theory of language, a generative (or phrase structure) grammar G is a 4-tuple (N, T, P, σ) consisting of:

(1) a finite set N of *nonterminals*;

(2) a finite nonempty set T of *terminals*, with $N \cap T = \emptyset$ and $N \cup T = A$ (the total *alphabet* of the grammar);

(3) a finite set of productions $P \subset ((N \cup T)^*\backslash T^*) \times (N \cup T)^*$ consisting of nonterminal arguments and their possibly terminal transforms;

(4) an element σ of N called the *starting symbol*.

The implementation of such a grammar is a deductive process leading from the general to the specific; starting from the most general symbol σ (which stands for "sentence"), increasingly specific productions lead to a terminal configuration. The production (x, y), often written $x \rightarrow y$, signifies replacement of x by y, or equivalently, the *substitution* of y for x. Where A^* denotes the set of all strings or "words" in A, and $A^*\backslash T^*$ denotes the complement of T^* in A^*, a word $w \in (A^*\backslash T^*)$ *generates* another word w′ if $w = w_1 X w_2$, $w' = w_1 X' w_2$, and $X \rightarrow X'$ is a production.

The theory of generative grammars classifies them according to the least powerful acceptor that can recognize the languages they generate. Type 0 grammars generate unrestricted languages requiring a universal computer (Turing machine) with unlimited memory; type 1 grammars generate context-sensitive languages requiring a linear-bounded automaton with memory proportional to word length; type 2 grammars generate context-free languages requiring a pushdown automaton with a memory stack in which a fixed number of elements are available at any point; and type 3 grammars generate regular languages requiring a finite deterministic automaton with no memory.

There is an obvious parallel between the states and state transitions of automata, and the strings and productions of a grammar. An automaton processes input strings through its internal states, expressing them in terms of its own "internal language". Indeed, a physical automaton in the act of processing an input string can be seen as a dynamic linguistic stratification incorporating the input language, the mutable programming of the automaton (including assembly and machine code), its hard-wired architecture, the nomological language consisting of the laws of physics according to which the hardware functions, and any "metaphysical" level of language necessary to define and maintain the laws of physics themselves. Since each language in this sequence is expressed in terms of the next one after it, the languages form a "descriptive nesting" in which the syntax of each distributes over all of those preceding it.

The *syntax* of a language consists of its grammar and the structure of its expressions. That is, a syntax is a compilation of the spatial (structural) and temporal (grammatical, transformational) rules of the associated language; its rules are invariant, general, and distributive with respect to the entire set of expressions comprising the language. This concept is as meaningful for automata as it is for the languages they process, applying to every level of the linguistic stratification just described. For example, where the concept of general covariance expresses the general and distributive nature of the laws of physics, these laws can be regarded as a "syntax" unto themselves, and so can the more general mathematical laws applying to the various mathematical structures to which the laws of physics implicitly refer.

Physics and mathematics are usually viewed not as languages, but as theories. Even though they are necessarily expressed in terms of language, they are usually considered "more specific". But like automata, they too meet linguistic criteria. For instance, mathematical theories have syntaxes consisting of axioms and rules of inference, and various derived expressions such as definitions, theorems and corollaries. More generally, a theory is simply an informational construct that plays a direct definitive, descriptive or explanatory role with respect to something that needs to be defined, described or explained. Because theories consist of recognizable strings of symbols taking the form of statements and equations and obey "syntaxes" consisting of axioms, principles, hunches or rules of thumb, and in fact share their syntaxes with the objects of theorization up to descriptive isomorphism, they are languages. Indeed, the very requisites of theorization, namely perception and cognition, are languages in the sense that they consist of sensory or conceptual "expressions" and conform to logical and nonlogical syntaxes consisting of general rules of structure and operation, including (but not necessarily limited to) the physical structures and dynamics of our brains and nervous systems.

Let us quickly review some of the technical details of theoretical languages. A mathematical theory consists of propositions containing basic predicates and functions representing fundamental concepts. For example, set theory is based on the concept of *membership* (∈); geometry is strongly dependent on primitive concepts like *angle* and *distance*; and elementary arithmetic incorporates the more or less basic concepts of *addition*, *multiplication* and *order* (<). The symbols that stand for these concepts, sometimes called *nonlogical constants*, form the signature Σ of the theory. This signature includes symbols for relations, functions and individuals, each with an "arity" or valence. Beyond its signature, a mathematical theory contains *variables* (preassigned symbols denoting objects from a previously specified domain) for individuals represented by symbols like (x, y, z, ...), logical symbols (logical constants) like (~, ∧, ∨, →, ↔, =, ∃, ∀), and auxiliary technical symbols (Gellert *et al.*, 1977).

An *elementary language* L_Σ (language of the predicate calculus) can be defined as a set of expressions or propositional forms incorporating these logical and nonlogical symbols and their syntactic rules. The terms of L_Σ are defined as follows:

(1) Variables and constants for individuals are terms;

(2) If F is an n-ary function symbol and $t_1, \ldots, t_n$ are terms, then $Ft_1, \ldots, t_n$ is a term;

(3) A sequence of symbols is a term *iff* it conforms to 1 and 2;

The expressions of L_Σ are characterized thusly:

(1) Where R is an n-ary function symbol and $t_1, \ldots, t_n$ are terms, then $Rt_1, \ldots, t_n$ is an *atomic expression* (i.e., an instance of attribution);

(2) If A and B are expressions, then so are ~A, (A ∧ B), (A ∨ B), (A → B), and (A ↔ B);

(3) If A(x) is an expression containing the variable x, but not ∃ x or ∀ x, then so are ∃ x A(x) and ∀ x A(x);

(4) A sequence of symbols is an expression only if formed in accordance with 1-3.

By nature, the languages of predicate logic are descriptive; their expressions describe relationships holding within various mathematical structures.[17]

It should now be clear that in order to define a theoretical language, one simply creates a syntax for it by extending the syntax of logic to accommodate the spatial and temporal relationships necessarily holding among its nonlogical

[17] Not all expressions are propositions; propositions are expressions with no free (unquantified) variables. Permitting the quantification of unary (one-place) predicates and treating them as individuals results in a generalization of the elementary languages to a more expressive class of languages, the monadic languages of the second order. Further generalizations are possible along these lines.

constants under all (or at least most) logical circumstances within the intended scope of the theory. Like the syntax of logic itself, which is a theory of descriptive inclusion or "truth" based on the cognitive and perceptual need to distinguish *that which is* from *that which is not*, the syntax of a theoretical language is also based on ingredients and imperatives of cognitive and perceptual reality. We have already remarked on the equivalence of automata and languages; the states and state transitions of automata parallel the strings and productions of a grammar. To find the theoretical language describing any class of transducers, we need merely adjoin to the syntax of logic the nonlogical descriptors of their transductive structures and processes.

The primary transducers of the overall language of science are scientists, and their transductive syntax consists of the syntax of generalized scientific observation and theorization, i.e. perception and cognition. We may therefore partition or stratify this syntax according to the nature of the logical and nonlogical elements incorporated in syntactic rules. For example, we might develop four classes corresponding to the fundamental trio *space*, *time* and *object*, a class containing the rules of logic and mathematics, a class consisting of the perceptual qualia in terms of which we define and extract experience, meaning and utility from perceptual and cognitive reality, and a class accounting for more nebulous feelings and emotions integral to the determination of utility for qualic relationships.[18] For now, we might as well call these classes STOS, LMS, QPS and ETS, respectively standing for *space-time-object syntax*, *logico-mathematical syntax*, *qualio-perceptual syntax*, and *emo-telic syntax*, along with a high-level interrelationship of these components to the structure of which all or some of them ultimately contribute. Together, these ingredients comprise the Human Cognitive-Perceptual Syntax or HCS.[19]

---

[18] Lest the inclusion of utility, qualia or feelings seem "unscientific", we need merely observe that it would be vastly more unscientific to ignore things that are subjectively known to exist on the wishful and rationally unjustifiable assumption that subjectivity and subjective predicates play no part in the self-definition of reality. Insofar as subjectivity merely refers to the coherent intrinsic identities of the elements of objective relationships, this would be logically absurd. But in any case, our aim at this point is merely to classify the basic elements in terms of which we view the world, whether or not they have thus far proven accessible to standard empirical methodology, and this means recognizing the reality of qualic and emotional predicates and adjoining the corresponding nonlogical constants to SCSPL syntax. If QPS and ETS predicates turn out to be reducible to more fundamental STOS/LMS predicates, then very well; it will permit a convenient reduction of the syntax. But this is certainly not something that can be decided in advance.

[19] Cognitive-perceptual syntax consists of (1) sets, posets or tosets of attributes (telons), (2) perceptual rules of external attribution for mapping external relationships into telons, (3) cognitive rules of internal attribution for cognitive (internal, non-perceptual) state-transition, and (4) laws of dependency and conjugacy according to which perceptual or cognitive rules of external or internal attribution may or may not act in a particular order or in simultaneity.

As every language user is aware, there is more to language processing than recognition and transduction. There is also *communication*. The concepts of language and communication are inseparable; languages are abstract structures through which communication is effected, and communication involves exchange of input and output by language users. Any time we have a syntax, a set of representative expressions conforming to it, and a set of language processors, we have a language and the possibility of meaningful communication among its processors. Where communication conveys information and information represents relationships, communication is about more than just the states of communicators; it can carry any information representing any kind of relationship. Accordingly, communicators not only accept and transduce language, but use it to represent to each other their views of the world.

The communication paradigm is perfectly general and holds on all scales. It applies not merely on the level of expressions exchanged by communicators, but even on the level of basic status information exchanged by the interactive processing elements of an automaton. That is, language processing itself can be regarded as a form of communication; in order to actualize a language through some form of processing, it must be intersected with a processing system that functions through operational communication among its parts. A universal machine, for example, is a "self-communication system" relaying information among its abstract components (indeed, communication is the very embodiment of the three basic operations of a universal computer, namely *read, write* and *relocation*). The structural and dynamical rules of language processors thus correspond directly to the syntaxes of languages; both kinds of system evolve by communicative intersection of syntax and state (or content). It follows that languages and automata can be treated on a par, and that since automata can be treated as self-communication systems, so can languages.

In the technical descriptions of automata and languages outlined above, a certain model of the world is implicit. Automata accept input from the outside world, transform it through their internal states, and (sometimes) return the result to the outside world when they are finished. Language and information are either *inside* or *outside* the automata, and they are inside only part of the time. The rest of the time, the information is presumably either lying around someplace else in the environment or *en route* from one place to another. But where communication happens on all scales, the distinction between *inside* and *outside* is not so clear. The languages communicated among language users and processors, and the languages embodied by users and processors themselves, occupy an overall medium with a unified communicative syntax largely indifferent to the distinction.

The laws that govern a system may be reposed in the space that contains its objects, or in the objects themselves. Classical physics reposes everything in space, applying spatial concepts like vectors and tensors to fields outside the objects. However, it is possible to apply a logical transformation which inverts this picture, turning it "outside-in". This results in a "distributed subjectivization" in which everything occurs *inside* the objects; the objects are simply defined to consistently internalize their interactions, effectively putting every object "inside" every other one in a generalized way and thereby placing the *contents* of space on the same footing as that formerly occupied by the containing space itself. Vectors and tensors then become descriptors of the internal syntactic properties and states of objects. In effect, the universe becomes a "self-simulation" running *inside* its own contents.

This view, which is complementary to the conventional geometric one, is called *transductive algebra*. The "dual" relationship between geometry and transductive algebra is called *conspansive duality*. In conjunction with other principles including hology and SCSPL-infocognitive-telic reducibility, conspansive duality can afford fresh insight on the nature of reality and the physical world. One simply takes the conventional picture, turns it outside-in, puts the two pictures together, and extracts the implications. The relationship of this new picture to the old one is extensional rather than competitive, embedding the geometric model in a larger and more complete conspansive model uniting it with its dual model. In any case, all equivalent models are on an equal footing, and the only scientific hypotheses on which doubt could be shed as a result of this extension are those based on the fallacious assumption that the geometric model is the "whole story".

Introduction to SCSPL

According to the Reality Principle, the universe is self-contained, and according to infocognitive monism, it regresses to a realm of nil constraint (*unbound telesis* or UBT) from which it must refine itself. According to the Telic Principle, which states that the universe must provide itself with the means to do this, it must make and realize its own "choice to exist"; by reason of its absolute priority, this act of choice is identical to that which is chosen, i.e. the universe itself, and thus reflexive. I.e., "existence is everywhere the choice to exist." Accordingly, the universe must adopt a reflexive form in which it can "select itself" for self-defined existence, with the selection function identical to that which is selected. This means that it must take a certain general or "initial" form, the *MU form*, which contains all of the requisites for generating the contents of reality. Due to hology, whereby the self-contained universe has nothing but itself of which to consist, this form is self-distributed.

One might at first be tempted to object that there is no reason to believe that the universe does not simply "exist", and thus that self-selection is unnecessary. However, this is not a valid position. First, it involves a more or less subtle appeal to something external to the universe, namely a prior / external informational medium or "syntax" of existence; if such a syntax were sufficiently relevant to this reality, i.e. sufficiently real, to support its existence, then it would be analytically included in reality (as defined up to perceptual relevance). Second, active self-selection is indeed necessary, for existence is not merely a state but a process; the universe must internally distinguish *that which it is* from *that which it is not*, and passivity is ruled out because it would again imply the involvement of a complementary active principle of external origin.

By the Principle of Linguistic Reducibility, reality is a language. Because it is self-contained with respect to processing as well as configuration, it is a *Self-Configuring Self-Processing Language* or SCSPL whose general spatio-temporal structure is hologically replicated everywhere within it as self-transductive syntax. This reduces the generative phase of reality, including physical cosmogony, to the generative grammar of SCSPL. This reality-generative grammar is called *Γ grammar* and the MU form, being the most general or prior form of reality, is its basis. By the Principle of Infocognitive Monism and the hology of MU, SCSPL consists of MU-configured infocognition, and Γ grammar describes the generation and transformation of this universal constituent.

SCSPL is not an ordinary language, and Γ grammar is not an ordinary generative grammar. The reasons come down to the inherent limitations of computational language theory. In standard computation theory, a language consists of the set of strings accepted by a given automaton or class of automata; e.g., a language L is called "regular" if there is a finite-state automaton that accepts it. However, this approach is inadequate for SCSPL. First, it centers on *computation*, a general type of information processing associated with an abstract automaton, the Turing machine or "universal computer", that could never have generated the informational structure of the real universe. Being an informational and metainformational (syntactic) construct, the universal computer can itself account for the genesis of neither syntax nor information. Second, unlike ordinary languages, the reality-language cannot rely on an external mind or automaton or preexisting hardware substrate for recognition and processing. Since any processor real enough to recognize and process reality is necessarily a *part* of reality, the language-processor distinction is without ontological force.

Thus, while ordinary discrete models of reality rely heavily on the language-processor distinction, SCSPL incurs no such debt. For example, cellular automaton models typically distinguish between a spatial array, the informational objects existing therein, and the distributed set of temporal state-transition rules by which the array and its contents are regulated. In contrast, SCSPL regards language and processor as aspects of an underlying infocognitive unity. By conspansive (ectomorphism-endomorphism) duality, SCSPL objects contain space and time in as real a sense as that in which spacetime contains the objects, resulting in a partial identification of space, time and matter. SCSPL is more than a reflexive programming language endowed with the capacity for computational self-execution; it is a protocomputational entity capable of telic recursion, and thus of generating its own informational and syntactic structure and dynamics.

Whereas ordinary computational models are informational and syntactic in character, the protocomputational nature of SCSPL requires a generalization of information and syntax. With respect to the origin or ultimate nature of perceptual reality, explanation is a reductive / inductive process that regressively unbinds constraints in order to lay bare those of highest priority and generality. This process eventually leads to the most basic intelligible descriptor that can be formulated, beyond which lies only the unintelligible. This marks the transition from information and syntax to a convergent reductive generalization, *telesis*.[20]

This points to a universal property of language: it is dynamic. While it is often conveniently assumed that languages are timeless Platonic ideals that waft around waiting to be processed by external acceptors, they can be treated in terms of static information only by users or processors that provide them with an external transductive syntax, and only then by neglecting certain underlying necessities. For example, to physically realize the informational structure of a language in a printed or electronic medium, it must be expressed in terms of physical particles that dynamically recognize each other's locations to the extent required to maintain the spatial relationships comprising its informational structure. This is a general requirement, extending from the physical and concrete to the mathematical and abstract.

Thus, languages are ultimately self-processing; they must either contain their processors in their expressions, or be expressed in terms of a more basic language fulfilling this requirement. Accordingly, the expressions of SCSPL are dynamic informational configurations of information-processors, implying that SCSPL everywhere consists of information and acceptive-transductive syntax in a state of logical intersection. Together, information and syntax comprise *infocognition*, self-transducing information in the form of SCSPL *syntactic operators* that cross-absorptively "communicate" by acquiring each other's informational states as cognitive-syntactic content. It is to the common basis of these two components that information may be reduced in the SCSPL context. Where the term *telesis* denotes this common component of information and syntax, SCSPL grammar refines infocognition by binding or constraining telesis as infocognition.

---

[20] To see how information can be beneficially reduced when all but information is uninformative by definition, one need merely recognize that information is not a stand-alone proposition; it is never found apart from syntax. Indeed, it is only by syntactic acquisition that anything is ever "found" at all. That which is recognizable only as syntactic content requires syntactic containment, becoming meaningful only as acquired by a syntactic operator able to sustain its relational structure; without attributive transduction, a bit of information has nothing to quantify. This implies that information can be generalized in terms of "what it has in common with syntax", namely the syndiffeonic relationship between information and syntax.

To the extent that any grammar functions by the recursive application of syntactic rules, SCSPL grammar is recursive ("self-calling"). However, SCSPL grammar is not merely deterministically or nondeterministically recursive, but telic-recursive. While an ordinary grammar recursively processes information or binds informational potential to an invariant syntax that distributes over its products, Γ grammar binds *telesis*, infocognitive potential ranging over possible relationships of syntax and state, by cross-refining syntax and its informational content through telic recursion. Telic recursion is the process responsible for configuring the syntax-content relationships on which standard informational recursion is based; its existence is an ontological requirement of reality. The telic-recursive cross-refinement of syntax and content is implicit in the "seed" of Γ grammar, the MU form, which embodies the potential for perfect complementarity of syntax and state, law and matter.

Since this potential can only be specifically realized through the infocognitive binding of telesis, and localized telic binding is freely and independently effected by localized, mutually decoherent telic operators, deviations from perfect complementarity are ubiquitous. SCSPL evolution, which can be viewed as an attempt to help this complementarity emerge from its potential status in MU, incorporates a global (syntactic) invariant that works to minimize the total deviation from perfect complementarity of syntax and state as syntactic operators freely and independently bind telesis. This primary SCSPL invariant, the *Telic Principle*, takes the form of a selection function with a quantitative parameter, *generalized utility*, related to the deviation. The Telic Principle can be regarded as the primary component of SCSPL syntax ... the spatiotemporally distributed self-selective "choice to exist" coinciding with MU.

SCSPL incorporates the concepts of syntactic stratification and syntactic distribution. For example, because the laws of mathematics everywhere apply with respect to the laws of physics, the former distribute over the latter in the syntactic sense. Thus, where the laws of mathematics and physics are denoted by S1=LMS and S2 respectively, S1 *distributes over* S2, i.e. forms a *syntactic covering* for S2. Essentially, this means that the laws of mathematics are everywhere a required syntactic component of the language of physics. With S2 is associated an SCSPL "sublanguage" called L_O (with a letter O subscript). L_O constitutes the world of perception, the classical objective universe of sense data traditionally studied by science. L_O is contained in the telic-recursive, pre-informational phase of SCSPL, L_S, which encompasses the cross-refinement of L_O syntax and L_O content from the pre-infocognitive aspect of SCSPL. The part of SCSPL grammar confined to L_O incorporates certain restrictions to which L_S is not subject; e.g., the grammatical portion of L_O (S2) is fixed, distributed and supposedly continuous, while that of L_S can also be mutable, local and discrete ... in a word, telic.

Γ grammar is the generative grammar of SCSPL = ($L_S \supset L_O$). Γ grammar is unlike an ordinary grammar in that its processors, products and productions coincide and are mutually formed by telic recursion. Syntax and state, loosely analogous to *form* and *content* (or *productions* and *products*), are mutually refined from telesis through telic recursion by infocognitive processors. Production rules include the Telic Principle, distributed elements of syntax formed in the primary phase of telic recursion, and more or less polymorphic telons formed by agent-level telors. The corresponding modes of production are global telic recursion, informational recursion by distributed syntax, and local telic recursion.

The "words" produced by Γ grammar are not strings of symbols, but L_O spatial relationships among parallel processors that can read and write to each other's states. In effect, the states of its processors are roughly analogous to the symbols and strings of an ordinary language. The processors of Γ grammar thus function not only as transducers but as symbolic placeholders for observables and values, while their external states correspond to products and their state transitions realize the productions of the grammar. In other words, the states and state transitions of the processors of Γ grammar comprise a representation of Γ grammar, rendering SCSPL a dynamic self-modeling language or "interactive self-simulation".

In the following description, products are limited to the physical end products associated with L_O. This is in no way to be interpreted to mean that the only SCSPL products are physical products; in fact, everything in SCSPL is to some extent a telic-recursive "product" of everything else. However, where the primary explanandum is physical (scientifically observable) reality, L_O is an appropriate choice.

$\Gamma = (O, R, P, \mu)$ contains:

(1) A set O of active reflexive objects including Γ itself, the *processors* (producers-reducers)[21] of Γ. All processors are capable of and responsive to informational (deterministic) recursion. O includes a distinguished set $\Sigma = \{\Gamma, A, Q\}$ of *syntactic operators*, coherent processors capable of or responsive to telic recursion. In general, the parts of SCSPL syntax active within a given operator depend on its specific type. Σ includes the global processor Γ, the set $Q = \{q_i\}$ of reducible and irreducible stable particles that are responsive to telic recursion in a degree proportional to their freedom and coherence, and the set A of telic *agents*, active telic-recursive operators or *telors* capable of expressing teleology on the local level. Elements of A need not occupy L_O, but may exist in L_S. Where υ denotes the generalized self-selection parameter of Γ, the elements of A are required by Γ as internal υ-responsive "sensor-controllers".

(2) A set $R = (R_o, R_s)$ of products containing a subset $R_O = (\sigma_O, \tau_O, \pi_O)$ of L_O *product states* including all of the relations of L_O, and a subset R_s of *telons*

[21] For present purposes, reduction may be understood as reverse production. As such, it is essential to grammatical recognition.

or "pre-products" generated in L_S by telic recursion associated with global or agent-level syntactic operators. The elements of σ_O are *spatial* relations, those of τ_O are *temporal* relations, and those of π_O are *spatiotemporal* relations containing both σ_O and τ_O relations. σ_O consists of relations of states regarding which the corresponding transducers have no mutual input; τ_O consists of sequences of consecutive states of single objects; and π_O consists of *processes*, or combinations of both. These states are not in general fully determined by L_O rules of state transformation, but require telic augmentation for actualization.

(3) A set P = (N, T) consisting of the *productions* of Γ. N consists of the distributed descriptors of π_O, including the spatiotemporally-distributed Markovian and conservative descriptors called "laws of physics", while T consists of active L_S entities analogous to those of N but associated with specific telons (pre-products) from R_S and subject to coordinated local variation in the context of higher-order telonic relationships. The elements of N are determined in the primary stage of telic recursion, while those of T are determined in the secondary stage of telic recursion.

(4) A starting configuration, the *MU form* μ, which is identical to the telic recursion event which creates it (the inception of SCSPL and Γ grammar is a telic recursion, not an informational-algorithmic recursion). It is this identity of event and outcome that determines the innate spatial and temporal characteristics of spacetime, for which μ is the "seed". The MU form can be regarded as an "intrinsic perturbation" or "intrinsic asymmetry" in UBT. The MU form is distributed over SCSPL.

A processor of a grammar G is any natural or artificial dynamical system that operates, changes state or processes information in conformance to the rules of G. Unlike ordinary generative grammars, Γ grammar requires no external processors; its processors and productions are identical. Thus, Γ grammar is executed by its own productions in levels of syntactic distribution ranging from the global to the object level. In fact, O, R and P – processors, products (states) and production events – all coincide and thus exhibit a form of triality. This three-way coincidence is characteristic of Γ grammar and captures many of its essential features.

O-R-P coincidence is already to some extent realized in the standard language-grammar-processor model of computation theory, but only inadvertently. While linguistic processing is dynamically paralleled by changes in the internal and external states of processors, the processors are still considered separate from the language and grammar being processed. Moreover, the basic medium of processing is not considered, the model is not self-sufficient, and recursion is merely informational and computational; there is no allowance for infocognition or telic recursion. SCSPL shares none of these limitations.

Γ grammar generates SCSPL according to the utility of its sentient processors, including the self-utility of Γ and the utility of its L_O relations to telors in A. Γ and A generate telons on the global and local level respectively; thus, they must be capable of recognizing and maximizing the selection parameter υ (in the case of human telors, for example, this requires the QPS and ETS components of the HCS). As such, they are responsible for telic recursion and may be regarded as the "generators" of Γ grammar, while the set Q of elementary physical objects are freely and competitively acquired by telons and thus occupy an ontologically secondary position.

Γ grammar is conspansive. Non-global processors alternate between the generation and selective actualization of possible productions, and thus between the generative and selective (inner expansive and requantizative) phases of conspansion. The selective phase of an operator coincides with interactive mutual-acquisition events, while the generative phase coincides with the generation and selective actualization of possible productions through hological multiplexing. In conjunction with extended spatiotemporal superposition, conspansion provides the means of local (telic and informational) recursion.

Conspansion is a global process which cannot be locally differentiated as to rate. Its rate is thus a globally invariant "time-space conversion factor", and because all changes of local state must be expressed in terms of it, it is maximal. This invariant maximal rate of production is referred to as the *rate of conspansion* c and can be physically identified with the speed of light *in vacuo*.[22] The implications of the constancy of c and N in light of MAP have already been noted with respect to internally-apparent accelerating expansion of the global operator.

It is instructive to experiment with the various constructions that may be placed on L_S and L_O. For example, one can think of L_S as "L-sim", reflecting its self-simulative, telic-recursive aspect, and of L_O as "L-out", the output of this self-simulation. One can associate L_O with observable states and distributed-deterministic state-transition syntax, and L_S with the metasyntactic Telic Principle. One can even think of L_S and L_O as respectively *internal* and (partially) *external* to SCSPL syntactic operators, and thus as loosely correspondent to the *subjective* and *objective* aspects of reality. Where L_S and L_O are associated with the coherent inner expansion and decoherent requantization phases of conspansion, so then are subjective and objective reality, simulation and output, "wave and particle". In other words, the subjective-objective distinction, along with complementarity, can be viewed as functions of conspansive duality.

The fact that L_O has a foliated structure consisting of spacelike sheets, with temporal rules confined to the operators embedded in the sheets, suggests that

---

[22] Because the propagation of light is a conspansion-dependent phenomenon, c cannot be affected by impedance or manipulation of photons.

its inter-operator (state-wise, ectosyntactic) level of structure be regarded as essentially spatial in character. Thus, where *space* denotes the external relationships among operators and *time* denotes their internal self-relationships, one might also think of L_S and L_O as corresponding approximately to time and space. (The correspondence is "approximate" because L_S and L_O are mutually inclusive, reflecting the logical coupling of space and time; L_O topologically contains (L_S, L_O)-structured operators, while the operators descriptively contain L_O.) Where space and time respectively correspond to information and a combination of generalized cognition and telic recursion, one may therefore conclude that the conspansive evolution of spacetime is an alternation of teleo-cognitive and informational phases cross-refined by telic recursion involving extended, trans-Markovian telonic relationships.

Although it contains the observable aspect of SCSPL, L_O may in principle contain hidden (inobservable) parameters implicated in nomological relationships and therefore relevant to physical state. I.e., in addition to the standard dynamical variables of physics, L_O may contain additional dynamical variables that cannot be directly observed, but only theoretically inferred on the basis of more or less general observations. For example, string theorists, M-theorists and others often conjecture that the universe may be floating in some sort of external embedding space, its relationship to which yields explanatory value regarding the internal state or condition of the universe itself. SCSPL conspansive duality suggests that the putative "externality" of such a space is not a meaningful property; if such a space exists and is sufficiently relevant to the universe in which we live to be of explanatory value to us, then it is by definition an ingredient of SCSPL syntax and can therefore be regarded as residing within SCSPL syntactic operators. In fact, this is a direct consequence of the Reality Principle.

Although it would be possible to go on at length, this paper is intended to present a general outline of the theory rather than an exhaustive development. In conjunction with the principles and features enumerated above, the given framework should suffice to characterize SCSPL on an introductory basis and distinguish it in flavor and content from other theories.

SCSPL as the Self-Excited Circuit

We are now in a position to draw a few parallels between Wheeler's vision of reality theory and the CTMU.

The **Self-Excited Circuit**, the informational logic loop through which physics engenders observer participation, which engenders information, which engenders physics, is a tight characterization of SCSPL ... so tight that it would be difficult if not impossible to replace SCSPL with anything else and neither violate nor fall short of Wheeler's description. SCSPL is logical in construction, has a loop-like dynamic, and creates information and syntax, including the laws of physics, through telic recursion generated by agent-level syntactic operators whose acts of observer-participation are essential to the self-configuration of the **Participatory Universe**. These acts are linked by telic recursion to the generalized cognitive-perceptual interactions of quantum-level syntactic operators, the minimal events comprising the fabric of spacetime.

Through telic feedback, state and syntax are cross-refined from unbound telesis or UBT, a zero-information domain of ontological potential, under the guidance of a higher-order law called the Telic Principle ... a protean **Law Without Law** through which order is extracted from disorder as laws are configured according to the generalized utility of state-syntax relationships for agent-level operators or observer-participants. The binary yes-or-no indications prescribed by **It from Bit** are demanded by infocognitive monism and the fundamental status of two-valued sentential logic in SCSPL syntax. The world is not merely a cybernetic monstrosity, a "giant machine ruled by preestablished law", but a metacybernetic system with logical priority over machines and the laws they obey.

How come existence? is answered by the fact that the universe is a global SCSPL operator amounting to one vast, self-selective, self-expressive act of reflexive observer-participation, while **How come the quantum?** is answered by the hological self-replication of the universe in each one of its microscopic syntactic operators and agent-level telors. **Many observer-participants** yield **one coherent world** because, through MU, the universe relates to its contents as a homogeneous distributed syntax that syndiffeonically supports and expresses their distinctions even as they help it evolve through observer-participation and telic recursion. Individual solipsism becomes *distributed solipsism* through the mutual absorption of SCSPL syntactic operators, made possible by a combination of distributed SCSPL syntax and shared teleology.

The Reality Principle, along with MAP, M=R and other logical guarantors of cosmic self-containment, shows that the syntactic stability of reality rules out any infinite reptilian regress of **turtle on turtle**, while the familiar **continuum** of classical physics corresponds to a syntactic (LMS) interpolation of the conspansive manifold generated by discrete SCSPL grammatical operations. Where **space and time** correspond to *information* and *generalized cognition* respectively, and where information and cognition are logically entwined in infocognitive SCSPL syntactic operators intersecting in states and state-transition events, space and time are entwined in a conspansive event-lattice connected by syntax and evolving through mutual absorption events among syntactic operators, symmetric instances of generalized observation influenced by telic recursion. Thus, time is not "fed into" the explanation of existence, but is a function of conspansive, telic-recursive SCSPL grammar.

The ultimate **boundary of the boundary** of the universe is UBT, a realm of zero constraint and infinite possibility where neither boundary nor content exists. The supertautologically-closed universe buys internal diffeonesis only at the price of global synesis, purchasing its informational distinctions only at the price of coherence. **No question? No answer!** reflects the fact that reality consists not of mere information, but *infocognition*, and that information on state is crucially linked to and dependent on syntax ... the syntax of the "questions" asked of itself by the self-configuring universe. Due to the self-configurative freedom inherited by reality from UBT, the dynamically self-configuring universe displays uncertainty and complementarity and thus cannot be locked into locally-determinate answers for all possible questions at once, while the extended self-connectivity of conspansive spacetime unavoidably implicates the environment in the Q&A.

The **Super-Copernican Principle** reflects the distribution of the creation event over every point of spacetime according to the Extended Superposition Principle, which describes the way consequent states and events are atemporally superposed in their antecedent states and events. **Generalized consciousness**, the unitary reflexivity of the universe, is a fair description of the self-configuration and self-processing capacity of SCSPL as captured by the Telic and M=R Principles, while conspansive spacetime links spatially and temporally distant objects in a web of contact and communication exceeding even the neural connectivity of a human brain. And the CTMU describes the universe as just the sort of complex, teleologically self-variegating, self-synthesized information system prescribed by **More is different**, telic-recursively explicating multiplicity and diffeonesis from the unity and synesis of distributed SCSPL syntax, the (unique) CTMU counterpart of what has sometimes been called "the Implicate Order".[23]

The above analogy is only partial, and it is nowhere implied that the current presentation is without its gaps. But in fairness, several crucial points can be made in favor of the CTMU even at this early stage of exposition. First, it is supertautological; being constructed to mirror logical tautology up to the level of model theory and beyond, it is true in much the same way that a theory of pure mathematics would be true, but with reference to an expanded universe consisting of both mathematical and physical reality. Indeed, the CTMU can be viewed as a theory of the mathematical structure of a new mathematical object, SCSPL. Second, it has considerably more explanatory scope than other theories, providing a framework that is capable of accommodating the logic of self-

---

[23] The Implicate Order is a concept championed by the late physicist David Bohm in connection with his interpretation of quantum mechanics. Using the analogy of a hologram, Bohm proposed a species of self-similarity whereby "everything is enfolded into everything". According to Bohm, this "implicate order" is complementary to an "explicate order", the result of unfolding it through a process called the "holomovement". This bears mention because of its superficial resemblance to the conspansive unfolding of state from syntax in SCSPL, which displays its own version of logical hyperconnectivity, and the similarity between hology and holography. However, since the CTMU intersects only partially with Bohm's theory, the analogy should not be overextended.

determinative cosmogony without prejudicially excluding (e.g.) subjective and non-material aspects of reality. Third, it largely embeds current models of reality, at least to the extent that these models have not already been pushed beyond their explanatory capacities. And fourth, it has what appear to be empirically valid implications absent from other theories except as assumptions or unexpected observations, e.g. accelerating cosmic expansion. But perhaps the most important thing at this point is that in principle, any apparent explanatory gaps can be filled. That is, if something can be explained within the realm of standard science, then it can be even better explained in an inclusive model fortified with conspansive duality.

The ramifications of the CTMU are sufficiently extensive that their issuance from a single theory almost demands an explanation of its own. The scientific, mathematical and philosophical implications of the CTMU are many and varied, running the gamut from basic physics and cosmology to evolutionary biology, the theory of cognition, the foundations of mathematics and the philosophies of language and consciousness. But to be fair, nothing less is to be expected of a true "reality theory", particularly one that takes the form of a description of the relationship between mind and the universe. After all, the CTMU is so-named because it is a symmetric cross-interpretation of mental and physical reality, logically mapping the concrete universe into an abstract theory of generalized cognition and vice versa according to the M=R Principle. Were its implications anything less than profound, it would be miscategorized and misnamed.

The CTMU says that reality is a language ... a self-explanatory, self-theorizing, self-modeling structure identical to its universe. As such, it can be regarded as a limit or "contraction" of model theory in which theory and universe coincide on the syntactic level of theoretical structure. Whereas most scientific theories are hopefully mapped into or onto the universe across an unbridgeable dualistic gulf, the CTMU is a monic theory of perception that simply eliminates the gulf by tautologically injecting logic in its entirety, including logical extensions of model theory and the logic of formalized theories, into reality as distributed self-transductive syntax. The CTMU is able to do this because it is a hard mathematical fact that anything which does not conform to the syntax of logic is inconsistent and therefore not a part of any stable, coherent reality. Because the reality we inhabit is visibly stable and coherent, the correctness of this move is assured. By eliminating the theory-universe gulf on the level of cognitive and perceptual syntax, the CTMU admirably fulfills the criterion of theoretic economy ... and supertautologically at that.

Does the CTMU qualify as a realization of Wheeler's vision, and is it alone in this distinction? While one naturally hesitates to put words into the mouth of an icon, one or both of two things seems to be true: either relatively few reality theorists are inclined to share Wheeler's far-reaching worldview, or relatively few reality theorists are able to understand this worldview and pursue its implications. Consequently, despite Wheeler's eminence as a physicist, his has almost seemed a voice in the wilderness, leaving some with the impression that his deepest speculations have more the ring of distant prophecy than immediate theoretical practicality. But while the questions, "no's" and clues in terms of which Wheeler describes his vision may occasionally appear heavier on intuition and creativity than on logical and mathematical rigor, they are just the sort of intuitive distillations of mathematical insight that one might expect from one so experienced in logical and quantitative reasoning. In conjunction with other necessary principles, they can be transformed into logico-mathematical properties of reality with enough restrictive power among them to determine a new mathematical structure called SCSPL, the ultimate "intrinsic language". When all is said and done, there are excellent reasons to believe that this structure is unique, and that any additional principles that Professor Wheeler might have in mind can be accommodated by the CTMU as surely as by logic itself.

The CTMU and Intelligent Design

Design theory, which traces its origins to traditional theological "arguments from design" holding that nature was more or less obviously designed by a preexisting intelligence, maintains that the observed complexity of biological structures implies the involvement of empirically detectable intelligent causes in nature. *Intelligent Design*, the most recent scientific outgrowth of Design Theory, is a scientific research program based on a more philosophically neutral, and therefore scientific, search for instances of a clear, objective, standard form of biological complexity. According to William Dembski (1997), one of the movement's leading spokesmen, this has led to "a theory of biological origins and development" according to which "intelligent [and empirically detectable] causes are necessary to explain the complex, information-rich structures of biology." In view of the informational nature of complexity, Dembski observes that "information is not reducible to natural causes ... the origin of information is best sought in intelligent causes. Intelligent design thereby becomes a theory for detecting and measuring information, explaining its origin, and tracing its flow."

One of the first things to note about the above definition is that it couples the implied definitions of *intelligence, causation* and *information* to a greater extent than do most dictionaries, pointing in principle to a joint definition of all of them. Since any good definition requires a model, one might be strongly tempted to infer on this basis that ID, as here defined, has a well-defined model in which all of its constituent concepts are related. It may therefore come as a surprise to many that perhaps the most frequent, or at any rate the most general, objection to ID in the wider intellectual community is that it "has no model". According to its critics, it lacks any real-world interpretation specifying a fundamental medium able to support it or a means by which to realize it.

Furthermore, its critics claim, its central hypothesis is not only beyond proof, but unrealistic and not amenable to empirical confirmation.

In all fairness, it must be noted that insofar as science has itself spectacularly failed to agree on a global model of reality, this is really nothing more than an exercise in hypocrisy. Science observes, relates and extrapolates from observations with what often turns out to be great efficiency, but has time and time again proven unable to completely justify its reductions or the correspondences between its theories and the real universe as a whole. Although some critics claim that beyond a certain point, explanation is pointless and futile, they do not speak for science; the entire purpose of science is explanation, not rationally unsubstantiated assertions to the effect that a closed-form explanation is "unavailable" or "unnecessary". In seeking a coherent explanation for existence – an explanation incorporating an ontological design phase that is rational, coherent and therefore intelligent – the ID program is in fact perfectly consistent with science.

However, being perfectly consistent with science means merely that something is in line for a model, not that it already has one. It has thus been possible for dedicated critics of ID to create the illusion, at least for sympathetic audiences, that they have it at a critical disadvantage. They contend that while science must be instrumental to society, yield specific predictions, and thus cite specific structural and dynamical laws that nontrivially explain its contexts of application, ID is nothing more than a Trojan horse for religious ideology, makes no nontrivial predictions, and is devoid of theoretical structure. Due to the number of sympathetic ears that such claims have found in Academia, this illusion has all but promoted itself to the status of a self-reinforcing mass delusion in certain closed-minded sectors of the intellectual community. Obviously, it would be to the advantage of the ID movement, and society as a whole, to end this contagion by putting forth something clearly recognizable as a model.

The problem, of course, is that as long as science in general lacks a fundamental model, so do all particular strains of science *including* Intelligent Design. Due to the close connection between fundamentality and generality, ID or any other field of scientific inquiry would ultimately have to provide *science in general* with a fundamental model in order to provide one for itself. This might have led some people, in particular those who doubt the existence of a stable fundamental model of reality, to suppose that the ID controversy would remain strictly within the realm of philosophy until the end of time. But this is not the case, for if there were really no fundamental model – if there were no way to map theoretic cognition onto reality *in its entirety* – perception itself would lack a stable foundation. Perception, after all, can be described as the modeling of objective reality in cognition, and the modeling of cognition in objective reality. The self-evident perceptual stability of reality, on which the existence and efficacy of science and scientific methodology absolutely depend, bear unshakable testimony to the existence of a fundamental model of the real universe.

The general nature of this model can be glimpsed merely by considering the tautological reflexivity of the term "self-evident". Anything that is self-evident *proves* (or *evidences*) *itself*, and any construct that is implicated in its own proof is tautological. Indeed, insofar as observers are real, perception amounts to reality tautologically perceiving itself. The logical ramifications of this statement are developed in the supertautological CTMU, according to which the model in question coincides logically and geometrically, syntactically and informationally, with the process of generating the model, i.e. with generalized cognition and perception. Information thus coincides with information transduction, and reality is a tautological self-interpretative process evolving through SCSPL grammar.

The CTMU has a meta-Darwinian message: the universe evolves by hological self-replication and self-selection. Furthermore, because the universe is natural, its self-selection amounts to a cosmic form of *natural* selection. But by the nature of this selection process, it also bears description as *intelligent self-design* (the universe is "intelligent" because this is precisely what it must be in order to solve the problem of self-selection, the master-problem in terms of which all lesser problems are necessarily formulated). This is unsurprising, for intelligence itself is a natural phenomenon that could never have emerged in humans and animals were it not already a latent property of the medium of emergence. An object does not displace its medium, but embodies it and thus serves as an expression of its underlying syntactic properties. What is far more surprising, and far more disappointing, is the ideological conflict to which this has led. It seems that one group likes the term "intelligent" but is indifferent or hostile to the term "natural", while the other likes "natural" but abhors "intelligent". In some strange way, the whole controversy seems to hinge on terminology.

Of course, it can be credibly argued that the argument actually goes far deeper than semantics ... that there are substantive differences between the two positions. For example, some proponents of the radical Darwinian version of natural selection insist on randomness rather than design as an explanation for how new mutations are generated prior to the restrictive action of natural selection itself. But this is untenable, for in any traditional scientific context, "randomness" is synonymous with "indeterminacy" or "acausality", and when all is said and done, acausality means just what it always has: *magic*. That is, something which exists without external or intrinsic cause has been selected for and brought into existence by nothing at all of a causal nature, and is thus the sort of something-from-nothing proposition favored, usually through voluntary suspension of disbelief, by frequenters of magic shows.

Inexplicably, some of those taking this position nevertheless accuse of magical thinking anyone proposing to introduce an element of teleological volition to fill the causal gap. Such parties might object that by "randomness", they mean not acausality but merely causal ignorance. However, if by taking this position they mean to belatedly invoke causality, then they are initiating a causal regress. Such a regress can take one of three forms: it can be infinite and open, it can terminate at a Prime Mover which itself has no causal explanation, or it can form some sort of closed cycle doubling as Prime Mover and that which is moved. But a Prime Mover has seemingly been ruled out by assumption, and an infinite open regress can be ruled out because its lack of a stable recursive syntax would make it impossible to form stable informational boundaries in terms of which to perceive and conceive of reality.

What about the cyclical solution? If one uses laws to explain states, then one is obliged to explain the laws themselves. Standard scientific methodology requires that natural laws be defined on observations of state. If it is then claimed that all states are by definition caused by natural laws, then this constitutes a circularity necessarily devolving to a mutual definition of law and state. If it is then objected that this circularity characterizes only the *process* of science, but not the objective universe that science studies, and that laws in fact have absolute priority over states, then the laws themselves require an explanation by something *other than* state. But this would effectively rule out the only remaining alternative, namely the closed-cycle configuration, and we would again arrive at ... magic.

It follows that the inherently subjective process of science cannot ultimately be separated from the objective universe; the universe must be self-defining by cross-refinement of syntax and state. This brings us back to the CTMU, which says that the universe and everything in it ultimately evolves by self-multiplexing and self-selection. In the CTMU, design and selection, generative and restrictive sides of the same coin, are dual concepts associated with the alternating stages of conspansion. The self-selection of reality is inextricably coupled to self-design, and it is this two-phase process that results in *nature*. Biological evolution is simply a reflection of the evolution of reality itself, a process of telic recursion mirroring that of the universe as a whole. Thus, when computations of evolutionary probability are regressively extrapolated to the distributed instant of creation, they inevitably arrive at a logical and therefore meaningful foundation.

The CTMU says that on logical grounds, reality has generative and restrictive phases, and that evolution has generative and restrictive phases that are necessarily expressed in terms of those of reality. It asserts that the meta-cybernetic mechanism of evolution is telic recursion, an atemporal process which sets up a stratified dialectic between syntax and state, organism and environment, with mutually consistent mutable and invariant levels. It says that this process, though subject to various forms of noise, interference and competition predicated on the internal freedom of reality, tends to maximize the utility of the universe and its inhabitants. And it thus says that evolution is much more than a mere environmental dictatorship in which inexplicable laws of nature call the tune as biology slavishly dances the jig of life and death.

The CTMU says that by its self-generative, self-selective nature, which follows directly from the analytic requirement of self-containment, reality is its own "designer". Other features of the generative grammar of reality imply that reality possesses certain logical properties traditionally regarded as theological or spiritual, and that to this extent, the self-designing aspect of reality is open to a theological or spiritual interpretation. The CTMU, being a logical theory, does not attempt to force such an interpretation down anyone's throat; not all semantic permutations need affect theoretical structure. What it does do, however, is render any anti-theological interpretation *a priori* false, and ensures that whatever interpretation one chooses accommodates the existence of an "intelligent designer" ... namely, reality itself. In light of the CTMU, this is now a matter more of logic than of taste.

In any case, it should be clear that the CTMU yields new ways of looking at both evolution and teleology. Just as it is distinguished from other theories of cosmic evolution by its level of self-containment, particularly with regard to its preference for self-determinacy rather than external determinacy or indeterminacy, so for its approach to biological evolution. Unlike other theories, the CTMU places evolutionary biology squarely in the context of a fundamental, self-contained model of reality, thus furnishing it with an explanation and foundation of its own instead of irresponsibly passing the explanatory buck to some future reduction; instead of counting it sufficient to model its evolutionary implications in the biological world, the CTMU establishes model-theoretic symmetry by providing a seamless blend of theory and universe in which the biological world can itself be "modeled" by physical embedment. This alone entitles it to a place in the evolutionary debate.

Acknowledgments

The author wishes to thank Gina Lynne LoSasso for creating the diagrams that appear in this paper and for many stimulating discussions and feedback regarding the presentation of this work. Many thanks are also in order for a number of individuals who have supported and followed my work and contributed stimulating questions that helped me to explicate the CTMU in its current level of detail. These include, among others, Robert North Seitz, John Chapman, Angell O. de la Sierra, Mike Doyle, Phillip Hopp, Quinn Tyler Jackson, Evangelos Katsiolis, Paul Kisak and Anthony Nolan.

References

Behe, M. J. (1998) *Darwin's Black Box: The Biochemical Challenge to Evolution*. New York: Simon & Schuster.

Berlinski, D. (2001) What Brings a World into Being? *Commentary*, Vol. 111, № 4, p. 17.

Dembski, W. A. (1997) Intelligent Design as a Theory of Information. *Perspectives on Science and Christian Faith*, Vol. 49, № 3, pp. 180–190.

Dembski, W. A. (1998) *The Design Inference: Eliminating Chance through Small Probabilities*. Cambridge: Cambridge University Press.

Eddington, A. (1933) *The Expanding Universe*. Cambridge: Cambridge University Press.

Gellert, W., Kustner, H., Hellwich, M., Kastner, H. (Eds.) (1977) *The VNR Concise Encyclopedia of Mathematics*. New York: Van Nostrand Reinhold Company.

Gödel, K. (1962) *On Formally Undecidable Propositions of Principia Mathematica and Related Systems*. Translated by B. Meltzer. New York: Basic Books.

Langan, C. M. (1999) Introduction to the CTMU. *Ubiquity*, Vol. 1, № 1.

Langan, C. M. (2001a) Flash from Deep Space: Supernews on Supernovas. *Noesis-E*, Vol. 1, № 3. Originally published in 1998. Available online at megafoundation.substack.com/p/flash-from-deep-space

Langan, C. M. (2001b) Of Trees, Quads, and God. *Noesis-E*, Vol. 1, № 4.

Langan, C. M. (2002a) *The Art of Knowing: Expositions on Free Will and Selected Essays*. Eastport, NY: Mega Foundation Press.

Langan, C. M. (2002b) The Theory of Theories. *Noesis-E*, Vol. 2, № 1.

Russell, B. (1979) *History of Western Philosophy and its Connection with Political and Social Circumstances from the Earliest Times to the Present Day*. London: Allen & Unwin Ltd.

Shannon, C. E. (1948) A Mathematical Theory of Communication. *Bell System Technical Journal*, Vol. XXVII, № 3, pp. 379-423 and 623-656.

Wheeler, J. A. (1979) From the Big Bang to the Big Crunch. *Cosmic Search Magazine*, Vol. 1, Issue 4, pp. 2–8.

Wheeler, J. A. (1980) Beyond the Black Hole. In H. Woolf (Ed.) *Some Strangeness in the Proportion: A Centennial Symposium to Celebrate the Achievements of Albert Einstein*, pp. 341–375. Addison-Wesley Educational Publishing Company.

Wheeler, J. A. (1990a) Information, Physics, Quantum: The Search for Links. In W. H. Zurek (Ed.) *Complexity, Entropy and the Physics of Information, SFI Studies in the Sciences of Complexity*, Vol. VIII, pp. 3–28. Addison-Wesley Educational Publishing Company.

Wheeler, J. A. (1990b) *A Journey into Gravity and Spacetime*. New York: Scientific American Library.

Wozniak, R. H. (1992) *Mind and Body: René Descartes to William James*. National Library of Medicine (U.S.), American Psychological Association.

Cheating the Millennium: The Mounting Explanatory Debts of Scientific Naturalism

Introduction: Thesis + Antithesis = Synthesis

In agreeing to write this essay, I have promised to explain why I find Darwinism unconvincing. In order to keep this promise, I will be compelled to acknowledge the apparently paradoxical fact that I find it convincing as well. I find it convincing because it is in certain respects correct, and in fact tautologically so in the logical sense; I find it unconvincing because it is based on a weak and superficial understanding of causality and is therefore incomplete. Explaining why this is so will require a rather deep investigation of the nature of causality. It will also require not only that a direction of progress be indicated, but that a new synthesis embracing the seemingly antithetical notions of teleology and natural selection be outlined. But first, some essential background.

It would be hard to imagine philosophical issues bearing more strongly on the human condition than the nature of life and the meaning of human existence, and it would be hard to imagine a scientific issue bearing more strongly on the nature and meaning of life than biological origins. Our view of evolutionary biology, whatever it happens to be at any particular juncture, tells us much of what we believe about who and what we are and why we are here, unavoidably affecting how we view (and ultimately, treat) ourselves and each other. Unfortunately, the prevailing theory of biological origins seems to be telling us that at least one of these questions, *why are we here?*, is meaningless[1] ... or at least this is the message that many of us, whether or not we are directly aware of it, seem to have received. As a result, the brightest hope of the new millennium, that we would see the dawn of a New Enlightenment in which the Meaning of it All would at last be revealed, already seems to have gone the way of an extravagant campaign promise at an inauguration ceremony.

The field of evolutionary biology is currently dominated by *neo-Darwinism*, a troubled marriage of convenience between post-Mendelian genetics and *natural selection*, a concept propounded by the naturalist Charles Darwin (1999) in his influential treatise *On the Origin of Species*. It has often been noted that the field and the theory appear to be inseparable; in many respects, it seems that evolutionary biology and Darwinism originated and evolve together, leading some to conclude that the field properly contains nothing that is not already accommodated by the theory.

Those attempting to justify this view frequently assert that the limitations of the theory are just the general limitations imposed on all scientific theories by standard scientific methodology, and that to exceed the expressive limitations of the theory is thus to transgress the boundaries of science. Others have noted that this seems to assume a prior justification of scientific methodology that does not in fact exist – merely that it works for certain purposes does not imply that it is optimal, particularly when it is evidently useless for others – and that in any case, the putative falsifiability of neo-Darwinism distinguishes it from any definition of *science* according to which the truth or falsity of such theories can be scientifically determined.[2] Nevertheless, neo-Darwinism continues to claim exclusive dominion over the "science" of evolutionary biology.

Until the latter part of the 18th century, the story was quite different. People tended to regard the matter of biological origins in a religious light. The universe was widely considered to have been freely and purposively designed and created by God as described in the Book of Genesis, and divine purpose was

---

[1] "Meaning" entails recognition, referring specifically to a recognizable and therefore informational relationship among related entities. Since information is abstract, so is recognition, and so is meaning (whether or not the related entities are themselves physical and concrete). *Naturalism*, of which the theory of evolution is an example, is an essentially materialistic viewpoint which denies or disregards abstract modes of existence, thus limiting meaning to "material" drives and instincts. But where the abstract contains the physical, capturing its structure in the form of meaningful informational patterns called "laws of nature", abstraction and meaning are plainly essential to both science and nature.

[2] Science is a two-step, two-level process concerned with (1) formulating hypotheses about nature, and (2) proving or disproving these hypotheses to some degree of confirmation. Relative to level 1, level 2 requires a higher level of discourse incorporating truth-functional criteria independent of any particular falsifiable hypothesis. Because maintaining this distinction helps to insure that false hypotheses do not figure in their own "validation", purportedly falsifiable (level 1) theories like neo-Darwinism should not be confused with the confirmational level of science.

thought to be immanent in nature and open to observation and study. This doctrine, called *teleology*, drew rational support from traditional theological "arguments from design" holding that nature could only have been designed and created by a supreme intelligence. But teleology began to wane with the rise of British empiricism, and by the time Darwin published his theory in 1859, the winds of change were howling his anthem. Since then, the decline of teleology has accelerated to a point at which every supposedly universal law of nature is confidently presented as "irrefutable evidence" that natural events unfold independently of intent, and that purpose, divine or otherwise, is irrelevant to natural causation.

The concept of teleology remains alive nonetheless, having recently been granted a scientific reprieve in the form of *Intelligent Design theory*. "ID theory" holds that the complexity of biological systems implies the involvement of empirically detectable intelligent causes in nature. Although the roots of ID theory can be traced back to theological arguments from design, it is explicitly scientific rather than theological in character, and has thus been presented on the same basis as any other scientific hypothesis awaiting scientific confirmation.[3]

Rather than confining itself to theological or teleological causation, ID theory technically allows for any kind of intelligent designer – a human being, an artificial intelligence, even sentient aliens. This reflects the idea that intelligence is a generic quality which leaves a signature identifiable by techniques already heavily employed in such fields as cryptography, anthropology, forensics and computer science. It remains only to note that while explaining the inherent complexity of such a material designer would launch an explanatory regress that could end only with some sort of Prime Mover, thus coming down to something very much like teleology after all, ID theory has thus far committed itself only to design inference. That is, it currently proposes only to explain complex biological phenomena in terms of design, not to explain the designer itself.[4] With regard to deeper levels of explanation, the field remains open.

Because neo-Darwinism is held forth as a "synthesis" of Darwinian natural selection and post-Mendelian genetics, it is sometimes referred to as the "Modern Synthesis". However, it appears to fall somewhat short of this title, for not only

---

[3] Properly speaking, science includes both the empirical and mathematical sciences. Most of those who call themselves "scientists", as well as many proponents of ID theory, assume that scientific confirmation can only be achieved by strict application of the scientific method and must thus be empirical. However, this is an oversimplification. The empirical sciences are not only mathematical in structure, but too heavily indebted to mathematical reasoning to exclude mathematical methods as possible means of confirming facts about nature. So with regard to the scientific status of ID theory, both empirical and mathematical methods of confirmation must be duly considered.

[4] It can also be cogently argued that the design inference requires the establishment of means, motive and opportunity for a designer to act, and that meeting these requirements amounts to explaining the designer.

is its basic approach to evolutionary biology no longer especially modern, but despite the fact that it is a minority viewpoint counterbalanced by cogent and far more popular alternatives including theistic evolution[5] and ID theory (Robinson, 1995) it actively resists meaningful extension. Many of its most influential proponents have dismissed ID theory virtually on sight, declaring themselves needless of justification or remedial dialectic despite the many points raised against them, and this is not something that the proponents of a "modern synthesis" would ordinarily have the privilege of doing. A synthesis is ordinarily expected to accommodate both sides of a controversy regarding its subject matter, not just the side favored by the synthesist.[6]

Given the dissonance of the neo-Darwinist and teleological viewpoints, it is hardly surprising that many modern authors and scientists regard the neo-Darwinian and teleological theories of biological evolution as mutually irreconcilable, dwelling on their differences and ignoring their commonalities. Each side of the debate seems intent on pointing out the real or imagined deficiencies of the other while resting its case on its own real or imagined virtues. This paper will take a road less traveled, treating the opposition of these views as a problem of reconciliation and seeking a consistent, comprehensive framework in which to combine their strengths, decide their differences, and unite them in synergy. To the extent that both theories can be interpreted in such a framework, any apparent points of contradiction would be separated by context, and irreconcilable differences thereby avoided.

The ideal reconciliatory framework would be self-contained but comprehensive, meaning that both theories could be truthfully interpreted within it to the maximum possible extent, and consistent, meaning that irreconcilable differences between the theories could not survive the interpretation process. It would also reveal any biconditionality between the two theories; were they in any way to imply each other, this would be made explicit. For example, were a logical extension of neo-Darwinism to somehow yield ID-related concepts such as teleological agency and teleological causation, these would be seen to emerge from neo-Darwinist premises; conversely, were ID-theoretic concepts to yield ingredients of neo-Darwinism, this too would be explicated. In any case, the result would wear the title of "synthesis" far more credibly than neo-Darwinism alone.

---

5 Theistic evolution is a simple conjunction of theism and Darwinism which pays no real attention to their mutual consistency or the model-theoretic implications of combining them.

6 "... neo-Darwinism exhibits a great power of assimilation, incorporating any opposing viewpoint as yet another 'mechanism' in the grand 'synthesis'. But a real synthesis should begin by identifying conflicting elements in the theory, rather than in accommodating contradictions as quickly as they arise." (Ho & Sanders, 1979, p. 574)

Two Theories of Biological Causality

In order to talk about origins and evolution, one must talk about causality, and because causality is a function of the system called "nature", one must talk about nature. Theories of biological origins and evolution like Neo-Darwinism and ID theory are both theories of causality restricted to the context of biological origins and evolution, and because causality is a function of nature, each points toward an underlying theory of nature incorporating an appropriate treatment of causality. That is, biological origins and evolution, being for scientific purposes instances of causation or the outcomes of causal processes, require definitions, theories and models of nature and causality. But these definitions, theories and models involve deeper and more complex criteria than meet the casual eye, and even to experts in science and philosophy, it is not entirely obvious how to satisfy them. This is why causality remains a controversial subject.

A *cause* is something that brings about an effect or result, and *causality* is the quality or agency relating cause and effect. Because there are different requirements for bringing about an event or situation, there are different kinds of causation. In common usage, a "cause" may be an event which causes another event, the reason or rationale for an event, an agent or the motive thereof, the means by which an event transpires, supporting conditions for an event, or in fact anything satisfying any logical or physical requirement of a resultant effect. Because causal relationships would seem to exist in a causal medium providing some sort of basic connection between cause and effect, the study of causation has typically focused on the medium and its connectivity ... i.e., on the "fabric of nature".

The kinds of causation that are required in order to explain natural changes or events were enumerated by Aristotle in the 4th century BC. He posed four questions involving four types of causes:

(1) What is changed to make the entity (Of what is it composed)?

(2) What makes the entity change, and how?

(3) What is the shape or pattern assumed by the entity as it changes?

(4) What is the goal toward which the change of the entity is directed?

He respectively defined the answers to these questions as the *material cause*, the *efficient cause*, the *formal cause*, and the *final cause*. With its explicit allowance for formal and final causation, Aristotle's classification ultimately implies the existence of a purposive, pattern-generating Prime Mover, and thus laid the groundwork for a teleological explanation of nature that went all but unchallenged for well over a millennium.

But when the Age of Reason (circa 1650–1800) had finished taking its toll on traditional Scholastic doctrines largely based on Aristotelian insight, only material and efficient causes retained a place in scientific reasoning ... and in the hands of philosophers like Hume and Kant, even these modes of causation were laid open to doubt. Hume (1975) claimed that causal relationships are nothing more than subjective expectations that certain sequences of events observed in the past will continue to be observed in the future, while Kant (1965) went on to assert that causality is a category of cognition and perception according to which the mind organizes its experience of basically unknowable objects. Nevertheless, contemporary science retains its concern for material and efficient causes while letting formal and final causes languish in a state of near-total neglect.[7]

Distilled to a single sentence, the prevailing scientific view of nature and causality is roughly this: "Nature is associated with a space, generalizable to a spacetime manifold, permeated by fields under the causal influence of which objects move and interact in space and time according to logico-arithmetical laws of nature." Despite its simplicity, this is a versatile causal framework with the power to express much of our scientific knowledge. But the questions to which it leads are as obvious as they are unanswered. For example, where do these laws reside? Of what are they composed? How and why did they originate? What are their properties? How do they function, and how are they sustained?[8]

In addition to generating questions about natural laws in general, the prevailing oversimplification of causality contains further gaps which have done as much to impede our understanding of nature as to further it.[9] The associated problems are numerous, and they lead to yet another set of questions. For example, is causality formally and dynamically contained, uncontained or self-contained? What is its source, on what does it function, and what additional structure does it predicate of that on which it functions? What is its substance –

[7] While material and efficient causation are superficially physical and can be described in more or less materialistic terms, formal and final causation are more abstract. Francis Bacon (1997), who strongly influenced scientific methodology, classified these abstract modes of causation as metaphysics rather than physics.

[8] These questions about *laws* of causality address the nature and origin of causality itself, and are thus metacausal analogues of Aristotle's questions about causality. The answers presented in this paper – roughly, that laws are elements of syntax of the language of nature, that they are composed of telesis and self-transducing metainformation, that they reside in syntactic (space-time-object) operators whose states they govern, that they arose through the metacausal self-configuration of the language of nature, that their properties include closure, comprehensiveness, consistency and coherence, and that their functionality and maintenance rely on intrinsic features of the language of nature – are thus metacausally analogous to Aristotle's modes of causation.

[9] For example, there is the gap between mind and matter; the gap between abstract and concrete existence; the gap between causality and generative cosmogony; the gap between classical and quantum mechanics, and so on. Because these gaps are serious, there is no reason to think that causality can be adequately explained as long as they exist.

is it mental, physical or both? How does it break down, and if it is stratified, then what are its levels? These questions lead in turn to further questions, and until all of these questions are answered at least in principle, no theory of biological causality stands on *terra firma*.

But before attempting to answer this question, let us have a look at the models of causality on which neo-Darwinism and ID theory are currently based.

Causality According to Intelligent Design Theory

Teleological causation is "top-down" causation in which the design and design imperative reside at the top, and the individual actualization events that realize the design reside at the bottom. The model universe required for teleological causality must therefore incorporate:

(1) a source and means of design, i.e. a designer or designing agency;

(2) a design stage in which designs are generated and/or selected;

(3) an actualization stage in which designs become physically real from the viewpoints of physical observers;

(4) a means or mechanism for passing from the design stage to the actualization stage.

If such a model universe permits these observers to empirically detect interesting instantiations of teleology, so much the better.

Particular teleological model universes that have been proposed include any number of celestial hierarchies and heavenly bureaucracies with God at the top giving the orders, angels of various ranks serving on intermediate levels as messengers and functionaries, and humans at or near the bottom; the Aristotelian universe, incorporating formal and final causation and embodying the *telos* of a Prime Mover; teleologically "front-loaded" mechanistic universes in which causation resembles clockwork that has been set in autonomous motion by a purposive, mechanically talented designer; and the panentheistic universe explicated by (among others) Alfred North Whitehead (1985), in which the teleological will of the designer is immanent in nature because in some sense, nature is properly contained *within* the designer. Although each has its strengths, these and other well-known teleological models are inadequate as formulated, failing to support various logical implications of requirements 1–4.

The model universe of ID theory, which can be regarded as a generalization of traditional teleological design theory with respect to causal agency, has essentially the same requirements. However, it also contains certain novel ingredients including a focus on intelligence, an emphasis on mathematical and information-theoretic concepts, and two novel ingredients called *irreducible complexity* and *specified complexity*.

Irreducible complexity, which is intended to describe biological systems and subsystems unlikely to have been produced by gradual (piece-by-piece) evolution, is by definition a property of any integrated functional system from which the removal of any one or more core components critically impairs its original function (Behe, 1998). Although proposed examples have drawn fire – such examples include the bacterial flagellum, the human eye, the blood clotting cascade, and even the conventional spring-loaded mousetrap – the concept has a valid basis with roots in logic, graph theory and other branches of mathematics and engineering.

Specified complexity, which is intended as a more general description of the products of intelligent causation, is by definition a property of anything that exhibits a recognizable pattern with a very low probability of occurring by chance. Whereas irreducible complexity is based on the sheer improbability of complex functionally-coherent systems, specified complexity adds an *intelligence* (rational pattern generation and recognition) criterion that lets functional complexity be generalized to a pattern-based form of complexity better suited to probabilistic and information-theoretic analysis (Dembski, 1998).

Specified complexity amounts to a relationship between three attributes: *contingency*, *complexity* and *specification*. Contingency corresponds to freedom and variety (as when there are many distinct possibilities that may be selectively actualized), complexity corresponds to improbability, and specification corresponds to the existence of a meaningful pattern which, in conjunction with the other two attributes in sufficient measure, indicates an application of intelligence. Wherever all three of these attributes are coinstantiated, specified complexity is present.

Contingency is associated with *specificational* and *replicational* probabilistic resources. Specificational resources consist of a set or class of distinct pre-specified target events, while replicational resources consist of chances for at least one of the specified target events to occur. The chance of occurrence of an instance of specified complexity is the chance that these two kinds of resource will intersect in light of total contingency.

For example, the total contingency of a 4-digit lottery consists of the set of all possible drawings over unlimited trials and is associated with the numbers from 0000 to 9999, the specificational resources consist of a subset of distinct pre-specified 4-digit winning numbers to be replicated (matched or predicted), and the replicational resources consist of the tickets purchased. The chance that the lottery will have at least one winner equals the probability of intersection of the set of winning numbers and the set of tickets, given that there are ten thousand distinctly-numbered tickets that might have been purchased.

More topically, the total contingency of a particular evolutionary context consists of all possible (productive or dead-end) lines of evolution that might occur therein, the specificational resources consist of instances of specified complexity or "intelligent design", and the replicational resources consist of all possible lines of evolution which can occur within some set of practical constraints imposed on the context, for example time or space constraints tending to limit replication. The chance that an instance of specified complexity will evolve equals the probability of intersection of the set of instances and the set of constrained lines of evolution, given the multiplicity of all of the possible lines of evolution that could occur. Where this probability is extremely low, some form of intelligent design is indicated.

Specified complexity is a powerful idea that yields insight crucial to the meaning and satisfaction of requirements 1–4. First, probability estimates for instances of specified complexity are so low as to require that specificational and replicational resources be linked in such a way that such events can actually occur, in effect raising their probability. It must therefore be determined whether the satisfaction of this requirement is consistent with the premise that low probabilities can actually be calculated for instances of specified complexity, and if so, how and why this can be reliably accomplished. And next, it must be shown that the required relationship implies intelligence and design.

Up to its current level of detail and coherence, the model universe of ID theory does not necessarily conflict with that of neo-Darwinism with respect to causality, but rather contains it, requiring only that causality be interpreted in light of this containment.

Causality According to Neo-Darwinism

Neo-Darwinism is the application of Darwinian natural selection to modern (post-Mendelian) genetics, which indifferently assumes that genetic mutations occur due to "random" DNA copying errors. This short but revealing description contains a certain amount of useful information. First, it reveals that causality is being at least partially reduced to some (ontic or epistemic) form of randomness. Even more revealingly, the phrase *natural selection* explicitly implies that nature is selective. Indeed, the term *natural* alone is instructive, for it reflects a naturalistic viewpoint according to which existence is ascribed exclusively to the natural world, i.e. "nature".

In practice, most scientists consider *nature* to consist of that which is physical, observable and amenable to empirical investigation as prescribed by the scientific method, in their adherence to which they see themselves as following a naturalistic agenda. This is in keeping with *scientific naturalism*, a worldview of which neo-Darwinism is considered representative. Scientific naturalism ascribes existence strictly to the physical or natural world consisting of space, time, matter and energy. Two strains of naturalism are sometimes distinguished, *philosophical* and *methodological*. While philosophical naturalism claims ontological force, methodological naturalism is epistemological in flavor, merely asserting that nature *might as well* equal the physical world for scientific purposes. But in either case, scientific naturalism effectively confines the scientific study of nature to the physical. So inasmuch as neo-Darwinism is exemplary of scientific naturalism, it is physical or materialistic in character (Pigliucci, 2000).

In the picture of causality embraced by scientific naturalism, processes are either *random* or *deterministic*. In deterministic processes, objects are affected by laws and forces external to them, while in random processes, determinacy is either absent or unknown. A process can be "random" due to ignorance, statistics or presumed acausality ... that is, because epistemological or observational limitations prevent identification of its hidden causal factors, because its causal outcomes are symmetrically or "randomly" distributed in the large, or because it is presumed to be nondeterministic. The first two of these possibilities are basically deterministic, while the last is (unverifiably) nondeterministic. So a neo-Darwinist either takes a deterministic view of causality or sees it in terms of the dichotomy between determinism and nondeterminism, in either case relying heavily on the theory of probability.

In fact, given that natural selection is based on the essentially trivial observation that nature imposes constraints on survival and reproduction,[10] neo-Darwinism boils down to little more than probability theory, genetics and a very simple abstract but nominally physical model of biological causality based on "survival and reproduction of the fittest" or some minor variant thereof. Thus, when its practitioners claim to have generated a prediction, it is generally not a deep secret of nature unearthed by means of advanced theoretical manipulation, but merely the result of applying what amounts to a principle of indifference[11] to some question about mutation, adaptation, selection or reproduction, running the numbers, and tracking the implications through its simplistic model universe. If there were no such "theory" as neo-Darwinism, the same conclusion might have been reached with a straightforward combination of biology, genetics, chemistry, physics, a statistics calculator and a bit of common sense. This is why neo-Darwinism is so astonishingly able to absorb new effects and mechanisms the minute they come out of the core sciences.

Something else that neo-Darwinism seems to do with astonishing ease is absorb what appear on their faces to be contradictions. For example, many people, some might say a large majority, find it to some degree incredible that what amounts to a principle of indifference can be seriously offered as a causal

---

10 At the time that Charles Darwin made this observation and formulated his natural selection thesis, it was still obscured by centuries of teleological dominance.

11 The canonical *principle of indifference* (or *insufficient reason*) states that where there is no positive reason for assigning different probabilities to competing statistical or predictive assertions, e.g. different possible mutations weighted by relative frequency, equal probabilities must be assigned to all. Since this is essentially how neo-Darwinism calculates its random distributions of mutations and other events, it is just a biological variant of the principle of indifference.

explanation for the amazing complexity of the biological world, or for that matter any other part of the world. The fact that a principle of indifference is essentially devoid of information implies that neo-Darwinism yields not a causal explanation of biological complexity, but merely an open-ended simulation in which every bit of complexity delivered as output must have been present as input, any appearances to the contrary notwithstanding. This implies that neo-Darwinism *per se*, as distinguished from the core sciences from which it routinely borrows, adds precisely nothing to our knowledge of biological complexity or its source.

In order to deal with this seemingly inescapable problem, the proponents of neo-Darwinism have eagerly adopted the two hottest slogans in the theory of complex systems, *self-organization* and *emergence*. Self-organization is a spontaneous, extrinsically unguided process by which a system develops an organized structure, while emergence refers to those global properties (functions, processes) of composite hierarchical systems that cannot be reduced to the properties of their component subsystems ... the properties in which they are more than the sums of their parts. But the fact that these terms have been superficially defined does not imply that they have been adequately explained. Actually, they remain as much of a mystery in complexity theory as they are in biology, and can do nothing for neo-Darwinism but spin the pointer toward another hapless and equally helpless field of inquiry.

Because scientific naturalism denies that existence of any kind is possessed by anything of a supernatural or metaphysical character, including a designing intelligence, the definitions, theories and models of nature and causality on which it implicitly relies must be "physical", at least in name. However, as we have already noted and will shortly explain in detail, what currently passes for an understanding of causality in the physical sciences leaves much to be desired. In particular, since the kind of causality treated in the physical sciences is ontologically and functionally dependent on the origin and evolution of the cosmos, scientific naturalists trying to answer questions about causality are obliged to consider *all* stages of causation and generation all the way back to the cosmic origin, constantly testing their answers to see if they continue to make sense when reformulated in more fundamental terms.

Unfortunately, this obligation is not being met. One reason is the reluctance of those who most need an understanding of causality to admit the extent of their ignorance. Another is the seeming intractability of certain problems associated with the causality concept itself.

A Deeper Look at Causality: The Connectivity Problem

Because causal relationships would seem to exist in a causal medium providing some sort of basic connection between cause and effect, the study of causation has typically focused on the medium and its connectivity ... i.e., on the "fabric of nature". How does this fabric permit different objects to interact, given that to *interact* is to *intersect* in the same events governed by the same laws and thus to possess a degree of sameness? How can multiple objects each simultaneously exhibit two opposite properties, sameness and difference, with respect to each other?

Equivalently, on what underlying form of connectivity is causality defined? When one asserts that one event "causes" another, what more general connection does this imply between the events? If there is no more general connection than the causal connection itself, then causality is underivable from any logically prior condition; it is something that happens *ex nihilo*, the sudden synthesis of a connection out of nothing. As Hume maintained, causal relationships are mere accidental correlations of subjectively-associated events.

But it can't be quite that simple. In fact, Hume's characterization of causality as mere random correlation presupposes the existence of a correlating agent who recognizes and unifies causal correlations through experience, and the abstractive, experiential coherence or *consciousness* of this correlation-inducing agent constitutes a prior connective medium. So in this case, explaining causality requires that the subjective medium of experience, complete with its correlative "laws of causality", be related to the objective world of real events.

Unfortunately, Hume's thesis includes a denial that any such objective world exists. In Hume's view, experience is all there is. And although Kant subsequently registered his qualified disagreement, asserting that there is indeed an objective outside world, he pronounced it unknowable, relegating causality to the status of a category of perception.[12] This, of course, perpetuated the idea of causal subjectivity by continuing to presuppose the existence of an *a priori* subjective medium.

How can the nature of subjective causality be understood? As Kant observed, perception and cognition are mutually necessary; concepts without percepts are empty, and percepts without concepts are blind.[13] It must therefore be asked to what extent perceptual reality might be an outward projection of cognitive processes, and natural processes the mirror images of mental processes.

This leads to another problem, that of *mind-matter dualism*.

---

12 Kant (1965) p. 83: "Things which we see are not by themselves what we see ... It remains completely unknown to us what the objects may be by themselves and apart from the receptivity of our senses. We know nothing but our manner of perceiving them."; p. 147: "We ourselves introduce that order and regularity in the appearance which we entitle 'nature'. We could never find them in appearances had we not ourselves, by the nature of our own mind, originally set them there."

13 Ibid., p. 93: "Thoughts without content are empty, intuitions without concepts are blind."

The Dualism Problem

The Kantian distinction between phenomenal and noumenal reality, respectively defined as those parts of reality[14] which are dependent on and independent of perception, mirrors a prior philosophical viewpoint known as *Cartesian (mind-matter) dualism*. Associated with René Descartes, the polymath mercenary who laid the groundwork for analytic geometry by helping to develop the concept of coordinate spaces, this is a form of *substance dualism* which asserts that reality consists of two immiscible "substances", mind and matter. Cartesian dualism characterizes a certain influential approach to the *problem of mental causation*: how does the mind influence the physical body?

Cartesian dualism leads to a problem associated with the connectivity problem we have just discussed: if reality consists of two different "substances", then what connects these substances in one unified "reality"? What is the medium which sustains their respective existences and the putative difference relationship between them? One possible (wrong) answer is that their relationship is merely abstract, and therefore irrelevant to material reality and devoid of material influence; another is that like the physical epiphenomenon of mind itself, it is essentially physical. But these positions, which are seen in association with a slew of related philosophical doctrines including *physicalism*, *materialism*, *naturalism*, *objectivism*, *epiphenomenalism* and *eliminativism*, merely beg the question that Cartesian dualism was intended to answer, namely the problem of mental causation.

Conveniently, modern logic affords a new level of analytical precision with respect to the Cartesian and Kantian dichotomies. Specifically, the branch of logic called *model theory* distinguishes *theories* from their *universes*, and considers the intervening semantic and interpretative mappings. Calling a theory an *object language* and its universe of discourse an *object universe*, it combines them in a *metaobject domain* consisting of the correspondences among their respective components and systems of components, and calls the theory or language in which this metaobject domain is analyzed a *metalanguage*. In like manner, the relationship between the metalanguage and the metaobject domain can be analyzed in a higher-level metalanguage, and so on. Because this situation can be recursively extended, level by level and metalanguage by metalanguage, in such a way that languages and their universes are conflated to an arbitrary degree, reality can with unlimited precision be characterized as a "metalinguistic metaobject".

In this setting, the philosophical dichotomies in question take on a distinctly mathematical hue. Because theories are abstract, subjectively-formed mental

[14] For cognitive (and thus for theoretical and scientific) purposes, *reality* consists of perception plus the abstract cognitive apparatus required to generate, support and sustain it.

constructs,[15] the mental, subjective side of reality can now be associated with the object language and metalanguage(s), while the physical, objective side of reality can be associated with the object universe and metauniverse(s), i.e. the metaobject domain(s). It takes very little effort to see that the mental / subjective and physical / objective sides of reality are now combined in the metaobjects, and that Cartesian and Kantian "substance dualism" have now been transformed to "property dualism"[16] or *dual-aspect monism*. That is, we are now talking, in mathematically precise terms, about a "universal substance" of which mind and matter, the abstract and the concrete, the cognitive-perceptual and the physical, are mere properties or aspects.

Translating this into the scientific *status quo* is not difficult. Science regards causality as "objective", taking its cues from observation while ignoring certain philosophical problems involving the nature of objectivity. But science also depends on theoretical reasoning, and this involves abstract analogues of causality to which science is equally indebted. To the extent that scientific theories accurately describe the universe, they are *isomorphic* to the universe; in order that nature be amenable to meaningful theorization, science must therefore assume that the basic cognitive ingredients of theories, and for that matter the perceptual ingredients of observations, mirror the corresponding ingredients of nature up to some minimal but assured level of isomorphism. Consistent theories of science thus require that physical and abstract causation be brought into basic correspondence as mandated by this necessity.

Abstract analogues of physical causation are already well-understood. Logically, causality is analogous to *implication*, an active or passive relationship between antecedents and consequents; theoretically, it is analogous to the application of rules of inference to expressions formulated within a theory; linguistically, it amounts to substitution or production according to the rules of a generative grammar; and mathematically, it amounts to the application of a rule, mapping, function, operation or transformation. In every case, the analogue is some form of recursive[17] or iterative morphism to which a nonlogical

---

15 It makes no difference that scientific theories are based on "objective" empirical observations; the key point is that scientific observation and theorization require subjectively conscious agents called "scientists", and that there exists no possible means of ruling out subjectivity on the part of any other kind of observer-theorist. Whatever reality "might have been without us", our presence immediately implies that it possesses a subjective dimension.

16 *Property dualism* asserts that the properties *mental* and *physical*, while essentially different, apply to the same objects. Dual aspect monism asserts that these two properties together characterize the fundamental "substance" of nature.

17 The *Church–Turing Thesis* asserts that the class of recursive functions and the class of effectively computable functions are the same. This is generally taken to imply an isomorphism between the formal, abstract realm of recursive functions and the physical, mechanical realm in which abstract Turing machines are instantiated. For theoretical purposes, this isomorphism must be taken for granted; without it,

interpretation may be attached.[18] The object is therefore to understand physical reality in terms of such operations defined on an appropriate form of dual-aspect monism.

This leads directly to the *structure problem.*

The Structure Problem

A description or explanation of causality can only be formulated with respect to a particular "model universe" in which space, time and matter are defined and related to each other in such a way as to support the description. This relationship must account for the laws of nature and their role in natural processes. A little reflection should reveal that both neo-Darwinism and ID theory, as well as all other scientific theories, are currently deficient in this regard. At best, scientists have a very limited idea where the laws of nature reside, how they came to be, and how they work, and due to the limitations of their empirical methodology,[19] they have no means of clarification.

We have already encountered Aristotle's four modes of causation: *material, efficient, formal* and *final.* These follow no special prescription, but are merely generic answers to questions about certain features of Aristotle's mental representation of nature ... his model universe. There are as many additional modes of causation as there are meaningful questions regarding the structure and dynamics of a given model universe. For example, in addition to Aristotle's questions *of what, who and how, what* and *why,* we could also ask *where* (positional causation), *when* (order or timing of causation), *by virtue of what* (facilitative causation), and so forth. Thus, we could say that something happened *because* it was positioned in a medium containing its material cause and supporting its efficient cause, *because* the time was right or certain prerequisites were in place, *because* certain conditions were present or certain tools were available, et cetera.

On what kinds of model universe can a causality function be defined? Among the mathematical structures which science has long favored are *coordinate*

---

theoretical instances of recursion could not be model-theoretically interpreted in physical reality, and physical reality could not be scientifically explained.

[18] ... even if what gets iterated is a "continuous" function representing motion in a differentiable manifold.

[19] Scientific methodology conforms to the *scientific method,* which prescribes that nature be treated as if it were everywhere both discernable and replicable, and the related doctrine of falsifiability, which asserts that science is concerned only with hypotheses that are conceivably false and susceptible to empirical disproof. However, *nature* cannot be meaningfully defined in such a way that these criteria always hold within it. For example, no full description of nature can exclude references to universal, unconditional and therefore unfalsifiable properties of nature, and such unfalsifiable properties need not be scientifically trivial.

spaces and *differentiable manifolds*. In differentiable coordinate spaces, laws of physics formulated as algebraic or differential equations may conveniently define smooth geometric curves which faithfully represent (e.g.) the trajectories of physical objects in motion. A model universe based on these constructs supports certain causal relationships to an impressive level of accuracy. However, it fails with respect to others, particularly those involving discrete or nonlocal[20] changes or requiring high levels of coherence. In particular, it is incapable of modeling certain generative processes, including any generative process that might have led to its own existence, and beyond a certain point, its "continuity" attribute has eluded a completely satisfactory explanation.[21]

These and other difficulties have prompted some theorists to suggest model universes based on other kinds of mathematical structure. These include a new class of models to which the concepts of *information* and *computation* are essential. Called "discrete models", they depict reality in terms of bits, quanta, quantum events, computational operations and other discrete, recursively-related units. Whereas continuum models are based on the notion of a *continuum*, a unified extensible whole that can be subdivided in such a way that any two distinct points are separated by an infinite number of intermediate points, discrete models reflect the fact that it is impossible to describe or define a change or separation in any way that does not involve a sudden finite jump in some parameter. Discrete models reflect the rising investment of the physical sciences in a quantum-theoretic view of reality, and the increasing dependence of science on computer simulation as an experimental tool.[22]

Discrete models have the advantage that they can more easily incorporate modern cybernetic concepts, including information, computation and feedback, which conduce to an understanding of reality as a control and communication system. In the context of such models, informational and computational reductionism is now pursued with a degree of enthusiasm formerly reserved for attempts to reduce the universe to matter and energy. However, certain

---

[20] In physics, spatial and spatiotemporal manifolds are usually constrained by the *locality principle*, according to which nothing travels faster than light. Locality can be more fundamentally defined as the condition that in relocating from one point to another in a metric space, an object must traverse the entire sequence of adjacent finite or infinitesimal intervals comprising some intervening path within the metric on which locality is being enforced. In other words, locality means "no sudden jumps from one point to another, through the space containing the points or any external space thereof." The bearing on causality is obvious.

[21] Continuity is understood in terms of infinitesimal displacements. Several approaches exist to the topic of infinitesimals, some more controversial than others. The most common is the Cauchy–Weierstrass epsilon-delta formalism; the most sophisticated is that of which A. Robinson's (1966) *non-standard analysis* is the earliest and most successful representative.

[22] Perhaps the most fashionable discrete model universe is explicitly based on a computational paradigm, the cellular automaton. An encyclopedic account of this paradigm can be found in Wolfram (2002).

difficulties persist. Discrete models remain dualistic, and they still cannot explain their own origins and existences. Nonlocality is still a problem for them, as are the general-relativistic spacetime deformations so easily formulated in continuum models. Because they allow the existence of discrete gaps between events, they tend to lack adequate connectivity. And in the shadow of these deficiencies, they can illuminate the interrelationship of space, time and object no more successfully than their continuum counterparts.

The unregenerate dualism of most discrete models demands particular attention. As we reasoned above, solving the problem of dualism requires that the mental and physical aspects of reality be brought into coincidence. Insofar as information and computation are essentially formal and abstract, reducing the material aspects of nature to information and computation should bring the concrete and the abstract, the material and the mental, into perfect coincidence. But because most discrete models treat information and computation as objective entities, tacitly incorporating the assumption that bits and computations are on the same ontological footing as particles and collisions, their mental dimension is overlooked. Because making no explicit provision for mind amounts to leaving it out of the mix, mind and matter remain separate, and dualism persists.

Is there another alternative? The model-theoretic perspective, which simultaneously juxtaposes and conflates subjective languages and their objective universes, suggests that reality embodies an ontic-nomothetic medium with abstract and physical aspects that are respectively related as syntax is related to language. For example, because scientific observation and theorization must be consistent, and logic is the backbone of consistency, the syntax of every scientific theory must incorporate logic. In the case of a geometric theory of physical reality like classical mechanics or relativity theory, this amounts (by model-theoretic implication) to the requirement that logic and geometry literally coincide. But where geometry is a property of "physical" spacetime, so then is logic, and if logic resides in spacetime, then so must logical grammar. This leads to the requirement that physical dynamics be *objectively* reconciled with the formal grammar of logic and logic-based theories, ultimately including abstract causality in its entirety.

Obviously, conventional continuum and discrete models of reality fail to meet this requirement. As far as they and those who embrace them are concerned, the physical world is simply not answerable to any theory whatsoever, even logic. According to the standard empirical doctrine of science, we may observe reality but never impose our conceptions upon it, and this means that theory – even a theory as necessary to cognition and perception as logic – is always the beggar and never the master at the scientific table. The reason for this situation is clear; scientists need a means of guarding against the human tendency to confuse their inner subjective worlds, replete with fantasy and prejudice, with the factual external world conventionally studied by science.

But there is a very clear difference between logic on one hand, and fantasy and prejudice on the other. While science never needs the latter, it *always* needs the former. By excluding logic from nature, mainstream science has nothing to gain and everything to lose; in not attributing its own most basic requirements to its subject matter, it is cheating itself in a crucial way. Whether or not a theory which fails to predicate on its universe the wherewithal of its own validity turns out to be valid, it can be neither more nor less so for its false and subtly pretentious humility. On the other hand, failing to attribute these requirements to its universe when its universe in fact exhibits them, and when its universe would in fact be unintelligible without them, can ultimately cost it every bit of truth that it might otherwise have had ... particularly if its methodology is inadequate to identify the problem and mandate a remedy.

Because they fail to provide definitive answers for questions about causality, conventional continuum and discrete models of reality devolve to acausality or infinite causal regression. No matter what causal explanations they seem to offer, one of two things is implied: (1) a cause prior to that which is cited in the explanation, or (2) random, spontaneous, acausal emergence from the void, no explanation supposedly required. Given the seeming absence of alternatives to determinism or randomness, or extrinsic[23] and null causation, how are meaningful causal explanations to be completed?

The Containment Problem

A certain philosophically controversial hypothesis about causality presently rules the scientific world by fiat. It asserts that physical reality is closed under causal regression: "no physical event has a cause outside the physical domain."[24] That is, if a physical event has a cause, then it has a *physical* cause. Obviously, the meaning of this principle is strongly dependent on the definition of *physical*, which is not as cut and dried as one might suppose.[25] It also

---

23 This means "extrinsic to the object affected by causality". For example, consider the problem of the origin of the real universe. Where the *real universe* is defined to contain all that is perceptible and/or of relevance to that which is perceptible, anything sufficiently real to have originated, caused or influenced it is contained within it by definition. Thus, extrinsic causality (standard determinacy) cannot be invoked to explain the origin of the real universe. Because every instance of causation *within* the real universe ultimately leads back to the origin of reality by causal regression, standard determinacy fails as a causal paradigm.

24 This particular formulation of the "physical causal closure thesis" is due to the contemporary philosopher Jaegwon Kim (2000). By the mathematical definition of closure, causal closure implies reflexive self-determinism. Because the physical causal closure thesis instead relies on standard determinism, it is conceptually deficient and powerless to effect causal closure.

25 *Physical* is a rather ambiguous term that currently means "of or relating to matter and energy or the sciences dealing with them, especially physics". It thus refers to a relationship of unspecified extent, namely the extended relational plexus generated

contradicts the obvious fact that causality is an abstraction, at best indirectly observable through its effects on matter, which functions independently of any specific item of material content. How, then, does this principle manage to maintain its hold on science? The answer: false parsimony and explanatory debt. Concisely, *false parsimony* is when a theory achieves deceptive simplicity in its native context by sweeping its unpaid *explanatory debts* (explanatory deficiencies) into unkempt piles located in or between other areas of science.

It is an ill-kept secret that the scientific community, far from being one big happy family of smoothly-connected neighborhoods, consists of isolated, highly-specialized enclaves that often tend toward mutual ignorance and xenophobia. Under these circumstances, it is only natural to expect that when caught between an observational rock and a theoretical hard place, some of these enclaves will take advantage of the situation and "pass the explanatory buck", neither knowing nor caring when or where it comes to rest as long as the maneuver takes some of the heat off them and frees them to conduct business as usual. While the explanatory buck-passing is almost never productive, this can be conveniently hidden in the deep, dark cracks and crevices between disciplines. As a result, many pressing explanatory obligations have been successfully exiled to interdisciplinary limbo, an intellectual dead zone from which they cannot threaten the dominance of the physical causal closure thesis.

However, this ploy does not always work. Due to the longstanding scientific trend toward physical reductionism, the buck often gets passed to physics, and because physics is widely considered more fundamental than any other scientific discipline, it has a hard time deferring explanatory debts mailed directly to its address. Some of the explanatory debts for which physics is holding the bag are labeled "causality", and some of these bags were sent to the physics department from the evolutionary biology department. These debt-filled bags were sent because the evolutionary biology department lacked the explanatory resources to pay them for itself. Unfortunately, physics can't pay them either.

The reason that physics cannot pay explanatory debts generated by various causal hypotheses is that it does not itself possess an adequate understanding of causality. This is evident from the fact that in physics, events are assumed to be either *deterministic* or *nondeterministic* in origin. Given an object, event, set or process, it is usually assumed to have come about in one of just two possible ways: either it was brought about by something prior and external to it, or it sprang forth spontaneously as if by magic. The prevalence of this dichotomy,

---

by the concepts of matter and energy. While causality does indeed relate to *matter* and *energy*, it can be neither held in the hand nor converted to heat, and because it thus bears description as neither matter nor energy, it resides elsewhere in this extended relationship. It follows that causality is more than physical. Where *physical* is further defined as "belonging to the class of phenomena accessible to the scientific method", only those levels of causality which are both discernable and replicable may be called "physical".

determinacy versus randomness, amounts to an unspoken scientific axiom asserting that everything in the universe is ultimately either a function of causes external to the determined entity (up to and including the universe itself), or no function of anything whatsoever. In the former case there is a known or unknown explanation, albeit external; in the latter case, there is no explanation at all. In neither case can the universe be regarded as causally self-contained.

To a person unused to questioning this dichotomy, there may seem to be no middle ground. It may indeed seem that where events are not actively and connectively produced according to laws of nature, there is nothing to connect them, and thus that their distribution can only be random, patternless and meaningless. But there is another possibility after all: *self-determinacy*. Self-determinacy involves a higher-order generative process that yields not only the physical states of entities, but the entities themselves, the abstract laws that govern them, and the entire system which contains and coherently relates them. Self-determinism is the causal dynamic of any system that generates its own components and properties independently of prior laws or external structures. Because self-determinacy involves nothing of a preexisting or external nature, it is the only type of causal relationship suitable for a causally self-contained system.

In a self-deterministic system, causal regression leads to a completely intrinsic self-generative process. In any system that is *not* ultimately self-deterministic, including any system that is either random or deterministic in the standard extrinsic sense, causal regression terminates at null causality or does not terminate. In either of the latter two cases, science can fully explain nothing; in the absence of a final cause, even material and efficient causes are subject to causal regression toward ever more basic (prior and embedding) substances and processes, or if random in origin, toward primitive acausality. So given that explanation is largely what science is all about, science would seem to have no choice but to treat the universe as a self-deterministic, causally self-contained system.[26]

And thus do questions about evolution become questions about the self-generation of causally self-contained, self-emergent systems. In particular, *how* and *why* does such a system self-generate?

The Utility (Selection) Problem

As we have just noted, deterministic causality transforms the states of preexisting objects according to preexisting laws associated with an external medium.

[26] In any case, the self-containment of the real universe is implied by the following contradiction: if there were any *external* entity or influence that were sufficiently real to affect the real universe, then by virtue of its reality, it would by definition be *internal* to the real universe.

Where this involves or produces feedback, the feedback is of the conventional cybernetic variety; it transports information through the medium from one location to another and then back again, with transformations at each end of the loop. But where objects, laws and media do not yet exist, this kind of feedback is not yet possible. Accordingly, causality must be reformulated so that it can not only transform the states of natural systems, but account for self-deterministic relationships between states and laws of nature. In short, causality must become metacausality.[27]

Self-determination involves a generalized atemporal[28] kind of feedback between physical states and the abstract laws that govern them. Whereas ordinary cybernetic feedback consists of information passed back and forth among controllers and regulated entities through a preexisting conductive or transmissive medium according to ambient sensory and actuative protocols – one may think of the Internet, with its closed informational loops and preexisting material processing nodes and communication channels, as a ready example – self-generative feedback must be ontological and *telic* rather than strictly physical in character.[29] That is, it must be defined in such a way as to "metatemporally" bring the formal structure of cybernetics and its physical content into joint existence from a primitive, undifferentiated ontological groundstate. To pursue our example, the Internet, beginning as a timeless self-potential, would have to self-actualize, in the process generating time and causality.

But what is this ontological groundstate, and what is a "self-potential"? For that matter, what are the means and goal of cosmic self-actualization? The ontological groundstate may be somewhat simplistically characterized as a complete abeyance of binding ontological constraint, a sea of pure telic potential or "unbound telesis". Self-potential can then be seen as a telic relationship of two lower kinds of potential: potential *states*, the possible sets of definitive properties possessed by an entity along with their possible values, and potential *laws* (nomological syntax) according to which states are defined, recognized and transformed.[30] Thus, the ontological groundstate can for most

---

[27] Metacausality is the causal principle or agency responsible for the origin or "causation" of causality itself (in conjunction with state). This makes it responsible for its own origin as well, ultimately demanding that it self-actualize from an ontological groundstate consisting of unbound ontic potential.

[28] Where time is defined on physical change, metacausal processes that affect potentials without causing actual physical changes are by definition atemporal.

[29] *Telesis* is a convergent metacausal generalization of law and state, where law relates to state roughly as the syntax of a language relates to its expressions through generative grammar ... but with the additional stipulation that as a part of syntax, generative grammar must in this case generate itself along with state. Feedback between syntax and state may thus be called *telic feedback*.

[30] Beyond a certain level of specificity, no detailed knowledge of state or law is required in order to undertake a generic logical analysis of telesis.

purposes be equated with all possible state-syntax relationships or "self-potentials", and the means of self-actualization is simply a telic, metacausal mode of recursion through which telic potentials are refined into specific state-syntax configurations. The particulars of this process depend on the specific model universe – and in light of dual-aspect monism, the *real self-modeling* universe – in which the telic potential is actualized.

And now we come to what might be seen as the pivotal question: what is the *goal* of self-actualization? Conveniently enough, this question contains its own answer: self-actualization, a generic analogue of Aristotelian final causation and thus of teleology, is *its own* inevitable outcome and thus its own goal.[31] Whatever its specific details may be, they are actualized by the universe alone, and this means that they are mere special instances of cosmic self-actualization. Although the word "goal" has subjective connotations – for example, some definitions stipulate that a goal must be the object of an instinctual drive or other subjective impulse – we could easily adopt a reductive or functionalist approach to such terms, taking them to reduce or refer to objective features of reality. Similarly, if the term "goal" implies some measure of design or pre-formulation, then we could easily observe that natural selection does so as well, for nature has already largely determined what "designs" it will accept for survival and thereby render fit.

Given that the self-containment of nature implies causal closure implies self-determinism implies self-actualization, how is self-actualization to be achieved? Obviously, nature must select some possible form in which to self-actualize. Since a self-contained, causally closed universe does not have the luxury of external guidance, it needs to generate an *intrinsic* self-selection criterion in order to do this. Since *utility* is the name already given to the attribute which is maximized by any rational choice function, and since a totally self-actualizing system has the privilege of defining its own standard of rationality,[32] we may as well speak of this self-selection criterion in terms of global or generic self-utility. That is, the self-actualizing universe must generate

[31] To achieve causal closure with respect to final causation, a metacausal agency must self-configure in such a way that it relates to itself as the ultimate utility, making it the *agency*, *act* and *product* of its own self-configuration. This 3-way coincidence, called *triality*, follows from self-containment and implies that self-configuration is intrinsically utile, thus explaining its occurrence in terms of intrinsic utility.

[32] It might be objected that the term "rationality" has no place in the discussion ... that there is no reason to assume that the universe has sufficient self-recognitional coherence or "consciousness" to be "rational". However, since the universe does indeed manage to consistently self-recognize and self-actualize in a certain objective sense, and these processes are to some extent functionally analogous to human self-recognition and self-actualization, we can in this sense and to this extent justify the use of terms like "consciousness" and "rationality" to describe them. This is very much in the spirit of such doctrines as physical reductionism, functionalism and eliminativism, which assert that such terms devolve or refer to objective physical or functional relationships. Much the same reasoning applies to the term *utility*.

and retrieve information on the intrinsic utility content of various possible forms that it might take.

The utility concept bears more inspection than it ordinarily gets. Utility often entails a subject-object distinction; for example, the utility of an apple in a pantry is biologically and psychologically generated by a more or less conscious subject of whom its existence is ostensibly independent, and it thus makes little sense to speak of its "intrinsic utility". While it might be asserted that an apple or some other relatively non-conscious material object is "good for its own sake" and thus in possession of intrinsic utility, attributing self-interest to something implies that it is a subject as well as an object, and thus that it is capable of subjective self-recognition.[33] To the extent that the universe is at once an *object* of selection and a self-selective *subject* capable of some degree of self-recognition, it supports intrinsic utility (as does any coherent state-syntax relationship). An apple, on the other hand, does not seem at first glance to meet this criterion.

But a closer look again turns out to be warranted. Since an apple is a part of the universe and therefore embodies its intrinsic self-utility, and since the various causes of the apple (material, efficient and so on) can be traced back along their causal chains to the intrinsic causation and utility of the universe, the apple has a certain amount of intrinsic utility after all. This is confirmed when we consider that its taste and nutritional value, wherein reside its utility for the person who eats it, further its genetic utility by encouraging its widespread cultivation and dissemination. In fact, this line of reasoning can be extended beyond the biological realm to the world of inert objects, for in a sense, they too are naturally selected for existence. Potentials that obey the laws of nature are permitted to exist in nature and are thereby rendered "fit", while potentials that do not are excluded.[34] So it seems that in principle, natural selection determines the survival of not just actualities but potentials, and in either case it does so according to an intrinsic utility criterion ultimately based on global self-utility.

It is important to be clear on the relationship between utility and causality. Utility is simply a generic selection criterion essential to the only cosmologically acceptable form of causality, namely self-determinism. The subjective gratification associated with positive utility in the biological and psychological realms is ultimately beside the point. No longer need natural processes be

---

33 In computation theory, *recognition* denotes the acceptance of a language by a transducer according to its programming or "transductive syntax". Because the universe is a self-accepting transducer, this concept has physical bearing and implications.

34 The concept of potential is an essential ingredient of physical reasoning. Where a *potential* is a set of possibilities from which something is *actualized*, potential is necessary to explain the existence of anything in particular (as opposed to some other partially equivalent possibility).

explained under suspicion of anthropomorphism; causal explanations need no longer implicitly refer to instinctive drives and subjective motivations. Instead, they can refer directly to a generic objective "drive", namely intrinsic causality ... the "drive" of the universe to maximize an intrinsic self-selection criterion over various relational strata within the bounds of its internal constraints.[35] Teleology and scientific naturalism are equally satisfied; the global self-selection imperative to which causality necessarily devolves is a generic property of nature to which subjective drives and motivations necessarily "reduce", for it distributes by embedment over the intrinsic utility of every natural system.

Intrinsic utility and natural selection relate to each other as both reason and outcome. When an evolutionary biologist extols the elegance or effectiveness of a given biological "design" with respect to a given function, as in "the wings of a bird are beautifully *designed* for flight", he is really talking about intrinsic utility, with which biological fitness is thus entirely synonymous. Survival and its requisites have intrinsic utility for that which survives, be it an organism or a species; that which survives derives utility from its environment *in order to* survive and *as a result of* its survival. It follows that neo-Darwinism, a theory of biological causation whose proponents have tried to restrict it to determinism and randomness, is properly a theory of intrinsic utility and thus of self-determinism. Although neo-Darwinists claim that the kind of utility driving natural selection is non-teleological and unique to the particular independent systems being naturally selected, this claim is logically insupportable. Causality ultimately boils down to the tautological fact that on all possible scales, nature is both *that which selects* and *that which is selected*, and this means that natural selection is ultimately based on the intrinsic utility of nature at large.

But in light of causal self-containment, so is teleology. Why, then, do so many supporters of teleology and neo-Darwinism seem to think them mutually exclusive?

The Stratification Problem

It is frequently taken for granted that neo-Darwinism and ID theory are mutually incompatible, and that if one is true, then the other must be false. But while this assessment may be accurate with regard to certain inessential propositions attached to the core theories like pork-barrel riders on congressional bills,[36] it is not so obvious with regard to the core theories themselves. In fact, these theories are dealing with different levels of causality.

---

35 Possible constraints include locality, uncertainty, blockage, noise, interference, undecidability and other intrinsic features of the natural world.

36 Examples include the *atheism* and *materialism* riders often attached to neo-Darwinism, and the *Biblical Creationism* rider often mistakenly attached to ID theory.

The scientific method says that experiments must be replicable, and this means that the same laws must govern the same kinds of events under the same conditions throughout nature. So where possible, the laws of nature are scientifically formulated in such a way that they distribute over space and time, the same laws applying under similar conditions at all times and places. Science also requires that the laws of nature be formulated in such a way that the next state of an object depends only on its present state, including all of the forces impinging on it at the present moment, with no memory of prior states required. Little wonder that science enforces these two conditions with extreme prejudice wherever possible, for in principle, they guarantee its ability to predict the future of any physical system from a mere knowledge of its current state and the distributed laws of nature.[37]

Science imposes yet further constraints on causality. One, the empirical discernability criterion of the scientific method,[38] guarantees the recognizability of physical states by insisting that they be formulated in terms of first-order properties[39] called *observables* that can be unambiguously measured in conjunction with physical objects. Another, which we have already encountered, is the locality principle, which says that there can be no "nonlocal" jumps from

---

37 This view was captured by the French astronomer and mathematician Pierre Simon Laplace (1749–1827) in his 1814 *Philosophical Essay on Probabilities* (1902, p. 4): "Given for one instant an intelligence which could comprehend all the forces by which nature is animated and the respective situation of the beings who compose it—an intelligence sufficiently vast to submit these data to analysis—it would embrace in the same formula the movements of the greatest bodies of the universe and those of the lightest atom; for it, nothing would be uncertain and the future, as the past, would be present to its eyes." This view, called *Laplacian determinism*, went virtually unchallenged until the first half of the 20th century, when it was undermined by such new concepts as quantum uncertainty and theoretic undecidability. But even though such problems seem to rule out an explicit calculation of the sort that Laplace envisioned, his ideal is still very much a driving force in science.

38 The scientific method mandates a constructive relationship between empirical observation and rational theorization that is designed for the investigation of phenomena possessing two criteria, *discernability* and *replicability*. That is, it confines scientific attention to that which can be exclusively and repeatedly observed under similar conditions anywhere in time or space; it does not cover any indiscernible or localized natural influence that is not conditionally (and thus falsifiably) distributed over space and time. Yet, indiscernables and unconditional universals must exist in order for nature to be stable – e.g., the universal, unconditional and intangible logical syntax which enforces consistency throughout the universe – and the exclusion of localized causal influences from nature is rationally insupportable.

39 In addition to other criteria, relations and properties are distinguished by *arity* and *order*. The arity (adicity, cardinality) of a relation is just the number of *relands* or things related, while its order depends on whether its relands are individual elements, relations of elements, relations of relations of elements, or so on. Similarly, a property (attribute, predicate) is distinguished by whether it is attributed to individual elements, properties of elements, properties of properties of elements, or et cetera.

one point in a physical manifold to another non-adjacent point.[40] This adds an adjacency or continuity constraint to the Laplacian ideal; the laws of nature must not only be formulated in such a way that the next state of an object depends only on its present state, but in such a way that successive states are "near" each other, i.e. so that smaller amounts of time and energy correspond to smaller distances. This proportionality of distance and effect permits the laws of causality to be consistently applied on the macroscopic and microscopic scales.

Of all the preconceived restrictions and unnecessary demands imposed on causality by science, the least questioned is the requirement that the relationship between physical states and laws of nature be one-way, with states depending on laws but not vice versa. Science regards the laws of nature as immutable, states as existing and transforming at their beck and call, and the directional dependency relationship between laws and states as something that has existed for all time. When the laws dictate that an event should happen, it happens; on the other hand, any event chancing to occur without their guidance is uncaused and totally "random". This leads to the determinacy-versus-randomness dichotomy already discussed in connection with the containment and utility problems.

Due to these criteria, what science calls a "law of nature" is typically an autonomous relationship of first-order properties of physical objects, and so for the laws of state transformation that govern causation. There can be little doubt that science has succeeded in identifying a useful set of such laws. Whether or not they suffice for a full description of nature and causality (and they do not), they are an important part of the total picture, and wherever possible, they should indeed be tracked down and exploited to their full descriptive and prescriptive potential. But at least one caveat is in order: they should be regarded as explaining *only that which they can be empirically and/or rationally shown to explain*. As with any other scientific assertion, they must be kept pure of any metaphysical prejudice tending to artificially inflate their scope or explanatory weight.

It is thus a matter of no small concern that in pursuing its policy of causal simplification, the scientific mainstream seems to have smuggled into its baggage compartment a certain piece of contraband which appears, despite its extreme resistance to rational or empirical justification, to be masquerading as a tacit "meta-law" of nature. It states that every higher-order relationship of objects and events in nature, regardless of complexity or level of dynamic integration, *must* be strictly determined by distributed laws of nature acting independently on each of its individual components. Along with the other

---

[40] Some manifolds come with special provisions for motion and causality, e.g. *metrics* defining the notion of distance, *derivatives* defining the notion of movement, and *affine connections* permitting the parallel transport of vectors through space and thereby supporting the concept of *fields*.

items on the neo-Laplacian wish-list of causal conveniences to which the scientific mainstream insists that nature be held, this criterion betrays a marked preference for a "bottom-up" approach to causation, suggesting that it be called the *bottom-up thesis*.[41]

The bottom-up thesis merely underscores something that we already know about the scientific mainstream: it wants with all of its might to believe that in principle, the whole destiny of the natural world and everything in it can be exhaustively predicted and explained on the basis of (1) a Laplacian snapshot of its current details, and (2) a few distributed laws of nature from which to exhaustively develop the implications. So irresistible is this desire that some of those caught in its grip are willing to make a pair of extraordinary claims. The first is that science has completely explained some of nature's most complex systems in terms of microscopic random events *simply by generically classifying the microscopic events that might possibly have been involved in their realization*. The second is that observed distributions of such events, which they again call "random", prove that no system in nature, regardless of its complexity, has ever come into being from the top down.

The genotype-to-phenotype mapping is a case in point. Many neo-Darwinists seem to have inferred that what happens near the endpoints of this mapping – the seemingly random mutation of genotypes and the brutal, deterministic competition among phenotypes – offers more insight regarding nature and causality than does the delicate, exquisitely complex ontogenic symphony performed by the mapping itself. In response to the observation that the theoretical emphasis has been lopsided, one hears that *of course* neo-Darwinists acknowledge the involvement of intermediate processes in the emergence of biological complexity from strings of DNA. For are not genes converted to proteins, which fold into functional forms and interact with other molecules to alter the timing of gene expression, which can lead to cytodifferentiation, pattern formation, morphogenesis and so on, and is this whole self-organizational process not highly sensitive to developmental interactions with the environment?

Unfortunately, where the acknowledged processes and interactions are still assumed to be micro-causal and deterministic, the acknowledgement is meaningless. In fact, the higher-order structure and processing of complex biological systems has only been shoveled into an unkempt pile sexily labeled "emergent phenomena" and bulldozed across the interdisciplinary divide into complex systems theory. And thus begins a ramose paper trail supposedly leading to the final owners of the explanatory debt, but instead looping, dead-ending or petering out in interdisciplinary limbo. The explanatory buck is thereby passed

[41] The bottom-up thesis is insidious in the way it carries the apparent randomness of experimental distributions of mutation events upward from low-order to high-order relationships, all the way to the phenotypic and social realms. This is what encourages many neo-Darwinists (and those whom they influence) to view mankind, and life in general, as "random" and "purposeless".

into oblivion, and the bottom-up thesis rolls like righteous thunder over any voice daring to question it.

In fact, the top-down and bottom-up approaches to causality are not as antithetical as they might seem. In the bottom-up view of causality, states evolve according to laws of nature in a temporal direction preferred by the second law of thermodynamics, which holds under the assumption that physical states are governed by laws of nature independent of state. But this assumption can hold only up to a point, for while the prevailing model universe supports only bottom-up causation, the situation is dramatically reversed with respect to cosmology. Because cosmological causal regression terminates with an ancestral cosmic singularity representing the whole of nature while omitting all of its details, standard cosmology ultimately supports *only* a top-down approach. The natural affinity of the cosmos for top-down causation – the fact that it is *itself* an instance of top-down causation – effectively relegates bottom-up causation to secondary status, ruling out the bottom-up thesis and thus making room for a new model universe supporting and reconciling both approaches.

It turns out that in a certain kind of model universe, the top-down and bottom-up approaches are to some degree mutually transparent.[42] Two necessary features of such a model universe are (1) sufficient causal freedom to yield probabilistic resources in useful amounts, and (2) structural support for metacausal access to those resources. As it happens, a well-known ingredient of nature, *quantum uncertainty*, provides the required sort of causal freedom. But while nature exhibits quantum uncertainty in abundance and can thus generate probabilistic resources at a certain respectable rate, the prevailing model universe supports neither metacausal relationships nor sufficient access to these resources. In fact, it fails to adequately support even quantum mechanics itself.

The new model universe must remedy these shortcomings ... but how?

---

[42] Within a given set of constraints, many possible future states of a physical system may be causally compatible with a single present state, and many alternative present states may be causally compatible with a single future state. Thus, higher-order and lower-order causal relationships describing the same system need not uniquely determine each other by top-down and bottom-up causation respectively. The physical situation is suggestive of formal model-theoretic ambiguity as captured by (e.g.) the *Duhem–Quine thesis*, according to which a given set of observations may be consistent with multiple theories of causation, and a single Laplacian snapshot can result in many possible predictions or retrodictions depending on the causal influences that are physically active or theoretically presumed to be active. Dual-aspect monism ultimately transforms model-theoretic ambiguity into causal freedom, revealing nature as its own creative theoretician and physical modeler and thereby effecting causal closure.

Synthesis: Some Essential Features of a Unifying Model of Nature and Causality

Classical mechanics, inarguably one of the most successful theories in history, is often cited as a model of theoretical progress in the sciences. When certain problems arose that could not be solved within its conceptual framework, it was extended to create a *metatheory* in which it exists as a "limiting case". In fact, this was done thrice in fairly rapid succession. The first extension created the Special Theory of Relativity, in which classical mechanics holds as a low-to-medium velocity limit. The second created the General Theory of Relativity, in the curved spacetime manifold of which the flat Minkowskian manifold of Special Relativity holds as a local limit. And the third created quantum mechanics, in which classical mechanics holds as a "decoherence limit".[43] Indeed, whenever a theory is extended by adjoining to it one or more new concepts, this creates a metatheory expressing the relationship between the adjoint concept(s) and the original theory.

The model universe of neo-Darwinism is just a special-purpose refinement of the continuous coordinate spaces of classical mechanics, and its causal limitations are shared by neo-Darwinism and most other scientific theories. This is because most sciences, not including certain branches of physics and engineering, have been unable to absorb and utilize the relativistic and quantum extensions of the classical model, each of which suffers in any event from many of the same difficulties with causality. It follows that another extension is required, and since neo-Darwinism holds true within a limited causal domain, it must hold in this extension as a limiting case (minus its inessential philosophical baggage). In other words, causality must become the objective, distributive limit of metacausality.

Such an extension has already been described (Langan 2002), and it embodies solutions for all of the problems discussed in this paper. Concisely, it embeds physical reality in an extended logico-algebraic structure, a *Self-Configuring Self-Processing Language* or SCSPL. SCSPL incorporates a pregeometric[44] *conspansive manifold* in which the classical spacetime manifold is embedded as a limiting configuration. SCSPL brings formal and physical causality into seamless conjunction by generically equating the laws of nature with SCSPL syntax, and then contracting the semantic, model-theoretic correspondence

---

[43] These extensions are to some extent mutually incompatible. In order to reconcile the outstanding conflicts and conceptual dissonances between General Relativity and quantum mechanics, yet another metatheoretic extension is now required.

[44] Patton & Wheeler (1975) use the term *pregeometry* in reference to "... something deeper than geometry, that underlies both geometry and particles ... no perspective seems more promising than the view that it must provide the Universe with a way to come into being." The SCSPL extension of physical reality fits this description.

between syntax and state (or laws and observables)[45] so that they coincide in *syntactic operators*, physical quanta of self-transducing information. Through properties called *hology* (syntactic self-similarity) and *triality* (space-time-object conflation), total systemic self-containment is achieved. In particular, the system is self-deterministically closed under causation.

SCSPL evolves by *telic recursion*, a higher-order process[46] of which causality is the physical limit (as required). In standard causality, physical states evolve according to laws of nature; in telic recursion, syntax-state relationships evolve by maximization of intrinsic utility. The temporal phase of telic recursion is *conspansion*, a dual-aspect process coordinating formal/telic and physical modes of evolution. By virtue of *conspansive duality*, SCSPL simultaneously evolves like a (metacausal, telic-recursive) generative grammar and a physical dynamical system, at once implementing top-down and bottom-up causation. Conspansion involves an alternation between *self-replication* and *self-selection*, thus constituting a generalization of Darwinian evolution in which specificational and replicational probabilistic resources are rationally linked. In this way, neo-Darwinist and design-theoretic (bottom-up and top-down) modes of causality become recognizable as complementary aspects of a single comprehensive evolutionary process.

From a formal standpoint, SCSPL has several unique and interesting features. Being based on logic,[47] it identifies itself with the logical syntax of its perceptual universe on grounds of logical-perceptual isomorphism. This eliminates the conventional model-theoretic distinction among theory, universe and theory-universe correspondence, contracting the problematic mapping between abstract and concrete reality on the syntactic (nomological) level. This brings the physical world into coincidence with its logical counterpart, effecting dual-aspect monism and putting logical attributes on the same explanatory footing as physical attributes. SCSPL thus adjoins logic to nature, injecting[48] nature

---

45 Where laws of nature incorporate not only observables but the abstractions relating them, bringing physical states and natural laws into coincidence reduces the set of physical (observable) properties to a subset of the set of abstract properties. Thus, the abstract is recognized as a *natural* generalization of the concrete.

46 *Telic recursion* is quantum metaprocess based on a generalized form of recursion maximizing intrinsic utility over entire (pregeometric) regions of spacetime through telic feedback under the guidance of coherent metacausal invariants called *telons*.

47 SCSPL is developed by adjoining to (propositional and predicate) logic a limiting form of model theory from which it acquires certain necessary high-level properties of any possible valid theory of reality at large. Thus, its syntactic and semantic validity can be logically established. By its method of construction, SCSPL is classified as a metaphysical tautology or *supertautology*.

48 Inflationary cosmology, membrane theory and various other theories have been assumed to require extensions external to physical reality. In contrast, SCSPL conspansive duality permits the extension mandated by SCSPL, as well as all other valid extensions of physical reality, to be physically internalized in a certain specific sense relating to conspansive duality.

with the abstract logical infrastructure of perception and theorization and endowing physical reality with the remedial conceptual apparatus demanded by the problems, paradoxes and explanatory deficiencies straining its classical descriptions. At the same time, it adjoins nature to logic in the form of perceptual categories and necessary high-level properties including closure, comprehensiveness, consistency and teleo-nomological coherence, thus opening logical routes to physical insight.

SCSPL offers yet further advantages. In defining nature to include logic and cognition, it relates physics and mathematics on a basic level, thus merging the rational foundations of mathematics with the perceptual foundations of physics and letting each provide crucial support for the other. By affording an integrated conceptual framework for prior conflicting extensions of classical reality, it sets the stage for their ultimate reconciliation. And its cross-interpretation of the cognitive and physical aspects of nature renders the universe *self-explaining* and *self-modeling*, thus effecting self-containment on the theoretic and model-theoretic levels. That is, SCSPL self-containment effects not just causal and generative closure, but closure under the inverse operations of explanation and interpretation, thus permitting nature to physically model and teleo-nomologically justify its own self-configurative determinations. In SCSPL, natural laws and physical states are seen as expressions of the intrinsic utility *of* nature *by* and *for* nature.

The reflexive self-processing and (telic) self-configuration functions of SCSPL imply that nature possesses generalized functional analogues of human self-awareness and volition, and thus a generalized capacity for utilitarian self-design. The self-design and self-modeling capacity of nature suggests that the universe is a kind of stratified "self-simulation" in which the physical and logico-telic aspects of reality can be regarded as respectively "simulated" and "simulative" in a generalized quantum-computational sense. This makes SCSPL relevant to self-organization, emergence and other complexity-theoretic phenomena increasingly attractive to the proponents of neo-Darwinism and other causally-challenged theories. At the same time, the fact that SCSPL evolution is both nomologically coherent and subject to a rational intrinsic utility criterion implies that the universe possesses properties equivalent to generalized intelligence, suggesting the possibility of an integrated SCSPL approach to the problems of consciousness and evolution.

The overall theory which logically extends the concepts of nature and causality to SCSPL and telic recursion, thereby merging the perceptual manifold with its cognitive and telic infrastructure, is known as the *Cognitive-Theoretic Model of the Universe* or CTMU, and its approach to biological origins and evolution is called *Teleologic Evolution*.[49] Based on the concept of telic-recursive meta-causation, Teleologic Evolution is a dynamic interplay of replication and

[49] See www.teleologic.org for a description of Teleologic Evolution.

selection through which the universe creates itself and the life it contains. Teleologic Evolution is a stratified process which occurs on levels respectively associated with the evolution of the cosmos and the evolution of life, thus permitting organic evolution to mirror that of the universe in which it occurs. It improves on traditional approaches to teleology by extending the concept of *nature* in a way eliminating any need for "supernatural" intervention, and it improves on neo-Darwinism by addressing the full extent of nature and its causal dynamics.

Due to their implicit reliance on different notions of causality, teleology and evolution were once considered mutually exclusory. While teleology appears to require a looping kind of causality consistent with the idea that ends are immanent in nature (even in beginnings), evolution seems to require that mutation and natural selection exhibit some combination of nondeterminacy and linear determinacy. In contrast, the phrase *Teleologic Evolution* reflects their complementarity within a coherent self-configurative ensemble identifying nature with its own utilitarian self-actualization imperative. In the associated metacausal extension of physical reality, the two central processes of evolution, replication and selection, are seen to occur on at least two mutually-facilitative levels respectively associated with the evolution of the universe and that of organic life.[50] Meanwhile, the intrinsic utility criterion of self-selection implies that nature, as rationally defined in the CTMU, possesses a generalized form of intelligence by which all levels of evolution are driven and directed, equating selection with specification and metacausally relating it to replication.[51] Reality is united with its generative principle by the rational linkage between the domain and codomain of the teleological, meta-Darwinian level of natural selection.

Because nature consists of all that is logically relevant to perception, and logic consists of the rules of thought and therefore comprises an essential theory of cognition, the CTMU couples mind and nature in a way suggestive of Ouroboros divided and reunited ... two intimately entwined constrictors, estranged centuries ago by mind-body dualism but now locked in a renewed embrace, each swallowing the other's entailments. Perhaps this reunion will deter the militant torch-bearers of scientific naturalism from further reneging on their explanatory debts and fleecing mankind of its millennial hopes and dreams after all. And if so, then perhaps mankind can snuff the rapidly dwindling fuse of its insidious ontological identity crisis while these hopes and dreams still have a fighting chance of realization, and the intrinsic utility of mankind is still salvageable.

---

[50] Human psycho-intellectual, sociopolitical and technological modes of evolution may also be distinguished on various levels of aggregation.

[51] In the CTMU, instances of irreducible and specified complexity are metacausally generalized to dynamic syntax-state relationships called *telons* which self-actualize by telic recursion.

References

Bacon, F. (1997) *Thoughts on the Nature of Things*. Kila, MT: Kessinger Publishing. Reprinted excerpt. Originally published in 1824 as Miscellaneous Tracts. In B. Montagu (Ed.) *The Works of Francis Bacon, Lord Chancellor of England*, pp. 406–455. London: William Pickering.

Behe, M. J. (1998) *Darwin's Black Box: The Biochemical Challenge to Evolution*. New York: Simon & Schuster.

Darwin, C. (1999) *On the Origin of Species*. New York: Bantam Classics. Original work published in 1859.

Dembski, W. A. (1998) *The Design Inference: Eliminating Chance through Small Probabilities*. Cambridge: Cambridge University Press.

Ho, M. W. & Saunders, P. T. (1979) Beyond neo-Darwinism—An Epigenetic Approach to Evolution. *Journal of Theoretical Biology*, Vol. 78, Issue 4, pp. 573–591.

Hume, D. (1975) *Enquiries Concerning Human Understanding and Concerning the Principles of Morals*. 3rd Edition. Edited by L. A. Selby-Bigge & P. H. Nidditch. Oxford: Oxford University Press. Reprinted from the posthumous edition of 1777. Original work published in 1748 and 1751.

Kant, I. (1965) *The Critique of Pure Reason*. Translated by N. K. Smith. New York: St. Martin's Press. Original work published in 1781.

Kim, J. (2000) *Mind in a Physical World: An Essay on the Mind-Body Problem and Mental Causation*. Cambridge, MA: MIT Press.

Langan, C. M. (2002) *The Cognitive-Theoretic Model of the Universe: A New Kind of Reality Theory*. Princeton, MO: Mega Foundation Press. Originally published in Progress in Complexity, Information, and Design, Double Issue, Vols. 1.2-3.

Laplace, P. S. (1902) *A Philosophical Essay on Probabilities*. New York: John Wiley & Sons. Original work published in 1814.

Patton, C. M. & Wheeler, J. A. (1975) Is Physics Legislated by Cosmogony? In C. J. Isham, R. Penrose, D. W. Sciama (Eds.) *Quantum Gravity: An Oxford Symposium*, pp. 538–605. Oxford: Clarendon Press.

Pigliucci, M. (2000) Methodological vs. Philosophical Naturalism, or Why We Should Be Skeptical of Religion. In *Tales of the Rational: Skeptical Essays About Nature and Science*. Smyrna, GA: Freethought Press.

Robinson, B. A. (1995) Public Beliefs about Evolution and Creation. *Ontario Consultants on Religious Tolerance*.

Whitehead, A. N. (1975) *Process and Reality*. New York: The Free Press. Originally published in 1929.

Wolfram, S. (2002) *A New Kind of Science*. Champaign, IL: Wolfram Media.

COSMOS AND HISTORY

An Introduction to Mathematical Metaphysics

Abstract

Since the time of Aristotle, metaphysics has been an ill-defined term. This paper defines it as a logically idempotent metalinguistic identity of reality which couples the two initial ingredients of awareness: perceptual reality (the basis of physics), and cognitive-perceptual syntax, a formalization of mind. The explanation has been reduced to a few very simple, clearly explained mathematical ingredients. This paper contains no assumptions or arguable assertions, and is therefore presented as an advanced formulation of logic which has been updated for meaningful reference to the structure of reality at large. This structure, called the Cognitive-Theoretic Model of the Universe or CTMU (Langan, 2002), resolves the problems attending Cartesian dualism by replacing dualism with the mathematical property of self-duality, meaning (for reality-theoretic purposes) the quantum-level invariance of identity under permutation of objective and spatiotemporal data types. The CTMU takes the form of a global coupling or superposition of mind and physical reality in a self-dual metaphysical identity $M: L \leftrightarrow U$, which can be intrinsically developed into a logico-geometrically self-dual, ontologically self-contained language incorporating its own medium of existence and comprising its own model therein.

Keywords

Metaphysics; Mathematics; Mathematical Metaphysics

Introduction

In these early years of the new millennium, there has been considerable academic concern over the stubbornness of Cartesian dualism and the conceptual difficulty of uniting the mental and physical sides of reality. Meanwhile, it has gone all but unnoticed in academic circles that the formal aspect of this challenge was met decades ago, and in the most effective possible way. This paper contains a brief introductory account of the work in question, including a highly simplified description of the logico-mathematical reasoning needed for the proper high-level theoretical description of reality. Everything follows from the requirements of this objective.

Aside from the author's own writings, there is as yet no well-defined field of mathematical metaphysics. Being an explanation of the author's previous work, this paper is largely an exercise in self-reference. The same applies to its bibliography. Conveniently enough, this is well in accord with a major theme herein emphasized, namely, the self-referential nature of reality.

The Necessity of Metaphysics

Whether or not any given scientist or philosopher of science chooses to admit it, science is heavily invested in the idea that there exists a valid, comprehensive theory of reality or "theory of everything" (ToE), and indeed there must. For reality as a whole has structure and history, and in principle, that structure and history are isomorphic to a valid comprehensive theory of reality that can be written in a language of sufficient expressive power, the existence of which would seem to be implied by the fact that reality is accessible to the mind and senses. In a pinch, one need merely wave expansively at the universe and declare that the theory shares its structure.

Unfortunately, getting any farther than this entails a few difficulties. For example, we must decide how reality should be studied in order to discover its structure, and by whom. A closely related issue, which we encounter before even getting around to establishing a valid correspondence between theory and content, is how to identify a language of sufficient expressive power to contain a valid, comprehensive description of reality.

If there were a scientific consensus on these issues, it might well go something like this:

> Reality should be studied by physical scientists using the tools and methods of the physical sciences, and there can be no better language for expressing their findings than the language of physics. For after all, physics is the most fundamental science, that to which all other sciences can supposedly be reduced.

But this gives rise to yet another problem: physics is not self-explanatory. If physics is regarded as an expression of the structure of reality, then clearly it is real, and a comprehensive theory of reality must explain its every part and aspect. But then in order to qualify as a comprehensive theory of reality, physics must explain itself, its correspondence to reality, and arguably the biological origins and mental activity of physicists in whose minds it exists. In fact, physics as we know it can explain none of these things, or even the natural law of which it supposedly consists. Such explanations are prohibited by its methodology as encoded in the scientific method, according to which it must constantly be tested by observation and experimentation.

Technically, it is not permitted to depart from the world of the senses.

As a matter of logic, the task of explaining such things as physics, the possibility of physics, and the relationship of physics to the physical universe requires a metalanguage of physics, a higher-order language in which the "object language" of physics can be an object of reference. Let a metalanguage capable of these functions be called "metaphysical". In order to properly refer to its content, the required metaphysical metalanguage must include physics, understood as the theoretical aggregate of physical insight, as a sublanguage, along with distinctions and classifiers suitable for distinguishing between reality and its complement or negation.

Moreover, because its metaphysical content must be mapped into reality along with its physical sublanguage, it must describe reality in such a way that it contains the codomain of this mapping, thereby establishing that reality has a metaphysical aspect.

Because it contains physics as a sublanguage, and physics is largely mathematical, the required metaphysical metalanguage must be mathematical as well. But ToE mathematical structure must be mathematics of a higher order, literally embedding and distributing over the mathematical structures employed in standard physics ... as does the language of mathematical logic in particular.

Relational Structure

The relational level of mathematical structure is always of crucial importance, especially with regard to connectivity and coherence.

Historically, the coherence and connectivity of reality have been impaired by dualism. The parallel distinctions between mind and matter, form and content, attribute and argument, property and substance, language and universe, and so on are conventionally understood as dualistic, positing a fundamental dissimilarity or separation between two entities whose separation is problematical. This can often be addressed by mapping the dualistic distinction to discernible but coincidental aspects of a single coherent entity, thereby making it a duality and therefore dualic. A similar meta-distinction exists for a three-way distinction such as that between space, time, and object (or matter); it can be regarded as either trialistic or trialic.

Science proceeds by formal cognitive abstraction of perceptual content, for which it requires just two initial ingredients: perceptual reality (our perceptual relationship with the world of percepts or observable states), and the cognitive apparatus through which perceptual content, as well as the perceptual relationship between percept and percipient, is recognized and processed. Indeed, science can be regarded as the coupling of these two ingredients, and inasmuch as the ingredients are coupled, as an initial ingredient in its own right (later on, we'll be calling it "M"). Hence, the above dualisms and their remedial dualities bear heavily on scientific epistemology.

The cognitive and perceptual ingredients of reality have structure, and like all relationships, their relationship has structure as well. This structure is that of a syndiffeonic relation.

Syndiffeonesis, The Relational Structure of Reality

Any relationship of scientific interest is recognizable as a relationship and therefore intelligible, and every intelligible relationship involves the attribution of a property to its instances; i.e., the relationship is attributed to the things related by it. Other ways of saying this are that every intelligible relationship is a syndiffeonic relationship, or equivalently, that syndiffeonesis, the definitive property of syndiffeonic relationships, is the universal relational structure of reality as it is knowable to science.

Syndiffeonesis can be intuitively understood as "difference in sameness"; it is what occurs when an observer simultaneously recognizes any set of discernible objects in juxtaposition to their logical complements, or when a thinker simultaneously recognizes a set of concepts or values as similar or congruent. It captures the relational structure of attribution and the conceptual aggregation of discernibles, describing the situation in which a set of discernible instances is recognized or defined through a common syntax through which they are aggregated. Syntax consists of structural and dynamical invariants which support and constrain a language on all formal scales; they are distributed over the language intrinsically or by language-users. The concept can be generalized to any attribute or structure shared by a set of instances to be identified through it. Accepting syntax is that part or aspect of the mechanical structure or programming of a computational automaton which enables it to accept or "recognize" input; the concept can be generalized to non-mechanistic transduction.

Syndiffeonesis breaks down into sameness or synesis, which is stratified by order of relation or attribute, and difference or diffeonesis, which is a differentiative function of arity or cardinality. Every relationship, even a unary or nullary relationship, has two levels, synetic and diffeonic, with the synetic level consisting of common attributes or accepting-syntactic invariants, and the diffeonic level consisting of some number of discernible instances thereof (this number being the arity of the relationship). The generic syndiffeonic relation is "self-dual" in the sense that its synetic and diffeonic levels are dual to each other; they are respectively associated with orthogonal axes, the synetic ordinal axis and the diffeonic arity axis.

A syndiffeonic relation can be expressed in a number of ways, one of which is:

$$s(d_1, d_2, d_3, \ldots)$$

The diffeonic extension of the relation, being just a set of things that are simultaneously recognized as different instances of the property or syntax s which describes them, is inside the parentheses; the synetic intension which describes or distributes over the extension is to the left. This notation is designed to resemble an arithmetical expression like $n(a_1 + a_2 + a_3 + \ldots) = na_1 + na_2 + na_3 + \ldots$, in which multiplication distributes over addition. The d_i inside the parentheses are the "diffeonic relands", arbitrary things that are observed or conceived to differ from each other, while s is just the synetic intension or syntax which describes or defines them.

The generic syndiffeonic relation is "self-dual" in another sense as well: it has dual active and passive interpretations. It can be understood as the passive parallel or sequential assimilation of perceptual content through accepting syntax, or dually as the active, generative distribution of syntax over defined content (generative in the sense that the content does not technically exist prior to its formative syntax; pure cognition is virtually always generative, as it is inherently internal, subjective, and without external percepts by which the subject can be passively imprinted). It thus describes not only passive perception, with the synetic level equating to accepting syntax through which the diffeonic instances are recognized, but the generative conatus through which instances or values are brought into being.

This leads us to a third sense in which a syndiffeonic relation is self-dual: it relates cognition and perception, the subjective and objective aspects of reality. Where it is perceptually interpreted, its synetic level is associated with the subject as accepting syntax or generative conatus, while its diffeonic level is associated with percepts or objects of perception. The distribution of the synetic level of a syndiffeonic relation generates a medium. (A medium is a pointwise distribution of syntax. When a syndiffeonic relation is understood as a distribution of synetic syntax over diffeonic points, it functions as a logical "spacetime diagram"; its arity axis is spatial, while its ordinal axis is temporal. By logico-geometric duality, logical binding is dual to topological bounding, and syntax functions as a metric for the medium over which it distributes.) Where holological means "having the property that overall structure is distributed on all scales", the holological distribution of syntax throughout this medium makes it a connection among the diffeonic elements and subsets thereof. The intension "connects" the extension, imparting to the extension a degree of connectivity and coherence.

To sum it up: whenever, in the course of perception and/or cognition, one simultaneously recognizes any number of things as different from each other (discernible), one is necessarily bringing to bear a cognitive-perceptual syntax in order to accept both of them as cognitive or perceptual input, which amounts to distributing over them a common property. This common distribution belies any assumption that they are utterly different and therefore subject to dualism. Instead, they are syndiffeonically related, differing only against a homogeneous background of common syntax or cognitive sameness.

Because syndiffeonesis relates cognition and perception, our only means of knowing reality, it comprises the universal relational structure of the identity of reality, characterizing sameness and difference, generality and specificity, intension and extension, invariance and variation, set-theoretic aggregation and discernment, and so on. In effect, it brings the apparatus of apprehension and attribution into every relationship that is thought or perceived, permitting it to model an attributive coupling in a self-contained fashion.

Science as the Identification of Reality

The objective of science is the identification of reality, i.e., to express the identity of reality as clearly and succinctly as possible. "The identity of reality" means "that as which reality can be coherently identified, or apprehended and locally discerned from nonreality"; it identifies reality as a discernible cohesive entity.

An identity is a coherent, stratified, self-dual syndiffeonic relationship. Stratified means that the synetic level distributes over the diffeonic level as a common property, while self-dual means that both levels coincide in a single coherent entity (the syndiffeonic relationship itself). Where the synetic level is regarded as intensional and the diffeonic level as extensional, the identity is just an attributive coupling of intension and extension, i.e., of a label, attribute, or description with the set or other structure which it describes. An identity, or intension | extension coupling, is self-dual in the sense that the coupling distributes over both coupled entities; on this level of meaning, the coupling relation is symmetric regardless of any asymmetry on other levels.

Algebraically, an identity represents an entire algebraic system of which it may or may not be a distinguished element. In either case, it is trialic in the sense that the system it represents contains elements, relations, and operations, which means that the identity plays all three of these roles. Mathematically, this can be expressed as follows:

$$i = i \, R \, i = i * i , \tag{E.1}$$

where i is an identity, R is a generic relation, and * is a generic operation. This means that where i is the identity of a system S which it therefore represents, any relation or operation under which S is closed instantiates the generic identity relation i R i and/or the generic identity operation i * i, and can therefore replace it under substitution, denoted ⟶. Thus, where a and b are two related elements of S, and R′ and *′ are a specific relation and a specific operation on S, we have the following substitutions:

$$
\begin{aligned}
\mathrm{i\,R\,i} &\rightarrow \mathrm{a\,R'\,b} \\
\mathrm{i} * \mathrm{i} &\rightarrow \mathrm{a} *' \mathrm{b} \\
&\rightarrow \mathrm{a} *' \mathrm{b}
\end{aligned}
$$

These simple examples illustrate the syndiffeonic aspect of identity. The identity itself, along with its generic R and *, reside on the synetic level, while its possible specifications i ⟶ (a or b or ...), R ⟶ R′, and * ⟶ *′ reside on the diffeonic level. Note, however, that as conventionally understood, the operational and relational closure of an algebraic system does not imply that its identity is completely self-contained. Instead, the mathematician and any required display, storage, or processing media are typically missing from its formal idealization, and it is usually considered to exist in a Platonic realm beyond which no explanation is required. In a ToE, this is unacceptable.

The Metalinguistic Identity of Reality: Structure and Dynamics

The maximal generality (universality, comprehensiveness) criterion of a reality-theoretic identity, or ontologically necessary and sufficient "Theory of Everything", means that a fully general formal structure must be selected as the skeletal identity or ToE framework of reality. This structure is distributive by definition, applying hologically to every point and part of reality. Conversely, because this structure represents the entirety of the ontologically self-contained system from which it is induced, it is the ontic identity of that system.

Reality is formulated in mathematical terms, and mathematical structures, along with the various cognitive and perceptual structures they describe, can be scientifically formalized only as languages. Even the theory of languages is itself a language. This rule applies without exception, covering all quantitative equations and/or "master equations" purporting to describe reality; they too are languages, formalistic and meaningless without embedment in a ToE metalanguage formulated as an overall description of reality. It follows that the ToE identity must incorporate not only syndiffeonic structure, but the formal structure of a language, refined to whatever degree may be necessary. As always, the structure of the language (and syntactic and semantic consistency) is preserved by its model-theoretic mapping onto and into its universe, which means that the universe shares the linguistic structure and dynamics of the theory and is therefore also a language.

Language is often mistakenly conceived as the opposite of mathematics. It is thought to be structurally loose and semantically nebulous, whereas mathematics is thought to be tight, clear, and unequivocal. In fact, not only is language a mathematical structure in its own right, but it is the most general mathematical structure of all, subsuming every other. If language were not mathematical in every sense, and if it were not capable of concise and unequivocal expression, then not only would mathematics (and therefore mathematical physics) be inadequate to characterize reality, but there would be no such things as mathematics and physics at all.

However, reality is not "just a language". It is a very special kind of language which possesses properties never before seen in a language outside the work of this author. These are the properties without which the mathematical structure of language cannot serve as the structure of reality, and without which reality theory would be futile. It would be futile because the word theory is just an abbreviation of theoretic language, and any theory at all must impute its own linguistic structure to reality as its content ... and in the case of a ToE, as its form and content alike.

Dynamically, the most interesting feature of the ToE master-language is that it evolves in two orthogonal directions. One mode of evolution occurs in the familiar linear read-write direction; for users of the English language, this is horizontally from left to right (we ignore the downward succession of lines on a page; the important thing is that each line is written and read from left to right). The other, which occurs in the perpendicular direction, can be variously exemplified as, e.g., (1) a mathematical calculation or axiomatic derivation in which one starts at the top of a page with a horizontally written initial statement and uses axioms and rules of inference to work downward through a series of modified statements, finally terminating on the bottom line or conclusion; (2) a cellular automaton which begins as a horizontal line of contiguous colored squares or cells, and evolves downward line by line as chromatic transformation rules are applied to the cells of each line in succession; and (3) a grammatical derivation in which, by a series of substitutions, a start symbol evolves through a derivational series of nonterminal expressions to a communicable terminal expression.

As a language couples to its content on the syntactic and semantic levels, orthogonal evolutionary processes analogous to those of the language are induced in the content. Where the content of the language is an appropriately structured manifold, the manifold in question also evolves in two orthogonal dimensions. One of these dimensions is the terminal dimension; this dimension is timelike in the sense that it takes time to read or write a horizontal line of print, with read and write being timelike processes. The vertical spacelike dimension is then correspondent to axiomatic or grammatical derivation, comprising an orthogonal dimension of causation with orthogonal sequences of pregeometric nonterminals. Where the manifold is spatiotemporal, and the terminal dimension of its evolution is understood as timelike, causal, and physical, the grammatical dimension of evolution is metacausal, metatemporal, and metaphysical. It coherently transforms entire spatiotemporally extended systems rather than local states; it is not subject to local confinement as is terminal causation, which consists of many separate threads, but has nonlocal coherence and combinatorial (relational and operational) degrees of freedom which let it take the physical states of spatiotemporally extended systems as input and yield extended complexes of physical events as metasimultaneous output (where *metasimultaneous* means "occurring within the same null grammatical process").

Metacausation and other metaphysical criteria require that the standard "physical" conception of spacetime be superseded by a more advanced metaphysical conceptualization that is logico-geometrically dual to the linguistic structure of the trialic identity. Called the conspansive manifold, it is self-generative and requires three levels of topology and three corresponding levels of quantization. The conspansive manifold is dynamically self-contained; in coupling with the linguistic identity, it evolves by generative self-modeling, embedding conventional spacetime as a linear-ectomorphic semimodel corresponding to the semilanguage Lo of the ToE identity.

Properties of the Metalinguistic Identity of Reality

The ToE must be intelligible, and must render reality intelligible as its content. This does not mean simply that the theory can be understood by its readers; it means that as a formal representation of the universal structure of reality, it is the distributive syntax of reality, and must enable reality to recognize itself through its elements and subsystems. It must comprise a distributive identity through which reality can not only be recognized by sentient entities, but which supports the states, state transformations, and interactions of inert entities, with different kinds of entity responding (only) to the appropriate syntactic invariants.

Explanation is the identification of prior cause or reason, an identificative regress intended to distinguish the explanandum from its negation. A global theoretical language L, being totally self-contained and thus self-explanatory, exhibits the property of inductive idempotence (ontic closure, existential self-sufficiency), amounting to the self-verifiability of reality as demanded by its self-containment (which is self-evident insofar as anything capable of supporting, constraining, or affecting reality is real by definition).

To say that a language or system is inductively idempotent means that under the operation E of explanation or explanatory regression, the operand or explanandum L is unchanged:

$$E(L) = L \qquad (E.2)$$

where arbitrary n-ary operations on a set or structure S can be represented as mappings from S^n to S. Being existentially complete, L returns only itself when E operates on it.

In other words, explanatory induction is a recursive trialic identity mapping on L, with L playing the roles of explanandum, explanation function, and explanation. In short, for an inductively idempotent system, explanation is equivalent to identification; the system is its own identity, and the system is dually (formally and topologically) self-distributed over its own fundamental constituents or points (distribution of syntax over a point denotes the logical binding and/or topological bounding of a coherent element by the syntax; the point and its structure and dynamics must conform to all applicable syntactic forms and constraints). The origin or first cause of the trialic metaphysical metalanguage L can be apprehended only through L-syntax, which is invariant by definition and changeless under self-iteration. At the explanatory limit where explanation is global self-identification by the inductively idempotent identity-operator, it is called logical induction (which this paper exemplifies as both a description and an application). Logical induction equates to a form of deduction in which global syntax, including its own quantization mapping onto reality, is generically educed from the initial ingredients of reality, namely perceptual reality and accepting syntax (respectively, the objective and subjective aspects of reality, of whose elementary coupling science and experience undeniably consist).

The ToE identity must be logically consistent so that no irresolvable paradoxes destroy its intelligibility (or that of reality at large). Where physics is idealized as a first-order language capable of expressing all object-level truths about physical or perceptual reality, the ToE must be a metaphysical metalanguage of physics which supports the verification or falsification of physical attributions by attaching one of the truth values (T, F) to each of them. For this, the 2-valued (T|F) distinction of propositional logic is required, and because propositional logic is just a special sublanguage of predicate logic (as is model theory), predicate logic as an (operationally defined) whole is required. As a requirement of intelligibility and thus of science, logic comprises the top-level structure of the identity of scientific reality, distributing over every point of its medium (i.e., every coherent, discernible diffeonic reland). Where the meaning of logic boils down to "the rules of valid thought and perception", logical structure doubles as cognitive-perceptual syntax. It is through this logical syntax that a self-modeling (conscious) entity couples "anthropically" to its universe and "meta-tautologically" identifies itself in contradistinction from its environment and vice versa (for an explanation of the metatautology concept, keep reading).

In order to have any degree of certainty, the ToE cannot rely on assumption, or empirical or probabilistic induction. What is required is logical-inductive reasoning from known specifics to generalities by tautological necessity. These initial specifics are just what we are given without inference, namely perceptual reality and cognitive-perceptual accepting / generative syntax (including what

Kantians sometimes call the "categories of thought and perception", and what philosophers of mind call qualia). The ingredients of this reality-syntax need not be enumerated in order for it to serve as a basis for logical induction; if syntax is instead functionally or operationally defined and distributed over reality through its universal syndiffeonic relational structure – i.e., as the synetic level of a global syndiffeonic relation – the implications are already far from trivial.

One very important application of logical induction relates the two initial ingredients of reality just mentioned: the external world of objective perceptual content, and the internal world of cognition and the syntactic and semantic structures to which it conforms. These two ingredients, which are given and need not be assumed, are related by a mapping that can be simply expressed as:

$$M : L \leftrightarrow U$$

where L is a language of cognition and perception along with the formal, ideal, or abstract ingredients of that language (includes cognitive-perceptual accepting syntax in its entirety), U is the universe of the language, e.g. the perceptual universe, and M is the metalanguage expressing the mapping of L to U and U to L via the two-headed arrow $\leftrightarrow$, which can represent cognition, perception, attribution, recognition, information, or anything else that bridges the dualistic gulf between form and content.

The mapping M may look unimpressive, but it is deceptively powerful. It is a minimal model of cognition and perception (and attribution and information as well), and is capable of exhibiting all of the essential ToE properties mentioned above. It is also just another way of looking at a syndiffeonic relation, and is therefore a model of relational structure; U is just a set of discernible entities or diffeonic relands, while L incorporates the synetic intension of any given subset of U, with M taking responsibility for the distribution of L over U. And last but not least, it is also a way of characterizing the global identity of reality, a compact model of thought and experience which forms the nucleus of advanced reality theory.

A few observations regarding the above mapping are in order. First, the whole thing can be contracted to M alone. For as we have already mentioned, a metalanguage like M must include its object-language L, and L must syntactically embed its content. So:

$$M = (M \supset L \supset U) \qquad \text{(E.3)}$$

In other words, reality can be entirely packaged in the ToE metalanguage M. However, this requires a bit more explanation. For example, (1) the intensional language L is operator-quantized; (2) M is "meta-surjectively" mapped onto the L-quanta with a special holological kind of one-to-many mapping (M is coherent and therefore "singular", whereas L-quantization produces many L-operators); (3) the two ⊃ signs carry different senses of inclusion; and (4) L "metrizes" U, inducing on it a dual geometry.

Secondly, there is an aspect of syndiffeonesis that we have not yet discussed: a syndiffeonic relation amounts to something called a metatautology or "self-dual tautology". Whereas a propositional tautology generally amounts to an element of logical syntax roughly equivalent to the self-identity of a sentential variable (e.g., "X = X", X an arbitrary attribution), and a semantic tautology is a similar redundancy within a specific attribution (e.g., "a horse is a horse" or "a rose is a flower"), a metatautology is a self-dual "input-acceptor redundancy" whereby an input datum doubles as an element of the accepting syntax of the acceptor (or generic observer or thinker). The shift of data type (of an identity) between percept and percipient, the thing observed and the accepting/generative syntax of the observer, is a dualization operation, i.e., the symmetric rotation of a self-dual identity between logical data types, in this case perceptual input and cognitive-perceptual syntax.

Having defined a metatautology, we can now expand the concept by making use of our two initial ontological ingredients, the two things that we are absolutely given without assumption upon birth into this reality, the *esse est percipi* and *cogito ergo sum* of reality theory: perceptual reality, or the objective environment consisting of external states, and cognitive-perceptual syntax or mind, consisting of the subjective inner world of mental internal states and state transitions.

First, note that in addition to the two initial ingredients L and U, we are also given the coupling between them; this is M, and we are implicated in M as entities in which the L | U coupling occurs. As an identity, M can be considered one coherent entity which self-differentiates by syndiffeonic self-stratification through the cumulative factorization of telesis, a dual generalization of energy properly defined to serve as the ultimate "stuff" of reality. Thus, rather than going through a sequence of steps to construct M, we are simply explaining how the aspects of M: L ↔ U interrelate, breaking up the explanation into parts and steps for the sake of clarity. The steps do not occur as a temporal sequence, but metasimultaneously; the ingredients M, L, and U are interdependent and mutually recursively defined with respect to function (although M can be considered to stand alone as a self-potentializing, self-actualizing global identity creating L and U in the course of self-identification).

(a) U is the entire universe of replicable perceptions, or the object-level universe of the first-order language L. Conversely, L is the comprehensive first-order intension of U (which is dual to its extension or object-universe U). This is an irreducible mutual recursive definition of L and U:

$$L = F_L(U)\ ;\ U = F_U(L) \qquad \text{(E.4)}$$

where F_L and F_U are definition functions with their parameters (definientia) in parentheses. In other words, the definition of each definiendum (defined term) L and U necessarily takes the other term as a definiens (definitional parameter). This definitional linkage is ontological; it does not say merely that percepts must be perceived and properties substantively instantiated, but that states in U are meaningless without attributes in L, and attributes in L are unassignable without attributive operators in U. L and U cannot be separated, and U is therefore equivalent to a particular L|U coupling within the metaobject domain {L|U} consisting of all possible couplings of entities from L and U respectively.

(b) As related entities, L and U comprise a relational identity $M = M : L \leftrightarrow U$ which equates to a logically idempotent form of model theory, the CTMU, that can not only couple L and U, but can also couple itself to L, U, and the L-U coupling L|U, and so on to arbitrarily high order. Where observable states of the perceptual universe U are couplings of attributes from L with values from U:

$$M{:}\,L \leftrightarrow U = M{:}\,L \leftrightarrow (L \mid U) = M{:}\,L \leftrightarrow \{L \mid U\} \qquad \text{(E.5)}$$

Anything not sharing the structure of L is inexpressible in L, and because L amounts to cognitive-perceptual syntax and thus limits perception and conceptualization on U, nothing is perceptible or conceivable which is not expressible in L. That is, the perceptual universe U, which coincides with the (time-dependent) coupling L|U and the inclusive metaobject domain {L|U} forming the universe of M (the syntactic metaverse), contains nothing which is not in structural correspondence with cognitive-perceptual syntax L; U is entirely embedded in L as perceptual content, and secondarily, as inferred properties and relationships of perceptual content. The mapping may thus be contracted, retracting U = L|U = {L|U} entirely into L and leaving M : L.

(c) By duality of the syntactic distribution of the metalanguage M over its object language L, L retracts into M, so that everything resides in the global identity:

$$\begin{aligned} M{:}\,L \rightarrow M[\rightarrow M'{:}\,M \leftrightarrow \{L \mid U\} = M''{:}\,M' \leftrightarrow \{M \mid LU\} = \\ = \cdots, \text{by the logical idempotence of } M] \end{aligned} \qquad \text{(E.6)}$$

Saying that M distributes over L amounts to saying that it distributes over points of L, i.e., syntactic operators and their coupled state-transitional identification events, which are in turn locally embedded in M. The syntactic self-distribution of M involves self-similarity mapping called d-endomorphism, in which M "self-quantizes" as trialic self-operators (acceptors and/or generators) coinciding with relationships and identification events ("measurements"). Basically, this creates localized, attenuated copies of M immersed in a hological medium everywhere identical to M. D-endomorphism describes the past-to-future direction of time (and the inward direction of space); the dual mapping, evolution or d-ectomorphism, has the opposite spatiotemporal directionality. These mappings occur only together in a spatiotemporally symmetric metaprocess called conspansion, where they define primary (potentiative) and secondary (actualizative) stages of causation respectively associated with wavefunction propagation and collapse. This requires orthogonal (causal and metacausal, mechanical and telic) modes of quantization associated with not only physical state and state-transition, but adaptive syntax-state (medium-content) relationships of arbitrary order.

The end result is a self-quantized, self-stratified trialic identity M = (M ⊃ L ⊃ U) which is simultaneously equivalent to the perceptual universe U, the cognitive-perceptual language L, and an idempotent metalogical metalanguage which models itself by iteratively coupling L and U while coupling itself to the metaobject domain {L | U} and its inductive successors to arbitrarily high order. According to its structure, M is a supertautology, that is, a comprehensive, ontically self-contained metalanguage of reality which is equivalent to reality itself ... a reflexive self-recognition/self-generation operator whose reflexiveness amounts to a generalized form of consciousness heritable by its quanta. Nothing which violates the supertautology can ever break into it, or maintain a relationship with it at or below the level of the violation. A local identity in conflict with the global identity M is ultimately annihilated.

A supertautology is self-verificative, as it injects logic – including the tautological propositional level of predicate logic and the supertautology itself – into its own quanta as accepting and generative syntax, which are supported and constrained by it. A supertautology thus limits reality to just that over which it distributes (including itself); anything over which it does not distribute is neither supported nor constrained by it, and is therefore unintelligible and tautologically excluded from the reality of which it is the hological identity and unique source of coherence and intelligibility. In short, M is a self-identification operator which not only identifies and verifies itself, but excludes its negation; that which fails to conform to the most general level of accepting syntax, namely logic, cannot be accepted as cognitive-perceptual input, or for that matter as an ingredient of reality theory.

As observed above, logical induction is just an identity mapping on an inductively idempotent logical identity-operator. Self-identification by the reality-theoretic supertautology equates to a universal trialic identity mapping which plays the roles of explanandum, explanation, and explanation-function (a function which accepts the explanandum as input and yields its explanantia as output). Shifting the terminology to the model- theoretic level of predicate logic, we now see that the identity mapping plays the roles of universe, language, and model.

The trialic supertautological identity M can be developed into the Cognitive-Theoretic Model of the Universe or CTMU, a reality-theoretic structure with formal-linguistic, geometric, and model-theoretic aspects. Because its formal aspect is called *Self-Configuring Self-Processing Language* or SCSPL while its geometric aspect is called the *Conspansive Manifold* or CM, it is sometimes written "SCSPL | CM" to convey its global logico-geometric (form | content, language | universe) duality. Due to the way the geometry of reality couples with the identity-metalanguage M = SCSPL, physical (perceptible) reality is everywhere directly immersed in metaphysical (global-identic) reality.

In the first decade of this millennium, the CTMU was succinctly described as follows:

> **Interviewer**: Can you sketch the CTMU – in plain English – for our readers?
>
> **CML**: The name literally says it all. The phrase "Cognitive-Theoretic Model of the Universe" contains three main ingredients: *cognitive theory*, *model*, and *universe*. Cognitive theory refers to a general language of cognition (the structural and transitional rules of cognition); universe refers to the content of that language, or that to which the language refers; and model refers to the mapping which carries the content into the language, thus creating information. The way in which the title brings these three ingredients together, or "contracts" their relationship to the point of merging, reflects their perfect coincidence in that to which the title implicitly refers, i.e., reality (the physical universe plus all that is required to support its perception and existence). Thus, the CTMU is a theory which says that reality is a self-modeling universal language, or if one prefers, that *the universe is a self-modeling language*.
>
> The operation of combining language, universe, and model to create a perfectly self-contained metalanguage results in SCSPL, short for Self-Configuring Self-Processing Language. This language is "self-similar" in the sense that it is generated within a formal identity to which every part of it is mapped as content; its initial form, or grammatical "start symbol", everywhere describes it on all scales. My use of grammatical terminology is intentional; in the CTMU, the conventional notion of physical causality is superseded by "telic causation", which resembles generative grammar and approaches teleology as a natural limit. In telic causation, ordinary events are predicated on the generation of closed causal loops distributing over time and space. This loop-structure reflects the fact that time, and the spatial expansion of the cosmos as a function of time, flow in both directions – forward and backward, outward and inward – in a dual formulation of causality characterizing a new conceptualization of nature embodied in a new kind of medium or "manifold" (Langan, 2010).

Summation

In sum: there are two kinds of causality in the CTMU, mechanical and telic. Telic causation is associated with conscious self-modeling entities, i.e., entities with sufficient mental coherence and complexity to internally model the relationship between self and environment (this being the kind of entity which can meaningfully participate in the self-identificative evolution of M by self-modeling, which happens automatically with embedment in M). Reality is usually understood to evolve along an ordinal parameter, time, as a sequence of spatial distributions of matter and energy, with allowance for perspectival variation. Treating reality as a unique kind of language, the CTMU shows that it is instead generated or grammatically produced by a telic (self-generative) grammar which operates dually to standard terminal causation and orthogonally to time.

SCSPL can be factored into a pair of semilanguages Ls and Lo, which are respectively dynamical and static and associated with the conspansive model and the linear-ectomorphic semimodel of the CTMU. Standard physics is largely confined to the linear-ectomorphic semimodel, which is retroscopic. In the CTMU, (meta)causation and dynamics reside solely in SCSPL. Ordinarily, the geometry of reality is described as a spatial or spatiotemporal manifold, the structure of which is usually explicated as a fiber bundle. However, a fiber bundle resides in the linear-ectomorphic semimodel and thus lacks support for its own evolution. The CTMU describes the geometry of reality using a more sophisticated structure associated with the self-dual conspansive model, which evolves metacausally and orthogonally to the terminal (temporal, ordinal) axis. Unlike the linear-ectomorphic kind of manifold usually employed in physics, the conspansive manifold has the characteristics required of a true universal medium, accommodating causal quantization requirements while incorporating General Relativity as a linear-ectomorphic terminal limit.

In the CTMU, the Lagrangian formalism of mainstream physics is merely the timelike image of a more general CTMU process, telic recursion, which accommodates a generalization of so-called physical causation called telic causation. The CTMU effects this accommodation by incorporating full support for the metacausal-grammatical dimension of causation. Its "spacelike" telic feedback, orthogonal to time, is what other theories mistake for "causation and retrocausation", which are confined to terminal event-sequences in the linear-ectomorphic semimodel. In particular, any quantum Lagrangian fits into the conspansive manifold as a conspansive cycle embedded in a conspansive metacycle or quantum of telic metacausation. The CTMU interpretation and extension of quantum mechanics, Sum Over Futures (SOF), is a generative "quantum metamechanics" supporting this embedment.

In the CTMU, quantum mechanics is modeled rather than merely interpreted. Unlike a mere interpretation, a model is logically valid; unlike a conventional "interpretation" of quantum mechanics, a QM model has a definite codomain. Without a verifiable "pre-quantum" picture of reality, the codomain of a QM-interpretative mapping is indefinite, precluding its validation as a real "model" (the absence of a model alone ensures the invariance of quantum mechanics under interpretative variation). Where a valid ToE is lacking, it cannot be decided which (if any) QM interpretations are correct, or how they relate to each other. In fact, QM itself defies verification on its own merits, and the field of QM interpretation is arguably pointless except insofar as it points toward the deeper structure of reality.

Independently of arguments for the Holographic Principle in string theory, the CTMU logically establishes that the universe is indeed a holographic self-simulation. The difference between the stock cosmic hologram and the CTMU "hologram" is that the latter is a trialic hological language which, because it is a metaevent and metaprocess as well as a metaobject and metarelation (the metacausal analogues of event, process, object, and relation), displays spatio-temporal closure. As required by the self-distribution of M (and arguably by the definition of "hologram"), the cosmic hologram is replicated in every identification event or internal point of the conspansive manifold. In the CTMU hologram, reality everywhere "simulates itself" through, and for, its own constituents.

As might be expected, there is much more to this story. The CTMU theory-universe-model dates from the mid-1980s, and has since been extensively developed in nearly total isolation from the academic community.

Please bear in mind that this paper contains only a minimal introduction.

References

Langan, C. M. (1989) The Resolution of Newcomb's Paradox. *Noesis*, № 44.

Langan, C. M. (2002) *The Cognitive-Theoretic Model of the Universe: A New Kind of Reality Theory*. Princeton, MO: Mega Foundation Press. Originally published in *Progress in Complexity, Information, and Design*, Double Issue, Vols. 1.2-3.

Langan, C. M. (2002) The Theory of Theories. *Noesis-E*, Vol. 2, № 1.

Langan, C. M. (2010) Interviewed by *Super Scholar*. The interview is available at https://superscholar.org/interviews/christopher-michael-langan/

Metareligion as the Human Singularity

Abstract

Based on the author's notes for a presentation to the Foundations of Mind[1] group in October of 2017, this paper examines the role of metareligion in obtaining a favorable outcome for the human species as it approaches a Singularity with both Human and Technological aspects. For limited technical background, please consult two of the author's previous papers (Langan, 2002, 2017).

Keywords

Cognitive-Theoretic Model of the Universe; CTMU; Dualism; Metareligion; Singularity

The Meaning and Purpose of Spirituality and Religion

Man has always craved knowledge of his own identity. Human identity is coherent, existing as a unified whole, without gaps or breaks that might interrupt the connection between its various parts and aspects. If human identity were ultimately pluralistic – e.g., dualistic, existing in two essential or fundamental

[1] **Editor's note** *Foundations of Mind*, the independent research group that has provided the papers for this special edition, has never taken either corporate or state money and is financed entirely by donations. Authors keep copyright without paying. The typical fee for this charged by open-access journals such as those published by PLOS, is around $2k. If you value this project, and wish to see further such proceedings from this group, we ask you to consider donating to Foundations of Mind – as little as $5 per download, through their website: http:// www.foundationsofmind.org/donate. This will ensure there will be further published proceedings on the foundations of mind like this one for you and others to enjoy free.

parts – then the equally fundamental gap between the parts would destroy its continuity, rendering it dissociative and pathological.

To understand his own identity, man requires a coherent and therefore monic self- model reflecting its psychological coherence and relating it to all levels of reality. That is, man requires a valid interpretation of the human individual in society, and of the individual and society in reality at large. This interpretation must take the form of an unbroken correspondence spanning the extended relationship between man, as an inhabitant of reality, and reality in its most basic and universal form; man must see himself as an integral part of reality, and reality as an extension of his own being within a single unified ontology or metaphysics. In short, man and reality must share a common metaphysical identity.

Where metaphysics is a language expressing the relationship between mental and physical reality, spirituality can be understood as the metaphysical essence of human identity, and religion as its organizational manifestation. In its various benign forms, religion provides man with self-understanding and a sense of community ... a model of the individual and his or her relationship to other people, society, and reality at large. Religion tells people who they are, and mankind what it is, by establishing their relationship to the global environment on the spiritual level; it is a binary relationship of man to his real environment, and where the global environment of each human being includes all others, the relationship of mankind to itself.

The spiritual model of self, the extended man-reality relationship required by religion, is thus a stratification of human identity from the individual to ultimate reality, the level of reality that cannot be explained in terms of anything prior to itself or any sort of exterior embedment. This follows from the fact that man is embedded in reality and thus shares all of its most general and ubiquitous properties, up to human limitations of structure and dynamics. Parallel to this degree of extension is the outward extension of self that is sought in certain Asian religious traditions; the self becomes ever more expansive as its hidden depths are plumbed.

But here we must note that the phrase "ultimate reality" is necessarily a partial description of God, incorporated in the (otherwise variously defined) identity of all viable monotheistic religions. Any God *not* incorporating ultimate reality could exist only in a properly inclusive reality partially beyond His influence and creative power, and would thus come up short in virtually every major strain of monotheism. On the other hand, this description holds regardless of any more specific properties incorporated in various definitions of God.

Dualism: Reality Torn in Two

In mainstream social and economic theory, a human being is understood as a mechanistic automaton driven by individual self-interest and governed by impersonal laws of nature and rules of behaviorism. Human automata are subject to conditioning on the basis of individual self-interest, which is a function of the individual's pleasure and happiness, freedom from want, pain, and sadness, and standards of biological fitness including survival and reproduction, all of which inhabit a standardized economy with a monetary metric. Man is thus simplistically viewed as an economic agent subject to monetary control, through centralization of which the entire future of mankind can in principle be mechanistically determined by the calculated pushing of buttons. Obviously, this dualistic view of man represents a complete negation of human dignity and sovereignty, reducing the human race to cattle. It is also incompatible with any kind of religion other than that referred to by Marx as an "opiate of the masses".

Dualism is usually associated with the French philosopher René Descartes. Cartesian dualism, which has several more or less equivalent formulations, simply asserts the mutual exclusion of mind and matter on the strength of an apparent absence of a connection or overlap between them. On the other hand, avoiding it is not quite as simple, for this would seem to require a conceptual framework which differs in certain counterintuitive ways from our usual picture of the world. In order to avoid having to cope with these differences, it is much easier to simply detour around them, which puts one back on the main road and saves one from having to do any steep or lonely climbing.

The historical persistence of Cartesian dualism may have something to do with another innovation of Descartes, Cartesian analytic geometry (independently invented by Fermat), which utterly permeates the scientific landscape. Analytic geometry defines and depicts space in such a way that it excludes anything which cannot be completely represented as a set of coordinates defined along a linear trajectory, with or without various objective properties attached. In particular, the mental and spiritual aspects of conscious subjects are excluded from physical spaces and thus separated from the objective, observable contents thereof.

Whereas religion must connect human beings to all levels of human identity up to and including reality at large (the entirety of which is understood to be spanned by the presence, knowledge, and creative power of God), thus in effect defining the human soul by way of the connection, Cartesian dualism cuts the soul in two, setting man adrift from physical reality and pitting science and religion against each other. Dualism fundamentally divides the world, leading to a seemingly endless sequence of futile attempts to patch it back together.

In short, Cartesian dualism is inconsistent with the spiritual connectivity demanded of religion, and thus makes religion inconsistent with empirical and mathematical science. Neither can be expressed in terms of the other, and we arrive at the variant of Cartesian dualism referred to (first by Stephen Jay Gould in 1997) as "non-overlapping magisteria" (NOMA). Yet science and religion both claim truth, and therefore have a common requirement: conformance to logic, i.e., to the structure of truth. Thus, restoring the coherence and consistency of religion requires its embedment in a high-level formulation of logic (truth-structure) which also accommodates the theoretical, observational, and methodological aspects of science.

Spirituality and the Crisis of Human Identity

Dualism has precipitated a dualistic and therefore dissociative crisis of human identity which, by impeding the spiritual level of human self-identification, amounts to a crisis of spirituality. This crisis has metastasized into the socio-political and economic realms, where philosophical dualism has been cynically and sometimes catastrophically exploited. Even as people are assured that their interests are of primary importance by government, corporate, and institutional leaders, they are more often viewed and treated as common and therefore disposable stimulus-response machines, of no greater value than other readily available forms of matter.

Societies are cybernetic; they are control-and-communication feedback loops in which top-down control tends to increase over time. Absent religious opposition, governments increasingly behave like malevolent egregores, using secrecy, disinformation, indoctrination, coercion, and conditioning to control their citizens. Even religion itself may be used to such ends; for example, ecumenical and interfaith conferences and councils nominally dedicated to a millennial religious reformation or synthesis may be controlled by those preferring to *shape* human identity rather than *comprehend* it ... to "remodel" man at their administrative or personal convenience. Secular control, facilitated by mutual suppression of divergent ideologies, masquerades as religious unification in the absence of coherent scientific, philosophical, or theological opposition.

Unfortunately, organized religion finds itself without a counterargument. Present approaches to religion and religious scripture instantiate dualism on the level of both theory and model; a scriptural language, subjectively internalized by adherents, is set apart from its objective content. The minds which endow the language with meaning are set apart from material reality, and somewhat oxymoronically, the intervening gap is *spanned* by a *separative* model or interpretative mapping. That the model itself is implicitly associated with mind, and that mind is thus present on both ends of the mapping, goes unrealized and unacknowledged.

As one might expect by its dependence on those who fund and control it, academia has been recruited to their cause. The dominant worldview in academia is another form of dualism called *naturalism*, a term which is synonymous in some contexts with *atheism* and *secularism*. *Methodological* naturalism excludes the supernatural or metaphysical from the content and methodology of science, while *metaphysical* naturalism oxymoronically excludes metaphysics from all of existence, presumably on the strength of some unspecified (and in fact, logically impossible) "physical ontology" which lets existence be confined to the physical or observable realm.

Obviously, neither kind of naturalism is consistent with spirituality and religion. Human beings invest religion with positive feelings, emotions, and expectations based on truth-claims which require it to have real-world content related to the formation and actualization of human purpose and destiny. Religion must therefore describe not a mere *object* of hope, but a verifiable *basis* of hope. Naturalism, being dualistic, excludes human identity from reality, depriving the spiritual imperative of coherent self-identification.

As metaphysical naturalism (physical monism) undermines spiritual cohesion, religion declines. The separation of church and state, originally meant to safeguard the freedom of the people to worship as desired, now often seems to be misinterpreted as freedom of the state from religious competition for the hearts and minds of the populace. In other words, Jefferson's "wall of separation" seems to have been misconstrued as yet another expression of Cartesian dualism, portending and falsely justifying an anti-religious secular pseudo-theocracy which threatens to displace true spirituality from its previous role in human self-identification and the formation of human destiny.

Dual Singularities

For present purposes, a "singularity" is a point at which a system must undergo a directional break, jump through a limit, or be redefined in order to survive regardless of how it may evolve before or after. Accordingly, it can be understood as a kind of systemic destiny, an inevitable convergence of possible paths or trajectories of systemic evolution. Paths converge on points, and where such a point marks a sharp change in the smooth overall trajectory of a system, it comprises a kind of systemic "metapoint" which can be seen as marking a systemic mutation or change of inertia. This provides a tentative mathematical conceptualization of "singularity" for social systems.

The related forms of dualism thus far discussed – Cartesian dualism, naturalism, NOMA, and so on – are opposed to the human need for a coherent spiritual identity. This implies a bifurcation or divergence, a human evolutionary choice between two possible adaptations or destinies respectively corresponding to the anthropic and technological aspects of an impending "singular" transformation. Each possible destiny corresponds to the dominance of one aspect over the other, and may be associated with its own conventional type of singularity.

On one side is the **Human Singularity**, a mass realization of the expansive spiritual identity of the human species. Basically, this is the mass spiritual awakening that we have been led to expect by, e.g., certain currents in "New Age" thought. The prototype for this kind of singularity is the Omega Point of Pierre Teilhard de Chardin, representing an evolutionary terminus and divine spiritual unification event through which mankind, and reality itself, will achieve "Christ-Consciousness" and be forever transformed. (After its introduction by Teilhard, the concept was developed by various authors, ironically including a few science-oriented, ostensibly tech-minded authors like Frank Tipler and David Deutsch.)

On the other side is the **Tech Singularity**, seminally formulated by the celebrated mathematician John von Neumann as the approaching juncture at which "technological progress will become incomprehensively rapid and complicated", prior to which "the ever-accelerating progress of technology ... gives the appearance of approaching some essential *singularity* [italicized for emphasis] in the history of the race beyond which human affairs, as we know them, could not continue" (Ulam, 1958). In short, von Neumann foresaw an uncontrollable technological quickening, a sudden acceleration of complexity followed by the transformation (or extinction) of humanity.

Most discussions of the Tech Singularity have been naive to the point of disingenuity, boiling down to starry-eyed encomiums to the power of human intelligence to inventively couple with reality on the physical level of being using technological marvels both real and imagined, including implants, prosthetics, genetic engineering, virtual realities, and above all, a merging of human intelligence with AI. The problem with such discussions is that they seem to inhabit a socioeconomic and political vacuum, whereas in fact, the singularity concept is fraught with worrisome complications involving economic and sociopolitical factors apart from which it cannot be properly evaluated.

The Human and Tech Singularities relate to each other by a kind of duality; the former is extended and spacelike, representing the even distribution of spiritual and intellectual resources over the whole of mankind, while the latter is a compact, pointlike concentration of all resources in the hands of just those who can afford full access to the best and most advanced technology. Being opposed to each other with respect to the distribution of the resources of social evolution, they are also opposed with respect to the structure of society; symmetric distribution of the capacity for effective governance corresponds to a social order based on individual freedom and responsibility, while extreme concentration of the means of governance leads to a centralized, hive-like system at the center of which resides an oligarchic concentration of wealth and power, with increasing scarcity elsewhere due to the addictive, self-reinforcing nature of privilege. (Note that this differs from the usual understanding of individualism, which is ordinarily associated with capitalism and juxtaposed with collectivism; in fact, both capitalism and collectivism, as they are monopolistically practiced on the national and global scales, lead to oligarchy and a loss of individuality for the vast majority of people. A Human Singularity is something else entirely, empowering individuals rather than facilitating their disempowerment.)

The existence of two possible singularities presupposes a point of bifurcation or divergence beyond which the evolutionary momentum of mankind must carry it. Presently, all of the momentum belongs to the Tech Singularity; it is preferred by the financial, corporate, and governmental interests which drive the general economy. This momentum is reinforced by the seeming unavailability of alternatives, i.e., the nonexistence of any other track onto which society might be steered in order to escape an oligarchical AI lockdown. It is one thing for humankind to awaken *en masse* to its impending enslavement through a seemingly inevitable Tech Singularity; it is quite another to have a superior alternative clearly in view.

In order to reach any alternate destination whatsoever, humanity must understand what has been driving it toward the Tech Singularity. At this point, the reason is clear: the virtually automatic concentration of wealth and power, which has been observed to occur under both capitalism and socialism, fractionates humanity into an overclass and an underclass between which all else is crushed out of existence as though by the jaws of a vise. That is, the top and bottom levels of society become the jaws of a vise which, due to the screwing down of the upper jaw against the anvil-like lower jaw, crushes the middle class and all meaningful competition out of existence, thus normalizing the hive through the economic, physical, and psychological standardization of its drones and workers.

For reasons that should by now be evident, let us call this process a "parasitic divergence" – i.e., an organized divergence of humanity into a parasitic overclass and a relatively impoverished underclass serving as its mind-controlled host, mirroring the gruesome effects of certain obligate parasites on the organisms they attack – and acknowledge that it is driven by the self-reinforcing and therefore accelerating acquisition of wealth, power, and technological control by the rich. Left to run away with itself, this process ultimately leads to a "singular" concentration of wealth and power ... a kind of sociopolitical-economic "black hole" that never stops gravitating. As the top jaw of the vise grows smaller, denser, and stronger, the bottom jaw grows larger and weaker; and as human utility becomes increasingly concentrated, every significant increase in the wealth of the overclass translates into a greater amount of misery for the underclass, arbitrarily diminishing the net utility of mankind.

Parasitic divergences have occurred many times in history, but the present one is different. Due to the double whammy of globalization and powerful surveillance and coercion technology, the one now in progress is geographically ubiquitous and quite possibly irreversible. If humanity is to save itself from the insectile, hive-like future associated with a Tech Singularity, the Human Singularity must prevail, empowering mankind to exert sufficient control over the production, distribution, and application of technology to prevent its unlimited oligarchical abuse. To bring this about, it is not enough to merely distribute a cognitive avoidance mechanism out of which the moneyed elite can buy and bribe their way as usual, given the absence of a well-defined alternative direction in which humanity can proceed; rather, an alternative direction must be defined and universally distributed in cognitive and attitudinal form.

In short, in order to have a meaningful mass awakening, the content of the awakening must be defined and distributed to the members of humanity, thus immunizing them against parasitic mind control. Because this content must be spiritual, the involvement of religion is unavoidable.

Twenty First Century Metaphysics as Metareligion

Thus are we presented with an existential choice of singularities ... a choice on which our future certainly depends. These dual singularities, Human and Tech, respectively correspond to spiritual monism, the self-dual unification of humanity in a common spiritual identity spanning all of reality, and physical or material monism, the terminal mechanization and de-spiritualization of human identity. For the sake of humanity, spirituality must triumph; yet religion as we know it is not only under withering attack by the forces of secularism, but lacks the conceptual wherewithal to overcome Cartesian dualism and reunite with the scientific and technological aspect of reality.

Religions evolve in coupling with cultures, conventions, and morals that come into real-world conflict with each other and/or the state, with the result that religions themselves conflict in all of these ways. Obviously, a mass spiritual awakening requires a means of resolving these conflicts. Suggestions include segregating religions, allowing or encouraging a single religion to become everywhere dominant, playing different religions off against each other in a strategy of mutual containment, merging religions by syncretism (collecting their respective beliefs and rituals under one aegis), and eliminating religion entirely. But there is another more promising option: unifying internally consistent religions in a well-structured metareligion, i.e., a theological relationship among religions which provides their valid truth claims with logical support.

By virtue of the functionality of this relationship, its structure is that of a language. Like theories of science, religions themselves are languages; specifically, they are scriptural and doctrinal languages taking as their content a single shared reality including not only the physical universe, but subjective beliefs, codes of behavior, and notions of the sacred arising therein. This implies that a metareligion is a comprehensive metalanguage of religious languages. However, to exceed the limitations of dualism and couple with science, this religious metalanguage must also be scientific in bearing. In fact, it must bring spirituality and science into contact on all scales of reference. How is this to be accomplished?

All intelligible languages include logic as a syntactic ingredient, which means that ostensibly independent languages share common syntax and are merely parallel aspects of one all-encompassing language, namely logic itself. This implies that a sufficiently powerful formulation of logic comprises a common language for science and religion alike, and can thus function as a bridge between them and their respective symbolic and semiotic aspects. But standard predicate logic is itself understood as a dualistic language; although it links attributes and objects together in attributions, it does so on a weak and tentative basis. Whatever it attributes to real-world content can be ascertained only with respect to more or less arbitrary axioms and/or by empirical confirmation. What can be done to defeat this requirement, thus transforming linguistic dualism into linguistic self-duality?

Languages are conventionally defined as complex attributes of their universes of discourse. By definition, they are dualistically separated from their universes just as an attribute is abstractly separated from that to which it is assigned by predicate-logical quantification. Yet languages take their universes as content, which contradicts the dualistic premise that they are separate. Like all conventional languages, religious (scriptural and doctrinal) languages are dualistic and thus conflicted; they are implicitly held apart from their universes even while taking them as content. Similarly, standard theology – a metalanguage for the analysis and comparison of religions and religious conceptions of God – is dualistic as well, standing apart from the religious languages which form its content (and for that matter, from God). This implicit dualism renders standard religious and theological languages fundamentally incapable of expressing the spiritual unity of man and nature. To eliminate the dualistic separation of languages and their universes, dualism must be formally eliminated from the intrinsic structure of language.

The fact that all intelligible languages include logic as a syntactic ingredient implies that a sufficiently powerful formulation of logic comprises a common language for both science and religion, where "a sufficiently powerful formulation of logic" expresses logic on a metaphysical level appropriate to its application to reality as a whole. This amounts to the requirement that it be a *supertautology*, i.e., a metaphysical analogue of logical tautology which employs logic as the identificative syntax of a generic reflexive identification operator. Supertautology describes the structure of an ontological and epistemological metalanguage bearing description as "metaphysics" in the sense required for a true understanding of spirituality and religion.

As usually considered, there are just two basic kinds of science: empirical science, relying on causal inference described by general "laws of nature", and mathematical (logical, linguistic) science, using substitution in formal (axiomatic or grammatical) systems. Although they operate in ways that are superficially dissimilar, they are nonetheless inseparable; the theoretical aspect of empirical science depends on formal inference, a *mental process* occurring in the minds of mathematicians and scientists who actualize it using brains which obey *laws of causation*. In order to deal with the mutual dependency relationship between mathematical and observational reality, we require a higher level of science which includes both empirical and mathematical science, but is more powerful than either in how it relates them.

Causal inference is interactive and linear; events are predicted from prior events occurring along timelines. Formal inference is derivational; specific relationships are substituted for more general ones (or vice versa) irrespective of temporal priority. A supertautology evolves self-dually or "metaformally", in a way that couples formal and causal evolution. This is the proper mode of evolution of an ontological metalanguage capable of not only supporting causation, but of justifying existence, including its own existence, without the help of any other language. Metaformal inference, elsewhere referred to as "telic causation", properly includes both formal and causal inference as aspects; thus, it supports both generative origination and causal evolution, and can be understood as a higher mode of inference embracing logical deduction, empirical induction, and metalogical reasoning about reality as a whole.

There already exists a branch of logic, *model theory*, that deals with the interpretation of empirical phenomena in theories and the mathematical structures of which they consist, but its standard formulation is dualistic. It now has a reflexive self-dual extension called the *Cognitive-Theoretic Model of the Universe* (CTMU; Langan, 2002), or as some have called it in more traditional and religiously loaded terms, *Logos*, mainly in recognition of its status as a metaphysical formulation of logic. Technically, the CTMU is a reflexive, high-level kind of model theory designed to support the description of reality on the ontological level of discourse ... the level on which reality *exists* independently of anything external.

On this level and all of those beneath, the supertautological structure of the CTMU is virtually unassailable. Just as standard logic requires no assumptions, neither does the CTMU; it requires only the cognitive and perceptual faculties that we are given from birth. And because the CTMU is intrinsically valid in a way that empirical science alone is not, it supports the expression and development of scientific truth in a self-dual (nondualistic) conceptual environment. As a metaformal ontology which replaces dualism with self-duality, the CTMU can be viewed as the outcome of what might be called the "Metaformalist Program" in the joint foundations of science and mathematics.

To model religious languages on the appropriate metaphysical level of logic and consistently express their interrelationships, the CTMU employs a *trialic* metalogical language which constitutes its own universe and its own model, and is thus capable of autonomously validating certain religious claims of truth and consistency. In effect, this language comprises the "metascripture" of a verific and potentially unificative metareligion. Its supertautological structure is that of a *Self-Configuring Self-Processing Language* (SCSPL) exhibiting referential closure and thus reflecting the structure of the self-contained, self-sufficient reality in which we live. Encoding the relationship between man and Deity, humankind and the metaphysical structure of reality, it is the only valid basis for eliminating the existential confusion and religious conflict that threatens our world without sacrificing that which makes us human.

Conclusion

Despite the fact that an elementary formulation of the CTMU has elicited relatively little academic interest after nearly three decades since its introduction, it is natural to ask how it is likely to impact the intellectual environment. Perhaps the most profound change in our worldview will come from learning that living, breathing human beings are essential and logically necessary ingredients of reality, not just "emergent phenomena" which "supervene" on brute physical processes. In the CTMU, human beings comprise a class of entities with a very specific mathematical formulation and an essential role in the structure and dynamics of reality. Once this role is properly understood, the spiritual and scientific realms fall back together of their own gravity.

In academia – which suffers from closure and a preference for intellectual orthodoxy over profound conceptual innovation – this realization will be characteristically retarded. For mathematics and the hard sciences, it will probably be mostly business as usual, especially at first; this is because in their current forms, both already have places in the CTMU. That is, pure mathematics inhabits SCSPL syntax, while science inhabits the *linear-ectomorphic semimodel* of the CTMU as a physical limit (ignoring for now the relationship between the syntax and the limit). For "softer" and more amorphous sciences which do not enjoy rigorous mathematical theories that compensate for their lack of solid conceptual foundations, the benefits may be more readily felt, at least among those who have not been locked into academic naturalism. Philosophy and theology have the greatest potential to undergo more immediate change; at any rate, it will no longer be possible to rationally dismiss the metaphysical aspect of reality or its implications, or to concoct *ad hoc* rationalizations based on relativism and existential ambiguity. Meanwhile, the emergence of a common foundational language for all of these disciplines will probably be only gradually realized.

As for religion, believers of which often languish under the crushing weight of orthodoxy and peer pressure, a whole new level of courage and open-mindedness will be required. But fortunately, perhaps as an unintended consequence of the steady erosion of religious dogma, many minds have already opened enough to accommodate a greatly enriched understanding of spirituality. Let us hope that there are enough of them to help us attain the Human Singularity, and redeem mankind from its otherwise gloomy and potentially catastrophic future.

References

Gould, S. J. (1997) Nonoverlapping Magisteria. *Natural History*, Vol. 106, pp. 16–22 and 60-62.

Langan, C. M. (2002) *The Cognitive-Theoretic Model of the Universe: A New Kind of Reality Theory*. Princeton, MO: Mega Foundation Press. Originally published in *Progress in Complexity, Information, and Design*, Double Issue, Vols. 1.2-3.

Langan, C. M. (2017) An Introduction to Mathematical Metaphysics. *Cosmos and History: The Journal of Natural and Social Philosophy*, Vol. 13, № 2, pp. 313–330.

Teilhard de Chardin, P. (1959) *The Phenomenon of Man*. Translated by J. Huxley, introduction by B. Wall. New York: Harper.

Ulam, S. (1958) John von Neumann 1903–1957. *Bulletin of the American Mathematical Society*, Vol. 64, Issue 3, pp. 1–49

The Metaformal System: Completing the Theory of Language

Abstract

The standard theory of languages has two levels, one centering on the study, teaching, and application of natural languages, and the other on formal languages and formal systems as applied throughout the mathematical and empirical sciences, in analytic philosophy, and for computer programming, software engineering, artificial intelligence, and related technologies. On both of these levels, standard language theory is *dualistic*, defining languages in isolation from their domains of discourse and treating attributes in isolation from their objective instances while omitting important properties and functions ordinarily provided or executed by language users, automata, or physical systems on which they appear to supervene. This decoupling of languages from their universes, and from necessary linguistic functions such as *display*, *processing*, *interpretation*, and *communication*, has profound epistemological bearing, limiting scientific knowledge by precluding the linguistic formulation of any verifiable comprehensive description of reality. This paper proposes that in addition to the two existing levels of standard language theory involving natural and formal languages and systems, the theory of language be recognized to possess a third "metaformal" level on which languages and their universes are "wrapped" in a uniquely structured, totally self-contained metalanguage, the *Metaformal System*, which restores missing linguistic functionality while using a supertautological intelligibility criterion to generically couple languages with their universes on a fundamental level of shared structure and dynamics, thereby restoring the potential for a verifiable comprehensive and fully connected understanding of the reality we share.

Keywords

Language Theory; Generative Grammar; Metaformal; Formal Language; Formal System; Metalanguage; CTMU

The Nature of Language

Language, which is widely regarded as an artificial construction of human beings, is conventionally defined in a variety of ways associated with various perspectives on philosophy, psychology, sociology, and pedagogy, often with an emphasis on culture and politics, and with various technical prerogatives having mostly to do with the study and employment of automata. This has led to an emphasis on linguistic variety and contextual specificity, with variations cancelling out rather than summing to a meaningful overall characterization. Thus, the "bare bones" definition of language omits any reference to content, application, or structure, reducing it to an abstract set of "strings" or linear sequences of sounds or symbols devoid of meaning.

All existing definitions of language – even those which delve deeply into its structure, production, genesis, and evolution, or have to do with its formalization or computational application – sell it short by restricting its definition and ignoring its ubiquity. In reality, language is not dependent on the vagaries of human culture and psychology, the preferences of academic specialists, or the capacity, functionality, or architecture of computational machinery. In fact, language is the most general and necessary abstract structure in mathematics, the protean abstract structure of which all other structures are instances by virtue of their effability and mathematical descriptions alone. Its generality is reflected in the fact that it can be equally well described as an *object*, *attribute*, *process*, or *operational medium*, and in fact as all of these in seamless combination.

From a technical standpoint, one of the most important features of language is that its study is necessarily self-referential. That is, any attempt to define, describe, analyze, or construct language uses language itself to do so. Self-referentiality is a powerful and underexplored feature of language, one which has puzzled and fascinated mankind for centuries. But at the juncture where language meets cognition, it is also a source of logical and epistemological complications. Due in part to these complications, language has met with many apparent failures, ranging from piles of self-contradictory and seemingly unproductive philosophical verbiage to the undefinability and undecidability problems of formal language-systems.

There are many who seem to believe that this is the end of the story, and that language-using creatures such as we will never understand or agree on the nature of the reality we inhabit. But fortunately for science, philosophy, and the intellectual destiny of mankind, complete surrender to epistemological relativism turns out to be premature.

The Stratified Theory of Language

The standard theory of language recognizes two main classes of language, *natural* and *formal*, each occupying its own compartment. These compartments may be respectively described as:

(1) The mostly academic study of human language acquisition and usage and the structure and grammar of various natural languages. A *natural language* is any human language, whether spoken, written, or manually signed, that appears to have arisen spontaneously and evolved naturally, through repetition with variation, usually along with the native culture of a specific population. Natural languages are used for routine self-expression and communication, public speaking, literary production, governance, commerce, and other purposes to which ease and freedom of expression and understanding are important.

(2) The (mostly academic and industrial) study and application of formal languages and systems and computational languages. Formal languages are often just natural languages that have undergone some degree of formalization, usually with convenient simplification and technical standardization. They are used wherever clarity, precision, and lack of ambiguity are required. Applications include mathematics, the empirical sciences, computer science, engineering, and other fields where exact technical specifications must be expressed and communicated with maximum precision and minimal semantic ambiguity.

Just as it would be hard to overstate the importance of natural language to (e.g.) human psychology, social organization, and psychosocial evolution, it would be hard to overstate the importance of formal language to science and technology. The mathematical sciences overwhelmingly rely on formal systems, especially "proof systems", to achieve their results. The empirical sciences are equally reliant on formal systems, which are theoretically coupled with observation and experimentation as prescribed by the Scientific Method. As for computer science and the information technology that supports modern civilization, they clearly rely as heavily on language as they do on computers.

If we regard natural and formal languages as the first two levels of modern language theory, it is possible to identify a third. Whereas the first two levels are dualistic, taking languages in isolation from their universes, the third is monic and self-dual, structurally fusing language and universe into a single coherent identity on the highest possible level of discourse. Featuring an intrinsic definition of language which is analogous to intrinsic geometry in its self-containment and external independence, it consists entirely of identities, or coherent self-dual language-universe couplings, and the operators and operations which generate and act on them.

This logico-metaphysical level of language theory is designed to formulate and address fundamental metaphysical and metalogical questions regarding a comprehensive and therefore totally self-contained reality with respect to which nothing deeper or more extensive exists, regarding which no external references are possible, and in which literally everything comes down to the self-reference of a self-defined language-universe complex that has no background and to which nothing intelligible is external. This is the level on which language merges with universe, and on which signs and symbols merge with their associated objects, interpretations, and interpreters. Or in more pedestrian terms:

(3) The level on which language can be productively employed in the conclusive identification and analysis of reality in part and in whole, i.e., on which language and universe are generically coupled in terms of common structure and intelligibility. It is on this level that the true and ultimate nature of reality – that which we can know of it strictly on the basis of our ability to sense it and draw logical inferences about it – can be fruitfully explored.

Where the deceptive dualistic isolation of language from universe obscures their true relationship and precludes certainty, conclusive identification is impossible. But as physical reality can in fact be clearly identified through direct observation and logical deduction, without which neither science nor human experience could exist, and as deductive logic is a *formal language* inherent in cognition, this assumed linguistic incapacity is unfounded and illusory. Hence, the third level of language theory herein proposed already has an irrefutable basis.

Let us more closely examine these levels of linguistic theorization.

Level 1: Natural Languages

A *natural language* is conventionally understood as the system and method of spoken or written expression and communication among human beings to whose various communities, nations, and cultures it is particular. Natural languages "evolve naturally" in real-world communities of language users, often in coupling with their cultures and conventions.

Natural languages consist of the structured and conventional use of *words*, i.e., meaningful elements of speech or writing that (where not used alone) are organized to form *sentences*. A sentence is a complete sequence of words typically containing a *subject* and a *predicate* which assigns some attribute, often involving an action, to the subject, thus constituting an attributive statement. A sentence includes a main clause which can stand alone but may have one or more subordinate clauses attached, and can take the form of a question, command, or exclamation without losing attributive functionality.

A given natural language $L = (\Sigma, \Gamma, S_\Sigma)$ can be specified in terms of:

(1) A finite *alphabet* $\Sigma = (s_1, s_2, ..., s_n)$ of uninterpreted *terminal symbols* s_i that can be linearly combined to form written words and expressions and are often associated with sounds that can be spoken in the same order, along with any transitory nonterminal symbols.

(2) A *grammar* or set of structural and transformational rules Γ which determine which *strings* or sequences of alphabetic symbols and words are "well-formed", i.e., which strings are potentially meaningful expressions of the language.

(3) The set S_Σ of strings determined by Γ.

Whereas expressions contain only terminal symbols of Σ, Γ may use any nonterminal symbols in Σ as means to its ends, inserting and then replacing them with other symbols in order to produce or "derive" terminal expressions as the final output of derivation trees orthogonal to the expressions themselves in the ordering of their derivational steps. In this case, the alphabet Σ may be divided into two sets N and T containing nonterminal and terminal symbols respectively: $\Sigma = (N, T)$.

There are two possible ways to construct a string or expression of L. One can write the expression as a linear sequence of terminal symbols and words by using a set of orthographic rules according to which one symbol or word follows another by concatenation or composition. Alternatively, one can rely on rules of production or generation consisting of syntactic forms and rules of substitution. Production works orthogonally to strings themselves, deriving a string by replacing nonterminal symbols with cumulative refinements along its entire length. Generative grammar begins with the universal start symbol of L, a compact way of representing L as a complete expressive potential, and progressively replaces nonterminal symbols with other nonterminal symbols and finally with the terminal symbols of the final expression.

The grammar of a natural language can be complex and unpredictable, and is typically discovered through intensive empirical investigation; until it is "formalized" by exhaustively enumerating its rules, the language generated by the grammar cannot be formalized, and the formal definition $L = (\Sigma, \Gamma, S_\Sigma)$ remains useless for reconstructing the language. The formalization criteria described and illustrated above are due to Zellig Harris (1951), who first described generative grammar as currently understood, and his student Noam Chomsky (1956), who subsequently named and explicated the concept.

Natural language is fundamentally conceptual; its words represent concepts or mental abstractions. The words and signs attached to the concepts are more or less arbitrary; they can vary as long as the concepts themselves retain their integrity. However, there is some amount of controversy regarding the relationship between language structure and cognition. Theories of this relationship range from *universalism* (cognition determines language), through mutual independence, all the way to *strong linguistic determinism* (language determines cognition), with intermediate degrees of dependency considered.

Note that while this account of natural language refers to the representation of *concepts*, nowhere does it refer to the *objective content*, *meaning*, or *interpretation* of its expressions. Yet even though one of the primary functions of language is to represent objective facts about the real world – facts which naturally reflect its structure – the natural language concept is defined in complete abstraction from the external universe, which is commonly treated as though it were completely "independent" of language and cognition. This putative mutual independence of language and universe – the way $L = (\Sigma, \Gamma, S_\Sigma)$ omits the symbol U for "universe", thus isolating L from its domain of discourse – amounts to a pernicious form of dualism which abstractly, artificially, and wrongly separates the elements of an inseparable pair.

In short, the capacity of language to accurately represent the external universe is far too reliable and extensive for this unequivocal kind of dualism to hold. It is not that we are always successful in predicting factual observations from language alone; rather, it is *the very possibility of bringing language into conformance with reality* (and vice versa) that dualism cannot explain. If dualism were justified, and language and reality were truly separate and independent, then there would be no basis for bringing them together, and neither science nor any form of human experience would be possible. The above definition, $L = (\Sigma, \Gamma, S_\Sigma)$, is therefore misleading in the way that its parentheses, like impermeable walls, isolate human cognitive language from the universe it so effectively but improbably represents.

Some readers will recognize that this is simply a formal generalization of the oft-cited "unreasonable effectiveness of mathematics in the natural sciences" to nonmathematical languages and reality at large. It is also a statement to the effect that any language which truly supports the identification of its universe must everywhere couple to that universe through classical 2-valued logic, which provides the 1 | 0 distinction in terms of which U and its contents may be distinguished. This coupling is an untapped well that goes very deep indeed.

Level 2: Formal Languages and Systems

Any natural language can be formalized to the extent that its alphabet is known and its grammar has been formalized, i.e., fully determined by empirical investigation and exhaustively enumerated. That is, a formal language is a natural language $L = (\Sigma, \Gamma, S_\Sigma)$ for which the elements of the signature and the rules of grammar are exhaustively enumerated.

A *formal language* L usually comes implicitly packaged with a *metalanguage* L′ which supports the description and analysis of its structure, functionality, and application. It also comes with predicate and propositional logic, which it requires for its intelligibility and functionality. While logic stands alone, being formulated independently and invariantly with respect to the other languages to which it is applied, it is itself a formal language conforming to the same general description. In addition, L must incorporate such abstract qualitative and quantitative concepts and relationships as may be necessary to specifically identify its content.

In contrast to the often-complex grammar of a natural language, the grammar of an artificial formal language like propositional logic (PL) is simple by design. Its properties are not empirically discovered but chosen for precise expression and analysis. In particular, PL is designed to analyze and describe the relationships among conjunctions, disjunctions, conditionals, and negations (*and, or, if-then, if and only if, not*) linking arbitrary sentences (sentential variables) to each other, i.e., to precisely model key properties of the logical operators and connectives naturally inherent in human cognition so that statements can be reliably and consistently formulated and reliably classified as *true* or *false*.

Any sufficiently expressive formal language L with an easily recognized (or at least computable) grammar can be incorporated in a *formal system* T which adjoins to L a *deductive apparatus*, or a set of axioms and rules of inference designed for the systematic investigation of some specific universe U (domain of discourse or field of knowledge). Data from U are plugged into T, and conclusions are deduced in T regarding U.

A formal system T can be defined as $T = (L, A, I)$, where L is a formal language, A is a set of axioms, and I is a set of rules of inference. The deductive apparatus $\{A, I\}$ can be viewed as an expanded version of the grammar Γ of L, somewhat blurring the distinction between a formal system and a formal language. However, L and T can often be distinguished by two related facts:

(1) The syntactic forms of L usually contain nonterminal symbols representing word-classes or "parts of speech", whereas the axioms of T usually do not.

(2) The production of expressions by the rules of Γ usually involves nonterminals, whereas the derivation of consequences from axioms using the rules of inference of T usually does not.

Predicate logic, including sentential or propositional logic, is implicit in any formal system T. This is because logic must distribute over every part and every element of T no matter how T is partitioned and quantized. It takes predicate logic to form attributions, and propositional logic to identify them from their logical complements; without logic, there can be no discernibility. Also implicit in any formal system T is a metalanguage defining and relating its components.

The second level of the standard theory of languages, studied mainly in mathematical logic and computer science, includes the study and application of these formal languages and systems, the means of proving theorems from their axioms, and their valid interpretations or *models* in their respective domains of discourse U.

The Limitations of Formal Systems

Despite their inadequacies, formal systems are powerful tools of (partial) understanding. However, a formal language omits its array, its processors, its universe, its models, the capacity to determine how it evolves (telesis), and even the capacity to perform its own linguistic operations such as *read, write,* and *erase*, let alone more advanced operations like the expression and communication of meaning. It is thus completely unable to stand on its own, instead relying on language users and sometimes on computers. It is therefore unable to model (share structure with) a perfectly self-contained universe. To do that, it must assimilate that on which it implicitly depends.

Note that formal systems are limited in the same way as language in general; just as for $L = (\Sigma, \Gamma, S_\Sigma)$, the formal system $T = (L, A, I)$ contains no symbol U representing the external universe. Again, the parentheses effectively wall off L and T, effectively isolating them and offering no assurance that T can be successfully interpreted in any given U. Thus, a given U need not preserve the structure of T, provide T with a model, or conform to T as semantic content; until it has been fully explored and put into exhaustive correspondence with T, the actual relationship between T and U is uncertain. In short, a formal system comes with what philosophers have called the *problem of induction*.

This model-theoretic formulation of the problem of induction implies that as long as theories of science, philosophy, and mathematics are regarded as formal systems (or something even less reliable), they can afford no certainty with respect to the real universe, and therefore cannot yield a verifiable understanding of reality. Indeed, the scientific investigation of reality is limited in just this way. Beholden as it is to the scientific method, the sciences have been limited to establishing tentative correspondences between formal systems and sets of physical observations (or subjective conceptualizations), and thus doomed to non-validation and epistemological inadequacy.

While it can be hard to establish that a given language has a model in a given universe, especially when the correspondence must be monitored and maintained "in real time" as U evolves, there are cases in which a model-theoretic correspondence can be taken for granted. For example, we can trivially define U on L; given the internal consistency of L, we can simply define U to be any universe that conforms to L. This is how mathematical theories and their underlying conceptual languages are often treated; their domains of discourse are simply defined as any universe U, i.e., any set of objects, relations, and functions, in which L can be modeled by virtue of common structure, i.e., by virtue of the nonempty structural intersect of L and U.

On the other hand, the structure of U must be known in order to define L on U, and U may not be completely knowable (e.g., in the empirical sciences, nothing about the observable universe can be taken for granted). In this case, it is natural to ask whether there exists any formal language L whose compatibility with any given U can be taken for granted.

Provided that U and its contents are intelligible or identifiable, the answer is yes. Any form of identification, conceptual, perceptual or otherwise, amounts to distinguishing something from its logical complement. This requires the formal language of logic, especially 2-valued propositional logic, which functions in effect as an "identification language" of U. By virtue of the very possibility of apprehending a given universe – by virtue of its *intelligibility* – it must everywhere incorporate the formal system "logic". Conversely, any supposed part or aspect of U devoid of logical structure cannot be identified, and is therefore not recognized as a part or aspect of U.

In short, logic evidently functions as a kind of "identity language" for U – something like the set of criteria through which a computer identifies input, or dually, an attribute that U displays in order to be identified. But it cannot do this alone, for whatever logic distinguishes from its logical complement must exhibit nonlogical properties – size, color, duration, and so on – of which it is an instance. So it appears that we must add these logical and nonlogical ingredients together to get a functional identity language L. But we still have a problem, as propositional logic can only identify full attributions. There must also be something capable of coupling the attributes in L to the objects in U, and while predicate logic comes with means of quantification, it has no way of producing or selecting such nonlogical attributes in order to couple them with the contents of U.

Throughout the empirical sciences (and most other sciences and the humanities as well), it is roundly denied that the physical universe and its internal content, or states, arise in coupling with L. Instead, it is held that physical reality evolves according to its own internal causal processes, mechanisms, and laws, despite the obvious fact that "laws" are formal (abstract, qualitative) entities. As usually conceived, causation is dualistic; it requires that reality evolve independently of any language, with causal functions that are metrically parameterized exclusively by the objects and independent medium on which they act. But this leads to a contradiction: it implicitly presupposes, once again, that U is already coupled with an intrinsic self-identity language that provides it with attributes and values from which to form the states which comprise the input and output of causal functions. In short, dualistic, language-independent physical causation implies that U is already inseparably coupled with its very own identity-language, which just happens to be the same as ours.

Finally, we realize that we might as well cut to the chase: Somehow, an intelligible universe, one that we can actually identify, must already be coupled with an identity-language. The simple fact is that where any degree of intelligibility is given, structure must already be shared between mind and external reality; mind must be equipped with a cognitive identity language L through which U can be recognized, and in a complementary way, U must have the capacity to display content recognizable to L. Unless these dual requirements are filled, the mind has no way to reach out and acquire external reality, and external reality has no way to send itself into the mind. And that, of course, means that the question is really just this: What are the implications of the fact that mind and reality already share structure which includes logic and various nonlogical attributes?

This is not the only issue in overcoming linguistic dualism. Standard dualistic languages lack a number of self-containment properties. They lack closure with respect to not just modeling and the acquisition of content (*reference*), but such functions as *read, write, production, processing, display, communication*, and the selection or execution of any constructive or restrictive process or mapping that might generate expressions or restrict the exhaustive linguistic potential represented by $\{S_\Sigma\}$, the exhaustive set of all grammatically well-formed strings. In order to determine and perform these operations, languages require ingredients that are not contained in their standard dualistic definitions. Dualistic language theory excludes everything that makes language telic or dynamic, waving it off to some language-user or programmed automaton, or supervening it on some random or determinate physical process that supposedly has no need of it.

The Metaformal System: An Informal Introduction

Due to limitations of time and space, this account is brief and non-justificative. We cannot hope to give anything but the most summary and superficial account of the Metaformal System in this paper.

First, it might be wondered why this paper is entitled "*The* Metaformal System" rather than "Metaformal Systems". This is because the only metaformal system known to exist is that which corresponds to reality as we know it, and as far as we know, there are no other realities to be considered (this system has been known for the last several decades as the *Cognitive-Theoretic Model of the*

Universe or CTMU; Langan, 2002). There are other essentially metaphysical constructs, including the Everett (1956) *Many-Worlds Interpretation of Quantum Mechanics* and the Tegmark (2008) "Mathematical Universe Hypothesis", which speculatively refer to other possible realities. But such acts of reference connect these "other" possible realities to this one, implying that the resulting web of interconnections constitutes one overarching "ultimate reality" after all. It is this ultimate reality that the Metaformal System represents.

The Metaformal System is an "extended formal system" which is perfectly self-contained, i.e., which has all of the properties and carries all of the ingredients required for its existence and functionality, including those just mentioned (*read, write, reference, interpretation, production, processing, display, communication, potentiation, determination*, and so on). In short, all of these functions are executed by objects of M (syntactors, telors) using the grammar of M.

Even though the Metaformal System can hold an arbitrary number of formal systems in its framework, it can be represented in or described by a formal system of its own. This is how it is represented even here, on paper and on monitor screens controlled by patterns of bits in electronic computers. However, despite the fact that the Metaformal System can be *described* by a formal system, i.e., embedded in the formal system as descriptive content, it turns the tables, mapping into its own structure the very formal system which describes it along with any language-user reading the formal description. In relating itself to its own ingredients, M places itself in its own universe by a self-dual grammatical inclusion mapping called *involution*, and is thus a metalanguage of not only the formal systems in question, but itself.

The Metaformal System can be thought of as a kind of "closed wrapper" for formal systems. The wrapper is a supertautological metalanguage (Langan, 2002, 2017) in which formal systems are "located" or assigned a place in M as referential content on the basis of shared structure. Specifically, they are located in L_M, the formal or linguistic aspect of the Metaformal System M, the *identity-language* of U_M which contains all of the abstract properties which suffice to identify not only U_M, but all that can be identified in the ontic identity $M = L_M | U_M$ including all of its internal subidentities or discernible attribute|value couplings. Some of these identities and subidentities are *telors*, which are defined in terms of functional capacity and include human beings; M thus absorbs human language-users and the linguistic functions they perform, firmly embedding them in the metaformal structure of reality.

The **Metaformalist Program**, as it has previously been called (Langan, 2018a), requires that language function as an *identity*, specifically the identity of a universe. That the universe we inhabit is intelligible to us means that our internal cognitive-perceptual language couples (interfaces or intersects) with it, allowing us to identify it by direct perception. It is in this sense that we function as "identification operators", using our cognitive language as an *identity language* to identify the universe and its contents including ourselves. Because identification operators use the syntax (formal structure) of an identity language to identify their operands, they are called *syntactic operators* or *syntactors*. Telors are structurally complex syntactors which can "factorize telesis" or actualize ontic potential, and have sufficient complexity to consciously generate internal representations of themselves and their relationships with the external environment.

The grammatical language | universe coupling effected by M is *self-dual* in the sense that language and universe are just different aspects of shared structure in which they everywhere intersect. This implies that they are the source and target, or (dually) the target and source, of a mutual *dualization* mapping (involution) which generically "models" one in the other. The coincidence of language, universe, and model makes the entire three-way coincidence not just self-dual, but *trialic*. (Triality was adumbrated by the American logician Charles Sanders Peirce [1976], whose semiotic approach to logic characterized signs in terms of three classes: *signifier*, *object*, and *interpretant*.) Because language is one of its coinciding aspects (no matter what its particulars may be and regardless of the degree to which they are specified in advance), this relationship has linguistic structure. Specifically, it is a new kind of "language" called an *intrinsic language*.

An intrinsic language can be understood as a trialic (three-aspect) self-identification language which does triple duty as language, universe, and model while functioning as a global primary syntactor, of which internally localized secondary and tertiary syntactors are *points of involution* (targets of the grammatical involution mapping). The primary syntactor, which is also the Primary Telor or universal start symbol of M-grammar, spans the entire involution mapping from L_M to U_M, while tertiary syntactors are confined to the U_M (terminal, physical) end and include fermionic (mutually exclusionary) point-like particles. Secondary syntactors occupy the terminal region of the mapping to variable depth, where – in their dual role as secondary telors – they fill causal deficits due to the underdetermination of tertiary syntactors by the Primary Telor, which is invariant with respect to them and of which they are images by μ-morphic involution (the grammatical involution mapping is also called the *μ-morphism*).

At the trialic limit where the semiotic representation relationship between intension and extension contracts to form self-dual intension | extension identities as required by systemic self-duality, information "about" the state and evolution of the system on all scales equates to the evolution of the system itself; in evolving, the system is "informing itself" of its own evolution. Primary, secondary, and tertiary syntactic operators generate three levels of information. Primary information is associated with the primary syntactor and best described as *teleology*; its function is to provide for the requirements of self-identification on all scales down to that of tertiary syntactors. (Where generic utility is that which enhances identity or "selfhood", a utilitarian definition of teleology as the maximization of generic utility equates it to the Self-reinforcement of the identity of reality or Primary Telor, [Langan, 2003]) Tertiary syntactors are the quantum objects of which secondary syntactors are composed and in which their extended phenotypes materially intersect. Tertiary information, which is about the terminal "surface" level of evolution and equates to it when attributes and values are merged as terminal identities by full L_M-U_M coupling, is the only level studied in physics and other empirical (terminal, surface-level) sciences. Secondary information fills causal deficits which owe to the underdetermination of tertiary states and interactions by primary information, this being the local freedom of the system and its inhabitants.

The Metaformal System, a Bit Less Informally

In much the same way as a formal language, the Metaformal System can be formally expressed as:

$$M = L_{INT} = Ls \mid Lo = (\Sigma = \{N, T\}, \Gamma_{MU}, S_\Sigma)$$

M as a Whole

Moving from left to right, $M = L_{INT}$ says that the Metaformal System is an intrinsic language. Complete (ontic) self-containment is required of both M and L_{INT}, implying structural equivalence. It turns out that this intrinsic language doubles as a pair of complementary *semilanguages*, Ls and Lo, that are in fact dual to each other insofar as they comprise the intensional and extensional aspects of M.

Notice that $M = (\Sigma = \{N, T\}, \Gamma_{MU}, S_\Sigma)$ is very close to the way in which formal languages are defined, with a signature, a grammar, and a set of linear "strings". The meaning is almost as simple: The fundamental objects of M are active signs called *telors* (N) and *syntactors* (T), and its grammar Γ_{MU} is a self-identification operation which, just as in a standard language, "identifies" strings by writing or generatively producing them. In other words, M evolves by "identifying itself".

The generative derivation and orthographic dimensions of self-identification are orthogonal, but not entirely independent; N and T must be continuously coupled so that telors can use Γ_{MU} to acquire terminal resources from T for grammatical transformation in N.

The metaformal intrinsic language M is an ingredient of its own syntax on all orders of reference. That is, M is self-referential and logically idempotent (Langan, 2017).

Signature

In the signature $\Sigma = \{N, T\}$, the subsets N and T of Σ are sets of nonterminal and terminal identity operators respectively; N = {telors} and T = {syntactors}. These two kinds of "active sign" are distinguished by functionality as well as location; because production ("derivation") occurs in N while concatenation ("write") occurs in T, telors can generatively produce entire strings in a spacelike or "nonlocal" fashion, while non-teloric (tertiary) syntactors, being restricted by the locality, timelike linearity, and continuity of T, cannot.

The signs in Σ are all identity operators which use M as an identity language to identify the world; hence, all are syntactors. There are three levels of syntactor: primary, secondary, and tertiary. The primary and tertiary levels respectively correspond to the universe as a whole, and the smallest and most elementary objects therein; the secondary level is intermediate.

N and T, which can be respectively likened to the internal and external and/or the mental and physical aspects of reality – and at the risk of considerable oversimplification, to a programmable automaton and its display screen – have some amount of overlap; every telor is a syntactor, but not every syntactor is a telor. Primary and secondary syntactors are telors, but tertiary syntactors are not.

Telors can exist in both N and T; tertiary syntactors can exist only in T. Because grammar operates mainly in N, only telors can fully exploit its power, generating spacelike derivation trees which output terminal strings in parallel, across time. But in order to meaningfully do so, they must have sufficient complexity in T to model their own relationships with the external environment.

Grammar and Semimodels

Γ_{MU} is a self-dual distribution mapping, the *μ-morphism*, that is inwardly involutive and consists of spacelike metatemporal parse trees which produce terminal states and trajectories in T, but outwardly evolutive with respect to the timelike linear trajectories thus produced. μ-morphism is associated with a rescaling operation called *conspansion*, which occurs in a *conspansive manifold* which takes the form of a *syntactor bundle*, a self-dual analogue of the topological construct known as a "fiber bundle".

Conspansion is a coupling of spacelike dual processes, *inner expansion* or *d-ectomorphism*, a one-to-many mapping which outwardly distributes compact identities (states, points) to potentials, and *collapse* or *d-endomorphism*, a one-to-many dual mapping which inwardly distributes attributes to states or points. (The *d-* stands for "distributive".) The self-dual conspansion morphism corresponds to the *conspansive semimodel* of M, which models derivation and is associated with the nonterminal semilanguage L_S; it evolves discretely and in just the way that reality appears to evolve in the quantum-theoretic picture.

The terminal analogue and dual model of the conspansive semimodel, the *linear-ectomorphic* (or *l-ectomorphic*) *semimodel* of M associated with Lo, is based on dual (inward and outward) conspansive gradients and yields a very different picture of M, namely the classical one that we see and hear, in which terminal objects follow linear trajectories through "spacetime", which – together with its physical content – amounts to the surface structure of the conspansive medium. The superposition of these dual semimodels reflect the fact that reality evolves by self-dually "modeling itself" in two complementary ways, only one of which we actually see when we look outward, but in the other of which true Γ_{MU} dynamics reside.

Semilanguages

Ls and Lo are dual semilanguages into which L_{INT} is "factorized" by Γ_{MU}. Ls and Lo correspond to N and T respectively.

Ls and Lo correspond to the linguistic structure superimposed by M on the sets N and T respectively, providing the elements of N and T (telors and syntactors) with enough structure and organization to productively interact and fill their roles as identification operators. Respectively corresponding to the intensional and extensional aspects of M, they remain in superposition as Γ_{MU} locks them together in order to form the self-dual identities that comprise the terminal states of M.

Simplistically, Ls exists "inside" secondary telors of N, while Lo consists of the external states and trajectories or state-transitions of their constituent tertiary syntactors. As Ls transforms and is dually transformed by Lo through the action of Γ_{MU}, Ls is conspansively "involuted" in metatime, and Lo "evolves" linearly in time along dual conspansive gradients.

Strings

$S_\Sigma \supset T$, the set of "terminal strings" of M, consists of the external states and trajectories of syntactors. S_Σ, associated with the semilanguage Lo, occupies the "surface structure" of M.

The Relationship Between Levels 1, 2, and 3

Level 1: Natural Languages

In a natural language, expressions are produced by writing or speaking linear sequences of words (or signing in a sequence of gestures). Empirical investigation of natural language grammars reveals that beneath their *surface structure*, they have *deep structure* which can be explicated in terms of word-classes ("nonterminal symbols") and phrase markers (parse trees, derivation trees). These *phrase markers* have derivational structure orthogonal to the linear expressions on which they terminate, suggesting that expressions are actually generated along two axes.

Natural language grammars are sometimes classified in terms of the least powerful acceptor, or language-recognizing automaton, that can recognize the languages they generate. As defined in the Chomsky Hierarchy, type 0 grammars generate unrestricted languages requiring a universal computer (Turing machine) with unlimited memory; type 1 grammars generate context-sensitive languages requiring a linear-bounded automaton with memory proportional to word length; type 2 grammars generate context-free languages requiring a pushdown automaton with a memory stack in which a fixed number of elements are available at any point; and type 3 grammars generate regular languages requiring a finite deterministic automaton with no memory (Chomsky, 1956).

Note that where languages are defined on the grammars that generate them, and grammars are classified in terms of automata, languages themselves are implicitly classified in terms of automata. This language-automaton association has been in force since the mid-20th century, and it has only reinforced the unavoidable theoretical coupling of natural language with human cognition.

Level 2: Formal Languages and Systems

In a formal system T, the production grammar of the formal base language L refines nonterminal symbols to terminal symbols, progressively applying rules of substitution or transformation to a universal nonterminal "start symbol" representing the entirety or "identity" of L, terminating on fully formed expressions which contain nothing but terminal sentences consisting entirely of terminal symbols and words. To this, a formal system T adds a "terminal grammar" consisting of terminal forms called *axioms* which provide a general description of some universe or domain of discourse U, along with rules of inference.

The axioms – which are terminal with respect to grammatical production and constituent symbols, but initial with respect to the axiomatic description of their universe – comprise a boundary between the nonterminal and terminal realms. They come in two forms: one, often used in mathematics, continues derivation in the same metatemporal direction as production, while the other is typified by continuously iterative "laws of nature" or "laws of physics" which evaluate to numbers, accepting the values of prior states as input and yielding as output subsequent states which appear to arise "orthographically", in the temporal *write* direction of symbolic and lexical concatenation or composition. The axioms may be applied to known data or assumptions regarding U in order to derive consequences via rules of inference, usually in the hope and expectation of obtaining useful information – descriptions, predictions, or explanations – about U.

Because the formal system is merely a terminal extension of its formal language L, all of the above characterizations of language apply to it. Either no universe or model is explicitly included in the formal descriptions of L and T (including the case where U = L or U = T), or there is a particular application in mind, as when a formal language is structured for computer programming or control of a particular automated system. In either case, the association between language and automata persists, but all the more strongly due to the increased emphasis on mathematical logic and computational applications.

Level 3: The Metaformal System

The third level of language theory is that of the Metaformal System. It resembles any other language in that its grammar brings to bear both rules of production and orthographic rules of symbolic concatenation and lexical composition along orthogonal *generation* and *write* axes which map to a pair of orthogonal axes in the conspansive medium of U_M. This medium, the *conspansive manifold*, evolves through a stratified, generative, self-dual rescaling operation associated with metaformal grammar Γ_{MU}, namely *conspansion*, which generates a logico-linguistic "boundary algebra" in which syntactic distinctions are mapped to closed static surfaces and hypersurfaces, a general concept associated with such logicians and mathematicians as Venn, Peirce, Frege, and most recently George Spencer-Brown (1969). It incorporates formal languages and systems, nonterminal forms and terminal strings, and is therefore linguistic in structure and operation. Nevertheless, it includes unique explanations for gravity, cosmic expansion, quantum phenomena, consciousness, and even spirituality. It is in every important way linguistic in structure and operation.

As for the trademark linkage of language and information processing, the Metaformal System M takes it to its logical conclusion by uniting language with the largest and most powerful "information processing system" conceivable, namely, reality at large. Over the last several years, there has been considerable buzz regarding the so-called "simulation hypothesis" or "simulation theory", which centers on the possibility that despite its propensity to fool us all, reality is an artificial simulation running on a vast cosmic computer, astral automaton, or some other kind of host. Some versions of this hypothesis seem reminiscent of science fiction movies; others involve ideas emerging from ongoing research into "virtual reality". But the most logically well-founded and therefore credible version of this idea is the unique logico-linguistic approach of the CTMU, with references as far back as the late 1980's (Langan, 1989). The CTMU shows that no matter how the issue is slanted, reality can indeed be categorized as a "self-simulation" with a nonterminal grammar and a terminal display space of which we are all the occupants, the spectators, and to a limited extent, the programmers.

Self-simulation aside, there is a certain logical inevitability to the proposed extension. Not only are fields like physics, cosmology, biology, consciousness, AI, and philosophy beating their heads against conceptual brick walls when they try to answer metaphysical questions for which standard theoretical languages are fundamentally inadequate – itself a powerful argument for rummaging around the language department for missing conceptual infrastructure – but the questions themselves are often garbled. If I might be permitted to employ the third level of language herein proposed, modern scientists and philosophers are telors in a state of total denial who – while blithely using their own technical languages to pose profound questions and seek deep answers – can often do nothing but look about them at the terminal domain $T \subset \Sigma$, peer at it with all their might, and fail to recognize even the linguistic surface structure of the associated M-semilanguage Lo Little wonder that nothing ever gives.

But still, there are some very good, very logical reasons to think that any attempt to avoid the proposed extension would be futile. Let's look at it from a metalogical vantage. A metatheory is a theory about theories. Theories of science and mathematics are formal systems; thus, the capacity in which M incorporates formal languages and systems, including those employed throughout the mathematical and empirical sciences, is that of a metatheory. An axiom of a metatheory that deals with its object-theories and/or their axioms is called a **meta-axiom**, and likewise for any axiom referring autologically to itself.

The CTMU metaformal system can be brought to rest on a single "closure meta-axiom", the *Analytic Reality Closure* Principle (ARC), which is also called the "CTMU Identification (or Intelligibility) Axiom". It can be expressed as the CTMU "master-equation" (which, interestingly, can be regarded as a logical generalization of Einstein's Equation, with a medium on one side and its content on the other):

$$\mathbf{reality_{INT} =^* reality_{EXT}}$$

This axiomatic equation, which for various reasons cannot be mistaken for a mere assumption, expresses the innate self-duality of M. Notice that in addition to being a global meta-axiom, it has on its left side an attribute or *descriptor*, and on its right side a collection of *operators* that are μ-morphic Self-images of a single *Global Self-Operator*. Thus, without too much blame, it might fairly be abbreviated as "the GOD Equation", where "GOD" is acronymic for "Global Operator-Descriptor".

This axiomatic equation translates as follows:

> *The intension of reality* (the global property L_M = reality_{INT}) *is dually equivalent to the extension of reality* (the physical universe U_M = reality_{EXT}),

where *dual equivalence* describes the situation in which dual entities coincide as aspects of a self-dual identity which "glues them together" using mutually distributed structure in which they everywhere intersect.

ARC can also be rendered as the "CTMU Reality Principle":

> Reality contains all and only that which is real.

This can be reformulated as a self-contained proof by contradiction:

> If there were anything *outside* reality on which reality were in any way dependent – e.g., for origination, causation, support, maintenance, or anything else – then the entire dependency relationship would be real and therefore *inside* reality. It follows that reality is self-contained with respect to physical and metaphysical functions like origination, causation, support, maintenance, etc.

This proof by contradiction immediately invalidates the premise of external relevance, implying the ontic closure of M = reality. It also means that the level of "reality" we are considering is the deepest level possible, namely the CTMU level. This is because "reality is ontically self-contained" doubles as a definition of reality under which it is idempotent under inclusion (or if one likes, "self-inclusive", with all the bells and whistles required to support this property; Langan, 2017).

The intelligibility criterion of the Metaformal System M overrides seemingly preclusive metalogical results such as the Tarski Undefinability Theorem and the Gödel Undecidability Theorem. In a nutshell, arguments based on such results, which usually apply explicitly to standard formal systems, can only undermine their own intelligibility (and that of the formal systems to which they are attached) by undermining M. Similarly, insofar as intelligibility relies on classical 2-valued propositional logic, objections based on the existence of alternative logical systems, e.g. constructive logic, are without force. Other logics may be relevant within bounded contexts, but criticisms which rely on them can only compromise their own intelligibility by opposing M.

With any luck, this has brought the reader at least a little closer to a very important realization about the CTMU Metaformal System: as asserted in previous papers, it bids fair to be recognized as an integral part of the universal foundational language of science and philosophy, and thus an integral part of the scaffolding for the long-awaited bridge between science and theology, physics and metaphysics, and physical and mental reality. Indeed, it should now be easy to see why it would be very difficult indeed to erect any other kind of scaffolding for this purpose.

References

Chomsky, N. (1956) Three models for the description of language. *IRE Transactions on Information Theory*, Vol. 2, Issue 3, pp. 113–124.

Chomsky, N. (1965) *Aspects of the Theory of Syntax*. Cambridge, Massachusetts: MIT Press.

Everett, H. (1956) *The Many-Worlds Interpretation of Quantum Mechanics*. Princeton University, PhD dissertation.

Harris, Z. S. (1951) *Methods in structural linguistics*. Chicago: University of Chicago Press.

Langan, C. M. (1989) The Resolution of Newcomb's Paradox. *Noesis*, № 44.

Langan, C. M. (2002) *The Cognitive-Theoretic Model of the Universe: A New Kind of Reality Theory*. Princeton, MO: Mega Foundation Press. Originally published in *Progress in Complexity, Information, and Design*, Double Issue, Vols. 1.2–3.

Langan, C. M. (2003) Cheating the Millennium: The Mounting Explanatory Debts of Scientific Naturalism. In W. A. Dembski (Ed.) *Uncommon Dissent: Intellectuals Who Find Darwinism Unconvincing*. Wilmington, DE: ISI Books.

Langan, C. M. (2017) An Introduction to Mathematical Metaphysics. *Cosmos and History: The Journal of Natural and Social Philosophy*, Vol. 13, № 2, pp. 313–330.

Langan, C. M. (2018) Metareligion as the Human Singularity. *Cosmos and History: The Journal of Natural and Social Philosophy*, Vol. 14, № 1, pp. 321–332.

Peirce, C. S. (1976) *The New Elements of Mathematics*. Edited by C. Eisele. Atlantic Highlands, NJ: Humanities Press.

Spencer-Brown, G. (1969) *Laws of Form*. London: Allen & Unwin Ltd.

Tegmark, M. (2008) The Mathematical Universe. *Foundations of Physics*, Vol. 38, Issue 2, pp. 101–150.

Introduction to Quantum Metamechanics (QMM)

To John Archibald Wheeler and Martin Gardner, who encouraged me when others did not

Abstract

Solutions for problems arising at the limits of science and philosophy require ontological grounding. Quantum Mechanics (QM) is increasingly called upon as a source of insight regarding such problems, but is not itself well-understood. The fact that QM has many conflicting interpretations for which ontological status is claimed demands a "post-quantum" theory which clarifies its meaning, settles the differences among its interpretations, and facilitates the analysis and solution of otherwise intractable problems. Herein described as Quantum Metamechanics (QMM), this theory is a "meta-interpretative" mapping of QM and its various interpretations into a supertautological description of reality, the CTMU Metaformal System. By incorporating the CTMU, a true ontic identity supporting the self-identification and self-existence of reality,

QMM provides QM with a valid ontology in terms of which its various interpretations can be evaluated and synergized.

Keywords

Cognitive-Theoretic Model of the Universe; Quantum Mechanics; Quantum Metamechanics; QM; QMM; Ontology; Quantum Ontology; Metaformalization; Formal Quantization; Metacausation; Retrocausation; Interpretation of Quantum Mechanics

Introduction

As never before, scientists and philosophers are trying to solve "big questions" having to do with such imponderable concerns as the nature and extent of reality, the origin and nature of life, the nature of mind and consciousness, the origin of the cosmos, the nature of space, time, and causality, the essence of human existence and spirituality, so-called paranormal phenomena, and other matters seemingly resistant to mechanical, material, or physical explanation. Accordingly, science and philosophy have been gravitating toward the broad and highly successful theory of quantum mechanics (QM) as a source of insight. But despite its great theoretical and methodological utility, QM is as much a mystery as the questions themselves, and that of which the meaning is unclear is not itself a credible source of meaning. This has led to the search for a "post-QM" theory that properly explains QM itself and is thus better equipped to deal with metaphysical issues.

The purpose at hand is to identify the requirements of such a post-QM theory and then describe it in logical terms. Because this theory is necessarily a metatheory (or theoretical metalanguage) of QM, it is called **Quantum Metamechanics** or **QMM**.

Its purpose is to map QM, along with any valid hypothetical correlates designed to obviate or accommodate its apparently problematical features, into the CTMU Metaformal System (Langan, 2018), a comprehensive high-level formulation of the structure of reality independent of QM itself, and then to explicate their relationship and thereby synergistically relate the microscopic and macroscopic scales of reality to each other. Because the Metaformal System is a supertautological (intrinsically valid) reflexive model of reality predicated on its manifest intelligibility, QMM can be described as a reflexive application of model theory which reliably locates QM within the theater of being.

Opinions to the contrary notwithstanding, QM itself is not an ontology. QM is a formal system standing apart from its universe, a mathematical apparatus incorporating such ingredients as linear algebra, Fourier analysis, and probability theory. Given the existence of certain measurements, QM merely yields statistical predictions of their outcomes. QM does not include definitions or attributions of being, existence, or reality. Assertions relating these concepts to QM reside elsewhere, usually in a more or less speculative interpretation of QM in an imperfect description of an incomplete set of observations labeled "physical reality".

That ontological status has nevertheless been claimed for various interpretations of QM – that they are called "quantum ontologies" – reflects a widespread misunderstanding of the word "ontology". In the minds of most scientists and philosophers, ontology consists of "claims about existence", e.g., the kinds of object, relation, operation, and process that exist in the world, and related epistemological claims about the nature and limits of knowledge, e.g., what kinds of knowledge are possible under what conditions. But insofar as anyone can make any claim at all about anything one likes, this is a trivialization. If existence can be meaningfully attributed to anything at all, then a valid ontological language must exist, and it must consist of actual knowledge rather than mere "claims".

Concisely, an *ontology* is a theoretical language that accounts for the nature and content of being (reality, existence) and logically supports its attribution on all scales and all levels of discourse. This carries certain requirements that QM cannot fulfill. *Being* is not an ordinary attribute, but the highest attribute of all; no lesser attribute can be meaningfully attributed to anything of which some kind or level of being, even if "purely conceptual", is not already a property. Moreover, just as QM suggests, ontology is intimately related to epistemology, which deals with the nature and limits of knowledge. Because something must *exist* in order to be known or identified, while that which exists must be *identifiable* as a value or instance of the attribute "existence", identifiability and existence must coincide.

While QM is considered by some to define the limits of physical measurement and thus of empirical identification, there are other things to be identified in the name of science – ideas, concepts, sensations, feelings, judgments, intentions, intuitions, and theories like QM itself, for example. It is simply not the case that abstract and subjective forms of existence and identity can be wholly supervened on "physical" objects and processes.

Overview of QM

Informally, a *quantum* of X is the smallest particle or indivisible instance of X, while mechanics is the branch of physics dealing with the motion of bodies and the energy and forces producing motion, including statics, dynamics, and kinematics. It follows that quantum mechanics is the study of how energy and forces relate to the motion of particles. But this is a bit deceptive, as the elementary particles studied in physics do not actually "move" in the usual sense of moving bodies.

In fact, the elementary particles of physics are observed only when they are measured, and what they do between measurements is never witnessed. To call it "motion" in the usual sense is an assumption. Moreover, to measure them is to cause them to change state, which means that they are seen only in conjunction with state-transition events. Experimental data suggest that between these events, they become waves. This is the easiest conclusion to draw from (e.g.) the famous double-slit experiment, in which sending particles through a pair of slits in a partition produces a distinctly wavelike interference pattern on a screen.

The interference pattern produced in the double-slit experiment is considered strange because it occurs even when the particles are sent through the slits at widely spaced intervals. Thus, their respective "waves" cannot be interfering with each other in real time. The waves can only be interfering with each other in association with *each individual particle*, implying that each particle is somehow equivalent to coherent set of "probability waves" that superpose on and interfere with each other in association with it, influencing the motion of the particle if not actually guiding it to its point of impact. In other words, between the emission event and its impact on the screen, each particle behaves like a coherent superposition of waveforms. Accordingly, each particle or physical system capable of quantum coherence is associated with a "quantum wave function".

Why a wave, and why a function? First, we have the wavelike behavior of light, extended to particles of matter by de Broglie. Secondly, that only one of multiple possibilities is actualized as the outcome for a given quantum event requires a many-to-one function to select the one from the many (or, given the fact that quantum experiments can be formulated as yes-no questions, to select one of two possibilities). Thirdly, that we have an apparent superposition of possibilities means that we require something that obeys the superposition principle (or property), which says that for a linear function or system, the net output is the sum of the individual inputs. (That is, if input b produces output x and input c produces output y, then input b + c produces output x + y.) Although the world contains many nonlinear phenomena, its linear aspects are what make possible a reasonable notion of causality whereby cause and effect are "in proportion".

Linear systems include both wave media and vector spaces; a superposition of waves is just the sum of their amplitudes at each point, while a superposition of vectors is just a vector sum. Because (by Fourier's Theorem) any wave, classical or quantum, can be expressed as a unique sum of sine waves in superposition, and because the Hilbert space of quantum states is both a vector space and an inner product space in which vectors can be superposed, added together, multiplied by scalars, and multiplied by each other, the superposition of two possible states is again a state of the system. That is, if $|\psi_1\rangle$ and $|\psi_2\rangle$ are possible states of a system, then so is $|\psi\rangle = a_1|\psi\rangle + a_2|\psi\rangle$.

This can be interpreted to mean that the system is "in both states at once". When the system is in a superposition of possible states, its wave function is said to be *coherent*. On the other hand, when it is measured, its possible states suddenly decohere and its wave function "collapses", or at least *appears* to have collapsed, into a single definite state.

In classical mechanics, the state of a physical system consists of values for all of its observable attributes or "observables". In contrast, a quantum state consists of values for a "complete set of commuting observables" that can be measured one after another without disturbing the rest. The commutativity restriction owes to the fact that some observables are "conjugate" and therefore do not commute; measuring one can disrupt the other, putting it into a superposition of different possibilities and thus destroying the information available on it.

This relationship between conjugate observables defines an epistemological limit of QM. It is called the **Heisenberg Uncertainty Principle** (Messiah, 1999) and is written:

$$\Delta x \Delta p \geq \frac{\hbar}{2}$$

where the symbol Δ denotes the "spread" or loss of information on the associated variable. Here, the noncommuting observables are position x and momentum p. The principle could also be expressed, for example, in terms of energy E and time T as:

$$\Delta E \Delta t \geq \frac{\hbar}{2}$$

In the view of Bohr, Heisenberg, and others, the epistemological uncertainty principle has ontological bearing. According to Bohr's Quantum Postulate, reality is naturally quantized, or discretely partitioned into measurable stationary states with nothing intelligible inside or between them, and according to his Correspondence Principle, quantum mechanics must reproduce classical physics in the macroscopic limit of large quantum numbers. In other words, QM must "scale" from discrete microscopic to continuous macroscopic physics (Bohr, 1928).

Insofar as the focus of QM is the limiting scale of physical measurement on which classical mechanics breaks down, Bohr and Heisenberg saw QM as a terminal theory of reality of which classical physics is just the macroscopic limit. Yet at the same time, Bohr insisted that only the classical conceptual language of the macroscopic, fully observable physical domain be used to express knowledge of quantum objects and processes, and that all scientific investigation must rest on a concrete foundation.

> [An] unambiguous communication of physical evidence demands that the experimental arrangement as well as the recording of the observations be expressed in common language, suitably refined by the vocabulary of classical physics. [...] In all actual experimentation this demand is fulfilled by using as measuring instruments bodies like diaphragms, lenses and photographic plates so large and heavy that, notwithstanding the decisive role of the quantum of action for the stability and properties of such bodies, all quantum effects can be disregarded in the account of their position and motion (Bohr, 1963, p. 91)

In short, QM not only characterizes the microscopic limits of measurement, but also marks the limits of reality. In a way reminiscent of Wittgenstein's observation that "the limits of my language are the limits of my world", the descriptive limits of QM also seem to limit reality itself, compelling the use of classical language to describe it. This, basically, is the theme of the Copenhagen Interpretation.

The Copenhagen Interpretation as "Quantum Ontology"

The Copenhagen interpretation originated in the mid-1920's as a product of expert collaboration involving Niels Bohr, Werner Heisenberg, and others (Herbert, 1985). Following close on the heels of QM itself, it is considered the original and paradigmatic interpretation of quantum mechanics. In keeping with the uncertainty principle, it asserts that physical systems lack definite properties prior to being measured, and that quantum mechanics can only predict the probability distribution of the possible results of a measurement. In effect, the measurement "collapses" the probability distribution to just one possible value, which did not previously exist but has come into existence due to the measurement (we will later refer to this property as "generativity"). In short, the real, physical state of the measured entity relies on measurement itself, which relies on those who do the measuring, and therefore cannot be separated from them. Because the probability distribution is described by a wave function, this is called the *reduction* or *collapse* of the wave function.

Because the Copenhagen interpretation asserts the *nonexistence* of physical properties, values, and states between measurement events, but asserts that they come into *existence* upon measurement, it has ontological bearing. Hence, it is widely considered the original "quantum ontology", a phrase which requires explanation. Conventionally, quantum ontology does not address the ontological or existential status of QM itself. Rather, it takes QM for granted and addresses its ontological implications for physical reality and sometimes reality in general. In this context, *reality* is synonymous with *being* or *existence*; if something is *real*, then it *exists* and has *being*. But as for the meaning of these synonyms, they are regarded as either primitive (i.e., associated with direct physical observation) or defined; and where they are deemed needful of definition, they are defined on either QM and the rest of physics (as in the

Copenhagen Interpretation), or on the very interpretation of QM in question. In any case, it has become the height of fashion for anyone with an "interpretation of quantum mechanics" to declare it a self-contained "quantum ontology".

Unfortunately for such declarations, one cannot properly interpret a theory without having something definite to interpret, and something definite in which to interpret it; and one cannot have an ontology which fails to incorporate a metalanguage which is so defined as to support attributions of *reality*, *being*, and *existence* on all scales, from the quantum scale up to the entire cosmos. Quantum mechanics only partially meets only the first of these criteria, and makes no pretense of meeting the second. Aside from statements dealing specifically with physical measurements, most of it resists interpretation in concrete reality, and those interpretations of QM which receive the most attention do nothing to improve the situation. In fact, most of them are primarily concerned with getting around what many consider the biggest problem in quantum theory, the collapse of the wave function, and are thus reactions against Copenhagen.

The Measurement (Collapse) Problem of QM

The reduction of the wave function can be described as follows: the time evolution of a quantum state or wave function $|\psi\rangle$ is given by the nonrelativistic Schrödinger wave equation:

$$i\hbar \frac{\partial}{\partial t}|\psi(t)\rangle = H(t)|\psi(t)\rangle$$

where $i = \sqrt{-1}$, $\hbar$ is Planck's reduced constant $h/2\pi$, t is the time parameter with respect to which the wave function $|\psi(t)\rangle$ is differentiated, and H is the Hamiltonian operator representing the total energy of the system. The Schrödinger Equation is linear in several important respects. For example, it can be simplified to unitary transformations which preserve inner products, and it has linear operators acting on a linear state vector (or wave function) in such a way that linearity holds among its solutions: if ψ_1 and ψ_2 are solutions, then so is $a_1\psi_1 + a_2\psi_2$. This restricts the entire description to linear spaces and linear geometry, excluding anything that cannot be expressed in a linear continuum. From the instant that $|\psi\rangle$ arises as a solution of the Schrödinger equation to its transformation into a new state, its dynamic is assumed (but not observed) to be both linear and wavelike.

Adapting the notion of linear causation to the quantum scale, QM drops the classical assumptions of determinacy and locality and makes the best of quantum uncertainty by replacing the notion of causal determinacy with inductive probability. Its success in describing microscopic reality thus comes at the cost of statistical deresolution, a trade-off which reflects its inadequacy for describing the deep structure of reality. The best that QM can provide under these circumstances is a statistical approximation of acausal "wave function collapse".

Wave function collapse occurs as follows. At the point of measurement, $|\psi\rangle$ becomes a superposition of eigenstates of the quantity being measured and immediately collapses: $A|\psi\rangle \rightarrow a_n|\psi_n\rangle$. Mathematically, $|\psi\rangle$ has "expanded in" (been filtered or partitioned into) the eigenbasis of the operator (A) associated with the measured quantity or "observable", and then for all practical purposes instantaneously transformed from a superposition of many possible states into a single eigenstate.

The essence of the measurement problem is that the projection postulate and its measurement-induced reduction (collapse) of the wave function conflicts with Schrödinger's equation, which prescribes the continuous deterministic unitary evolution of states rather than sudden, seemingly inexplicable punctuations. As mandated by the uncertainty principle, QM is probabilistic; between measurements, physical systems "exist" not as definite, directly observable states and state-transition events, but as undetectable and therefore unphysical superpositions of *possible* states.

These probabilistic superpositions, which are not even described by classical probability theory but require a "quantum" theory of probability all their own, must alternate with the definite outcomes of physical measurement events themselves. This implies that vast, spatially extended potentials coinciding with distant points of space, possible future states of a physical system which have not yet been actualized, are carried outward from quantum events at the speed of light only to be instantaneously confined to more or less precise locations.

This problem, which is connected to complementarity and wave-particle duality, is viewed as the central problem of QM. More than anything else, the measurement problem drives the "QM interpretation industry" of modern physics and philosophy, the successor of QM itself with respect to a considerable number of academic publications.

The QM Interpretation Problem

QM began as a mathematical theory of physical measurement. Considered in isolation, it is a rather skeletal affair consisting primarily of linear algebra along with other kinds of mathematics related to it by inspiration and convenience. Like all mathematics, its formal expression is abstract and symbolic. It can appear stark and intimidating when considered apart from the many experimental contexts to which it is applied, but for mathematics this can hardly be considered damning. The real problem has to do with the fact that it is difficult to motivate in terms of the small-scale measurements it describes. Such measurements require QM precisely because more intuitive macroscopic descriptions fail to work.

This suggests a sartorial analogy. On the bare-bones algebraic mannequin of QM, layer upon layer of clothing has been draped and piled by designers who differ strongly in their opinions of what looks good on it. To put it mildly, their visions clash with even less appeal than the underlying skeleton, tending to cancel each other. Consequently, even after their disorganized attempts to bury it under a mound of conceptual raiment, its bones poke through as starkly as ever, dangling and jutting like the girders and cranes of an unfinished skyscraper. Thus, while grudgingly praised for its spectacular empirical success, it continues to be roundly panned for its abominable aesthetics, prompting various efforts to "dress it up" in new interpretations that on close examination turn out to be equally counterintuitive and unappealing.

An interpretation of a theory is a "structure-preserving" correspondence or mapping between a theory regarded as the formal domain of the mapping, and a range or codomain consisting of a universe in which the theory is instantiated (i.e., in which it has instances consisting of specific objects, relationships, and processes that conform to it). The oxymoronic concept of "literal interpretation" notwithstanding, interpretation is a necessary stage in the recognition of any theory. No theory can be understood in terms of uninterpreted symbols, and the meanings of its constituents are dependent on the interpretative context in which it is defined. That such context must be provided in order to determine the "intrinsic structure" of a theory suggests that even when it is assumed that a given interpretative mapping will not change the theory to which it is applied, there is in fact a potential for theoretical structure to be changed.

The intrinsic structure of a theory can be at least partially clarified by *formalization*. A theory can be formalized by interpreting it in one or more well-defined structures including the language in which it is expressed along with any axioms and rules of inference supporting its descriptive or normative functionality. But while this definitely helps limit ambiguity, these formal structures – languages and axiomatic systems – may contain ambiguities of their own. Try as we might to nail everything down, theories do not always uniquely determine their universes, models, or interpretations, and universes do not always uniquely instantiate theories. (For purposes of orientation, the logical principles usually associated with interpretative variability include the *Löwenheim–Skolem theorem* and the *Duhem–Quine thesis*.)

Technically, interpretations of QM are correspondence mappings of which the formal domain is always the theory of QM, and the universe is always empirical or "physical" reality including the set of microscopic measurement events. As for the domain, QM is a theory that comes with principles and postulates that make it a formal system. However, like any formal system, it does not come packaged with a model, i.e., a valid correspondence between itself and any particular universe or set of instances. It merely refers generically to empirical reality, providing a general prescription for the execution and analysis of measurements of submicroscopic phenomena. Nevertheless, QM may be considered to intersect with empirical reality in precisely the measurement outcomes that it correctly (statistically) predicts.

The problem is that this intersection is only partial. Although the degree of correspondence between QM and the measurement events in which it is interpreted is often impressive, it is merely probabilistic; QM underdetermines its content and thus exhibits causal deficits. In particular, QM does not include the level of causation that predicts the occurrence of measurement events or determines their specific outcomes. Even worse, most of the complex mathematical apparatus of QM has nowhere to go; the empirical universe contains nothing that obviously corresponds to it and thus has "nowhere to put it". There appears to be nothing in empirical (observable) reality capable of supporting such things as probability waves and the equations that govern them.

In keeping with the scientific method, a scientific theory and its empirical universe absorb each other through their structural correspondence in an inevitable process of mutual contextualization and accommodation. Whereas a formal system isolates theory from universe, a theory cannot be isolated from any universe to which it is actually applied. This is especially true for any scientific theory intended to describe the empirical universe as it is progressively revealed to the human mind and senses. In establishing a "structure-preserving" correspondence between theory and universe, we are forced to deal with a joint structure that evolves as they feed back on each other as prescribed by the scientific method. For QM, this joint structure is limited to bare measurement events and their QM-predicted observable consequences; the rest of QM is either excluded, or empirical reality is interpretatively embellished with extra ingredients designed to accommodate it.

In other words, the description of the empirical universe changes as new observations are made, and the scientific method requires constant feedback between theory and universe in order to ensure a good descriptive / instantial fit. However, explanatory gaps and creative slack in this process of mutual accommodation can create holes through which one could drive a truck, so to speak. When this occurs, physicists are often completely unconstrained in making up new, empirically untestable "physical" structures and processes to which problematical QM ingredients can be mapped, and/or adding to or subtracting ingredients to/from QM in order to fit it to their preferred descriptions of the universe. (E.g., von Neumann implicitly adds to reality the physically undefined concepts of mind and consciousness in order to explain wave function collapse, Everett adds an unobservable cosmic wave function that generates countless totally unobservable alternate universes, de Broglie and Bohm add unobservable "pilot waves" that guide particles along linear localistic trajectories through nonlocal pilot fields, and so on.)

While QM is formulated in hard mathematical language, the empirical universe it describes is known strictly by direct observation and logical deduction. The only sure way to characterize the physical universe is by the minimal description of physical observations with no inferences but those obtained by plugging bare observational data into deductive logic. With respect to QM, this leads to a problem: there is no sure way to describe the physical universe that can fully accommodate QM, which contains ingredients of which the empirical universe appears to contain no observable instances.

If there were such a description, then QM could be at least partially coupled with it in an appropriate language, pinning down both ends of the model-theoretic correspondence and thereby restricting the interpretation of QM.

Formal Quantization

Physics involves an operation called "quantization" which involves the division of physical properties and substances into their smallest possible discrete units or instances. For example, chemical compounds can be quantized in terms of molecules, molecules can be quantized as atoms, atoms can be quantized as protons, neutrons, and electrons, and the frequency of a standing wave like an atomic orbital can be quantized in terms of its ability to accommodate only a whole number of wavelengths without self-destructive interference. In quantum mechanics, the property which is quantized is called "action", defined as energy multiplied by time (making it both energetic and time-like) and roughly synonymous with "physical change". Many other physical properties can be quantized in terms of it.

However, quantization applies just as well to other kinds of property, including those which are purely mathematical. As physics is expressed in terms of various mathematical formalisms, it relies on the quantization of mathematical concepts well before it arrives at quantum mechanics. Of course, formal quantization is a well-recognized mathematical necessity. Sets are quantized as elements, topological spaces are quantized as points, geometry is quantized as lines, which are quantized as points and units of length, and angles, which are quantized as radians or degrees. More generally, any kind of formal system is quantized as symbols representing objects, relationships, functions, and operations. These symbolic "quanta" characterize the signature of the system; every symbol in the system must conform to one of these descriptors (including typographical symbols including empty spaces), and to each a degree of coherence is assigned.

The coherence of a symbol is what enables it to have a definite meaning and to be treated as a single unified entity. The coherence property is crucial; it means that anything possessing it can be treated as a unitary entity which behaves or transforms in a unified and regular way under certain mathematical or physical operations. Defined in terms of the coherence property, quantization means "division of a coherent identity into coherent subidentities which act as unitary entities and thus behave coherently". (This holds true in QM, where the coherence of a quantum wave function means that all of the possible states of the associated physical system cohere in mutual superposition and evolve in phase with each other.)

Unfortunately, there are problems with mathematical quantization, and they carry over into physics. This is easy to see in the case of a classical manifold, basically a space consisting of zero-dimensional (0D) points (real numbers, elements of the real continuum R^n) and equipped with a metric that is "locally Euclidean", permitting a reasonable in-frame notion of distance and locality.

Immediately we detect a paradox: "of zero extent in a given space" means "nonexistent in that space" – *existence* in a space means taking up space in it – and we cannot assert the existence of a space consisting of *nonexistent* points that take up no space at all. Even if we could, there would be yet another problem associated with continuity, the *adjacency paradox*. An infinitesimal line element or increment of linear motion must relocate a point-object from one point to an adjacent point. But where points are zero-dimensional as continuity demands, adjacency or "being in mutual contact" effectively identifies them. Adjacent points simply merge, and no relocation can occur. Linear motion is out of the question.

It follows that continuous motion requires finite termination or bounding of the interval in order to scale and sum infinitesimal increments, produce a definite integral, and assign a length to the interval. But this still leaves non-zero (albeit sub-finite) separations between each pair of successive points along a path or physical trajectory, and an object must "jump out of" the manifold or space in order to get from one point to the next. Motion then becomes a series of infinitesimal "quantum jumps" through hyperspace. This, of course, is not what mathematicians and physicists typically have in mind when they talk about "the continuous (differentiable, smooth) motion of objects or waves through the continuum". (We ignore for now the various workarounds that have been proposed for these problems, at least one of which – usually the Cauchy–Weierstrass epsilon-delta formalism based on infinite converging sequences – is generally invoked in introductory calculus courses in order to deflect the "problem / paradox of infinitesimals", which is never satisfactorily resolved with respect to the existence or nonexistence of 0D points and infinitesimal intervals between them.)

Naturally, this problem, being deeply rooted in the foundations of mathematics, was bound to emerge in the course of formulating and following mathematical procedures in physical calculations. Specifically, it was bound to emerge in connection with submicroscopic measurement procedures.

QMM Metaformalization

Quantum Metamechanics or **QMM** is a "meta-mapping" that maps entire QM interpretations into the absolute structure of reality, and depending on their logical consistency, embeds them there. Only thus can the most valuable features of various QM interpretations be merged in a single coherent overall description of reality.

The formalization of a theory T amounts to the addition of axioms and rules of inference to a formal language L accommodating the expression and development of T, thus embedding T in L. Where the syntax and grammar of L amount to the "axioms and rules of inference" of L, the axioms and rules of T amount to an extension of the syntax and grammar of L, making T a special-purpose "sublanguage" of L. Where T = QMM and L = the supertautological Metaformal System (Langan, 2018), this process amounts to "QMM Metaformalization". Because the intrinsic-language structure of the Metaformal System suffices to determine QMM, additional (QMM) axioms and rules of inference are unnecessary.

QMM metaformalization requires metaformalization of the generic QM-interpretative (QMI) mapping QMI: QM $\leftrightarrow$ U, the domain and codomain of which are QM and empirical reality respectively. The problem with the generic QMI mapping is that thanks to the fertile imaginations of various scientists and philosophers, empirical reality has "overrun its buffers" in various conflicting ways and thus lacks a coherent formulation of its own, with the result that interpreters of QM are basically willing their preferred versions of reality into hypothetical existence as they please and calling it "ontological". No QM interpretation incorporating what may be an incoherent description of empirical reality can be judged trustworthy.

In contrast, QMM maps QM into empirical reality as represented by the CTMU. Thus, QMM is not just another interpretation of QM, but a metaformal extension of QM which specifies the absolute structure of its codomain and is itself embedded therein. Whereas ordinary QM interpretations speculatively modify the "reality" concept in order to solve or circumvent the measurement problem, QMM interprets QM in the supertautological Metaformal System.

Having itself been metaformalized, QMM can thus be described as the "metaformalization" of QM. This amounts to a true and inevitable extension of the QM formalism to incorporate a true metaformal ontology. It is in this extended structure that ordinary interpretations of QM, including their "ontological" claims, must themselves be interpreted.

The codomain and domain of QMM mapping may be described as follows.

Specifying the QMM Domain

Just as for ordinary QM interpretations, the formal domain of the higher-level QMM mapping includes QM. However, this requires a qualification: the co-domain now consists of the CTMU supertautology, a metaformal ontic identity that equates existence to its own self-identification and thus couples ontology and epistemology. This affords a crucial advantage: being logically induced from the intelligibility of reality, the supertautology requires no additional proof and immediately qualifies the mapping as a true "quantum ontology". But it also comes with what may seem a disadvantage: the codomain can no longer unconditionally accommodate mere speculations, instead holding them in the domain along with QM. The domain now contains the entire QMI mapping, no extra features of which can accompany QM into the codomain unless they are consistent with the embedment of QM itself in the CTMU supertautology.

As for the usual formalism of QM, it can be described in terms of a variable axiomatic structure which is to some extent a matter of preference. But while the core concepts of QM can be differently organized, certain mathematical ingredients are indispensable. This allows us to describe the formal structure of QM with a set of postulates similar to the following:

1. The state (vector) of a quantum mechanical system, including all the information that can be known about it, is represented mathematically by a normalized ket $|\psi\rangle$. At each instant, this ket represents the state of a physical system in the space of states, a vector space called *Hilbert space*. Associated with the system is a *wave function* which, unlike a particle, extends throughout space and consists of possibilities in mutual superposition.

2. A physical observable is represented mathematically by an operator A that acts on kets. When operating on a wavefunction with a definite value a_n of that observable, it yields that value times the wavefunction: $\{A|\psi\rangle = a_n|\psi\rangle$.

3. The only possible result of a measurement of an observable (physical) property is one of the eigenvalues a_n of the corresponding operator A.

4. The probability of obtaining the eigenvalue a_n in a measurement of the observable A on the system in the state $|\psi\rangle$ is $\text{Prob}(a_n) = |\langle a_n|\psi\rangle|^2$, where $|a_n\rangle$ is the normalized eigenvector of A corresponding to the eigenvalue a_n.

5. After a measurement of A that yields the result (eigenvalue) a_n, the quantum system is in a new state that is the normalized projection of the original system ket onto the ket (or kets) corresponding to the result of the measurement:

$$|\psi'\rangle = \frac{P_n|\psi\rangle}{\sqrt{\langle\psi|P_n|\psi\rangle}}$$

[This is a particular formulation of the *Projection Postulate*, which is central to the measurement problem.]

6. The time evolution of a quantum system is determined by the Hamiltonian or total energy operator H(t) through the time-dependent Schrödinger equation $i\hbar \frac{\partial}{\partial t}|\psi(t)\rangle = H(t)|\psi(t)\rangle$

Note that the mathematical definitions and formulae in these postulates are formal and "syntactic" in character, conforming to generic and often mathematical rules which exist in the cognitive-perceptual syntax of observers ... the accepting syntax through which they recognize and absorb cognitive and observational input, supporting and constraining cognition and perception. As these rules are being projected on that part of physical reality consisting of submicroscopic events which would otherwise be unidentifiable as physical phenomena, they amount to features of the physical medium itself. Moreover, because physical reality exists on the limiting submicroscopic scale and all coarser scales, they are being ascribed to physical reality, and where physical reality is an aspect of reality at large, to reality as a whole.

Standard interpretations of QM are targeted on the same reality as that to which QM itself applies. Hence, their "extra" ingredients map to the same overall structure as does QM itself.

Specifying the QMM Codomain

Reality, the target of the partial interpretative mapping of QM which various interpretations supposedly "complete", has always been hard to define. From academically cloistered quasi-robotic gray men pushing symbols from equation to equation on dusty blackboards to Buddhist monks chanting together in their saffron robes, the vast majority who have tried have come up short. Reality displays immense complexity, and one tends to either become lost in the tangle or throw up one's hands in surrender. As much insight as QM brings to the reality-theoretic table, it is also very strongly affected by this problem. One cannot establish the true correspondence of QM with reality when, having used calculus and linear algebra and Fourier analysis to put the source of the mapping in the crosshairs, one still cannot positively identify the target.

Before QM can be used in the solution of metaphysical problems, it must be integrated with reality on the highest, most general, most inarguable level of structure and dynamics. For this to happen, reality must first be conclusively identified on the required level of description. Popular ignorance notwithstanding, this mission has been accomplished; reality has been succinctly described as a primal self-identification operator which reflexively identifies itself and thereby attributes existence to itself, functioning "trialically" as an object, a self-relationship, and a self-operation. In order to do this, the operator – herein denoted by the acronym G.O.D. or "Global-Operator-Descriptor" – uses the most generic level of its own self-attributed being as ontic potential, generatively actualizing itself from that potential in the form required for existential self-identification and scientific intelligibility. Specifically, it takes the form of a super-tautological intrinsic language which has been described in these Proceedings as the **Metaformal System**.

The Metaformal System $M = \{\Sigma (N, T), \Gamma_\mu, S_\Sigma\}$ is a supertautological *intrinsic language* characterized by complete self-containment. That is, its self-dual signature Σ consists of a nonterminal metaphysical domain N and a physically emergent terminal domain T, and its generative grammar Γ_μ controls the N | T relationship, producing the timelike strings (histories, paths, trajectories) S_Σ of T from the nonterminals of N. As an ontic identity language, M factors into dual *semilanguages* Ls and Lo which comprise its intensional and extensional aspects respectively. Although the definition of M strongly resembles that of an ordinary formal language with a signature, a grammar, and a set of linear "strings", it is a trialic metaformal language which evolves by modeling itself in its own intrinsic universe.

The fundamental objects of M are active signs called *telors* (telic identification operators), secondary quanta existing in N along with the primary quantum or G.O.D. representing the entire system, and *syntactors* (syntactic identification operators), including tertiary syntactors comprising the subatomic particles of the physical domain T, which depend on more complex but nonetheless coherent self-modeling telors for their regenerative existence.

Its μ-morphic grammar Γ_μ is a self-identification operation which produces $S_\Sigma \supset T$, the set of terminal strings of M – the external states and linear trajectories of tertiary syntactors in the terminal domain T comprising the "surface structure" of M – by generating them in the pregeometric domain of N and projecting them onto physical timelines. Thus, M evolves by generatively identifying itself.

Conveniently enough for present purposes, the supertautology M is *self-quantizing*. It is an ontic identity – a coherent, intension | extension coupling formulated with respect to the attribution of being and identity – which naturally partitions and "multiplexes" itself by its intrinsic μ-morphic grammar Γ_μ into coherent subidentities (active signs) that mirror its essential structure and can thus function coherently as its generative operators and/or physical content. Its self-quantization is both syntactic, applying to the formal level of being that controls identification, grammar, and orthography or well-formedness in T, and extensional or "semantic" (substantive, physical), forming meaningful configurations consistent with not just the syntax of intelligibility, i.e., the self-distributed rules of G.O.D. self-identification, but with the state of the external world. As a self-defining entity, the ontic identity must abut its logical complement *in situ*, permitting it to distinguish self from nonself.

The metaformal trialic self-quantization of the G.O.D. devolves to an *identity system* consisting of states coinciding with self-identification events that distinguish between self and nonself. The system generates at least three levels of identity: the primary or global level (the G.O.D.), the tertiary level (the level of ultimate and near-ultimate constituents of matter as already partially addressed by QM), and an intermediate level resolving the causal deficits of the primary and tertiary levels. This is the secondary mesoscopic level of identity, the classical scale occupied by various kinds and levels of secondary telor including human beings. Thus, just as the G.O.D. is the primary metaformal quantum of reality and elementary particles are tertiary quanta, human beings are "secondary quanta" with a combination of coherence and complexity that allows them to freely and meaningfully "self-model", configuring reality by configuring themselves. In other words, in the Metaformal System, life and consciousness are specifically quantized as innately coherent secondary telors whose coherent existence surpasses their physical emergence.

Like tertiary id-quanta – the submicroscopic, physically irreducible identities whose state transitions are quantized as action – secondary quanta are coherent identities mirroring the structure of the G.O.D. But unlike tertiary quanta, which are detectably coordinated by nothing but localistic fundamental forces and are otherwise seemingly probabilistic, they have sufficient complexity and nonlocal integrity to generatively "self-model" on behalf of the larger reality they populate. Just as envisioned by theorists like von Neumann, Wheeler, and Stapp, the self-realizing "conscious minds" of secondary quanta play a crucial role in filling QM causal deficits. They are among the key missing ingredients of standard quantum mechanics ... primitive elements of the Metaformal System directly and indirectly responsible for the coordination and collapse of quantum wave functions, and thus for completing physical identification processes that cannot be completed by QM alone.

This being understood, QMM maps QM directly into the structure and dynamics of the ontic identity in a most unequivocal way.

The QMM ↔ Reality Correspondence

As a trialic intrinsic language, the CTMU Metaformal System M combines *language*, *universe*, and *model* to create a perfectly self-contained metaphysical identity. The intensional aspect of M is a self-configuring self-processing language, and the extensional aspect couples to this language as a pointwise distribution of its syntax which provides the language with instances. This intrinsic language is self-similar in the sense that it is generated within a formal identity to which every part of it is mapped as content; its initial form, or grammatical "start symbol" – herein discussed at various levels of resolution as the ontic identity, the G.O.D., and the Metaformal System – everywhere describes it on all scales. In this system, time, causation, and the spatial expansion of the cosmos as a function of time flow in both directions – inward and outward, forward and backward – in a dual formulation of causality characterizing a new conceptualization of nature embodied in a new kind of medium or "manifold".

The CTMU conspansive manifold, conceived as a joint medium for QM and General Relativity, differs from a classical manifold in important respects, several of which are geometrically transparent.

1. Whereas a classical manifold consists of points which exist independently of their transient (moving) content – the points are static parameters rather than the states they contain – the points of a conspansive manifold are trialic, which means that they are content. The state of a particle (tertiary syntactor) in the manifold "inner expands" to become an open potential which engenders and contains the next state. (Insofar as these syntactors are coupled in mutual state-transition events, the conspansive manifold superficially resembles the conventional pseudo-Riemannian spacetime manifold which has "events" as its points.)
2. The conspansive manifold contains three topologically nested levels of "point" corresponding to three levels of active sign in the signature of M: the primary telor, secondary telors, and tertiary syntactors, each instantiating its own level of metaformal quantization. Each kind of point has internal and external states. Points are rescalable by inner expansion and collapse.
3. Simplistically (nonrelativistically), the conspansive manifold can be "layered" in terms of the terminal and nonterminal subsignatures T and N of the signature Σ of M, consisting respectively of (a) current states, including those in T itself, which have occurred but have not yet been succeeded by newer states; and (b) past states deeper in N which have already been succeeded by newer states. The current layer includes the set of newly collapsed external states paired in mutual identification events that are occurring "right now/in the present" (ignoring for now the matter of simultaneity of events), and the set of "open" (still current) states which have not yet collapsed. The second, deeper layer consists of states which have already been replaced by newer states. Because states can never leave the overall manifold or primary point (as there is nowhere else for them to go), they remain in a state of *extended superposition* and are outwardly rescaled even after they have been replaced by new states and receded into the past, nesting and progressively interpenetrating as the manifold evolves (in contrast to the static extension of a "block world"). This "inner expansion" of the manifold, equating to the relative shrinkage of its content, is to be viewed in terms of rescaling morphisms rather than ordinary expansion and shrinkage. Thus, the entire manifold and its points are dynamically and in fact generatively coupled, changing in unison.

The generative dynamic of the conspansive manifold has primary and secondary stages. The primary stage consists of a self-dual n-ary (n-fold unary, $n \geq 2$) operation, *conspansion*, with two alternating phases, *inner expansion* and *collapse*. Inner expansion *potentializes* the state transitions of the syntactors coupled in events – the points and events are brought into intersection by internal rescaling through the primary quantum (point) so that they "overlap" – while collapse re-actualizes the inner-expanded syntactors as compact objects in new interactions. Conspansion requires the progressive rescaling of points with respect to the manifold as a whole, leading to the apparent overall expansion of the manifold with respect to its content. The conspansive manifold thus evolves analogously to the physical cosmos, with intrinsic effects analogous to cosmic expansion, propagation of the wave function, and quantum wave function collapse.

The conspansive manifold evolves by way of generative information mappings that supersede standard physical causation occurring along timelike (or null) worldlines. Inner-expansive potentialization "opens" a tertiary identity or point of the manifold to N while collapsative actualization "closes" it in T; potentialization is null, while collapse is spacelike. Together, inner expansion and collapse comprise conspansive potentialization-actualization cycles which form self-dual information mappings, each of which initiates and actualizes a potential by restricting it to a specific outcome for a net gain of information. It is through these generative, metaphysical prereality-to-physical reality information mappings that the Metaformal System models and defines itself.

The evolution of the conspansive manifold is that of M itself. By way of conspansion, the generative grammar Γ_μ of M produces the extensional (physical, geometric, topological) semilanguage Lo from its dual intensional semilanguage Ls. As QMM maps QM into the conspansive manifold, it thus induces an extension of QM into Γ_μ and Ls. Driving Γ_μ is a secondary stage of evolution of the conspansive manifold associated with primary and secondary quanta (the coherent higher-order points of the manifold) and called *telic recursion*. By telic recursion, adaptive exploratory feedback between Ls and Lo generates the terminal expressions of M, coupling and collapsing syntactors in interactive mutual identification events including the measurement events of QM.

QM is thus mapped to the open top layer of the conspansive manifold, where inner expansion is approximated by the symmetric time-like propagation of the wave function according to the time-dependent Schrödinger equation (bearing in mind that $\Psi(x)$ and the Schrödinger equation must ultimately be formulated in terms of the point-structure and topological dynamic of the manifold), and conspansive collapse is thus analogous to wave function reduction (collapse, projection). The irreducible characterization of a QM measurement like $\langle\psi_i|\psi\rangle$, in which a present or prepared state ψ "casts its shadow" on vectors representing possible next states, implies that collapse is just the conspansive rescaling of inner-expanded static potentials: $A|\psi\rangle \rightarrow a_i|\psi_i\rangle$. This reflects the topological dynamic of the conspansive manifold, in which both phases of conspansion regulate and transform its point-structure. Adding a metaphorical touch to the implicit anthropic coupling of observer and universe, one might observe that this causes the manifold to "breathe". The conspansive manifold thus mirrors quantum dynamics, rescaling, combining, and collapsing quantum wave functions along with its own points in a fully coordinated fashion.

Self-Modeling

Like M itself, the conspansive manifold evolves by "self-modeling", i.e., by letting active Ls semantic potentials "inside the points", their internal states, guide and coordinate wave function collapse to produce the physical content of Lo "outside the syntactors" as external states (this can be likened to a directed form of "decoherence" replacing the more familiar random kind). In terms of information: where the manifold is a pointwise distribution of M-syntax, collapse is a spacelike information mapping which achieves specificity by the semantic reduction of inner-expansive scope or extent. QM indeterminacy and probabilism allows this process the freedom required by its innate generativity.

In the conspansive manifold, the coordinated self-modeling of primary and secondary quanta flows "retrocausally" outward and pastward from the advanced semilanguage Ls into the retarded semilanguage Lo. The content of the flow is determined by the telic-recursive feedback of Ls, the dynamical semilanguage of M, and Lo the "static" semilanguage of M which is already bound in the past and thus parameterizes Ls. But even as determinacy flows from future to past, the objects through which it flows are propelled along timelike gradients from past to future by retarded causation, a linearized approximation of the true metacausal dynamic. The true dynamic is not located in T at all; T is merely what the true dynamic produces as output, and reality can thus be described as a kind of "self-simulation".

The linear semimodel of T – the collapsed "pixels" of the terminal "display space" to which classical physics is confined – affords access to only a localistic, linearized correlate of true causation, a mere projection of the real metacausal dynamics of M. It is on this surficially displayed causal simulation, the M-subsignature $T \subset \Sigma$, that true quantum reality is projected from the non-terminal domain $N \subset \Sigma$, thereby literally casting the shadows which seem to move across the walls of Plato's dank and dimly lit cave.

Telesis

Strictly speaking, the "action" of Γ_μ is not energetic, but telic and informational in nature; it effects the punctuated redistribution of energy as medium-content relationships are generatively redefined and emerge into the terminal domain

T. Where energy is confined and conserved in T by definition, its longstanding function as the ultimate reduction of reality – a role it often fills in modern physics, and arguably in the history of physics as *vis viva* – requires some amount of adjustment, especially when moving away from T and deeper into the nonterminal domain. In the deep structure of M, the physical quantity *energy* must be reductively generalized to a protean "metasubstance", *telesis*, which intrinsically determines its own properties and composition by unbinding and binding itself in conspansive potentialization-actualization cycles, and of which physical energy is just the conservative orthogonal restriction to T.

Telesis, a metaphysical generalization of both energy and volition, can be described as "self-actualizing self-potential" which generates medium-content relationships. According to Heisenberg, quantum objects are not so much hard bits matter as "probabilistic tendencies" or *potentiae* which "stand in the middle between the idea of an event and the actual event, a strange kind of physical reality just in the middle between possibility and reality" (Heisenberg, 1958). By process of elimination, Heisenberg could only have been talking about telesis

In CTMU terms, Heisenberg's quantum potentiae coincide with tertiary syntactors (point-quanta) of the conspansive manifold that are inner expanding due to the generative self-action of telesis while interfacing between the nonterminal and terminal domains N and T. Thus, their actualization, amounting to quantum wave function collapse, is the mechanism of emergence of T from N, and of Lo from Ls.

Generativity

The Metaformal System M is an identification system consisting of an ontic identity with many subidentities which all identify each other, passively recognizing and actively transforming each other. By definition, generativity characterizes the evolution of any identification system not governed by timelike laws of causation relating a preexisting array or background medium to its content. Law, medium, and content must all be determined together in order for identification to occur, and because this amounts to a determination of causation itself, the process is by definition "metacausal" in nature.

In classical mechanics, a closure principle is assumed under which causation must be defined on the medium and content of nature, with outside factors excluded. For example, (1) a fixed law is induced – e.g., Newton's second law F=ma – which is intended to capture some aspect of the generic medium-content relationship, (2) a specific distribution of content in the medium is fed to this law as input (the "cause"), and (3) the law converts the causal input to output (the "effect"). Classical mechanics is thus ruled by "causal efficacy", in which cause determines effect by timelike laws which relate the medium of nature to its physical content. (Although we have deliberately chosen a very simple example in F=ma, it characterizes physical reasoning in general.)

The CTMU replaces the classical closure assumption with an ontic closure principle (ARC, or Analytic Reality Closure) under which the entire relationship between medium and output is self-selected from unlimited possibilities in the self-potentialization of the ontic identity and its subidentities. Rather than merely relating cause and effect (causal input and output states) using a timelike law within a fixed array, metacausal functions relate entire medium-content relationships within the conspansive manifold. Not only are cause and effect coupled in mutual dependency, but so is the law by which their mutual determination emerges. This mirrors the intrinsic dynamic of the generative grammar of M (and for that matter generative grammar in general), wherein any given expression and its grammatical abbreviation, its "start symbol", must be determined before the grammatical derivation of the expression can proceed.

In generative evolution, the exact relationship between a fixed array and its content is fundamentally indeterminate precisely because there is no fixed medium independent of its content, and paths cannot be defined until the dependency relationship has been generated. Inner expansion turns terminal states in T into a "typographical array" in which the next states can be written (Langan, 2017, 2018), unbinding and opening the quantum metric. For each new event, the entire relationship must be regenerated along with the "law" describing that relationship before a trajectory can be determined. Such a relationship is called a *telon*, the self-configuration of an active sign of Σ called a *telor* (self-configurable telic identification operator), i.e., a secondary and/or the primary quantum of QMM. Thus, primary and secondary telic quanta are the true source of quantum dynamics.

Conspansive Duality: Distributed and Linear Morphisms

The inner expansion-collapse cycles of the conspansive manifold are "distributed endomorphic" and "distributed ectomorphic". A distributed endomorphism can be approximately visualized as a sphere collapsing to an interior point, while a distributed ectomorphism can be approximately visualized as a point expanding into a sphere. (Such morphisms are defined as "hological", preserving essential structure on all scales.) On the other hand, temporal sequences of the successive collapsed states of objects – the strings of S_Σ – are "linear ectomorphic" in both directions, with objects traveling along linear paths or trajectories. This is called the *linear ectomorphic semimodel* of the conspansive manifold, associated with "display space" T.

Standard physics, including spacetime, is linear-ectomorphic. In a linear ectomorphism, an object leaves or enters a point of a manifold along a linear trajectory. (We have already seen that this leads directly to the adjacency paradox, forcing objects in motion to exit the manifold.) In the inner expansion-collapse cycles of the conspansive manifold, on the other hand, a telic point repeatedly "self-factorizes" into a self-nonself content-medium relationship under the guidance of one or more telons in such a way that an incremental linear displacement amounts to a conspansive cycle consisting of the inner-expansion of an initial state (the "medium") and collapse of its successor therein. Terminal content never leaves any point in its history, but merely *contracts within it*, projecting onto a timeline to define an ectomorphic interval.

Spacetime too is ectomorphic, consisting of points called "events" which are specified by four coordinates, three of space and one of time, that are separated by spacetime geodesics called *worldlines*. Spacetime can be overlaid on a continuously collapsed idealization of the terminal point-set T of the conspansive manifold, the points of which are fully collapsed tertiary syntactors already conveniently coupled in mutual identification events. Just "below" T, and interspersed with T, is the top level of N, in which inner-expanded but as-yet uncollapsed points of the manifold lie open, comprising what appears to be "empty space". This is where QM belongs, in the open states of the conspansive manifold. Below this open top stratum of N lies the deep structure of M including secondary quanta, the metacausal influence of which resolves the causal deficits associated with QM potentiae (open tertiary syntactors).

In short, spacetime is just a kind of "ectomorphic dual" of the conspansive manifold. Spacetime approximates T in the sense that objects "move" by skipping along timelike gradients like stones on the surface of a pond, their paths effectively interpolated between points generated on the surface. But unlike spacetime, the *surface itself* is regenerated with each skip of the "stone" or tertiary identity, and while spacetime can only confine its evolution to an ectomorphic scenario devoid of any extrinsic pregeometric background, T resides on an *intrinsic* background, the nonterminal domain N. T is thus adjoined to deeper structure supporting teleodynamics, which cannot reside on the surface of the manifold and is not actually supported there.

Due to these and other limitations of spacetime and the classical (Cartesian coordinate-space) model of classical physics from which spacetime evolved, QM cannot be fully modeled there, and its resulting homelessness has engendered various ontologically unsound interpretative modifications of empirical reality. QMM therefore maps potentiae to the open points of the top nonterminal stratum of the conspansive manifold, thus providing QM with a home at last.

Examples of QMM Meta-interpretation

Here we give a very brief account of the application of QMM to several well-known QM interpretations, particularly those which have been mentioned in the forum discussions of *Foundations of Mind*. Please bear in mind that all mainstream interpretations of quantum mechanics are confined to the linear semimodel of M and require adjustment in order to conform to the conspansive manifold. These interpretations will be treated as tersely described in Nick

Herbert's respected and highly readable classic *Quantum Reality: Beyond the New Physics* (Herbert, 1985).

Copenhagen (Bohr and Heisenberg): *There is no deep reality.* While quantum indeterminacy leaves ample room for generativity, making use of it would require acknowledgement of the nonterminal (and non-physical) domain N and the deep structure of M, which this interpretation explicitly denies. This is unfortunate, as Heisenberg erred in placing his potentiae "between the idea of an event and the actual event, [in] a strange kind of *physical* reality just in the middle between possibility and reality" (Heisenberg 1958, page 41), thus explicitly calling ideas and possibilities "physical" even when they are *physically unrealized* and therefore *not* physically real. This, of course, is semantically inconsistent; if a phrase like "physically unrealized ideas and possibilities" has any content at all, it cannot be physical in nature. We may thus infer, for the sake of consistency, that potentiae are metaphysical, which means that QM is either metaphysical and thus reliant on the metaphysical structure of reality, or merely physical and therefore needful of augmentation by deep reality in order to explain how reality identifies itself. Mere probabilistic tendencies are by definition inadequate to determine individual state transitions, and even if they are included in QM, something more is required in order to account for the self-identification of reality. It follows that the statement "there is no deeper reality than QM itself" cannot be mapped into the supertautology, and must therefore be excluded from our understanding of reality.

Observer-Created Reality (John Wheeler): *Reality comes into being through the observations of observer-participants.* Reality corresponds to the metaformal supertautology (or to its physical domain $T \subset \Sigma$), creation is mapped to generative-grammatical production by Γ_μ, and observation is mapped to secondary and tertiary identification events (i.e., to quantum measurements, and generically, to bare tertiary interactions). To the extent of its description, Wheeler's interpretation meets the standards of QMM; reality can indeed be generated by teloric observers.

Consciousness-Dependent Reality (von Neumann–Wigner–Stapp): *Consciousness creates reality.* Insofar as the generic definition of "consciousness" overlaps with the self-identification of the ontic identity and its internal self-images, it passes the test of metaformal consistency for basically the same reasons as does Observer-Created Reality.

This approach merits additional explanation. Associated with the vNWS interpretation is a "process description" of quantum dynamics. Dirac originally noted that there are two ways in which a quantum system evolves: (1) the wave function deterministically explores all possible interactions as it propagates, and (2) a single possibility is randomly actualized. In his book *Mathematical Foundations of Quantum Mechanics* (1932), John von Neumann elaborated on these modes of evolution, observing that two distinct alternating processes are transpiring: Process 1, a non-causal, nondeterministic process in which a measured particle randomly assumes one of the possible eigenstates of an observable property determined by the relationship between the particle and a measuring apparatus, and Process 2, a causal, deterministic process in which the wave function evolves between measurements according to the Schrödinger wave equation.

Process 0 "Some process that is not described by contemporary quantum theory, but that determines what the so-called 'free choice' of the experimenter will actually be."

Process 1 "The basic probing action that partitions a potential continuum of physically described possibilities into a (countable) set of empirically recognizable alternative possibilities."

Process 2 "The orderly mechanically controlled evolution that occurs between interventions."

Process 3 "The process that *selects the outcome*, 'Yes' or 'No', of the probing action. "The choice of nature."

QMM maps Stapp Process 0 to a generative event associated with a secondary telor endowed with free will (generative capacity); Process 1 is mapped to the expansion of the measured system in an eigenbasis of the telonic observable (syntactic or semantic property) in the generative syntax of the telor, priming the system's wave function for collapse; Process 2 is mapped to the underlying telic-recursive process approximated by the Schrödinger equation under ectomorphic confinement to the surface of the conspansive manifold; and Process 3 is mapped to the combined action of the primary and secondary telors on the system, which triggers the collapse. In short, QMM maps all four of Stapp's processes into the conspansive manifold.

Bohmian Mechanics (early David Bohm): *Quantum particles are ordinary objects steered by guide waves in a nonlocal pilot field.* Bohmian mechanics is disallowed by QMM for the following reasons: (1) It is a so-called "realistic" interpretation of QM which holds that reality exists independently of the observer, precluding the crucial dynamical functionality of primary and secondary quanta (including conscious human telors); (2) It is deterministic, thus violating generativity; (3) it is fundamentally dualistic, holding the particle apart from its pilot wave (function) in such a way as to imply ontic inequivalence; and (4) It is often considered to violate the locality principle of classical physics (no superluminal influences) by requiring that the force on a point-particle instantaneously depend on the precise location of many other particles in the universe. Yet it is also widely considered to violate Bell's theorem by incorporating this nonlocal information as "hidden variables" which account for its determinism (d'Espagnat, 1979, 1989).

In the conspansive manifold, problem 4 is at least partially obviated by extended superposition, which distributes information on distant particles to every location within range of their wave functions (the scope of their inner-expanded states). Extended superposition means that no violation of locality is necessary. However, while the pilot field to some extent approximates the extended-superpositional structure of the conspansive manifold, it falsely objectivizes particles by turning them into ordinary objects which compactly persist between linear-ectomorphic state transitions, and thus commits to a form of dualism fundamentally separating medium (the pilot field) from content (the particle). Pilot waves supposedly guide particles, but in order to do so, must be determined in advance of the states of the particles themselves. Thus, field and particle are dynamically as well as structurally distinct. This is inconsistent with conspansion, whereby points of the conspansive manifold inner-expand to become their own media. Triality demands that the particle and its wave field coexist within a single identity in conspansive alternation.

The Implicate Order (late David Bohm): *Reality is an undivided wholeness.* This interpretation is rather nebulous, but if held apart from the insupportable aspects of Bohmian Mechanics, it passes in several important respects. For example, it is explicitly generative; the implicate and explicate orders correspond to Ls and Lo respectively; its wave function entanglements mirror the extended-superpositional structure of the conspansive manifold; and the holomovement (Bohm, 1980) resembles the conspansive evolution of the manifold, including telic-recursive Ls|Lo feedback over a metaformal "semantic network" of wave function entanglements and the terminally restricted flow of advanced metacausal data from Ls to Lo. While it lacks an ontology of its own due to insufficient logical support for its conceptual ingredients, Bohm's "undivided wholeness" qualifies as a limited success by QMM standards (Bohm & Hiley, 1993).

Many worlds (Everett): *Reality is a multiverse of many alternate universes.* In order to circumvent the reduction of the wave function, this interpretation reifies a higher-order wave function, the so-called *universal wave function*, and associates it with physical reality as a whole. Then it lets this vast "meta-quantum" evolve "deterministically", splitting into separate universes at each quantum event.

First, the good news: Everett racked up an impressive QMM score just by proposing the existence of a cosmic wave function. With some adjustment, QMM can map it to the primary quantum of the Metaformal System, i.e., to the ontic identity as a whole. But unfortunately, this is where the correspondence ends, for wave function collapse is a basic feature of the conspansive manifold and cannot be avoided. QMM maps wave functions in general to identities consisting of superpositions of the M-semilanguages Ls and Lo, and these are not merely optional. Secondly, the evolution of semilanguage superpositions is neither temporal nor deterministic but generative, and when it comes to generative events, there is no way for the cosmic superposition of possible universes to see them coming. It could only wait for them to occur and then pretend that elsewhere they didn't, splitting the metaverse into the universe where the event and its outcome have occurred, and an "alternate universe" where they have not. Moreover, it takes more than a cosmic wave function to make an ontology, and now that a proper ontology has been discovered, it is evident that there are ontological criteria that Everett failed to take into account. At the very least, the coming-into-existence of any given alternate universe depends on these neglected criteria.

Quantum field theory (QFT): While QFT is a complex and powerful extension of the formalism of QM, its inclusion here is justified by the fact that it incorporates QM and thus involves some amount of QM interpretation. Simplistically, QFT replaces particles and wave functions with quantum fields as primitive objects, defining the particles as "excitations" of the fields which emerge analogously to wave function collapse while permitting variation in particle numbers. This much is consistent with the structure of the conspansive manifold, at least where tertiary syntactors need persist only for the duration of a single state terminated by generative or annihilative events.

But as the conspansive manifold evolves by conspansive self-dualization in generative potentialization-actualization cycles terminating on interactions of its quantum point-identities, it symmetrically dualizes the relationship, making the field internal to the points just as the points are internal to the field. The quantum fields of QFT, like the elementary particles they replace as fundamental entities, are thereby identified with the points of the manifold, i.e., with tertiary syntactors, and thus equivalently reduced to pointwise syntactic distributions. But while both QM and QFT are confined to the open top layer of N and thus superficially excluded from the deep structure of M, they are now embedded in the conspansive manifold, a "quantum metafield" where physical systems superpose directly on deeper levels of metaphysical structure and dynamics.

Lastly, it is perhaps appropriate to mention the existence of a handful of interpretations involving Lagrangianism (e.g., **the Path Integral formalism** of Richard Feynman), advanced causation (e.g., the **Transactional Interpretation** of John Cramer and its PTI variant by Ruth Kastner), or both at once.

The coupling of Lagrangianism and advanced causation may seem quite natural in the linear-ectomorphic spacetime "dynamic". While Lagrangian mechanics requires that the initial and final states of a moving object be determined before a definite linear path can be calculated using the stationary action principle, an advanced cause serves nicely as a final state. However, in the absence of determinism, it is given by assumption rather than any sort of explanation, and so for the initial state as well. This is no less true in the quantum realm, where classical determinism is out of the question; the initial and final states of a particle must be determined before its path can be determined and used for the ectomorphic transmission of causal influences. This is the case whether causation is considered to run forward or backward, and whether the initial or final point is designated as the cause or effect ... i.e., whether determinacy is retarded or advanced. This spells trouble for retrocausal theories which rely on the linear-ectomorphic transmission of advanced influences along definite linear paths, even if this is wishfully attributed to, e.g., the "back-action" of mind on a pilot field. The first requisite is to establish a generative relationship between minds and fields, and this something that only the conspansive manifold can provide.

In the CTMU, determinacy in either direction of time is superseded by generative metacausation, in which the generative action of telic identity operators gives both the initial state and final state by projecting them onto a timeline from the depths of the conspansive manifold (and which, if desired, can be factored into advanced and retarded components in spacetime despite the inadequacy of spacetime alone for mental causation).

All QM interpretations are subject to this kind of analysis, some for better and others for worse. Because QMM stands on the CTMU Metaformal System and its logical-inductive groundwork, it is unconditionally validated by the intelligibility of reality and cannot be falsified by empirical induction. In short, just as with the Metaformal System itself, there is no way out of it.

Summary

We have explained how QMM maps certain key concepts of the QM formalism, along with several popular interpretations of QM, into the supertautological CTMU Metaformal System, a true ontological metalanguage formulated in such a way as to logically support reflexive attributions of existence associated with the high-level self-identification of reality. By virtue of this mapping, QM itself, along with certain hypothetical ingredients attached to it by various interpretations dubiously claiming to provide it with an "ontology", finally have a true ontology against which they can be tested for relevance and consistency, and where those passing the standards of QMM may come to rest.

Concisely, the development can be described as follows. Any "interpretation" of A in B (or vice-versa) is a correspondence C between A and B: C: A $\leftrightarrow$ B. Obviously, both A and B must be defined with some amount of coherence and precision before the correspondence C can be mapped. In any interpretation of QM, A = QM and B = reality as a whole. The problem has been that while the mathematical formalism of QM was well-defined, "reality as a whole" was not, in part because its description changes or has the potential to change with each new scientific theory or experiment. This is no longer the case; reality can now be unequivocally characterized as the supertautological and therefore rationally undeniable CTMU Metaformal System, and the true correspondence can therefore be established. By definition, QMM is that correspondence. All that remains is to apply QMM to the beasts of the quantum jungle, namely, the often strange and apparently wildly conflicting QM interpretations that have been freely proliferating at the frontier where physics meets metaphysics. It has just been demonstrated how this is done, with examples.

Nevertheless, let us explain once more *how* it was done in the clearest and briefest possible way. The CTMU Metaformal System is an intrinsic language, the involutional coupling of a language with a manifold whose points are the elements of the signature of the language, or in semiotic terminology, its "signs". This manifold is dynamic, with dual outward and inward gradients accounting for gravity and the relative linear motions of physical objects (the curvature of spacetime equates to the inner expansive gradient of the conspansive manifold, which is dual to the timelike collapse gradients of T). Its evolution can be described in terms of two operations, *conspansion* and *telic recursion*. By virtue of conspansion, the evolving manifold closely resembles the existing formalism of quantum mechanics; and by virtue of telic recursion, this formalism can be carried into the linguistic aspect of the manifold and the deep structure of the Metaformal System. (For now, we refrain from asserting that the Metaformal System is a "theory of quantum gravity", but this will no doubt eventually emerge.)

Thus, QMM is not to be confused with any mere interpretation of QM, examples of which relate to QMM as input. Rather, QMM is a "metamodel" which carries QM and its various prospective models (speculative interpretations) into the supertautological ontic identity. In short, QMM is what QM must become if it is ever to qualify as a truly reliable source of insight and authority regarding deep metaphysical questions of which the answers require certainty, generativity, and true ontological support.

Of course, there is much more to this story. But owing to space limitations, further details must be set aside for later presentation.

References

Bohm, D. (1980) *Wholeness and the Implicate Order*. London, New York: Routledge.

Bohm, D. & Hiley, B. J. (1993) *The Undivided Universe*. London, New York: Routledge.

Bohr, N. (1928) The Quantum Postulate and the Recent Development of Atomic Theory. *Nature*, Vol. 121, pp. 580–590. From Bohr's Lecture at the Volta Conference at Lake Como, Italy, on September 16, 1927.

Bohr, N. (1963) The Solvay Meeting and the Development of Quantum Physics, 1962. *Essays 1958-1962 on Atomic Physics and Human Knowledge*, pp. 79–100. New York: Interscience Publishers.

d'Espagnat, B. (1979) The Quantum Theory and Reality. *Scientific American*, Vol. 241, № 5, pp. 158–181.

d'Espagnat, B. (1989) *Conceptual Foundations of Quantum Mechanics.* 2nd Edition. New Jersey: Addison-Wesley.

Heisenberg, W. Z. (1927) Über den anschaulichen Inhalt der quantentheoretischen Kinematik und Mechanik. *Zeitschrift für Physik*, Vol. 43, № 3-4, p. 172-198. English translation published on December 1, 1983. The Actual Content of Quantum Theoretical Kinematics and Mechanics.

Heisenberg, W. Z. (1958) *Physics and Philosophy.* New York: Harper.

Herbert, N. (1985) *Quantum Reality: Beyond the New Physics.* Garden City, New York: Anchor Press/Doubleday.

Langan, C. M. (2017) An Introduction to Mathematical Metaphysics. *Cosmos and History: The Journal of Natural and Social Philosophy*, Vol. 13, № 2, pp. 313-330.

Langan, C. M. (2018) The Metaformal System: Completing the Theory of Language. *Cosmos and History: The Journal of Natural and Social Philosophy*, Vol. 14, № 2, pp. 207-227.

Messiah, A. (1999) *Quantum Mechanics.* Unabridged Dover republication of the English translation originally published in two volumes by John Wiley & Sons, 1958. Mineola, New York: Dover Publications.

Stapp, H. (2007) *Mindful Universe.* Berlin: Springer-Verlag.

Von Neumann, J. (1955) *Mathematical Foundations of Quantum Mechanics*, New Edition. Princeton: Princeton University Press

The Reality Self-Simulation Principle: Reality is a Self-Simulation

> Behind it all is surely an idea so simple, so beautiful, that when we grasp it – in a decade, a century, or a millennium – we will all say to each other, how could it have been otherwise.
>
> – JOHN ARCHIBALD WHEELER, 1986

Abstract

The Simulation Hypothesis is the widely discussed conjecture that we inhabit a directly experienced but nevertheless artificial reality which, while supporting human consciousness and perception, is produced and displayed by a host system occupying a higher-level reality unseen from below. Reality is thus implicitly defined to have at least two levels, the one in which we seem to exist, and another associated with the host system. As the term "reality" is undefined beyond this hypothetical relationship, the Simulation Hypothesis is indifferent to the details, e.g., where the host system is located, how the host system works, who or what created and/or controls the host system, and in what respects the simulation resembles the higher reality containing it. But in any case, there must be an ultimate all-inclusive reality or "ontic ground state" that contains and supports whatever reality-simulations may exist, and it is natural to ask whether some aspects of the simulation concept may apply to it. The Reality Self-Simulation Principle states that ultimate reality is itself a natural reflexive self-simulation in which all intelligible levels of reality must exist whether simulated or not. Where ultimate reality is a global self-identification operator configured as the CTMU Metaformal System (Langan, 2018), i.e., the identity-language of intelligible reality, the Reality Self-Simulation (RSS) can be identified with the Cognitive-Theoretic Model of the Universe (CTMU; Langan, 2002), which thus describes reality as a self-simulating identity operator $R_{INT} | R_{EXT}$ and details its structure and dynamics, showing that it possesses its own universal form of consciousness (coherent self-identification and self-modeling capacity), an unbreakable quantum ontology, and a new paradigm for self-organization and emergence.

Keywords

Simulation, Self-Simulation, Self-Simulation Theory, Self-Simulation Principle, Self-Simulation Hypothesis, Reality Self-Simulation, Cognitive-Theoretic Model of the Universe, CTMU, Metaformal System

Introduction

In science and technology as well as popular culture, simulation is ubiquitous. A wide variety of processes are now being computationally simulated for purposes ranging from entertainment and scientific exploration to urban planning, drug design, and military and commercial flight training. These days, many apparently sane people are even looking about them with suspicion and wondering whether reality itself could be a simulation in which they have somehow been trapped, and if so, whether there is any possibility of escape.

By definition, a *simulation* is a model or imitation of an actual situation or process that fools an observer into thinking it is real given limited suspension of judgment. That is, it is similar enough to the original system or process to be "realistic" even while differing from it in some respects, preferably not so many as to destroy the illusion. But where the simulation concept is taken to the global limit – where we are considering the simulation of reality at large – illusion takes on a whole new meaning, converging on fundamental sensory and epistemological limitations.

Where it is the entire universe that is simulated, strange complexities and unexpected simplifications may arise. For example, although the simulation must differ from true reality in order to maintain its definitive artificiality, it is ideally indistinguishable from genuine reality for its inhabitants. In this sense, "artificially simulated reality" is an oxymoron; a "realistic" (convincing) artificially simulated reality must be real enough to make its resident observers regard it as genuine after all. Yet insofar as any simulated reality is embedded in all higher levels of reality, it may still be possible to discern ultimate reality within it.

The Simulation Hypothesis

The "Simulation Hypothesis" or "simulation theory" is a relatively new hypothesis with ancient roots, a modernization of the old idea that the physical world is merely a perceptual representation of a deeper level of reality. It posits that we inhabit an artificial system, e.g., a simulation programmed and running on a computer or other advanced digital construct possibly overseen by a higher intelligence (God, the devil, aliens, post-humans, etc.). Some variants involve consideration of how and when the technological capacity for reality-simulation evolves or emerges from base-level reality, and the assignment of likelihoods to the associated possibilities.

In the Western philosophical tradition, the simulation hypothesis can be traced back through the "evil demon" of René Descartes (Descartes, R., 1996), which he described as presenting "a complete illusion of an external world ... devised to ensnare my judgement", to Plato's Allegory of the Cave (Plato, n.d.) and beyond. In the Eastern tradition, it is foreshadowed in certain strains of Vedic and Buddhist philosophy and literature. But perhaps the first technical application of the modern version of the Simulation Hypothesis to an outstanding philosophical problem appeared in the paper "The Resolution of Newcomb's Paradox" (Langan, 1989), which features a reality-simulator capable of simulating the deterministic and nondeterministic aspects of real processes, including volition and cognition, in an otherwise paradoxical decision-theoretic scenario.

In addition to the Simulation Hypothesis, there exists a related trilemma with which it might sometimes be confused. This trilemma, called the "Simulation Argument", is expressed as follows (Bostrom, 2003):

> At least one of the following propositions is true: (1) The human species is very likely to go extinct before reaching a "posthuman" stage; (2) Any posthuman civilization is extremely unlikely to run a significant number of simulations of their evolutionary history (or variations thereof); and (3) We are almost certainly living in a computer simulation.

Whereas the Simulation Hypothesis forthrightly asserts that we live in an artificially simulated reality (subject to proof), the Simulation Argument is about the *likelihood* of living in an artificially simulated reality under specific evolutionary circumstances involving implicit technical assumptions. According to Bostrom (2008), "one can accept the simulation argument and reject the simulation hypothesis". Thus, the Hypothesis and the Argument should not be confused.

However, the Hypothesis and the Argument share something in common: both are about *artificial* reality-simulations, and are thus distinguished from the true (artificially unsimulated) or *ultimate* reality in which they are implicitly assumed to exist. This raises a pair of questions. One question is ontological, having to do with the relationship between true reality and artificially simulated pseudo-reality, while the other question is epistemological, having to do with the possibility and means of distinguishing one from the other:

(1) How do true reality and artificial simulation differ, and what if anything do they have in common?

(2) Can the inhabitants of an artificial simulation distinguish it from true reality and *vice versa* (i.e., can real physically-embodied human beings distinguish the true reality they inhabit from an artificial one)?

To answer one or both of these questions, it seems that we require a sound definition of "true reality", which to some extent entails a *verifiable theory* of reality. This "reality theory" must contain the required distinctions and a reliable means of applying them.

Concisely Formulating Simulation

First, let's see if we can find a simple way to look at simulations as usually conceived.

A simulation, being defined as the imitation or imaging of a situation or process, is an input-to-output transformation that converts the situation or process to a more or less faithful image thereof.

In other words, a simulation is a kind of mapping, which we can write as:

$$\text{Sim} : \text{input} \rightarrow \text{output} \tag{E.1}$$

where "Sim" stands for a simulative mapping operator or a natural or artificial processor which operates as programmed on input from some combination of internal and external sources. The output, consisting of images of the input data under the mapping, "simulates" the input to the extent that it resembles the input or shares its properties. (At this stage of simplification, the observers and programmers for whom the input is "simulated", and by whom the resemblance will be judged, are implicit.)

Equivalently, a reality-simulation can be expressed as follows:

$$\textbf{Input} \rightarrow \textbf{processor} \rightarrow \textbf{output} \tag{E.2}$$

That is, a simulation occurs when a processor "models" input as output. Basically, the **processor** (simulator, host system) runs a program which is a **model** of a **source system** (a real or abstract system to be simulated, of which a model, representation, or formal equivalent has been programmed into the processor). Many seemingly ordinary computer programs contain hidden models of real or artificial systems and processes which are activated when the program is run, generating a "simulated miniverse" instantiating them on input data from the user.

In the standard simulation hypothesis, the host system (a connected processor-display ensemble) is a computer programmed with a model of the source system, where the source system may exist only formally in the model itself. In any case, the host system is assumed to be a machine embedded in a higher, "truer" level of reality from which the simulation in some way deviates for better or worse (as otherwise it would be pointless).

To simulate X is to provide instances for a characteristic invariant formal description of X that remains unchanged with respect to variation among possible instances. The instances "deceive the observer" by supplanting a generic formal description, the model of X, with specific instances, physically imitating mental properties and formal patterns thereof. Only the instances, not the programming, are visible inside the simulation (in the "display"). The subjective nature of models, programs, and formal properties are hidden, either in an external processor or deep within the observer himself.

Where X = reality, the simulation is a reality-simulation. Where the processor is artificial – e.g., a mechanical or electronic construct – the reality simulation is artificial; where the processor is natural (involving no artificial construct), the reality simulation is natural, and may be considered a part of nature or reality itself. If it can be considered a simulation at all, it is a self-simulation of, by, and for nature itself.

Real simulations of reality exist in reality, and in principle may be "nested" in simulations of simulations ... of simulations. Thus, we have the following diagram:

(1) Ultimate Reality → [nested subrealities] → (2) source system (to be simulated) → (3) host system with model of source system → (4) displayed target system (E.3)

The universal relational structure of reality is *syndiffeonic*. A syndiffeonic relation consists of two levels, one *synetic* and singular and the other *diffeonic* and (usually) plural. The synetic level distributes over the diffeonic level, thus lending the diffeonic (individually discernible) relata a basis for cognitive or perceptual coherence. In other words, whenever one perceives or conceives of many things at once, it is by virtue of the fact that they have a common synetic property: they are conceivable or perceivable (and have other more specific properties in terms of which to be distinguished). This property relates to one's ability to discern the relata, i.e., one's ability to tell them apart while nevertheless seeing them all at once. Syndiffeonesis is clearly present in every property or attribute with multiple objects, instances, or values, and in every object that displays multiple properties or attributes. Every relation, even a simple identity relation, is syndiffeonic; a property associated with the relational symbol is distributing over whatever is to either side of it.

Ultimate (true, base-level) reality refers to a level of reality that is perfectly self-contained; it must conform to the CTMU Metaformal System, which distributes a synetic "intelligibility property" or universal syntax of identification – an instance-free, semantically unbound *universal distributed form* or UDF supporting and constraining all possible recognizable input – over its localized (semantically bound) instances within a universal syndiffeonic identity, thus providing itself with the internal coherence that allows bound states to be coherently assembled in the building of artificial processors. (Although every reality-simulation is in a sense "ultimate reality simulating itself", its displayed output is extensional and localized to particular locations or regions within a display medium which may, if artificial, be localized within ultimate reality and thus unable to display ultimate reality in its entirety. This localization implies that there is a higher external level of reality on which it depends, implying that it is not self-contained and is therefore non-ultimate.) *Nested subrealities* are reality simulations within reality simulations within ... within ultimate reality.

Host systems simulate the host systems of any lower-level (nested) simulations which are run on them. Host systems may thus be described as "internally emulating" the host systems of lower-level simulations.

We can now refine steps 2-4 in the above diagram. First, note that the target of any structure-preserving mapping "simulates" the source – i.e., that the homomorphism $m: x \rightarrow y$ *simulates* x with y. We can thus define a **simulation** as an iterative composition of structure-preserving state-transformation mappings, usually executed by a processor P programmed with an internal model M^X of a specific system X to be simulated by successive simulated states (marked with a prime symbol) $X' = \{X'_0, X'_1, X'_2, \ldots\}$ on input consisting of some initial state X'_0, where time-indexed simulated states X'_T of the simulated system X are expressed in a display D at time T (the present moment), advances step by step, state by state, and change by change:

$$P(M^X, X'_T): D(X'_{T=n}) \rightarrow D(X'_{T=n+1}) \qquad \text{(E.4)}$$

In plain language: the processor P, operating on a model M^X of system X to be simulated and a simulated state X'_T at time T in the display D, transforms the input state $D(X'_{T=n})$ at time T = n in D to the output state $D(X'_{T=n+1})$ at time T = n + 1 in D. (Note that this mapping is a loop; where n stands for "now", n + 1 immediately becomes n again as time advances, output is recycled as input, and the function P iterates.)

$D(X'_n)$ means that the display D is displaying ("performing a display operation on") the simulated states X'_n of the simulated system X′ as they are symbolically communicated to it by the processor P as language or code. In other words, P is being treated as a function converting input to output, and D serves as its domain and codomain. Neither P itself nor its various subroutines and individual processing operations are distinguishable in D, and insofar as it determines the content of D, P exists and functions in a "pre-display" capacity.

In a **self-simulation**, the model is a description of the processor and display themselves:

$$P\left[M^{(P,D)}, (P,D)'_T\right]: D(P,D)'_{T=n} \rightarrow D(P,D)'_{T=n+1} \tag{E.5}$$

Here, "self-simulation" is interpreted in such a way that there is a real display with a simulated processor and simulated display as its content.

But alternatively, because the processor contains a model of both itself and the display, and could simply be internally updating its internal model, everything could be inside the processor. When the simulation is confined to P, we have:

$$P\left[M^{(P,D)}, (P,D)'_T\right]: (P,D)'_{T=n} \rightarrow (P,D)'_{T=n+1} \tag{E.6}$$

or simply:

$$P\left[M^{(P,D)}, M_T^{(P,D)'}\right]: M_{T=n}^{(P,D)'} \rightarrow M_{T=n+1}^{(P,D)'} \tag{E.7}$$

with the processor internally running its "self-modeling program".

In short, while we cannot put P in D – a processor can process states, but a display cannot display processing – we can put D in P provided that P alternates between process and display functions. This amounts to putting both P and D in the self-dual processing-display element P, thus avoiding P-D (time-space) dualism. P simply alternates between display and processing stages, or state and state-transition. In any case, as we explain in the next section, the processing and display functions are fundamentally incompatible; at best, they can alternate within dual-purpose operators handling both functions. The processing stage is separate from the display stage whether external or internal to P.

D, being incapable of displaying processing intervals, merely displays output from P with retrodictive interpolations between states. (Physically, this is how we observe the "display" we call spacetime; all processing is retrodicted from the current state.) As D does not affect the output it receives from P, P can simply ignore D along with the rest of its output-to-input loop (which in this case is not external but hardwired or simulated within P). P is thus internally emulating itself on the initial state P'_T.

This makes P its own mapping or state-transition function P: $P'_n \rightarrow P'_{n+1}$. P need produce no actual output at all; everything remains internal. Only the internal state of P actually changes; no change is visible from outside (i.e., no external output or input signals are present on any external wires connecting it to an external display). So immediately we get:

$$P(M^P, P'_T): P'_{T=n} \rightarrow P'_{T=n+1} \tag{E.8}$$

where the P'_T are the self-simulated states of P itself.

A **reality simulation** is just a simulation in which the modeled system is called a "reality" on the strength of whatever criteria may be associated with the term. (People often disagree on how reality should be defined, so its definitive criteria vary widely.) A commonly used criterion: it must be sufficiently realistic, or faithful to one's expectations of what reality should be – e.g., immersive, replicable, and logically consistent – to convince an observer. (However, one should not fool oneself about what it actually takes for this "anthropic" criterion to be fulfilled.)

A more interesting case is that in which reality simulates itself. In a **reality self-simulation**, reality is self-contained, reflexively incorporating its own processor and display functions. In effect, reality becomes its own processor and display, reducing the "simulation" to reality itself (Langan, 2002). This situation can be expressed by simply replacing the processor in E.8 with reality:

$$R(M^R, R_T): R_{T=n} \rightarrow R_{T=n+1} \tag{E.9}$$

where R is reality, M^R is a self-model of reality, and the R_T are the time-dependent states of R (the prime symbols are now superfluous, as states are no longer confined to artificial machinery but instead belong to reality itself.) That is, reality R runs a self-modeling program M^R on a given state R_T thus "simulating" the structure and dynamical evolution of R.

Where *telesis* is a convergent generalization of intension and extension coinciding with the ontic ground state, and R* is an ultimate (self-contained, self-defining, ontically closed) telic identity operator that couples (and dually, separates into) its own intensional and extensional aspects, a reality self-simulation can simply be written:

$$R^*: R_{INT} \leftrightarrow R_{EXT} \tag{E.10}$$

where the subscripts mean *intensional* and *extensional* respectively. In this context, intensional means "processor/model/program-like", while extensional means "display-like". As R* is an ontic identity or irreducible self-dual telic identity operator, intension and extension are codependent, and a 2-headed arrow is required.

Reality Self-Simulation

Reality and *simulation* have nearly opposite meanings; one is real, and one is counterfeit. It is thus appropriate to ask whether the phrase "Reality Self-Simulation" is meaningful.

Where a system produces and displays its own content, there arises a question of whether it can still be described as a reflexive "simulation". Whether it retains the structure and functionality of a simulation depends on the extent to which it inherently supports the distinctions on which *simulation* is defined, especially the distinction between the *processor* and *display* aspects of the system, and between the system model and the system itself.

The definitive criterion: Reality is self-contained and has a "display stage" or extensional medium in which substantive content is perceptibly displayed, and an intensional "processing stage" which is not immediately displayed, but generates an advance model of its future evolution which is transferred to the display stage for actualization.

Here are the possible kinds of "reality simulation" usually considered:

(1) **Central processing – discrete display** (the computer-simulation paradigm): Reality resembles a standard computer simulation displayed on the discretely pixelated screen of a computer monitor controlled by a remote CPU which schedules the action on the screen using its central clock.

(2) **Central processing – continuum display** (classical physics, with disjoint processor and display): Reality, consisting of a separate processor and continuous display, displays itself on an infinitesimally (sub-finitely) pixelated 3D "monitor screen" to which the processor is external but to which it is somehow connected.

Except for the infinitesimal "pixels" of the display (as required for continuity), this configuration resembles case 1. The display screen resembles a classical void or spatial medium containing observers and serving as their common environment, with time amounting to clock pulses distributed throughout the display from the CPU.

(3) **Distributed processing – discrete display** (cellular automaton model/digital physics): Reality is confined to a finitely partitioned display resembling the discrete array of a cellular automaton. Its "cells" are either dual-purpose processing-and-display elements, or two kinds of element are somehow packed together without disrupting observation. The display is equipped with its own clock to coordinate the activity of its cells, or the cells somehow coordinate with each other automatically.

In this configuration, the display evolves by way of distributed, locally parameterized, locally processed instructions and thus obeys a locality principle. This kind of "digital physics" is supposed to allow for the reduction of reality to information, or bits and computational operations thereon. But among its weaknesses is the fact that informational reductionism applies only to the content of the display; the display itself must be taken for granted as an informational medium (Langan, 2002; Wolfram, 2002).

(4) **Distributed processing – continuum display** (modern physics and cosmology): Processing, and thus the evolution of the system, are intrinsic to the geometry of the display. Processing is either associated with points of the spacetime manifold (just as it is associated with the display elements of a cellular automaton in case 3), or points have dual processing and display functionality. In either case, the laws of physics are pointwise-distributed over the spacetime manifold or some other manifold, possibly with "extra dimensions" that might accommodate hidden processing. This conforms to the trend in modern physics which confines reality to a distribution of matter and energy in a "physical" continuum, with quantum mechanics demanding more or less problematical accommodations.

In all of these configurations (1–4), the distinctions among processor, display, display content, programming, and conscious observers are either nebulous or dualistic. By association, so are the distinctions among time, space, matter, causation, and life and consciousness. By the time we reach configuration 4, the GR relationship holds for space and time, and the QFT relationship holds for matter and quantum fields (spatial distributions of energy of which elementary particles are energetic excitations). But we still have Cartesian mind-matter duality, the continuum vs. discrete array problem, causal deficits associated with quantum indeterminacy, and the apparent absence of anything holding processor and display together.

Why should it be necessary to "hold processor and display together"? The reason, which will be explained in more detail below, is basically an inexorable dualism. While configurations 1–4 reflect the historical physicalism (or "metaphysical naturalism") of mainstream science in general, thus mandating the confinement of reality to the display (which is where perceptible physical states reside), processing is a *function* of states which is not amenable to display. To an extent, processing and display are incompatible functions; processing must be halted in order to display the states it produces. It follows that processing – the dynamics of reality – cannot inhabit the display and must therefore reside elsewhere, specifically in a location adjoining the display and separated from it by an interface that allows it to receive input from and deliver output to the display.

In the sciences and beyond, reality is usually considered to be confined to space and time, which are now understood to be everywhere coupled in a hybrid medium called "spacetime". But the nature of this linkage is not yet fully understood, and there are many outstanding questions. In particular, what is time, and if it is a physical process, where is it hiding? Empirical science cannot tell, because time is not empirical – one sees only its correlates, its input and output states. Because science wants so badly for everything to be "physical" and thus directly observable as states in the display, the structure and dynamics of reality are pigheadedly reduced to structured sets and temporal sequences of states and state-transition events visible in space – i.e., of display and processing respectively – and it is thus assumed that reality consists entirely of time and space as a joint medium for matter, energy, and life. Aside from empirically induced transformation equations, it is not usually considered what might be holding space and time together.

While each of these configurations (1–4) has its own problems, all of them must accommodate the exclusory relationship between processing and display, or time and space. While it is sometimes held that certain mathematical equations, e.g., the Lorentz transformations, explain the relationship between time and space, such mathematical niceties are more descriptive than explanatory. The perception of space – the apprehension of state – displaces the direct perception of intervals between states. The only way to perceive an interval is to terminate it on both ends, and this can only be accomplished once the leading (advanced) terminus has already appeared, forcing the reconstruction of the processing stage. But in the end, just one thing is certain: states and virtually instantaneous state-transitions are what display media actually display. Obviously, once a medium has been strictly defined on its capacity to display states (by any name including physicality, materiality, or observability), that which cannot be displayed cannot inhabit the medium, and what transpires between the display level and that which cannot be displayed must transpire beneath the surface through an appropriate interface.

Let's look at the problem more closely. The processing of one state into the next occurs between states – e.g., to produce a state s_2 of X from a prior state s_1 of X, processing must occur between s_1 and s_2 in a processing (state-transition) interval s_1s_2. But the processing interval s_1s_2 can have no duration and therefore must remain hidden, as otherwise there is a time lag during which X has no state at all and thus drops out of the display medium. Thus, what get displayed "in real time" are instantaneous state-transition events coinciding with states themselves; in the display, "real time" becomes discontinuous, and the transition of one state s_1 into its successor s_2 coincides with display of the single state s_2. Thus, processing and display are necessarily distinct functions and must alternate. The display must therefore be stratified, with the focus bouncing between a hidden processing layer and a visible display layer. In other words, avoiding perceptual or observational disruption requires that processing be hidden so as not to occasion discontinuities between displayed states. But this contradicts the supposition that the display "has both processing and display functionality". To put it as succinctly as possible, **processing is not displayed**.

This is perfectly evident on the quantum scale of processing. First, let the "display" be defined as that which is inhabited by all and only that which it actually displays (we ignore for now the question of whether display requires observation, assuming that reality "observes itself" by the interaction or "mutual display" of its contents). Between measurement events (state transitions) in the display, matter particles exist as *quantum wave functions* containing information on their states, evolving as prescribed by the Schrödinger equation

(Schrödinger, 1926). But quantum wave functions are not displayed, and neither are the "probability waves" of which they supposedly consist. Although many people have seen effects and causal correlates of wave functions (e.g., double-slit interference patterns), no one has ever seen a wave function or a probability wave directly. It follows that quantum wave functions do not inhabit the display, and neither do the particles whose states they represent. It follows that particles are not actually present in the display until they have "collapsed" there from another place, and that inasmuch as they continue to exist at all, they must go "somewhere else" for processing.

This implies that processing and display are connected by a background identity that bridges the alternating *display* and *process* functions. Fully satisfying this connectivity criterion requires an ultimate level of reality that is totally self-contained and requires no external background; its background is purely intrinsic and connects all levels of reality and their associated "simulations".

This ultimate identity is the ontically closed Metaformal System (Langan, 2018), on which the following configuration is defined. Readers are referred to previous papers on the CTMU (Langan, 2002, 2017) for further information.

(5) **Generative (conspansive hological) processing and display** (CTMU): Ultimate reality is an ontically self-contained telic self-identification operator $R^*: R_{INT} \leftrightarrow R_{EXT}$ that simulates itself in a generative multi-level processor-display feedback loop driven by the self-dual grammar Γ_μ of the **CTMU Metaformal System** M describing the structure and dynamics of the operator.

$$R^*: R_{INT} \leftrightarrow R_{EXT} [= N \leftrightarrow T] = \text{processor} \leftrightarrow \text{display} \qquad \text{(E.11)}$$

Note the 2-headed arrow, which replaces the 1-headed arrows of the last section. It denotes feedback between processor and display. This feedback reflects a terminal form of time symmetry associated with the advanced (retrocausal) and retarded (causal) semimodels of the CTMU, amounting to the "display confinement" of metacausal CTMU grammar (Langan, 2019).

Before continuing this description in Section VI, here is a simplified review of the Metaformal System (Langan, 2018).

CTMU/The Metaformal System

The CTMU Made Simple

The CTMU can be succinctly described as a language that is structured in such a way as to talk *to* itself *about* itself (which actually makes it a "metalanguage of itself"). This is how it exists and evolves. This reality-language is of a novel form required by its functionality, but still bears description as a language. The form required is that of the Metaformal System M, the supertautological intrinsic identity-language of reality. M is the linguistic self-configuration through which reality identifies itself, and which allows us to identify reality with certainty by direct replicated perception. Its existence is implied because this is what we, and reality, actually do.

When we say that reality "talks to itself", this means that its ingredients, syntactic identification operators, are communicating and interacting with each other, "talking and listening" for purposes of self- and mutual identification. (Identity operators are active objects which identify or determine the states of other identity operators.) Reality uses internal images of itself to self-communicate and self-transform. When we say that reality "talks about itself", it means that the formal and substantive aspects of identification operators are inseparably coupled within them; the formal intension of each operator couples to its extensional substance in the (external and internal) states of the operators. Thus, what reality "says to itself about itself" is truthfully bound to the actual content of reality.

If one finds it hard to see how language and reality could have anything in common even though everyone uses language to describe various parts of reality every day, then one can focus on the identification operators themselves, and view reality as a kind of "operator algebra". This is an *identity system* consisting of identity operators that identify the states of other identity operators. Each identity operator identifies itself by self-modeling, and by the virtual synonymy of modeling and simulation, "self-simulates" in the bargain. One can think of these operators, albeit simplistically, as a swarm of interpenetrating spheres which intersect and absorb each other – or internally "simulate" each other – in space and time, but extend into "metaphysical pre-reality" or the nonterminal domain N in order to do so. They interpenetrate through a grammatical operation called *conspansion*, and couple with each other and transform their states through a "protocomputational" self-modeling operation called *telic recursion* which collapses and displays their wave functions in terminal measurement events (where protocomputation is a primitive generalization of computation appropriate to CTMU structure). Thus, not only is reality a self-modeling identity operator and therefore a global self-simulation, but it everywhere consists of conspanding identity operators and is therefore "quantized" in terms of self-simulative conspansive cycles as well. That is, it is *everywhere* a self-simulation (Langan, 2002). The conspansive cycles of identity operators quantize reality with respect to not just self-simulation, but consciousness and emergent quantum information.

The metaformal structure of the RSS leaves no doubt about the self-simulative character of reality. Unlike the artificial reality-simulations depicted in movies like *Tron* and *The Matrix* (which the CTMU predated by over a decade), the reflexive reality simulation of the CTMU is natural, reflexive, and without need for an artificial host system. It is its own host system, with all functions distributed among the id-operators of M and the points of its manifold. This manifold has both processing layers and a top spatiotemporal layer which serves as its "display medium". In its nonterminal processing stage, the self-distributed host system of ultimate reality self-simulates by (1) using its metacausal grammar to "pre-simulate" its future states in terms of its past states in nonterminal pre-reality N; (2) advancing to the display stage and collapsing nonterminal *telons* (metacausal relationships) to the terminal content of its display T; and (3) iterating over and over again in a vast, highly orchestrated multilevel feedback loop, generating all of the action visible on its spatiotemporal "display screen".

The Metaformal System can be dually expressed as follows:

$$(\mathrm{M}\colon \mathrm{L} \leftrightarrow \mathrm{U}) = \left(\Sigma_{\mathrm{M}}\{\mathrm{N}, \mathrm{T}\}, \Gamma_{\mu}, \mathrm{S}_{\Sigma}\right) \tag{E.12}$$

This equation identifies M as an ontic identity operator and a trialic intrinsic language.

On the left side of equation E.12, which reads M: L ↔ U are the following symbols:

M The active identity (metaformal linguistic id-operator) of reality as we know and perceive it. Its structure and dynamics are those of the CTMU Metaformal System, the intrinsic metalanguage of reality whose signs are active identities (intension | extension coupling operators, identity operators, id-operators). M is logically induced from reality as it is known through human cognition and perception of its pattern and substance, where logical induction is the inference of necessary and sufficient support for the self-identification of reality, i.e., inference of the requirements of existence and intelligibility.

M is also referred to as the "global operator-descriptor" or G.O.D., so called because it is a global identification operator which distributes itself over reality as a universal syntax or descriptor in terms of which it identifies reality both directly and through secondary and tertiary id-operators. M may be called R* when the self-simulative aspect of M is emphasized, or symbolized as Γ when the emphasis is on its grammatical aspects. Where the emphasis is on cosmology and/or quantum theory, M can be taken to stand for multiverse, metaverse, metareality, or metaphysics (words which are conventionally ambiguous, but amenable to precise definition in the metaformal context).

M is a metaphysical structure in which existence is coupled with identification and intelligibility, which amounts to a coupling of ontology and epistemology. (This coupling should in any case be obvious; something must be identified before existence can be attributed to it, and only where it exists as a concept or percept can it be identified as such.) Owing to the dual role of its id-operators as the signs of its signature and the points of its inherent manifold, M is a supertautology and thus a logical necessity. Much can be known of the structure of reality from its functional properties even without enumerating its nonlogical ingredients.

M is a metalanguage of the object-language L of cognition and perception. Accordingly, the signature of M can be regarded as a "metasignature" relative to the signature of the formal object-language L, as it "wraps" or packages L in syntactic operators forming the "active signs" of M and the dynamic points of the conspansive manifold. In effect, the metasignature of M "converts form into substance" by packaging the generic formal identities of L in id-operators instantiated by their states, filling them with content and thus coupling L and U. Without this coupling, reality would be neither self-intelligible nor intelligible to us as its internal self-images, over whom its generic structure is distributed as the syntax through which we recognize the world and the grammar through which we help create it.

L The purely intensional (uninstantiated/non-extensional) aspect of the active identity (metaformal id-operator) M; the universal descriptor. L can also be regarded as the object-language of M, coupled by M with specific instances in U. It is associated with M-syntax or the universal distributed form (UDF), which includes all properties and formal or abstract structures in terms of which all real objects, relations, and processes are identified in M. The UDF consists of syntactic forms (grammatical nonterminals), formal definitions (formal / generic identities with empty extensions), and formal expressions.

$\longleftrightarrow$ The generic structural and grammatical coupling of L and U, refined by identity operators via the CDF to the content of the terminal domain. The CDF is the conditional distributed form, a kind of "post-semantic syntax" or metrical pattern of extended superposition in N which records and potentializes the progressive semantic binding and localization of id-operators leading to the terminal distribution of mass and energy in spacetime. The CDF can be thought of as connecting the UDF, the "universal accepting | generative syntax" or generic intensional self-model of reality, to the terminal domain or extensional "display". It can be dually associated with the grammatical factorization of telesis by id-operators, thus generating conspansive medium | content relationships in T (the self-simulations of id-operators), and the terminal interactions of physical objects in state-transition events.

U The extensional aspect of the active identity (metaformal id-operator) M. U, standing for the universe or domain of discourse of the formal language L, consists of specific directly discernible instances of L, i.e., fully localized tertiary id-operators obeying an appropriate exclusion principle. The discernibility criterion amounts to a "display criterion";

discernible instances are limited to fully localized states in U. U thus coincides with the physical content of the terminal domain ... the matter distribution of spacetime. The spacetime metric, on the other hand, occupies the CDF, which interconnects the contents of U and connects U to L, scaling and localizing L.

On the right side of equation E.12, which reads ($\Sigma_M = \{N, T\}, \Gamma_\mu, S_\Sigma$) we have:

$\Sigma_M = \{N, T\}$ The **signature** of M with nonterminal and terminal subsignatures (or "stages") N and T. Σ_M contains the "components" of M, including the self-dual primary (global-teleological), secondary (organo-telic), and tertiary (micro-physical) identity operators; their structural and functional similarity defines the self-similar mereology of M. The identity operators are both the elements of the id-operator algebra or "id-system" of M and the points of its intrinsic manifold, supertautologically binding algebra and geometry together in M. This relationship is not determined on the physical (tertiary) level alone, but on the primary and secondary levels associated with teleology and organic telesis. Each id-operator alternates between the terminal and nonterminal domains by conspansion; thus, N and T are best regarded as "stages" of the operators. Σ_M spans L and U, N and T, and all mereological scales of M (primary, secondary, tertiary).

Γ_μ The mu-morphic **grammar** of M. It is generative, factoring telesis into medium|content relationships. It involves two operations, *conspansion* and *telic recursion*, by id-operators. These operations are respectively analogous to generic processing and specific programming, where "programming" is the self-modeling of Σ_M and its secondary telic id-operators. Γ_μ determines bundles of intertwined and intersecting terminal strings or event-sequences in T, imparting to system dynamics hidden levels of causal organization which are not locally evident in T itself. Although the events thereby related seem to be independently (and often "randomly") determined by local causation, they may – given sufficient freedom provided by the generativity of Γ_μ – be determined instead by nonlocal dependencies among distant points of T which intersect and couple grammatically in N without actually violating locality.

Grammar evolves orthogonally (projectively) with respect to the timelike evolution of spacetime. The localistic causal evolution of spacetime is thus largely illusory except in classical (non-quantum) contexts, being projected from the nonterminal domain N onto the terminal domain T by Γ_μ as standard "physical causation". In other words, thanks to Γ_μ-mediated nonlocal feedback and telic recursion (adaptive grammatical self-organization) in N, terminal strings or temporal event-sequences effectively emerge into spacetime along their entire lengths. Together with other aspects of Σ_M, Γ_μ is thus the basis of a new approach to self-organization and emergence in complexity theory.

S_Σ **Strings** are timelike sequences of terminal-stage identity operators (active signs or symbols) from Σ_M, specifically from the terminal subsignature T, with their fully resolved external states. They are subject to rules of syntax and orthography, which include the "laws of nature" or "laws of physics" controlling sequences of states and state-transition events in T. Strings describe the apparent time evolution of the tertiary id-operators in T as they interact in mutual state-transition events.

The Metaformal System as Reality Self-Simulation

Here again is the RSS configuration presented at the end of Section IV.

(5) **Generative (conspansive hological) processing and display** (CTMU): Ultimate reality is an ontically self-contained telic self-identification operator $R^*: R_{INT} \leftrightarrow R_{EXT}$ that simulates itself in a generative multi-level processor-display feedback loop driven by the self-dual grammar Γ_μ of the **CTMU Metaformal System** M describing the structure and dynamics of the operator.

$$R^*: R_{INT} \leftrightarrow R_{EXT} [= N \leftrightarrow T] = \text{processor} \leftrightarrow \text{display} \qquad (E.11)$$

N and T are (respectively) the nonterminal and terminal levels of the signature Σ_M of M and its dual stratified (conspansive) manifold. The elements of Σ_M are id-operators occupying primary (cosmic), secondary (mesoscopic / biological), and tertiary (submicroscopic / quantum) levels, with higher-level elements coordinating those below. N and T are not disjoint, but describe alternating functions of the id-operators; i.e., the elements alternate between N and T in the conspansive cycles of Γ_μ. In N they are active signs subject to conspansion and telic recursion (the operations of Γ_μ); in T, they are the fully localized points of the "physical" spacetime manifold. Thus, the active (operational) signs of the signature are the dynamic (conspanding) points of its corresponding medium.

The Reality Self-Simulation (RSS) can be identified with the Metaformal System M as follows. The nonterminal aspect N of its signature Σ_M is the processing stage of the RSS, and the terminal aspect is its display stage T. The strings S_Σ of M are the spacetime histories of the tertiary identity operators (identification operators, id-operators) of Σ_M, i.e., the static and dynamical content of the display. M-grammar Γ_μ, a "metaprocess" orthogonal to the time parameter of T, is the self-dual process by which identity operators self-model or "program N" with input from T, and by which N adaptively processes the input and delivers output back to T. Secondary id-operators existing in both N and T thus function as observers, programmers, and processors. The primary or "teleological" id-operator M coordinates it all.

Because the processing and display functions are associated with each identity operator of M – because the display elements conspansively feed back on themselves in grammatical processing-display cycles – processing is distributed alongside display. T is the display, consisting of display-stage elements, while N is the processor, consisting of processing-stage elements, with everything coordinated by the primary id-operator R* = M using the same grammar and syntax of which more or less restricted versions are used by secondary and tertiary operators on biological and quantum scales, endowing the system with profound structural and dynamical self-similarity). The tertiary elements of N are in the processing stage of their conspansive cycles, while the tertiary elements of T are the display stage. N and T coexist as levels of a manifold over which the generic structure of M pointwise-distributes as grammar and syntax.

Identity operators, each of which has dual processing and display functionality in N and T respectively, repetitively collapse from N into T and then recede from T back into N by inner expansion. As identity operators play dual roles, being the signs of the signature and the points of its dual manifold, the conspansive evolution of the extensional manifold of M mirrors the grammatical evolution of the linguistic intension of M by factorization of nonterminals into their derivational successors. Thus, the points themselves furnish the processor-display N|T interface. Unlike the dualistic remote-processing configurations 1 and 2 in Section IV, processor and display are not separate, but dual aspects of a single unified logico-geometric manifold; unlike the distributed-processing configurations 3 and 4, processing does not occur in the display T, but in deeper layers of N (over which it remains distributed).

Due to the generativity of the system, whereby points inner expand into N and recollapse in T to form new states within their predecessors and are thus factored into medium and content, tertiary points of the conspansive manifold are subsimulations unto themselves. For any pair of successive states s_1 and s_2 of any given pixel, s_2 is displayed within its inner-expanded predecessor s_1 (the "display medium") by collapse. Secondary (self-modeling, self-organizing) meta-points connect the tertiary subsimulations to form telonic subsimulations of higher complexity, driving the emergence of higher-order information and filling causal deficits associated with quantum indeterminacy while functioning as observers with a generic "consciousness" attribute proportional to self-modeling or "programming" capacity.

Thus, the secondary and tertiary id-operators of the CTMU Metaformal System M serve as processing elements in N, display elements in T, the states transformed in N and displayed in T, self-modeling programmers, and conscious observers. M itself, the primary id-operator of reality, provides the distributed structure and overall coherence to fully support these functions.

Summary

The RSS, while not without complexity, may be compactly described as an ontically and epistemologically closed ("ultimate") telic identity operator R* with the unique structure of the CTMU Metaformal System. This telic identity operator can be conceptualized as a *multiverse*; this is due to the nature of generativity, whereby the generative grammar of M selects each new state from an uncountable range of possible next states, projectively coupling the invariant UDF with any terminal configuration supported by it ("coupling alpha with omega").

The Metaformal System is a supertautology, i.e., an intrinsic (perfectly self-contained) identity language that is trialic (comprises its own universe and model), with the identity operators of its signature being the points of its universe and the substantive content of the points while also serving every other essential function of the system. It is an ontological and epistemological necessity of the coherent existence and evolution of reality as it is experientially and scientifically known to its secondary telors, i.e., its resident observers including human beings, for whom it functions as a universal identity-language. Its generic implications are inarguable, revealing the nature of reality even without exhaustive enumeration of its nonlogical ingredients.

The ingredients of the Metaformal System exist in perfect correspondence with the functional components of the RSS, which are as follows:

(1) Processors and processing elements;

(2) Display and display elements;

(3) Processor-display connections or communication channels;

(4) Modeling or programming operators;

(5) Observers or observation operators;

(6) Display content, including states, events, and processes.

Respectively, the above components correspond to the following ingredients of M:

(1) Telic identity operators in N (nonterminal prereality);

(2) Tertiary identity operators in T (terminal or physical reality);

(3) Γ_μ conspansive $N \leftrightarrow T$ cycles involving all levels of identity operator;

(4) Telic identity operators configuring in N for actualization in T;

(5) Telic identity operators reading display content in T;

(6) Strings, or processes, consisting of the successive external states and interactive state-transition events or "worldlines" of tertiary identity operators.

As previously noted, the Γ_μ conspansive cycles of the identity operators in Σ_M and S_Σ quantize reality with respect to not just self-simulation, but consciousness (coherent self-identification and self-modeling capacity) and emergent quantum information. It follows that reality is ultimately a self-simulation possessing not only an unbreakable quantum ontology, but its own primal form of consciousness and a new paradigm for self-organization and emergence.

FAQs About Simulated Reality

1. Are there any theoretical universal models that are not causal by nature? For instance, if this was a simulation as suggested, then how might the rules be different?

Yes. The definitive theory is called the CTMU, short for *Cognitive-Theoretic Model of the Universe*. Publications date from the late 1980's through the present, most recently in the peer-reviewed academic journal *Cosmos and History*. Technically, it is a theory of mathematical metaphysics.

Basically, the CTMU says that reality is an ontologically closed "self-simulation" in the precise sense that it contains empirically undecidable "programming" as well as "displayed" observational content, and that its logico-mathematical structure is that of a self-structuring, self-axiomatizing "SCSPL", short for *Self-Configuring Self-Processing Language*. This language has a unique form called a *supertautology*, according to which its evolution is *metacausal* rather than strictly causal.

In addition, the CTMU embodies a logical form of self-similarity analogous to the so-called "holographic principle" as later developed in connection with string theory. Accordingly, it is *trialic*, effecting a distributive coincidence of theory, universe, and model in a coherent "metaformal system".

Last but not least, the CTMU was authored by a man reputed, at least in some circles, to be the most intelligent in the world.

2. Within the context of Christopher Langan's Cognitive-Theoretic Model of the Universe (CTMU), could we technically be in a computer simulation of sorts, with "God" being the programmer?

Yes. According to the CTMU, reality can be understood as a "self-simulation", at least in the sense that what we see outside of ourselves – the physical universe – conceals the process which actually produces it, and cannot account for the entirety of what producing it actually requires.

However, the simulation is not merely computational in the mechanical sense, but (in CTMU terminology) "protocomputational" or "precomputational". That is, it utilizes a kind of "metaprocessing" called *telic recursion*, which generates **entire timelines** instead of individual events.

God is indeed the Programmer-in-Chief, but does His "programming" largely through secondary sensor-controllers, including human beings, which locally inhabit the simulation. As for God Himself, He distributes over the entire simulation and is therefore omnipresent.

The entire simulation can be reduced to a master "programming language" sometimes referred to as *Logos*, which is *trialic*, serving as its own universe and model. On a more technical level, it is known as the CTMU.

3. Is the CTMU the first simulation theory?

Yes, the CTMU is the first detailed theory of reality as a self-simulation, having been used to resolve Newcomb's Paradox in 1989. (Note that reality must be a "*self*-simulation" insofar as that which does the simulating must itself be real.)

This in itself was a milestone; it seems to have been the first time that a major philosophical problem was solved in the context of a computationally simulated reality. While it had previously been suggested that our world is somehow embedded in a wider insensible reality – e.g., Plato's Allegory of the Cave – this was the first time that an existing philosophical problem was solved in an explicitly computational setting.

Note that the CTMU has a unique theoretic structure that is absolutely necessary for this purpose, and that its functionality exceeds that of ordinary computers (it was, and is, described as "protocomputational"). No other so-called "simulation theory" shares this property, ostensibly because the authors of such theories fail to understand the deep structure of reality and the limitations of standard automata.

4. What is the NST, and how does it resolve Newcomb's Paradox?

The NST, or Nested Simulation Tableau, can for present purposes be described as a computational hierarchy of virtual realities, each one nested within a higher reality surrounding it. The NST resolves the paradox by building an inclusive computative reality around the physical reality we take for granted. In this extended universe, time is directionally unrestricted, but can still be treated as a one-way linear dimension within the physical level. Thus, it is a *model* in which expected utility and dominance may coexist in the Newcomb context.

An easy way to envision the NST is to imagine that physical reality is a "program" running on a vast, ultrahigh-resolution 3-dimensional "monitor" consisting of one or more sub-monitors corresponding to physical cognitive agents. To whatever extent the sub-monitors interact on a physical level, their contents intersect at a mathematical interface. Thus, sub-monitors represent subjective frames of reference, whereas the interface represents objective Minkowski spacetime. The setup allows arbitrary localized access to the physical level, letting the higher-level programmer-controller avail himself of distributed and non-distributed programming at his convenience.

The NST model contains at least two possible mechanisms for creating the Newcomb scenario, each one supporting a distinct kind of "omniscience". In one, the programmer simply controls the thought and behavior of the physical subject, effecting omniscience by means of omnipotence. In the other, the programmer allows the physical subject to make all his own decisions, but "random-accesses" certain key physical-time junctures in order to control the effects of those decisions once they have been made. In the latter case, the predicted subject has free will; in the former, he does not.

5. Are you sure that nobody thought of the virtual reality scenario first with regard to Newcomb's problem?

I first encountered Newcomb's paradox in the early-to-mid-1980's, and as I recall, thought up the NST (Nested Simulation Tableau) shortly thereafter. For all I know, somebody else may have considered it even before that. But if so, it is still far from certain that he developed it to the required extent or was able to explain in logical terms why it constituted a resolution of the paradox. For example, did he describe an actual mechanism for omniscience, define the NST in logical terms, or recognize it as an extended hierarchy analogous to the theory of metalanguages and thus as a model for all relevant arguments including dominance and expected utility? Probably not. Did he pursue its logical implications, as I did with the CTMU? Again, probably not. Thus, even with credentials to ease his way, he may not have been able to convince the editor of any academic journal that his idea was publishable. And even if such a suggestion were published, then judging from the speed and thoroughness with which it seems to have been buried and forgotten, it was never properly justified or defended.

6. Haven't you been accused of "making metaphysical assumptions" in your resolution of Newcomb's paradox?

Such an accusation has been made, but it is baseless. The NST is amenable to a number of possible constructions. The one used to resolve Newcomb's problem – a computer within a computer ... running a simulation within a simulation ... – is closely analogous to the *theory of metalanguages*, already an essential ingredient of logic. Experimenting with various possible interpretations of this ingredient is not the same thing as making factual assumptions.

After all, physics already has a bidirectional conception of time. Granted, time usually behaves in what seems to be a directional way, and scientists have had a great deal of success in treating time as a one-way linear dimension. In the branch of physics called thermodynamics, the one-way linearity of time is known as *entropy*. On the other hand, particle physics treats time as a two-way symmetry, tachyonic time reversal emerges from relativity theory, and in quantum theory, measurements can have implications that may affect the past. Times have changed in physics, and so has time's directionality.

Looking at the matter from a cybernetic viewpoint, physics has already prescinded from one-way time and introduced a degree of time-directional *freedom*. The dominance argument then reduces this freedom by introducing a *constraint* to the effect that time is exclusively one-way ... a constraint that introduces *information* with respect to an extant pair of directional possibilities. Since only the constraint embodies new information, only the constraint qualifies as an "assumption". In other words, by prescinding to the NST and thereby suspending the one-way constraint, I avoided the risk of introducing spurious information to the Newcomb context and thus did the diametric opposite of "making an assumption". And in the process – hard though it may be to grasp – I remained truer to the spirit of modern science than did my critic(s).

A good case can be made that this accusation, coming from the source from which it came – the publisher of *Noesis*, the journal in which it appeared – has robbed the solution of the attention it deserved for the last seven years. In any case, the fact that it was made so carelessly, and has been maintained so obstinately, represents a singular injustice not only to me, but to the Mega Society and the readers of this journal.

7. What is the theory of metalanguages, why is the NST a metaphysical analogue of it, and what does the analogy imply?

Any language is used in reference to something of an abstract or concrete nature. With respect to this relationship, the language is called an *object language* and the referent is called its *object universe*. We can now define a higher-level language expressing the details of this relationship, a so-called *metalanguage*. The metalanguage can be used to selectively posit, derive and compare various possible rules of object-language syntax and semantics (something which cannot be done in the object-language itself, which begins with and can never deviate from its own syntax). Given such a metalanguage, we can then define a yet-higher metalanguage expressing the relationship of the first metalanguage to *its* universe, i.e., to the relationship of object-language to object-universe, including the rules of object-level syntax and semantics. Inductively extended, this logical relationship constitutes the "theory of metalanguages". This theory is an established ingredient of logic.

It has been realized by a long line of great minds, from ancient Ionian philosophers to scientists from Lavoisier to Freud to Chomsky, that cognition is a form of language. This is true not only in terms of content – rational cognition is generally reducible to a sequence of conventionally linguistic expressions – but in terms of form, as becomes evident when we represent thought as a mathematical sequence of neural state-transitions. If we accordingly replace the terms "object universe", "object-language" and "metalanguage" with "physical universe", "object-level cognition" and "higher-level cognition" respectively, and then add a cybernetic control parameter, the NST is the result. The NST is thus a well-defined mathematical object homomorphic to a necessary and well-studied ingredient of logic, the theory of metalanguages.

The analogy between metalanguages and NST "stratified virtual reality" implies that for logical purposes, any TOE (Theory of Everything), considered as a language, is required (a) to be its own regressive metalanguage, i.e., to explain how and why its own axioms and theorems apply to reality, how and why the explanation applies to reality, and so on *ad infinitum*; and (b) in placing reality within the framework of logic for analytical purposes, to explicitly incorporate the entire metalinguistic component of logic and a mapping of reality thereto. In other words, no NST, no TOE. Just as in Aristotle's day, cosmology is a branch of metaphysics and requires a logical framework of metaphysical scope. That framework begins with the NST and "ends" with the CTMU, which terminates the most general metaphysical stage of theorization and functions as a new beginning with respect to science.

8. No matter how well the NST can be logically justified as a conceptual tool, VR still seems highly unnatural as an explanation for reality. Does it become more "natural" as its implications are developed?

Yes. The CTMU, which is what the NST ultimately becomes, is a very natural theory well in keeping with the history of world philosophy and science. The NST is merely a starting point for working cognition into reality theory, which turns out to be a necessary step towards theoretical unification. This is unsurprising when we realize that the dependency of physical reality on observation and measurement, both of which are cognitive in a general sense, respectively characterize relativity and quantum theory, the dominant theories of physics on large and small scales.

9. Is it possible to distinguish dreams or hallucinations from reality?

A very deep question, given that reality is already a self-simulation. You can test the integrity of the suspected dream-world in several ways.

For example, can you will yourself to awaken from it? Does it last longer than a normal sleep cycle? Are all of your memories intact and consistent? Are you aware of any possible pharmacological or other abnormal effects? Does the dream-reality exhibit high definition, logical consistency, stability, continuity, coherence, replicability (confirming distributed solipsism), and all other properties familiar to you through your experiences of what you consider to be your primary reality?

If your suspected dream-world passes all of these tests, then you are in a closed, internally consistent self-simulation of ultimate reality ... one of the infinite possible self-simulations potentially accommodated by the CTMU logico-syntactic metaverse. (If you still have doubts, you can try watching the very last scene from the original movie *Total Recall* – it might help, at least on an emotional level.)

10. Applying the metaphysics of virtual reality to a longstanding philosophical conundrum seems like a radical and possibly momentous departure in the field of analytical philosophy. How important is it?

Very important indeed. The field of metaphysics qua philosophy has been stagnating for years. Professional philosophers, cowed by the success of the physical sciences, have all but abandoned cosmology to the physicists. Some philosophers still hope to advance the field, but – let's face it – the physicists are walking all over them. Regarding the ultimate nature of reality, a Hawking, Lederman, Penrose or Tipler can outsell books by factors of many, and if one actually does bother to look at what the philosophers are putting out, disappointment generally follows. Part of the reason: logic and mathematics are our most powerful tools for organizing abstractions, and most philosophers lack the mathematical creativity and discipline required by modern cosmology. Instead, they indulge in physics envy, exegesis, pointless criticism, or fussy games of pick-and-choose from the dusty scrolls of philosophical patriarchs. What philosophy needs is a new paradigm.

The NST and its logical development, the CTMU, are ideas whose times have come.

11. This whole matter seems to verge on the topic of religion. Can we draw any clear parallels between the two?

Yes. Western religion in particular has tended to separate God from the physical universe in a way requiring the full stratified NST scenario. And oriental religions, by regarding physical reality as an "illusion" concealing a deeper level of being, tacitly embrace the potential for multiple levels of ontological simulation. Once two distinct NST strata are posited, the rest follow by induction. Any attempt to unify science and religion, or the physical and spiritual universe, must therefore construct a mapping between the NST and scientific theory. Given the pressing social, political, and psychological need for such a unification, the importance and conceptual utility of the NST – and its full logical development, the CTMU – are undeniable. That's why the CTMU doubles as a "logical religion" of its own. This was first pointed out in my original paper, "The Resolution of Newcomb's Paradox".

Supplemental Glossary

CAMU / Cellular Automaton Model of the Universe

The CAMU represents a conceptual regression of spacetime in which space and time are re-separated and described in the context of a cellular automaton. Concisely, space is represented by (e.g.) a rectilinear array of computational cells, and time by a perfectly distributed state transformation rule uniformly governing cellular behavior. Because automata and computational procedures are inherently quantized, this leads to a natural quantization of space and time. Yet another apparent benefit of the CAMU is that if it can be made equivalent to a universal computer, then by definition it can realistically simulate anything that a consistent and continually evolving physical theory might call for, at least on the scale of its own universality.

But the CAMU, which many complexity theorists and their sympathizers in the physics community have taken quite seriously, places problematic constraints on universality. E.g., it is not universal on all computational scales, does not allow for subjective cognition except as an emergent property of its (assumedly objective) dynamic, and turns out to be an unmitigated failure when it comes to accounting for relativistic phenomena. Moreover, it cannot account for the origin of its own cellular array and is therefore severely handicapped from the standpoint of cosmology, which seeks to explain not only the composition but the origin of the universe. Although the CAMU array can internally accommodate the simulations of many physical observables, thus allowing the CAMU's proponents to intriguingly describe the universe as a "self-simulation", its inability to simulate *the array itself* precludes the adequate representation of higher-order physical predicates with a self-referential dimension.

(...) In other words, the reason that the Cellular Automaton Model and other so-called "computational models of reality" fail as models of evolution is that they cannot account for the evolution of computation itself, the array or machine or circuit in which computation occurs, or even the relationship between the array and its contents. This leads naturally to the topic of manifolds. A cellular array is a discrete manifold in which neighboring cells are separated and connected by boundaries. In language theory, the concept of a cellular array corresponds to structural syntax, but the correspondence breaks down where the grammatical or generative part of syntax requires more of a setting than a mere discrete manifold can provide. Similar objections apply to any manifold in which discrete fitness landscapes are regarded as "points".

Consciousness

Telesis factorizes itself into dual aspects of reality: form and substance, which given their dynamical oscillation, link mind and physical reality in conspansive cycles. Consciousness is that which is quantized to make the "components" of reality, and through which reality "self-simulates".

(...) The digital physics crowd has been orbiting the CTMU for decades, seemingly without a scintilla of awareness. But in the CTMU, reality would be more accurately described as a "self-protocomputing consciousness". In order to self-compute, reality would have to be, or be simulated in, a computer; but then the existence of the computer must be explained, and this takes more than the process to which we refer as "computation". Specifically, it takes a deeper metaphysical process, an ontological generalization of computation previously defined as "protocomputation". Calling reality a "self-computation" turns out to be insupportable without this generalization.

Conspansion

Conspansion consists of two alternating phases of matter and energy, inner expansion as an IED (Inner Expansive Domain) at the rate **c**, and requantization as an event-component. Requantization is just the collapse or decoherence of the quantum wavefunction, as computed by the simulative computation going on in the (coherent) IED, with a reduction in scale that follows from the absence of an external metric against which to obtain an "objective measurement" of the size of the universe. This implies that the size of the universe is for all practical purposes invariant, and its apparent "expansion" (from an interior vantage) is actually a relative shrinkage of content (from a global vantage). Eddington knew about this global-local duality, even realizing that as objects shrink, time scales would have to shrink apace, i.e., time would have to accelerate or contract. But he didn't know how to do the logical setup for the "shrinking atom" scenario. The CTMU does the setup, in the process solving problems Eddington could only have dreamt.

Conspansive Duality

The laws that govern a system may be reposed in the space that contains its objects, or in the objects themselves. Classical physics reposes everything in space, applying spatial concepts like vectors and tensors to fields outside the objects. However, it is possible to apply a logical transformation which inverts this picture, turning it "outside-in". This results in a "distributed subjectivization" in which everything occurs *inside* the objects; the objects are simply defined to consistently internalize their interactions, effectively putting every object "inside" every other one in a generalized way and thereby placing the *contents* of space on the same footing as that formerly occupied by the containing space itself. Vectors and tensors then become descriptors of the internal syntactic properties and states of objects. In effect, the universe becomes a "self-simulation" running *inside* its own contents.

This view, which is complementary to the conventional geometric one, is called *transductive algebra*. The "dual" relationship between geometry and transductive algebra is called *conspansive duality*. In conjunction with other principles including hology and SCSPL-infocognitive-telic reducibility, conspansive duality can afford fresh insight on the nature of reality and the physical world. One simply takes the conventional picture, turns it outside-in, puts the two pictures together, and extracts the implications. The relationship of this new picture to the old one is extensional rather than competitive, embedding the geometric model in a larger and more complete conspansive model uniting it with its dual model. In any case, all equivalent models are on an equal footing, and the only scientific hypotheses on which doubt could be shed as a result of this extension are those based on the fallacious assumption that the geometric model is the "whole story".

Dual-Aspect Monism

The CTMU treats mind and matter as coinciding aspects of a coherent monic substrate in a kind of framework known generically as "dual-aspect monism". Instead of being held separate, or one being gratuitously painted onto the surface of the other, they coincide in a telic "metasubstance" from which mental and physical processes emerge together in a way reflecting its self-referential, self-simulating, teleological nature. Mind is thus no longer trivialized, but meaningfully coupled with physical reality.

ERSU and USRE

In addition to generalized utility and generalized volition (teleology), the universe also possesses *generalized cognition* (coherent self-recognition). By any reasonable definition of the term "mental", this makes the universe *mental* in a generalized sense, where "generalized" means that these attributes conform to general functional descriptions of what humans do in the process of volition, cognition and mentation. The "coherent self-recognition" feature of reality appears as an explicit feature of conspansive spacetime, a model-theoretic dual of the expanding cosmos. Whereas the expanding cosmos is simplistically depicted in terms of a model called ERSU, short for *Expanding Rubber-Sheet Universe*, conspansive spacetime is depicted by a model-theoretic dual of ERSU called USRE, short for the *Universe as a Self-Representational Entity*. While ERSU is a product of Cartesian mind-matter dualism that effectively excludes mind in favor of matter, USRE, which portrays the universe as a "self-simulation", is a form of dual-aspect monism according to which reality is distributively informational and cognitive in nature.

(...) The cosmos can't be expanding in any absolute sense, because there's nothing for it to be expanding into. Therefore, we must invert the model in a way that "conserves spacetime"; the total "amount" of spacetime must remain constant. When we do so, the cosmos ceases to resemble a balloon inflating (extending outward) over time, and instead becomes an inward superposition of sequentially related states. The best way to think of it is in terms of a cumulative embedment of Venn diagrams (of state) on the inside surface of a sphere of extrinsically indeterminate size. As matter shrinks relative to the sphere, time speeds up. In Eddington's words: "Smaller and smaller. Faster and faster. ..."

Eddington's phrase for his primitive glimpse of conspansion was the "theory of the 'shrinking atom'."

(...) In the expansive model, photons and planets alike move by rotation and translation through a background called "space", which, while possessing sophisticated properties like Lorentz invariance, is still considered an independent medium containing particles and fields. In the conspansive model, on the other hand, motion is an endomorphism defined with respect to those nested Venn diagrams we mentioned, with outer circles playing the part of quantum wavefunctions with respect to inner ones, and inner ones playing the role of "moving particles" with respect to outer ones.

Because the circles in the Venn diagrams are just lightcone cross-sections corresponding to wavefunctions – because of the conspansive identification of lightcones with wave functions and worldlines with particle trajectories – the CTMU defines a general relationship between relativity and quantum mechanics. We end up with two kinds of motion: a primary "inner expansive" kind representing the growth of Venn diagrams and the interpenetration of their spherical wave functions, and a secondary "requantizative" kind representing the relative shifting of diagrams as particles and galaxies undergo "incidental" spatial rotation and translation. But this implies a logical distinction; simplistically, the first kind of motion characterizes a "simulative" phase of matter, and the second kind characterizes a "simulated" phase. Time, calibrated in terms of primary motion, is a process of dualization between these logically distinct levels. The universe becomes a "self-simulation" ... a Self-Representational Entity.

Hology

Hology, a logical analogue of holography characterizing the most general relationship between reality and its contents, is a form of self-similarity whereby the overall structure of the universe is everywhere distributed within it as accepting and transductive syntax, resulting in a homogeneous syntactic medium.

(...) Independently of arguments for the Holographic Principle in string theory, the CTMU logically establishes that the universe is indeed a holographic self-simulation. The difference between the stock cosmic hologram and the CTMU "hologram" is that the latter is a trialic hological language which, because it is a metaevent and metaprocess as well as a metaobject and metarelation (the metacausal analogues of event, process, object, and relation), displays spatiotemporal closure. As required by the self-distribution of M (and arguably by the definition of "hologram"), the cosmic hologram is replicated in every identification event or internal point of the conspansive manifold. In the CTMU hologram, reality everywhere "simulates itself" through, and for, its own constituents.

IED/Inner Expansive Domain

Spacetime is "qualitative" in the sense that space is like a layered sequence of Venn diagrams (spacelike hypersurfaces to which cosmic time is normal) in which circles represent predicates defined on physical observables, and time is like logical substitution within those domains according to various rules of implication. The domains or Venn circles are called "IEDs" and are associated with events from which objects are interpolated (to illustratively adopt an oversimplistic graph-theoretic model, object worldlines amount to edges in a spacetime graph whose nodes are events). These "qualitative" IEDs expand at the rate **c** and then collapse into a new event within their mutual intersects. Inside the IED you have simulative computation or "cognition"; in the compact output event to which the IED collapses in cosmic time, you have the mutual perception (interaction) of two or more objects. The interior of the IED is wavelike, cognitive and simulative; the output event is particulate, informational and objective (that's "wave-particle duality", actually a temporal alternation between spatiotemporal and objective states of material being). In this way, spacetime is identified with the mathematical and physical laws that govern it. This spares us the absurd necessity of having a nonphysical "storage space" for the laws of physics (or by extension, a nonmathematical storage space for the laws of mathematics).

Metatime

Reality can be envisioned as a self-simulation or metaverse-universe coupling. Time and causation are display functions; metatime and metacausation (which is really "why things happen") are processor functions. Metatime is orthogonal (perpendicular) to time. God distributes over time and space via metatime, which makes God "atemporal", or in the vernacular, "eternal".

Protocomputation

Whereas computation is a process defined with respect to a specific machine model, e.g. a Turing machine, protocomputation is logically "pre-mechanical". That is, before computation can occur, there must (in principle) be a physically realizable machine to host it. But in discussing the origins of the physical universe, the prior existence of a physical machine cannot be assumed. Instead, we must consider a process capable of giving rise to physical reality itself ... a process capable of not only implementing a computational syntax, but of serving as its own computational syntax by self-filtration from a realm of syntactic potential. When the word "computation" appears in the CTMU, it is usually to protocomputation that reference is being made.

SCSPL / Self-Configuring Self-Processing Language

In the theory of computation, a "language" is anything fed to and processed by a computer; thus, if we imagine that reality is in certain respects like a computer simulation, it is a language. But where no computer exists (because there is not yet a universe in which it *can* exist), there is no "hardware" to process the language, or for that matter the metalanguage simulating the creation of hardware and language themselves. So with respect to the origin of the universe, language and hardware must somehow emerge as one; instead of engaging in a chicken-or-egg regress involving their recursive relationship, we must consider a self-contained, dual-aspect entity functioning simultaneously as both. By definition, this entity is a *Self-Configuring Self-Processing Language* or SCSPL. Whereas ordinary computation involves a language, proto-computation involves SCSPL.

(...) Fundamental to this new model (CTMU) are revisions of basic physical concepts including space, time, matter and motion. Space, once a featureless medium aimlessly proliferating through cosmic expansion, becomes a distributed syntactic structure iteratively reborn of matter and subject to conspansive evacuation and rescaling. Time, previously envisioned as a quasi-spatial linear dimension along which the cosmos hurtles like a runaway locomotive, becomes the means by which the universe self-configures ... an SCSPL-grammatical symphony of logico-linguistic transformations played by the self-creating cosmos. Lumps of matter, no longer the inert pawns of external laws of physics, become SCSPL syntactic operators containing within themselves the syntactic rules by which they internally process each other to create new states of physical reality. And motion, once seen as the passage of material attribute-ensembles through adjacent infinitesimal cells of empty space displaying them as content, becomes an iterative, self-simulative sequence of endomorphic self-projections by moving bodies themselves.

References

Bostrom, N. (2003) Are You Living in a Computer Simulation? *Philosophical Quarterly*, Vol. 53, № 211, pp. 243–255.

Bostrom, N. (2008) The Simulation Argument FAQ (Updated 2011), www.simulation-argument.com

Chomsky, N. (1956) Three models for the description of language. *IRE Transactions on Information Theory*, Vol. 2, Issue 3, pp. 113–124.

Descartes, R. (1996) *Meditations on First Philosophy with Selections from the Objections and Replies*. Edited by J. Cottingham. Cambridge: Cambridge University Press.

Langan, C. M. (1989) The Resolution of Newcomb's Paradox. *Noesis*, № 44.

Langan, C. M. (1997) Some Q&A on "The Resolution of Newcomb's Paradox". *Noesis*, № 130.

Langan, C. M. (2001) Flash from Deep Space: Supernews on Supernovas. *Noesis-E*, Vol. 1, № 3. Originally published in 1998. Available online at megafoundation.substack.com/p/flash-from-deep-space

Langan, C. M. (2002) *The Cognitive-Theoretic Model of the Universe: A New Kind of Reality Theory*. Princeton, MO: Mega Foundation Press. Originally published in *Progress in Complexity, Information, and Design*, Double Issue, Vols. 1.2-3.

Langan, C. M. (2017) An Introduction to Mathematical Metaphysics. *Cosmos and History: The Journal of Natural and Social Philosophy*, Vol. 13, № 2, pp. 313–330.

Langan, C. M. (2018) The Metaformal System: Completing the Theory of Language. *Cosmos and History: The Journal of Natural and Social Philosophy*, Vol. 14, № 2, pp. 207–227.

Langan, C. M. (2019) Introduction to Quantum Metamechanics (QMM). *Cosmos and History: The Journal of Natural and Social Philosophy*, Vol. 15, № 1, pp. 265–300.

Plato (n.d.) The Allegory of the Cave. *Republic*, Book VII, 514 a2 to 517 a7. Translated by T. Sheehan.

Schrödinger, E. (1926) An Undulatory Theory of the Mechanics of Atoms and Molecules. *Physical Review*, Vol. 28, № 6, pp. 1049–1070.

Wheeler, J. A. (1986) How Come the Quantum? *Annals of the New York Academy of Sciences*, Vol. 480, Issue 1: New Techniques and Ideas in Quantum Measurement Theory, pp. 304–316.

Wolfram, S. (2002) *A New Kind of Science*. Champaign, IL: Wolfram Media.

FAQs

Questions 1–3, 9: 2018, Quora and Facebook.

Questions 4–8, 10–11: 1997, *Noesis*, № 130 (*see* References).

Glossary Sources

CAMU: 2001, *Noesis-E*, Vol. 1, № 3 (*see* References); 2003, Critical Questions in forum Brainstorms at ISCID Forums.

Consciousness: 2017, 2020, Facebook.

Conspansion: 1999–2002, Dialogue on the CTMU between Christopher Michael Langan and UltraHIQ community member, Russell Fred Vaughan Ph.D. (Engineering); reprinted in *Noesis-E.*

Conspansive Duality: 2002, *Progress in Complexity, Information, and Design*, Double Issue, Vols. 1.2-3 (*see* References).

CTMU: 1998, The CTMU as the Nexus of Spirituality and Cosmology: An Interdisciplinary Approach to Reality; 2001, *Noesis-E*, Vol. 1, № 3 (*see* References); CTMU Diagrams.

Dual-Aspect Monism: 2018, Facebook.

ERSU and USRE: 1998 (*see* CTMU); 2001, Discussion with a Physicist and Mega Society East Member.

Hology: 2002 (*see* Conspansive Duality); 2017, *Cosmos and History: The Journal of Natural and Social Philosophy*, Vol. 13, № 2, pp. 313-330 (*see* References).

IED: 1999–2002 (*see* Conspansion).

Metatime: 2020, Facebook.

Protocomputation: 2001 (*see* CAMU).

SCSPL: 2001 (*see* CAMU); 1998, The CTMU as the Nexus of Spirituality and Cosmology: An Interdisciplinary Approach to Reality.

FAQs About Reality:
Chris Langan's Social Media Posts
Book One: Quora

To the Members of the CTMU Community who faithfully follow my work

R

S

T

W

Publisher's Note

FAQs About Reality contains Chris Langan's answers to questions posted on Quora during the years he was active, 2016-2019. Langan was banned from Quora for his conservative worldview and uncompromising debate style. Although Langan never broke Quora's rules or violated its terms of service, he was summarily deplatformed without warning. All of his posts were deleted or orphaned. With the help of members of the Mega Foundation and CTMU Community, the majority of his work on Quora was preserved in this volume.

A

Academia

Academic admission

1. Is there a "back door" to top universities?

Absolutely not, unless you have personal connections who can wire you into the system (or a fashionable complexion or sexual orientation which "merits" an exception).

I happen to be in a position to know this for a fact. I've been featured in a NY Times number 1 best-seller as someone who fell through the cracks of the education system, but was nevertheless intellectually well-suited for a faculty position at a top institution like Harvard. Naturally, Harvard ignored this suggestion (apparently, Larry Summers – or should that be Drew Gilpin Faust – was too busy to notice the book and/or too important to care). I was subsequently contacted by a student at Syracuse who thought I should be welcomed onto the faculty; no cigar. Other similar initiatives failed as well.

Prior to that, my wife – who happens to have a very respectable PhD – had scoured academia for PhD programs willing to admit students directly, regardless of prior credentials. She found just a couple, albeit not in the US: they were located in Wales and South Africa. Neither wanted me even as a grad student, despite my evident ability to intellectually annihilate most members of their respective faculties in their chosen disciplines. Finally, we were forced to realize that it was all a facade – at best, these places had merely established a pretext for admitting supposed "outsiders" through their nepotistic, crony-based, insiders-only friends-and-relatives programs. It was a complete joke.

In case this series of unfortunate events seems vanishingly improbable, consider what academia is actually about these days. First, it comprises a diffuse multinational metacorporation which is unabashedly in business for profit. Tuition, government grants, research funding, sports income, endowments, investments – everything with monetary value feeds the bottom line, and this is ultimately all that counts. Secondly, its educational mandate has steadily given way to a mandate for ideological standardization, indoctrination, and socialization, putting it in the business of psychological mass production and social engineering. Unless one appears receptive to its standardized psychological conditioning, one's personal qualifications are all but irrelevant to admission. Thirdly, academia is an intellectual trade union which, while admirably inclusive when the money and credit flow inward, turns highly exclusive when the money threatens to flow outward. Defy it or even try to jump the queue, and unless you're one of the chosen few, it will make an example of you. Best get it straight as quickly as possible: in Academia, Inc., it's strictly pay to play.

But despair ye not, for there is still the conventional route to academic certification: run the admissions gauntlet, sit there like a good little boy or girl for 4–8 years or so, soak up the ignorance and opinionation of your instructors while latching onto any crumbs of actual knowledge that may accidentally exit their mouths as they scramble to advance their academic careers, and incur the customary mountain of debt. Then maybe, *just maybe*, you'll be grudgingly admitted to the club.

(Alas, I no longer have that kind of spare time on my hands. How lucky for me that I can nevertheless make nattering monkeys out of most professional academics, and refrain from doing so more often only because it feels like taking candy from babies. ;)

January 17, 2018

Academic credentials

2. Is it true that "anyone with the slightest experience with academia would laugh at the very notion of some journal editor even looking at a submission from someone without academic credentials, even if that person really was an Einstein"?

Over the years, I've asked roughly the same question of many academics on a more or less casual basis. The answers fell into two categories which may be illustrated as follows:

(A) "Unsolicited papers may be submitted to most academic journals without any mention of credentials and are usually subjected to blind review. Most of the time, the editor and reviewers either know or can figure out with whom they're dealing, and have an incentive to obtain such information insofar as their own reputations are riding on the papers that they accept. In any case, a few papers have indeed been accepted from unknown and uncredentialed people over the years. My advice would be to first submit two or three papers that are more or less obvious and clear-cut – just this side of trivial, but with a little something of merit to perk up their ears –

and only then one's more profound work, which might still be rejected. You might also try convincing a known academic to co-author with you – of course, he or she will probably want first-author credit, so that's something to bear in mind."

(B) "If you're the least bit out of the mainstream, don't bother. 99.9 percent of the time, journal editors know exactly with whom they're dealing, and if you're an unknown with anything important to say, they'll find out about you with some browsing and a couple of phone calls. For a journal editor, the down side of accepting bad work greatly outweighs the up side of accepting passable work, so your identity and credentials *will* be looked into, and you'll never get the benefit of the doubt. Because most work by uncredentialed unknowns is deficient, the assumption will immediately be that your work isn't worth anything; on the other hand, any good ideas in your paper can be misappropriated by your 'blind reviewers' with complete impunity, and as an unknown, you won't be able to get a word in edgewise. My advice is this: pay your quarter or half-million dollars for an advanced degree from a top institution, make a few friends in the field, and then try your luck – you'll get much better results."

So unfortunately, barring exceptional circumstances and solid connections in one's field, the answer would seem to be that yes, the vast majority of academic journal editors do *not* seriously consider submissions from authors without academic credentials. This is especially likely for potentially dangerous or disruptive work that approaches the originality and importance of Einstein's, especially if it is hard to understand or more than a little interdisciplinary (most academics are too busy to spend much time trying to understand what they don't already know, and too specialized to sign off on interdisciplinary content).

Unexceptional papers by unknown authors have a slight but definite advantage over more profound papers, which are nearly certain to be rejected due to the author's low perceived status or "lack of standing in the field". If one's name happens to be publicly known, and one has been subjected to trolling or half-baked amateur criticism which makes one seem "controversial" (as might have been the case with Einstein had the Internet existed in his day), one might as well forget about it – even sterling academic credentials might not have helped.

Here's a similar response that I gave to an interviewer a few years back:

> The academic world is far too wrapped up in itself to make time for anything or anybody else. The situation is exacerbated by a tendency among academics to assume that if it wasn't coughed up from the belly of the academic beast, it can't be worth a glance. The prospects are especially dim for potentially game-changing work that is perceived to run afoul of academic orthodoxy, threaten the economic status quo, or have difficult social or political ramifications.

As I've already mentioned, academia has all but monopolized gainful intellectual activity through its stranglehold on intellectual certification. The dependence of economic opportunity on academic certification is impossible to miss; it should be no less obvious that this dependency relationship extends to the intellectual advancement of mankind. Woe to any would-be contributor who parts ways with the academic machine, for intellectual commerce is governed by a publish-or-perish economy of publication and citation in academic journals, wherein the obstacles to publication and proper attribution are proportional to the obscurity of the author, the importance and controversiality of the topic, and the level and prominence of the periodical. As a result, important technical works by authors without academic credentials and affiliations are unlikely to be published or cited, and even if these authors beat the odds against them, they cannot benefit in any meaningful way from their effort.

There are at least two obvious reasons for this situation.

(1) Much like an intellectual trade union, academia reserves all of its benefits for members, affording nonmembers no standing, no incentives, and no avenue of recourse should others use their ideas without credit. By making sure that people cannot get along without academic credentials, academia assures itself of a continued influx of paying clients and thus feeds its pyramidal growth economy without necessarily delivering all of the educational services it owes the public.

(2) Professional academics consider it risky to associate with those who may be perceived as "unqualified", preferring to cite well-credentialed, well-connected authors likely to reflect well on their academic reputations. This aversion to risk applies *a fortiori* to journal editors and reviewers, and even to the commercial publishing houses specializing in scholarly nonfiction. This greatly increases the likelihood that meaningful contributions by academic outsiders will not be published, and if they are, that they will be used without credit.

Unfortunately, this means that people without academic credentials have nothing to gain, but everything to lose, by attempting to publish in academic journals, and thus that academic journals do not qualify as a rational venue for their ideas. Just as one need not step off a succession of cliffs to understand what will happen if one steps off the next cliff, one need not repeatedly hurl oneself at the closed shop door of academia to know that one will simply bounce off, the sound of impact echoing without citation or appreciation if at all.

March 30, 2018

Academic indoctrination

3. What would be your advice for highly intelligent youth who have lost academic motivation as a result of being unchallenged?

First, let me say that I sympathize with what would appear to be the motives behind this kind of question, which no professional educator is in a position to answer honestly. But that being understood, you should probably resign yourself to the following facts:

(1) The main purpose of modern education is not to prepare you for life as a scholar. In fact, it is not for worthwhile education of any kind. It is for indoctrination, socialization, and classifying you as either (a) a good little drone who can be put to work and squeezed interminably for revenue by your "betters" in the power structure, or (b) someone who should be marginalized and economically disenfranchised *en route* to possible incarceration by the penal system, at which point your "betters" will have their way with you anyway. All pretensions to the contrary exist strictly in furtherance of this goal. You may find a few instructors who delude themselves about this and actually try to help their students learn, but this is no longer the norm.

(2) Intelligence has been completely replaced by academic credentials as an employment criterion. You do not inhabit a meritocracy; the world you inhabit is run by only moderately intelligent, mostly filthy-rich sociopaths who place no value on truth or knowledge for the betterment of mankind or for their own sake. Thus, your prospects of social and economic advancement are not enhanced by extraordinary intelligence. They have far more to do with your credentials, your connections, and sadly enough, your complexion (the more non-White you are, the better you can avail yourself of affirmative action, take employment from better-qualified people, and avoid termination due to any bad attitude you may exhibit in negative appreciation for the affirmative action you received). Remember, capitalizing on one's intelligence requires material resources, and if those resources are denied you, your intelligence will not help you.

(3) If you flounder, do not expect your teachers to help you unless you have the proper complexion for it. If you are nonwhite and thus fashionably complexioned, educators can get bureaucratic brownie points by treating your mind as "a terrible thing to waste"; bending over backwards to help you makes them look good to their superiors in the bureaucracy. But if you are White, then you are highly likely to be deemed (a) automatically guilty of "White privilege", (b) an undeserving beneficiary of US colonialism who needs to be punished for the unspeakable effrontery of your existence to the rest of mankind, and (c) justly destined for demographic genocide as "more deserving" immigrants from overpopulated Third World cesspools, transported here using what little money you may have succeeded in accumulating, rightfully displace you from your ancestral homeland and the human gene pool.

(4) If you are rejected by the "education" system for any reason, you become ineligible for participation in the modern economy. Without connections, you probably won't be able to find work as anything more than a janitor (if

that), and your life will be rendered economically worthless. It is the possible avoidance of this plight for which you are paying when you buy a college "education". It used to be that a college education was a virtual guarantee of decent employment; this is no longer true, at least for White people. It is only the fear of complete, permanent economic exclusion that induces modern students to pay up to 20 (twenty) times for academic credentials what they cost in the 1970's (that extra money is being stuffed in the pockets of an ever-expanding pool of corporate-administrative yuppies, by the way – think of an insatiable army of Larry Summers clones). As soon as you are rejected by the education system, you effectively become ineligible for gainful employment; you probably won't be able to find a decent job of any kind, and your life will be rendered economically worthless as a result. I'm sure you can imagine the attending hardships.

(5) In a totally materialistic society, intelligence is of value only when it is turned to exclusively material pursuits ... that is, to the acquisition of money and what it buys. Your worth as an individual is measured strictly by the amount of money your socio-economic betters think they can make from you, or in rare cases, the amount of public credit they think they can get for helping you. Yes, you may be allowed to console yourself with feel-good platitudes like "money isn't everything" and "the best things in life are free"; talk, as they say, is cheap. But let's face it, you won't have much time for idealistic soliloquies as you scramble, against all the odds, to survive.

Incidentally, I haven't just rendered an opinion here out of prejudice; I've been invited to answer this question for a reason – i.e., because of who and what I am – and am merely explaining what I've learned from my own life experiences. That is, encouraged by the fact that I've never met a college professor I couldn't crush intellectually like an aluminum beer can, I went the hard way and did what I felt called to do in spite of everything you've just read. Given the will to do the same, you can succeed on your own terms even after being thrown away by the venal and corrupt indoctrination apparatus that passes for an education system in this country. You'll probably remain poor, unfortunately, but forewarned is forearmed.

Doubt any or all of this if you like, but don't say that no one tried to set you straight. It's simply not what many of us think it is out there, and the sooner you realize it, the better off you'll probably be.

December 26, 2017

4. Does Chris Langan realize that his brash, impertinent rants against academics are likely contributing to his theory not being taken seriously by many people who have heard of him?

Yes, I realize that openly criticizing academia is likely to precipitate a certain insufferable subset of academics and academic administrators into a permanent sulk. But I doubt that anything is lost by it, as at this point, I have ample reason to think that I'd never have gotten a fair shake from that crowd anyway. Accordingly, when people ask questions about academia and I choose to answer them, I'll continue to do so honestly.

Secondly, I don't "rant" against academia *as it should be*; I deplore *what it has become* in spite of the best intentions of its progenitors. Academia, having turned away from its educational mandate to pursue alternative missions including disinformative mass indoctrination, social engineering, shameless profiteering, and pandering like $20 hookers to the military-industrial-security complex and the purveyors of identity politics, is currently destroying Western Civilization, an institution that is far greater than academia and which must be preserved for the sake of mankind. Criticism, and more if necessary, is therefore exactly what academia needs.

Thirdly, I was ranted at, and against – often behind my back, I later learned – by academics at times in my life when it could (and did) hurt me the most, which of course had the cumulative effect of permanently depriving academia of any immunity from criticism that I may previously have granted it. Academia employs an embarrassingly large percentage of mean, nasty, sneaky people of Machiavellian bent, who thrive by positioning themselves advantageously in bureaucratic webs and anonymously pulling strands to the detriment of their chosen enemies. Not good at all, especially when they target those with important contributions of which they are themselves collectively incapable.

It's like this. If you're an organizational pickpocket, and you get caught lifting someone's wallet, and you really want forgiveness, you don't blame the victim. You suck it up, admit what you did, own up to your moral flaws, give back the wallet, and promise to do better. Why does everyone seem to assume that academia is above any need to do the same? Think about it a little.

[**Note**: This question closely resembles one which was erased by the questioner after I had already responded to it, and after my response had received well over 20 upvotes in the course of two hours this morning. For future reference, I don't appreciate having my written material destroyed by indignant trolls after their trolling backfires on them – if you want to be a smart ass, learn how to take your lumps. Thanks for your attention.]

December 29, 2017

Academic jacket

5. What is it like to be fired as a PhD student for not performing well?

One of the most-viewed answers to this question includes the sentence "The University is actually a rather forgiving institution." This may be true on some occasions, but a qualification is in order: as a rule, the university will treat you "forgivingly" only in proportion to the amount of money, prestige, and approval you can bring it. If it sees no prospect of gain from you, and if you can't even help fill an ethnic quota for it, then once something too pejorative goes into your "jacket", your academic career is quite likely to be over.

In case the meaning of "jacket" is unclear, most of us know that grade school and high school teachers everywhere are fond of threatening problem students with critical notations on their "permanent records". "Jacket" is synonymous with "permanent record". This general phenomenon is not confined to academia; for example, those who are sent to prison quickly learn that there is a "jacket" which follows (or more accurately, precedes) them as they make their ways through the penal system. Even when they are moved from one prison to another, its contents determine how they are treated. **As it is for convicts, so it is for students.**

One almost never sees the contents of one's own jacket. Consequently, even if it is packed with sheer libel motivated by the personal animosity and venom of those with access to it, no one but its subject is ever held to account for it. It typically emerges from its locked file cabinet only when being secretly updated or reviewed by parole boards or academic star chambers. Once a vindictive academic bureaucrat has gotten into a snit over one's "bad attitude" and written something awful in it – and academia notoriously teems with vindictive backstabbers who thrive and proliferate under layers of bureaucratic secrecy – one might as well call it a day (or, given the dire economic cost of not having academic credentials, a life).

In this unfortunate event, academia expects one to blame everything on oneself. "Oh woe is me!" one is expected to wail. "I screwed up! My folly knew no bounds!" And indeed, the student sometimes bears a large measure of responsibility for his/her own failure. But sometimes, grave injustices are done by academic miscreants who, while remaining safely hidden in the shadows, snicker into their palms with sadistic glee at the pain they have succeeded in causing. The more personal interaction a student must have with professional academics, the greater the risk. The risk becomes higher than ever in graduate school, where the wrong word to some professorial intellectual narcissist could spell total disaster.

We all know that academia likes to portray itself as spotless and above reproach; this is only natural given its need to justify its accelerating consumption of societal resources. But sadly, this image is increasingly undeserved. When it comes down to the nitty-gritty, academia is exactly what it appears to be: a venal bureaucratic rat's nest seething with petty rivalry, intellectual jealousy, Machiavellian intrigue, and internecine political conflict. At this point, no one should make the mistake of assuming that those who have found their callings within it are immune to its rot, and no one who finds oneself on the wrong side of it should rush to blame it all on oneself.

January 24, 2017

Academic reception

6. Is Chris Langan probably the greatest philosopher and mathematician of all time?

Let's just put it like this: if my work were understood in academia, which jealously controls the distribution of credit for advanced intellectual productions and has at its disposal a vast army of enforcers including proud certificants, adoring groupies, and Internet trolls who all suffer from something like the Stockholm syndrome and therefore treat it as their Mother Church, some people would probably be making some very positive statements along those lines.

However, in addition to having parted ways with academia, I suffer from two related problems: (1) many people evidently cannot understand what I write about the CTMU, and (2) instead of blaming any part of this on themselves, they blame it entirely on me. These problems can be expressed a bit more formally as, respectively, the Hollingworth "30 point rule" and the Dunning–Kruger effect. The former says that even people who are measurably smarter than average will find it hard or impossible to share their deeper thoughts with those whose IQs are more than about two standard deviations lower, and the latter says that less intelligent people tend to overrate their own intelligence relative to that of more intelligent people.

Both of these problems have been vastly amplified by decades of intensive social engineering designed to shift the intellectual bell curve objectively downward while shifting the subjective intellectual self-regard of the majority upward. The main tool of the social engineers is the indoctrination engine of Academia, Inc., an engine so efficient that it is in fact self-indoctrinating; its members actually believe the orthodoxies and political doctrine they dispense. Academia constitutes a pyramidal growth economy which has now passed the saturation point, but quite possibly has entered its decline too late to do society any good.

Obviously, as academia has contrived to establish a near-monopoly on advance certification for any form of employment more desirable than ditch-digger and toilet-cleaner, academic rejection costs people like me dearly. However, although I was initially willing to take some of the responsibility for my style of expression and lack of formal education, I have since learned in a variety of ways that this was never actually justified. That is, I've been forced to conclude that the nonrecognition of me and my work is academia's fault, academia's responsibility, and academia's future problem. (We could add "atheistic trolls" to the mix, but this should be enough on which to chew for now.)

I hope this helps clear things up a little.

July 26, 2018

Tuition inflation

7. You are given 20 million dollars up front in cash, or you are given access to higher education of any sort for life. What would be your choice?

Take the twenty and spend it quick. Tuition inflation guaranties that in 10–15 years, $20 million won't be enough to pry any self-respecting corporate administrator of Academia, Inc. out of his diamond-encrusted gold-and-ebony faculty-powered litter, and the deal would be broken. Plan ahead, for Pete's sake!

February 04, 2017

Afterlife

Possible outcomes

8. How many kinds of after-life existences are there? Or do we simply merge with God and have our soul/personality dissipate into one consciousness?

On the most basic level, just two: one's soul, or connection to God, is intact (good news), or it is corrupt and the connection is severed (bad news).

Without a soul that is at least partially intact, embedment in God cannot be maintained, and one is cut off from the source of existence. In this case, "the afterlife" consists of the disintegration, reduction, and recycling of personal identity, which is desperately resisted and thus a source of unimaginable despair (some call it "hell"). On the other hand, if the soul remains intact, the identity can persist in any of a number of specific ways depending on its strength and configuration.

Due to the possibility of a negative "life-after-death" outcome, this may not be the most popular answer you receive for this question. However, it is the *real* answer with a real basis in metaphysical logic.

May 15, 2017

Atheism

Atheism or agnosticism

9. How is atheism different from agnosticism?

An "agnostic atheist" is merely an agnostic, unless someone can explain how one differs from the other.

A case can be made that atheism has two definitions, because it can be broken down into lexical constituents in two ways: (1) *athe(o) + ism* (belief in the nonexistence of God) and (2) *a + theism* (nonbelief in the existence of God). As an atheist, one can be either 1 and 2 (1 implies 2), or just 2. "Agnostic atheists" seem to have chosen 2 alone.

Here's their problem: if one chooses definition 2 alone, this implies that one is not described by definition 1, which makes one an agnostic plain and simple (one believes in neither the existence nor the nonexistence of God). Only if one subscribes to both definitions 1 and 2 is one a real atheist. Take your pick.

Quora has many people claiming to be "agnostic atheists". It almost looks like none of them have thought it through as anything but a rather sneaky way to swell the ranks of atheism by reclassifying agnostics as atheists with a trick of etymology. Unless, of course, some "agnostic atheist" can satisfactorily explain the nomenclature.

I suspect that some might explain it as follows: "An agnostic atheist is one who believes that *there exists no evidence* for the existence of God" (an epistemological rather than an ontological claim). But aside from the fact that this has never been shown, many agnostics believe it as well, so again there is no real distinction; every "agnostic atheist" can be more economically described as an agnostic.

In other words, the ruse has failed.

June 26, 2017

10. Do atheists solely rely on logic?

There can be little doubt that atheists, as diverse as they may be, often seem to *think* that they "rely on logic". Many appear to pride themselves on it.

However, when one closely examines their arguments, it often turns out that their ideas of logic leave something to be desired. For example, according to the kind of "logic" on which many of them depend, there is no difference between an agnostic and the kind of "atheist" calling himself an "agnostic atheist".

To wit, agnostics and "agnostic atheists" both answer NO to both of the following questions:

(1) "Do you believe in the *existence* of God?"

(2) "Do you believe in the *nonexistence* of God?"

Previously, a real atheist was conventionally defined as someone who answers NO to the first question and YES to the second. This is quite a change!

The idea behind this semantic switcheroo seems to be that an "agnostic atheist" who denies that he actively believes in the nonexistence of God has no burden of proof. But clearly, a self-styled "atheist" who does not actively believe in the nonexistence of God is just another agnostic!

Obviously, a group which has trouble properly distinguishing itself from another group with which it has historically been contrasted may not be the best group on which to rely for a definition of "logic".

June 10, 2018

11. What is the evidence in support of atheism?

There are two definitions of atheism. It can mean either (1) the belief that God does not exist [*athe(o) + ism*], or (2) a lack of belief in the existence of God [*a + theism*, which, when accompanied by a lack of belief in the nonexistence of God, is identical to agnosticism]. There is no evidence for atheism under either definition.

Bear in mind that in light of the law of excluded middle, those "atheists" who are simply agnostic are merely begging the question; they cannot maintain their position indefinitely unless there is no evidence either for or against the existence of God ... but there is no evidence that this is the case. In fact, there may be evidence that God exists, e.g., the existence of reality itself, with which God may coincide in whole or in part. So unlike atheism (1), atheism (2) is not really a definite theological position.

Note that many of the self-styled "atheists" answering this question are merely agnostics who falsely believe that they have permanent grounds for their lack of belief. The proper question to ask such an "atheist" is this: "What is your evidence that the existence of God cannot be proven?" An "atheist" who cannot provide such evidence might as well get off the bandwagon, admit that he doesn't really know anything of theological import, and call himself an "agnostic".

June 17, 2017

Belief

12. Is atheism a belief or a conclusion?

Atheism is definitely a belief.

The word *atheism* can be broken down into lexical constituents in two ways: (1) *athe(o) + ism* (belief in the nonexistence of God) and (2) *a + theism* (nonbelief in the existence of God). An atheist must conform to both definitions (1 implies 2), not just definition 2.

Why? If one chooses definition 2 alone, this implies that one is not described by definition 1, which makes one an *agnostic* as conventionally defined in juxtaposition to theism and atheism. That is, lacking knowledge supporting either the existence or the nonexistence of God, one believes in neither.

Many atheists, some calling themselves "agnostic atheists", have taken to claiming that they are covered by definition 2 alone. Why? Because it seems to excuse them from any burden of proof. However, this renders their position theologically indefinite and therefore irrelevant. They merely lack belief in God *without having any good reason* for not believing in God, and thus have no definite theological position. (If one has a good reason for not believing in the existence of God, then one conforms to definition 1 after all.) Only if one subscribes to both definitions 1 and 2 is one a real atheist.

Any atheist who responds by saying that his "good reason" for not believing in God is that "there is no evidence for the existence of God" merely incurs another burden of proof, this one epistemological: he needs to *prove* (or at least confirm) that there is no evidence for the existence of God. But as no atheist is in a position to attest to anything but his own subjective knowledge state regarding such matters, this is out of the question.

So the bottom line is that if you want to call yourself an atheist, then you must believe in the nonexistence of God, period ... and once again, you have a burden of proof to that effect. Otherwise, you're just a garden variety agnostic, and in that case, you shouldn't be butting into theological discussions as though you have anything definite to say. Nor should you be claiming the rational high ground without rational justification of your own.

As for atheism being a "conclusion", that depends on how it is inferred and from what premises. Unfortunately, the best theoretical framework in which to derive such a conclusion is "methodological naturalism", which is simply the assumption or *belief* that only the content of the natural (empirical) sciences is real. In other words, the conclusion is derived from a mere belief, and we have arrived back at the start of the discussion.

July 18, 2017

Blasphemy

13. Atheists, if tomorrow it were proven beyond a shadow of a doubt that gods exist, would you want to remain an atheist? Why or why not?

Despite what one reads here, a large percentage of atheists will choose to remain atheists. This is because, being proselytizers for atheism (blasphemers) or supporters thereof, they sense that they are already lost and have nothing to lose by continuing to evince their contempt for, and implacable hatred of, the Almighty. (If equating "rationality" to hatred seems unjustified, the constantly fuming Richie Dawkins provides a clear example of antitheistic hatred masquerading as rationality.)

When the proof has been spelled out in a way that makes it absolutely inescapable for anyone with the high level of intelligence claimed by most atheists (who generally see themselves as rational skeptics and "Brights"), it will be impossible for them to conceal their blasphemous intent even from themselves. Beholding themselves in the convoluted funhouse mirrors of their own disordered minds, transfixed by a final blinding burst of metacognitive presence restoring the true geometry of nature and exposing their implacable self-hatred, they will seal their own metaphysical fate.

But all is not yet lost. The good news is that self-styled atheists still capable of repentance can always mend their ways by their own volition, provided that they are willing to part ways with those who lack the sense to do so. And in any event, those who remain courageous and pure of heart need never despair, for the fires of hell warm the Kingdom of Heaven.

In the end, the books will balance. Atheists, including most of the Powers that Be, have a severe red-ink problem that only they can eliminate before the bill comes due, and this cannot be accomplished by disingenuous posturing or conforming to PC doctrine or writing checks to fashionable causes.

A word to the wise.

January 10, 2017

14. Are there serious metaphysical consequences for unreasonably defaming the CTMU on Quora?

For those who have already severed their souls, amputating themselves from the identity of reality through blasphemy and moral atrophy, there are no further "metaphysical consequences". Barring an utterly heroic attempt at spiritual reconnection – the worse you are, the more heroic and costly the effort must be – the matter is completely settled, and divine "vindictiveness" has nothing to do with it.

I understand that the typical atheist, being a master thespian when it comes to empty bravado, will giggle like a tickled baby at the very thought. Oh, the sheer entertainment of it all!

But you know what they say: there are no atheists on deathbeds, at least who still have a choice about it.

So I'm afraid that it's laugh now, cry later ... or perhaps that should be, wail and gnash your teeth later.

(Now, how will *that* be for entertainment? ;)

July 10, 2018

15. Are some atheists afraid of the CTMU?

Of course some atheists are afraid of the CTMU.

Which atheists, exactly? The more intelligent ones who have actually read and understood some appreciable part of it.

Their terror owes to the penalty that awaits them for the sacrilege and blasphemy with which they have systematically polluted the minds of entire generations, thereby parting them from the source of being and damning them to perdition.

September 23, 2018

CTMU attacks

16. Why are there atheists who criticize the CTMU without understanding anything of it?

For essentially the same reason that someone might refuse to accept an apple as a piece of fruit because it fails to resemble an orange, banana, or kumquat.

As I've explained elsewhere on this site, true atheists actively disbelieve in God. (Despite the efforts of some to characterize themselves as "agnostic atheists", this phrase is nothing but an oxymoron demonstrating that they misunderstand not only the standard definitions of its constituent terms, but the relationship between faith and knowledge.) In many cases, this disbelief is so active that it becomes indistinguishable from virulent hatred, the nemesis of intellectual clarity and objectivity.

Let's have a closer look. Most atheists subscribe to (methodological or metaphysical) "naturalism", the dominant perspective in academia. For some of them, naturalism is more than just a perspective on reality; they see it as an *explanation* of reality. But as naturalism is fundamentally dualistic, destroying the connection between abstract and concrete reality, it doesn't work above a certain level of explanation; no coherent high-level explanation of reality can be fully interpreted in a naturalistic framework. Trying to do so anyway represents the height of confusion.

In contrast to naturalism, the CTMU is monic. When one tries to interpret a monic structure in a dualistic framework, one merely ends up ripping it in half, at which point one may fall through the gap that one has idiotically created between the resulting pieces, and then compounds the situation by complaining bitterly about it as though it's someone else's fault. This is the sad yet rather comical fate of many atheists who have tried and failed to understand the CTMU. They simply don't have a clue.

Obviously, I need not take responsibility for the intellectual shortcomings of atheists, which in some cases are both profound and numerous. They need to make at least a little effort to address their own conceptual deficiencies before running around making a lot of extremely unpleasant noise about things that they are incapable of comprehending.

February 12, 2018

17. Why do people still have atheistic leanings when the CTMU clearly proves that atheism is *a priori* false?

This recently posted question presently has 54 answers including this one, and another 38 which have been collapsed. I've gone through a few of these answers, and in addition to detecting quite a bit of nastiness, I've found that almost none of the authors displays the slightest knowledge of the theory they purport to be criticizing. Zero ... zilch ... nada.

Not much can be said on behalf of people who operate this way. One can't paint a smiley face on it. It's not as though these people are motivated by a love of truth and the advancement of knowledge; people who conform to that description at least acknowledge the content of that which they criticize.

The only explanation that seems reasonable at this point is that Quora is a dark carnival teeming with virulent atheistic-materialistic troll-clowns, and having driven off other potential targets with the insults and defamation for which they are famous, they are increasingly desperate for something, anything, to attack *en masse*. Hence, they've latched onto the CTMU like leeches, catapulting it to the status of premier troll-magnet at an already degenerate trollfest.

As I've pointed out before, I've never posted a question here (except to replace one that was erased after I'd already answered it), and I've never asked anyone else to post a question here. I've tried my best to provide sincere and informative answers for the questions I've encountered, that's all. But the troll attacks had begun before I even arrived (some even involved "Quora Top Writers"), they've only gotten worse as time has passed, and I'm starting to feel like the only surviving human in Zombie City. It's depressing.

Just in case any innocent, well-meaning naïf happens to bump into this feeding frenzy and mistakes it for a source of information, I'm advising you to avert your eyes. Most of it amounts to pseudointellectual hate-porn, and no decent person wants to dirty his or her eyes with it.

Thanks for your attention.

Addendum: There's something called an "Answer Wiki" at the top of this page. It is not accepting my edits, which is pretty much par for the course on this site, which is largely what makes this site a joke for purposes of getting anything properly established. My attempted addition read as follows:

> **Disputed by the author of the CTMU**. The undeniable fact is that those who express strong dislike of the theory typically fail to understand a single word of it – or at least so it appears from their failure to address its actual content – and the incomprehension often appears feigned rather than honest. Unfortunately, this is highly unlikely to change. Thanks for your attention.

As I say, this place seems to be nothing but an atheist-materialist trollfest. What a shame.

July 9, 2018

Intolerance

18. Is atheism a hoax?

Although atheism is non-factual, it is not a hoax. It exists as a (highly negative) theological doctrine, and many people actually believe in it despite the absence of anything approaching a proof that it is correct. That many atheists now occupy high places in the power structure, or exert pressure on those who do, has allowed them to do very real damage to the cultures and future prospects of Western nations and Western Civilization as a whole.

In contrast to many self-styled Christians who are passive to a fault, and who practice blanket tolerance of atheism on the absurd premise that the Bible somehow instructs them to do so, atheists tend to be highly intolerant, swarming outspoken theists and evincing the most vicious kinds of pack behavior, often attacking the good names, professional reputations, and livelihoods of their enemies. Yet despite their notoriety for this kind of mischief, atheists often characterize themselves as "moral" and "ethical", claiming that morality is independent of God. As this makes morality relative and thus neutralizes it – that's what it takes to explain why atheists embrace many different "moral codes" and definitions of morality – such assertions are preposterous.

This is why big-name atheists like Richard Dawkins and Daniel Dennett long ago settled on a policy of avoiding open dialogue with capable theists. Their stated rationale was transparently disingenuous: "Engaging with theists amounts to legitimizing them!" But of course, their real concern is that engaging with capable theists entails an unacceptably high risk of defeat and could eventually result in a resurgence of theism, an eventuality with which atheistic contempt and/or hatred for God is incompatible.

May 15, 2017

IQ level

19. Why is it that atheism is prevalent amongst high IQ individuals?

The meaning of this question depends on that of "high IQ". Very high IQ individuals – those near the top of the scale – have not usually been atheistic, historically speaking. Atheism has usually been espoused by those whose levels of intelligence range from average to moderately above average, perhaps a bit higher in a few cases (e.g., Voltaire, who has often been claimed by atheists due to his attacks on religious authority, but would be better described as a deist who leaned toward the idea that the universe was intelligently designed).

July 4, 2018

20. How do I stop being an atheist?

Spiritual understanding has two complementary aspects, faith and knowledge. The only real issue is which comes first, as you'll eventually need both. That is, to the extent that faith is true, it entails knowledge, and to the extent that someone who is rational must believe in that which one knows to be true, knowledge entails belief. (Recent atheistic attempts to couple and conflate atheism and agnosticism, e.g., the oxymoronic phrase "agnostic atheism" which seems to drive a wedge between knowledge and belief, are just a transparent ruse for which only fools can fall.)

Many atheists will tell you that there is "no logical or scientific reason" to believe in the existence of God, but they are mistaken. So the first thing you must do in your search for God is to disregard their opinions, which are in any case devoid of logical support. Remember, the vaunted correlation between atheism and IQ peters out above the "danger zone", where a danger-zone intellect is one that is sharp enough to be dangerous to ordinary people, but too dull to recognize and acknowledge its own limitations. (In contrast, when a very intelligent person has no support for his/her opinion, he/she wisely suspends judgment.)

Congratulations for realizing that atheism is something from which you need to escape. It's the first step toward spiritual understanding and redemption.

July 13, 2017

Irrationality

21. Is atheism more logical than religion? If God doesn't exist, theism doesn't "lose", but if God in fact exists, atheist will suffer.

Spiritual understanding has two complementary aspects, faith and knowledge. The only real issue is which comes first, as you'll eventually need both. That is, to the extent that faith is true, it entails knowledge, and to the extent that someone who is rational must believe in that which one knows to be true, knowledge entails belief. (Recent atheistic attempts to couple and conflate atheism and agnosticism, e.g., the oxymoronic phrase "agnostic atheism" which seems to drive a wedge between knowledge and belief, are just a transparent ruse for which only fools can fall.)

Many atheists will tell you that there is "no logical or scientific reason" to believe in the existence of God, but they are mistaken. So the first thing you must do in your search for God is to disregard their opinions, which are in any case devoid of logical support. Remember, the vaunted correlation between atheism and IQ peters out above the "danger zone", where a danger-zone intellect is one that is sharp enough to be dangerous to ordinary people, but too dull to recognize and acknowledge its own limitations. (In contrast, when a *very* intelligent person has no support for his/her opinion, he/she wisely suspends judgment.)

Congratulations for realizing that atheism is something from which you need to escape. It's the first step toward spiritual understanding and redemption.

January 21, 2017

22. How can someone be an atheist, if there is a chance there is a god? Because you can't prove there isn't, then when he dies he will find out he is doomed for eternity.

Being an atheist given the Biblical penalty for blasphemy is indeed very unwise. In fact, given that there is no proof of the nonexistence of God, it is irrational; professing atheism in spite of one's ignorance is like unnecessarily traversing a minefield in pitch blackness.

This irrationality is precisely how we know that atheism is not a mere abstract belief, instead amounting to active hatred of God. It is not for a mere lack of conviction that atheists are sentenced to hell; it is for actively, irrationally negating the existence of their own highest level of being. Obviously, those who are determined to negate their own being on the highest level thereof might as well forget about lower levels, which are automatically extinguished as a result. (Sad, isn't it.)

December 28, 2017

Irresolvable paradox

23. What important part of you was unlocked after becoming an atheist?

Becoming an atheist "unlocks" your capacity to embrace paradox. That is, it breaks you free of your psychological need for logical consistency, enabling you to embrace inconsistency as an integral component of your own psyche. This is not usually a good thing. In the context of God, it is in fact a very *bad* thing, as it deprives you of any sound basis for morality and disrupts your connection to reality.

As for the ridiculous idea that nothing is actually unlocked because atheism does not imply active belief in anything, the simple fact is that atheism *does* imply belief, specifically belief in the nonexistence of God. In other words, atheism means "actively disbelieving in God". "Not believing in God", on the other hand, is simply "non-theism".

Here's what amounts to a mathematical proof that atheists actively disbelieve in God. The word *atheism* can be etymologically deconstructed in two ways: *athe(o) + ism* and *a + theism*. The first literally means "belief in the nonexistence of God" (active disbelief in God), while the second means "nonbelief in the existence of God". But the second definition is not yet complete, because it fails to specify whether or not one believes in the nonexistence of God. If one does not believe in the nonexistence of God, then this amounts to plain old agnosticism, i.e., believing in neither the existence nor the *nonexistence* of God. Only if one actively believes in the nonexistence of God does the second definition cease to equate to agnosticism. In this case, one actively opposes God on the ontological level, and is that to which scripture refers as a "blasphemer".

Here's the bottom line: If you're a real atheist, then you actively disbelieve in God, and are therefore a blasphemer. Otherwise, you're just a plain old garden variety agnostic, in which case you shouldn't run around absurdly calling yourself an "atheist" or "agnostic atheist", as it would only cause others to question your intelligence (and eventually cause you even worse problems than that. ;)

I've already discussed this in several answers on this site. Feel free to look them up if this is still unclear to you.

October 21, 2017

24. Is Atheism as indefensible a position as believing in God is from a purely logical perspective? What do you think?

Atheism – the oxymoronic metaphysical statement that reality lacks a coherent metaphysical identity, which equates to "God" under certain definitions – is an irresolvable paradox. That is, it implicitly *affirms* its own truth while *denying* its own truth. A statement which is metaphysical by construction – i.e., which refers to nature or physical reality from above, occupying a higher level of reference – cannot be true where it explicitly denies its own metaphysical support.

In other words, atheism is too general in the kind of "God" it denies. Where a proper definition of God consists of logical necessities, a blanket denial of God renders itself paradoxical by negating these logical necessities. These necessities can be encapsulated in the necessary existence of a metaphysical identity (of reality), without which there can be no metaphysics, without which the metaphysical position of atheism – not to mention reality itself – cannot exist (because *existence* is a metaphysical and specifically an ontological function). On the other hand, theists who believe in "God" under this definition remain within the bounds of logic.

Many atheists are presently scrambling to sidestep logical objections to atheism by calling themselves "agnostic atheists". However, they might just as well drop the *atheist* from "agnostic atheist" and refer to themselves as *agnostics*. Given the way "agnostic atheism" is being promoted here on Quora (and elsewhere), it appears to be a disingenuous attempt to change the language so as to blur the distinction between atheists and agnostics.

What is the difference between atheism and agnosticism? Obviously, in order to distinguish between an *atheist* and a *non-atheist*, two questions must be answered: (1) "Do you believe in the **existence** of God?"; (2) "Do you believe in the **nonexistence** of God?" An agnostic answers NO to both questions; an atheist answers NO to 1 and YES to 2.

Failing to answer YES to question 2 while calling oneself an "atheist" amounts to declaring oneself *both* an atheist *and* an agnostic; the terms become effectively redundant. This fails to provide enough information to determine one's theological stance vis-a-vis the existence of God.

The only way around this is to claim that an atheist differs *epistemologically* from an agnostic, somehow "knowing" that *neither* the existence *nor* the nonexistence of God can be conclusively established. However, this too entails a burden of proof, and we don't see this burden being met.

So logically speaking, the bottom line is that where God is properly defined as the metaphysical identity of reality – and this can be done with a very high degree of logical precision – **theism beats atheism hands–down**.

May 30, 2018

Metaphysical naturalism

25. Why haven't atheists started a belief system yet?

Because even the very best and most capable of them can be easily crushed by anyone who knows what he's doing. As any high-profile theist can attest, the closest the atheists have ever come to a coherent "theory of atheism" is the "theory" of insult, defamation, and dirty polemics to which so many of them resort after coming up short against someone more skilled at theological argumentation.

Their best "serious" attempt at a distinctively atheistic philosophical theory is called "metaphysical naturalism", which characteristically turns out to be an oxymoron. It holds that nothing exists but the "natural" (physical) objects, relations, and processes which form the content of the natural sciences. In other words, the metaphysical or "supernatural" is completely excluded from reality. Their decision to call what amounts to their ideological flagship "metaphysical naturalism" despite its exclusion of metaphysics stands as an everlasting testament to their collective theoretical incompetence.

Of course, it would be a mistake to suppose that all atheists are unintelligent. It's just that those who do most of the talking typically have psychological issues which occlude their powers of comprehension, sometimes including guilt and a fear of punishment, extreme intellectual narcissism, poor impulse control, emotional infantilism, and a defiant personality accompanied by a deep hatred of authority (especially religious authority).

July 8, 2017

26. Does atheism entail metaphysical naturalism?

Superficially, atheism and metaphysical naturalism differ. The former merely excludes God from reality, while the latter excludes metaphysics, AKA "the supernatural", in its entirety (metaphysics is supernatural insofar as it refers to nature as a whole, and thus exists on a level above that of nature itself). But now let's see what logic has to say about it.

As far as logic is concerned, both metaphysical naturalism (the oxymoronic metaphysical statement that nature has no metaphysical aspect) and atheism (the ultimately equivalent and equally oxymoronic metaphysical statement that reality lacks a coherent metaphysical identity, which equates to "God" under pretty much any reasonable definition) are irresolvable paradoxes.[1] That is, they implicitly *affirm* their own truth while *denying* their own truth (because a statement which is metaphysical by construction – i.e., which refers to nature or physical reality from above – cannot be true where it explicitly denies its own metaphysical structure).

[1] Atheism is paradoxical because it is too general in the kinds of "gods" it denies. In particular, where a proper definition of God consists of logical necessities, a blanket denial of God renders itself paradoxical by negating these logical necessities. These necessities can be encapsulated in the existence of a metaphysical identity of reality, without which there can be no metaphysics, without which the metaphysical position of atheism cannot exist.

All irresolvable paradoxes are equivalent in the sense that they imply "True = False"; and where True = False, anything is true. In this sense, atheism entails metaphysical naturalism and vice versa.

Of course, any irresolvable paradox implies anything whatsoever. But what makes atheism and metaphysical naturalism so obvious as irresolvable paradoxes is the fact that they are oxymorons, i.e., paradoxical on their faces. Their inconsistency is not hidden, but as plain as day.

Only a fool could fall for either of them.

[**Note**: I see several answers here which claim that an atheist is simply a person who answers "no" to question (1): "Do you believe in the existence of god(s)?" This is disingenuous. There's another question that must be answered: (2) "Do you believe in the *nonexistence* of god(s)?" An *atheist* is someone who answers "no" to the first question and "yes" to the second.

Anyone who answers "no" to both questions is merely an *agnostic*. Claiming that only question 1 need be answered to distinguish an atheist from a non-atheist amounts to claiming that *atheism* and *agnosticism* mean the same thing. In fact, they have never meant the same thing and never will. It would be nice if atheists would stop trying to fool everyone about this.]

May 29, 2018

Social injustice

27. If you are not accountable after death, you can do whatever you want to and get away with it. How do atheists defend this void of justice?

Most atheists conflate justice with legality. They maintain that evil is punished in the physical world "with a few exceptions" – after all, some criminals are sent to jail – and that deviations from justice are unavoidable but negligible.

However, in a world run by sociopaths, this is absurd. Sociopathic ruthlessness and dishonesty are systemic, being more often rewarded than punished (the more vile and audacious the better). We have war criminals in high office, mass murderers running the military-industrial-security complex, and galaxy-class thieves and extortionists running the banks and controlling the money supply. Their corruption and injustice are contagious; using government as a means of control, they write laws under which anyone who refuses to go along is economically disenfranchised, incarcerated, or killed. Injustice is thus self-reinforcing, eventually becoming the norm.

It follows that an atheist who considers himself "rational" has no choice but to deny that justice exists in any necessary sense, thus voiding any responsibility to explain its absence. And as the concept of justice flies out the window, so does any enforceable basis for morality, thus reducing phrases like "moral atheist" and "atheistic morality" to meaninglessness.

July 17, 2017

B

BELIEFS

Non-believing

28. What do you call someone when they don't believe in God, Satan, Jesus and religion?

This question is all over the place. If one does not believe in the existence of Satan, then one can be called (e.g.) "a non-follower of any religion incorporating the concept of Satan". If one does not believe in the sayings of Jesus, then one can be called (e.g.) a "non-Christian" and a "non-Muslim". If one does not believe in the existence of Jesus, then one can be called (e.g.) a "non-follower of any Abrahamic religion". If one does not believe in the existence of God *qua* Logos (essentially, logic), then one can be called (e.g.) "illogical" or "irrational"; if one does not believe in God under any definition, then one can be called (e.g.) "atheistic" or "agnostic". If one does not believe in any religion at all, then one can be called (e.g.) "irreligious". If one does not believe in the very existence of religion, then one can be called (e.g.) "delusional". If one does not believe in any of these things, then one can be described as all of the above (and more, very little of it complimentary. ;)

June 11, 2017

BIOLOGY

Evolutionary biologists

29. Why do some evolutionary biologists belittle supporters of Intelligent Design and Creationism?

For several reasons. First, many of them – despite being relatively knowledgeable about some of the mechanisms involved in evolutionary biology – do not understand that the issue of ultimate origins is not strictly scientific, but metaphysical by definition. Some of them even dispute the very existence of metaphysics, subscribing to an oxymoronic philosophy called "metaphysical naturalism" which, despite its name, effectively denies that nature has a metaphysical aspect. In order to get away with it, they simply replace the word *metaphysical* with *supernatural*, and under what they imagine to be the deep cover of this transparent semantic switcheroo, absurdly issue what amounts to a metaphysical injunction against "the supernatural".

Secondly, many evolutionary biologists are hostile toward religion, and even though ID is not defined in religious terms – it is innocuously defined as "the theory that life, or the universe, cannot have arisen by chance and somehow involves the application of intelligence" – everything looks like a nail to people with nothing but a hammer. I learned this the hard way many years ago, when I was hounded all over the Internet by trolls accusing me of being an "Intelligent Design Creationist" when I actually had a preexisting theory that bears no resemblance to the stock ID theory promoted by the ID Movement. Once a surfeit of these unpleasant creatures fixate on you, crouch like skunks, and let fly in your direction with page after page of insult, venom, and illogical nonsense, they don't stop. The spray continues to exit their reeking rear ends until their hostility is accidentally refocused by, e.g., a major natural disaster occurring in their immediate vicinity.

Thirdly, some people get their jollies by putting on exhibitions of spite, and certain less talented academic scientists have learned how to attract unmerited attention and approval by engaging in pack behavior, mocking and belittling those attacked by their more eminent colleagues in the hope that a bit of prestige will trickle down on them. No, it's not nice, and it's not pretty to watch. But anyone who thinks that those who specialize in "debunking ID" are driven or inhibited by insight, compassion, or cold scientific rationality needs to pry the tube of model airplane glue from one's nose and suck down a hot cup of coffee.

This comedy will continue until the dialogue is shifted into its proper context, which is presently understood by neither side of the debate. But to judge from the incredible tenacity of those who latched onto "ID Creationism" twenty years ago like rabid badgers and have never since managed to let go of it or even gain an inch of perspective on it, it's probably not a good idea to hold one's breath.

May 10, 2018

C

Communication

Barriers of communication

30. What are some examples of the barriers of communication?

The first step is to classify the requirements of communication and the possible causes of failure.

Communication is the transfer of information (a message) from a sender to one or more receivers; hence, any problem with the form of the message (encryption, improper syntax, semantic degeneracy, etc.), or the mental or computational state, structure, or dynamics of sender or receiver, impairs it. One also needs a medium and channel for the transfer; if the medium is deficient or too high in entropy, or the channel is blocked or subject to excessive noise or interference, communication is impaired. The content of any message must be placed in a properly layered context, up to and including the language in which it expressed, which may be limited in its expressive capacity or designed and utilized for deception and control rather than honest communication. Obviously, if the message contains lies (content is corrupt), or if it must be understood in an ambiguous context or in the context of a false narrative or body of false doctrine (context is corrupt), a communication problem exists.

Now here's the really bad part. (1) Largely because governments, corporations, and social engineers (including academia and the mass media), who are self-serving and ethically confused or uninhibited, work under the (often accurate) assumption that people and society can be controlled by twisting and manipulating information and communication, we are presently dealing with all of these problems in spades. (2) These problems render democracy, scientific progress, and a proper understanding of reality all but impossible, guaranteeing the continued political, socio-economic, and psychological degeneration of mankind.

[Of course, there is a more "touchy-feely" interpretation of this question which might be better expressed as "What are the psycho-social barriers between human beings which can impair their ability to express or understand each other's thoughts, feelings, and intentions?" The answer would obviously include every source of emotional conflict and personal confusion known to man, complicated by every personality disorder and combination of psychological traits and circumstances under the sun.]

May 24, 2017

Computer Science

Self-simulation

31. What does the CTMU contribute to the field of computer science, especially artificial intelligence?

Regarding computer science, CTMU originally stood for "Computation-Theoretic Model of the Universe" before it was extended to encompass the entirety of cognition. It is by far the most advanced theory of reality as a self-simulation; other so-called "simulation theories" are beneath comparison.

Regarding AI, the CTMU is not about "artificial" intelligence. It is about *real* intelligence, including *synthetic* intelligence. Anything less is just the race to make money by selling rope to the would-be hangmen of mankind in the form of mechanical golems that can be used to subvert, suppress, and manipulate human cognition, communication, and action.

September 5, 2018

Consciousness

Consciousness and life

32. What do you think came first, consciousness or life?

Neither; they come as one. Attempting to define either without using the other as a definitional parameter is meaningless; they have an irreducible semantic overlap.

Of course, the linguistic operation of definition leaves much room for creativity. If one likes, one can simply define these two terms, *consciousness* and *life*, in any of four ways: (1) they are mutually independent; (2) life depends on consciousness, but not vice versa; (3) consciousness depends on life, but not vice versa; (4) they come bound together in mutual dependency. But then one faces the problem of mapping one's definition to reality, and that's where one can run into trouble.

Obviously, condition 1 equates to **Cartesian mind-body dualism**, with consciousness standing for *mind* and life standing for *body*. That's a no-go, as it precludes any meaningful relationship between the two. There's simply no way to hold them together in a single unified reality such as the one we inhabit.

Condition 2 is almost right, but there's a problem: consciousness – understood simplistically as "self-awareness" – immediately requires that an entity have dual aspects: (a) that aspect which is subjectively aware, and (b) that aspect which is the object of subjective awareness. This self-duality immediately implies a view of reality (the CTMU) in which life and consciousness are "metasimultaneous" in the sense of condition 4.

Condition 3, the "emergence" of consciousness from certain physical phenomena to which "life" has been attributed by definition, is the dominant view in academia and elsewhere. Unfortunately, it is absurd insofar as it merely amounts to the supervenience of mind on physical substance and process, and thus equates to physicalism – or if one prefers the conventional oxymoron, "metaphysical naturalism" – which in effect denies the distinction between mind and matter, or consciousness and life, while implicitly claiming that the latter (physicality) has priority.

Finally, we have condition 4, which is true by implication (condition 2) and elimination of conditions 1 and 3. As previously mentioned, this requires a new paradigm called the CTMU (*Cognitive-Theoretic Model of the Universe*) in which *life* and *consciousness* have a higher form of simultaneity, *metasimultaneity*, which is technically defined within the theory.

[**Note regarding the standard physicalistic view of emergence** (condition 3): Obviously, substance does not exist apart from its properties. Properties *characterize* substance, which *instantiates* properties. Given any substance, its properties are what lend it form and make it discernible from the general environment; therefore, it cannot stand alone as the source from which its properties emerge. Its properties cannot "supervene" on it, or "emerge" from it, unless the dependency can be reversed so that the substance is regarded as supervening on/emerging from its properties.

This amounts to saying that any process of emergence requires a "syntax" (formal structure, generic metaproperty) implicitly containing all of the possible properties of any substance that can instantiate them. In other words, for a property like *consciousness* to emerge from living matter (the substance which instantiates it), that property must "already" be implicit in syntax (as must the properties *living* and *materiality*), in which case it is just as true that living matter is a function of consciousness.

Therefore, condition 3 – as understood in the context of physicalism – is ultimately absurd. Stock physicalism ("mind supervenes on matter, but not vice versa") is ontologically inconsistent, and *emergence* must be defined as it is in the CTMU (condition 4).]

May 22, 2018

Hard problem of consciousness

33. How does Chris Langan's CTMU (Cognitive-Theoretic Model of the Universe) address the "hard problem of consciousness"?

By formulating the structure of reality in a monic, self-dual way.

Remember, the "Hard Problem" is really just a formulation of Cartesian mind-matter dualism. What's "hard" (in fact, impossible) about it is the insistence that it can be properly resolved only while leaving its central assumption intact, i.e., the irrational insistence that it be answered entirely within the constraints of dualism, or somewhat equivalently, physicalism.

The CTMU "short-circuits" the problem by expressing the structure of reality in a self-dual (dual-aspect monic) ontological framework. This causes it to vanish by denying it the conventional dualistic framework in which it is improperly formulated.

February 4, 2018

Conspiracy Theories

Coudenhove-Kalergi Plan

34. Is the Kalergi plan a conspiracy theory?

Of course it is, on two counts.

First, the Coudenhove-Kalergi Plan is a conspiracy *involving* the European ruling class and the international bankers on whom its members have always relied for money to fund their wars, colonial adventures, and so on.

Secondly, the Coudenhove-Kalergi Plan is a conspiracy *theory* because it has rightly been called a "conspiracy" by others. In other words, it is the object of a *theory* externally formed *about* a conspiracy.

A little background. War loans are a very profitable business for international bankers, who get to fund, and collect, from both sides of every conflict. The winner pays willingly because it extorts reparations from the loser, and the loser pays because those reparations have been doubled in order to ensure it (and because losers can also be dunned directly by their "loan officers").

And of course, wars and colonial adventures can also be very good for heads of state, high-ranking military officers, and those who profit from the manufacture of weaponry and munitions (eventually dubbed "the Military-Industrial Complex" by US President and former General Dwight D. Eisenhower). And let's not forget multinational corporations supplying the military and engaging in international trade.

It follows that these groups have a *motive* for conspiracy, namely, prospective gains of money and power. It also follows that they have *means* and *opportunity*, which follows immediately from *already having* most of the wealth and power. It even follows that they must lie about their actual motives, lest the populace rise up against them and take their money and power away from them. Their collusion and secrecy, often rationalized in terms of "national security", establishes that they are implicated in a *conspiracy* for which they possess means, motive, and opportunity.

As explained by Coudenhove-Kalergi himself in *Praktischer Idealismus* (1925), and even considerably earlier in the associated movement, the idea was to:

(1) Blame the indigenous people of Europe for European war and colonization ("Indigenous Europeans are congenitally greedy, quarrelsome, and violent!") despite the fact that unlike European royalty and the bankers who funded their various wars, the vast majority of Europeans had never seen a bit of the money or territory gained in such adventures at the cost of their lives, limbs, and happiness in their otherwise peaceful and productive societies, using this false pretext to justify the elimination of European cultural and ethnic (genetic) identity by means of politics and mass immigration and its unavoidable consequences including transformative effects on European population genetics;

(2) Intermix and communize the resulting deracinated population, depriving them of their ethnic and cultural identities in order to thoroughly domesticate them and prepare them for global governance under the enlightened leadership of the bankers and despots who had actually been guilty of starting, funding, and profiting from all of the wars and colonial misadventures for which the innocent masses of Europe had been falsely blamed;

(3) Justify permanent control of the resulting divided-and-conquered population, a so-called "Eurasian-Negroid Race of the Future", by a "Master Race" consisting of, you guessed it, a combination of European royalty and international bankers and their friends and progeny (which, owing to its supposed moral and intellectual superiority, would not be subject to genetic dilution).

This is all in print, in the abovementioned book and elsewhere. There's simply no denying it, at least if one has any respect for the truth. The movement was handsomely funded by famous international bankers including Warburg and Rothschild and eagerly embraced by European royalty ... i.e., by the other (non-banking) constituents of the "Master Race", which today controls the EU economy and EU immigration policy.

Notice that this policy is exactly, precisely what Coudenhove-Kalergi prescribed in *Praktischer Idealismus*, and that the despotic, undemocratic bureaucrats of the EU take all of their orders strictly from above, i.e., from members of the EU "Master Race". So there is no question that what we see today is a perfect reflection of what the movement prescribed.

R.N. Coudenhove-Kalergi, founder of the Pan-European movement and often called the "Father of the European Union", was the winner of the EU's first Charlemagne Prize. He even chose the EU anthem. There is no question whatsoever that those who run the EU embraced his program completely and continue to do so.

Note that the ultimate success of this program would result in a 2-tier society consisting of an inbred, filthy-rich overclass and an impoverished, genetically homogenized underclass inhabiting what amounts to a "world-hive". In such a hivelike arrangement, war and colonization would no longer be necessary for the royalty-banker "Master Race" to remain in control. The hive would contain workers, drones, and soldiers to police the workers and drones 24/7 and squeeze them for every last drop of blood on behalf of their masters, no wars necessary except for purposes of population control.

[If you want to argue about this, please don't expect me to engage with you on it. As far as I'm concerned – and many others share my opinion – these are well-documented facts and thus bear no debate.]

September 21, 2018

Powers that be

35. What is the most ludicrous scientific or political conspiracy theory?

The idea that legions of wingnuts, fruitcakes, and outhouse rats are "out to get" the powers-that-be by accusing them of collusion and conspiracy when in reality, the powers-that-be are enlightened and uniformly benign entities who always tell the truth, obey the law, and want what's best for everyone else even if they must nobly sacrifice themselves on our behalf.

A bit of reflection should quickly reveal who the real "wingnuts, fruitcakes, and outhouse rats" are here. The fact is that whenever a potential conspiracy has positive mathematical expectation, those in a position to execute it and reap its rewards will almost always attempt to do so. (It's called "human nature".) Obviously, those most often in this position are the powers-that-be.

Beware the indoctrination victim who denies this aspect of reality. Such people are dangerous in the extreme.

May 11, 2018

Secret organizations

36. Could secret organizations exist?

Yes. Otherwise, why would there be a legal concept called "criminal conspiracy", defined as an agreement between two or more parties to commit an unlawful act? As the act is unlawful, the conspirators generally keep their agreement secret to avoid being arrested. So by definition, almost any criminal conspiracy qualifies as a "secret organization". (Conversely, many secret organizations amount to criminal conspiracies.)

Nevertheless, many people would answer this question negatively for fear of harassment. In this day and age, daring to call out the members of a conspiracy is likely to get one mocked and ostracized as a "conspiracy theorist". But this is absurd, for well-designed conspiracies occasion more gain than risk for those in a position to keep their secrets. Such people often find conspiracy economically and/or politically rational, a fact which virtually guarantees its widespread existence.

One example of a gainful but very secret organization was the Italian Mafia, whose existence was admitted only relatively recently. There have been countless others, up to and including hidden layers of government.

March 21, 2018

CONSTRUCTION

Cold joint

37. What are the effects of a concrete cold joint?

The effects may be classified as structural, functional, and cosmetic. Structurally, a cold joint or cold seam is weaker than the surrounding concrete and thus a potential plane of fracture. Functionally, it can result in difficulty working the concrete, causing waviness of the surface, and as a result, difficulty in closing seams between floors and walls and (in extreme cases) potential loss of footing. Cosmetically, a cold joint can result in an unsightly, highly visible linear discoloration which will not stain properly, and the extra difficulty of working the concrete due to reduced plasticity of the older section can lead to visible surface imperfections (rips, tears, bubbles, pits, grooves, scuffs, and so on along the joint) which may require expensive sanding and polishing to repair.

In short, cold joints are bad news, and are only weakly mitigated by dense steel reinforcement (plenty of rebar in the concrete). A concrete supplier whose delivery intervals are known to be so long that they could produce cold joints should not be hired. Adding retardant to the concrete can help, but only up to a point.

May 13, 2017

CTMU

Academic reception

38. Could Chris Langan's CTMU just be one big ploy to demonstrate the gullibility of academics?

Hmmm ... I wonder. After all, we have the Sokal debacle to consider. But if so, then to judge from the number of academics who have expressed interest over the years, one would think that academia had "passed the test". On the other hand, one might also think that some courageous academic big shot who actually knows what he's doing in the field would have stepped forward to either hit the CTMU with a knockout punch, or take a hard shot on the chin and get put on his smug, self-satisfied ass by it. (My money's on the CTMU. ;)

December 28, 2017

Algebraic algorithms

39. What would Christopher Langan's Cognitive-Theoretic Model of the Universe look like if it was reduced to an algebraic algorithm?

The phrase "algebraic algorithm" is perhaps inappropriate in this context, if only because it is too limiting. (Algorithms are problem-solving procedures which always have algebraic aspects, but are generally of limited applicability.) On the other hand, as a mathematical structure which "solves the problem of describing and explaining reality", the CTMU looks like any number of abstract structures which have been explicated in numerous essays, web pages, and academic papers on the CTMU (complex problems often require complex solutions).

February 14, 2018

Comprehending CTMU

40. Why don't people understand the CTMU?

Some people have difficulty understanding the CTMU because they spend no significant amount of time trying to absorb the maximally simplified language in terms of which the theory has been described. While it contains neologisms (as does almost any new and original theory), standard terminology is nowhere redefined in it, so those confused by it cannot point to linguistic usage as a reason to dismiss it.

As the CTMU has been designed as a comprehensive theoretical framework for reality at large, it cannot be reduced or condensed in any way that would diminish its intended scope. By the same token, introducing such a theory requires that one explain why the overall worldview it replaces is deficient, which unfortunately can easily be mistaken for the claim that "all that has gone before it is wrong" (*deficient* need not mean "wrong"; it can also mean "incomplete").

In fact, the CTMU has been designed so that all valid scientific theories can be modeled within it.

May 20, 2018

CTMU vs atheistic globalists

41. When will the CTMU finally take off?

In all likelihood, the CTMU will take off when the avenues of intellectual commerce are no longer monopolized and dominated by people who perceive the CTMU as antithetical to their interests. Unfortunately, this dominance has been globally bureaucratized through Academia, Inc., which is itself thoroughly dominated by atheistic progressive globalists. At this point in history, the global bureaucracy has a great deal of momentum; the ownership and enslavement of the planet has been its primary incentive for quite some time, and the resulting damage to society is cumulative and ongoing. Obviously, it is not up to me alone to break this momentum.

I suggest that people who want the CTMU to "take off" should consider doing a little something to help this happen, as opposed to sitting around, carping about it, and overloading fora like this one with *ad hominem* attacks on its author (me). I do more for humanity every day by upholding and developing the structure of reality than the entire set of CTMU critics could do for humanity in a lifetime of bitter, confused bellyaching about it.

January 12, 2018

Defining usefulness

42. What's the most useful and tangible consequence of the CTMU (Cognitive-Theoretic Model of the Universe)? What is an example of a way the CTMU has changed or altered one's life in a real way?

"Useful" to whom, and for what?

Some people consider a basic understanding of reality psychologically and socially useful and even indispensable for the health of individuals and societies. (They're right.) Once people are equipped with such knowledge, they can choose to apply it in more or less obvious ways, using it to guide and parameterize ethical judgments and personal, political, educational, and administrative decisions. (Of course, problems can arise for those who are completely uncreative, and need certified acadummies to explain to them, for up to several hundred thousand dollars over the course of a university education, how to think. Such people naturally have our sympathies.)

Others want new gadgets or substances that they can use for various more specific purposes. The "cooler" the substance or device and its applications, the better. (It's a shame when people find themselves unable to evaluate utility in any other way than by using teenage vernacular like "cool", but that's the world we inhabit.) Yet others want to run the companies that manufacture such substances and devices in order to make "buttloads" of money and be *really* cool. (Nothing like a bit of super-cool "disruptive tech" to become filthy rich and preside over your own personal empire even if you're just a greedy, self-serving sociopath who had to steal it!) Others, already rich and powerful, may want new tech just for the purpose of sitting on it so as not to disrupt their own income streams (or major rivers), which they have monopolized and which may involve inferior tech with a great deal of economic inertia. (Talk about greedy, self-serving sociopaths!)

But here's the real problem with gadgets and disruptive tech: if there is anything about any given device or substance or technology that can be weaponized or abused for profit or pleasure, it will be. This is a virtual certainty in today's shallow, hedonistic, and highly competitive world, and what it comes down to is this: if there's any way at all for one's "cool" inventions to circle around and bite one in the derriere, or bite society in the derriere, one can start the countdown and time it on one's watch!

Fortunately, a proper understanding of reality comes without this major drawback. Some people would call that a very good thing. (They're right.)

[**Note**: I just received an automatic notification advising me to "add a credential" in order to help my readers. Rushing to comply, I entered "Sole Author of the CTMU", which is absolutely true about me and potentially very helpful for those seeking correct information on the CTMU, which after all is clearly mentioned in the question. But this "credential" is not being displayed. The reason given by Quora: "This credential is hidden and needs editing to be helpful to readers. Please edit and improve it to have it appear next to your answers." I have no idea what this seemingly idiotic instruction is supposed to mean. Therefore, no credential. Thank you for your attention.]

March 20, 2018

Evaluating CTMU

43. Why are many physicists incompetent when it comes to evaluating Christopher Langan's Cognitive-Theoretic Model of the Universe?

There are a number of CTMU papers online, every one of which clearly states one or more problems and then solves or explores them.

Perhaps the most general encapsulation would be: "The CTMU solves the scientific, mathematical, and philosophical problems attending Cartesian dualism by instead placing these disciplines within a form of dual-aspect monism, thus supplanting dualism with self-duality."

Most physicists are trained to make sense of neither this problem nor its solution, both of which are very important to physics as a whole. Depending on their specialties, this does not necessarily make them "incompetent" as physicists. However, it obviously makes them incompetent to criticize the CTMU, especially in a casual or peremptory way.

Fortunately, a great many physicists have the general intellectual competence to refrain from opposing what they are unequipped to comprehend. This works greatly in their favor, as they avoid denying crucial implications of physical science.

February 11, 2018

Logical integrity

44. To all those with IQ scores above 150, what is your take on the CTMU?

I'm fairly well-known for having an IQ well above 150, and I also have the advantage of being the author of the CTMU. Not only is the CTMU the only theory of reality with complete logical integrity, but any other theory of reality which cannot be wholly interpreted in the CTMU – i.e., mapped into it on all orders with complete preservation of structure – is erroneous.

No one has ever succeeded in both exhibiting any understanding of what the CTMU actually says, and finding an error in it. This is because the CTMU has been structured in such a way that no errors are logically possible. (Bear in mind that the CTMU does not rely on empirical methodology, from which nothing can be definitively established beyond the level of direct perception, but instead utilizes a new kind of metalogical inference to justify itself.)

January 6, 2018

Meaning of life

45. Does your CTMU include a theory for the meaning of life? If it does, what does it suggest? Is it for us to create ourselves, or are you seeking the fundamental reason for existence?

The CTMU is a theory of the meaning of life. "Meaning" is how things relate to each other; a bit more abstractly, it is the existence and evolution of relational structure and structure-preserving morphisms. Where relational structure is referred to as "identity", it is thus about the expression and evolution of identity on all levels of reality at large. For a human individual, the general idea is to exist and evolve consistently on all levels of identity, through the family, group, social, species, planetary, and cosmic levels all the way up to the spiritual level.

Note that those who deny the spiritual (top) level of identity might as well not even think about real meaning, as it is simply not in their prospectus (except insofar as it might be imposed on them from without, e.g., as when their employers deem them tools and wage-slaves, politicians deem them voters and useful idiots, the government deems them cannon-fodder and sources of tax revenue, and their bankers deem them chattel or "useless eaters"). Sadly, the dominant ideology among such people – be it called naturalism, physicalism, secularism, atheism, or whatever they like – fails to incorporate a coherent top-level identity, which means that when it comes down to lower dependent levels of identity, they might as well hang it up.

It thus goes without saying that if you want your life to have meaning, you'd better get straight with the metaphysical structure of reality and your own ultimate spiritual identity. The CTMU exists to help you do that.

(Incidentally, I'm the author of the CTMU. Accept no substitutes.)

February 10, 2018

Metaformal system

46. Why is Langan's CTMU not a real theory?

This question is not valid; it is not meant to inform, but to impugn. There is no one anywhere who can show that the CTMU is "not a theory", and anyone who makes this claim is a fraud.

The CTMU is a *new and very advanced* kind of theory, but the differences between the CTMU and ordinary theories make it stronger rather than weaker. Instead of being a mere body of empirical-inductive speculation haphazardly glued onto one or more formal systems which serve as its theoretical backbone, the CTMU is a metaformal system in which language and universe are combined using the property of intelligibility. It is the only theory of its kind, and deserves applause rather than unpleasant insinuations.

(One obviously biased CTMU question posted for purposes of harassment; six pejorative and misinformative answers posted by people who, to judge by what they have written, know absolutely nothing about the theory itself. What a site!)

September 28, 2018

Metalogical system

47. In which sense is Christopher Langan's Cognitive-Theoretic Model of the Universe mathematical, meta-mathematical, logical and metalogical?

The CTMU is logical and mathematical because it explicitly contains logical and mathematical structures, concepts, and reasoning. The CTMU is metalogical and metamathematical because it is reflexively self-modeling or distributively closed under valid interpretation, mapping the correspondence between itself and its universe of discourse and describing the relationship between logico-mathematical syntax and the real phenomena to which it applies.

March 7, 2018

Metaphysical formulation of logic

48. Is CTMU an introduction to metaphysics?

No, the CTMU is not formulated on a mere "introductory" level. It is metaphysics at its most advanced, with a powerful and unique design that only trolls and metaphysical ignoramuses are foolish enough to contradict. It has been repeatedly published in reputable academic journals, and has withstood every informed critique ever leveled against it. Its structure is that of a metaphysical formulation of truth and logic themselves; hence, it is not open to doubt and will never be superseded by a "better" theory.

March 28, 2018

Metaphysical metalanguage

49. What is the relationship between science, religion, and logic?

It is called the **CTMU** (*Cognitive-Theoretic Model of the Universe*), it has been around since the 1980's, it was authored by me (Christopher Langan), it is absolutely superior in every way to every other theory of its kind, and it is written in stone (very *hard* stone).

The CTMU comprises a metaphysical formulation of *logic* which has the form of a special scientific and theological metalanguage that takes both scientific theories and bodies of religious belief as object languages, and in which *science* and *religion* require interpretation for mutual and self-consistency. Modeling each of these classes of object language in the CTMU leads to their convergence, eliminating much of the conceptual distance between them.

In other words, the relationship in question takes the form of a logical language in which scientific and religious truth can be expressed and related, and the structure of this language is already a well-solved problem.

May 24, 2018

50. Does the CTMU show that some religious ideas can be critically examined with metaphysical logic?

Yes, it does. That's because it is a metaphysical metalanguage with unique logical support for the expression of religious and spiritual concepts such as God, the human soul, the nature of good and evil, alternate realms of spiritual existence, and the relationship between life and death.

Lesser languages – e.g., physicalism, metaphysical naturalism, and comparable pseudo-philosophical conceptual frameworks – lack such support. This is why people who insist that reality is all and only "physical" often reject spiritual and religious concepts across the board.

Where the limits of one's language are also the limits of one's powers of comprehension, spiritual and religious concepts are utterly inconceivable to those from whose cognitive languages metaphysics has been excluded.

July 14, 2018

Next calculus

51. What is the "next calculus," meaning what is, if there is one, its 21st century analog?

The CTMU (Cognitive-Theoretic Model of the Universe). Based on a powerful, highly sophisticated, and unique mathematical structure, it's a way to link mathematical formalisms with their contents (the things which model them in various ways). It has been out there for around thirty (30) years. Its author was introduced at the turn of the millennium as "the smartest man in America/the world" on every major US television network, along with many other major media outlets abroad. It has never been successfully challenged.

Sadly for those who depend on Academia for mathematical and empirical-scientific insight, academic interest in the CTMU remains minimal to nonexistent. It's not that Academia has anything better or more important or useful on which to concentrate; it's just that Academia has been corporatized, has established a virtual monopoly on scientific communication and intellectual commerce, and no longer has any incentive whatsoever to entertain contributions from outside sources.

June 3, 2017

Nonspeculative character

52. Is the CTMU theory of Chris Langan provable or a speculation?

There's nothing speculative about the CTMU. (For the answers to questions like this one, feel free to go through the many CTMU-related answers I've already posted here.)

May 10, 2018

Paradigm shift

53. It has been stated that "if you can't explain it simply, you don't understand it well enough." Can Christopher Langan explain his Cognitive-Theoretic Model of the Universe (CTMU) simply?

Obviously, "simple" is a relative term. Something which may be unfathomably complex to a monkey – integral calculus, for example – might be very easy for a more intelligent creature, e.g. a moderately intelligent human being, to grasp. This general situation applies throughout the entire range of variation of intelligence and complexity. In short, the meaning of "simple" depends on intelligence and intelligence differentials.

Just as obviously to anyone who is both intelligent and well-motivated enough to have familiarized himself / herself with my published work before criticizing it, I have given what I consider very simple, very straightforward explanations of the CTMU. Unfortunately, certain people persist in claiming that they are too complex to be understood ... and just to make sure that no one misses their displeasure, these people run around posting their disappointment to various social media, sometimes seeming to cross the boundary which separates honest, constructive criticism from trolling.

My honest impression is that either my simplifications have remained too complex for certain people to understand given their best efforts to do so, OR certain people didn't really make their best effort to make sense of them. For example, perhaps certain people rejected my explanations because accepting them would have forced them to let go of something dear to them, e.g., atheism.

(I need scarcely mention that it would be inappropriate for me to apologize to such people, as in this case, the fault lies entirely with them.)

This being understood, the CTMU represents a major paradigm shift. The thing about paradigms is that they have a great deal of inertia, comprising underlying models in which people implicitly interpret what they read and hear. Thus, paradigm shifts are obstructed by the conceptual inertia of existing paradigms. In academia, the prevailing paradigm is "naturalism", which in various contexts is synonymous with physicalism, materialism, and even secularism. I think the problem here may be that some people are so cognitively invested in this paradigm that they unwittingly attempt to interpret proposed alternatives in the very cognitive framework which those alternatives are designed to replace, thus sabotaging their own comprehension.

I hope this helps, if only a little. Rest assured that I intend to keep trying – even when a person who is drowning goes into an irrational panic and ignores the life preserver he/she is thrown, the rescuer still has the option of jumping into the water, physically subduing him, and making him wear it. (It's either that, or the victim might actually drown. ;)

January 7, 2018

54. Why do people have such a hard time understanding the Cognitive-Theoretic Model of the Universe by Chris Langan?

Different people no doubt have different reasons for their confusion regarding the CTMU. Some lack the background to understand it, and are unwilling to put forth the time and effort needed to remedy this situation; others feel threatened by the prospect that the CTMU will displace some cherished belief such as atheism or materialism.

But perhaps the most general reason is that the CTMU represents a new paradigm for the human conception of reality, and new paradigms cannot be fully understood in terms of the old paradigms which they replace. It follows that paradigm shifts are resisted by people who, unwittingly or by preference, cling to old paradigms.

January 12, 2018

55. Will the CTMU of Chris Langan become the most important theory of humankind?

There are many important theories which deservedly occupy places in the sun. But as time inevitably passes and minds evolve, the weather changes. At this point, the once-productive "naturalistic" worldview on which most of our dominant theories are based has gone no-frills, stone-cold, pockets-turned-inside-out, can't-squeeze-blood-from-a-turnip bankrupt.

In short, the formerly successful theories that scientists and philosophers are still trying to use in order to answer their most profound questions have proven to be abject failures for that purpose. (They're still quite useful for some problems, mind you, but not for the biggest ones.) We're not just talking "the jury isn't in yet"; we're talking "not even within broadcast range of anything approaching a real answer." It has been that way for quite a while now, and only the infants among us are still denying it.

When such a situation arises – as it has arisen at several past junctures in human intellectual history – the only possible remedy is a paradigm shift. The CTMU is indeed a paradigm shift, and it does indeed resolve many of the issues plaguing the dominant worldview. We may thus infer that the CTMU is unsurpassed in its importance to the continued progress and future history of science and philosophy.

Of course, the problem with any paradigm shift is that almost everyone invested in the current paradigm idiotically tries to cram the new one into it, and when it proves a bad fit, they squawk like singed boobies. Attempting to shoehorn a new paradigm into the old one it is designed to replace is generally an exercise in futility, potentially leading to a Galileo-versus-the-Church situation.

Just be patient. Even infants must one day become adults.

June 23, 2018

Predictions

56. Does Chris Langan's CTMU give rise to any kind of physical predictions that could be experimentally tested? Is any part of it potentially falsifiable?

Yes, the CTMU is as empirically falsifiable as any scientific theory. For example, it predicts the apparent accelerating expansion of the universe, and is therefore confirmed by the corresponding "observation statement" (a statement which affirms the empirical content of an observation).

However, owing to model-theoretic ambiguity, it is not always possible to empirically distinguish between two explanatory theories with respect to a given observation statement or prediction. In such cases, it is necessary to replace empirical confirmation with a more unequivocal standard, that of logical verification. This is where the CTMU shines, putting all competition out of the running (except insofar as "the competition" is really just a recapitulation of the CTMU after all).

This may be hard to understand without a little background. First, with respect to anything but directly observed data and their logical implications, the phrase "empirical falsifiability" is an oxymoron. No theory with nontrivial content is either empirically verifiable or falsifiable unless it is either limited to a universe that is already completely known (in which case it is scientifically redundant and superfluous), or formulated in such a way that it is possible, at least in principle, to observe a direct counterexample or make an observation from which a contradiction of the theory can be logically inferred.

For example, say that you have a theory which asserts that all crows are black. (Note the involvement here of what is called *universal quantification* in predicate logic.) It is possible to empirically falsify this theory by observing a single non-black crow, e.g., an albino crow, or perhaps a crow with feathers that have been bleached and dyed. But unless one's theory deals with some aspect of reality that is both observable and absolute or universal, it is generally impossible to make an observation that directly contradicts it or implies a contradiction.

The logical verification of a theory requires that it have logically necessary (or otherwise verifiable) axioms or first principles, and a means of deriving theorems or consequential statements under which truth is heritable. This is what the CTMU, and only the CTMU, brings to the table.

September 16, 2017

57. Does the CTMU actually predict anything? If so, what? If not, doesn't that disqualify it as not being theorem, (due to the fact that it couldn't be falsifiable)?

Of course the CTMU makes predictions, e.g., the accelerating expansion of the cosmos, the invariance of the speed of light, biological evolution, and so on.

The problem is, so do a number of reverse-engineered theories which, instead of coming at these issues from first principles, proceed from a set of convenient empirically induced assumptions designed to support the "deduction" of such observable facts, often interspersed with similarly back-engineered quantitative equations so that they can pass as "mathematics". These speculative *ad hoc* theories get more airtime than those of independent researchers because their authors are members of Academia, Inc., a sprawling, monopolistic, incredibly profitable intellectual trade organization which promotes its own members ahead of anyone else regardless of minor concerns like truth, honesty, and intelligence.

It seems that once someone pays hundreds of thousand dollars into this glorified trade union and undergoes a bit of circular "peer review" by (*quelle* surprise) other academics, one is as good as gold, gaining automatic respectability for one's supposed "qualifications" to pontificate on one's favorite topics even at the expense of better minds and ideas.

March 20, 2018

Prerequisites

58. What mathematics is needed to understand Christopher Langan's Cognitive-Theoretic Model of the Universe?

A legitimate question, but first I think we'd better get something straight: there exists no coherent argument to the effect that new theories always require only existing mathematics. Some theories require that one learn new mathematics which are needed for their concise expression. This includes the CTMU, in the source material of which various new mathematical concepts have been introduced.

As for the existing fields of mathematics with which one should have a bit of familiarity before expecting to get the gist of the CTMU, one should definitely understand a little predicate and propositional logic along with rudimentary model theory and a bit of set theory. One should also know a bit about algebra and language theory. A tad of computation theory wouldn't hurt, and neither would a little calculus. (Later on, you'll need much more than that, but first things first.)

Caution: despite their explicit or implicit claims to the contrary, at least a couple of those who have attempted to answer this question obviously lack even the limited mathematical knowledge I've just mentioned. In particular, answers that come down to "None, because the CTMU doesn't have any real mathematical structure!" are absurd and misleading, and should be disregarded.

April 5, 2018

59. What assumed knowledge does one need in order to understand Chris Langan's CTMU?

The CTMU involves no assumptions whatsoever. It requires just two things that each of us is given from birth: The world, and one's ability to perceive and conceive it.

August 16, 2018

60. How much prerequisite physics knowledge is required to understand the CTMU?

An initial grasp of the theory requires only the most basic knowledge of mainstream physics. This includes a bit of classical mechanics, a little Special and General Relativity, a smattering of (standard) quantum mechanics, some introductory cosmology, and so on.

August 16, 2018

Principles

61. (1) Can Chris Langan explain his CTMU M=R principle? (2) Why is it necessarily true?

(1) Very easily. As explained from the time it was first mentioned, the expression "M=R" expresses the duality (*not* dualism) of internal and external state.

(2) It is necessary as a requirement of intelligibility, without which there is no discernible reality to be scientifically explained.

Some people don't like this principle, claiming that it's not a "real mathematical equation". But it certainly is. The problem is that the contentious little geniuses who reject it understand neither the full range of modern mathematical usage of the "=" sign, nor the perils of showing one's ignorance by criticizing what one fails to understand.

February 15, 2018

62. Why should we believe the three fundamental axioms (MAP, M=R, and MU) of Christopher Langan's Cognitive-Theoretic Model of the Universe (CTMU)?

One should believe these three principles because they are true. They reflect dire ontological necessities including the closure and self-duality of reality, neither of which can be rationally characterized as an "assumption".

February 18, 2018

Resistance towards CTMU

63. Is Chris Langan's CTMU being resisted by people lacking virtue?

Over the years, most of the resistance to the CTMU has been found to emanate from people who:

1. Dislike it because they associate it with God, the very idea of Whom they hate;
2. Misunderstand it because they are ignorant or otherwise mentally limited;
3. Oppose it because they envy or resent its author, whom they regard with jealousy or hostility;
4. Hold it in contempt because it was not authored by a professional academic, whereas they indulge academic snobbery;
5. Reject it because they feel that it threatens them and their opinions and preconceptions;
6. Trivialize it because they are lazy, and it looks too much like work.

In other words, they lack some combination of spiritual awareness, reverence, gratitude, knowledge, intelligence, generosity, benevolence, open-mindedness, humility, courage, self-confidence, initiative, and the desire to learn. And because virtually all of them falsely blame their negativity toward the CTMU on the CTMU itself, they also lack honesty.

So regrettably, it would appear that the answer is yes.

May 20, 2017

Responding to criticism

64. Chris Langan, why haven't you written your CTMU more broken up in such a way that our thought patterns can follow and understand?

Your evident difficulty in understanding the CTMU suggests that your thought patterns are not sufficiently compatible with CTMU structure to comprehend the simplified accounts of it which already exist. In that case, bringing your patterns of thought into conformance with the CTMU may take a bit more effort on your own part. I'm sure that if you put forth this effort, you'll find it well worth your trouble.

I've written about this before, and I don't want to beat a dead horse. But if intellectual commerce and communication were not dominated by an academic monopoly, the CTMU would have "taken off" a long time ago. By now, there would be a huge amount of expository material available on it. Instead, we have a pile of academics arguing so noisily about their own speculative theories, and so thoroughly dominating the media with their ridiculous disputes, that nothing else can be heard above the din.

If one didn't know better, one might almost surmise that mankind doesn't really want to understand reality or reap the benefits of such understanding, but prefers to praise and pamper its academics like precious lapdogs, sacrificing everything to their insatiable desire for attention and pretensions to intellectual greatness.

In which case, all that one can really say is "good luck with that." ;)

June 22, 2017

65. When will a relatively known academic openly and publicly criticize specific contents of Christopher Langan's Cognitive-Theoretic Model of the Universe (CTMU), putting his reputation at risk because the CTMU is absolutely true?

It could happen any day, although I wouldn't give you a fig for the academic's chances of making his or her "specific criticisms" stick. As I've pointed out repeatedly, the CTMU is logically unassailable, and this is not subject to spontaneous change.

Incidentally, the last paper published on the CTMU ("An Introduction to Mathematical Metaphysics") has reportedly been downloaded many thousands of times, perhaps even setting some kind of record for the academic journal in which it appeared.

January 11, 2018

66. Can critics of Christopher Langan's Cognitive-Theoretic Model of the Universe (CTMU) offer specific criticism that identifies which specific aspects of the CTMU are false?

CTMU critics occupy two general categories: those who at least partially understand the CTMU, and those who do not.

Those who understand the first thing about the CTMU realize that by virtue of its structure, its content is invulnerable to attack; their critiques thus tend to focus on presentation and emphasis.

On the other hand, this invulnerability to substantive criticism naturally limits those who *do not* understand the CTMU to various indirect and illogical forms of argumentation including straw man arguments, *ad hominem* polemics, stylistic critiques, and irrational outbursts sometimes including insult, profanity, and outright libel (this can be verified right here on Quora, where one malicious but nonetheless popular anti-CTMU screed stridently declares it to be the manure of two different species of large farm animal).

As for the hypothetical possibility of "specifically identifying false aspects" of the CTMU, it would be irrational to assign it a probability greater than zero.

February 10, 2018

67. What is wrong with Christopher Langan's Cognitive-Theoretic Model of the Universe (CTMU)?

Absolutely nothing ... nothing at all.

In order to make sense of this question, it must be modified to read:

"What has gone wrong with the communication of the CTMU?"

This form of the question is very easy to answer. Every communication process involves a sender, a receiver, and a channel, and all three must be functioning properly for the message to be properly communicated. Unfortunately, only one of these three components has been functioning properly with respect to the CTMU communication process: the sender.

As for the other two components, the channel is not functioning properly because it is under the monopolistic control of Academia, Inc., which denies full access to anyone lacking its certification, and most of those at the receiving end of the transmission are too disinterested and/or intellectually impaired to comprehend the message even when they manage to hear it.

Should any professional academic with a reputation to protect want to take issue with this, he or she is welcome to try. I'll be frank: there is no professional academic on this planet who is even remotely knowledgeable enough to attack the CTMU in the full light of day and avoid being crushed by me.

(This offer applies only to known figures who are both reputable and well-qualified in fields related to the CTMU. For others, the penalty for specious argumentation is insufficient to "keep them honest", so to speak. They stand to gain too much more than they risk by engaging with known figures, and are too likely to engage in specious argumentation and dirty polemics.)

[**Note 1**: Another answer for this question, evidently with comments disabled, criticizes the CTMU on the alleged grounds that "a panentheistic god is outside of time and space, somewhat like the Abrahamic God, thus beyond reality and not merely generated by reality." This is mistaken. A panentheistic God, being metaphysical, properly includes physical reality and is thus omnipresent therein. The CTMU is precisely what it takes to fully realize this criterion.]

[**Note 2**: Yet another answer is followed by comments from one "David Moore" (again with replies disabled), who criticizes the CTMU on the alleged grounds that "educe" means "infer", and that the CTMU therefore runs afoul of Gödel's theorems. In fact, the first definition of "educe" is "to bring out or develop", which is consistent with undecidability. Mr. Moore also states that "tautologies imply nothing except themselves...". However, while this is superficially true for propositional *tautologies*, tautology is in fact a much more general concept, and the CTMU applies it in a whole new way.]

February 24, 2018

68. What do other profoundly gifted individuals think of Christopher Langan's Cognitive-Theoretic Model of the Universe (CTMU)?

To my knowledge, no profoundly gifted individual has ever found fault with the CTMU. No *real, verified* one, that is.

As for the nasty answers posted here, I'm afraid they're more likely to have come from resentful mediocrities than "profoundly gifted individuals". I've been dealing with the profoundly gifted for decades, and none of these people sounds anything like them.

August 15, 2018

69. Why do people claim that the conclusions of the CTMU are "unjustified assertions" when they are all results of rigorous deductions (including logical induction)?

There are several possible reasons. E.g., they don't like theories said to be related to God; they don't like people who are said to be smarter than they are; they don't like theories related to God and written by people who are smarter than they are, and so on.

But perhaps the most important reason is that of those who claim to understand logic, few actually do. They merely *pretend* to understand logic, and in order to console themselves for their ignorance of logic, use their misunderstandings of logic to attack theories about which they have yet other misunderstandings.

In short, it's all one big misunderstanding. ;)

August 21, 2018

70. Why does Chris Langan receive so much unreasonable hostility from otherwise reasonable people just for saying what he thinks like everyone else does? Is it that people are threatened by his high IQ, causing them to feel a need to assert themselves?

Permission to speak frankly? It's because, when one is just a cheesy, resentful, pseudonymous little Internet troll with a hard-on for God, the CTMU, and anyone reputed to be more intelligent than oneself, one has nothing of value to lose by shooting off one's vile, foaming mouth like a messy, smelly vinegar-and-baking soda Vesuvius whenever one gets the chance.

August 28, 2018

Scientists and CTMU

71. What do scientists think of Chris Langan's CTMU?

There are two kinds of science: empirical, and logico-mathematical or linguistic (logic and mathematics are just special kinds of language). They are inseparable in the sense that while mathematics exists in the minds of observable (physical) entities including mathematicians, the theoretical aspect of empirical science is dependent on logic and mathematics along with other ingredients of language.

In order to deal with the mutual dependency relationship between mathematical and observational reality, we require a level of science which spans both. This is the sense in which the CTMU is "scientific"; it is science of a higher and more powerful kind than empirical or mathematical science alone. In the vast majority of cases, neither mathematical nor empirical scientists are properly trained for this level of scientific discourse.

Model theory is a branch of logic having to do with the interpretation of empirical phenomena in theories and the mathematical structures of which they consist. Technically, the CTMU is a special kind of model theory designed to support the description of reality on the ontological level of discourse ... the level on which reality fundamentally "exists". On this level of science, the CTMU is absolutely impervious to attack. Just as logic itself contains no assumptions, neither does the CTMU. No one has ever gotten to first base against it, and no one ever will.

While many scientists have been misled into believing that their academic training should allow them to understand and even pass judgment on a high-level scientific theory like the CTMU, nothing could be farther from the truth. Nevertheless, some of the more incautious among them – especially those who fear its apparent theological implications – have expressed extremely negative opinions of it and its author (me). It is important to remember that none of these people has ever succeeded in putting a dent in it, and that every one of them could be easily crushed by anyone with sufficient understanding.

Readers are advised to bear this in mind as they read some of the other answers in this thread.

May 12, 2017

Syntax and state

72. What do "syntax" and "state" mean in Christopher Langan's Cognitive-Theoretic Model of the Universe (CTMU)?

In the CTMU, these terms refer to properties and valuations thereof.

[Please note: The level of detail in my answers regarding the CTMU, which was authored by me around three decades ago, is being limited by the apparent efforts of another Quora participant to appropriate them. In particular, Quora seems to have become a major outlet for a self-described 22 year-old philosophy undergraduate who is trying to put together a graduate thesis, and is apparently bent on doing it by feigning originality for certain aspects of my work. Although he already claims to have a "more elegant, precise and complete" "theory of metaphysics" which he tries to pass off as "naturalistic" (a contradiction in terms), I have seen no evidence whatsoever of scientific, mathematical, or philosophical competence in anything he has written. This person's constant efforts to misinform people about the CTMU are harmful to the veracity of Quora. Thanks for your attention.]

February 14, 2018

Understanding the importance of CTMU

73. Do people realize how important it is to understand the CTMU and save billions of souls and for the sake of one's neighbor to truly change the course of mankind which is only possible with a new worldview and not with more technology?

Some do and some don't.

To some, a theory must be "useful" in order to be worthwhile, and no theory is "useful" which does not immediately lead to the manufacture of one or more physical substances or devices to satisfy their material needs or serve their convenience, amusement, or pleasure.

Obviously, these people are incapable of forming meaningful judgments regarding theories of which the primary (explicit) applications are psychological, social, political, spiritual, or comparable. Give them a theory with the potential to reveal man's inner nature, bring people together, and save the world, and they'll simply complain, in effect, that it cannot be eaten, worn, driven, or used to play videos and send email.

On the other hand, some people rationally admit that human utility exists in non-commercial areas as well as commercial ones, and display a refreshing measure of intellectual competence in recognizing the characteristics that a viable worldview or ideology must exhibit. It is these people among whom the CTMU finds its most ardent and discerning followers, and not unrelatedly, on whose shoulders the fate of humanity rests.

(Of course, this is not to be construed to mean that the CTMU lacks scientific and technological applications. But the clear existence of psychological, social, political, spiritual, and other cognitive or "anthropic" applications suffice for the dismissal of those who irrationally deny their importance.)

February 5, 2018

Uniqueness

74. What is the world's current, in use, equivalent to Christopher Langan's Cognitive-Theoretic Model of the Universe (CTMU)?

The CTMU is unique, and has the conceptual structure to prove it. In fact, it can be described as an ontological characterization of self-uniqueness, and as such, it occupies a higher level of discourse than any other human conceptual framework.

However, with certain exceptions which vary from example to example, the best mainstream approximations of the CTMU might include (1) certain strains of advanced mainstream mathematics, particularly at the foundational level;

(2) certain novel scientific constructs like the Holographic Universe; (3) certain works of mainstream philosophy like A.N. Whitehead's "panentheism"; and (4) the world's great religions, each of which instantiates certain aspects of it.

Bear in mind that none of these examples "duplicates" the CTMU to any large extent, and that all include statements that must be modified or eliminated for complete CTMU consistency (if only as doctrinal baggage). The CTMU accepts none of them *a priori*, and brings to bear certain insights that all of them lack.

February 1, 2018

D

DM THEORY

Acutt, Matt

75. Matt Acutt seems to understand Christopher Langan's Cognitive-Theoretic Model of the Universe (CTMU) well enough to criticize it. Can he explain what exactly are its fundamental assumptions, possibly in common English, thus making the theory unsound?

I'm the author of the CTMU.

On the basis of the limited material we've seen from Mr. Acutt regarding his aptly-named (DMT) "theory", I do not accept the premise that he "understands the CTMU well enough to criticize it" effectively. It seems to me that where he does not simply misunderstand what the CTMU actually says, he argues in favor of the illogical or logically untenable.

If Mr. Acutt's CTMU comprehension ever improves to a level at which I find that his concerns make sense to me, I may decide to address them. Meanwhile, I would merely observe that anyone who wants to initiate a debate on philosophy should choose a topic of which one has a thorough understanding.

As a general rule, I have more pressing things to do with my time than serve as an unpaid CTMU tutor for self-styled metaphysicians who criticize my work even when they exhibit little or no understanding of it.

Thank you for your attention.

January 9, 2018

DMT vs CTMU

76. Is Chris Langan's CTMU superior to Matt Acutt's DMT?

Yes, vastly. The CTMU is an exact, original, and tightly integrated theory of mathematical metaphysics introduced by me in the 1980's and mentioned by numerous sources in the mainstream media. Due to certain unique properties, it is demonstrably superior to any other theory of metaphysics.

In contrast, "DMTheory", as it is somewhat comically named (note the unwitting double-entendre), is a jumbled hodgepodge of disconnected pieces – some from the CTMU itself – which has been slapped together by some number of absurdly ambitious college undergraduates with nothing better to do, and presented here on Quora as though there were something to it. My best guess is that these kids are shooting for the position of "world's foremost naturalistic metaphysicians" (an oxymoron), buoyed to fame and fortune by their virulent opposition to the CTMU. One almost admires their cheek. But while it never looks good to take candy from babies, their bubble begs to be popped ... if only to spare Quora readers from being inundated with philosophical misinformation.

[**Note (from the Comments)**: The problem with "DMTheory" is very easy to state. Naturalism says that everything arises from natural objects, properties, and causes – i.e., from *nature*, which in the broadest sense is the physical or material world or universe (Wikipedia) – with no metaphysical, supernatural, or spiritual explanation required. This means unequivocally that "naturalistic metaphysics" comes down to "physicalistic or materialistic metaphysics", which is an oxymoron plain and simple. This oxymoron is the very starting point of "DMTheory", which purportedly combines physical data with "logic" while excluding any ontological support for the latter due to its non-physicality. This means that "DMTheory" is pure nonsense, and that discussing it is a complete waste of everybody's time. (Note also that I've just deleted an incoherent attempt to justify this nonsense using "the metaverse" along with yet more nonsense including "absolute materialism". Pardon me, but I've had it with this garbage – at worst it's gobbledygook, and at best it's just a CTMU rip-off undertaken thirty years too late.)]

February 21, 2018

77. Which is correct: DataLogical Metaphysical Theory or Christopher Langan's Cognitive-Theoretic Model of the Universe (CTMU)?

The CTMU is a theory of mathematical metaphysics with a unique logical structure which ensures its correctness. It has had this structure ever since it was conceived decades ago, before the promoters of "DataLogical Metaphysical Theory" were born.

The CTMU is tight and technically precise. It has been published in several academic journals. It has been mentioned on several major news networks and is known to hundreds of thousands if not millions of people throughout the world. It has profound applications from science and mathematics to philosophy and theology. Its unique mathematical structure is duplicated by no other theory in existence, including "DataLogical Metaphysical Theory" (the promoters of which are evidently incapable of understanding it, given their confused, irrelevant, incessant CTMU critiques here on Quora). Its only problem, in fact, is being a magnet for ignorant and obsessive trolls, who are seemingly drawn to it as flies to honey.

As for "DMTheory", this is from my answer to a previous similar question here on Quora (of which there have been several):

> "DMTheory", as it is somewhat comically named (note the unwitting double-entendre), is a jumbled hodgepodge of disconnected pieces – some from the CTMU itself – which has been slapped together by some number of absurdly ambitious college undergraduates with nothing better to do, and presented here on Quora as though there were something to it. My best guess is that these kids are shooting for the position of "world's foremost naturalistic metaphysicians" (an oxymoron), buoyed to fame and fortune by their virulent opposition to the CTMU. One almost admires their cheek. But while it never looks good to take candy from babies, their bubble begs to be popped ... if only to spare Quora readers from being inundated with philosophical misinformation.
>
> The problem with "DMTheory" is very easy to state. Naturalism says that everything arises from natural objects, properties, and causes – i.e., from *nature*, which in the broadest sense is the physical or material world or universe (Wikipedia) – with no metaphysical, supernatural, or spiritual explanation required. This means unequivocally that "naturalistic metaphysics" comes down to "physicalistic or materialistic metaphysics", which is an oxymoron plain and simple. This oxymoron is the very starting point of "DMTheory", which purportedly combines physical data with "logic" while excluding any ontological support for the latter due to its nonphysicality. This means that "DMTheory" is pure nonsense, and that discussing it is a complete waste of everybody's time.

March 2, 2018

E

Ethics

Christian morality

78. What does Christopher Langan think about Christian morality?

I embrace it. However, it must be noted that as most people understand it, "Christian morality" is in some ways a rather ill-defined concept. While I agree with morality based on the underlying logic of the Golden Rule and the rest of the New Testament, many sectarian versions of so-called "Christian morality" are based on nonlogical interpretations of Christian scripture associated with various bodies of doctrine authored long after the New Testament was written. Just as Christian scripture is independent of sectarian doctrine, so is the morality which can be directly derived from it on a purely logical basis.

March 11, 2018

Ethical system and abortion

79. Would the existence of artificial wombs change the abortion debate?

Not really. Without a verifiable set of ethical principles covering this issue (and other abortion-related issues), such questions qualify as little more than baseless opinion magnets. A valid ethical system covering all aspects of abortion would have to be sociobiologically parameterized, and could not avoid the reasoned consideration of eugenics ("genetic hygiene", in the socially and biologically rational sense of the concept). Unfortunately, this is diametrically opposed by "political correctness".

May 12, 2017

Eugenics as genetic hygiene

80. Is Chris Langan a racist by implication or otherwise of his support for eugenics?

This question seems to come down to "Does support for eugenics (under its original and valid definition) imply racism?"

The answer is very clear: NO. Eugenics, defined as the maintenance or improvement of human genetic quality and integrity, can be pursued without respect to race, or even simultaneously within each existing racial group.

However, there seem to be some unspoken assumptions here ... for example, that Christopher Langan is a "eugenicist" under some unspecified definition of the term. Where the definition is allowed to vary, the accuracy of this assumption would depend on the particular definition that is chosen.

For example, where eugenics is defined as reasonable attention to genetic hygiene in order to prevent genomic degradation and reverse evolution attending the artificial suspension of environmental constraints and limitations – i.e., the suspension of "natural selection" using medical and other kinds of technology – the assumption would be accurate. (Show me someone who does not share this opinion, and I'll show you someone who is profoundly irrational on the species level of human identity.)

But where eugenics describes the current effort of the parasitic oligarchy to monopolize adaptive traits by "dumbing down", suicidally indoctrinating, and genetically homogenizing various sectors of the human population without respect to fitness (as defined with respect to a sustainable refinement of modern civilization), the assumption would be false. In particular, I am against the use of weaponized mass immigration to commit demographic genocide against the indigenous populations of all and only Western (White majority) nations, as well as the rapid ongoing destruction of Western culture, family values, educational integrity, population genetics, and patterns of reproduction in the name of multiculturalism, diversity, and social justice.

[There is an answer below which accuses me of wanting to "sterilize people with low IQs, average IQs, *and* moderately high IQs", and even to drag young girls away from their parents in order to perform "invasive surgical procedures" on them. Nowhere have I ever advocated anything of the kind. The answer in question is factually incorrect, misleading, and defamatory.]

January 11, 2018

F

FACEBOOK

Discussion group

81. Does Chris Langan read the CTMU FB Discussion page?

I've visited the page only a couple of times. But then again, I just managed to get subscribed. (Please bear in mind that if I were to answer every question that anyone feels like asking me, I'd get very little done on the CTMU *or* the ranch. The US economy is degenerate, and good help is getting very hard to find these days. You can't even find a high school kid willing to mow the lawn –

apparently, most of them are either lazy or get too much allowance. Or maybe it's that they're "above all that", being precious snowflakes who prefer to reserve their time for the properly vetted PC/ SJW campaigns of their choosing.)

UPDATE (June 03, 2017): A very queer situation has arisen at Facebook, apparently due to policies instituted by its proprietor Mark Zuckerberg. Although I'm given to understand that Facebook has several accounts for "Christopher Langan" and even "Christopher Michael Langan", the *real* "Christopher Michael Langan", namely yours truly, was recently delisted from Facebook on the absurd premise, advanced by an obvious troll, that I was "using a false identity" (the name under which I've gone since I was 6 years old) to contribute to the Facebook CTMU Group. In short, all the fake and/or lesser-known "Christopher Langans" are still on Facebook; only the real one, from whom people really want to hear, has been rejected.

Naturally, I appealed this decision, sending in two (2) forms of identification listed in the Facebook document "What types of ID does Facebook accept?" Breaking its own rules, Facebook declined to restore my account! I hate to have to say it, but it seems that Facebook rules are strictly for those with whom Mark Zuckerberg, a notorious atheist, happens to agree. Unfortunately, Christopher Michael Langan does not appear to be a person with whom Mr. Zuckerberg agrees.

As long-time CTMU fans are well aware, the CTMU and I have been subject to vicious attack by atheistic trolls for the last three decades. This is just the most recent proof that there is something concerted, coherent, and to put it bluntly, nasty and counterproductive going on with the "New Atheism" (i.e., the Old Atheism dressed up in the emperor's new clothing) ... which is why, for example, there is no CTMU article on Wikipedia to this very day.

It's been dirty pool all the way with CTMU/Langan critics in the social media. It's the only kind of pool these people seem to know.

June 3, 2017

FREE WILL

Free will vs predestination

82. How does Christopher Langan's Cognitive-Theoretic Model of the Universe (CTMU) resolve the conflict between predestination and free will?

By simultaneously and consistently interpreting both of the concepts "free will" and "predestination" in the structure of the CTMU. (In any other conceptual setting, these concepts are too ill-defined to conflict as usually assumed; when properly defined and interpreted in the CTMU, they do not conflict.)

February 26, 2018

Humans as telors

83. Can Chris Langan prove that free will exists?

Yes, I can. It's a matter of properly characterizing the freedom of reality itself, and then showing that this freedom is internally replicated for systems meeting the appropriate criteria *within* reality (in the CTMU, these special subsystems of reality are called *telors*).

Such a proof is possible only within the CTMU. Others have advanced their own arguments for free will, or pointed to what they assume is "evidence" for free will. However, the arguments and evidence in question are selectively interpreted in light of supporting assumptions which are themselves both ill-defined and devoid of any proof whatsoever. As the CTMU requires no assumptions, it avoids this shortcoming.

In principle, human beings are telors and therefore possess free will. However, many are degenerate, having surrendered or sold their freedom in return for various amenities and necessities of survival. This is largely why the world often seems to be hurtling toward self-destruction like a runaway freight train.

February 13, 2018

Meaning of life

84. Does Chris Langan believe in free will?

Yes. My belief in free will stems from my knowledge concerning it. This knowledge relates to a theory of mathematical metaphysics called the **CTMU (Cognitive-Theoretic Model of the Universe)**. The CTMU is the only intrinsically viable, self-contained ontology in existence. In fact, it is the only one possible.

[Note: Another answer to this question, emanating from a troll account of which Quora Moderation has been repeatedly informed but to no avail, asserts that free will is pointless if there are penalties attached to its improper use. However, the point of free will is not to use it improperly, but to use it properly in service of teleology, thus avoiding the metaphysical penalties attached to its chronic abuse. One is allowed to make the occasional honest mistake, but not to latch onto the same tired mistake over and over again like a mad dog, growling and gnawing and foaming from the mouth. (Free will is a requisite of meaning in our lives, but can be a very dangerous thing. It's a pass-fail test of sorts, but with very appreciable risks and rewards attached to one's score. Obviously, it is best that everyone understand this, and not end up having to whine about it as though one had no choice between good and evil.)]

June 23, 2018

G

God

Conformance to logic

85. God is who He says He is. He created the universe and us. What right do we have to expect Him to conform to our ideas of who He is or should be?

We may expect God to conform to our ideas of Who and What God should be precisely when they have achieved cognitive emergence in logical form, thus reversing the mapping by which God originally implanted them in the human mind. (Despite its concision and explicit invocation of logic, this answer is fully consistent with the New Testament and compatible scriptures.)

In other words, God does indeed tell us Who and What He is, and He does it with logic, which provides the only general means of verifying the message. This convergence of theological information and verification is no accident; unverifiable theological information is ultimately pointless, and God always has a point.

Obviously, any denial of what God tells us with logic is trivially false, and the deniers are illogical by definition. Such people and their opinions may be safely ignored for theological purposes.

May 9, 2017

Defining God

86. What is God according to Chris Langan? What does he think about the established religions such as Christianity or Islam?

God, best understood as Ultimate Reality, has (supertautological) structure which implies certain properties which are consistent with the definitions of God that occur in major world religions like Christianity, Islam, and Judaism.

Simultaneously modeling the main scriptural components of these religions in this structure makes them theologically and ontologically consistent.

Theological and ontological consistency is limited to just those religions which can be modeled in this way.

January 12, 2018

Endomorphic images

87. In the CTMU, human beings are seen as endomorphic images of the mind of God. Can this mapping be described? We are very constrained local entities, so how does it work?

In the CTMU, the global identity of reality is explicitly stratified, and distributed morphisms are utilized. So (1) one needs to be very clear on the level of identity to which any given statement or mathematical symbol refers, and (2) one needs to allow for the distributed nature of certain morphisms, where by *distributed* we mean that the morphism applies to *each point* in a topological point set.

Notice the implication: the use of distributed morphisms implies that each point of any topological space subject to distributed morphism has internal structure. In the CTMU, this internal structure is called *syntax*; hence, the points are called *syntactic operators*.

Some syntactic operators cannot support full isomorphism to the global identity; hence, they are subject to syntactic restriction, and isomorphism is limited to some proper part or aspect of global syntax. But the points in which the syntax of the global identity is *fully* imaged are not restricted in this way. It follows that their syntax is isomorphic to global syntax. (This does *not* imply isomorphism with respect to external state.)

And now for a bit of advice. Typically, questions like this come from people who have not bothered to read available material on the CTMU. They are looking for fast answers, they are often confused, and many of them dislike work so much that they don't even review the answers I've already posted on Quora. Unfortunately, this gives others the opportunity to post false or misleading answers, and Quora is sometimes very lax in letting such people misinform its readers.

Readers are therefore cautioned against blindly accepting any answer for any CTMU question on its face unless it comes from me or someone else whose knowledge of the theory can be verified by checking it against the available material, especially if it emanates from someone who has been guilty of misinterpreting and/or badmouthing the CTMU in the past.

September 28, 2018

Logical theology

88. What is logical theology? How does it relate to Chris Langan and the CTMU?

Logical theology consists of the theological (God-related) implications of the CTMU, a metaphysical formulation of logic.

Others have used this term to describe strains of theology that conform to their ideas of logic. However, (1) "ideas of logic" often deviate from any well-defined logico-mathematical structure known to logicians and mathematicians; (2) it is hard to derive theological implications from standard logic; and (3) problems arise when trying to employ standard logic on the metaphysical level of discourse required by theology.

Properly applied to theology, the CTMU solves problems 1–3.

May 23, 2018

Omnipotence

89. Do you accept that any god could not be omnipotent due to the logical paradoxes this creates?

In order to infer divine non-omnipotence from, for example, the paradoxical assertion that an omnipotent God could create a rock too heavy for God to lift, one must be able to show that God Himself is not responsible for the operative constraints on His power, as these take precedence over God's self-imposed "inability" to lift the rock.

In other words, if God has tied His own hands by establishing prior constraints or commitments that prevent Him from lifting a given rock (or doing anything else), then the real measure of His power is the establishment of these prior constraints, and the real measure of His will is His own refusal to break them.

Human examples are easy to find. E.g., it would be misleading and ridiculous to accuse a smoker who successfully forbids himself to smoke one more cigarette of "lacking willpower". In fact, willpower is exactly what such a person demonstrates.

As long as God Himself is responsible for any limitations on His own power, including even the structure of logic, this remains entirely consistent with divine omnipotence.

August 15, 2018

Proving God's existence

90. Why is the physical world the best proof of God?

If you're talking about the physical world alone, it is not the "best proof" of the existence of God. The reason is very simple but often overlooked: proof is a two-way street with a sender on one end and receivers on the other, and many receivers are defective. In particular, those who disbelieve in God typically attribute the physical world to a cosmogonic form of "randomness" (with respect to origination) and cite "logic" as their pretext for rejecting any claim to the contrary. They can receive no information through this blockage.

The physical world is indeed the (partial) basis of proof, but must be combined with logic and mathematics to constitute a proof that defeats the fake, hand-waving sort of "logic" brought to bear against it by atheists and materialists.

[date unknown]

91. Can Christopher Langan's Cognitive-Theoretic Model of the Universe (CTMU) prove God's existence by logic?

The answer is *yes*, and it is completely unequivocal. This is obviously not the forum in which to argue about it, as out of the 10 answers which have thus far been posted, at least 8 are saying *no* for entirely the wrong reasons.

If you want a meaningful debate about this, my advice would be to find a widely recognized authority on the subject and get him to express an opinion on the matter under his real name in the full light of day, with his reputation on the line just as mine is. That way, he or she has something to lose as well as something to gain, and I have something to gain as well as something to lose (that's how fair debates are conducted).

Examples: Richie Dawkins, Daniel Dennett, Sam Harris, or someone else of that ilk. (Your problem, of course, will be that these people have been ducking me for years, and don't want me to publicly mop the floor with them. Which I can certainly do.)

May 7, 2017

92. How does Christopher Langan's Cognitive-Theoretic Model of the Universe (CTMU) prove the existence of a god?

The CTMU proves the existence of God by (1) explicating the deep structure of reality, and (2) showing that this structure possesses attributes conventionally ascribed to God.

No qualified and well-reputed physicist, philosopher, mathematician, linguist, theologian, or other professional has ever demonstrated that the CTMU contains a single error, or that it is not what I've just said it is, or that it fails to do what I've just said it does. In all likelihood, at least part of the reason for this is that no qualified, well-reputed professional academic in any relevant field is intellectually capable of doing so and making it stick, at least under his or her real name in the full light of day.

After nearly 30 years since the first papers were published on this subject, sincere and well-motivated readers may consider the CTMU and its implications to be written in stone. Further clarifications will be forthcoming.

(Papers on the CTMU have, by the way, been published in peer-reviewed journals. It is hard to say why any CTMU critic would deny this. But two fairly obvious reasons are that the critic recognizes only a select subset of journals in specific fields – in a word, "academic snobbery" – or that the critic is simply trying to conceal his or her inability to understand the content, in which case he or she should not be making negative comments about it on social media sites.)

May 11, 2017

93. Would an atheist believe in God if he/she found proofs beyond a doubt of God's existence?

No, and the reason is very clear: disbelief is a cognitive filter. In order to recognize a proof of the existence of God, one's mind must be open to the possibility, and true atheists have closed their minds to any such intellectual content. Thus, they cannot "find" (or recognize) such a proof in the first place.

I have thirty years of relatively high-profile experience with this. The minds of most atheists are cognitive wind-up toys which absolutely cannot be swayed by any theory, model, or form of reasoning which even smells like it might confute their (anti)theological preconceptions. Wind them up by mentioning the possible existence of God, and their minds instantly spiral into an infinite loop.

The underlying conceptual mechanics are also very clear. Real atheists (as opposed to so-called "agnostic atheists" or agnostics with mere atheistic leanings) implicitly or explicitly subscribe to "metaphysical naturalism", a self-reinforcing (and therefore academically dominant) worldview which oxymoronically excludes the metaphysical from its arbitrary physicalistic conception of reality and thus precludes any meaningful (metaphysical) definition of God.

Obviously, the existence of X cannot be proven to anyone who refuses to let X be properly defined.

So you see, reasoning with true atheists on this matter is simply not in the cards. The only "atheist" who could possibly be convinced of the existence of God is one who unwittingly falls somewhere short of true atheism.

July 19, 2017

94. Is God really there? How can you prove it?

Proof is a logical operation. It follows that there is just one way to prove the existence of God: using logic. Unfortunately, the things with which standard logic ordinarily deals do not include God, Who resists being captured by it (because, in effect, God is by definition the One Who spans all that exists and thus does all the "capturing"). Therefore, logic must be adapted to the task in question, and reformulated on the metaphysical level of discourse. Proving the existence of God thus requires a system called the CTMU, which incorporates the required metaphysical formulation of logic.

[date unknown]

95. How does CTMU prove that God exists?

The CTMU proves that God exists by providing a valid framework for the overall structure of reality, and showing that this structure exhibits properties traditionally attributed to God.

Note: I see that someone is requesting a definition of God. In the CTMU, God is defined as *Ultimate Reality*, with a recursive base consisting of *perceptual reality* (the physical world) plus its regularities (patterns). "Ultimate" means, roughly, "idempotent with respect to containment and explanation". This has been the CTMU definition of God for decades.

(Those with questions about the theory might try reading up a little, especially if you're capable of handling appreciable abstraction – abstraction is the name of the game when it comes to metaphysical reasoning.)

It may be hard to believe, but some people really can't handle much of it at all. This evidently includes many people who use Quora, even some who claim to respect "science".)

July 11, 2018

96. Is it possible to use science to prove or disprove the existence of God?

There are two kinds of science: mathematical and empirical.

The only statements that can be proven in the empirical sciences are (1) observation statements about things that can be directly observed with replication; and (2) statements that can be logically deduced from replicable observation statements.

Where *God* is defined as *omnipresent*, He cannot be directly observed in His entirety; and because nothing that is not omnipresent qualifies as God, God's existence cannot be empirically verified using limited powers of observation. Because our powers of observation are limited not only by the cosmic horizon, but by the limits of quantum measurement expressed in terms of Heisenberg uncertainty, God's presence goes beyond our observational limitations on the macroscopic and microscopic scales; His extent exceeds our ability to perceive it. Due to Hume's Problem of Induction, which is related to the horizon and uncertainty problems, His existence cannot be logically deduced from limited observation statements alone.

This rules out the empirical sciences, leaving only the mathematical sciences to consider. But here we arrive at another set of problems, including the form of dualism whereby statements deduced from mathematical axioms cannot be globally attributed to any incompletely observable external universe to which those axioms might hypothetically apply. Again, the problem of induction rears its head; that the axioms of a theoretical language are found to apply in any given region of its universe does not mean that they necessarily apply everywhere.

It turns out that what is actually required is a higher-level proof system capable of generically relating theories, their universes, and their model-theoretic relationships, and thus of uniting the mathematical and empirical sciences. Called a "metaformal system", it has a unique structure that was explicated decades ago as the CTMU, short for *Cognitive-Theoretic Model of the Universe.*

In other words, this is a well-solved problem, and has been for quite some time. If one likes, one can simply forget about the dozens of conjectural answers to this question and concentrate on this one alone. On the other hand, anyone who is unable to understand what I've just written, or doesn't like the conclusions to which my answer leads – e.g., "God exists" – is free to keep on groping.

May 3, 2018

97. Is Chris Langan's CTMU the first intellectual production to incontrovertibly prove God's existence? Didn't any earlier philosophers (e.g. Kant, Hegel) make cogent cases before him?

While many well-known philosophers have attempted, over the centuries, to "make cogent cases" for the existence of God, they all lacked a coherent and comprehensive metaphysical framework in which to do so.

As the CTMU is inarguably the required metaphysical framework, it alone can support the rehabilitation or restructuring of their arguments and carry the issue to its proper conclusion.

June 13, 2018

98. How can I explain the CTMU, the proof of God and the afterlife with metaphysical logic, to an atheist?

Science and human experience depend on the intelligibility of what science studies, and what people experience. This means that the integrity of science depends on an explanation for the intelligibility of physical reality.

The CTMU is a high-level formulation of logic designed to provide such an explanation. It answers questions like the following: What is it that makes reality identifiable? What features does reality possess that make science and human experience possible – what forms do they assume in the overall structure of the cosmos? How do these features manifest on all scales?

The nonoccurrence of irresolvable paradox in physical reality is a scientific fact. No scientist has ever managed to observe an irresolvable paradox; somehow, reality maintains its consistency under all circumstances. When a scientist sees what looks like an inconsistency in the structure of reality, the usual paradox-avoidance strategy is simply to make the paradox vanish by adjusting an existing theory, or finding a new one. *But why should this work?*

The undeniable fact is that science has no answers for questions like these, and mainstream philosophy has been of no help. Science lacks any cognitive framework in which the actual relationship between theory and content, knowledge and reality, epistemology and ontology, can be expressed, and its standard methodology does not permit this deficiency to be meaningfully addressed. In order to remedy this situation, it must be shown how and why reality is able to maintain its self-consistency and ontological integrity. This is what the CTMU does, *and more.*

I've scanned some of the answers below. Most are full of ignorance and derision, contain almost zero information on the CTMU, and are profoundly misleading. Readers are hereby advised to be on guard against the efforts of Quora's ever-expanding troll population to mislead the unwary about a theory that its members, for all intents and purposes, are intellectually incapable of understanding. That a few of these people seem to know a smattering of scientific and/or philosophical terminology is beside the point, as they evidently have no idea how to properly apply it.

[The other parts of this question, which is actually several questions in one, are addressed in other answers I've posted on this site. A bit less misinformation out of you trolls, please.]

July 21, 2018

99. Can the CTMU prove God with roughly 80 symbols?

Of course, and easily. However, the definitions of those symbols might exceed that number, and even the term "symbol" might not be defined in quite the usual way.

August 21, 2018

Self-configuration

100. What created God according to Chris Langan?

God is eternal. He exists without respect to any external clock; clocks exist *within* God, not outside of him (basically, that's why time is relative in General Relativity; this relates to something called "background freedom", specifically from any external standard or metric). On His most general level of Being, God simply *exists*. All change in or evolution of the structure of God, known in the CTMU as "Self-configuration" (often through secondary telors), is strictly internal to God Himself.

February 28, 2018

Self-integrity

101. Does Rabbi Harold Kushner correctly imply that God cannot violate physics because God invented physics, and God does not destroy his creations usually?

Yes, given a sufficiently broad definition of "physics" which recognizes its necessary embedment in metaphysics. (With this qualification, the Rabbi is not the only one to have acknowledged such an implication.)

On the global level of nomological invariance, God does not break His own laws, for this would be to paradoxically violate His own will. The law that reality does not contain irresolvable paradoxes is certainly inviolable, as any violation would render reality self-inconsistent and therefore unintelligible, when in fact, we know reality to be intelligible by our very ability to identify it through direct observation with mutual perceptual corroboration.

The problem is to determine the exact form of the invariants that God, defined as the Ultimate Reality, must preserve for the sake of His own Self-integrity. It is important to note that these are not necessarily "laws of physics" as *physics* is ordinarily understood.

Fortunately, there is a *metaphysical* theory which permits such determinations, namely, the Cognitive-Theoretic Model of the Universe (CTMU).

[**Note**: The original form of this question was "Does (—) correctly imply that God cannot violate physics because God invented physics?" As for the part which has been added to the original question, I do not necessarily agree that God "usually does not destroy His creations". He destroys them, with nonviolation of high-level Self-consistency, precisely when they persistently fail to serve His purposes, with the understanding that on the level of metaphysical invariance, His purposes are *always* served.]

October 21, 2017

Ultimate Reality

102. Is God ultimate reality?

Yes, beyond any shadow of doubt.

This answer can be logically justified with complete certainty in a unique "metamodel" of reality called the CTMU. It can be gainsaid by no other belief system (including Zen and other forms of Buddhism), as the CTMU is a reflexive model-theoretic extension of logic in which every intelligible belief system is by definition embedded, and which is as inviolable as predicate and propositional logic themselves. (Note that *śūnyatā*, whether regarded as an ontological feature of reality, a meditation state, or a phenomenological analysis of experience, has conceptual structure and is therefore subject to the demands of logic.)

In short, just as there can be no escape from ultimate reality, there can be no escape from God. Make peace with Him, or pay the price: dissolution and reabsorption, not necessarily without a very great deal of pain and regret in the "dissolution" phase (as deserved).

Good luck on the road to enlightenment regarding the true nature of being.

February 13, 2018

Worshipping God

103. How can it be wrong to worship God in other forms?

Not to worship God in totality is in effect to deny some aspects of God, which amounts to denying God, period.

October 31, 2017

I

Infocognition

Definition

104. What is the definition of "infocognition" by Chris Langan?

Infocognition is a convergent generalization of information and cognition based on the fact that *cognition* is a process which, by definition, *informs* (generates *information* for) the cognitive processor.

That is, when you (let's call you "X") cognitively apprehend Y, you are *informed of* Y by the act of cognition. Conversely, when anything X is *informed of* (or affected by) something Y which causes a state transition in X, this amounts to what can be described as a *generalized cognition* event, namely the (re)cognition of Y by X.

Thus, by the universality of information, it is possible to generalize cognition in such a way that the coupling is universal.

January 13, 2018

Intelligence

Brain

105. How is a person's brain with a high IQ physically different from a person with a low IQ?

Well, I suppose I'm the highest-profile living specimen with relevance to this question, having been mentioned in this regard by multiple sources in the mass media.

My cranial volume, computed by subjecting *in vivo* caliper measurements to several estimation formulae including the Lee-Pearson formula and the spheroid formula, is in the neighborhood of 50 percent greater than that of the average human male, almost enough to be considered a species-level difference. (I'd be considered large-framed, but I'm really not far above the human average in this regard.)

Not too much should be made of this, as there are many other morphological, physiological, and especially neurological criteria to consider. However, one of them is *not* environment – my brothers and I all suffered from quite a bit of environmental impoverishment and poor nutrition as children, but ended up with high IQs anyway.

I suggest that we all face it: the fact that human cranial volume has been shrinking over the last few millennia is not, I repeat *not*, a good sign, evolutionarily speaking.

In fact, it will probably turn out to have been a very *bad* sign.

September 23, 2018

Chris Langan and John von Neumann

106. Who was/is smarter, John Von Neumann or Chris Langan?

I've always admired John von Neumann and greatly respected his intelligence. But still, a couple of things need to be said here.

Obviously, answering this kind of question requires a suitable measure of intelligence. Failing to specify such a measure, and for that matter a suitable definition of "intelligence", amounts to throwing open the door to any measure or definition at all, including measures and definitions which eventually turn out to be unseemly mixtures of bias and opinion.

Secondly, while I agree that John von Neumann was a very interesting person, his life was largely determined by the fact that he was a relatively coddled child prodigy who was fast-tracked every inch of the way through it. My life has been lived much "closer to the bone", so to speak. (Yes, von Neumann was a genius, but there are many geniuses out there who didn't get all of his opportunities.)

Thirdly, in order to make an authoritative comparison between two people, one must know an appreciable amount about both of them. But while many people know a great deal about John von Neumann, relatively few know anything at all about me, and even fewer know anything about my work (a topic of great confusion among so-called "experts" in science and philosophy).

January 6, 2018

Danger zone

107. Why do some people with slightly high IQs (120-135) think they are so great?

People with slightly or moderately above-average intelligence may fancy themselves possessed of intellectual greatness because of something very much like the Dunning–Kruger effect, which is not merely limited to people of low ability but scales upward into what is sometimes called the "IQ danger zone".

The Dunning–Kruger effect is usually understood as a cognitive bias in which people of low intellectual ability, due to their own metacognitive incapacity, imagine that they are intellectually superior. But the meaning of "low intellectual ability" is relative, which suggests that it exhibits no absolute IQ cut-off below the level at which metacognitive competence is unavoidable.

In fact, the Dunning–Kruger Effect scales up through what is sometimes called "the danger zone", approximately IQ 135–150, and sometimes higher. The danger zone syndrome is typified by those who, being experientially convinced of their own relatively high intelligence, are further encouraged by egotism, narcissism, and/or other personality disorder(s) to forget or willfully disregard their own intellectual limitations, thus displaying what amounts to metacognitive incapacity.

The danger-zone syndrome is very common among people in positions of wealth, power, and leadership, particularly those who confuse their own lack of moral inhibition with intelligence. Psychopaths and sociopaths tend to be ethically uninhibited, as it conduces to the acquisition of wealth and power at the expense of others, including those who are more capable. Foolishly equating morality to stupidity, they come to believe that ethically inhibited people are their intellectual inferiors – after all, have the morally inhibited not let morality tie their hands, thus showing themselves to be "losers" and therefore "stupid"? – and fall into a spiral of excessive self-esteem which the approval or adulation of others can only reinforce.

This is how the minds of danger-zone psychopaths and sociopaths often work, leading them to severely underestimate their betters. A great many of the world's problems can be traced to such people, who frequently manage to overtake and displace their moral and intellectual superiors from positions of responsibility and reward for which the latter are objectively much better-qualified.

Obviously, this represents a serious net loss of social and intellectual efficiency to mankind.

October 12, 2017

Economic opportunity

108. Assuming that intelligence is defined by either I.Q. or crystallized knowledge, do you know any really smart people who work dumb jobs?

I'm sorry to burst anyone's bubble, but there is no direct relationship between IQ and economic opportunity. In the supposed interests of fairness and "social justice", the natural relationship has been all but obliterated.

Consider the first necessity of employment, filling out a job application. A generic job application does not ask for information on IQ. If such information is volunteered, this is likely to be interpreted as boastful exaggeration, narcissism, excessive entitlement, exceptionalism (you're a prima donna, think you're "special", and may break the rules), and/or a lack of team spirit. None of these interpretations is likely to get you hired.

Instead, the application contains questions about job experience and educational background, neither of which necessarily has anything to do with IQ. Universities are in business for profit; they are run like companies, seek as many paying clients as they can get, and therefore routinely accept people with lukewarm IQ's, especially if they fill a slot in some quota system (in which case they will often be allowed to stay despite substandard performance). Regarding the quotas themselves, these may in fact turn the tables, advantaging members of groups with lower mean IQ's than other groups. Literally anyone can get a higher education these days if one is in a position to pay the rising price despite the diminishing returns, and sometimes, people with lower IQ's are expressly advantaged in more ways than one.

These days, most decent jobs require a college education. Academia has worked relentlessly to bring this about, as it gains money and power by monopolizing the employment market across the spectrum. Because there is a glut of college-educated applicants for high-paying jobs, there is usually no need for an employer to deviate from general policy and hire an applicant with no degree. What about the civil service? While the civil service was once mostly open to people without college educations, this is no longer the case, and quotas make a very big difference in who gets hired. Back when I was in the New York job market, "minorities" (actually, worldwide majorities) were being spotted 30 (thirty) points on the civil service exam; for example, a Black person with a score as low as 70 was hired ahead of a White person with a score of 100. Obviously, any prior positive correlation between IQ and civil service employment has been reversed.

Add to this the fact that many people, including employers, resent or feel threatened by intelligent people – especially ones who resemble men instead of pasty little geeks – and the IQ-parameterized employment function is no longer what it was once cracked up to be. If you doubt it, just look at the people running things these days. They may run a little above average, but you'd better not be expecting to find any Aristotles or Newtons among them. Intelligence has been replaced in the job market with an increasingly poor substitute, possession of a college degree, and given that education has steadily given way to indoctrination and socialization as academic priorities, it would be naive to suppose that this is not dragging down the overall efficiency of society.

In short, there are presently many highly intelligent people working very "dumb" jobs, and conversely, many less intelligent people working jobs that would once have been filled by their intellectual superiors. Those sad stories about physics PhD's flipping burgers at McDonald's are no longer so exceptional.

Sorry, folks, but this is not your grandfather's meritocracy any more.

June 25, 2017

IQ distribution map

[source map unavailable]

109. How do you interpret this map of IQ scores?

Permission to speak frankly? Because IQ is mostly a matter of nature rather than nurture, different genetic groups – groups that evolved in different parts of the world, facing different environmental (natural and social) stressors requiring different adaptations – have different mean IQ scores.

Note that China, which is (relatively) genetically homogeneous, has a mean IQ over 105. But surprisingly, the nations whose majority inhabitants are descended from those responsible for building Western Civilization from the ground up, in whom we would expect to see the highest mean IQ's, seem to have slightly lower mean IQ's, e.g., 95 or 100. This is because

(1) Western nations are no longer genetically homogeneous, but have admitted groups with lower mean IQ's in very large numbers (often irrationally preferring those groups with the lowest mean IQ's);

(2) Western Civilization is dominated by an extremely disloyal and pathologically self-interested overclass which wants to lord it over a tractable, easily governed underclass that cannot think and cohere well enough to challenge the power structure or question the "wisdom" of its superiors therein.

In other words, the Western parasitic elite, having coopted "liberalism" and "progressivism" using the power of money, have systematically destroyed the genetic uniformity, family values, and public education systems of Western nations, "dumbing down" their populations in comparison with certain other IQ-adapted nations which do not subscribe to multiculturalism.

I know that this isn't what many people want to hear, but it's as simple as that.

January 4, 2018

In marriage

110. How does someone with a genius-level IQ avoid making his/her spouse feel inferior? To me, intelligence has no bearing on the worth or value of a person, how can I make him/her understand this?

Find and marry an intelligent spouse, and then sincerely compliment his/her intelligence occasionally.

One need not be measurably as intelligent as one's partner. Throw a little love, creativity, and mutual respect into the mix, and intellectual disparities cannot detract from anyone's self-esteem. After all, it takes someone special to satisfy someone more intelligent, and that's not just lip service.

January 28, 2017

Race

111. If a scientific study supports the intelligence and superiority of white man, would it be published? And why?

No study purported to show that one racial group is intellectually superior to another can be published in the present ideological climate of Academia, Inc.. The reasons are at least twofold:

(1) Academia must pursue a "big tent" admissions policy in order to maximize its tuition profit. (Regarding Whites in particular: as people of European descent are a worldwide minority relative to other "major" racial groups, they have less money to offer in the long run than other more populous groups, and are therefore increasingly likely to find themselves holding the short end of the academic stick.)

(2) In order to stay in business at all, academia must stay on the right side of the entities which regulate it and/or ultimately provide most of its additional funding. These, of course, are global / multinational corporations and the governments with which they "partner", both of which find it advantageous to exploit modern "identity politics" as a part of their overall divide-and-conquer strategy for world domination. (We're not talking about "world democracy", by the way, but about something more akin to the EU, which has reposed all of its political power in a group of unelected bureaucrats operating on behalf of anonymous oligarchs who prefer to remain behind the curtain as a matter of personal privacy and security.)

For these reasons, and of course the avoidance of interracial strife, radical egalitarianism is a main feature of the standard ideology promoted in academia. Obviously, testing the elitist and egalitarian hypotheses would require unbiased research, and this is simply impossible at the present time.

January 19, 2017

In society

112. How would the world react to a man who seems to combine Von Neumann's brilliance, Tesla's ingenuity and Einstein's/Gödel's depth times ten?

That all depends on what the world, and specifically those who run the world, think they can get out of such a man.

We all like to think of ourselves as uncompromising individuals. But as a bit of honest reflection will reveal, one "sells oneself to the rich" whenever one accepts a bank loan, a position of corporate, political, or bureaucratic responsibility, or for that matter a paid job of any kind from them. A man willing to sell himself and his potential to the wealthy and powerful and to dedicate himself primarily to their interests will do fine, provided that they fairly reward him for his services and make him rich like themselves. But where egocentrism, narcissism, and deceit outweigh empathy, compassion, and loyalty in climbing the ladder of material success, this is an increasingly improbable outcome. When people sell their souls to the devil, as most people do without even realizing it, the devil has every intention of cheating them blind, and he makes no exceptions for the highly intelligent. As always, the object is to steal their patents, break contracts or promises, offshore their jobs, pass them over for well-deserved promotions in favor of relatives and sycophants, make them train cheaper replacements before laying them off, steal their pensions, tax and gouge them silly, and so on *ad nauseam*.

That's the down side of our modern pseudo-meritocracy. The up side is that if one defies the odds and manages to capitalize on one's talents without being robbed and cheated, then the world – which is in perennial bondage to the elite and suffers from the global equivalent of the Stockholm syndrome – will kiss one's rear end just for being rich! After all, one has "made the most of one's opportunities" and "made something of oneself", financially speaking (as success is usually measured, its ultimate currency is currency, period). In this event, one ends up sitting on top of the world as a "winner", as opposed to having the world write one off as a "loser" before skidding its cold, runny nose, like that of a freezing, starving orphan, across the shop window of the global economy to gaze in wonderment and envy at the next smug, sociopathic, and above all lucky and entitled social-media whiz-kid billionaire to streak across the sky ... who will, in the splendor of his riches, be compared to Einstein and Tesla whether he deserves it or not. (And he usually does not, by a very wide margin.)

On the other hand, a truly brilliant man who prefers not to spend his life chasing money had best become a hermit, hide behind the walls of an institution like a church or a university (which will claim part or all of his soul in return for its shelter), or learn to dissemble, pursuing a double-life strategy and passing himself off as relatively normal. This will help him blend in with the rest of society, whose regard for money is the only thing that protects the rich from its implacable envy and resentment. The last thing he wants is for the public to regard him as super-intelligent, as this may not only trigger hostility, but may cause people to assume that he is wealthy, and thus to double or triple the prices of even the most basic goods and services that they sell him. (Think of it as a crippling form of taxation on those perceived as wealthy, but who lack the actual wealth to retaliate legally against those who cheat them.)

Of course, such a man should also resign himself to being parasitized by the elite in any event, as this is simply what they do to everyone who can't stop them (it's a frog-and-scorpion thing. ;)

December 25, 2016

University admissions

113. Despite having an IQ of 160, I did not get into Harvard. I even submitted proof of being in Mensa. How can I cope with this and move on with my life?

The first step toward reconciliation is to realize that it is more honorable these days to be in Mensa than it is to be at Harvard. Anyone who scores above the threshold on the MENSA Test will be admitted to MENSA, no discrimination allowed. But if you're a US citizen with the bad luck to be a member of the ethnic majority (White and of European Christian ancestry), like the Founding Fathers of the United States or Harvard itself, for example, then your chances of being admitted to Harvard are far lower than they should be.

The student body of Harvard was recently found to have an American majority enrollment of a bit under 20%, far lower than one would expect on the basis of academic ability. There is simply no way to explain this except as a function of racial, national, and/or religious discrimination. Quite simply, the people who run Harvard despise people of White European ancestry, especially if there happens to be some Christianity in the mix. Evidently, this group is not sufficiently PC and "diverse" enough for them. Multiculturalism, not merit, is the primary consideration.

It's hardly an honor to be accepted by an institution which boasts of being America's most elite while discriminating viciously against the American majority. No matter what your own ethnicity may be, thank your lucky stars that you haven't been implicated in its vile Cultural Marxist social engineering program, which amounts to a criminal conspiracy to destroy the United States from within and deliver its people into slavery. Better to have clean hands even if they're not full of money.

[**Addendum** (from Comments): "I have heard Canadian Psychologist Jordan Peterson making the same point as you do in your last paragraph re social constructivism and cultural Marxism. What makes you think that this is done with the intent of bringing down the US, rather than their purported motive of compassion for marginalised groups?"

In Marxism, "the end justifies the means" is axiomatic. Marxists are allowed and in fact encouraged to lie – it's explicit in their playbook. For a Marxist/communist, literally anything is justified to bring about the Dictatorship of the Proletariat ... i.e., the dictatorship of Marxist activists and ideologues. (Yes, right-wingers lie too, but at least some of them actually feel guilty about it.)

Marxist strategists long ago announced their intention to subvert and control America by mass deception and the systematic infiltration of its power structure, including academia and the mass media. And what do we now see? Extreme leftward bias in the corporate media, monopolization of intellectual commerce and high-level employment by leftist indoctrination mills/credentials factories like Harvard, politicians openly funded and controlled by Marxist organizations and front groups, Marxist platitudes and policies coming out of their mouths and pens and legislative deliberations in great profusion, and the openly hostile presence of certain problematical "marginalized" groups which did not arise in the US, but are being imported by the millions with full governmental knowledge of their unassimilability and the danger they pose to American unity and security.

This unmistakably implies the presence of enemies within, and our "friends" the Marxists are looking very good for it ... certainly better than anyone else. It remains only to note that as always, the Marxists are secretly in bed with the wealth-and-power elite, as we see by the fact that corrupt corporate fascists like George Soros are funding domestic socialist groups, paying rabble-rousers and demonstrators to riot, and generally fueling unrest with no resistance or explicit disapproval by the government. The ultimate goal has become obvious: a 2-tier neofeudal society with corporate fascism at the top as always, and communism for all the "little people" down at the bottom (that way, people won't ask why the rich have so much and they so little; they will concentrate on trying to stay alive while jealously watching each other to make sure that their rich masters have no annoying competition from lowly members of the proletariat).

Every time the Marxists take over and rule a nation, it's only a matter of time before its citizens are disarmed, relieved of their freedom, and standing in bread lines, where they talk in very low voices lest the wrong word send them to the gulag, reeducation camp, or killing field. Meanwhile, the party elite – and of course, those by whom they are financially leveraged – fatten on the blood of the people, gorging until their wealth, power, and privilege exceed all bounds. This is the way it went in Russia, China, Romania, Yugoslavia, and Cambodia ... and how it's going right here, right now, in the Americas. It's a massive, murderous, phenomenally abusive pseudo-utopian scam run by con artists with big ambitions and very little in the way of humanity, and it's strictly for dummies.

If that's really what you want, it can only be because you think you'll be part of the inner circle, the Politburo, the Supreme Soviet. But the problem is, so do you all ... and at best, only a handful of you will make it. So the question is, how much of your soul will you sell to make the cut? ;)]

October 20, 2017

Verbal IQ

114. What is this verbal IQ thing? I've never heard of it before.

Intelligence manifests in many ways. These have been reduced to a set of "intelligence factors" which can to some extent be separately tested. In addition, there is a general intelligence (meta)factor called "g" which shows up in composite scores based on multiple tests, and which in principle is expressed across all IQ factors and subtests.

Due to the cognitive universality of g, an IQ test can in principle test for just one factor (or irreducible complex of factors) in a given factorization of intelligence and still be reasonably accurate in providing an "estimate" of IQ. However, due to impairment, disability, or a developmental deficit, problems may occur which can lead to asymmetry among the factors (as when one or more factors are simultaneously tested at levels significantly below most of the others for a given test subject). Thus, while an IQ test including only a single factor is not always as accurate or reliable as a multi-factor test, it can provide a fairly good estimate of overall cognitive ability even where problems exist with regard to other factors.

An all-verbal IQ test typically measures such abilities as vocabulary, language comprehension, conceptual abstraction and discernment, and general knowledge or "crystallized intelligence". At one time, such tests were fairly common.

That being said, there is no getting away from language in any intelligence test whatsoever, as some form of language must be employed to communicate the tasks and deliver the responses.

February 16, 2018

Wealth

115. Are smartest people the richest too? Is there any correlation between intelligence and being rich?

"Wealth = power," yes. "Wealth = intelligence," never. The wealth = intelligence equation is absurd on its face.

One can be born rich, but due to upper class inbreeding and sheer mental laziness, quite slow. On the other hand, if one is born poor, one very probably lacks the wealth and connections that it certainly takes to accumulate large amounts of capital by modern standards. Due to the structure of the economic system, capital attracts capital, and where large concentrations of capital accumulate, connections with other rich people accumulate as well. Hence the old aphorism "It takes money to make money." Quite simply, when you're rich, other rich people, and people aspiring to be rich, want to be your friends. But if you're poor, good luck getting past security.

The matter is complicated by several other factors. Competitive advantages like ego, ambition, deceit, and a psychopathic willingness to manipulate others arguably outweigh intellectual brilliance on the scales of material success, and even where intellectual superiority exists, its competitive advantages may be neutralized by financially irrelevant traits like morality and compassion. And of course, if one has been raised to believe that rich people are inordinately selfish and cannot be trusted, one will tend to avoid them, which in most cases means avoiding their money as well.

Lastly, brilliant people often have things on their minds which they rightly consider to be more important than chasing money, especially given that the competition for money is so fierce that accumulating it from scratch tends to demand all of one's time and mental resources, pushing more important and/or socially responsible priorities aside.

December 21, 2016

116. Is there something wrong with the idea that if I'm not smart enough to change the world, I can be rich enough to hire smart people to the change world?

Yes, there is something very wrong with the premise that one who is not smart enough to change the world for the better can make up for it by getting rich enough to hire smart people to change the world for the better. The problem is that if one is not smart enough to know how to change the world for the better – and this takes more intelligence than one might think – then one is not necessarily smart enough to conclusively identify people who are smart enough to know how to change the world for the better, and will therefore be in no position to hire them (unless one gets lucky).

The intellectual requirements for being rich are far less stringent than those for knowing how to change the world for the better. Yet the rich, always craving praise and publicity, want to justify their materialism after the fact, and to expiate their (often spectacular) transgressions on the way to success by providing smarter and more dedicated people with financial resources to do what the rich lack the brainpower to do themselves. This sometimes yields limited success, but it cannot as a rule be done efficiently; a rich nincompoop can too easily make a bad situation worse by planting himself and his employees in the middle of it when it requires better minds, even displacing the very people who could otherwise have dealt with it in the right way.

Nevertheless, this get-rich-and-throw-money-at-it routine is the rule rather than the exception. Name a high-profile problem, and it's only a matter of time before some wealthy dunce decides to become a big star by hurling himself and his money at it, usually by scouring academia for highly credentialed academics who can point out other academics whom he can hire to make him the Savior of Mankind and get his mug on the cover of Time Magazine despite the fact that his own head contains nothing but rocks and dollar signs.

Unfortunately, it's not usually that simple. Truly profound solutions for extremely hard problems do not usually come about as a function of wishful thinking and financially stimulated academic nepotism. We know this because there are too many big, important problems that academics have chronically failed to solve no matter how much time and money they were given to do it.

However, a note of caution is appropriate. If the only alternative to misguided philanthropy and private enterprise is handing everything over to the government, then misguided philanthropy and private enterprise it will have to be. For you see, in this inverse meritocracy we inhabit, it's as Ronald Reagan used to say: "The nine most terrifying words in the English language are: 'I'm from the government, and I'm here to help.'" Help of the kind we're accustomed to getting from the government is often considerably worse than no help at all.

December 12, 2017

117. Is Chris Langan correct that we "have no business equating wealth with intelligence"? Some of my few Republican colleagues seem to do just that, and more.

Of course I'm right.

Granted, IQ and income are somewhat positively correlated; people with higher IQ's tend to make more money, at least below a certain limit. However, the strength of the correlation does not support the inference of a definite causal relationship. On the other hand, the relationship between IQ and wealth is all over the place, suggesting that there is no such thing as a social-Darwinistic wealth-based meritocracy. There's a touch of meritocracy in the distribution of income – it peters out very quickly as we go higher on the IQ scale – but not in the ownership of large mansions and fat investment portfolios. It turns out that the hereditary rich, lucky investors, etc., have nothing very special going on upstairs.

Unfortunately, this is a problem. Just as the power to affect the course of society is amplified by wealth, the ill effects of the relative moral and intellectual deficiencies of the rich are *also* amplified by wealth. In short, the rich are not in general sufficiently intelligent to properly use their wealth and influence to guide us toward a better world, especially in the presence of as many diversions and distractions as now assail us, and their wealth and power only amplify their mistakes. The common "meritocracy assumption" – that wealth correlates with intelligence and responsibility throughout the IQ scale – turns out to be so much hot air, and due to the inevitable "wealth = power" equation, systemic injustices of the real economy translate directly into socioeconomic degeneration.

In other words, it is simply not true that as the power of the wealthy to determine the evolution of society increases as a function of net worth, they are proportionally capable of manifesting the integrity and intelligence to provide society with optimum guidance and financial impetus. The meritocracy assumption is just a variant of the illusory "just world hypothesis" and the associated conception of "justice" according to which reward is always proportional to merit. We do not live in a just world, and the rich are not on top of society because God put them there.

Of course, there supposedly exists a patch for this situation. It's called "Academia, Inc.", from which the uber-wealthy and powerful are supposedly able to hire experts who know how to make judgments and solve problems that the rich lack the brainpower to handle on their own. But unfortunately, academia – in its own special bureaucratic way – is every bit as degenerate as the investment class; in effect, it is led around by its nose at the merest scent of money, whoring out like a drug-addicted hooker at any hint of the long green. (I'm terribly sorry if this offends anyone, but it is what it is.)

Add to this the fact that academia and its minions systematically exclude any outsider who tries to participate meaningfully in intellectual commerce – that is, anyone lacking not only wealth, but credentials and connections – and the problem becomes obvious to anyone with half a brain (which naturally excludes fat cats and acadummies. ;)

December 28, 2017

L

Langan, Chris

Intelligence

118. Are there really self-educated geniuses from the streets like in Good Will Hunting?

Of course there are. Back in the late 1990's, possibly inspired by the success of "Good Will Hunting", reporters began looking around for someone like the character. It seems that they found me, after which I appeared on every major television network in North America and many others abroad.

I've never met anyone involved with the movie, and I'm sure that none of the cast or producers had ever heard of me. So what made people think that I was something approaching a real-world template for the movie character? Well, I had a rough home life growing up. I barely have a high school diploma – they didn't even want to give me one of those – but I can make almost any professional academic look like a dunce at will. I've worked as a cowboy, construction worker, firefighter, and fisherman, and I was perhaps the best-known bar bouncer in the Greater New York area for 25 years. I was the first person ever to solve the venerable chicken-or-egg problem (my solution is the one for which somebody else is credited anew every two or three years). I'm even the author of the CTMU, the one and only viable theory of mathematical metaphysics.

Despite all that, I'm sure that most people have no idea who I am. I don't really care, as I've always been a private person (the press always contacted me, never vice versa). But if I had to guess at the reasons, there are at least three that I might consider. One is Academia, which runs a closed shop and monopolizes scientific communication and intellectual commerce to the complete exclusion of outsiders. Another is that IQ has been deconstructed and democratized until everyone and his dog thinks of himself or herself as a genius. (Everyone deserves to be respected for his or her special intellectual attributes, but I suspect that many people have gone a little overboard in the self-esteem department these days.) Yet another is that I'm a White male who dared to use the forbidden G-word (God) when questioned about what I believed, and was mercilessly disparaged and defamed by every half-baked atheistic troll on the planet in revenge for my theological opinions.

But enough. If you want to read more, you can try this link:

https://superscholar.org/interviews/christopher-michael-langan/

July 6, 2017

119. Is Christopher Langan the smartest person in the world?

I agree with the idea, more or less stated in another comment, that being smart and having a high IQ are not necessarily equivalent properties. But this makes judging intelligence all the more difficult. To be judged extraordinarily intelligent, one must communicate extraordinary thoughts to those doing the judging. But this raises a vexing issue: as any really intelligent person knows, communication is always a two-way street. This is not only because communication consists of at least two people exchanging information, but because each must understand what the other is saying. Unfortunately, extraordinary thoughts are often extraordinarily hard for most people to understand.

As relatively few people properly understand the CTMU, it is occasionally subject to erroneous opinionation. For example, it is not self-contradictory as claimed in this Q&A, but uniquely consistent ... provably so, in fact. And at this point, perhaps the main reason – aside from bias and obstinacy – that its critics don't better understand it is that the modern system of intellectual communication and commerce dominated by Academia, Inc. has been completely walled off from those who lack the academic stamp of approval (to which, if the truth be told, any reasonable concept of intellectual excellence is increasingly irrelevant). It's a circular enterprise, a corrupt intellectual trade union that would make the Mafia blush. I can't even post a paper to arXiv – one needs an institutional affiliation for that! (So much for the alleged neutrality and infallibility of "peer review".)

Just to dispel any remaining confusion, the CTMU is absolutely serious Furthermore, it is incontrovertible. What do I mean by that? I mean that if anyone were to gather the "best" philosophers, theologians, mathematicians, and physicists from the "best" universities in the world, they couldn't put a dent in it even if those criticizing it here were helping them. They could, and would, be squashed like bugs, as publicly as one might like. If no one else were able or disposed to perform this service, I could easily handle it myself. The most with which they could get away would be an admission that the theory is unclear to them, or that they don't find it "mathematical enough". But in fact, the CTMU is a theory of metaphysics that is based *entirely* on mathematics, albeit of kinds that are not taught to most students in these fields.

Although some have maintained that the CTMU is "information-free" despite the mathematics that it obviously contains, the real truth is that they are personally incapable of registering the level of information that the CTMU conveys. For this, I can take no more than partial responsibility.

December 25, 2016

120. Does Chris Langan really have an IQ of 200?

Perhaps. This is a psychometric extrapolation, i.e., an estimate. In principle, my IQ could be lower. On the other hand, it could be higher. (With all due respect to "the competition", I'm certainly not worried about getting shown up by anyone claiming higher intelligence, which doesn't necessarily equate to IQ anyway.)

Let's just say that there are several people for whom even higher numbers have been estimated, at least with regard to ratio as opposed to deviation IQ. So if one's idea is that actually racking up such high scores is impossible, one may as well start at the top and work one's way down.

[**Note**: This may well be a troll question, as implied by the fact that one of the two parties who requested this answer of me recently announced that he'd be trolling me. Yesterday, on February 24, 2018, in response to my answer to the question "What is Chris Langan up to these days?", in which I reported that I was being stalked by a couple of trolls, he commented as follows: "You just earned yourself another one. Have a nice day!" It isn't very often that a troll openly announces his evil intent, but when he does, his target can usually expect the trolling to begin immediately, and this A2A would seem to fit the mold. Just as suspiciously, another answer – pejorative as usual – has just been posted by one of the very stalkers to whom I'd referred. If this is about trolling, then what a shame that Quora is being used to chisel away at the credentials of notable participants. Not that it would be terribly surprising given the harassment to which I've been recurrently subjected here of late by the trolls in question.]

February 26, 2018

121. Why do so many people believe Chris Langan is super-intelligent when it is obvious to any actually intelligent person that he is putting up a façade?

I hate to disappoint any Quora trolls who may be on the lurk, but I don't rely on façades. I simply have no need of them. I can squash the vast majority of my intellectual opponents like insects, and the "tall tales" of my intelligence have been born out several times over on standard tests, experimental tests, and the body of work that carries my name.

On the other hand, the person who posted this question is evidently *very much into* façades. First, we have the fact that no name has been given; the poster is anonymous (the façade is so weak that one can't even put a name to it). Secondly, there's the pretense that whereas I (Chris Langan) am merely feigning intelligence, the poster of this question is an "*actually* intelligent person" who is peering scornfully down at me from the heights of Olympus.

Now, anyone who buys *that* kind of façade is *really* feigning intelligence! But not very convincingly, I'm afraid.

August 7, 2018

Quora

122. Why have we seen so little of Chris Langan?

I don't think it's quite fair to say that I've been "seen so little". I've actually posted quite a few answers here, mostly of a kind that no other participant could have duplicated.

It is possible that there would be more answers from me if:

(1) a number of my contributions hadn't been eliminated to placate various trolls from whom I'd tried to defend myself and my work;

(2) the page hadn't spontaneously reloaded during my attempt to correctively comment on an answer previously provided by one of Quora's most popular science contributors, at which point comments were suddenly disabled and my corrections vanished (I got the impression that someone at Quora was able to read what I was writing in real time, and had opted to protect the answer by resetting the page to disable comments and destroy what I'd written in response).

In addition, I've notice that my comments are typically viewed far less than those of certain prolific contributors here, suggesting to me that Quora may employ a rating system or viewing policy that somehow favors high-volume contributors at the expense of lower-volume ones irrespective of quality, and/or that the readership of Quora is strongly biased toward contributors with certain viewpoints that I find insupportable.

If at some point I have reason to change my mind regarding these concerns, I may contribute more often. Quora is basically a good idea; it's just the execution that worries me.

Thanks for your interest.

June 21, 2017

Typical day

123. What does Chris Langan's typical day look like?

Thanks for asking, but I'm afraid this may be a little boring. I live on a ranch in an isolated rural location, so there's very little in the way of nightlife and commercial entertainment. While it's very pretty and private, the virtual absence of a local service economy means that we're almost entirely on our own – for most of what needs to be done, there's no nearby tradesman able to do it competently and affordably. Self-sufficiency is an indispensable virtue.

On a typical day, I wake up and make coffee for me and my wife. We may talk for a while about what needs to be done. I sit down at my desk and work on intellectual matters, transcribing the answers to any questions that had been on my mind before I retired (I usually wake up with the answers). I check my email and catch up on any accumulated business that may have hit my inbox – I run a nonprofit foundation, and I'm sometimes included in email distributions for physicists, mathematicians, and cognitive scientists with whose technical problems I sometimes try to help.

After a while, I exit the house – a 1500 ft^2 farmhouse built at the end of the 19th century, which I've largely rebuilt over the last 12 years or so – and tend to any urgent ranch business that may have emerged during the night, doing a few chores. (There are almost no professional ranch hands anymore; even in agricultural areas such as this one, everyone wants to be a wealthy big shot like the gasbags and stuffed shirts on TV, or why work at all?) I return to the house, browse a couple of news sites to get some headlines – we have no live television connection here, just a television connected to a computer and a DVD player – and see to more business or do some more thinking and writing.

As midday approaches, my wife describes to me what she's contemplating for lunch; I tell her how good it sounds, or request a slight modification. We eat lunch. I return to my desk and do a bit more work ... how much depends on the lunch (too many carbs, and I might feel the urge to take a brief nap). I exit the house for another round of chores – feeding the animals, fortunately not an everyday necessity, consists of moving up to 20,000 pounds of hay into various pastures using either of two old but powerful JD diesel tractors. I come back inside and do some more thinking and writing. Dinner involves a slightly more elaborate repetition of the lunch routine. While the sun is still up, I go out and button everything down that needs to be secured before nightfall. I return to my desk and write until it's time to go to sleep. I might watch a movie with my wife. Reload and repeat.

Interspersed among regular activities are limited breaks consisting of recreational pastimes like reading, writing music, or pumping a little iron. Telephone usage is relatively infrequent; there's no real cell phone service in this remote location. When necessary, I make a trip to town (the two nearest towns are 8 and 10 miles away, but they lack real grocery stores, and the next-nearest towns are nearly 40 miles distant). If something requires me to catch a plane, I must leave 4-5 hours in advance because the nearest large airport is 120 miles away, and parking is often in limited supply. A recent unexpected trip to NYC for a television interview, for example, consumed two full days with almost no down time. It's hard to get away for much longer than that due to the nature of the operation.

Of course, there's plenty of *ad hoc* variation. Ranches typically have many mouths to feed, and sometimes emergencies arise. The most recent emergency, for example, was an amorous 2000 pound black Angus outside the fence ... one of my bulls, who had broken through the fence in order to court lonely cows across the road. Other kinds of emergency: an animal falls ill or gives birth; a stallion breaches the mares' pasture "without permission"; a lightning storm passes through and sends a limb or an entire tree where it shouldn't have gone, and so on. (It might even come to my attention that some logic-challenged troll has launched a polemical online attack on me or the CTMU, in which case I may attempt a little damage control ... or maybe not.)

This is all very flexible and approximate, but I'm sure you get the picture.

May 8, 2017

Logic

Absolute Truth

124. Do Christians like the CTMU, the proof of God?

That all depends on whether or not they're real Christians as opposed to fake "I believe everything my duly ordained priest/pastor says and nothing else" Christians.

People who believe in Christian Scripture *must* believe in the CTMU, as scripture must be interpreted in logic in order to be apprehended, and the CTMU is merely logic formulated on the required level of discourse.

On the other hand, fake Christians cannot apprehend scripture except by filtering it through layers of doctrine authored by fallible human beings pretending to have divined its proper interpretation. This means that they cannot directly apprehend scripture at all, hence cannot meet the most fundamental requirement of true Christianity.

There's a big difference between real and fake Christians. Christ himself would have told you to choose logic over doctrine, as it is immediate to God. In fact, God and Christ are scripturally identified with logic in the form of "Logos" or Absolute Truth [John 1:1].

(We'll keep it short due to the troll problem here at Quora – questions and answers have a strange way of simply disappearing, and I have only so much time to spend, and potentially to waste, on such a volatile medium.)

July 12, 2018

125. What do you consider the true religion and why?

If one wants a belief system based on Absolute Truth, then one needs to have identified Absolute Truth ... or, as it is called in Christianity, Logos. It has been scripturally approximated in Christianity itself, and is even personified by Christ. However, its mathematical structure is not identified in any known body of scripture.

There is just one well-defined comprehensive truth-structure, or knowledge-system, based on Absolute Truth. It has not been promoted as a religion, but merely described as a "metareligion" in which other religions are subject to interpretation. Technically, it would be classified as a metaphysical formulation of logic. It has been here for decades, but has never been embraced or even acknowledged by spokesmen for major organized religions (or minor disorganized ones, for that matter). Such people usually say what they must to ensure maximum popular and financial support for themselves and (secondarily) their own creeds; deny it though they might, logic and truth are strictly side-issues for them.

This truth-structure is called the CTMU, and it's been out there for decades. There is not a philosopher, theologian, mathematician, or scientist on Earth who is remotely knowledgeable or intelligent enough to gainsay a single word of it. (This is not to disparage anyone's intellectual ability; it's a function of how the system is structured. As its author, I know this with certainty. I'm relatively well-known, yet I have never seen a valid, meaningful, content-oriented critique of it from anyone with any weight in the field. Given that I'm better-known than most of those who would be doing the critiquing, the reason can only be that no real authority has the knowledge or confidence to take issue with it.)

Those who go on spiritual quests for the ultimate spiritual system should beware, as this is a very noisy field; there is a great deal of unwarranted competition for peoples' attention. The loudest and most insistent voices usually belong to the most venal and ambitious people. I would sincerely advise readers against embracing any other system to the exclusion of the CTMU, as there would be a spiritual cost associated with doing so.

I hope everyone has a great day.

October 8, 2017

Branch of philosophy

126. Why is logic a branch of philosophy?

Logic consists of the rules for attributing properties to objects (predicate logic), and for determining whether the resulting attributions are true or false (propositional or sentential logic). Thus, it is integral to ontology, which concerns attributions of existence, and epistemology, which concerns the ability to know attributions as true or false.

Ontology and epistemology are coinciding aspects of metaphysics, which is central to philosophy insofar as it spans all that exists, or – placing the emphasis on epistemology rather than ontology – of all that can be known.

Hence, logic has everything to do with philosophy on the deepest possible level, and vice versa.

July 28, 2018

Human Cognitive-Perceptual Syntax

127. Can something be conceived without any axiom?

Yes, to the extent that it instantiates the (self-instantiating) *a priori* HCPS (*Human Cognitive-Perceptual Syntax*). The HCPS, which can be understood as the accepting syntax of any human observer (or a generic observer defined thereon), is necessary and therefore universal because nothing can be conceived or perceived without it.

Note that it would be correct to substitute *logic* for "the HCPS" in the above paragraph. However, this would require that *logic* be defined to accommodate the CTMU, a truly universal and therefore metaphysical formulation of logic which applies not just to generic attributions, but to the entirety of logic and all that goes with it.

In contrast, theoretically induced axioms usually refer to hypothetical *a posteriori* content without which they cannot be fully instantiated, and their universality is not given for any universe not defined on them by fiat.

May 30, 2018

Intelligibility of reality

128. What if the CTMU were correct?

This question is equivalent to "What if *logic* were correct on the metaphysical (metalogical, metamathematical) level of reality and all levels beneath?"

The answer is very simple: Logic / the CTMU must be correct on all levels of reality, for if it were not, then reality would be unintelligible, in which case there would be no science, philosophy, mathematics, or reality as we know it.

So really, all that one need do is decide how much one likes science, philosophy, mathematics, and reality itself, and then answer accordingly. ;)

March 29, 2018

Interpretation of scripture and observations

129. Which tells the truth, Bible or science?

In a way, this is a meaningless question. Truth is a logical property; its structure can be considered that of logic in its entirety. Science and religion are true only insofar as they can be interpreted in logic, or shown to be logically consistent and thus to "model logic", and the jury remains out until this requirement is satisfied.

Science can be interpreted in logic with respect to direct replicable observation and the immediate logical entailments thereof; beyond that, it relies on empirical induction and does not support claims of truth or proof. Religion is even more problematic, especially when based on revelatory scripture; miracles are by definition non-replicable in any reliable sense, and must therefore be taken on faith by those who do not witness them directly.

Although the Bible is based on a powerful set of universal (and therefore unfalsifiable) truths, some of which are shared or "replicated" by other religions, its revelatory and universal contents make it impossible to completely verify using standard scientific methodology. However, **this does not rule out substantial verification by means of logic**, which in any event has always been the sole means of verification beyond the perceptual realm. Using a sufficiently advanced form of logic, it can be shown that Christianity is largely based on verifiable (true, logically self-consistent) principles.

It thus turns out that both science and the Bible are verifiable with respect to direct objective perception of the external world and all logical necessities thereof. Because they can both be largely interpreted in the same advanced logical system or "truth structure", they can in fact be shown to correspond with each other in certain key respects. It is thus meaningless to call either one of them "more true" than the other; they are simply concomitant aspects of reality.

While this may be reassuring, a word of caution is in order: neither science nor religion is immune to error and confusion. It is thus a very good thing that through their mutual dependency on logic, each is a potential source of improvement for the other.

February 10, 2017

Logic vs physical naturalism

130. What would require you to give up your naturalistic worldview in order to embrace the truth of Christopher Langan's Cognitive-Theoretic Model of the Universe?

I'll answer this one, if I may: as I've previously explained elsewhere on this site, **logic** is what requires the rejection of (naive physical) naturalism. The CTMU is logic in the sense that it is a globally distributed ingredient of reality which contains no assumptions or nontautological axioms, and it precludes naive naturalism due to its referential incapacity for self-explanation.

As the CTMU has been out there for thirty years or so, we can no longer say that the technical relationship between this universe and global reality is "in its infancy" (as another answer claims). In fact, it is quite evolved and robust. What remains in its infancy is the openness of academia and its journals to material emanating from anyone but certified, dues-paying academics, who have a mile-wide conflict of interest when it comes to serving truth over their own collective interests, and of course, over the interests of the shady, similarly conflicted proprietors of their glorified trade union / corporate monopoly.

Once that conflict of interest is resolved, and the honor and chastity of truth have been rightfully restored, the world can more easily take account of existing developments in the field of mathematical metaphysics.

February 13, 2018

Nature of tautology

131. Why do people assume that the CTMU contains assumptions when, in fact, it doesn't?

People erroneously assume that the CTMU contains assumptions because assumption is their intellectual default condition. They fall back on assumption because they are intellectually incapable of discerning the necessities on which human knowledge is actually based.

Knowledge entails certainty by definition, and certainty entails logic. Logic is based not on assumption, but on tautology (tautological forms and inferences). Without fully understanding the structural and dynamical properties and manifestations of tautology, they simply assume that nothing of scientific interest can ever be logically deduced from well-verified initial data.

This assumption is very common but ultimately mistaken. The irony is that because so many people stumble into it – including many highly trained people who are supposed to know better – those who rightly question it are the ones usually accused of "making assumptions".

Eventually, one learns to see this situation as rather comical, like a passel of clowns frantically unicycling around the center ring, bumping into each other and toppling over and generally creating mayhem under the Big Top while cacophonously squeezing the air out of their red rubber noses. ;)

April 26, 2018

Predicate logic

132. Is predicate logic dualistic as stated in the latest CTMU paper published in the academic, philosophical journal Cosmos and History?

That's what it says in the papers, right?

Predicate logic is dualistic because it considers attributes separately from the objects or arguments to which they are attached by (universal or existential) quantification.

Equivalently, predicate logic is a formal system which includes a formal language (including its formal grammar) and a deductive apparatus consisting of axioms and rules of inference. The external universe is excluded from the system, isolating the language from the rest of reality.

July 11, 2018

Properties of logic

133. Is logic learned? Is it absolute or relative? Is it influenced at all by society?

As the term is ordinarily used, logic is a fundamental requirement of human experience that is universal and unvarying. This must be the case insofar as it governs the coupling of objects and attributes (predicate logic), permits the discernment of logical complements (propositional logic), and controls implication and entailment (determines how one thing follows from another).

Logic is innate and *a priori* to the extent that its failure on the level of sensory perception and basic reasoning would preclude the individual and consensual identification of physical reality and its ingredients, preventing discernment of objects, relations, and events. This is confirmed by the fact that regardless of our logical training, we all perceive the same physical world and can agree on what it contains. The fact that we all inhabit the same world, consisting of the same physical objects, relationships, events, and processes, attests to the unconditional invariance and universality of logic.

The perfect distribution of logic over physical reality and valid cognition is a hard requirement of empirical science, which depends on the replicability of observation and the unbroken possibility of valid inference. Conversely, the existence of empirical science implies that given any apparent violation of logic, we can always resolve the paradox and restore consistency.

While pathological deviations from logic are not precluded, we are assured that when such deviations collide with reality, reality and logic will ultimately prevail.

April 17, 2018

Propositional logic

134. Why are propositional and predicate logic important in the CTMU?

For the same reason that they are important in science and philosophy in general: predicate logic is basic to attribution, while propositional logic is required for classifying attributions as true or false. (Technically, propositional logic is properly included in predicate logic; after all, truth values are attributes.) One cannot have a meaningful theory without them.

However, there are differences between the CTMU and standard theories with regard to where logic resides, the particular form it takes, and how it is applied.

June 25, 2018

Self-justifying nature of tautology

135. The CTMU says tautologies are self-justifying. How can this be true?

A statement can be proven true or false – in fact, it can be *identified* – only by using 2-valued propositional logic. Therefore, 2-valued propositional logic is verified by the identifiability or provability of anything at all.

A specific propositional tautology amounts to an axiom of 2-valued logic; thus, it is implicated in its own identification. As it can be identified only if it is true, we know that it *must* be true by the mere fact that we can identify it.

The CTMU develops this idea in order to elevate the field of mathematical logic to the metaphysical level of discourse, the better to answer profound questions that empirical scientists insist on asking, but lack the tools to answer.

And now a brief word regarding the anti-CTMU trolls which proliferate like a toxic fungus here at Quora. There is no one on Quora who is remotely smart enough, or good enough at math, logic, philosophy, physics, chemistry, biology, or anything else, to get over on the CTMU. In fact, the mere effort to get over on the CTMU, and for that matter on me, is incontrovertible, rock-hard proof that the person trying to do it is a logical ignoramus, a nincompoop, and a buffoon. The nastier he or she is about it, the more he or she is attempting to mislead the readership of Quora, and the more he or she deserves censure as an absolute gibbering idiot.

Let's be clear about this. There are people on Quora who claim to have advanced degrees in this or that from big-name universities; some are even lauded as "Quora Top Writers", whatever that's supposed to mean. So far, so good. But after tooting their own horns like the steam-powered pufferbellies of yore, some of them have revealed themselves as mean-spirited imbeciles by flying into protracted fits of projectile diarrhea of the mouth, calling the theory "bullshit", "horse shit", "crap", and "nonsense". Alleged upvotes notwithstanding, the readers of Quora would be vastly better off without a single one of them.

I'm tired of having to make a hobby out of digging everyone else out from under the piles of manure spewed by these people. It would be wonderful if they could simply run off and find another topic about which to blow smoke up everyone else's rectum while knowing not a damned thing about it.

Thanks for your attention.

June 29, 2018

Structure of truth

136. Are all truths logical?

Yes, as logic can be described as the "definition" or "structure" of truth.

However, the truth predicate takes attributions as its arguments, and the *signature* or "alphabet" of the language in which a given truth is expressed may contain symbols representing both *logical* and *nonlogical* elements, where by "logical elements" we mean the symbols and associated concepts of mathematical logic, and by "nonlogical elements" we mean (e.g.) symbols representing concepts like *point, line,* and *angle* in the signature of the language of geometry. So the truth need not be exclusively logical or even logically determined; it may involve nonlogical elements and be freely determined by nonlogical considerations.

That being understood, every nonlogical language must conform to logic in order to be intelligible – it must fit into appropriate logical expressions as instances thereof – and in this sense, logic and truth are inseparable.

October 11, 2017

137. Could God prove his existence?

Of course He could. He would simply show that His structure is that of truth and existence themselves.

It would take just two questions.

(1) What is the structure of truth and existence?

(2) What properties do you share with this structure ... i.e., how are you identified with it?

If the shared properties include those on which the "God" concept is conventionally defined, then God would have proven that He exists in the form of truth or Logos, just as stated in the New Testament. Basically, Truth is queried about itself, and divinity is recognized in its logical structure.

(I've already been through this in considerable detail, starting three or four decades ago. The relevant structure is called the CTMU, acronymic for *Cognitive-Theoretic Model of the Universe*.)

June 19, 2017

Tool for studying reality

138. Which field of science studies reality as a whole?

Technically, the answer would be "metaphysics", considered as an explanatory generalization of physics and other scientific disciplines. However, most people would argue that metaphysics is not a science, but merely an unkempt collection of abstruse philosophical ideas.

What about modern sciences? There are two general kinds of "science", *mathematical* and *empirical*. The empirical sciences are constrained by the Scientific Method and thereby limited to observation, logical deduction from observation, and empirical induction, which are inadequate for the study of reality as a whole. This suggests that the study of reality falls within the range of the mathematical sciences.

However, most mathematical theories are formulated with respect to abstract universes defined on sets of axioms which do not fully characterize empirical reality, let alone reality as a whole. They apply only to those parts of empirical reality which happen to obey their respective axioms. There is only one branch of mathematics which covers every aspect of reality without exception, including the empirical aspect: logic. This is because the truth predicate of two-valued logic can be used to denote inclusion in reality at large.

So the answer is logic, along with any other branch of mathematics, or bundle of branches, whose axioms separately or collectively cover every part of reality. A global reality theorist therefore needs, above all, to be a mathematical logician. He or she can then fit other, narrower branches of science into mathematical logic as auxiliary theories. (For example, physics is just a bundle of mathematical theories in which certain terms are observationally defined, and every mathematical theory can on some level be interpreted in a sufficiently advanced form of logic.)

Again, what I've just described is a framework for what "metaphysics" was originally supposed to be. Unfortunately, modern philosophy, which lays claim to metaphysics, has become too disconnected, disorganized, and equivocal to exhibit the coherence expected of a science. In short, too many people recognized as "philosophers" and "metaphysicians" have too long operated in too unscientific a way for their versions of "metaphysics" to qualify as scientific explanations of reality.

So for the time being, mathematical logic it is. This may not be what most people want to hear – they want something easier, or perhaps "sexier" – but it's the only discipline with an unlimited capacity for both ontological generality and scientific specificity, and thus able to accommodate all of reality without prejudice or preconception.

February 12, 2017

M

Mass Media

Indoctrination

139. Why are people with a low IQ getting more news coverage than Chris Langan and his ideas?

This is the case because the mass media and its mass audience contain many more people of ordinary intelligence than people of extraordinary intelligence, and there is money to be made by telling people just what they can understand and want to hear.

As it happens, many people don't want to hear that there is someone more knowledgeable than they on the nature of reality, that the academic degrees and doctrines they bought so dearly aren't really a substitute for intelligence and hard-earned knowledge, that there is a God independent of their preferred body of religious or antireligious doctrine Who may discard them if they insult, disobey, or actively oppose Him, et cetera. They want to hear the exact opposite, and the establishment is only too happy to sell them what they desire.

People of normal intelligence are far more common and easier to indoctrinate than those of extraordinary intelligence. The world is run by a small but powerful minority which tells the majority what to believe and thus what it eventually wants to hear; it indoctrinates the majority to adopt and prefer irrational beliefs that violate majority interests but conduce to its own. Being a nonmember of the controlling minority, I am in no position to take advantage of the mass media or the machinery of public indoctrination in order to promote work which deviates from standard doctrine.

The highest and deepest kinds of truth have no monetary value in a materialistic society, and may even be suppressed by the controlling minority as opposing its interests. Under these circumstances, the mass media naturally prefer more orthodox, widely palatable, and therefore lucrative content.

January 13, 2018

Mathematics

On conceptual abstractions

140. Is it possible to mathematically represent conceptual abstraction of Christopher Langan's Cognitive-Theoretic Model of the Universe? If so what is the proof?

Yes. In fact, it is possible to mathematically represent *any* coherent conceptual abstraction. All one needs to do is define one or more symbols to represent it and/or its components, and then specify a mathematical structure to which the symbols conform, usually something "algebraic". A language is an algebraic structure, and the CTMU is the start symbol of the grammar through which its detailed structure emerges. Now here's the cherry on the sundae: because the language in question is supertautological, its grammar is self-verifying, thus obviating the need for additional proof ... which, however, is by no means precluded. (See how easy that was?)

February 5, 2018

Illogical exclusion

141. What is the most illogical thing in mathematics?

The most illogical thing in mathematics is the exclusion of the mathematician from mathematical structures and reasoning, technically rendering mathematicians powerless to reliably understand or connect with the content of their field. This systematic exclusion has been prolonged by the exclusion from mathematical journals of anything resembling a valid correction.

April 14, 2018

On infinitesimals

142. Could we "invent" a number h = 1/0 similarly to the way we "invented" i = √-1? What would be the effect on modern math if we did so?

This number already exists in the form of "infinity" (∞). However, there is a cheat involved: "0" becomes the infinitesimal "1/∞", which is "of measure 0" in the real unit interval. That is, 0 is interpreted as a "measure" rather than a quantity, and a nonzero "infinitesimal" *quantity* is invented which coincides with *measure* 0. This can be expressed as follows: "An infinitesimal is a nonzero number that is less than any finite *quantity*, and is therefore quantitatively negligible, i.e., has sub-finite *measure* 0."

It's quite a trick; *quantity* has been disjoined from *measure* using the term *infinitesimal* and then crudely plastered back together with measure using the phrase "measure 0". An infinitesimal is a *nonzero* quantity representing *zero* progress along a line segment. The cost is that infinity "breaks" finite algebra, requiring that the rules be changed (as prescribed by Cantor and others). It is on the strength of this cheat that the infinitesimal calculus exists. For over a century, this has been officially rationalized to high school and college students using infinite converging series and the Cauchy–Weierstrass epsilon-delta technique. But on close examination, this fails to provide a satisfactory explanation of *either* infinitesimals *or* the calculus.

This situation has given rise to new mathematical perspectives and techniques including Lebesgue integration, nonstandard analysis, and pointless topology. The CTMU affords a new perspective of its own.

May 21, 2017

In natural sciences

143. Is mathematics unreasonably effective in the natural sciences?

Reality consists of patterned substance, i.e., substance which displays patterns through which it is recognized. "Pattern" is thus analogous to the syntax in terms of which a language is recognized by its users. Mathematics is a human formalization of the highest (most ubiquitous) level of this syntax, and thus describes reality and our recognition of it in a very general way.

As it presently exists, this formalization is not perfect; despite what one sometimes hears from self-congratulatory professional academics, the patterns which have thus far been formalized by human mathematicians are only imperfectly understood by them. However, mathematics can nonetheless be functionally defined as "the patterns through which we recognize our reality", thus circumventing the intellectual shortcomings of human mathematicians. This generic functional definition can then be trimmed using any special-purpose restrictions we might wish to apply in particular contexts.

When mathematics is defined in this way, its effectiveness in the natural sciences is no mystery at all; it is effective in the natural sciences precisely because it describes the syntax through which natural scientists identify reality. This relationship between reality and those who observe, study, and theorize about it is called a "supertautology", the structure of which requires that reality be modeled as a special kind of "language". This has all been formalized in a true "theory of everything" called the Cognitive-Theoretic Model of the Universe (CTMU). The CTMU can be described as a metatheory of science, or if one prefers, as "(meta)mathematical metaphysics".

The CTMU has been the best-kept secret in the history of science and philosophy for the last thirty years or so. This is because it amounts to metaphysics, which many so-called experts absurdly regard as impossible despite their manifest inability to argue coherently against it. Thus, even though the CTMU was authored by a man widely described as the smartest in the world (Christopher Langan), it has been roundly ignored in what can only be described as a collective fit of academic spite and incomprehension.

December 24, 2017

P vs NP problem

144. How close is Chris Langan to solving the P versus NP problem?

The matter has been satisfactorily resolved for my own purposes (which, incidentally, are the only purposes about which I care in this context).

There are two different solutions, one of which applies in a certain framework defined as "metacausal", "precomputational" or "protocomputational", and the other of which applies in a causal or computational limit thereof. This amounts to a relativization of the solution. Although I can't say for sure, I suspect that in the view of the "Millennium Prize" Committee, this falls short of an actual Y|N solution, leaving me nothing whatsoever to gain by submission.

(We also have the serious problem of bias on the Millennium Prize Committee, expressed by Andrew Wiles as their complete certainty that no non-academic is capable of solving any of the problems. This, of course, guarantees that the submissions of nonacademics will in effect be circular-filed, leaving them no rational choice but to disregard such prizes, and where necessary, the professional academics who award them to each other.)

February 14, 2018

Problem solving

145. Has it ever happened that an ordinary person found the solution to an unsolved mathematics problem?

Mathematically uneducated people have often found the solutions for previously unsolved logical and mathematical problems and puzzles. For example, the legendary Games columnist Martin Gardner credited many amateurs for the solutions of difficult math and logic puzzles published in *Scientific American*.

Although Academia, Inc. would like the public to believe otherwise – it is, after all, in business for profit – intelligence and ingenuity are innate human characteristics that cannot be bought and paid for at a university. A case can be made that some amount of academic training is all but necessary for the solution of certain very difficult mathematical problems, especially ones which cannot be understood (let alone solved) by the typical layman. But in the large, professional academics – who notoriously tend to fixate on just the problems salient within their personal specialties – do not have a monopoly on the selection or solution of "good" or interesting mathematical problems.

Historically, the intellectual penetration to formulate and solve really interesting problems of great import to the future intellectual and practical development of mankind came almost exclusively from formally untrained people, including those who laid the groundwork for modern disciplines which now reek of academic snobbery. Such people worked from first principles, an ability that professional academics, many of whom are weighted down with leaden orthodoxy, have forgotten how to use, recognize, and properly value.

September 7, 2018

Proving God's existence

146. Can mathematics tell if God exists?

Of course it can: the mere existence of mathematics suffices. Something is needed in order to distribute it as an ordered pattern over our entire domain of thought and experience, and it turns out that this "something" bears description as God. This can be couched in rock-hard mathematical language.

September 29, 2017

Pure mathematics

147. When will Chris Langan release a purely mathematical version of Christopher Langan's Cognitive-Theoretic Model of the Universe (CTMU)?

There is no "purely mathematical version" of anything with real-world meaning.

"Pure", uninterpreted, unapplied mathematics is devoid of external correspondence; no theory which consists exclusively of "pure mathematics" has real-world relevance. At the very least, definitions must be given which apply to the real world as well as to the mathematical structures themselves, along with an overall real-world correspondence (model).

As for the question "When will Chris Langan release a mathematical version of the CTMU?" (note the elimination of "purely"), I've already done this on several past occasions. Further mathematical descriptions are in the works, but must await proper venues for publication.

February 19, 2018

Mega Foundation

Operations

148. Is Chris Langan's Mega Foundation still active?

Absolutely. As a matter of fact, we have a large construction project in progress that is designed to facilitate and enhance our operations.

Unfortunately, the construction has been delayed by various errors and oversights on the parts of some of the contractors involved. We won't be able to make use of our new facilities until the problems are corrected, the construction project has been brought to a more advanced stage of completion, and we've brought in one or more volunteers to help with IT and physical maintenance.

Meanwhile, basic operations are continuous and ongoing, and we've tentatively scheduled a gathering for this autumn.

[By the way, the entire Mega Foundation website was placed on a secure platform two years ago in the expectation that the construction project would be complete and we could move forward again. The changes were quite extensive, but largely invisible to those visiting the site. As for the visible content of the site, it will change only if and when there is a need to change it. Additional content may of course be expected at some point.]

May 7, 2017

Metaphysics

Cosmology

149. What does the CTMU contribute to cosmology?

Cosmology, along with ontology and epistemology, began as a branch of metaphysics. Because the deepest issues in cosmology – the origin of the cosmos, accelerating cosmic expansion, the nature of causation, and so on – require a well-structured metaphysical metalanguage of physics for their proper formulation and solution, metaphysics is still where cosmology belongs.

This, of course, does not preclude the relevance of physical research to cosmology. Physics is more than welcome in cosmology, and has carved itself a very respectable niche in the field. Problems arise only when physics tries to steal the entire show, something which it is unequipped to do as a matter of logico-mathematical fact. The structure of physics simply doesn't allow it.

There is exactly one properly formulated theory of metaphysics: the CTMU. It follows that the CTMU must be credited with putting cosmology back where it belongs, and thereby making it possible to express and solve cosmological problems which do not admit of physical expression and solution.

July 14, 2018

Definition

150. What is metaphysics?

The proper definition of metaphysics has three escalating levels.

The body of theory which is "beyond physics". (This usage is due to Aristotle's "editor" Andronicus of Rhodes, who introduced the term to describe that part of Aristotle's explanation of reality which underlies physics.)

The field consisting of ontology, cosmology, and epistemology. In other words, the theory which deals with (a) the most general abstract properties, *existence* and *nonexistence*, and all lesser properties subsumed by them; (b) the set of all substantive referents of those properties, namely the cosmos including all of its parts and aspects; and (c) the epistemological (attributive, perceptual and/or cognitive) correspondence between the properties and their referents. (Note that these fields are not separable.)

A logically idempotent supertautological metalanguage of reality at large, defined in such a way as to coherently refer to reality, formulate questions regarding it, and express answers for the questions. This metalanguage is called the CTMU, short for *Cognitive-Theoretic Model of the Universe*. The CTMU, which functions as its own universe and model, and which may be considered identical to the high-level structure of reality itself, is described elsewhere on this site.

I sympathize with those who throw up their hands at the idea of describing an integrated discipline covering everything which has been called "metaphysics" over the years. In recognition of their difficulties, my own approach has been somewhat more limited: to educe a comprehensive reality-theoretic framework in which all valid physical and metaphysical concepts, relationships, and theories can be jointly interpreted or *modeled* in the logical sense. Due to this shift of focus, metaphysics now exists as a rigorous discipline which subsumes not only the empirical and mathematical sciences, but the humanities and religion as well.

February 13, 2018

Dyson, Freeman

151. What does Christopher Langan think about Freeman Dyson's metaphysics?

The CTMU is absolutely essential to the proper development of Freeman Dyson's "cosmic metaphysics of mind".

Dyson: "The universe shows evidence of the operations of mind on three levels. The first level is the level of elementary physical processes in quantum mechanics. Matter in quantum mechanics is [...] constantly making choices between alternative possibilities according to probabilistic laws. [...] The second level at which we detect the operations of mind is the level of direct human experience. [...] [I]t is reasonable to believe in the existence of a third level of mind, a mental component of the universe. If we believe in this mental component and call it God, then we can say that we are small pieces of God's mental apparatus" [***Infinite in All Directions*** (1988), p. 297]

The CTMU can be described as the complete explication of this viewpoint, and more. This is interesting, as Professor Dyson and I must have been thinking very similar thoughts during the 1980's. The difference: Professor Dyson, who was very capable as a physicist but lacked the specific kinds of knowledge needed for the development of such metaphysical ideas, was a professional academic with full access to the academic monopoly which controls scholarly communication and intellectual commerce (hard publishing, electronic media, and so on).

I, on the other hand, lacked this advantage. So although I was vastly better able to develop such ideas, I was completely ignored by Professor Dyson and his academic colleagues. There is simply no way to compel these people to pay the slightest attention to anything emerging from the real world outside their vast institutional bubble, and unfortunately, compulsion is exactly what would have been required.

So much for the efficiency and integrity of academia as the self-styled vanguard of human intellectual progress. It simply doesn't work as advertised. Let's hope that something can finally be done about it.

May 19, 2018

Epistemology and ontology

152. What are some differences between epistemology and metaphysics?

There aren't any, other than being "different" aspects of the same thing. Epistemology cannot be separated from ontology, and ontology is squarely situated in metaphysics.

Why can't epistemology, the study of *knowledge and knowing*, be separated from ontology? Because knowledge *exists*, and *existence* is the subject matter of ontology.

Can metaphysics be regarded as "more general" than epistemology? Not as long as metaphysics is about *knowing* and *being known as* "metaphysics".

Because each is an aspect or ingredient of the other, they are inseparable. This principle of inseparability amounts to *epistemological holism*, or if one prefers, *ontological* or *metaphysical* holism.

This kind of holism is central to the theory of mathematical metaphysics known as the **CTMU**, short for *Cognitive-Theoretic Model of the Universe.*

The CTMU is by far the most advanced theory of metaphysics in existence.

June 7, 2018

Framework of reality

153. What is the purpose of metaphysics?

Where metaphysics describes every aspect of the existence, intelligibility, origin, structure, and dynamics *of* reality *for* reality, it *equates to* reality, and its purpose is therefore that of reality itself. That is, its purpose is to provide reality with an ontological, epistemological, and cosmological framework supporting the existence, intelligibility, origin, structure, and dynamics of reality.

Where *science* is a real process that reveals the nature of reality to and for the inhabitants of reality itself, metaphysics includes science in its descriptive framework and thus qualifies as "metascience". In other words, it is the discipline by virtue of which science is rendered self-explanatory, and its purpose can be understood accordingly.

May 13, 2018

Metalogical system

154. What is the relationship of metaphysics to logic and epistemology?

Metaphysics is best understood as a metalogical system, the CTMU, which matches the overall structure of reality, and in which it is therefore possible to describe and model reality in its entirety. In other words, it equates to a comprehensive higher-level extension of logic, where *logic* is the syntax of 2-valued identification and therefore integral to epistemology. Thus, logic and epistemology are inseparable aspects of metaphysics. They're all part of an inseparable whole.

May 27, 2018

Metaphysical metalanguage

155. What is metaphysics according to Christopher Langan's Cognitive-Theoretic Model of the Universe (CTMU)?

According to the CTMU, metaphysics is a unique explanatory metalanguage of physics with certain mathematical properties that have been explored in various papers and essays which are readily available online.

February 9, 2018

Philosophy

156. What is the relationship between metaphysics and philosophy?

Traditionally, metaphysics has been regarded as a branch of philosophy which coexists with other branches of the philosophical tree. This amounts to a loose and simplistic definition of *metaphysics* as the application of *philosophical* methods – systematic inquiry through the use of reason, rational argumentation, and critical discussion – to fundamental *metaphysical issues* typically including first principles, ultimate origins and ends, being, knowledge, substance, causation, space, time, and identity.

However, a bit of reflection suffices to reveal a problem with the supposed proper inclusion of metaphysics in philosophy: just as philosophical methodology is so general that it distributes over (includes) metaphysics, the metaphysical issues enumerated above are so general that they apply to philosophy in its entirety. In short, philosophy and metaphysics are inseparable; the content of metaphysics cannot be rationally disentangled from the methodology of philosophy or vice versa.

Now here's another eye-opener: science and mathematics both began as philosophy, and the pairing of philosophy and metaphysics continues to apply to them with full force. So ultimately, all of these fields must be considered together.

The suppression of this epistemological holism by way of sloppy semantics and academic compartmentalization is artificial and strictly illusory.

May 5, 2018

Physics

157. What is the connection between physics and metaphysics?

Metaphysics is a metalanguage relating the theoretical object language of physics to the physical (empirical, observable) universe; it expresses a model-theoretic relationship between them. Owing to the mathematical structure of the metalanguage (implied by ontological closure and related criteria), physical reality – a coupling of physical theory and observational data – can be understood as the "observational limit" of an inclusive metaphysical reality described by the metalanguage. The metaphysical metalanguage which describes this all-inclusive metaphysical reality is a mathematical theory of metaphysics called the *Cognitive-Theoretic Model of the Universe (CTMU).*

May 2, 2018

Scope of metaphysics

158. Does metaphysics entail anything that is not empirical, and if so, is theoretical physics and for example the theory of relativity considered metaphysics?

Well, that all depends, doesn't it.

Let "physical reality" consist of that which is directly and replicably perceptible, plus all that can be directly logically inferred from it. This definition depends on the meaning of *logical inference*, with the burden on anyone claiming that inobservables have been "logically inferred" to specify the logical system in which the inferences have been made.

Almost every physical advance in history has adjoined to the existing version of "physics" new inobservable concepts and quantities that could not be strictly deduced within any system then recognizable as *logic*, or within an axiomatic system known to be universal. Instead, they were empirically induced from particular observations in order to relate and explain them *without certainty*. Where the burden of logical proof has never been met, such advances cannot yet be pronounced physical.

But are they metaphysical? That depends on how *metaphysics* is defined. Where it is defined as the "deep structure of reality" and formulated as a consistent comprehensive metalanguage in which all valid physical constructs and theories can be expressed, related, analyzed, compared, and modeled – as it is defined in the CTMU – the answer is *yes*, provided that the theoretical advance is itself correct and logically valid within its domain. But where the theory is not a valid limit or subtheory of metaphysics, or where the wrong definition of metaphysics is in force, the answer is *no*.

In other words, to the extent that physical theories with inobservable components, such as both GR and QM (General Relativity and Quantum Mechanics), are correct, they are metaphysics under the most powerful definition of metaphysics (that of the CTMU). But under a weaker definition incapable of actually containing reality, one may be limited to saying merely that they are "partially nonphysical".

In either case, such theories are not merely physical.

May 27, 2018

Multiplex Unity

Coherent interconnection

159. Can someone elaborate on how the MU principle relates to the ''problem'' of Unity and Multiplicity in the CTMU? How does MU defines the relationship between the two?

In the CTMU, the Principle of Multiplex Unity – which can be simplistically expressed, in the language of John Archibald Wheeler, as "one universe out of many" – explains how reality coheres despite consisting of many parts which appear to be separate and largely independent of each other. In the 2002 paper *The Cognitive-Theoretic Model of the Universe: A New Kind of Reality Theory*, it is a well-defined mathematical concept explicitly related to dual forms of inclusion. It is strongly related to the CTMU concepts *syndiffeonesis* and *supertautology*.

Without the MU Principle, there would be no way to coherently interconnect our many "subjective universes" – or if one prefers, the set of scientific observers and frames of reference – in one common reality.

February 18, 2018

Musk, Elon

Evaluating utility

160. Who is more valuable to humanity, Chris Langan, who claims proof of god, or Elon Musk, businessman, who tries to secure power by presenting a good image and expanding his portfolio?

Let's have a rational look at this question, shall we?

Just so no one gets confused, let's make it as simple as possible. Take a socio-economic system X, and two people A and B within it. Suppose that A is a corporate CEO who has never actually invented or discovered anything (but managed to attract a lot of favorable attention from venture capitalists and investment bankers), and that B is a person who has weathered a lot of adversity in order to bring humankind a new worldview (no one who actually understands anything about the CTMU denies that it is both original and insightful, and in any case, I could easily crush anyone who denies it were I so disposed).

As we all know, value and scarcity are very closely related. The scarcer something is, the more valuable it is. CEO's are a dime a dozen, and so are salesmen who can attract a lot of favorable attention from venture capitalists and investment bankers. In fact, they've become annoyingly common, and even worse, we seem to have more of them every day. Clearly, such people are only as valuable as they are irreplaceable. But if techie billionaires / corporate CEOs were ever irreplaceable, this is no longer the case. Any one of them can be replaced by any other; they all cherry-pick whatever experts they might need from Academia, Inc., and delegate all of the real intellectual responsibility to their employees. And as if that weren't enough, their positive value exists only with respect to the system X in which their social utility is defined; if X happens to be displacing a better system, then their true value is negative.

It follows that their comparative human worth is strictly limited, and the burden of proof is on anyone who claims that there aren't a million others out there who could step into their shoes and rake in the loot just as well as they do.

Do you see where this is going yet? ;)

January 6, 2018

Outlier or not

161. Based on the book written by Malcolm Gladwell, would Elon Musk also be defined as an Outlier?

Well, yes and no.

Any billionaire is an "outlier" with respect to the overall distribution of wealth. In addition, we can probably say that Elon began with somewhat less wealth, and more initiative, than did the average billionaire. We can even say that Elon is considerably "luckier" than others, insofar as certain things came together for Elon that would not necessarily have come together for another person, in another place, at another time.

However, those who insist that Elon is a "genius" have their work cut out for them. Elon has never been an inventor or discoverer of new concepts, and as far as most of us are aware, has never really exerted himself for any cause or reason that did not promise to result in the further bloating of his personal bank balance. Thus, he is not an outlier with respect to his motives, which are every bit as materialistic and self-interested as those of any other businessman.

January 14, 2018

Productivity

162. Will Elon Musk study the CTMU and thus turn into a productive member of society?

I have no idea whether Elon Musk is even aware of the CTMU. However, the fact that he is reputed to be a convicted atheist suggests that the CTMU is not on his list of preferred reading material.

I have never said, and would not say, that Elon Musk is not "a productive member of society". I would merely point out that the uniqueness and long-term value of his contributions has not yet been determined.

There is another answer on this page which, itself and in its comments section, contains several factually incorrect and/or pejorative statements or insinuations regarding me and the CTMU. As its comments are now disabled – which raises the question of how any comments got there in the first place – and lest the author's personal opinion be mistaken by the unwary for the official stance of Quora, I'll take this opportunity to include a few corrections.

If there is a CTMU/Langan "advertising ploy in action", I have nothing to do with it. I've posted only one question on Quora, and it was merely a duplicate of a previous question which had been erased by the OP along with my answer (this could ostensibly be verified by the Quora IT department, although I have no idea how one would make that happen). I have never instructed anyone, either publicly or privately, to post any questions whatsoever about me or my work on this site or any other.

I have never said that "If Elon Musk doesn't believe in this [CTMU] theory then he isn't a productive member of society, and ... is going to hell."

I have never said that "every single other person [out of "billions of people"] who doesn't believe in this theory" is unproductive and therefore going to hell. I have no idea where this strange notion originated.

It is erroneous of the author to assume that "if she continue[s] to see such questions, [she] will know [her] theory is correct — that this is nothing more than a cloaked advertising campaign going on." Such questions could just as well be generated by honest interest in the CTMU. (After all, several academic papers have been published on it.)

It is not necessarily true that "Something that is truth and sincere is not promoted Anonymously." In fact, the truth value of a theory or assertion is independent of who promotes it or how.

It is not true that "[The CTMU] never went anywhere for no one really took it seriously." My CTMU papers have been downloaded hundreds of thousands if not millions of times, and attracted many sincere followers.

The author links to an extremely contemptuous and pejorative screed by another Quora participant claiming to be a "physicist" or a student of physics. This screed was duly collapsed long ago by Quora Moderation. She has attempted to "uncollapse it" by linking to it in a comment, prompting others to read it. This clearly opposes the attempt of Quora Moderation to quarantine it.

[**Addendum**: The person responsible for generating all of the above errors has apparently modified her original answer to assert that Quora never collapsed the misleading and aggressively disparaging answer of the so-called "physicist" she cites. Yet I still find it to have been collapsed, as I have found it since around the time the alleged "physicist" lost an argument with me in its comments section and had all of my comments (and possibly also my own answer to the question) collapsed or totally removed. So either the "physicist's" answer was collapsed as I've stated, or Quora is collapsing different answers for different people.]

That should do it for now. I hope I've dispelled or at least reduced any confusion that may have existed with regard to the above issues.

January 27, 2018

P

Peterson, Jordan

Debate invitation

163. Does Jordan Peterson regard the CTMU as a metamathematical, logically rigorous scientific theory of metaphysics?

God help poor Jordan if he ever says otherwise! (I'm afraid he wouldn't stand much of a chance against me, and as a North American clinical psychologist, he probably knows it.)

Of course, maybe someone doubts this who actually knows Jordan. If so, then have Jordan give me a call and we'll see about setting something up. I'm awfully busy, but if he's polite and respectful, there won't be a problem.

September 30, 2018

Worldview assessment

164. What does Chris Langan think about Jordan B. Peterson?

I think that in some ways, Jordan's heart is in the right place – he displayed courage in taking his stance on the issue of compulsory speech (at least with respect to gender pronouns, if not necessarily all aspects of PC doctrine).

However, it occurs to me that although Jordan is clearly intelligent and well-spoken, his personal history as a lifelong professional academic is bound to have limited his experience outside the Ivory Tower while exposing him to intensive indoctrination, and this may have limited his ideological flexibility and real-world awareness. For example, he seems to labor under the delusion that we inhabit a real meritocracy in which strong, upstanding "alpha males" are rewarded for merit and always get the girl. (That's just not how it works out here in the real world.)

It also strikes me that Jordan may be overextending himself by staking claims on difficult intellectual territory, including theology, religion, morality, and other metaphysically loaded fields. As a North American clinical psychologist with much to say about intelligence and intelligence testing, he has in all likelihood encountered my name. Yet, like academics everywhere, he shows no sign of knowing that I and the CTMU exist, and thus seems to think that his rather pedestrian views on certain metaphysical issues are state-of-the-art.

This being understood, I wish Jordan well. I'd merely suggest that if he really wants to be on the bleeding edge, he'll need to widen his acknowledgement of expertise and authority beyond the opaque, sweating walls of the academic hothouse. As always, out here in reality-land is where it's really happening.

June 12, 2018

Philosophy

Arrogation by Academia

165. What are the shortcomings of philosophy?

The major shortcoming of philosophy is that it has been arrogated by Academia, Inc., which functions like an intellectual trade union / pyramidal growth economy / Ponzi-style credentialization scam with membership standards irrelevant to the content of philosophy and for that matter intellectual merit. Because Academia has been corporatized and therefore tightly standardized, and because its socialization and indoctrination functions have come to outweigh its traditional education function, it now enforces a rigid philosophical narrative based on PC doctrine. This kind of ideological standardization is diametrically opposed to free intellectual inquiry and intellectual balance and integrity. As a result, the phrase "modern philosopher" is now all but oxymoronic. The typical modern philosopher, or at least one who commands recognition as such, is a professional academic who cannot think independently, who lives in abject fear of losing his/her job for deviating from institutional orthodoxy (or saying boo to an administrator), and who wouldn't know a worthwhile philosophical framework if it crashed the stage on live television, bit him on the nose, and yanked his undies down around his knees. This is why physicists and other scientists have been eating philosophy's lunch for the last few decades with total impunity, "bitch-slapping" the discipline around the block virtually at will.

October 2, 2016

Cartesian mind-body dualism

166. Do you agree with Chris Langan that Cartesian mind-body dualism fails because both mind and body, in any such theoretical separation, are still being discussed through one syntax (i.e. English), thus showing that the two are in the same reality?

WARNING: This question is badly garbled. Specifically, the restriction of cognitive syntax to one particular natural language has nothing to do with the CTMU or anything else I've ever said or written.

In fact, there is no way to cogently disagree with the CTMU. In particular, the simultaneous recognition of mind and body implies that they share common structure, and this precludes absolute dualism. There is not a logician or philosopher on Earth who can coherently argue otherwise.

Now to the issue of why the wording of this question is so badly skewed. Questions like this are the work of trolls; that much is clear. The real issue is whether the readership of Quora is sufficiently stupid and malevolent to regularly turn such trolling into a pretext for badmouthing people and issues about which most of its members know precisely nothing, or merely into an opportunity to slip in a snide jab or two in order to boost personal self-esteem at the expense of truth.

Some of the answers here aren't that bad. But still, Quora has been failing this test, and those who have contributed pejorative answers here are among the reasons. There is nothing positive that can be said about such people; they are gulls and suckers, and they are evidently too dull and witless to stop themselves from being played like yoyos.

My advice to the gullible: don't make a fool of yourself. Even when a troll pushes your buttons because they stick up a mile, exercise a little self-control and refrain from slavishly indulging your impulse to spew resentment and illogic. Quora is not supposed to be anyone's personal hatefest.

Thanks for your attention.

August 16, 2018

Chris Langan and Immanuel Kant

167. Is Chris Langan the Immanuel Kant of our era?

Only in the sense that the CTMU, which has comprised the sum total of (competent) mathematical metaphysics for the last three or so decades, is the ultimate extension of some of Kant's more perspicuous reasoning, transplanted into a vastly richer and more sophisticated conceptual environment. However, Kant made certain errors which the CTMU does not replicate, e.g., regarding the noumenon-phenomenon distinction. Thus, the CTMU is precisely what Kantian philosophy has consistently failed to produce: a comprehensive, self-consistent ontology and global conceptual framework supporting the entirety of human knowledge.

February 18, 2018

Harris, Sam

168. Sam Harris says the sense of self is an illusion, that we're just our brains, the ego doesn't exist. Does the CTMU agree with this claim?

No, the CTMU does not support the claims that "the sense of self is an illusion, that we're just our brains, [and] the ego doesn't exist."

From a CTMU perspective, these claims are ridiculous, and in my opinion, Mr. Harris wouldn't have a chance of prevailing against me in an argument about it. (As this is one of those threads that was already packed with misinformation by the time I got here, I'll just leave it at that for now. Suffice it to say that of the answers below, only one – by Joe Moyenne – contains so much as a single accurate statement about my work.)

April 29, 2018

Hume, David

169. What is David Hume's analysis and criticism of causality?

Basically, that **one cannot in general directly observe or logically deduce a causal connection**. One can observe a sequence of events and hopefully label it a "cause-effect relationship", but without logical implication, the label is merely a token of optimism; mere statistical confirmation proves nothing. This is called the "problem of induction" and permeates the empirical sciences, theories of which usually fall well short of logical implication or entailment. No matter how loudly an empirical scientist may proclaim that "The empirical evidence proves theory X!", proof is simply not possible within the constraints of empirical methodology.

May 10, 2018

Philosophy of physics

170. What do physicists think of the philosophers of physics who philosophize about physics?

If we're talking about *competent* philosophers of physics (or of science in general), roughly the same thing that the foxes guarding the henhouse think of the farmer pointing a shotgun at them. ;)

May 26, 2018

Prevalence of physicalism

171. What do philosophers think of Chris Langan's CTMU?

That all depends on the philosopher.

Most modern philosophers, particularly those of the academic variety, have been abjectly cowed by the relative success of the hard sciences, which – although they have not imparted the deep understanding of the overall structure of reality that they once seemed to promise – have nevertheless resulted in quite a lot of impressive and often very profitable gadgetry. Accordingly, ever cognizant of what is good for their academic careers, they subscribe to (metaphysical or methodological) naturalism.

Naturalism, which amounts to physicalism or physical monism, is the dominant worldview among academicians, especially those in the hard sciences. Because the hard sciences are where the corporate profiteers of Academia, Inc. butter their bread, academic philosophers stick to it like glue, at least where the rubber meets the road. That is, they confine their philosophical meanderings within the boundaries of physicalism, merely fleshing out its details while studiously ignoring any academic outsider who might have a better way.

Obviously, physicalism – which is strongly associated with various so-called "naturalistic" ideologies like atheism, secularism, and communism – is not valid from a CTMU standpoint, and the CTMU is therefore summarily rejected by the proponents of these belief systems.

In contrast, some philosophers have managed to retain three things once considered essential to success in philosophy: an open mind, high intelligence, and an understanding of the technical aspects of metaphysical philosophy as it relates to conventional scientific theories as object-languages. Within this group, those familiar with the CTMU are almost universally in favor of it.

One can only wish that there were more of them around. ;)

[**Addendum**: One of the answers in this thread, written by someone absurdly claiming to have a degree in "philosophy", is highly erroneous. Unfortunately, when I posted a warning about this in its comments section, it was immediately deleted and comments were closed. Accordingly, I'm repeating the warning here.

"**WARNING**: I'm the author of the CTMU, and this answer contains nothing that can be recognized as having anything to do with it. In addition to ironic insults involving "word salad" and "masturbation", it consists of a quote of which the critic professes complete incomprehension; a handful of terms taken out of context and mistakenly interpreted as synonyms of "circular or self-referential"; and erroneous descriptions of the CTMU as "perception and thought is reality" and "the claims of the theory are proven by the claims of the theory." As these errors are misleading at best and malicious at worst, readers are hereby advised to ignore this answer in its entirety. Thank you for your attention."

Answers like the one which occasioned this warning do not come from philosophers, but from trolls, and any upvotes or endorsements that they receive are from other trolls. Remember, not everyone who represents himself as a "philosopher" has the intellectual wherewithal and understanding to engage in philosophy, and many of those who do not are given to venting their resentment of those who do in unkind and misleading ways. Forewarned is forearmed.]

August 9, 2018

Physics

Force of gravity

172. If you couldn't feel the force of gravity, would the directions of up, down, left, and right still exist?

Yes. There would still exist 3 linearly independent spatial dimensions which, when reversed, yield 6 directions (up, down; left, right; forward, backward). But rather than being defined with respect to an observer standing on the surface of a gravitating body or oriented with respect to an external center of mass, they would be defined with respect to the observer alone. Where you're the observer, up and down would run from the center of your body toward your head and your feet respectively; left and right would run outward through your left and right sides respectively; and forward and backward would run outward to your front and rear respectively. In other words, your body would define a 3D coordinate system with axes lying along the lines of intersection of three anatomical planes: sagittal, coronal, and transverse.

September 17, 2017

Hawking, Stephen

173. Will there be a discussion between old Stephen Hawking and Chris Langan anytime soon?

ABC News reportedly tried to set up a meeting between me and Hawking in the late 1990's. (They asked me and I agreed.) Unfortunately, it turned out that Hawking was surrounded by academic flunkies who carefully screened all of his invitations, and they (apparently) screened me out.

Oh, well. C'est la vie.

January 14, 2018

Indispensability of logic

174. Can you have absolute proof of anything in physics, or is it all a case of having theories?

Of course you can.

I see a great deal of insistent yammering here about how nothing can be proven in the empirical sciences. That's true for the empirical sciences *minus logic*; proof is a logical operation, and is therefore a matter of logic rather than empirical induction. The problem, of course, is that logic is a very high-level science in its own right, and minus logic, there is no such thing as science at all (including physics).

Logic is the onboard "scientific master theory" that you are given when you are placed into the world and issued a brain, and it permanently holds the upper hand over all other scientific theories. Logic distributes over the empirical and mathematical sciences in the sense that as soon as a logical inconsistency arises in any of them, logic demands either that it be immediately resolved, or that the operative theory be deemed incorrect and repaired no matter how long it may take. This applies no matter where or how the inconsistency arises. Thus, science is never in a position to reject that which can be logically derived from logic itself ... e.g., from propositional tautologies like "x or not-x" (and there are other kinds of logical tautology as well).

The question thus comes down to whether or not any observable fact can be tautologically deduced, given that logic applies on the level of observation itself, with observability critically depending on logical truth-valuation. In particular, given any scientific observation or experiment, it must be true rather than false that one is making a replicable, scientifically valid observation, with everyone agreeing on what has been observed (this scientific criterion is called *replicability*). As logic is just the mathematical structure of truth, this automatic invocation of truth amounts to an automatic invocation of logic.

There is a certain naive (and absurd) viewpoint according to which scientifically relevant tautological inference is forever impossible. However, none of those espousing this viewpoint can prove it. The amazing fact here – the scientific elephant in the room – is that although the Scientific Method is basically a model-theoretic application of logic, most scientists fail to understand the mathematical logic that governs model theory. In fact, they can't even shake a stick in its direction, largely because much of it is new. Rather, they merely understand certain assumptions that fed into the smattering of logic that they were forced to learn in school.

It follows that asking the typical scientist what can ultimately be proven in the sciences is at best a waste of time, and at worst highly misleading. The game that scientists are trained to play is empirical induction, not advanced logic, and this is a fact from which they cannot escape. Some of them are quite good at plain old empirical induction – that's why science works! – but this is not to be mistaken for deep logical penetration and self-awareness on the parts of most scientists. That, unfortunately, remains a faraway pipe dream.

January 28, 2018

Particles in CTMU

175. What are atoms and other particles in the CTMU theory of Chris Langan?

In the CTMU, fermions (particles obeying the Pauli Exclusion Principle) are *syntactic operators*, quantum identities capable of mutual interaction. (These identities are called "syntactic operators" because they operate on each other according to the syntax of a certain language that couples with the manifold in which they exist, and in which other particles are defined on them.)

January 14, 2018

Physicists and CTMU

176. Why don't physicists accept the CTMU? The CTMU is a far superior theory of reality than any physics theory.

What on Earth would give anyone the idea that "physicists don't accept the CTMU"? This assumption is simply false.

Admittedly, many physicists *haven't heard* of the CTMU, largely because it's outside their specialties. (Physics has been pretty well carved up over the years.) But in just the last week, I've corresponded with three major physicists, one a Nobel Prize winner. Yet another famous physicist, John Wheeler, inventor of the phrase "black hole", actually *liked* it – he told me so around twenty years ago.

You know, this rabid anti-CTMU trolling has to stop, if only because it's making this whole website look ten kinds of ridiculous. There's not a "physicist" on this site – or a physics *troll*, for that matter, which is what some Quora "physics experts" seem to be – whom I couldn't pretty much mangle, intellectually speaking, if he/she tried to debunk or discredit it.

A little less of this nonsense, if you wouldn't mind.

September 20, 2018

Wheeler, John

177. Does Chris Langan have any evidence for his claim that John Wheeler liked the CTMU?

Yes, I do. Specifically, I still have one of Wheeler's personalized illustrated postcards on which he had penned a request that I meet with him at Princeton (unfortunately, I was unable to get away from any of the three or so jobs I had at the time).

The next time I see it around the office, I'll scan it and post a copy somewhere. Until then, any doubters may feel free to doubt it as much as they like, as I really couldn't care less.

September 29, 2018

POLITICS

Gun control

178. What would likely happen if suddenly there were no more presumptive questions posed on Quora about the suspension of the 2nd amendment and suddenly everyone is happily turning in their guns? Would life on Quora be forever changed?

Governments are by far the most prolific mass murderers in history, amounting in many cases to an especially brutal form of organized crime. Other mass murderers pale into insignificance by comparison.

The Second Amendment of the US Constitution equates to a US citizen's right of self-defense against the government. Without it, the wholesale murder of citizens would begin as soon as the government, possessing a virtual monopoly on the efficient application of deadly force, deemed it advantageous.

In that case, Quora would indeed change. Everyone would be anxiously parroting the official narrative on everything, just in order to avoid that pounding on the door in the middle of the night.

Don't even ask why this is so hard for some people to understand – it would only get them started. Idiocy is never to be encouraged.

[**Addendum**: It has sometimes been argued that mere guns leave citizens powerless against a government with military weaponry at its disposal. But governments tend to avoid the use of tanks, planes, and bombs against their people because this would involve the destruction of property, and governments consist of venal and corrupt individuals who generally prefer theft to destruction. The object is usually to kill or incarcerate people while stealing their money and possessions, which requires that armed agents be sent out against them ... uniformed flunkies who must worry about being shot by those whose property they are stealing under cover of law.

As the idea is usually to make a few high-profile examples so that everyone else will meekly hand over their money and possessions on demand, official acts of theft tend to occur in isolation. Guns can be quite effective in deterring amoral government thugs who would otherwise confiscate one's private property under asset forfeiture laws, auction off one's home for nonpayment of oppressive taxes, appropriate one's land under "eminent domain", take one's children in retaliation for home-schooling or refusing to vaccinate them, arrest one for political crimes or "hate speech", etc. And of course, guns can also be effective in defending oneself and one's family against criminals and wild animals, a basic human right if ever there was one.

Government officials who steal with pen and paper are for the most part a cowardly lot who rightly fear revenge for their depredations. Maintaining an armed citizenry is an absolute necessity of keeping them honest, which would otherwise be impossible given their innate criminal propensities. These days, the option to abuse their power for personal advantage is what makes most bureaucrats and politicians tick – no matter how ethical and public-spirited they pretend to be, it's always in the back (if not the front) of their minds. When it comes down to them versus the little people, their prime imperative is always to come out on top.

Sadly, this is the nature of the governmental beast. It jealously exercises a virtual power monopoly not to keep us free and safe, but to protect its own privilege and secure itself from those at whose expense it exists. Forewarned is forearmed.]

October 11, 2017

179. Why are both sides of the gun issue so at odds with each other? Can't both sides come to table and compromise?

In the sentence "The right of the people to keep and bear Arms shall not be infringed," the phrase "shall not be infringed" greatly limits compromise. Beyond a certain point, gun control laws infringe on the right of private citizens to bear arms by delaying, obstructing, or denying their efforts to exercise that right, thus simply violating the Amendment.

Remember, a "right" is something that everyone either has from birth or has the opportunity to earn. Historically, the "opportunity to earn" this particular right – which the Founding Fathers considered an absolute necessity given the known tendency of governments to abuse the defenseless – has taken the form of, e.g., hunting safety courses and bans on gun ownership by those with criminal convictions involving gun violence against their fellow citizens. Aside from these rules, almost all proposed gun control legislation amounts to infringement.

History and current events show that any move to compromise with the political Left on the issue of gun control simply causes it to grab whatever minor short-term advantages might become available, and immediately issue new and more stringent demands. It has thus become increasingly clear that the Left has no intention of compromising on such matters, and that its anti-gun caterwauling is only reinforced and amplified by each small advantage that it gains.

As lasting, worthwhile agreement is clearly impossible to achieve under these circumstances, and legal attacks on the Second Amendment of the United States Constitution are extremely dangerous to the security and welfare of American citizens, politicians bent on attacking it should be democratically removed from office ... and so, unfortunately, should those who cave into their demands for the sake of "compromise".

April 5, 2018

Identity politics

180. What is Chris Langan's view on identity politics?

My view on identity politics is that it can be justified only if everyone of any ethnicity is entitled to participate, in which case it is necessary for all (because failing to assert it, as when White people of European ancestry fail to assert it lest they be branded as "racists", means leaving oneself and one's group defenseless against competition for resources and opportunity). Alternatively, lest any group be denied its identity while others assert their own, group identity must be equitably denied to everyone.

Human identity is stratified, and thus has both individual and group levels. Accordingly, we can (and sometimes must) reason in terms of group identity. But when group self-identification is officially granted to some groups yet denied to others against which they compete, this can only result in imbalance and injustice. For example, when some overpopulating groups which have overtaxed their own resources by reproductive incontinence and homegrown oligarchy are allowed to migrate into the sovereign territories of worldwide ethnic minorities – e.g., people of European descent – and enjoy special "oppressed" status whereby they reap special benefits such as free food, free housing, free education, free healthcare, affirmative action, reproductive subsidies, and special treatment under the law, and are even credited with moral superiority due to their alleged "oppression", this can result in the destruction of the national, cultural, and ethnic identity of the hosts, leading ultimately to their extinction. Incoming groups which assert their own collective identities while denying their hosts any reciprocal right of political group cohesion thus amount to noxious, invasive, and ultimately lethal socioeconomic parasites. Obviously, any governmental authority which enforces or encourages such asymmetry – e.g., the European Union – is illegitimate.

Bear in mind that once we cease to treat individuals as individuals *per se*, thus allowing members of their respective groups to assert their ethnic, cultural, or religious (etc.) identities against their "oppressors", their group properties and statistics are automatically opened to scrutiny and comparative analysis. For example, if after several generations of special treatment in the educational sphere (compulsory school integration, special programs, modifications of educational procedure, racially defined college admission preferences, etc.), a particular "oppressed" group fails as a whole to outgrow these measures, its members are no longer entitled to exemption from objective characterization in terms of associated group statistics; if one wants to enjoy the social benefits attending ethically loaded group-defined properties like "belonging to an oppressed group", one must submit to rational policies formed on the basis of not just individual assessment, but empirically confirmed group-defined properties such as "belonging to a group exhibiting a relatively low mean IQ and a tendency to violently disrupt the educational environment". Continuing to pursue racially parameterized measures of human worth and achievement can only lead to personal injustice, social degradation, and biological degeneration (because such measures inevitably supplant any rational form of social, economic, and reproductive selection).

In short, identity politics should either be shut down immediately, or the majority populations of Europe and North America should be encouraged to assert their own ethnic and cultural identities and group interests with full force. Any governmental, academic, religious, or media authority which tries to prevent it is clearly unworthy of respect and obedience.

[**Note**: This answer has been "collapsed" on the supposed grounds that it "needs attribution". However, it absolutely does *not need* attribution, because it consists of my opinion on identity politics expressed entirely in my own original words as formed on the basis of my own original reasoning. The "collapse" has been appealed, so again we have a test of Quora and its moderation staff. If you have found this answer in a state of "collapse", it follows that Quora has once again failed the test.]

May 29, 2018

PSYCHOLOGY

Psychological disorders

181. Why do psychopaths and narcissists manipulate differently?

Narcissism and psychopathy are personality disorders. There are many personality disorders, and they can and usually do overlap. Various disorders give rise to various structural and causational theories which purportedly describe them, and these theories may be mutually exclusive. But it would be a mistake to assume that the disorders themselves must necessarily follow suit. In principle, there is no reason why narcissism and psychopathy cannot overlap.

Often, the best that one can do after identifying a given subject's personal spectrum of psychological traits and disorders is to ask how much of each is exhibited, and how they might interact in various kinds of situation. After deciding on a specific proportion or psychological profile, one can then try to predict the person's cognitive and behavioral tendencies as a function of these traits and disorders in a given context. But of course, this can be hit or miss.

If one looks closely at the socioeconomic elite, for example, one finds a raft of characteristic personality disorders which free those who possess them from psychological constraints and inhibitions, like empathy, compassion, honesty, and morality, which might have impeded their progress toward material success. Obviously, the successful landlord cannot be so compassionate that he can't evict a poor family in the middle of winter if that's what it takes to save his business; the corporate climber cannot be so fair and honest that she never takes credit for the work of underlings (insofar as those lower down the ladder are often the ones doing all the intellectual and physical work); and the successful financier cannot be so empathic that he becomes paralyzed when he must yield to his "killer instinct" and knife his competition, or his clients, in the chest or in the back.

Narcissism and psychopathy both work to disinhibiting effect. For example, psychopathy helps suspend empathy so that one can objectify and manipulate those who might get in one's way, while narcissism frees one from compunction by inflating one's sense of entitlement and justification (one has succeeded because one is "really better" than those whom one has surpassed, or because one is so "lovable" that one's success is a matter of sheer popular consensus). Naturally, the winners of the race for wealth and power tend to enjoy a powerful combination of these "advantages". (We're not just picking on the wealthy and powerful alone, by the way; it just so happens that they lend themselves to the discussion by displaying a characteristic personality spectrum which has always been problematical for the rest of society precisely because its ill effects are amplified by the wealth and power with which it correlates. We could just as easily have focused on street criminals, whose ill effects on society tend to be more localized.)

Different personality disorders may be triggered by different situations in various combinations, and the emotions which arise and the strategies employed may vary accordingly. The notion that the set of narcissists and the set of psychopaths are necessarily disjoint, and that one can always tell to which set a person belongs by his/her attitudes, strategies, and tactics, is simply inaccurate. The same person can act like a narcissist one minute and a psychopath or sociopath the next, or even like both at once, mixing and switching behaviors in largely unpredictable ways.

Such is the marvelous complexity of human psychology.

January 8, 2017

182. How do you reconcile the loss of identity (dementia for instance) with the existence of a personal God within Christopher Langan's Cognitive-Theoretic Model of the Universe (CTMU)?

Dementia is a "loss of memory", not of identity. Its symptoms include memory difficulties, but not a change or lapse of identity (provided that "identity" is not superficially and improperly defined on memories alone).

Dementia patients can be lucid provided that they can engage with physical reality and are thereafter "held in the moment" – they can tell stories, play cards, and so on. But when the context changes, their memory problems prevent them from keeping track of what they were just doing and/or thinking, and this "discontinuity in the narrative" can be mistaken by others for discontinuity of identity.

Provided that their condition is not complicated by an independent dissociative disorder, dementia patients typically experience no lapse or transformation of conscious identity as their symptoms come and go. This is due to the fact that human identity exists on a level of coherence that is independent of any particular context or narrative.

This is fully consistent with the CTMU concept of identity as formulated within that of Ultimate Reality or "God".

January 31, 2018

183. Can SCSPL from Christopher Langan's Cognitive-Theoretic Model of the Universe (CTMU) explain or map psychosis / schizophrenia and other mental health phenomena?

Of course it can. The CTMU isn't called the "*Cognitive*-Theoretic Model of the Universe" for nothing.

The discipline of psychology began with metaphysics. Initially, before it was overpowered by physicalism and behaviorism, psychology permitted the consideration of mind without insisting that it be supervened entirely on matter. In contrast to the "naturalism" in terms of which modern psychology is understood, its older and more primitive theory even included mention of supernatural entities like angels and demons, and sometimes weird diagnoses and treatments to go along with them.

But then came the Age of Reason, and eventually, a complete reversal. Instead of remaining a spiritual being, man came to be interpreted in a mechanistic model of self according to which a human being is nothing but a machine ... a classical-mechanical automaton driven by self-interest and governed by impersonal laws of nature and rules of behaviorism, subject to conditioning on the basis of individual utility as defined on the pursuit of pleasure, the avoidance of pain, and biological standards of "fitness" including survival and reproduction. As for *mind*, it was pretty much reduced to *brain*.

But another reversal is now at hand, and everything is poised to come full circle. Physicalism has gone bankrupt, standing in embarrassed silence before the great metaphysical conundrums of science and philosophy – conundrums at which its orthodox practitioners foolishly insisted on tilting like champions – as a tired old hobo with holes in its shoes and its pockets turned inside-out. Just as physics must now yield once again to metaphysics, as it has done so many times before, so must the physicalistic abortion that currently passes for psychology.

The rehabilitation of psychology will be complete only when its partial physical interpretation has been extended to include the metaphysical ingredients of reality on which a meaningful definition of *mind* clearly depends ... in short, when it remembers its spiritual origins and takes its place as one of many integrated disciplines within the CTMU "master-language", SCSPL.

January 31, 2018

Q

Quantum Theory

QM and CTMU

184. What does Chris Langan think of quantum theory, and how does it relate to his Cognitive-Theoretic Model of the Universe (CTMU)?

Quantum theory is a quantum-probabilistic approximation of an overall theory of causation, the CTMU.

June 11, 2017

Quora

Appealing

185. How do I appeal a Quora Moderation decision?

My advice would be not to bother, unless you have a lot of time to waste.

On at least two occasions when I tried to reasonably and politely defend myself from attacks against me and my ideas on this site, I was accosted by Quora Moderation, which informed me that my supposed "violations" of "Be nice, be respectful!" policy had been summarily removed, but meanwhile allowed the initial attacks against me to stand despite the fact that there was nothing nice, respectful, or accurate about them. When I appealed, I was curtly informed – with no stated reason whatsoever – that the decision was final and would not change.

Despite the fact that this caused me to stop posting here, I'm still getting unwanted messages from these people, apparently due to belated squawking by my disgruntled detractors ... who, had they kept their unkind remarks to themselves in the first place, would have had nothing about which to run whining to the moderators.

I only know about my own situation, of course, and claim no knowledge of anyone else's. But for anyone in a situation like mine, this place is strictly no-win. If it starts looking the same from your own perspective, then perhaps a change of venue would be the easiest solution.

March 18, 2017

Debating

186. Why does Chris Langan disregard anyone who disagrees with him as intellectually inferior instead of addressing their argument and deconstructing it? Why is he always so skeptical of people being trolls, or have I misinterpreted their Quora activity?

Quora is not a debate club. I'm not paid to be here by Quora, and I don't have time to roll around on the floor wrestling with Quora trolls all day. (A Quora troll never admits it is wrong, cannot be educated, and only becomes more obnoxious when directly addressed by its target *du jour*.)

That being understood, I do not typically disregard coherent, meaningful arguments by informed, well-qualified, and well-vetted people who object to something in the CTMU. It's just that I never see any.

Rather than engage in debate with Quora trolls – an activity from which I have nothing whatsoever to gain – I've made a good deal of CTMU material freely available on the web.

Interested readers are invited to take full advantage of it.

August 17, 2018

Gould, John

187. What does John Gould think about CTMU?

Obviously, John Gould has done quite a lot of talking about the CTMU right here on Quora. One might be tempted to infer from this that he knows just about everything there is to know of it.

However, as the sole author of the CTMU, I can state with complete certainty that this is anything but the case. Despite the several attempts I've made to set Mr. Gould straight on various points of relevance, he continues to make precisely the same errors regarding it that he has made in the past. It's like being locked in an empty discotheque with the exits sealed, a skipping record on the platter, and the speakers cranked up to 150+ decibels.

I'm not trying to say that John Gould is an especially stupid person; that's not the case. He manages to say some relatively intelligent things on occasion. But that's what makes it all the more disappointing, and ultimately more irritating, when he makes glaring mistakes and then automatically defaults to "infinite loop" mode. It's like trying to sleep with a nuclear-powered cricket in the room.

What I'm saying is that Mr. Gould has an angry little bee in his bonnet regarding the CTMU, and that possibly due to a touch of OCD – we all have a little of that, by the way – he would probably be harder to detach from it than a five-pound lobster with one's fingers pinched between its claws.

So the long and short of it is simply this: John Gould is "attached" to the CTMU, and inasmuch as the attachment has nothing to do with an actual understanding of it, the attachment may contain a certain *emotional* element.

If only the emotion were of a positive and constructive nature.

September 28, 2018

Hostility

188. Why do comments which disagree with Chris Langan/CTMU keep disappearing from Quora?

Owing to the double whammy of being known for high intelligence and metaphysical (anti-physicalist) reasoning, I'm often stalked by pests. Or perhaps it's simply because I'm a little better-known out here in the real world than most Quora participants, and thus seem an attractive target for those who would take a shortcut to notoriety by forcing me to recognize their existence in order to siphon off a little public recognition.

Unfortunately, this army of pests includes several Quorans who have displayed a marked tendency to "specialize in my case". Not only have some of them posted misleading, prejudicial, impertinent, and sometimes downright toxic answers for questions about me and/or my work, especially the CTMU, but where permitted to do so, they've been observed to clog up the comments sections of my own answers with *ad hominem* nonsense which violates Quora's "be nice, be respectful" policy. In short, they are neither nice nor respectful.

Sometimes – but only in the worst cases – I delete their comments, thus saving Quora Moderation the necessity of removing them. I suggest that if one doesn't want one's comments removed, one *be nice and respectful*. (A bit of veracity wouldn't hurt either.)

[Note: I should also mention that I don't have time to play games here on Quora, especially games which have no entertainment value for me. Accordingly, I've deleted an exchange in the comments between "two" characters using what appear to be pictures of the same individual on "their" accounts, one of them vaguely resembling Jack Sparrow from "Pirates of the Caribbean" and the other merely foppish. No more of this nonsense, please.]

February 20, 2018

Trolling

189. What is Chris Langan up to these days?

Thank you for asking!

I'm running my 140 acre ranch. I'm seeing to Mega Foundation business. I'm working on the next round of CTMU material. And I've even found the time to answer a few questions here on Quora!

But speaking of which, I've also been forced to take up a new hobby: riding herd on a couple of tireless, utterly implacable, and often vicious little trolls who follow me around here like flapping, squeaking vampire bats with claws full of hungry, writhing leeches.

(One absurdly claims to have authored a "naturalistic" theory of "metaphysics" – roughly speaking, this is like going off one's meds, sticking one's snout high in the air, sliding one's hand under the lapel of one's smelly, brown-stained hospital gown, and imperiously proclaiming oneself to be Napoleon Bonaparte – while the other merely considers himself to be a victim of my alleged unceasing racial and cultural discrimination, crying out plaintively for justice.)

I keep hoping that I can dispense with the hobby.

February 24, 2018

190. Why does Chris Langan get so much praise on Quora?

Oh, please. The undeniable fact is that I attract far more trolling than praise on Quora. There are "Top Writers" here who have trolled me viciously in spiteful, misleading, profanity-laced tirades, received literally thousands of upvotes for doing so, and resisted all attempts to remove their defamation despite repeated complaints. Meanwhile, I get at most a few dozen upvotes for unique, informative, demonstrably correct answers that they couldn't have duplicated if they were given several centuries in which to do so, and have been subjected to the removal of dozens of my own well-reasoned comments defending myself and my work from troll attacks. If I were looking for "praise", this would be one of the last places in the world I'd come for it.

February 3, 2018

Veaux, Franklin

191. Is Franklin Veaux qualified to attack the CTMU by Chris M. Langan?

Absolutely not. Mr. Veaux has not only demonstrated what can only be called thorough incomprehension of the CTMU, but lacks the technical background that one would need in order to meaningfully criticize it or even comment intelligently on the field of metaphysics.

March 15, 2018

R

Rational Wiki

Pseudointellectual trap

192. How does Rational Wiki compare to Less Wrong?

"Less Wrong" appears to be a Q&A site, perhaps a bit like Quora itself. As can be expected of such sites, it is slanted and not without its share of uninformed and/or objectionable opinion, but occasionally affords a modicum of useful information.

In contrast, "Rational Wiki" is a website which is designed to look like Wikipedia, but which is actually more like a pseudointellectual tourist trap whose inhabitants are fiercely dedicated to the mockery and defamation of people and ideas to which its proprietor objects. (As nearly as anyone can determine, this is an obscure Canadian resident and "secular humanist" calling himself "Trent Toulouse".) It exhibits a strong and persistent bias against conservatives, roundly insulting and belittling non-leftists at every available opportunity. It is also noted for its boundless contempt for religion and metaphysics, dismissing anything its contributors find difficult to understand as "nonsense" or "woo". Unlike Wikipedia, the site appears to have no clean-up crew(s) responsible for toning down the vitriol of its editorializers.

As I know from long personal experience, Mr. Toulouse and his gaggle of mostly pseudonymous partisans go after people and ideas they dislike quite aggressively. They've been going after me for years, always ranking high on the Google pages returned on my name. They've even taken to posting pictures of me with silly captions, doing their very best to make sure that none of their readers takes anything I say the least bit seriously. Such monkeyshines are unconstrained and apparently receive in-house encouragement.

Unfortunately, although these juvenile tactics fool no one with any intelligence, the misinformation shows no sign of abating, and its purveyors show no sign of even trying to comprehend the ideas they criticize. In the case of my own work, this means that they exhibit no understanding of metaphysical philosophy or the logic on which it properly comes to rest.

Browsers beware. With sites like this all over the place, a search engine can be a very dangerous thing. ;)

July 18, 2017

REALITY

Accepting syntax

193. Do you agree with Chris Langan that all things share a common reality and are to that extent similar, and that the mere fact that two things can be discussed through the same syntax (e.g., the English language) shows their difference isn't absolute?

Yes, I'm Chris Langan, and I usually agree with myself. To that extent, this answer belabors the obvious. But if I may, let me explain something brand spanking new to those of you who have already answered this question in such a way as to betray severe confusion about really hard concepts like *reality*, *structure*, and *language*.

That thing you've been typing on, and at whose screen you've been staring until your beady little eyes cross, is called a *computer*, and computation theory is the branch of science which studies *what it does* ("compute"). You even do a little computation of your own, *mentally* speaking, and according to many big experts who claim to know what they're talking about, your body does likewise, *physically* speaking. So let's begin with a little elementary computation theory, shall we?

If you were the kind of abstract automaton called an *acceptor* – which, by the way, represents all kinds of actual concrete automata built on its template, and even biological systems including you – you would (do, in fact) have something called *accepting syntax*. Conformance to this syntax – and by the way, "conforming to" means "sharing structure with" – is the criterion which allows you to *accept* some things as input while rejecting others.

For those of you capable of understanding it, this establishes that you "share structure" with everything that you can accept as input, i.e., *recognize* or perceive. What structure do I mean? Why, accepting syntax, of course – that part of your programming and/or basic architecture that controls what you accept as "real" and what you cannot. (Yes, yes, I know – you're a big shot who can decide all by yourself what you do or do not "accept as real". But there are others to consider, and the *replicability criterion* of scientific observation implies that we all have a standard accepting syntax controlling our perception. In fact, even when you just imagine something, it is *recognizable* to you only because it superficially conforms to this syntax! But I digress.)

Now, are we all following along here? At this point, we should all understand that we have an accepting syntax, and that things we perceive must conform to it and in fact share structure with it.

Just to make sure that we all understand this, let's have a little pop quiz.

Doesn't this mean that whenever you see two things you regard as "different", both of them conform to (share structure with) accepting syntax, and therefore have something in common, namely, the generic structure of the aforementioned accepting syntax?

But doesn't this imply that all that you see, feel, touch, smell, and/or taste has a nonempty structural intersect?

But doesn't this mean that everything you can perceive is *the same* in this respect?

And doesn't that mean that it's *not totally different* from anything else you can see, feel, touch, smell, and/or taste?

ANSWER KEY: **Yes, it does.**

Any questions? (If so, that's too bad, because at this point, still having questions would mean that you've got so many loose screws that if you stop clutching your head in consternation, it just might roll off your shoulders and bounce away like a tennis ball.)

We all know that abstraction can be hard ... for some, *very, very* hard. But let's get real here – if you're the kind of person who has trouble with this level of abstraction, then why are you trying to critique a theory that requires the highest level of abstraction possible?

I mean, shouldn't you be playing with matches in a haymow or something?

August 10, 2018

First causation

194. Why does Chris Langan state in the CTMU that the randomness of DNA mutations in Darwinian evolution is indeterminate and thus "magic" when there are clear deterministic explanations of how this randomness arises?

Because "random determinacy" is an oxymoron. Even statistical determinacy and mathematical chaos merely consist of microscopic events which themselves have no ultimate explanation, and therefore do not qualify as fully determinate. Full determination requires not just the full expression of a standard causal function including all parameters and inputs, but ordinal completion ("first causation" or causal closure). [In fact, it requires even more than that, but this should be enough to chew for now.]

February 10, 2018

Indispensability of logic

195. What is CTMU? Is it worth my time to understand it?

One cannot understand the reality that one inhabits without the CTMU. Therefore, if one wishes to understand reality, oneself, and one's embedment in reality, one should at least try to come to grips with it.

Perhaps this will be easier to understand if we put it another way. One should try to understand the CTMU for the same reason that one should study logic, only more so. The CTMU can be characterized as logic formulated on a very high level of discourse, and specifically as the level of metalogic which identifies logic itself as a central ingredient of intelligible reality.

On the other hand, if one is a Quora troll – especially the kind which always leads off by identifying itself as an atheist or a skeptic – then one needn't bother, as in this case, one lacks the intellectual acuity to make any headway.

August 10, 2018

Methodological inadequacies

196. What do mathematicians / physicists out there think about the Cognitive-Theoretic Model of the Universe article by Chris Langan (attached) that suggests that our current methods for understanding reality are fundamentally flawed?

As the author of the CTMU, let me state for the record that I don't recall saying that "current methods for understanding reality are fundamentally flawed." What I've been saying, more or less, is that *certain* methods of understanding reality are fundamentally *inadequate to explore reality in its metaphysical entirety*. These are different statements.

I can also state that no one who fails to acknowledge the shortcomings of certain methods of understanding reality is competent as a philosopher, and that such a person may be less than fully competent as a scientist or mathematician as well.

March 1, 2018

Self-communication

197. How would someone explain the CTMU (Cognitive-Theoretic Model of the Universe) to a child?

"When we talk to each other like we are right now, we're using something called a *language*. A language is something we use to communicate, just like you and I are doing when we talk to each other. It has symbols just like the letters on your alphabet blocks, words like *cat*, *dog*, and *skittles*, and sentences like 'See Spot run!', all of which help us tell each other what we mean.

"Well, every part of the world talks to itself and to other parts, and the CTMU is the language that it uses! Right now, I'm using the language called 'English' (French, German, Russian, Chinese, ...) to talk to you and tell you what I mean. You can understand me because you understand English (etc.) the same way I do. When the world (cosmos, universe, ...) talks to itself, it uses the CTMU!

"Remember, we say that a language is 'mathematical' because it's a bit like the counting numbers you use when you count to ten. Not only can you use those numbers to count; you can also use them to add, subtract, multiply, and divide (those are called 'operations'). You can even tell when one number is bigger or smaller than another (those are called 'relations')!

"Mathematical languages like the CTMU also have mathematical relations and operations just like the counting numbers. So we can always be exact, and the other person – if he or she is paying proper attention – can always understand what we mean. When the universe talks to itself using the CTMU, it always knows what it means as well!"

(Notice the part about "paying attention". Nature always pays attention to itself. Only the strange and unnatural phenomena calling themselves "philosophical naturalists" seem completely unable to do so with any success. ;)

March 15, 2018

198. How can you explain CTMU to a layman?

Reality necessarily *communicates* (exchanges information) with itself; it sends information on states, events, and processes from place to place within itself by various means of transmission and induction. By definition, the medium of communication is *language*; that is, language can be generically defined as "the medium of information and communication". **It follows that reality has linguistic structure.**

Unfortunately for science, most scientific languages are structurally inadequate to perform linguistic functions like communication on reality's behalf.[2] However, the definition of language can be furnished with extra structure that overcomes these inadequacies. **The language defined to overcome these inadequacies, thus enabling the self-communication of reality, is called the CTMU.**

---

[2] For example, ordinary scientific languages are not self-contained; they have extrinsic requirements such as external processors, e.g., human minds or computers, to execute their operations. This prevents them from modeling their content, reality at large, well enough to generate and carry the information that reality needs in order to function. The CTMU, on the other hand, is ontologically self-contained with respect to all required objects, structures, and operations.

(Note that this definition is functional, logico-mathematical, and independent of empirical induction and special terminology.)

Because the existence and validity of the CTMU (as just defined) are logical, ontological, and epistemological necessities, there is no escape from its implications. Laymen – and for that matter, amateur and professional scientists, mathematicians, philosophers, and theologians – have no chance of legitimately prevailing in arguments against it, and should avoid baseless complaints about its allegedly "unscientific" or "conjectural" nature.

May 28, 2018

Self-containment

199. (1) What exactly is Chris Langan trying to convey in the CTMU (in layman's terms), and (2) what is his goal/expectation in doing so?

(1) I'm trying to convey the deepest and most general formal characterization of the structure of reality. Specifically, I'm trying to convey the fact that reality as a whole takes the form of a certain uniquely structured and profoundly self-contained language.

(2) My goal is to help mankind avoid the nasty but at least partially avoidable pain and misery that it recurrently brings upon itself due to its substantial ignorance of the structure and meaning of the reality that it inhabits.

February 18, 2018

200. A key principle of the CTMU is that reality is self-contained. Can you really debunk this position and formulate the alternative?

The self-containment of reality cannot be debunked. It's a logical necessity.

Even a drunken monkey could see it: if there were anything *outside reality* on which reality were in any way dependent (for origination, causation, support, maintenance, or anything else), then the entire dependency relationship would be *real* and therefore *inside reality*. Therefore, reality is self-contained with respect to physical and metaphysical functions like origination, causation, support, maintenance, etc. (QED).

This is what logicians call a *proof by contradiction*, and it invalidates the premise of external relevance. (Sorry, but there's simply nothing else to give way here.) It also means that the level of reality we're talking about is the deepest level possible ... the CTMU level. That's because "reality is self-contained" doubles as a *definition* of reality which makes it idempotent under inclusion (or if one likes, "self-inclusive", with all the bells and whistles required to make it work). The CTMU develops the implications of this and certain other facts from first principles.

One really needs to understand how logic works on the metaphysical level of discourse to reason properly about metaphysical questions like the origin of reality, "dark energy", and so on. On the other hand, if you're trying to explain things on a more superficial level, then your explanation is incomplete, and once again, you don't get to take credit for anything I've already figured out.

I thoroughly understood the tautological implications of self-containment thirty years ago, which is about how long the CTMU has been around. I hope no one thinks he's going to get over on me now.

July 11, 2018

Self-reference

201. Are Russell's Paradox, Gödel's Incompleteness Theorem, and the Halting Problem just different instantiations of the same underlying difficulty? (If not, what are interesting differences?)

The "underlying difficulty" you evidently mean is called "self-reference" in a formal context and "self-inclusion" in set theory. These concepts are not themselves inconsistent, but seem problematical because they permit the formulation of paradoxes ("this sentence is false", "this theorem is unprovable") and/or apparently inconsistent mathematical constructions (the set of all sets, Russell's set of all non-self-inclusive sets).

At one time, various formalisms were "immunized" against such constructions by simply forbidding self-reference across the board, or by changing the names of things so that they couldn't display it (basically: "If X looks like it might self-refer / self-include, reclassify it or change its name so that it can't!"). But this was entirely uncalled-for and misleading, as reality itself is self-referential in the sense that it has various reflexive properties and incorporates various operations which act on it and functions that "take it as an argument". Given that reality makes extensive use of self-reference, no scientific or mathematical case can be made against it.

The self-referential nature of reality has been thoroughly developed in a theory called the CTMU, around which various supposedly intelligent people have been completely unable to wrap their heads. Of course, most of them swear that it's not them, but the theory. However, their stupefaction at best reflects an unfortunate glitch in their own thought processes. Just as in Turing's solution of Hilbert's *Entscheidungsproblem*, their minds exhibit a tendency to spiral into infinite loops at any mention of certain "forbidden" ideas.

In any case, self-reference is something that no conscious human being can avoid, as it constitutes the essence of self-awareness.

October 8, 2017

Syntactic metaverse

202. Are there meta-simultaneous universes within reality, according to the CTMU theory of Chris Langan? How are they related to our world?

In the CTMU, alternate universes exist in syntax as semantic potential. (This does not equate to the full concrete existence ascribed to them by other theories; it is not automatically guaranteed that a given potential reality is anywhere realized.)

May 12, 2018

True structure

203. What can the simple man or woman expect from Christopher Langan's Cognitive-Theoretic Model of the Universe (CTMU)?

The CTMU affords a basic understanding of God, Man, the human soul (which connects God and Man), and the true metalogical structure of reality. All that its comprehension requires is moderate intelligence, basic literacy, an open mind, and an honest effort to understand a well-structured metaphysical alternative to the current naturalistic (physicalistic, materialistic, "secular") paradigm.

Regardless of the excuses thus far offered by the confused and disgruntled for their incomprehension – very few of which repose any blame whatsoever on those who concoct them – it is the conceptual inertia of the naturalistic paradigm, plus certain mostly negative emotions and biases, that presently stand in the way.

Let's hope that it is still possible to change this for the better.

January 9, 2018

Ultimate Reality

204. What is the ultimate reality?

Ultimate reality is described by a unique metaphysical formulation of logic called the *Cognitive-Theoretic Model of the Universe* (CTMU). This description is not a product of guesswork, but is as real and verifiable as it can possibly be. No other description is even remotely viable; any description which fails to conform to the CTMU is logically precluded.

February 13, 2018

205. What does "ultimate reality" mean to you?

"Ultimate reality" is the level of reality that is both comprehensive and irreducible to anything deeper or more fundamental. (There are many technical properties that go along with these criteria, but this is the gist of it.)

May 21, 2018

206. Can we ever know the ultimate reality?

Of course we can, at least to an extent.

Because ultimate reality is induced from reality as we know it as a terminal generalization thereof, it couples with reality as we know it, and can thus be known through the coupling (in particular, it must have structural properties which make it amenable to inductive coupling; *amenability to inductive coupling* is itself a logical property, and all necessary supporting properties are thereby implied). Conversely, if "ultimate reality" *cannot* be known in any way through the inductive coupling, then it does not couple logically with reality as we know it, and cannot be meaningfully called "reality" at all.

The unique metaphysical formulation of logic which elucidates the structure of ultimate reality, otherwise known as "God", is called the *Cognitive-Theoretic Model of the Universe* (CTMU). The CTMU is a true "theory of everything" in the logical sense, i.e., the only sense in which a true theory of everything can actually exist.

February 13, 2018

Reality Principle

Valid tautology

207. Do you agree with Chris Langan that reality contains all that is real and only that which is real, and that if there was anything real enough to in any way influence reality, it would have to be inside reality?

I'm afraid that there's no way to disagree. This has been correctly identified as a tautology, albeit *misidentified* as a mere *propositional* tautology which cannot be refined to obtain useful information.

I'm not quite sure where those who deny this assertion learned what they persist in calling "logic", but one thing is for sure: they need to get their money back. ;)

August 16, 2018

Recycling

Logical fallacies

208. What is an example of logical fallacy about recycling?

The main recycling fallacy is the assumption that things are directly "recycled", i.e. put to the same specific uses to which they were originally put in the form they originally took. This is not generally the case. In fact, they must usually be reduced to some prior form from which new products serving new purposes can be made in different sizes and shapes.

This is sometimes an extremely expensive process, and sometimes it leads to more pollution than the recycled materials themselves. It is typically claimed that "science will find a way". Unfortunately, this claim usually turns out to be a way for certain privileged parties to stuff their pockets, much in the way that "closing the nuclear fuel cycle" was nothing but a toxic flatus in the breeze intended to narcotize the public as certain unscrupulous parties did their business in the field of nuclear energy ... i.e., stuffed their pockets. (Big surprise, right?)

And let's not forget our wonderful political hacks, who make careers out of inflating pseudoscientific pipe dreams into gilt-edged "done deals". Yes, there's no easier route to political success than promising technological fixes for urgent problems while squandering every tax dime squeezed out of the "little people" on the military-industrial-security complex-cum-burgeoning welfare state. Come one, come all!

That being said, recycling isn't always a bad idea. But never buy a jalopy without kicking the tires and poking around under the hood.

October 6, 2017

Religion

Interpreting religions

209. What religions are not compatible with Christopher Langan's Cognitive-Theoretic Model of the Universe (CTMU)?

All logically consistent religions are compatible with the CTMU in the sense that they can be interpreted (or modeled) within it. This follows from the fact that the CTMU is a metaphysical extension of logic in which logic can metalogically refer to itself.

January 26, 2018

S

Science

Methodological inadequacies

210. Why do so many scientists (and other "educated" people) accept the "Big Bang" theory when it contradicts one of the fundamental axioms of all knowledge: cause and effect?

First, most scientists are not extensively trained to properly evaluate the logical integrity of scientific theories; surprisingly, this is not typically an ingredient of scientific curricula. By the nature of their training, scientists are often quite effective in working out the mathematical details of their fields, but the big picture sometimes escapes them without necessarily interfering with the solution of the specific problems in which they are interested.

Secondly, while there exists a valid causal framework which accommodates the Standard Model, most scientists are totally unfamiliar with it.

April 16, 2018

Problem of induction

211. Why do physicists ignore the problem of induction? Do they assume it has a solution?

They ignore it because they lack the kind of knowledge that would enable them to sort it out. The simple fact is that the problem of induction precludes certainty for theories derived exclusively by empirical induction and confirmed by observation, and this qualifies as a very inconvenient truth for not just physicists but the entire "hard science" community.

In my personal experience, not only are most physicists unequipped to deal with such irreducibly epistemological and therefore metaphysical problems; they are hostile to those who can. If there's one thing that typical academic physicists love to do, it is to disparage philosophy in general as extraneous gobbledygook, thus excusing themselves from having to pay attention to anyone but other physicists. It's the case-hardened lock on the clubhouse door.

On a very high level of reasoning, the problem of induction can be (and has been) solved. The solution is called the *CTMU*, short for *Cognitive-Theoretic Model of the Universe*. But while this solution is logically all but immune to criticism, no one should hold one's breath waiting for physicists – almost all of whom consider themselves natural philosophers *par excellence* – to come to grips with it.

Nothing against physicists, mind you. Some of them are downright brilliant. But they really need to stick to physics, and leave metaphysical thinking to those specifically equipped for it.

June 7, 2018

Proving God

212. Why do scientists remind us that there is no proof of God when the arguments for Big Bang or Evolution are only based on evidence and testability?

Because of wishful thinking. No scientist has any way whatsoever of knowing that "there is no proof of God"; it just so happens that many of them have been told this all their lives by people whom they (mistakenly) considered to be experts on the subject, and many others simply like the idea of ruling it out. In any case, if there were a proof that no proof of God's existence is possible, it would not utilize the scientific method; it would necessarily be logico-mathematical in character. Such reasoning is better left to mathematical logicians with considerable knowledge of philosophy and theology.

May 2, 2017

Scientific method

213. What is Chris Langan's analysis of the scientific method?

The Scientific Method, which is central to the traditional methodology of the empirical sciences, is a dualistic metalanguage which purports to govern the bidirectional mapping between theory and observation. It is dualistic in the sense that theory is made entirely dependent on empirical observation, with absolutely no dependency in the opposite direction aside from the choice of which experiments to run and which observations to make.

The Scientific Method can be helpful in shaping constructive feedback between theory and experimentation, but obviously breaks down at the level of scientific inquiry on which observed reality cannot be properly described by dualism (because absolute dualism precludes observation itself).

Reality is ultimately monic, as otherwise it would be disconnected and therefore two or more realities instead of one. Because these separate realities would be incommunicado (being cut off from each other by absolute dualism), they could not interact at all, and the inhabitants of one could not recognize another.

Addendum (originally in response to a troll who evidently had my response to his own comment deleted, and whose comment was therefore deleted in turn):

If theorization and observation were not coupled in a very specific way by the Scientific Method, one could simply dispense with it. In that case, no one would ever have made a big deal out of it. Instead, students would simply be told "Get to making observations and theorizing about them!"

But it's not as simple as that. The observations have to be selected in such a way that the associated observation statements confirm or falsify the theory – this often involves the careful design and execution of experiments – and the theory has to be developed in a way that explains existing observations and/or predicts others.

This coordination of theorization and observation is what the Scientific Method is all about. This is how it "mediates" between them.

July 8, 2018

Standard model vs CTMU

214. If scientists can't really explain what the state of the universe was in before the 'big bang' and how it came to be that way, why is it taken seriously?

As an answer has been requested of me, and as this question keeps showing up in my feed, I'll answer it. (This is not to be construed as an invitation to attack.)

The Standard Model is taken seriously because, as a partial model of physical causation which omits origination and falls short of complete determinacy, it works pretty well. That is, it is often (but not always) confirmed by prediction and experimentation.

The Standard Model comprises a theoretical language that is often generically referred to as "physics". However, in order to discuss the origin or "initial state" of reality, a higher-order "metaphysical" language is required ... a language in which the physical aspect of cosmic evolution can be analyzed and justified.

In the 1980's, I provided such a language. It's called the "CTMU", standing for the "Cognitive-Theoretic Model of the Universe". It can be described as a reflexive metaphysical metalanguage of the object-language "physics" and its observational universe of discourse.

In the CTMU, the physical aspect of reality occupies a certain limit. The nature of this limit has been described in several academic publications (although one would never know it by the troll attacks to which it has recurrently been subject due to its theological implications).

As the CTMU is supertautological – because it has a structure which amounts to a global, self-contained analogue of propositional tautology – it has a certain metalogical property which can best be described as "logical self-veracity".

It is therefore extremely safe to say that this is where metaphysical extensions of standard physics are headed, whether or not any particular physicist or philosopher happens to like it.

January 22, 2018

Simulation Theory

Self-simulation

215. Within the context of Christopher Langan's Cognitive-Theoretic Model of the Universe (CTMU), could we technically be in a computer simulation of sorts, with 'God' being the programmer?

Yes. According to the CTMU, reality can be understood as a "self-simulation", at least in the sense that what we see outside of ourselves – the physical universe – conceals the process which actually produces it, and cannot account for the entirety of what producing it actually requires.

However, the simulation is not merely computational in the mechanical sense, but (in CTMU terminology) "protocomputational" or "precomputational". That is, it utilizes a kind of "metaprocessing" called *telic recursion*, which generates ***entire timelines*** instead of individual events.

God is indeed the Programmer-in-Chief, but does His "programming" largely through secondary sensor-controllers, including human beings, which locally inhabit the simulation. As for God Himself, He distributes over the entire simulation and is therefore omnipresent.

The entire simulation can be reduced to a master "programming language" sometimes referred to as *Logos*, which is *trialic*, serving as its own universe and model. On a more technical level, it is known as the CTMU.

January 31, 2018

216. Are there any theoretical universal models that are not causal by nature? For instance, if this was a simulation as suggested, then how might the rules be different?

Yes. The definitive theory is called the CTMU, short for *Cognitive-Theoretic Model of the Universe*. Publications date from the late 1980's through the present, most recently in the peer-reviewed academic journal *Cosmos and History*. Technically, it is a theory of mathematical metaphysics.

Basically, the CTMU says that reality is an ontologically closed "self-simulation" in the precise sense that it contains empirically undecidable "programming" as well as "displayed" observational content, and that its logico-mathematical structure is that of a self-structuring, self-axiomatizing "SCSPL", short for *Self-*

Configuring Self-Processing Language. This language has a unique form called a *supertautology*, according to which its evolution is *metacausal* rather than strictly causal.

In addition, the CTMU embodies a logical form of self-similarity analogous to the so-called "holographic principle" as later developed in connection with string theory. Accordingly, it is *trialic*, effecting a distributive coincidence of theory, universe, and model in a coherent "metaformal system".

Last but not least, the CTMU was authored by a man reputed, at least in some circles, to be the most intelligent in the world.

April 26, 2018

217. What is Chris Langan's view on the simulation hypothesis?

The CTMU is the first and only viable ontological theory of reality as a self-simulation. I wrote the CTMU. Hence, I have a high opinion of it.

June 23, 2018

218. Is the CTMU the first simulation theory?

Yes, the CTMU is the first detailed theory of reality as a *self*-simulation, having been used to resolve Newcomb's Paradox in 1989. (Note that reality must be a "self-simulation" insofar as that which does the simulating must itself be real.)

This in itself was a milestone; it seems to have been the first time that a major philosophical problem was solved in the context of a computationally simulated reality. While it had previously been suggested that our world is somehow embedded in a wider insensible reality – e.g., Plato's Allegory of the Cave – this was the first time that an existing philosophical problem was solved in an explicitly computational setting.

Note that the CTMU has a unique theoretic structure that is absolutely necessary for this purpose, and that its functionality exceeds that of ordinary computers (it was, and is, described as "protocomputational"). No other so-called "simulation theory" shares this property, ostensibly because the authors of such theories fail to understand the deep structure of reality and the limitations of standard automata.

September 14, 2018

SOCIAL MEDIA

Stratified hive

219. What did you assume was exaggerated until you experienced it?

Counterfactual groupthink and Kafkaesque bureaucratic skullduggery within social media / online social networks.

One tends to assume that such networks are all about the communication and dissemination of real, valid information for the good of society and the people comprising it, and sometimes this is actually the way it works. But each network has a business model and a tiny handful of self-interested people riding it, and deviating from their POV nearly always results in censorship and other abuses. Being the champion of certain unpopular truths (which I have an established ability to cram down the throat of any opponent in fair debate, so to speak), I'm in a special position to know this. Every time I see myself and my ideas being run down by opinionated ignoramuses in this or that social medium, I show up to defend them ... only to have it once more brought to my attention that almost no one is on the side of truth these days.

I've generally tried to find the best I could in people, and this is how I initially approached social media. I heard people complaining about how they're often run autocratically, even whimsically or perversely, but I'd ask myself: given the First Amendment of the US Constitution, how bad could it possibly be? Unfortunately, it turns out that the answer is "pretty bad indeed". I recently had a good deal of truthful and potentially valuable dialogue erased by the moderators of one of these fora, apparently in order to protect a longstanding contributor who had initially attacked me, but was trying to cover his tracks after finding himself on the losing end of an exchange with me. He whined to the moderators, and I suddenly became the "aggressor". Just another bad show in an ever-longer run of them.

The Internet Hive Mind is defined as an entity consisting of numerous individuals whose exchange of knowledge and opinions leads ideally to emergent collective intelligence, but can also lead to deadening conformity. We have on the one hand a potential for group intelligence to emerge from a distributed, self-organizing "human computation" occurring on a given proprietary software platform, and on the other, the actual reality of hives in nature and the human economy: they naturally stratify into classes, one of which runs the show and the other of which provides myriad group-thinking drones and workers for the pleasure of those in the steering compartment. Where monopoly capitalism ices the socialist cake, the freedom of people to spontaneously come together to achieve a truthful consensus is nonviable. Instead, social media implicitly provide an ideologically constrained setting in which the most exigent contributors, having slavishly "gotten with the program" of those higher up in the hive, sneakily pull strings and push levers to game forum policies and manipulate bureaucratic processes to their advantage. The result generally has nothing whatsoever to do with truth.

Participants need to question who is running the hives to which they belong, and what they're actually about. Just because you haven't yet found yourself on the wrong side of the ideology prevailing in a given forum doesn't mean that you won't, should you ever appreciably deviate from it. Have a look around you. Are the proprietors avowed atheists? Religious fanatics? Singularity-preaching techie plutocrats? Do they prefer globalism to national sovereignty and local democratic self-determination? Is there a strange alignment of contributor opinion with the publicly stated views and narratives of the proprietors? Are there any improbable confluences of opinion that appear to have been engineered?

Needless to say, this trend is not pointed in the best direction for launching and propelling humankind into the New Millennium. But for the time being, on the increasingly slim chance of opening a few minds and doing a little good, I'll remain open to limited participation.

May 2, 2017

Society

Sustainability

220. How does a global society behave that finally comes to accept that Christopher Langan's Cognitive-Theoretic Model of the Universe (CTMU) is true?

This question can be rephrased as follows:

How does a society behave that accepts truth on the physical and metaphysical levels?

It behaves more sustainably and with greater resilience than a society anchored to a flawed, inexorably sinking worldview. It offers real meaning to its members, and is not disrupted by frequent descents into irrationality. And by encouraging intellectual competence and creativity, it has the potential to maintain a high general standard of living.

February 19, 2018

SUPERTAUTOLOGY

Absolutization of truth

221. Is the CTMU absolute truth? If so, why?

Yes, the CTMU is absolute truth in a sense roughly analogous to that in which a propositional tautology is absolute truth, but in the expanded context of a new metaformal linguistic construct called a *supertautology*. (This has been defined and explained at length in several peer-reviewed papers – don't be afraid to do a little homework.)

Just as a propositional tautology like "X or not-X" is true regardless of the truth values of its sentential variables, the CTMU supertautology is true regardless of the form taken by specific scientific theories therein, and can thus guide the formulation of scientific theories in general. Metalogically, it amounts to an advanced reflexive formulation of model theory which utilizes brand new mathematics in explaining the "absolute" (self-verifying) nature of logic with respect to reality at large on the basis of direct intelligibility.

Because logic is the structure of truth, the absolutization of logic as a requirement of intelligibility amounts to the absolutization of truth. Thanks to a property known as "logical idempotence", the supertautology spans arbitrary levels of verification all the way up to metaphysical ontology. It is thus applicable in every scientific, mathematical, and philosophical domain.

Addendum: A reader asks the following question:

Question: Given Wittgenstein's statement that tautologies tell you nothing relevant about reality, why do tautologies take absolute precedence in your theory, and why is it so important that they be used in this way?

Answer: Before one can formulate a meaningful question about the CTMU, one must take the trouble to understand what it says ... for example, by reading the above response, which states that the CTMU is not a mere tautology, but a supertautology.

Whereas a tautology is just a circular expression within a language, a supertautology is an entire language configured in a tautological or circular way. In order to take on this structure, a language must assume certain unique and interesting properties. It is from this structure and these properties, which must be shared by the content of the language as a condition of its intelligibility therein, that interesting conclusions can be drawn regarding the content.

The CTMU leaves Wittgenstein and every philosopher before or since in the dust. But before one can understand how and why, one must make an effort to understand what the CTMU says about its own structure and its own properties. Again, **do your homework**. I've made study materials freely available, so it's no one's fault but yours if you fail to use them.

[**Warning**: Because the CTMU has theological ramifications unbeloved of "the atheist community" (if such we may call it), it has many implacable enemies on this site, not one of whom has ever displayed the slightest understanding of it. It is strongly suggested that readers not allow these trolls to mislead them.

A common ploy of such trolls is to demand a "prediction" of the CTMU. Some of these are history; e.g., an early one was that of accelerating cosmic expansion, for which the CTMU explanation has never been duplicated. But the short answer is that the CTMU does not rely on prediction; it relies on the fact that it provides necessary infrastructure for science in general.

A few less stupid (but still intellectually challenged) trolls have responded that science needs no infrastructure, as science already exists and it already works. However, it certainly *does* need infrastructure, and it certainly *does not* work on all levels – it tops out well beneath the metaphysical level of discourse occupied by extremely important questions regarding (e.g.) the ultimate nature and origination of reality, including quantum cosmology – and no scientist can explain why not. Hence, the CTMU is an absolute requisite of science.]

Note: Another answer has been posted to this question by a frequent (and just as frequently misinformed) CTMU critic, who states that "a tautology is a tautology with respect to a class of models." Quite so. To be more specific, a propositional tautology is absolutely true with respect to all *intelligible* models. We can forget about models outside of this class – after all, they are *unintelligible*, which means that you couldn't describe one of them even if you tried.

September 28, 2018

Logical core of metaphysics

222. Why is there a spate of Anonymous questions about the fringe CTMU conjecture?

As the sole author of the CTMU, I can only say that *if* this were true, I don't know why it *would* be. However, I can state with absolute certainty that whether or not the CTMU is "fringe", it is not a mere "conjecture". It's a supertautology, and one cannot worm one's way out of it. It's the everlasting adamantine logical core of mathematical metaphysics, and if one doesn't like it, one really needs to go out and find oneself another field, the actual state of which is more to one's liking (not that one can extricate oneself from its domain in any event).

[**Note**: This may be a troll question. In any case, my answer was requested by someone who uses the approving phrase "a fair analysis" to describe another (defamatory, collapsed) answer which alleges that I'm "irresponsible ... [with] a complete lack of moral conscience ... a pseudointellectual ... and sorry excuse for a genius". One can't explicitly endorse this kind of defamatory ad hominem tripe and not be considered as much a troll as the one who wrote it.]

March 1, 2018

Logical structure

223. How often does a theologist read an evidence-based argument against theism?

Never. (There aren't any, aside from one which atheists are intellectually incapable of comprehending.)

Evidence for a statement or theory X is generally applicable (instantiates X) only within the scope of any axioms or principles incorporated by X or implicit in the underlying model of X. However, the scope of X cannot usually be established within X itself, unless X has a very special structure.

This structure has been described on Quora as that of the CTMU Supertautology. This is the only context in which the meaningful probative power of evidence is unlimited, making it useful in deciding ontological issues like the existence of God. In this context, the existence of God is absolutely inescapable.

October 18, 2017

New mathematical system

224. Why does Chris Langan use the term "super-tautology"?

Isn't is obvious? "Supertautology" is a new and very important mathematical concept to which everyone had best get accustomed if one wants to understand the first thing about the true nature of reality. To reject supertautology is to reject the reality that we inhabit.

January 12, 2018

225. Is the CTMU true?

Yes, absolutely. Its structure is that of a *supertautology*, a new mathematical system which incorporates logic as a whole and thus ensures its own veracity.

[**Warning**: This site teems with anti-CTMU trolls and their sockpuppets, and the Quora moderation staff is completely inadequate to deal with them. Not one of these trolls has ever displayed any understanding whatsoever of the theory or any of its components. Answers to CTMU-related questions which do not come from me, or from someone whose statements can be checked against CTMU material published elsewhere, should be regarded as unreliable, especially if they are both pejorative and free of actual CTMU content. Thanks for your attention.]

September 30, 2018

Syndiffeonesis

Common syntax requirement

226. What does concept of "syndiffeonesis" mean in Christopher Langan's Cognitive-Theoretic Model of the Universe (CTMU)?

Syndiffeonesis translates to "difference in sameness". It defines the generic relational structure of reality in terms of syntactic distribution and coherence: given any relationship of any arity and order, its synetic (invariant syntactic) level must distribute over its diffeonic (variable) level, which consists of things to be discerned or distinguished, thereby providing them with a unified basis of cognitive and/or perceptual coherence. Basically, it says that common (uniformly distributed) syntax is required in order for any number of things to be recognized, no matter how generically, as different or distinct, or to identify any one thing as different from its complement.

February 9, 2018

T

Teaching

Social engineering

227. What controversial advice have people given to their children?

"Trust your teachers. They want what's best for you."

Modern teachers, bless their hearts, are in the indoctrination-and-socialization business. These days, trusting them blindly amounts to swallowing the social engineering imperatives of the socioeconomic elite, who want a 2-tier society consisting of a hereditary upper class which embraces global supercapitalistic corporate fascism in partnership with host governments, and a culturally and racially homogenized ("multicultural") lower class which stays on the ground, mewling pathetically with their mouths open for any crumbs that happen to fall off the upper-class table.

This might not be totally unjustifiable if there were a real standard of merit involved in class stratification, or for that matter in the education system. But unfortunately, one can only strain one's eyes and credulity trying to find it.

May 12, 2017

TELEOLOGY

Embracing truth

228. Does donating to Mega Foundation increase one's teleological value?

Well, that's rather a strange question, isn't it. (I wonder if it's sincere.)

Obviously, positive teleological value is not contingent on having money to donate to charitable causes – poor people can have it too. However, it is also true that a person who supports teleology by embracing and supporting truth to the best of his or her ability usually has greater teleological value that one who ignores it, rejects it, or opposes it.

Unfortunately, as we all know, most people with appreciable money and power couldn't give a hoot less about anything but their own future prospects, the rest of the population be damned. They wax philanthropical only when it wins them an instant pat on the head by the media and/or other rich and powerful people, improving their public image in such a way as to fatten their own bottom lines.

For the record, the Mega Foundation does not pander to such people.

February 19, 2018

Metaphysical optimization

229. Does Christopher Langan's Cognitive-Theoretic Model of the Universe do away with Hume's principle that one cannot derive an "ought" from an "is"?

Of course. "No ought from is" (Hume's Law, Hume's Guillotine) was just Hume's self-contradictory and therefore doomed attempt to turn his exclusion of *a priori* metaphysics into metaphysical preclusion.

Whereas Hume's bare-bones conceptualization of reality had no room for a metaphysical domain over which *that which is* can (and must) in some sense be optimized (*should* inevitably refers to such an optimization), the CTMU is explicitly metaphysical and therefore does not share this limitation.

Out of respect for standard philosophical terminology, the aspect of reality which makes *should* out of *is* (and vice versa) continues to go by the name **teleology**.

February 18, 2018

Self-reinforcing conatus

230. Chris Langan (author of CTMU) claims that he's logically proved atheists go to hell. To clarify, do atheists stay in hell for eternity, or only until they stop rejecting the possibility of God?

I don't recall publishing a proof that atheists go to hell (whether I can is another matter). If you can direct me to the exact statement to which you refer, perhaps I could comment at slightly greater length.

Meanwhile, remember that hell is necessary for not only the efforts of the Christian Church to curb attrition, but the welfare of society and mankind. For example, once an ambitious sociopath becomes a serial killer, mass murderer, hanging judge, bloodthirsty warlord, tyrannical emperor, evil dictator, crazed oligarch, rapacious international banker, or atheist-materialist techie billionaire who, by stealth or manipulation, can abuse other people with legal impunity (as nearly all successful sociopaths eventually do), society and mankind need a higher form of deterrence to protect them, and this is the very important function of hell.

As for whether hell is "real", that's a matter of valid metaphysical reasoning. This reasoning has much to do with a concept historically referred to as *teleology*, a self-reinforcing conatus by and for which the universe must include all necessary deterrents and safeguards, and the mathematical structure of metaphysical reality at large. Also important are the concepts of *God*, *free will*, and the *human soul*, on all of which I've previously commented on this site.

(Yes, hell is in some sense "real". But of course, this says nothing about the exact form that it will take for a given unregenerate miscreant. ;)

[**Addendum**: To those who insist on making an issue of my logic, please at least make an effort to understand it. First, hell is not a linear extension of the axes of spacetime, but is contained in an orthogonal expansion of spacetime; thus, objections based on the fact that "the dead no longer exist in space and time" are invalid. Secondly, one is not sent to hell for honest doubt or agnosticism, but for active denial or antitheism, i.e., blasphemy (among other things). This really doesn't require any proof, as it is a matter of common sense – one cannot survive in a metaphysical medium that one actively rejects and has encouraged others to actively reject. Instead, one is sent "in the opposite direction", so to speak ... to a complementary degenerative medium where identity is destroyed rather than sustained, refined, or transformed.

Many atheists seem to think that upon death, their minds will dissolve painlessly or perhaps even ecstatically into blackness or white light as though they were *arhats* or *bodhisattvas*, or perhaps just Kiefer Sutherland in "Lost Boys". But sadly for those of this optimistic persuasion, existence and identity aren't quite as easy to surrender as they might imagine. In fact, the utter dissolution of identity can be a process that is full of unimaginable pain and despair and which seems to last forever (or "for eternity", as religious people like to say.)

From here out it gets a bit more technical, but this should at least help put you on the right track regarding something that could – i.e., *will definitely* – end up being very important to you. A word to the wise.]

March 24, 2018

Telesis

Ultimate reduction

231. What is the definition of "telesis" by Chris Langan in simple and sophisticated terms?

Telesis is the convergent generalization and ultimate reduction (product) of reality. Energy may be considered a limiting form of it. (If you don't understand what any of these words mean, please get a dictionary. You might also try reading a little something I've written.)

January 10, 2018

Telic Recursion

Metacausation

232. What does Telic Recursion mean in Christopher Langan's Cognitive-Theoretic Model of the Universe?

In the CTMU, telic recursion is "metacausation". That is, it is a level of the causal evolution of reality which distributes over standard causation and is orthogonal to it in the sense that it generates entire timelines (cause-effect event sequences, networks of quantum entanglement) as opposed to mere single events.

February 10, 2018

Telor

Self-dual operator

233. What is the definition of "telor" by Chris Langan?

In the CTMU, a telor is self-dual operator which binds telesis. Human beings are telors. (Quora is not the proper venue for a full technical description at this time, as the typical Quora participant lacks the background necessary to understand it.)

January 12, 2018

Theorizing

Theoretical principle

234. What is the meaning of a theoretical principle?

Let a theory or "theoretical language" **T** be defined as a set of strings {s*} of symbols s from the signature or "alphabet" S of **T**. (This is just the ordinary definition of "language" for certain general purposes, but with the label *theory* attached to make it a "*theoretical* language"; at its most general, a theory need be neither axiomatic nor even descriptive).

A theoretical principle P of **T** is just an expression of **T** (i.e., a member of the set of strings {s*} comprising **T**) to which an assignment of "importance" justifies the exalted title of "principle".

Whether P be an axiom, a definition, a heuristic, a piece of guesswork, or just an uninterpreted string, it has no meaning until it is interpretatively mapped into or onto a *universe* U in which meaning can be ascribed to it. Once this has been accomplished, we can then worry about providing the "importance" of P with a justification. E.g., we can require that P be especially useful or "instrumental" in U, or that P be true in *all* models (valid interpretations) of **T** in U, or that certain other expressions (well-formed strings of **T**) be derivable from P by substitution under the operative rules of deductive inference, or that P be an encapsulation of **T** into which a significant part of U and/or **T** itself can be inductively mapped, and so on.

In other words, the **meaning** of anything, including a principle of a theory **T**, resides in its **correspondence** to something, even if this is just the thing itself. Given this meaning, we can then decide whether it is sufficiently important to merit recognition as a "principle".

February 23, 2018

Theory of everything

235. Has Chris Langan published any other theory / theories other than the CTMU?

I define the CTMU as a true "theory of everything". So by definition, the CTMU absorbs all of my valid theoretical content.

(I seldom publish theoretical speculations, preferring to make sure that my ideas can be modeled in the CTMU before releasing them.)

February 9, 2018

TRIALITY

Cosmic self-containment

236. Can anyone give a quick summary of Christopher Langan's Cognitive-Theoretic Model of the Universe (CTMU)?

It couldn't possibly be simpler. The CTMU is a theory that is not *just* a theory, but also its own universe AND its own model therein. In other words, it is *trialic*. Triality is an implication of a mathematical property called *closure*, which is analogous to cosmic self-containment. Any so-called "Theory of Everything" (ToE) must exhibit this property. Among all of the ToE candidates out there, only the CTMU actually does so. There is only one way to construct a theory of this kind: as a *supertautology*. A supertautology is simply the model-theoretic analogue of a propositional tautology, i.e., a tautology with ontological force. The entirety of modern empirical and mathematical science can be seamlessly embedded in this structure as a kind of limit.

Because of what it does, the CTMU may well be the single most important work of philosophy, theology, science, and mathematics ever conceived. Inasmuch as we will ever be privileged to truly know reality, we must know it as specified by this theory. When it comes to epistemology, the CTMU has no meaningful competition. In over thirty years, no one – regardless of claimed academic credentials and affiliations – has been able to put a dent of any kind in it, and due to its supertautological formulation, no one ever will.

This gives rise to a very important question: How is it possible for something like the CTMU to be discovered by a person who is subsequently introduced on every major television network in North America (and many abroad) as the "smartest man in America/the world", and still draw no attention from Academia? The answer is simple but profoundly disturbing: Academia, and the wider economy which unwisely relies on it, *also* exhibits self-containment and the associated mathematical closure property. Its constant accumulation of power and influence has now given it a virtual monopoly on scientific communication, intellectual commerce, and even the right to obtain a decent job on the basis of personal merit and intelligence.

Because Academia is closed, outsiders – people who have not paid tens or hundreds of thousands of dollars for its increasingly watered-down credentials – have no access to it, its personnel, or its journals. This allows it to exclude anything for which its own members cannot take exclusive credit (in any other context, this would amount to a "conspiracy to commit theft of intellectual property"). In my own case, there was never any choice about whether or not to obtain academic credentials, as not only did I lack the money to obtain them, but by some combination of error and contrivance which it is either unable or unwilling to explain, Academia repeatedly blocked me from doing so.

Anyone who has visited "10 different websites" pertaining to the CTMU and still "can't make heads or tails" of it is not trying hard enough to understand it. In the last several decades, I have heard similar complaints too many times to count. For what it's worth, essentially the same "unintelligibility" complaints were often heard about relativity theory and quantum mechanics (both of which are naturally embedded in the CTMU); it took many decades for this to change. But in the end, they prevailed. This is not only because they are good theories, but because they were introduced by professional academics, and were therefore taken seriously in Academia. Otherwise, they would have had no chance whatsoever.

Instead of railing against the CTMU and its author, those who find this state of affairs untenable should demand a reasonable accounting of the global academic bureaucracy, which has fattened beyond measure on the claim that it dutifully seeks, preserves, and disseminates valuable knowledge regardless of its nature or its source. In light of my own very telling experiences, this claim is not merely unjustifiable, but false on its face.

Thanks for your interest in the CTMU!

June 3, 2017

Trump, Donald

Intelligence

237. President Trump is now claiming to have one of the highest IQs. What would you estimate it to be?

With all due respect, I suggest that some of the respondents here try to suck it up a little and come to grips with reality. In fact, Donald Trump would very likely qualify for Mensa. There's plenty of evidence for this, not least his success as a negotiator and businessman. Even with his (appreciable) head start in life, he could not have expanded his holdings in the various directions he chose with a "double-digit IQ". One need not be a sublime genius to juggle all of the irons that he has in the fire, but one can't exactly be a chimpanzee either. In Trump's world, smart people are always trying to take advantage of you and come out on top. You can't defend yourself from them without meeting a certain threshold of intellectual ability.

Ever hear of the "double-life strategy"? That's where a very intelligent person, realizing that he must communicate effectively with those around him in order to be successful (and avoid resentment and ostracism), adopts their verbal mannerisms and habits of expression so as to be accepted in their circles and smoothly interact with them. In New York (where I lived for over 25 years), that means Trump-style vernacular and plentiful hand gestures ... unless, of course, one has the luxury of existing in an isolated, privileged, snowflake-populated bubble-world of the kinds which proliferate in the academic, artistic, political, and corporate spheres.

Many people disagree with Mr. Trump's ideas and policy proposals on philosophical grounds. That's fine, as long as one does not confuse "philosophical grounds" with ad hominem attacks on his intelligence. Stop making yourselves look bad to people who actually have the brainpower to see through your emotional polemics. (Thanks.)

February 4, 2017

238. What is your honest estimation of Trump's IQ?

Trump's IQ is probably at least as high as that of the typical professional academic, and higher than that of the typical Harvard student.

When a sample of 86 Harvard undergraduates was given an abbreviated version of the WAIS, their mean IQ came up as 128.1, with a range of 97-148 (lower than would be estimated from the SAT scores of Harvard students, but the SAT is not an IQ test). The mean IQ of tenured college professors is probably around 135.

Anyone who thinks that Donald Trump couldn't easily work as a high-functioning professor at a top business school like the one from which he graduated (Wharton) hasn't been watching him very closely. Success of that magnitude doesn't lie, not even with four strategic bankruptcies (The Trump Taj Mahal, 1991; Trump Plaza Hotel, 1992; Trump Hotels and Casino Resorts, 2004; Trump Entertainment Resorts, 2009). Anyone who would walk down a university hallway and pass by a business management course taught by Trump in order to reach one taught by almost anyone else either has a bee in his bonnet, or rocks in his head.

One can disagree with Trump's politics and dislike his personal style all one likes, but his detractors need to stop fooling themselves about how sharp he is.

July 8, 2017

W

Wikipedia

CTMU page

239. Is the current Wikipedia summary of the CTMU (available on the page for the author) a good summary of the theory?

Not really. The absence of a dedicated CTMU article on Wikipedia is a travesty that was engineered by some very aggressive yet very deceptive people circa 2006, and is quite possibly the worst show in Wikipedia history.

My involvement began when I was informed by a concerned party – whom I had never met – that two articles on Wikipedia were embroiled in controversy. One was a biography article about me; one was an article on my theory of mathematical metaphysics, the Cognitive-Theoretic Model of the Universe (CTMU).

Apparently, some number of Wikipedia editors, administrators, and arbitrators had decided among themselves that neither I nor the CTMU – both of which had received considerable international publicity from, for example, ABC, NBC, and Popular Science (among other highly reputable sources) – were sufficiently notable for Wikipedia's elevated standards. In particular, they falsely but adamantly maintained that both articles had been created by me (they had not), were being jealously controlled by me despite alleged conflicts of interest (they were not), and that the CTMU was baseless and did not belong on Wikipedia (it is factual and logically verifiable, and even if it were not, it was still sufficiently notable to merit description). Both articles were nominated for deletion. (Another article on the Mega Foundation, a nonprofit 501.c.3 corporation in continuous operation since 1999, was created and deleted during this period.)

In the course of the ensuing proceedings, which can only be described as a witch hunt culminating in a kangaroo court, the CTMU was repeatedly misclassified as "Intelligent Design" when in fact, it predates and is independent of ID theory, and was snidely compared to nonsense like the "Time Cube" and other theoretical oddities often used as targets for the spitballs of those who fancy themselves "defenders of science". I finally traced the initial push to remove these articles to a website called "the brights.net", a pseudoskeptical anti-ID website run in keeping with the atheist-materialist convictions of such opinionated people as Richard Dawkins, Daniel Dennett, and James Randi. Unfortunately, this initiative – the true nature of which its promoters made every attempt to conceal – was immediately latched onto by a number of prominent Wikipedians, some of whom apparently still lurk behind the curtains to this very day.

Ultimately, the biography article was kept, but the CTMU article was deleted and the "earth was salted" against its recreation (a phrase I seem to recall being used by one or more of those clamoring for deletion). The experience left such a bad taste in my mouth that personally, I wouldn't care whether a CTMU article ever appears on Wikipedia were it not for others constantly using the platform to take credit for CTMU ideas without proper attribution. If experience be our guide, such an article would only be ruthlessly attacked and used as a platform for defamation and misattribution by certain nefarious parties still concealed in the Wikipedia woodwork. On the other hand, we might hope that Wikipedia is no longer such a nest of spiders, and that it is no longer controlled by sneaky little arachnids secretly communicating and coordinating with each other and using their mutual connections and experience in navigating the bureaucratic web to suppress those with whom they disagree.

In fairness to Wikipedia, which I and many others sometimes find very useful, a "peoples' encyclopedia" is by nature a very hard thing to run. Such a project naturally tends to be dominated by aggressive and opinionated people who operate largely behind the scenes, resulting in a "hive" whose inner mechanics are largely hidden as its drones and workers succumb to various forms of censorship and peer pressure and eventually veer away from sound encyclopedic principles of fairness, veracity, and scholarship.

Had I not ultimately appealed directly to the Head of the Wikimedia Foundation, Jimmy Wales (who evidently interceded even though I never received an official reply from him), there might be no mention of either me or the CTMU on Wikipedia to this day. In effect, we'd have been obliterated from the Internet "hive mind" by people whose motives may simply have been misguided, but whose tactics were inarguably as black and dirty as sin.

May 20, 2018

240. Why is there no Wikipedia page about the CTMU?

First, let me say that I think Wikipedia is a great idea, and I sometimes get some real use out of it. It has many fine and dedicated editors. But it also has its problems.

In particular, Wikipedia is notorious for its freewheeling bands of obsessive and seemingly jobless "vigilante trolls" who, after getting a bee in their bonnets about something they find annoying or offensive, stop at nothing to get it deleted. They usually begin by tampering with its article, pumping it full of defamation and misinformation; then, should someone else try to repair the damage, by complaining to administrators and nominating the article for deletion; and finally, when their case is heard, by engaging in nonstop lying and cheating while threatening any and all opponents with dire consequences until they get their way.

Then, after the article has been thoroughly nuked, they keep buzzing around the crater like flies around an outhouse to make sure it never comes back.

Some of these trolls are mere editors; some are administrators; in the past, some were arbitrators. They are overwhelmingly of a "secular" (atheist-materialist) persuasion; thus, if any of them encounters anything on the web which seems to suggest that a potential target mentions God, or spirituality ("woo"), or was written by a theist (particularly a Christian, AKA an "Intelligent Design Creationist"), that's usually the end of it. Whether or not what they've read about their target or victim is correct – and in the case of the CTMU, it was absolutely false – they immediately freak out *en masse* and throw a collective fit until the object of their displeasure has been red-inked, ripped out, and consigned to oblivion. They don't even bother to read the source material. The whole routine is approximately as sane and rational as a medieval witch hunt.

Unfortunately, I'm not exaggerating. The CTMU and I were attacked as being "non-notable" even after having been on every major news network in North America and Europe, with considerable mass media exposure in South America and Asia as well, and were insulted and harangued by self-styled Wikipedia "mathematicians" and "physicists" whose actual command of these disciplines couldn't have exceeded that of an average high schooler. It eventually emerged that the whole Wikipedia anti-CTMU movement had migrated from the website of "the Brights", a shrine to the dubious brand of "rationalism" promoted by various celebrity atheists who run around deploring "the supernatural" and claiming to love "science", but who often seem unable to distinguish worthwhile science, mathematics, and philosophy from soggy corn flakes.

Perhaps the CTMU article will come back some day. It certainly should, because the CTMU has nothing to do with "intelligent design", it has thousands of followers, it is backed up by several peer-reviewed publications, it has urgent social, scientific, and philosophical applications, and it is based on exciting new mathematics. The deletion of its article may have been Wikipedia's darkest day, and it should definitely be put right. But one thing is for sure: I'm not wasting any of my own time to make it happen.

September 20, 2018

SUBJECT INDEX

The Art of Knowing: Expositions on Free Will and Selected Essays Revised and Expanded Edition

PUBLISHER'S NOTE

Although best known as the creator of the Cognitive-Theoretic Model of the Universe, Christopher Michael Langan has been publishing philosophical essays on a wide variety of topics for more than 40 years. While reflecting his diverse interests, these essays tend to gravitate toward a recurrent theme: the intimate, multifaceted connection between physical reality and human cognitive processes. This is a revised and expanded edition of the popular 2002 collection of the same name.

Special thanks to the members of the Mega Foundation and CTMU Community who help preserve and present Chris Langan's work and ideas.

PART ONE

So You Want to Be a Millionaire? All It Will Cost You Is Your Freedom

Imagine that you snagged some tickets from a buddy at NBC and are now a lucky member of the studio audience on the celebrated *Tonight Show*. The Tonight Show Band fires up and Jay Leno emerges.

"Thank you ... thank you!" says Big Jay through the applause.

"I'm proud to announce that we have some very interesting guests on the show tonight. In fact, one of them is known for seemingly inhuman powers of prediction ..., powers that he will demonstrate on a member of this very audience!" (Kevin Eubanks spookily simulates a theremin on his electric guitar.)

"And thanks to the cooperation of my friend Regis Philbin, that lucky audience member will have the chance to walk away with *one million dollars*!" (You and I both know that a Leno–Philbin collaboration would cross the Major Network Divide and is thus forbidden. So just consider it a suspension-of-disbelief thing.)

Jay does his monologue, trots out his first couple of guests. Finally, the magic moment arrives.

"And now the event you've all been waiting for," crows Jay. "It is my pleasure to introduce a veritable modern-day Nostradamus, the most amazing mentalist in history, Sir Damon Nukem!"

Nukem, tall of stature and angelic of countenance, emerges to thunderous applause. Everyone in the audience, including you, already knows exactly who he is; a regular in the *National Enquirer*, he's reputed to be an alien, to wear contact lenses that conceal yellow irises with vertical pupils, and even to have a couple of goatish horns under his immaculately coiffed hair. (Fortunately, you're not the kind of person who believes everything you read.) One thing you know for sure, however, is that he's the world's foremost mentalist, having correctly predicted the behavior of thousands of people in hundreds of different contexts under a wide variety of conditions. He has never failed ... not once.

In Nukem's most notorious routine, he offers a randomly chosen contestant a choice between one or both of two boxes, one of which (Box A) is transparent and contains a thousand dollars, and one of which (Box B) is opaque and contains either a million dollars or nothing at all. That is, the contestant is allowed two possible choices: (1) to take just the opaque box B, or (2) to take both boxes (A and B). Here's the catch: a neutral referee, usually either professional debunker James Randi or an equally curmudgeonly associate of the *Skeptical Enquirer*, holds a single well-sealed envelope containing a single slip of paper on which Nukem has written a prediction regarding the contestant's choice. If the slip says "B", then Box B contains a million dollars; if the slip says "A and B", then Box B contains nothing. In other words, if Nukem has predicted that the contestant will leave the thousand on the table, then there is $1,001,000 on the table; if Nukem has predicted that the contestant will grab the "extra" thousand, then what you see is exactly what you get, and there is nothing more on the table than the price of a good office chair.

Common sense tells you that since the prediction is already in the sealed envelope at the moment the contestant makes his or her choice, he or she can lose nothing by taking both boxes and grabbing the extra thousand. No disrespect to Nukem – he does, after all, seem to be on a world-class lucky streak when it comes to prognostication – but what's done is done, and things will unfold according to the Grand Plan regardless of what anybody thinks. Even if Nukem somehow scans the brain or reads the mind of the contestant before making his prediction, and even if this causes a million dollars to be placed in Box B, there is obviously nothing stopping the contestant from changing his mind at the last minute and snagging that extra thousand. After all, that's what free will is all about. And an extra grand could be the difference between banking that million at the autoteller in Hampton Bays, and personally depositing it in a special account in the Bahamas during a cut-rate Princess Cruise with the Significant Other!

But the thing is, the prediction in the envelope always turns out to be correct, and no matter how hard Randi and his buddies have tried to catch Nukem in a mistake or a deception, the mentalist always comes up squeaky clean. Somehow, Nukem knows exactly how much confidence the contestant has in his abilities. Those who believe that he has predicted correctly and choose accordingly are always rewarded with a seven-figure retirement fund; those who think they can "pull a fast one" at the eleventh hour and belie his prediction are doomed to kick their own cans forever after for throwing away a certified ticket to Easy Street, the additional $999,000 they'd have won had they merely given predictive credit where credit was due. Furthermore, Nukem never seems to force the issue; when questioned, all of the contestants report that their choices were made "freely".

Lost in your reverie, you realize that someone with a distinctive nasal tenor is speaking into your ear. Fearing the worst, you look up just in time to see a gigantic boulder crashing down on you from above! Then you see what it really is: The Chin. It positively blocks the sky.

"Excuse me ... are we boring you here?"

"Er, not at all," you quickly respond.

Everybody laughs at you.

"I'm very happy to hear that," says Leno. "Because we were just about to call security!"

Everybody laughs at you again. You feel pathetic.

"Anyway, I'm pleased to inform you that you've been chosen as the lucky contestant who will have a chance to win One Million Dollars!"

He sticks the mike under your nose.

"Uh, er ... sure," you say, nodding.

"OK then," says Leno as he takes you by the elbow. "Let's go!"

Before you know it, you're up on the *Tonight Show* stage. The lights are blinding. With a flourish, Nukem draws a curtain, and there they sit: two boxes on a table. As expected, one of them is transparent and contains ten crisp $100 bills; the other is larger and opaque. The audience falls silent as Newcomb begins to speak, his voice at once hypnotic and resonant. Your ears buzz as he recites a game description that sounds like a magician's incantation. Finally, he finishes, gestures dramatically towards the table, and folds his arms across his chest. The fateful juncture has arrived: you must choose! Will it be the opaque box, or will it be both boxes? Will you embrace Nukem's predictive infallibility and place all your faith in his ability to predict your behavior, or will you use this moment in the spotlight to strike a blow for common sense, freedom of choice, and an extra G in your checking account?

Well – what'll it be?

Alright, alright, let's stop the clock and think about this for a minute. If the opaque box contains the million dollars, taking just that box will earn you a cool million. But why would you do that, when you could easily grab the transparent box as well and make an additional grand? After all, the contents of the opaque box won't suddenly change simply because you decide to take the other one too. This whole setup has been checked out by Randi and his antiparanormal goon squad, and they've made sure that the contents of the boxes are as safe and unchanging as the laws of physics can make them. On the other hand, if the opaque box is empty, then that thousand dollars in the transparent box is your only chance to walk away with anything at all! When you think about it in these terms, it really looks as though the issue is cut and dried. You might as well cover all bets and take both boxes.

But as long as we've already got a suspension of disbelief going, that's not the only way to look at it. Nukem has quite a track record, and it hardly seems rational to ignore it. The fact that he has played this game a thousand times and been right a thousand times – well, the odds against that are 2 to the thousandth power to one, and thus way beyond astronomical. This alone would seem to suggest that the laws of physics may not be everything they're cracked up to be. Doesn't particle physics recognize time as flowing in both directions? Nukem is either passively reading the future without affecting it, or actively controlling the thought and behavior of his contestants in such a way that they cannot directly detect his influence. In any case, statistical induction would seem to indicate that the contestant's chance of beating the renowned mentalist at his own game are around one half to the thousandth power, as close to nothing as one can get without falling through the zero. You might as well bet against death, taxes and 1 + 1 = 2!

Then again, there are a couple of philosophical kinks in the wire. The first one, *Goodman's Grue*, is about the limitations of induction. Suppose you have two hypotheses, namely "emeralds are green", and "emeralds are green and will so remain until New Year's Day 2050, at which point they will all turn blue." (This can also be expressed by saying that "emeralds are grue", where "grue" is defined as "green until 2050 and blue thereafter".) Statistically, both of these hypotheses are equally confirmed by current observations. This implies that Nukem's past successes can also be regarded as confirming two hypotheses, namely "Nukem is able to predict the future" and "Nukem was able to predict the future up to the last contest, but not thereafter." With loopholes like that, arguments from induction can be hard to trust.

On the other hand, laws of physics are themselves discovered by induction. This suggests that all of the observations and experiments confirming the currently accepted laws of physics also support the following hypothesis: "The currently-known laws of physics hold perfectly well until Damon Nukem decides to predict the future, at which point they bend subtly to his will." This, of course, raises an important question: in what kind of reality could a prognosticator like Nukem bend the laws of physics with the power of his will, yet remain undetected? Even if his powers are entirely passive, even if he only "watches" the future without ever taking a hand in it, how can he transcend the flow of time? Where time is understood to progress steadily from past to future, how can his mind leapfrog into the future and back again? How can information go from future to past?

Suddenly you remember a hit movie called *The Matrix*. What if the world is a kind of digital simulation running on a powerful computer and displayed on a vast quantum-pixelated 3D monitor? What if "people" are just software homunculi who consider themselves sentient, but cannot detect the higher level of reality containing the computer and its programmer? Perhaps the programmer of such a simulation could influence the behavior of these homunculi without being detected. In fact, maybe he could freeze the action, fast-forward the program, gather information on future events, back the program up to the "present moment" and resume the simulation precisely where it left off so that nobody inside would ever know. What if Nukem has a direct line to the Great Programmer in the sky, or maybe His competition down below? Where would the laws of physics, including the flow of time, be then?

Breaking surface from your deep concentration, you look around the studio. The audience is murmuring and shifting in their seats. Nukem, looking as much as possible like the Antichrist, is glaring at you with his arms still folded. Leno is looking conspicuously at his wristwatch and making silly faces in your direction. Obviously, your time is up. Your heart pounding in your chest, you make your choice and announce it. And when the opaque box is opened ...

Nukem the Magnificent is right again!

Now, among the questions you're probably asking are these: What is this story really about? Why is this guy writing about it? Where did it originally come from, and when? And how is anybody supposed to make sense of it, when it's obviously crazy? The answers, in order: This story is about free will. I'm writing about free will because I find the topic very interesting (as should you). The basic story, minus a few stylistic and atmospheric touches of my own, originally came from a physicist named William Newcomb and was popularized by the Harvard philosopher Robert Nozick circa 1969. (As we will see, it is no accident that the story originally came from a physicist.) And if you think it's crazy, it's not because you can *prove* it's crazy, but only because you *assume* it's crazy. For all you really know, it could actually happen.

In case you're wondering why I say that you "should be" interested in free will, I'll risk belaboring the obvious and point out that the free will question ranks right up there with "Is there a God?" and "What is the meaning of life?" among philosophical riddles. Moreover, although philosophers have long treated free will as an open question, more practical minds long ago jumped the gun on their lofty discourse and adopted free will as a foregone conclusion. Most of us mirror their position: because we feel as though we are exercising free will in order to make decisions and perform voluntary actions, we take its existence for granted. Our social, political and religious structures, as well as our ideas about human dignity, human rights, civil liberties, moral responsibility, crime and punishment, and good and evil are already heavily invested in the concept ... heavily enough that if the existence of free will were ever disproved, it would literally destroy everything we believe about ourselves and the world we inhabit.

The above scenario, along with the question of how the contestant should choose, is found in the literature under the names "Newcomb's Problem" and

"Newcomb's Paradox". Although one might be tempted to assume that they mean the same thing, there is a subtle difference between the "problem" and the "paradox". In order to solve Newcomb's Problem, one must explain how the contestant should choose and why.

Unfortunately, there are two possible ways to do this, and it is extremely difficult to say which one of them is right. In fact, it's so difficult that they are regarded as having equal claim on correctness. And that's where Newcomb's Paradox enters the picture: two "equally correct solutions" that contradict each other add up to a paradox, plain and simple. To resolve the paradox, one must do one of three things:

1. Decide the issue, showing why one solution is correct (and the other incorrect).
2. Describe a framework into which both solutions fit under different circumstances.
3. Prove the scenario unreal and the question meaningless.

How might one achieve a resolution of Newcomb's paradox? First, one would need to know more about the solutions of the problem on which it centers. The first, "take both boxes", is called the *dominance* solution. The second, "take only the opaque box", is called the *expected utility* solution. The dominance solution is supported by the idea that since time runs in only one direction, the player can lose nothing by taking the extra thousand dollars in the transparent box; the million is either in the opaque box or it is not, and this fact is not subject to last-minute change. The expected utility solution, on the other hand, is a straightforward calculation of mathematical expectation based on statistical induction from Nukem's long and unbroken string of successful predictions. One simply constructs a payoff matrix for the game defined by the Newcomb scenario, computes the expected value of each possible move, and chooses the move with the highest value.

So much for the current philosophical impasse regarding Newcomb's paradox. In order to progress beyond this point, we need to avail ourselves of a branch of mathematical logic called *model theory*. According to model theory, using a theory to correctly describe some part of the world is logically equivalent to specifying a valid *interpretation* of that theory, which means specifying a *model* of the theory. In this case, the theories are the rationales used to justify the solutions, and the models of those theories are their (superficially) valid interpretations in the real world ... assuming, of course, that the Newcomb scenario is possible in the real world, which is a bit of a stretch (in this context, "valid" means only "not yet invalidated"). The key ingredient of the dominance model is a one-way arrow of time, while the key ingredient of the expected utility model is a natural induction principle. While neither of these ingredients is yet a well-understood feature of reality, neither has yet been disproved. But the fact remains that they have sharply conflicting implications in the Newcomb scenario, and therein resides the paradox.

That Newcomb's Paradox can be reduced to conflicting elements in a hypothetical configuration of physical reality suggests that its proper resolution requires a theory and model of reality in which the relationship between these elements can be meaningfully evaluated. In other words, we need a theory of reality in which we can explore the nature of time in an inductive context. This is why, several paragraphs ago, I called it "no accident" that the paradox was originally cooked up by a physicist; physicists are among those most concerned with the nature of time and the inductive derivation of functional theories of reality. Insofar as weird or improbable scenarios have held the keys to new insight many times in the history of science, the far-out nature of the paradox is beside the point. Where the stakes include a better understanding of time, induction and free will, even a scenario as unlikely as Newcomb's is worth a gander.

The Buck Starts Here: Another Disquisition on Free Will

In the previous essay we talked about free will, and I asked you to pretend that you were playing an imaginary game taken from the hypothetical philosophical scenario called "Newcomb's Paradox". In this essay, we're still talking about free will. But this time, I'm asking you to imagine that you're part of a scientific experiment.

You're seated at a table with electrodes glued to your head. Across from you sits a man in a white lab coat with a clipboard and a stopwatch. There's a camera on the ceiling; mounted on the wall in front of you is some kind of fancy clock with a bright circle of light revolving rapidly around its face where a second hand should be. The man carefully explains what he wants you to do.

"Over the next few minutes, you will be expected to make spontaneous, freely-initiated movements of your right hand. The movements should consist of simple flexions of your right wrist. Any time you want to move your hand, you can and should do so of your own free will. All you have to do is keep your eye on the clock and report the position of the circle at the precise moment you decide to act."

You agree.

"Then let's begin," says the man.

Keeping your eye on the revolving circle, you mentally command your hand to move and note the position of the circle at the instant you make your decision. You report the circle's position: halfway between 9 and 10 o'clock. The man marks this down. You continue. Again and again you move your hand, each time reporting the position of the circle at the exact moment you decide to do so; each time, the man marks it down on his clipboard. Finally, he indicates that he is satisfied.

"Very good," he tells you. "That will suffice. In the name of science, we thank you."

"My pleasure," you say, getting up and unceremoniously unpeeling the electrodes from your scalp. "I hope this helps clarify what goes on in the human brain after a person decides to commit a voluntary act!"

"I'm certain your data will be of great help," says the man. "But since you've been so kind as to participate in the experiment, let me clarify one little detail for you. This experiment was not about what went on in your brain *after* you decided to act. Everything of interest, including the initial electrical brain potential precipitating each of your actions, occurred considerably *before* you consciously decided to act."

With a slight smirk, the scientist turns and walks out of the room.

Wait a minute here. The initial electrical brain potential precipitating each of your actions occurred *before* you decided to act? But then how could you have been acting of your own free will? What was that white-smocked lab geek talking about? Who did he think he was, amusing himself at the expense of an experimental subject – Doctor Giggles? What were they pumping through those electrodes? And who cares about a far-fetched hypothetical experiment like this one anyway?

Well, maybe not so hypothetical. In 1983, an experiment just like this one was performed at the Neurological Institute at San Francisco's Mount Zion Hospital and Medical Center, and reported in the journal *Brain* in a paper entitled "Time of Conscious Intention to Act in Relation to Onset of Cerebral Activity (Readiness-Potential)" by Libet, Gleason, Wright and Pearl. The paper reported that after subjects were invited to make freely-timed voluntary movements of their right hands, an electrical brain wave called the *Bereitschaftspotential* ("readiness-potential") appeared over the motor areas of their left cerebral hemispheres some 500 to 800 milliseconds (0.5–0.8 seconds) before they were aware of intending to act, with muscular electrical activity not appearing until its near-completion another 0.2 seconds later. On the basis of these data, the experimenters concluded that voluntary neural activity appears to be preceded by physiological neural activity, and that even though awareness of intent comes almost a second after the onset of the neural activity in question, it is interpreted by the subject as having come "first". In other words, a voluntary act may begin well in advance of the conscious intent to perform it, but the subject interprets things the other way around.

Where does this leave free will? Right about where causality would be if it were ever found that what we think are "causes" actually occur *after* their "effects": in the OUT basket. Could we really be so clueless that what we call our "intentions" are just delusions that occur only after our actions are first initiated by our subconscious minds, perhaps under the rigid guidance of deterministic mechanisms in light of which we amount to little more than self-deluding automata? What could possibly be going on here?

Let's begin by taking a look at bodily movements in general. A *reflex* is an automatic muscular reaction that does not directly involve the brain. Because reflexes have a protective function, speed is of the essence, and the resulting movements are general enough that no conscious action is required. An example is the patellar reflex, the familiar medical "knee-jerk test" in which the tendon just below the kneecap is tapped with a rubber hammer. In a healthy patient, this causes a quick, unintended outward jerk of the lower leg. Why? The tap of the hammer triggers sensory nerve cells that send a message to the spinal cord.

Ordinarily, sensory stimuli travel up the spinal cord to the brain for conscious processing, but not in this case. Instead, the spinal cord responds directly to motor neurons in the muscle, causing the muscle to contract and the lower leg to kick out. Although the brain seems to exercise a general one-way "gating effect" influencing reflexes throughout the body, the sense-response loop or "reflex arc" never gets far enough up the spinal cord for brain-based conscious volition to specifically intervene. A particular reflex can be surprising to one experiencing it for the first time; the idea of being able to act (or react) without willing it can be as hard to absorb as the idea of acting *before* willing it.

In contrast, a voluntary movement originates not with the sensation of heat on a fingertip or a hammer to the knee, but in the brain itself, the seat of consciousness. It has always been assumed that voluntary acts are just what they appear to be: actions resulting from, and therefore following, causative volition. The chain of events is supposed to be as follows:

1. We think about something we desire, possibly on the basis of sensory input.
2. We form the goal of having it and formulate a plan to get it, even if that plan consists of only a simple body movement.
3. We execute the plan.
4. We obtain or fail to obtain the desired object or state.

Step 1 takes place in the cerebral cortex and hypothalamus (the source of drives). Step 2 involves higher-order reasoning and judgment occurring mainly in the frontal lobes. Step 3 involves those parts of the cerebral cortex specialized for voluntary movement, including the primary motor cortex, its associated premotor areas, and other motion-specific neural aggregates. And step 4 feeds back into the limbic system, which registers the satisfaction or frustration of the original drive.

Now let's have a closer look at the distinction between a reflex and a voluntary act. A voluntary act is purposeful and directed towards the completion of a task; a reflex is merely a response to a stimulus. A voluntary act performed in response to a stimulus can vary according to the task being performed; a reflex obviously cannot (because no task is involved). A voluntary act is "endogenous" or internally generated; a reflex is externally stimulated. A voluntary act can be adapted to various circumstances, and its speed and accuracy improved through practice; a reflex is relatively immutable. And because higher levels of the motor system can dissociate the informational content of a stimulus from its capacity to trigger a movement, the information can be voluntarily processed and the movement either permitted or interdicted. With a reflex, there is no such opportunity.

However, the work of Libet *et al.* introduces a new degree of similarity to voluntary and reflex actions: both are to some extent beneath awareness and therefore "subconscious" in that sense of the term, which we shall henceforth replace with "unconscious" to avoid Freudian connotations. In the type of voluntary act studied by Libet and his colleagues, conscious awareness is still technically in control; it precedes the act itself by 0.2 seconds and can in principle "veto" the movement even after the readiness-potential has been initiated. But regarding this point of initiation, one kind of act is no more voluntary than the other. In neither case does the act begin with conscious awareness of the intention to act. The main difference is that while something belatedly purporting to be conscious intent finally appears half a second or more after the beginning of the so-called "voluntary" act, it never appears at all with the reflex. Or if one prefers, the reflex lacks the volitive equivalent of *esprit de l'escalier.*

How are these results to be interpreted? Libet himself feels that since the 200 ms separating conscious intent from muscular contraction is time enough to consciously veto or permit the movement precipitated by the readiness potential, free will remains intact. Thus, although we cannot be held accountable for our apparently unconscious impulses to act, we remain accountable for whether or not we yield to them. However, others have more extreme opinions. For example, C. M. Fisher (2001) speaks for many when he suggests that Libet's research points to a conclusion that "may be reached on the basis of rather elementary observation ... the behavior of humans represents the electrochemical activity of the brain, involuntary, devoid of will and wish. We are automatons." While Fisher admits that this bleak impression may somehow turn out to be incorrect, he holds it sufficiently likely to be worth serious exploration.

But as regards the question of free will, neither of these interpretations is entirely satisfactory. For while neither Libet nor Fisher holds that the data alone decide the issue, each nevertheless voices a bias regarding the existence of free will, Libet's being affirmative and Fisher's negative. Libet holds free will to be a matter of conscious clearance of prior impulse; Fisher holds free will to be less likely than servitude to deterministic brain chemistry, suggesting that we get used to the idea that we are "automatons". Thus, each one (1) admits that the data are insufficient to decide the free will issue, and (2) unabashedly proceeds to weigh in on the issue anyway, trying to use the data to support his personal opinion on the matter. And in each case, the opinion posits a definite restriction on what most of the human race considers "free will" to be.

In fact, there are other ways to explain the data. For example, the subject may unconsciously choose a target position for the circle and wait for the circle to reach this position before "deciding" to act. In this case, the readiness potential could be merely anticipatory, following the unconscious choice of a target position.

Along much the same lines, the subject may choose a task and formulate an appropriate general plan on a combination of conscious and unconscious levels, delegating certain parts of the execution of this plan, as well as details of scheduling, to the unconscious mind. The unconscious mind would then take responsibility for that which has been delegated to it, executing various subtasks when appropriate and "reminding" the conscious mind of its responsibility to "intend" those acts just before committing them. The conscious mind would automatically take these cues, thereby reserving the right of final authorization and selectively permitting or interdicting the associated unconscious impulses.

The latter explanation incorporates several key concepts. One such concept is the mind's supposed ability to delegate various responsibilities to conscious or unconscious levels of processing, implying that the unconscious mind can to some extent function autonomously, without benefit of direct, step-by-step conscious oversight. Another is the distinction between tasks and subtasks in goal-related processing and behavior. This distinction permits the distinction of an overall task-related decision, e.g. deciding to perform a *sequence* of voluntary hand movements, from constituent subtask-related decisions, e.g. deciding to perform *one* of the hand movements in question.

Yet another such concept is higher-order intentionality, or "intent to intend". For example, intending to perform a sequence of voluntary hand movements amounts to *intending to intend* to perform each of the hand movements in the sequence, and where the latter (lower-order) intentions are generated by the unconscious level of processing, they can in turn be regarded as a *unconscious intentions to consciously intend* to permit or veto the impulses associated with the unconscious intentions themselves. We thus have a kind of "volitional loop" involving two levels of processing, and two levels of intentionality, instead of the single level usually acknowledged ... a multilevel control loop in which the "higher" (conscious) level of volitional processing is insulated from the noise and complexity generated by the "lower", unconscious nuts-and-bolts level, which thus functions to some extent autonomously.

Does this new explanation of volition as a multilevel control loop have any weaknesses? One possible weakness is the fact that because we associate control with consciousness, the very idea of "unconscious volition" seems semantically inconsistent. Relegating any part of a volitional control function to a non-conscious level of mental processing seems to contradict the premise that we possess the freedom to control our actions.

However, a little reflection should reveal that the horse of cognition is already long gone from the barn of consciousness anyway. If the conscious mind, which has an innate need to function within a well-defined conceptual system in order to ensure its informational integrity, were ever made responsible for the details of the complex, tentative, rapid-fire neural dialogue that microscopically relates one well-defined state of consciousness to its successor, cognition would immediately break down like a tired old jalopy. With a catastrophic "kapow!" from its exhaust pipe and a sad sigh of defeat from beneath its hood, it would forcibly retreat into the wakeless sleep of unrealizability. One might as well demand that the output of a computer never be acknowledged until the user has accounted for each of the millions of logical operations by means of which it was generated. Such a demand cannot be met within the bounds of practicality.

The moral of this story should now be obvious. Above, we cited an assumed chain of events comprising a voluntary act. We can now refine that chain as follows:

1. We think about something we desire, possibly on the basis of sensory input. This step amounts to consciously setting general parameters for a supertask or task sequence.
2. We form the goal of possessing the object of desire and formulate a plan to get it. This step amounts to consciously defining the supertask using the parameters of step 1, and automatically delegating as much of it as possible to the unconscious mind.
3. We execute the plan. This step, which may consist of numerous individual tasks or substeps, is executed on a tentative basis by the unconscious mind, which generates impulses that are subjected to conscious oversight as they become sufficiently well-defined to emerge into consciousness.
4. We obtain, or fail to obtain, the desired object or state, at which point the limbic system registers satisfaction or frustration.

Notice that at no point does anything come into being without the involvement of the conscious mind. The conscious mind chooses the overall task and clears or vetoes each subtask as it emerges from the unconscious background to which it was delegated under conscious oversight. In interpreting the Libet experiment, we need simply remember that when the subject consciously accepts the task of generating a sequence of voluntary actions (subtasks) over a given period of time, the entire future contribution of the unconscious mind is automatically requisitioned. At all times, the subject is doing just what he or she has consciously agreed and decided to do, nothing more and nothing less, but with the indispensable help of unconscious faculties without which the conscious mind could not function.

So the real value of the Libet experiment is not that it answers the philosophical question of whether or not free will exists, but merely that it provides data that clarify the operation of free will on the assumption that it *does* exist by elucidating the cerebral dynamics of volition.

What if the brains of experimental subjects, including electrical potentials and conscious ideations, could be monitored without their knowledge by means of a remote scanning procedure, and the experiment were concealed to make them unaware of it? Then they would not be known by the experimenters to have agreed to a specific experimental task, and some of their actions might appear truly spontaneous. It would be interesting to see whether the Bereitschaftspotential still precedes conscious intent in all cases. But if so, could spontaneity be ascertained? It is hard to see how a voluntary act of any kind could be certified as "spontaneous" where responsibility for subtasks is *automatically* delegated to the unconscious mind. Any such act could be related to any purpose consciously adopted at any time in the past, and even the subject may be unable to identify the purpose in question. In principle, the unconscious mind could be involved in a complex, protracted, consciously-authorized sequence of sensorimotor transactions in which the role of a given "spontaneous" act may be quite inscrutable.

What would it take to decide the question of the existence of free will? Because the answer depends on whether the universe evolves in a deterministic or nondeterministic way, this is a metaphysical rather than a merely psychological question. Indeed, answering this question requires an understanding of not only psychology and reality at large, but their logical interface ... the relationship of mind and reality. In other words, it requires a comprehensive theory of reality uniting the subjective and objective sides of existence. Within the overarching framework of such a theory, psychological and neurological phenomena could finally be interpreted in a way that clarifies their deeper philosophical significance.

A Free Will Cyber-Synthesis

That's right, folks ... we're back to free will again. In my defense, I'll say only that if you already believe in free will, then you should want to know as much as possible about how it works, and if you don't, then you have no choice but to resign yourself to your fate and read up on it anyway.

At the start of this book, we discussed Newcomb's Paradox, a controversial prediction paradox that highlights an apparent conflict between two basic principles: the macroscopically apparent linearity of time, whereby time always flows from past to future, and the principle of induction, whereby we form conclusions about reality on the basis of what we observe. We concluded that the relationship between these principles depends on the model of reality in which we interpret them. This was really no surprise, since only certain models of reality support the possibility of the Newcomb scenario in the first place. The surprise, if one exists, is that science has thus far failed to decide which model is correct, and has thus failed to preclude the real existence of a Nukem-like predictor.

Then, in the last chapter, we analyzed a controversial scientific experiment which appears to show that the supposedly voluntary acts of human beings precede any conscious intent to perform them. Although Benjamin Libet, the lead scientist in that experiment, opined that the free will hypothesis survives this finding intact, others believe that it proves mankind a race of automated puppets driven by laws and circumstances beyond our control. We leaned decidedly in favor of Libet in that debate, but again concluded that a final resolution of the issue would require a conceptual framework illuminating the interface between objective reality and the subjective source of volition, namely the mind.

As we saw, it is possible to explain the Libet experiment by positing discrete levels of intentionality, with broader and less immediate levels being relegated to the unconscious mind. Where this generalized unconscious level of self-determination is responsible for the integration of specific acts, it must initiate the preliminary stages of any voluntary act and leave only final clearance to the conscious mind; logically, this is the only way that meaningful volition can work. Indeed, the role of the unconscious is simply to remind the conscious mind of its chosen goals and notify it of an impending opportunity to act in such a way that some long-term goal will be wholly or partially achieved. Unfortunately, it is not easy to see how this applies to Newcomb's paradox.

The Newcomb scenario, in which a predictor (Nukem) unerringly forecasts the behavior of human subjects, and the Libet experiment, in which the subject's own nervous system seems to play the role of "predictor" in beating consciousness to the punch, have more than a little in common: where it is assumed that a future decision process is being passively read rather than actively caused by the predictive agency, they both require that the predictor make a counterintuitive leap through time. In Newcomb's paradox, the predictor must leap into the future and back again with information about a future state or event, while in Libet *et al.*, the subject's unconscious mind must leap forward in time in order to "predict" an instance of conscious volition and its outcome (otherwise, it would have to usurp conscious volition entirely).

While the Libet experiment was explained in terms of orders of intentionality arranged in a neural control hierarchy, with higher-order (inclusive) neural relationships controlling lower-order (included) neural relationships, Newcomb's paradox seems to resist such an explanation. This is partially because it is not confined to the brain of a single subject, but deals with the interaction of two non-cooperating subjects in a general environment containing both of them ... an environment not so easily stratified. A major difference between the two scenarios is that in order to win the Newcomb game, Nukem must predict a specific act in advance, while ordinary volition requires merely that reality at large provide alternative outcomes for all possible acts and their timing ... provide "multiple alternative futures" in one of which a given act will automatically result. But to some extent, certain essential features apply to both scenarios. For volition to work in either case, the body and environment must predictably mirror the neural dynamics of the brain. In other words, the picture of reality in our heads must accurately depict reality as it really is, and reality must in turn allow us to form reliable pictures of it.

What "pictures of reality" are we required to have in our heads? First, we must have a picture that represents our bodies. At coarse resolution, this "picture" is a brain map called the *motor homunculus*. If the human body were a puppet and motor nerves its strings, the motor homunculus would be the internal wiring diagram that tells the brain which strings to "pull" to cause particular voluntary movements involving particular parts of the body. Second, we must have an immediate, real-time picture of the environment with which our bodies interact and in which our movements have meaning. Conveniently enough, we ordinarily get such a picture through our eyes and other sense organs. And third, we must have at least a rudimentary understanding of how the world works so that we can predict the outcomes of our actions. This picture is gradually learned by experience and augmented by theoretical reasoning.

Our success in negotiating the often-treacherous road of life depends largely on the accuracy of our pictures of reality and how closely our actions conform to them. To this extent, reality is like the Newcomb predictor. If it "predicts" that our mental pictures of it are faithful and that we will act in conformance with them, it rewards our volition, while if it predicts that our pictures are inadequate, it frustrates us instead. Newcomb's paradox involves an appreciable time interval between prediction and outcome, in the middle of which resides the act that must be predicted. But in the Libet scenario, and in fact for volition in general, the "prediction" is actually *exhaustive provision* for a full range of outcomes. No matter what we do, the world must have our reward or frustration waiting for us in advance; it can make it up on the spot no more than we can choose our goal-oriented actions without advance planning. The burning question is how the world makes this "exhaustive prediction" with such unerring accuracy.

One might object that since it is possible to build a machine that rewards those who use it correctly and frustrates those who do not, and since the universe may be exactly this kind of machine, there is no need for prediction or exhaustive provision. After all, machines function automatically. But this simply begs the question, for the "automatic" operation of any machine depends on preexisting laws of mechanics. Were the laws of mechanics to suddenly break down, so would the machine. So as a first step, the laws of mechanics must be reified and maintained. Unfortunately, reifying and maintaining complete laws of mechanics is a cosmological issue very different from building a machine. When one builds a machine, there is a preexisting nomological framework of which one may take advantage; in building the nomological framework itself, there is not (if there were, then instead of being excused from delivering an explanation, we would simply have to explain the more basic nomological framework).

So the world's exhaustive real-time "prediction" of our actions, and the way it rewards or frustrates us in response, is ultimately synonymous with the laws by which it enforces its own self-consistency. Prediction and control are passive and active sides of the same coin; the laws that allow the world to write itself, i.e. evolve, also allow it to read itself. Thus, the laws of physics double as the means by which the world detects, analyzes and evaluates our actions, and the means by which it "predicts" and readies the outcome. In providing general causal mechanisms by which the world can yield an appropriate result for every action, these laws also provide means by which the world can detect and distinguish one action from another. Without such means, the concept of "action" would have no meaning, and the results of particular actions would not even be regular, much less "appropriate".

In other words, maintaining a set of unbreakable laws, e.g. the laws of physics, amounts to predicting (or exhaustively "meta-predicting") that anyone who tries to break those laws will definitely end up a loser, while predicting different outcomes for different actions entails a set of laws that allow actions and outcomes to be distinguished. Our real-world faith in these laws is analogous to our faith in Nukem's predictive infallibility, and our willing obedience to them is analogous to our acceptance of his predictive powers as a strategic criterion. Perhaps the main difference is that while there may be doubt about whether we can beat Nukem at his own game – his existence may, after all, be incompatible with necessary features of reality – there is very little hope that we can beat reality itself. For while we are physically separate from Nukem the Magnificent, reality contains all living beings and does not permit them to violate its laws.

As we observed, Newcomb's paradox pits the principle of induction against the idea that time flows exclusively from past to future. Either we can reason from local reality to global reality, *or* we can continue to think of time as nothing but an orderly past-to-future sequence, but not both (as we can see from the way these principles support conflicting decisions in the Newcomb scenario). Because the principle of induction permits us to attribute generality to laws active in our local vicinity, it has a "meta-predictive" aspect; it lets us "predict" that observations distant in space or time, regardless of specific content, will be similar to local observations (although this principle is troubled by the aptly-named "problem of induction", we use it out of necessity anyway). But as we have just observed, prediction and control are passive and active sides of the same coin; in some respects, enforcing laws is equivalent to "proactively predicting" the outcomes of a set of possible actions. Thus, the predictive principle of induction is in some respects equivalent to a control function that distributes general laws over space and time.

Do science and mathematics offer us any insight on matters of control? As it happens, all of this is right up the alley of a relatively young field of mathematics called *cybernetics*. Cybernetics, billed by its creator MIT wunderkind Norbert Wiener as "the science of control and communication in animal and machine", pits *freedom* against *constraint* in biological and engineered systems consisting of multiple control strata, with higher strata dominating lower ones in overall self-regulatory relationships. The central dynamical concept is the *feedback loop*, whereby actions or instructions issued by a control object affect a controlled object, which sends information on its new state back to the controller, which then updates its routine to further maintain or adjust the state of the controllee (and round and round they go in more or less complex flow diagrams). Cybernetics, the mathematical bedrock of the computer age, is about as close as mainstream science has yet come to forming a mechanical picture of purposive, goal-oriented behavior.

However, cybernetics goes only so far in its explanations, confining itself to mechanical or biological systems more or less analogous to electronic computers. Regarding the pre-mechanical or metaphysical phase of control and communication, the phase concerned with selecting, enacting and distributing laws and forces that allow machines to be built, cybernetics yields only silence. Where a system is completely self-contained, cybernetics explains it intrinsically on a mechanical level through internal feedback; where a system is not self-contained, cybernetics pursues mechanical control and feedback past the systemic boundary into the environment. But beyond this, cybernetics makes the usual round of too-convenient assumptions, e.g. that it is possible to regress from one control level to another *ad infinitum*, and that because explanatory regression through successive control strata never terminates, pre-mechanical or metaphysical considerations need never be addressed. The system of interest is simply placed within a nested series of mechanical systems-within-systems through which control and feedback can be plotted.

But the universe is more than a mechanical system; as we have established, it is a nomological "predictor" of all that it contains. The universe is so powerful a predictor that unlike Nukem or one of his subjects, it does not even need to know what specific act a person will commit in advance; it can simply allow for all possible acts by providing a possible outcome for every one of them. As Newcomb's paradox reminds us, we usually think of time as one-dimensional and one-directional, a linear dimension along which the present moves from past to future. Unfortunately, this kind of time is simply too limited to support nomological meta-prediction, providing means and explanation for neither its own existence not that of volition. Such an explanation, along with that which is to be explained, requires the existence of a control level above that of time itself ... a level representing hidden temporal structure.

What is the nature of this stratification? Cybernetics can offer some pointers. First, let's try to get a cybernetic view of a generic competitive scenario. Consider a 1 or 2-player game based on a computer simulation or *virtual reality*. There are two control levels in this game, one simulated and one simulative, to which correspond the roles of "subject" and "programmer". The programmer, being in charge of a computer on whose monitor runs a simulation containing the subject, controls the subject and his fate. In fact, as far as the subject is concerned, the programmer is a veritable "master of time", having the ability to fast-forward, reverse and replay the simulation at will. Therefore, the object of the game is to occupy the role of programmer. In addition, there is a third "role" to consider: that of the game's Designer. Because the Designer makes the rules that the player(s) must obey, this is an important role indeed.

There are four possible positions in the 2-player version of this game:

1. Player A and player B are both in the role of subject. In this case, neither controls the game, and it's a toss-up.
2. A is in the role of programmer and B is in the role of subject. In this case, A controls the game and therefore wins.
3. B is in the role of programmer and A is in the role of subject. In this case, B controls the game and therefore wins.
4. A and B are both in the role of programmer. In this case, both A and B control the game, and because neither controls the other, they are forced to cooperate. This leads to a draw (or a mutual win or loss, depending on how the game is structured).

Now let's see who or what we can cast in these roles. First, consider Newcomb's paradox. Where Nukem is player A and his subject is player B, the Newcomb scenario corresponds to position 2 above. Nukem – who always wins – is definitely in the role of programmer, and because he determines the outcome whether or not the subject chooses to cooperate, he is always in a higher control stratum (this precludes position 4). What is it about Nukem that makes it possible for him to win this game 100% of the time? That, of course, is a question for science fiction authors. For present purposes, the important part is what cybernetics tells us about where he must be sitting: somewhere above the subject in a control hierarchy.

Now let's consider the Libet experiment. In this scenario, there is only one player, the experimental subject. In the *solitaire* version of this game, a lone player must play both roles, and the only possible object of the game is therefore self-control. But what is self-control? It is precisely that which enables voluntary action. That is, the experimental subject, in attempting to commit a string of voluntary acts, is trying to simultaneously put himself in the roles of programmer and subject so that he can control and perform his own movements. According to our explanation for the peculiar timing of the neural events involved in this endeavor, the subject executes his duties as "programmer" in two phases relegated to his unconscious and conscious mind respectively.

Notice what this says about the relationship between the Newcomb and Libet scenarios. In displacing the player from the role of programmer, Nukem is either usurping the player's self-control and thus depriving him of free will, or making sure that whatever self-control the player insists on retaining turns out to be disappointingly unprofitable. In the latter case, Nukem must do something more impressive than merely controlling the player; he must control the contents of a certain box to ensure that the player is robbed of success and thereby punished for his or her willfulness. In other words, in making his prediction, Nukem must control some part of the subject-environment complex in a way that violates our intuitive notions of time and causality.

What about the Designer of the game? Because the Designer makes the rules, and because rules are analogous to laws, the Designer's real-world analogue is the universal principle responsible for selecting, distributing and enforcing the laws of reality, and thereby setting the stage for volition. This represents a higher level of control than those occupied by programmer and subject; the programmer derives his "programmatic" ability to see into and/or control future states of reality, subjective and objective alike, wholly from the Designer. That this applies to volition in general should now be clear as a bell.

(Of course, there still remains a problem: we have not yet decided whether or not free will really exists. But if the set-up were all that easy, the free will question would have been definitively answered a long time before we ever got around to it.)

Where does it all lead? To exciting new extensions of cybernetics and other fields of mathematics? To new inroads in physics and cosmology? To new insights regarding theological concepts like God, the human soul, and ethical refinements based on an enhanced understanding of good and evil?

As a matter of fact, yes it does. But why not leave a little something for later?

Solutions for the Problems of Free Will, Good and Evil, Consciousness and God

For three chapters now, I've subjected you to merciless philosophical, neurobiological and mathematically-oriented disquisitions on free will, even arranging your involuntary participation in a virtual game show and a scientific experiment. As a reward for your long suffering, we'll now try to extract some momentous implications from the little thread we've got going. Specifically, we're going to show that free will definitely exists. Then we're going to explain that while this doesn't necessarily mean that human beings possess it in any immediate sense, there are good reasons to believe that they do. And then we're going to discuss a few important philosophical corollaries.

First, let's review the virtual reality game described in the last chapter. The point of introducing this 1- or 2-player Game was to provide a standard analytical framework for Newcomb's paradox and the Libet delayed-choice experiment. The Game has two control levels respectively inhabited by a "subject" and "programmer". The subject inhabits a simulated world under the control of the programmer, who is able to fast-forward, reverse and replay the simulation at will. The first step toward winning the Game is to occupy the role of programmer and thereby gain control of the outcome (sadly, there are not yet any detailed instructions for doing this). The *solitaire* version requires that a lone player play both roles, making self-control, i.e. personal volition, the only possible object of the Game.

In addition, we defined a third "role", that of the Game Designer who makes the rules that the player(s) must obey. These rules include the laws of causality that operate on each level, and the laws governing the relationships among control levels. Because the Designer in effect occupies a third control level, there are three control levels altogether. Control flows downward through these levels, from Designer to programmer to subject. Due to the quality and quantity of control associated with the role of Designer, the Designer has a clear theological analogue: God. In the context of the Game, denying the existence of the Designer would amount to asserting that despite the complex structural organization of the Game, it is random in origin.

Let's begin by reviewing a few historical viewpoints regarding the free will issue. First, we have the idea, dear to the hearts of most of us and especially to those who framed the U.S. Constitution and other rational codes of law and ethics, that human beings have free will. While we acknowledge that there are laws of causality that apply to everyone regardless of time or place – e.g. "for every action there is an equal and opposite reaction" – these are not sufficient to control our behavior, leaving us free to "make our own laws" when it comes to our personal preferences. This corresponds to the *solitaire* version of the Game, but with a tacit restriction on the fast-forward, reverse and replay options: they are usually considered to be exclusively "mental", i.e. confined to our respective thought processes.

Then we have *determinism*, the idea that the laws of causality determine our every move. Determinism holds that we resemble billiard balls rolling around and colliding on a barroom pool table, bouncing off each other under the impulses imparted to us by environmental cue sticks. Like the cue sticks themselves, the internal impulses triggered in our minds and bodies by external stimuli are completely motivated by the laws of physics, reductively including the laws of genetics, biology and psychology. There is no causal gap in which free will can find a toehold. This corresponds to a version of the Game in which the Designer leaves no open controls for the programmer to use; the structure of the Game, as fixed by the Designer, absolutely controls the simulation and the subject. (While Laplace, widely considered the father of determinism, once remarked to Napoleon that a deterministic universe "has no need" of a Designer, he thereby left himself with no way to account for the mathematical laws of causality offered as a replacement.)

Juxtaposed to determinism is indeterminism or *randomism*, the idea that causality is only an illusion and that everything, including human behavior, is utterly random. This represents a bizarre mutation of the Game in which one or both of the following conditions hold:

1. The Designer is absent, dead, or out to lunch; the Game, and its rules of causality and interlevel control, exist solely by "chance". An obvious drawback of this viewpoint is that it fails to account for the probabilistic laws of chance themselves.
2. The rules of causality do not exist at all; they are merely *supposed* to exist by the subject and break down where the subject's powers of self-delusion end, beyond which point chance rules openly. In other words, when a subject fancies that he is "making a decision", he is really rolling the dice, and somehow, his sequential delusions of intent, action and outcome perfectly match the dice roll. This, of course, seems to give a whole new meaning to the term "improbable".

In addition to free will, determinism and randomism, there is another doctrine worth considering: *predestination*. Predestination is the idea that God can look at a person, measure his or her tendencies towards goodness and/or badness, predict how the person will decide to behave in his or her particular setting, and consign that person's soul to heaven or hell on that basis. Where God's insight relies on the existence of laws that transform information on a person's innate tendencies and environmental conditions to information on behavior – these laws are ostensibly what He uses make His prediction – predestination amounts to determinism with a theological twist. Where God determines both the personality of the subject and the laws that determine his behavior, this boils down to the following scenario: God makes a person, sizes up His handiwork, says either (a) "Man, am I good!" or (b) "Screwed up again!", and tosses His newest creation onto the pile marked "Heaven Bound" or "Born To Lose". If you happen to be a Calvinist, this is your lot – or should that be lotto? – in life.

However, there is another strain of predestination, favored by (e.g.) the Roman Catholic Church, that does not rely on determinism. According to this version of the doctrine, God might instead do one of the following:

1. Scan the subject for freely-formed higher-order intent. As we recall, persistent higher-order intent is what the experimental subject generates in agreeing to participate in the Libet experiment and move his hand repeat-

edly at will. To capture the theological flavor of predestination, compare this to "what the subject generates in freely selling his soul to the devil and agreeing to commit a lifelong string of felonies and misdemeanors".

2. Let the subject play out and determine his own simulation, then use the rewind control to back up to the subject's moment of birth and invisibly stamp him "Accepted" or "Rejected". In this case, God is simply jumping forward in time to observe the subject's timeline, and then jumping back through time to affix the proper invisible label to its origin. This kind of predestination leaves room for free will.

Notice that where free will is absent, the mind is reduced to a mere byproduct of deterministic material reality. Consciousness, including the psychological sensation of intentionality and self-awareness in general, becomes a kind of meaningless sideshow compulsively played by our irrelevant "minds", as we are forced to call them, to interpret that which they cannot affect. The idea that the mind is just a side effect of objective physical processes is called *epiphenomenalism*. One can, of course, split hairs over the question of whether the mind is truly intrinsic to our material brains and bodies, or just extra metaphysical baggage that is somehow tied to them by some kind of ethereal thread. But either way, if intentionality has nothing to do with the structure of reality, then matter and the laws of physics are all that really count. Our "minds" are just along for the ride.

We've enumerated the above scenarios to show that the cybernetic paradigm, involving feedback among layers of control, clarifies the distinctions among traditional approaches to volition. This is a bit surprising; while cybernetics is a respected branch of modern science, the position of the scientific establishment as a whole is evidently nowhere near as sophisticated. It is based on a naive understanding of causality which holds that with respect to anything not locked into a determinative causal relationship, randomness prevails. For example, modern physics characterizes reality on three scales: the overall cosmic scale, the macroscopic scale of ordinary objects, and the ultramicroscopic scale of atoms and subatomic particles, to which it applies the rules of relativity, classical mechanics and quantum mechanics respectively (where relativity is understood as a classical theory). Unfortunately, the dichotomy between "classical" and "quantum" reality is essentially the same as that between causality and randomness or determinacy and indeterminacy. It thus effectively excludes all of the above scenarios except determinism and randomism, leaving the existence of free will with no apparent possibility of scientific explanation.

Determinacy and indeterminacy ... at first glance, there seems to be no middle ground. Events are either causally connected or they are not, and if they are not, then the future would seem to be utterly independent of the past. Either we use causality to connect the dots and draw a coherent picture of time, or we settle for a random scattering of independent dots without spatial or temporal pattern and thus without meaning. But there is another possibility after all:

self-determinacy. One can be the author of one's own personal "causality", setting up a volitional feedback loop between one's will and one's goals, where "goals" include everything from spontaneous bodily movements to the fruits of long labor and well-laid, well-executed plans.

Self-determinism is a rather subtle concept. It means that one determines one's own path *independently* of general laws of causality, where "independence" describes a situation in which one is constrained by, but also free to exploit at will, the laws of physics and biology. It therefore implies that neither randomness nor the laws of causality have the final word in the determination of reality; there is enough room between them for human beings, who are themselves parts of reality, to "get a word in edgewise". But how is this possible, given that the general laws of causality, linear and chaotic, are the only things short of "quantum randomness" that can possibly account for the structure of reality as a whole? It is possible only if there is more to the overall structure of reality than randomness or the laws of causality can fully determine, and only if human beings possess a means of shaping this extra structure at will.

To expand further on the concept of volition, we must take a closer look at its meaning. When we do, we find that "meaning" is precisely what volition entails. Concisely, to "have meaning" is to play a critical role, to make a crucial difference, in something of value, where *value* is reckoned in practical, emotional, intellectual or spiritual (edificative) terms. So to be meaningful, volition must enable one to advance, in a way transcending the laws of causality and imparting value beyond that of causality alone, toward freely-chosen goals which themselves possess meaning or value. But for goals to have meaning, *the system in which they reside* must have meaning. If reality were meaningless, there would be no basis on which to ascribe meaning to anything in it; "valueless value" is no value at all. So if meaningful volition exists, reality itself must have meaning, and there must be an objective, meaningful scale on which its value can be gauged.

But now there arises a problem: in order to be "objective", something must be *real*. On what real, objective scale can reality be measured? If we define "reality" in the most straightforward and tautological way, namely as "that which contains all and only that which is real", no such thing as an external scale is possible. For by our definition, anything real enough to quantify reality from the *outside* is already included in reality and therefore on the *inside*! Since this is a logical contradiction, the idea that reality can be externally evaluated is logically invalid. This implies that in order to possess meaningful existence and thus be capable of supporting meaningful volition, reality must be "self-evaluating". All of its value scales must be internal to it, and its meaning must be intrinsic.

In fact, we can go somewhat farther. In the traditional view of science, that which is not random must be created or selected according to ambient (externally extended) principles of causality. But where external causality requires external scales against which to be defined, it can no more possess real existence than the external scales themselves. Again, anything real enough to affect reality from the outside is already included in reality and therefore on the inside, and since this is a logical contradiction, the idea of external causality is *a priori* invalid with respect to reality. This implies that in order to possess meaningful existence and therefore support meaningful volition, reality must be "self-caused". But this comprehensive kind of self-causation entails an even deeper form of self-determination than that usually associated with volition; reality must not only determine its own structure independently of external causality, but must do so starting with nothing but itself! In other words, the universe must possess a global analogue of free will that lets it internally define and calibrate the very scale on which its intrinsic value is internally measured.

But wait a minute. Aren't we ignoring the possibility that the universe is simply "random", i.e. uncaused? Not really, for externally speaking, that's exactly what it is! In deducing that the universe is unaffected by external causality, we find that it is externally acausal or "random" in that specific sense of the term. The operative distinction, of course, is that which holds between internal and external causality. And since our observations of a coherent, well-regulated, profoundly ordered universe rule out the possibility of internal randomness from the start – there is no way that a system as coherent and complex as the real universe can be accurately described as "randomly disordered" – we've got the issue of cosmic randomness sufficiently well-covered for present purposes.

So reality, being self-determinative but externally unconstrained, possesses a global, self-enabling analogue of free will that generates its own means of realization. If these means can be utilized by human beings *within* reality, then human beings possess free will, and because they are included in reality, they can use it to contribute to the realization of the global Self-structuring imperative. On the other hand, even if human beings cannot avail themselves of free will, they can still be used *by* it for the same purpose. In either case, human beings are integral parts of reality that contribute to its structure, and must either be using the inherent freedom of reality to do so, or be freely used as tools by some higher level of realty to the same end. So while the using-versus-used question remains up in the air, one fact has nevertheless been rationally established: whether it belongs exclusively to the universe or to man as well, *free will exists*. (Q.E.D.)

Now let's look at some convincing cybernetic evidence that we participate in the self-creation of reality. Because intentional self-creation entails an internal stimulus-response dynamic consisting of feedback, any self-configuring system needs internal sensors (agents, internal self-proxies) capable of not only recognizing and affecting its state from local internal vantages, but of responding to higher-level instructions tending to enforce global structural criteria. Moreover, the system must possess a stratified utility function allowing it and its agents to prefer one possible future over another. Human beings and other intelligent life forms are useful to reality on both of these counts. So the first criterion of reality is the possibility, and in fact the inevitability, of the existence of "sensors" just like us ... sensors with an advanced capacity to recognize, evaluate and respond to internal states of the system.

How, in general, would the universe self-configure? It would select itself from a set of internally-generated, internally-refined structural possibilities in order to maximize its self-defined value. In the (somewhat inadequate) terminology of quantum mechanics, this set of possibilities is called its *quantum wave function* or QWF, and the utility-maximizing self-selection principle is traditionally called *teleology*. In exploiting this self-actualization mechanism, human beings would select their specific goals from the global QWF according to their own specific self-selection principles or "teleses". In the course of being realized, these individual teleses would interfere with teleology (and each other) in a constructive or destructive way, depending on whether they and their specific methods of implementation (modes of interference) are teleologically consistent or inconsistent. In this way, the "good", or teleologically constructive, may be distinguished from the "bad", or teleologically destructive. I.e., free will would give human beings a real choice between good and evil ... a choice like that which we already seem to possess.

For example, say that you choose great personal wealth as your goal. Whether this goal is inherently good, bad or indifferent depends on whether or not its realization would necessarily lead to a net increase or decrease in global utility or value, and thus interfere constructively or destructively with teleology (in which cases it is good or bad respectively). Suppose that it is not inherently bad. Then whether its *specific realization* is bad depends on whether your means of realizing it will be prohibitively costly to the aggregate consisting of you, those near you, mankind as a whole and the entire Planet Earth (if they are, then the ends do not quantitatively justify the means). As we recall from the last two installments, any goal that can only be reached through a more or less complex sequence of actions is assigned to a higher, long-term level of intentionality that suggests and coordinates one's actions in light of it, with the subject retaining "power of veto". To qualify as "good" or at least "indifferent", you must use your power of veto to ensure that your individual actions meet the same teleological consistency criterion that applies to your overall goal.

But hold on there – doesn't the question of good and evil automatically evoke the topic of theology? Of course it does. "God" is just the name we give to the protean creative principle that generates and enforces teleology, or equivalently, that Entity whose "Will" equals teleology. Since this is the same entity whose self-structuring operations are guided by teleology, namely reality, we have a measure of equivalence between reality and God. This makes sense for the following reason: reality can be mathematically modeled as a *Self-Configuring Self-Processing Language* (SCSPL), a dynamic system possessing not just a global analogue of volition, i.e. teleology, but a global analogue of self-awareness as well, i.e. SCSPL self-processing. So where "consciousness" is an attribute with active and passive dimensions amounting to volition and self-awareness, "God" and "reality" are equally good answers for the question "What entity possesses the global form of consciousness described by SCSPL?" In other words, it's just as we might have expected: God is the ultimate reality.

This, of course, brings us back to the notion that "God" is a real-world analogue of the Designer in our cybernetic Game. But now the game is reality, its object is the Designer's own Self-creation, and any number of players occupying any number of lower control strata are just participants in the Self-configuration of the reality they inhabit (for now, we'll ignore the long-running debate over whether God can inject Himself into the bottom-level simulation as a "messiah", simply answering "yes" and leaving it at that). When we use the freedom we inherit from reality at large to properly choose our teleses and exercise our powers of veto, we favorably tilt the balance between teleology and anti-teleology, good and evil, and thus inherit meaning in the bargain. This meaning enriches our lives and sustains the conscious identities with which reality has endowed us as its internal proxies.

Notice that our answer for the question of free will has brought with it the answers for some bonus questions: "What are good and evil?", "What is consciousness?" and "What is God?" (We've left out a few technical details, of course, but that's only because we need to reserve a few surprises for future essays.) Given that you could spend your entire life's salary on popular and academic literature and still not get a tentative logic-driven answer for even one of these questions, how's *that* as a payoff for four straight chapters on the science and philosophy of volition?

By any reasonable measure, you've done quite well for yourself.

PART TWO

On the Perils of Metaphysical Skysurfing Without a Parachute

or "Je n'avais pas besoin de cette hypothèse-là"

Although we often forget to act like it, human beings are indisputably the brightest bulbs on the evolutionary tree. So in describing where we stand in relation to the rest of the animal kingdom, we tend to stress our intellects, asserting that we are on top of the food chain, the clothes chain and the housing chain because virtually every other species that crawls, hops, swims or flies is a dim bulb by comparison. No matter how much stronger, faster, or keener of eye, ear and nose other species may be – and pound for pound, they are virtually all much stronger, faster or sharper-sensed than we – we are the ones who climbed to the top of the evolutionary ladder ... spectacularly outsmarted the biological competition ... cleaned up in the All-Time Grey Matter Sweepstakes.

So much for the creatures that preceded us on the timeline of creation. What of those which came after us ... the creatures of metal, plastic and silicon that we birthed to aid us in our physical and mental labors? Here, the same applies *a fortiori*. Regardless of horsepower, clock rate or memory capacity, none of our machines can hold a candle to even the weakest among us. For machines merely obey the laws of physics, while we alone can study, formulate and manipulate those laws to our own advantage. Ours are the hands on switch and dial, on mouse and keyboard, on tiller and wheel, on gas hose and power plug. Without our protection and guidance, even our most durable machines would be naturally selected for fast extinction. We are human not just because of our intelligence, but because of our will to control ... our will to supremacy.

So far, so good. But then we collectively make a daring leap. Not satisfied with our quantitative superiority on the scales of intelligence and control, we philosophically transform it into a qualitative difference setting us forever apart from other animate and inanimate entities. Historically, this difference was boldly expressed in statements like "human beings have souls; machines and the lower animals do not." Because this is an age of science, and science pretends to have no use for that which cannot be empirically demonstrated, such statements are now heard a bit less frequently. But the assumed difference lingers on in our minds, implicit but ineffable, a tacit but unbreakable confidence in the unique nature of human consciousness and its sovereign place in the eternal hierarchy of being.

This putative qualitative difference is central to our ethos. It justifies the lion's share of earthly resources that we take, use and waste. It is our philosophical "license to kill" ... to overpopulate the planet, to extinguish other forms of life, to pollute the environment. It is central to what we call the "meaning of life" and to our own meanings as individuals. And because it is absolutely essential to our collective self-importance, we like to think of it as anything but accidental. Indeed, if it has any real significance, it cannot be a quirk of fate, a chance pattern of spots on the cosmic dice. Our claim to preeminence must be based not on mere physical dominion, but on *metaphysical distinctiveness*. We are what and who we are because this is how it was *meant to be*. Despite having disposed of pre-Copernican anthropocentric cosmology several centuries ago, we yet embrace a metaphysical analogue of it that refuses to die.

Unfortunately, certain trends in science threaten to kill it nonetheless. Take, for example, our vaunted intelligence. Dolphins, porpoises and other cetaceans display playfulness, task learning ability and mental agility impressive even to humans. A strain of lab mice called "Doogie mice" – remember Doogie Hauser? – are bred for high densities of brain proteins called NR2B receptors that enhance their performance on task-based animal intelligence tests, narrowing the IQ gap between humans and rodents. Researchers on human intelligence are even now locating human genes associated with high IQ, thus accumulating the knowledge to genetically engineer super-smart children like the precocious

Doogie himself. And pharmaceutical companies are already analyzing the effects of such genes to develop "smart drugs" expected to cause IQ spikes of up to 50 extra points. This gives rise to a disturbing question: how special is intelligence when it can be dispensed from a bottle or syringe to humans and animals alike?

But wait! Even if task performance has a biological basis, could there not be something special in our use of *language*? Well, not necessarily. Human linguistic ability has been localized to specific brain modules, like Broca's and Wernicke's areas, that seem to have counterparts in the brains of some animals. For example, a monkey's premotor cortex contains an area known as F5 that is considered analogous to Broca's area and contains special cells, called "mirror neurons", that appear to be present in human brains as well. Meanwhile, certain African grey parrots can seemingly hold rudimentary but meaningful conversations involving both quality and quantity. And great apes like gorillas and chimpanzees, with whom we share up to 98% of our DNA, regularly communicate using sign language vocabularies of thousands of words ... if certain studies are correct, enough to rival the functional vocabularies of many humans.

But does primate language use not differ from ours in its lack of *empathy*, and is empathy not a distinctively human "spiritual attribute"? Not by the look of certain recent findings. Having noted that the brain of a human observer generates excitatory patterns that mirror patterns in the brain of someone performing a task, a team of researchers has examined monkey brains and found a new class of brain cells in which such mirrored patterns reside. (Can you guess their location? That's right ... F5.) The sympathetic firing of these "mirror neurons" bids fair to be the neural equivalent of empathy. No longer is empathy a moral and spiritual attribute of personhood, a *je ne sais quoi* of human social interaction. Like the jerk of a tapped knee and the dilation of a pupil in sudden darkness, empathy appears to be just another biological phenomenon reducible to the ubiquitous *pas de trois* of proton, neutron and electron in random chunks of matter.

So much for intelligence and empathy as evidence for human spiritual ascendancy. What of spirituality itself? As it happens, world literature affords several telling adumbrations of modern scientific research on this topic. In his tragicomic novel *The Idiot*, Fyodor Dostoyevsky described the expansive feeling of beatitude experienced by his protagonist Prince Myshkin just prior to the onset of epileptic seizures. Insofar as Dostoievsky himself suffered from epilepsy, these fictionalized accounts seem to have been taken directly from the author's personal experience ... and the great Russian novelist was far from alone. Many patients afflicted with temporal lobe epilepsy have reported that when a spell overtakes them, they "see God" or experience "enlightenment". Searching for the explanation, neurologists were finally able to find a connection between temporal lobe epilepsy and spontaneous religious feelings: epileptic seizures can stimulate a structure called the "God module" (!) in the temporal-limbic region of the human brain.

Inspired by such successes, science is now looking at the neurological correlates of religious experience in general. This emerging field already has a name, *neuro-theology*, and a central premise: that feelings of spirituality are due to the genetic wiring of the human nervous system. To illustrate, neuro-theological researchers recently conducted an unusual experiment in which a Tibetan Buddhist tugged on a strand of twine as he entered a state of meditative transcendence. Upon this prearranged signal, a researcher injected a radioactive tracer into an IV line attached to the subject's arm and scanned his brain with a sophisticated machine known as SPECT. SPECT's circuits registered a dramatic reduction in neural activity within a specialized region at the top rear of his brain ... an area known to be responsible for translating sensory data into comprehension of the boundary between self and environment. Having shut down his "self-delimiter module" by meditative sensory deprivation, the subject had destroyed the subjective distinction between internal and external reality and become "one with the universe".

The effect is nonsectarian. According to brain scans, praying makes Franciscan nuns feel "oneness with God" by suppressing the activity of their self-delimiter modules. And the neural aspects of religion are by no means confined to meditation and prayer. A 1997 Japanese study showed that the rhythmic repetition of ritual chanting and dancing causes a certain brain component, the hypothalamus, to generate feelings of serenity or arousal. For example, while repeatedly chanting a mantra transports a yogi to a state of deep inner calm, Sufi mystics feverishly dance themselves to states of excitement in which they feel like "live wires" crackling with the infinite energy of the universe. But does this provide evidence for anything of an objectively transcendental nature? The response of neuro-theology to this question is just that of Laplace to Napoleon regarding the existence of God: *"Je n'avais pas besoin de cette hypothèse-là."* Given the physical nature of that which is to be explained, there is no need for such a hypothesis.

The elevated status of human consciousness grows ever more precarious upon the swaying tightrope of metaphysical distinctiveness. First came evidence for the physical basis of consciousness in the various aphasias and apraxias associated with brain lesions; then came the frontal assault of operant conditioning on free will; now we have brain-scanned nuns and the possibility of genetically engineered intellectual and spiritual "geniuses". Intelligence, language, empathy, spirituality ... all of our claims to metaphysical distinctiveness seem to be falling beneath the heartless scythe of physical reductionism. So where will we go now, at the dawn of the New Millennium, for renewed confirmation of our "specialness" in the scheme of things?

Fortunately, we are not yet facing an ideological dead end. For it seems that just ahead on the intellectual horizon looms a new science of metaphysics ... a logical framework in which the importance of humankind is unthreatened by reductionism, and in which the significance of human feelings and emotions is uncompromised by their correlation with lowly biological processes. Rather than declaring us the abject slaves of natural laws beyond our control, this framework yields a new understanding of space and time in which the very laws of physics can be viewed as an expression of our minds. Granted, this framework remains hidden despite its portentous approach. But if its ongoing delay contributes to our collective store of humility, perhaps this is not entirely a bad thing.

For now, we may simply observe that science consists of building accurate conceptual models of the world, and that the theory of models has thus become a crucial ingredient of scientific reality. Perhaps the most distinctive characteristic of model theory is its use of "metalanguages", or languages that talk about languages, to analyze the mappings, or correspondences, between scientific theories and their universes of discourse. Logically, such metalanguages amount to relationships between theoretical cognition and perceptual reality... or with a slight frame shift, between the mind and the real world of scientific observations. Because these relationships do more than just describe reality – because they turn out to be primary conditions of its existence – they distribute over it on all scales, uniting the ubiquitous subjective and objective sides of its nature. The universal distribution of these relationships implies the existence of a homogeneous cognitive and perceptual medium embracing the real universe and propagating the influence of our minds throughout the cosmos.

Examined within the distinctive logical structure of this medium, time and causality turn out not to be confined to the familiar past-to-future direction. Instead, the future-to-past direction becomes important as well. The bidirectionality of time implies that the universe is in a state of extended spatio-temporal self-superposition, each of its serial configurations, as defined at each successive moment of time, in contact with all others. At first glance, this seems to imply that the universe is completely determined ... that as Laplace believed, *every* state of the universe is implicit in *any* state. But not only is such an assumption unnecessary, it would ultimately lead to intractable inconsistencies. In fact, the logical structure of spacetime provides the universe with the wherewithal of being, endowing it with self-creative freedom and permitting it to rise from a sea of undifferentiated ontological potential.

What does this have to do with neuro-theology? Insofar as God is generally defined as the Prime Mover or first principle of reality, God is by definition real, and thus a part of the real universe. And because this first principle is by definition primary – because there is by definition no preexisting reality to contain it – reality cannot properly include it. So the real universe must coincide with God; scientific knowledge is theological knowledge, and science a form of theology. It follows that our ability to use science to better the plight of our species – our capacity to know God and to help others know God – is part of what makes us "special". And where true science is a key ingredient of our claim to metaphysical distinctiveness, our place in the hierarchy of being cannot be threatened by science done thoroughly, correctly, and without *a priori* restrictions on our rôle in the scheme of things.

But what is that rôle? Where does humanity fit into the scheme even now being revealed to us by science ... we, who have lately found ourselves reduced to the wearing of metaphysical sackcloth? It turns out that in a certain precise sense, we are microcosms, images of the universe within the universe. And because of the symmetric connection between source and image, cosmos and microcosm, we function as agents through whom the universe realizes its being. We are its children and heirs, poised on the threshold of adulthood and charged with shaping Destiny itself, with helping the living universe choose its form and content from a background of undifferentiated potential. Though eons removed from the moment of creation, we actively retrodict it, sending the power of our minds back through time to help the Prime Mover, our parent and provider, self-creatively embody the universe we inhabit.

Let us do well our filial duty. For by the partial identity of parent and child, we serve ourselves and each other in the bargain.

A *Very* Brief History of Time

I just had a chance encounter with a garden slug, and it got me thinking about time.

In this ostensibly inanimate, impersonal universe, a garden is a miracle. All the more so is a *garden* slug, an animal that can extract sufficient energy from the garden's vegetable matter to move from place to place under its own power. When one is in the right mood, watching the shimmering spotted slug slide over the mulch evokes the miracle of biology in all its splendor; the creature's pulsating aliveness is hypnotic. But then one recovers his bearings and realizes that this is only, after all, a garden slug, and that the ladder of biology goes much higher. The miracle of life has culminated in one's own species, *man*. Unlike the slug, whose nervous system has barely enough complexity to let it interface with the environment, a man's nervous system, nucleated by the adaptive and inventive human brain, can abstractly model its surroundings and project itself consciously and creatively through *time*.

A slug can learn. The small neural network that serves as its brain can be modified by sensory input from its environment, and the slug's behavior modified accordingly. To this extent, the slug "remembers" the input. But because its simple brain cannot form an internal model of its changing relationship with the garden, the slug cannot recognize its memories as "changes"; the state of its nervous system at any given moment can pass for all that it has ever known. Because the neural function by which the slug identifies *self* is instinctual and perceptual as opposed to cognitive – because the slug "defines itself" strictly by nonreflective instinctual processing of environmental stimuli – the dependent neural function *time* is limited to here-and-now. The slug recognizes no past self or future self on which to define an extended temporal relationship.

As the slug's primitive example shows, our awareness of time depends on the extent to which our mental models of reality reflect *change*. To see an object change, one must recall its former state for comparison to its present state, and to do that, one must recall one's former perception of it. Because perception is an interaction between self and environment, this amounts to bringing one's *former self* into conjunction with one's *present self*. That past and present selves can be brought into conjunction across a temporal interval implies that momentary selves remain sufficiently alike to be conjoined; that they can intersect at any given moment to compare content means that the intersection is changeless. So when self is generalized as *the intersection of all momentary selves*, it acquires a property called *time invariance*. It is the rock of perception, the unchanging observation post from which the net of temporal connections is cast and to which it remains anchored. Indeed, it is the fabric from which the net is woven, its relationship with the environment serving as the universal template for all temporal relationships.

Through learning, mental models *of* time evolve *in* time. As the brain's neural connections are modified and the strengths of existing connections are adjusted to account for new information regarding both self and environment – as it *learns* – its model of time changes as a *function* of time. In other words, the model changes with that which is modeled. If the brain is smart enough, then it can model itself in the process of being changed, and depict its own learning process as a higher level of time. But even as the self absorbs its educational history and deepens its reflexive understanding, it remains static at its core. Otherwise, it would lose temporal cohesion and fall apart. Since self is static, time too should possess a static description that does not change in the temporal flow it describes (if time were the water flowing in a river, then a static description of time would be analogous to the rocky banks that determine the river's course).

Such a description arises by *abstraction*. As cognitive models become more sophisticated, cognition becomes increasingly abstract; concepts become increasingly independent of the particular objects they describe. Among the first things to be abstracted are space and time. The most general abstract system incorporating both is a *language*. Although the term "language" usually refers to a natural language like English, it is actually more general. Mathematically, a *formal language* consists of three ingredients: a set of elements to be combined as strings (e.g., symbols, memes), a set of structural rules governing their arrangement in space, and a set of grammatical rules governing their transformations in time. Together, the latter two ingredients form the *syntax* of the language. It follows that neural, cognitive-perceptual, and physical systems can be described as languages, and the laws which govern them as their syntaxes. On a subjective level, time itself can be abstractly characterized as the *grammar* of the joint language of cognition and perception. The rules of this grammar are the general ingredients of subjective time.

Because time is defined in terms of transformations among spatial arrangements of objects, it is conceptually entwined with space. Thus, it is actually part of a linguistic complex called *spacetime*. Spatiotemporal relations exist on many levels; if level one consists of simple relationships of objects in space and time, then level two consists of relationships of such relationships, and so on. Because logic is *stratified* in much the same way, one can say that time is stratified in a manner corresponding to predicate logic. This must be true in any case, since any meaningful description of time is logically formulated. Spatiotemporal stratification allows time to be viewed on various scales corresponding to ascending series of contexts: e.g., personal awareness, interpersonal relationships, social evolution, evolutionary biology, and so on. The histories of people, institutions, cultures, and species are nested like Chinese boxes, with the abstract principles of each history occupying a level of temporal grammar corresponding to an order of predicate logic.

Because of the relation between self-awareness and temporal awareness, temporal stratification induces a stratification of self. What we have already described as the static intersect of momentary selves becomes a stratified relationship ... a terrace of temporal vantages conducing to long-term self-integration. As the self becomes stratified, the principles abstracted from higher orders of experience tend to be *objectivized* due to their generality, with science and philosophy among the results. Thus, the subjective and objective sides of reality – the self and the environment – tend to merge in a symmetric way. On one hand, the environment is absorbed by the self through experience, and the laws of nature are thereby abstracted; on the other hand, the self is projected onto the environment in such a way that it "selects" the laws of nature by analogy to its own internal laws. Either way, the core self tends to intersect with the environment as momentary selves are intersected within *it*. This brings the subjective and objective phases of reality – and time – into closer correspondence, blurring the distinction between them from an analytic standpoint.

As time grows more abstract, ways are sought to measure it, diagram it and analyze it numerically. This requires a universal depiction of space and time against which arbitrary processes can be differentially graphed and metered. Such a depiction was introduced by the Frenchman René Descartes in the first half of the 17th century. It was called *analytic geometry*, and it depicted time and the dimensions of space as straight, mutually perpendicular axes. In analytic geometry, any set of numerically-scaled space and time axes associated with any set of properties or attributes defines a *coordinate system* for assigning numbers to points, and simple processes appear as the graphs of algebraic functions. A few decades later, Newton and Leibniz independently discovered a new kind of mathematics, the *infinitesimal calculus*, by which to numerically quantify the rates of such processes. These innovations, which laid the foundations of modern science and engineering, suffice to this day in many practical contexts. Even though garden-variety analytic geometry was technically superseded by the Theory of Relativity – which was itself constructed on an analytic-geometric foundation – it gives a very close approximation of relativity in most situations.

Unfortunately, the conveniences of analytic geometry came at the price of *mind-body dualism*. This was Descartes' idea that the self, or "mind", was a nonphysical substance that could be left out of physical reasoning with impunity. For some purposes, this was true. But as we saw in the next-to-last paragraph, the *relationship* of mind to reality is not that simple. While the temporal grammar of physics determines the neural laws of cognition, cognitive grammar projects itself onto physical reality in such a way as to determine the form that physical grammar must assume. Because the *form* of physical grammar limits the *content* of physical grammar, this makes cognition a potential factor in determining the laws of nature. In principle, cognitive and physical grammars may influence each other symmetrically.

The symmetric influence of cognitive and physical grammars implies a directional symmetry of time. Although time is usually seen as a one-way street, it need not be; the mere fact that a street is marked "one way" does not stop it from being easily traveled in the unauthorized direction. Indeed, two-way time shows up in both quantum physics and relativity theory, the primary mainstays of modern physics. Thus, it is not physically warranted to say that cognition cannot influence the laws of physics because the laws of physics "precede cognition in time". If we look at the situation from the other direction, we can as easily say that cognition "precedes" the laws of physics in *reverse* time ... and point to the strange bidirectional laws of particle physics to justify our position. These laws are of such a nature that they can as well be called laws of *perception* as laws of *physics*.

Before we get to the final word on time, there is one more aspect of physical grammar that must be considered. Physical reasoning sometimes requires a distinction between two kinds of time: ordinary time and cosmic time. With respect to observations made at normal velocities, ordinary time behaves in a way described by Newtonian analytic geometry; at higher velocities, and in the presence of strong gravitational fields, it behaves according to Einstein's Special and General Theories of Relativity. But not long after Einstein formulated his General Theory, it was discovered that the universe, AKA spacetime, was *expanding*. Because cosmic expansion seems to imply that the universe began as a dimensionless point, the universe must have been *created*, and the creation event must have occurred on a higher level of time: cosmic time. Whereas ordinary time accommodates changes occurring within the spacetime manifold, this is obviously not so for the kind of time in which the manifold itself changes.

Now for the fly in the cosmological ointment. As we have seen, it is the nature of the cognitive self to formulate models incorporating ever-higher levels of change (or time). Obviously, the highest level of change is that characterizing the creation of reality. Prior to the moment of creation, the universe was *not* there; afterwards, the universe *was* there. This represents a sizable change indeed! Unfortunately, it also constitutes a sizable paradox. If the creation of reality was a real event, and if this event occurred in cosmic time, then cosmic time itself is real. But then cosmic time is an aspect of reality and can only have been created *with* reality. This implies that cosmic time, and in fact reality, must have created *themselves*!

The idea that the universe created itself brings a whole new meaning to bidirectional time, and thus to the idea that cognition may play a role in the creation of reality. As a self-creative mechanism for the universe is sought, it becomes apparent that cognition is the only process lending itself to plausible interpretation as a means of temporal feedback from present to past. Were cognition to play such a role, then in a literal sense, its most universal models of temporal reality would become identical to the reality being modeled. Time would become cognition, and space would become a system of geometric relations that evolves by distributed cognitive processing.

Here comes the surprise: such a model exists. Appropriately enough, it is called the *Cognitive-Theoretic Model of the Universe*, or CTMU for short (Langan, 2002). A cross between John Archibald Wheeler's Participatory Universe and the Stephen Hawking–James Hartle "imaginary time" theory of cosmology proposed in Hawking's phenomenal book *A Brief History of Time*, the CTMU resolves many of the most intractable paradoxes known to physical science while explaining recent data which indicate that the universe is expanding at an accelerating rate. Better yet, it bestows on human consciousness a level of meaning that was previously approached only by religion and mysticism. If it passes the test of time – and there are many good reasons to think that it will – then it will be the greatest step that humanity has yet taken towards a real understanding of its most (or least?) timeless mystery.

And so the circle closes. Time becomes a cosmogonic loop whereby the universe creates itself. The origin of our time concept, the self, becomes the origin of time itself. Our cognitive models of time become a model of time-as-cognition. And the languages of cognition and physics become one self-configuring, self-processing language of which time is the unified grammar. Talk about "time out of mind"!

And all this because of a little garden slug.

Which Came First?

Some people believe that children should be allowed to use their minds as freely and imaginatively as possible, without attention to the tedious laws of rationality. Others think that a child is never too young to get his or her first dose of logical and scientific reasoning. But in any case, a child with the intellectual maturity to ask a question like "which came first, the chicken or the egg?" is probably ready for a valuable lesson in logic and biology ... more of a lesson, perhaps, than many of us are ready to give. This little essay aims to change all that, and thereby protect you and your pint-size inquisitors from the perils of ignorance (and specifically, being recognized as an incurable case thereof!).

The question "which came first, the chicken or the egg?" looks at first glance like a matter of straightforward reproductive biology. But before we can even begin to answer this question, we must define our terms. So actually, it is a classic case of semantic ambiguity ... a problem of meaning and interpretation. Specifically, while the term "chicken" is biologically unambiguous – we all know what a chicken looks, sounds and tastes like – the term "egg" is somewhat more general and is therefore a possible source of ambiguity. Do we mean (1) just any egg, or (2) a *chicken* egg? And if we're talking about a chicken egg, then is a "chicken egg" (2a) an egg laid by a chicken, (2b) an egg containing a chicken, or (2c) both? Reformulating the question to reflect each possible meaning of "egg" leads to four distinct versions of the chicken-or-egg question:

1. Which came first, the chicken or (just any old) egg?

2a. Which came first, the chicken or an egg *laid by* a chicken?

2b. Which came first, the chicken or an egg *containing* a chicken?

2c. Which came first, the chicken, or an egg laid by *and* containing a chicken?

Contrary to popular belief, there is indeed a definite answer to each of these questions. Specifically, the answers are: (1) The egg. (2a) The chicken. (2b) The egg. (2c) The chicken. Given some knowledge of logic and biology, these answers are not hard to verify. To get this show on – or should that be across? – the road, let's go through them in order.

First, consider question 1: which came first, the chicken or (just any old) egg? This question is answered "the egg" because species that lay eggs have been around a lot longer than modern chickens. For example, we have plenty of fossil evidence that dinosaurs laid eggs from which baby dinosaurs hatched, and dinosaurs predate chickens by millions of years. Indeed, a growing body of research indicates that dinosaurs were among the biological *ancestors* of chickens!

Now let's look at question 2a: which came first, the chicken or an egg *laid by* a chicken? The answer to this question is "the chicken" on semantic grounds alone. That is, if a chicken egg must be laid by a chicken, then before a chicken egg can exist, there must *by definition* be a chicken around to lay it. And question 2c – which came first, the chicken or an egg laid by *and* containing a chicken? – is answered the same way on the same grounds; logically, the fact that a chicken egg must be laid by a chicken precedes and therefore "dominates" the (biologically subsequent) requirement that it contain a chicken. So whereas we needed paleozoological evidence to answer question 1, questions 2a and 2c require practically no biological knowledge at all!

Having saved the best for last, let us finally consider the most interesting version, 2b: which came first, the chicken or an egg *containing* a chicken? This version is interesting because an egg containing a chicken might have been laid by a chicken *or* a non-chicken, which of course affects the answer. Thanks to modern genetic science, we can now be sure that the egg came first. This is because reproductive mutations separating a new species from its progenitor generally occur in reproductive rather than somatic DNA and are thus expressed in differences between successive generations, but not in the parent organisms themselves. While the *somatic* (body) cells of the parents – e.g. wing cells, drumstick cells and wishbone cells – usually contain only the DNA with which they were conceived, *germ* (reproductive) cells like ova and spermatozoa contain non-somatic DNA that may have been changed before or during mating by accidental deletion, insertion, substitution, duplication or translocation of nucleotide sequences. This is what causes the mutation that results in the new species.

Where an animal qualifies as a member of a given species only if its somatic DNA (as opposed to its reproductive DNA) conforms to the genotype of the species, the parents of the first member of a new species are not members of that new species. At the same time, all the biological evidence says that the ancestors of modern chickens were already oviparous or egg-laying ... that a male and a female member of the ancestral species of the modern chicken, call this species "protochicken", mated with each other and created an egg. (Could the first chicken have evolved from a viviparous or live-bearing species, and after being born alive, have started laying eggs? All the biological evidence says "no".) But because their act of mating involved a shuffling of reproductive genes that were not expressed in the body of either parent – if they *had* been expressed there, the parents would themselves have been members of the new species – the fetus inside the egg was not like them. Instead, it was a mutant ... a modern chicken!

Only two loose ends remain: the "gradual" and "sudden" extremes of the evolutionary spectrum. These extremes are *evolutionary gradualism* – Darwin's original slow-paced timetable for natural selection – and *punctuated evolution*, as advocated more recently by evolutionary theorists including the controversial Stephen J. Gould.

Gradualism says that mutations are biologically random, but subject to a selection process determined by environmental (external) conditions to which species must *adapt* over the course of many generations. Taken to the limit, it implies either that each minor mutation that occurs during the evolutionary change of one species into another is random and independent of any other mutation, in which case a useful combination of mutations is highly improbable, or that each individual mutation confers a selective advantage on the mutant ... that every evolutionary advantage of a new species over its precursor decomposes into smaller advantages combined in a more or less linear way. Unfortunately, this makes it almost impossible to explain complex biological structures that do not break down into smaller structures useful in their own right ... structures like bacterial cilia and flagella, and even the human eye.

The hypothetical gradualistic evolution of one species into another via mutations accumulated over many generations leads to the following question: when does the quality and quantity of mutations justify a distinction between "species" ... when does a protochicken become a chicken? It's a good question, but our chicken-or-egg answers remain valid no matter how we answer it.

At the other extreme, evolution sometimes appears to progress by leaps and bounds, moving directly from the old to the new in "punctuated" fashion. And to complicate matters, this sometimes seems to happen across the board, affecting many species at once. The most oft-cited example of punctuated evolution is the *Cambrian Explosion*. Whereas sedimentary rocks that formed more than about 600 million years ago are poor in fossils of multicellular organisms, slightly younger rocks contain a profusion of such fossils conforming to many different structural templates. The duration of the so-called "explosion", a mere geological eyeblink of no more than 10 million years or so, is inconsistent with gradualism; new organs and appendages must have been popping out faster than the environment alone could have selected them from a field of random mutations. Clearly, the sudden appearance of a new appendage would leave little doubt about the evolutionary demarcation of ancestral and descendant species.

But the kind of punctuated evolution that occurs between generations is not the end of the line in sheer biological acceleration. Sometimes, an evolutionary change seems to occur within the lifespan of a single organism! For example, in the spirit of "ontogeny recapitulates phylogeny", insect metamorphosis almost seems to hint at an evolutionary process in which an ancient grub or caterpillar underwent a sudden transformation to something with wings and an exoskeleton ... or alternatively, in which a hard-shelled flying bug suddenly gave birth to an egg containing a soft and wormy larva. While that's not what really happened – as is so often the case, the truth lies somewhere in the middle – what occurred was just as marvelous and just as punctuated.

What seems to have happened was this. Due to a reproductive mutation, a whole sequence of evolutionary changes originally expressed in the fetal development of an ancestral arthropod, and originally recapitulated within the womb and egg it inhabited, were suddenly exposed to the environment, or at least to the hive, in a case of "ovum interruptus". A fetal stage of morphogenesis that formerly occurred within womb and egg was interrupted when the egg hatched "prematurely", making the soft fetus into an equally soft larva and giving it a valuable opportunity to seek crucial nourishment from external sources before being enclosed in a *pupa*, a second egg-like casing from which it later hatched again in its final exoskeletal form. So metamorphosis turns out to be a case of biological common sense, providing the fetus-cum-larva with an opportunity to acquire the nourishment required for the energy-consuming leap into adulthood.

Does this affect our answer to the chicken-or-egg question? Not really. For even where the life cycle of an organism includes distinct morphological stages, the DNA of egg-laying insects does not change after conception. And since it is reproductive and not somatic DNA modification that distinguishes one species from the next in line, our answers stand firm. (Of course, this says nothing of science fiction movies in which something bizarre and insidious causes runaway mutations in the somatic DNA of hapless humans, causing them to evolve into monsters before our very eyes! Such humans have either undergone a random or radiation-induced "meta-mutation" whereby their genetic code suddenly rearranged itself to incorporate a self-modification routine that is executed somatically, within their own cells, or they are the victims of a space virus which inserted such a routine into their DNA for its own nefarious purposes.)

OK ... perhaps there's yet another loose end. Asking which of two things came first implies that time flows in a straight line from past to future (those are the "loose ends"). But what if time were to flow in either direction, or even to loop around, flowing in what amounts to a circle? No more loose ends. In fact, loops have no ends at all! But in this case, the answer depends on whether we're on the forward or reverse side of the loop, heading towards the future or the past. Another way to formulate this question: does the cause lead to the effect, or is there a sense in which the effect leads to the cause? Suffice it to say that no matter which way we choose to go, the original answers to the four versions (1, 2a, 2b and 2c) of the chicken-or-egg question are all affected the same way. They are either all unchanged or all reversed, with no additional ambiguity save that pertaining to the direction of time (not a problem for most non-physicists and non-cosmologists).

Now that we've tied up every last loose end, what about the most important question of all, namely what to tell a curious child? The answer: take your pick of versions. Some kids will prefer the dinosaur angle of version 1; some kids will prefer the "birds and bees" reproductive biology lesson of version 2b. In my opinion, if we limit ourselves to one version only, the most valuable explanation is probably that of 2b; but due to its relative complexity, a younger child can probably derive greater benefit from a T. Rex-versus-Triceratops embellishment of version 1. To exhaust the golden opportunities for logical and scientific instruction, one should of course answer all four versions. But no matter which way you go, make sure the child knows exactly which version(s) of the question you're answering. If you leave out the one he or she had in mind, you'll no doubt be egged on until it gets answered!

Of Trees, Quads and God

Ever hear this old Ronald Knox limerick about a lonely tree?

> There was a young man who said, "God
> Must think it exceedingly odd
> If he finds that this tree
> Continues to be
> When there's no one about in the Quad."
>
> REPLY
>
> Dear Sir, Your astonishment's odd:
> I am always about in the Quad.
> And that's why the tree
> Will continue to be
> Since observed by, Yours faithfully, God."

No? Then perhaps you've heard the old koan-like question that seems to be every schoolchild's introduction to philosophy (it's still about a lonely tree, but at least this one is doing something): "If a tree falls in the forest and no one is there to hear it, does it make a sound?"

No, you say again? Then for the consolation prize, maybe you'd like to guess what the limerick and the question have in common.

The answer is twofold. They're both about the empirical philosophy of the celebrated 18th century philosopher Bishop George Berkeley. (That's right – Berkeley was a bishop as well as a philosopher. How else could God have gotten into a philosopher's limerick? Oh, yes ... Knox was a cleric too!) And they're both about the relationship between reality and perception, or more generally, the relationship between reality and mind.

As with most philosophers, the nature of reality was something that preoccupied Berkeley. The nature of reality ... now, there's a difficult topic. The first thing we notice about reality is that it's everywhere. And the next thing we notice about it is that because it's everywhere, there's nothing around to contrast it with or compare it to in order to define it. There are always dreams, fantasies and illusions, of course, but what are they? And if we don't know what *they* are, then how can we use them to define reality?

The problem gets worse. An understanding of reality as the medium of existence is necessary for understanding anything that exists. We're parts of reality. So if we don't know what reality is, how do we know what *we* are? And if we don't know what we are, then how do we know what anything else is? Indeed, how do we know anything at all? (I'm sorry to say that this is not a joke, but a sad reality.)

Science doesn't help. When we use science, all we're doing is using various parts of reality to define other parts, not defining reality with respect to its complement. We can't just point to physical reality and say, "there it is ... that's what reality consists of!" while banging our knuckles or heads against a massive object, e.g. a table or wall, to drive home the sheer solidity of our conviction. For physical reality is not a stand-alone proposition; it requires universal rules in order to exist and maintain its consistency. And from a logical standpoint, those rules are more than merely physical.

Let's go back to every schoolchild's introduction to philosophy for a moment. "If a tree falls in the forest and no one is there to hear it, does it make a sound?" It is no accident that people have come to regard this as a western version of "what is the sound of one hand clapping?", for both can be reduced to the notion of unrequited duality.

In the case of one hand clapping, the duality is the concept of (two-handed) clapping, and the missing part of the duality is the other (non-clapping) hand, which is either absent or motionless depending on the point you want to make. In the case of the sound made by the tree falling in the forest, the duality is the concept of a sound, and the missing part is the listener. For in contrast to the objective physical disturbance called a "sound wave", a sound is a *perception* that requires both an object to make it and a subject to hear it.

The lesson is clear: when it comes to perception, it takes two to tango. Perception is a duality involving both object and subject; for every perception, there is something perceived and someone who perceives it. But here the clarity ends, for the lesson has more than one possible interpretation. In fact, there are two, each one predicated on a distinct hypothetical relationship between perception and reality.

The first of these relational hypotheses goes like this: perception and reality are separate. Perhaps the foremost exponent of this viewpoint was the French mercenary, philosopher and mathematician René Descartes (1596-1650). At this point, we need to know just two things about Descartes, the first being that he was a *rationalist*. Rationalists believe that the most basic concepts in terms of which we understand the world – e.g. self and causality – are known intuitively rather than through experience. Descartes held that we can deduce truths with absolute certainty from such "innate ideas" in the same way that theorems are deduced from axioms in geometry. In fact, rationalists see mathematical demonstration as the perfect means of establishing truth, and thus as a general model for the pursuit of knowledge.

The second thing we need to know about Descartes is that his philosophy was dualistic, positing an absolute separation between mind and matter ... and specifically, the human mind and body. To put it another way, Descartes said that reality consists of two independent substances, mind and matter. Both are real, Descartes assured us, but never the twain shall meet except through the act of perception, which must itself have a dual nature in order to bridge the gap. (In contrast to *dualism*, which reduces reality to two essential, irreducible substances, *pluralism* reduces it to many and *monism* reduces it to just one. The problem with any -ism other than monism, however, is that one gets stuck trying to explain the composition of the medium in which the various essential substances are related to each other.)

The second relational hypothesis is that perception and reality are intimately related. The first philosopher usually associated with this viewpoint was the Englishman John Locke (1632-1704). Locke argued against the rationalist belief in innate ideas, holding that the mind is a *tabula rasa* or "blank slate" on which all knowledge is imprinted by experience. He distinguished primary objective qualities like number, extension and solidity, which mechanically affect the sense organs through perception, from secondary subjective qualities like color, smell and sound, which are produced by the direct impact of the world on the sense organs. Once past the sense organs, Locke theorized, these secondary experiential qualities combined to form ideas that mirror reality and constitute the basis of science.

But what kind of science? Although Locke believed that we know the world only through experience, he also believed in the existence of a world apart from the mind, with perception and experience mediating between the mind and the world around it. It thus seems that Locke's apple fell not too far from Descartes' tree after all. For his theory of primary and secondary qualities, whereby the mind knows only secondhand experiences impressed upon it by perception, implies that scientific knowledge is itself a secondhand affair ... that the true essences of physical objects cannot be scientifically known. So Locke agreed with Descartes not only about mind-body duality – that the conceptual, subjective world of the mind is forever separate from the concrete, objective world it perceives – but that mathematics is the only route to certainty.

So far, the only real difference between Descartes and Locke is ... well, not much. Descartes says that the mind possesses innate knowledge; Locke says that the mind is a blank slate on which reality leaves its imprint. But in the final analysis, the intuition-versus-*tabula rasa* distinction turns out not to make a lot of difference. For Descartes and Locke are both dualists, and their respective solutions of the tree-in-the-quad problem are exactly the same: since the material existence of trees is independent of perception, they exist whether or not anyone is watching them. This agreement is quite strange in view of the fact that Descartes and Locke are respectively considered to be the founders of *rationalism* and *empiricism*, two opposing philosophies about the source of human knowledge. So where do these philosophies diverge?

Enter George "What-you-see-is-what-you-get" Berkeley (1685-1753), an Anglo-Irish philosopher and clergyman who went Locke one better. Whereas Locke had allowed the independent existence of matter, Berkeley held that reality and perception are one and the same, and thus that matter cannot exist without perception. (In the intellectual style of the day, he dressed his thesis up in Latin: *esse est percipi*, "to be is to be perceived.") Technically, this made him an adherent of *idealism*, a strain of philosophy which holds that nothing exists apart from minds and their contents. In contrast to the mind-body dualism of Descartes, Berkeley's perceptual monism reduced reality to a single "substance", perception.

Whereas Locke had believed that real objects possess their own innate qualities along with mental qualities stimulated by perception, Berkeley held that all qualities are mental or ideal in nature. But he also acknowledged a key distinction: whereas ideas are mutable and evanescent, material objects are stable and persistent. From this, he reasoned that the mind on which material objects depend must be divine rather than human. That is, from the common-sense notion that physical objects exist even when no one perceives them, Berkeley inferred that the Official Perceiver of trees and other objects must be God Himself. Where Locke's philosophy had opened the epistemological door to the demons of skepticism and atheism, Berkeley aimed to close it and thereby restore faith in God.

At this point, we should note that the God-in-the-quad answer to the tree-in-the-quad question is in some ways an oversimplification. In fact, Berkeley would have said that the material quad does not exist at all except as an archetype in God's mind, God's own subjective description of the quad's physical appearance. According to Berkeley, people who enter the quad will have the appropriate mental experiences, i.e. will "perceive" the quad, only because these experiences are produced by reference to God's own archetype. Just as a musician produces music by reference to the composer's musical score, we perceive quads by referring to the contents of the Mind of God; our perception follows God's cognition. To this extent, Berkeley's perceptual monism is *cognitive* monism.

If Berkeley's reality-equals-perception thesis seems radical, then watch out for the Scottish philosopher and historian David Hume (1711-76). Hume carried the empiricism of Locke and Berkeley to the logical extreme of *radical skepticism*, repudiating the possibility of certain knowledge. According to Hume, the mind is nothing but a series of sensations, and causality is merely an illusion that happens when one impression follows another. Even though cause precedes effect, says Hume, there is no proof that the cause is responsible for the effect's occurrence. Unfortunately, because Hume also disagreed with Berkeley on the subject of God and flatly rejected theology, he took the tree-in-the-quad problem out of God's perceptual lap and put it back in man's (not an entirely fortuitous move, given the increasing prevalence of attention deficit disorder).

The vulnerability of Hume's reality to ADD is not the only problem associated with his philosophy. Hume is also credited with another problem, the so-called *problem of induction*: how do human beings form beliefs about unobserved matters of fact, and are these beliefs justified? That is, do our powers of reasoning permit us to form valid conclusions about parts of nature we have not observed? The bearing of this problem on that of the tree-in-the-quad is obvious: in principle, a strong constructive solution for the problem of induction would let us determine the existence of any tree anywhere from our perception of any subset of trees. If human reasoning were sufficiently powerful, and human powers of generalization sufficiently strong, then to perceive any subset of trees would be as good as perceiving them all. In effect, human cognition (and not divine) would suffice to relieve perception of much of its ontological responsibility.

But that would be optimistic, and Hume was not an optimist. Indeed, his solution to the problem of induction was downright gloomy. First, he defined induction as the construction of a general theory from limited data by applying a principle asserting the uniformity of nature. That is, induction generalizes from the part to the whole on the assumption that nature is everywhere the same. From this it followed that all beliefs about unobserved facts are derived by induction from experience. But while Hume considered the inductive schema (pattern of reasoning) logically valid, he denied the validity of the uniformity premise. For while the validity of the premise can only be established by induction, every inductive argument necessarily employs it. Since the premise is "circular" or self-justifying, said Hume, it cannot be proven.

After Hume did his bit, things looked bad indeed for any tree unlucky enough to be in an empty quad; neither God nor induction could philosophically warrant its existence. But even as the gloom of radical skepticism cast its pall over much of Europe, all was not lost. For if Hume was the philosophical Little Engine That Could, he and his followers were about to learn the meaning of Kant.

Immanuel Kant (1724-1804) was a German philosopher who complained of being awakened from his "dogmatic slumber" by the Little Engine's whistle, a shock which prompted him to synthesize the skepticism of Hume with the rationalism of Descartes' fellow rationalist Gottfried Wilhelm von Leibniz. Kant's central thesis was this: we can know objective reality only insofar as it conforms to the structure of our minds. Beyond this point, Kant deemed reality unknowable. Kant called knowable objects of experience *phenomena*, and the unknowable objects underlying these phenomena *noumena* or "things-in-themselves" (think "objects independent of perception"). By definition, noumena are immune to affirmation, denial or scientific confirmation. Unfortunately, there is a problem with the Kantian relationship between phenomena and noumena: it is bedeviled by a particularly troublesome form of dualism that resists transformation into a unified (monic) explanation of reality. Whereas Descartes asserts that mind and matter occupy a divinely-mandated similarity relationship, Kant denies that the parallel relationship between phenomena and noumena can be known at all, even to the extent of calling it a "similarity relationship". From this follows some bad news: noumenal trees are untouchable by phenomenal perception. But there is some good news as well: phenomenal trees must conform to the structure of our minds. In principle, this yields a solution to the problem of induction with respect to phenomenal reality.

Although it may seem hard to believe, this brings us right up to the present in terms of "metaphysics", or our knowledge of deep reality. Indeed, since Kant's linkage between phenomenal reality and the structure of our minds invites further exploration, it propels us towards the future. Even now, science struggles in vain with the problem of induction, waiting for philosophy to tell it how to construct a general theory of reality from the limited data accessible to its instruments; floundering in a confusing welter of dualism, rationalism and empiricism, it cries out for a metaphysical resolution. But analytic philosophy, long ago cowed by the brilliant successes of its scientific offspring, remains preoccupied with emulating the hard sciences, and thus finds little room for metaphysical thought. As each side looks vainly to the other for guidance, it sees only its own shadow.

But the tree-in-the-quad saga is not yet over. Using ideas from each of the above philosophers along with some newer concepts, it is possible to synthesize a new mind-matter, mind-body connection that leaves reality neither irrevocably split, nor hanging in the ontological limbo of a quad at once atheistic, noumenal and empty. Look to the future. For when science and philosophy remarry at last, their progeny will populate and enrich our intellectual world.

The Pros and Cons of "Machine Intelligence"

If you want to be intelligent, you have to be smart about it. And the first thing you need to know about intelligence is that it comes in two varieties: natural and artificial. Let's start with the shallow, artificial kind and save the deep stuff for later. (Then let's show that the two cannot be objectively distinguished and pretend they're the same.)

AI, oh magical acronym! Artificial Intelligence, perhaps the hottest topic of the digital era, has captivated the imaginations of computer scientists and screenwriters since the days of vacuum tubes and punch cards. But as any star buzzword should know, it's a long fall from the top, and being overhyped is little better than not being hyped at all. AI, having failed repeatedly to emerge on schedule, now finds itself with a bad record, a bad name, and increasingly bad credit at the Bank of Big Expectations (in fact, the alternate title for this piece was "The Gurus of AI – Pros or Cons?").

No big surprise, really. It was all right there in the name. "Artificial" and "intelligence" go together like *jumbo* and *shrimp*, *communist* and *party*, or *educational* and *television*. Indeed, insofar as *artificial* is nothing but an antonym of *real*, AI seems to have shot itself point-blank in the foot with Webster's Collegiate Dictionary. No wonder, then, that it seems all but destined to go the way of genuine *pleather*, real imitation fur and bona fide faux pearls. As it now stands, AI is the stuff of late-night TV ... of hardcore sci-fi and breathless blurbs for various direct-marketed techno-toys.

However, there are a couple of new and disturbing trends on the horizon. In fact, they started at the horizon some time ago and have since gotten considerably closer. Do they involve startling breakthroughs in computer science? Well, yes and no. They *do* involve computers. But the "science" has at least as much to do with biology – or as they say in the AI biz, "wetware" – as with hardware and software, and it opens up a number of disturbing possibilities reminiscent of movies like *Colossus: The Forbin Project*, *Demon Seed*, *War Games*, *Terminator* and *The Matrix*. Suffice it to say that if you still take solace from the trite belief that "computers can't do anything we humans don't program them to do!", you need to ... well, re-compute your position.

But before we go there, let's answer the obvious question: how did it all begin? That's easy – it began with Alan Turing. The subject of a play called *Breaking The Code* – he was famous for cracking Germany's "uncrackable" Enigma code in WW2 – Turing was a British mathematician whose ideas sowed the seeds of modern computer science. No sooner had scientists wishfully added electronic computers to the List of Neat Things We Want for Christmas, than their friends the mathematicians began to use abstract models to explore their capabilities. As the world waited for a real computer to be built, Turing's *universal machine* became the abstract model computer of choice. A harbinger of disappointments to come, Turing used it to show that there are problems that no computer will ever be able to solve (in effect, Turing mathematically fed a symbolic universal machine to itself and then stepped back to avoid the flying nuts-and-bolts symbols).

Not one to give up, Turing nevertheless went on to become the world's first AI theorist. First, he reasoned that a computer would some day become powerful enough to duplicate some of the functions of the human mind. Would this make the computer intelligent? To answer this question would require a test, but of what kind? Since subjective attributes like intelligence do not readily lend themselves to the scientific method, the test should measure the objective correlates of intelligence rather than intelligence itself. But against what scale? Since people introspectively know how it feels to be intelligent, and know intelligence when they encounter it in others, human judgment should serve as the scale. So the best test, reasoned Turing, would be to put a computer and a person on opposite sides of an opaque partition and let them interact, e.g. engage in a written dialogue. If the person is fooled into thinking that another person is behind the partition, then the computer passes the *Turing Test* for artificial intelligence.

Unfortunately, a machine that fools a human proves itself no more than a convincing fake. After all, the computer can do nothing but what it is programmed to do ... or can it? Granted, a computer is mechanical. But human beings can themselves be described as biomechanical systems. Granted, a computer is deterministically programmed and therefore devoid of free will. But there remains a philosophical controversy over the possession of free will by humans, whose own "programming" involves a mixture of genetics and environmental exposure over which they arguably have little control. In fact, Turing even described abstract machines capable of making intelligent guesses, which amounts to a form of free will and obscures the distinction completely. So where is the functional difference between computers and people? The superficiality of the Turing Test – its failure to probe beneath surface appearances in search of the "true essence" of intelligence, human or otherwise – reflects the difficulty of pinning down this distinction, at least for a mathematician.

The bottom line: our best working definition of intelligence is "problem-solving ability". But then the meaning of *intelligence* devolves to that of *problem*, and where a "problem" is whatever you want it to be, intelligence is whatever you want it to be as well. Which, of course, recreates the original problem, namely how to find a meaningful definition of intelligence. This takes us right back to the Turing Test, which – if we sensibly go along with Turing regarding the limitations of the scientific method with respect to subjective predicates like intelligence – is an optimal test of intelligence in man *or* machine. After all, when one person sits behind a partition and gives another person an IQ test, this is merely a quantitative refinement of the Turing Test involving no *a priori* distinction between a human being and a test-taking computer program. This means that when it comes to intelligence, we might as well talk about it in terms of machines as human beings. So let's just pick up where Turing left off.

Not too long after Turing invented his eponymous abstract model of a computer, real mechanical computers had been built and were ready to be put to the Turing Test for machine intelligence. When programmers first considered ways of doing this, it occurred to them that there were a couple of obvious methods at their disposal. They could program a computer to play a game against a human and either beat the human or come close to it. Or they could program a computer to convincingly answer a series of questions ordinarily posed to a human expert like a doctor or a lawyer. Either way, they reasoned, a human participant behind a partition would have a hard time figuring out that he was being bamboozled by a bag of hot circuits. And thus originated two of the main currents in AI, gaming programs and *expert systems*.

The most famous gaming program of all time is *Deep Blue*, the wood-pushing automaton that beat World Chess Champion Garry Kasparov in 3.5 out of 6 games in May, 1997 (in a draw, each player gets ½ game). Billed by *Newsweek* as "The Brain's Last Stand", it was ballyhooed as a huge victory for AI. Unfortunately, if there was ever a Turing-style partition between the players, it was anything but opaque. In response to each of Kasparov's moves, the machine was known to be rapidly executing thousands upon thousands of lines of code containing cumulative "if ... then" statements, something of which a human mind would be incapable even if it seemed like a good idea. Partition or not, it was all too clear that the combination of heuristics and brute force employed by Deep Blue was not what most of us really mean by "intelligence".

What about expert systems? To build an expert system, a "knowledge engineer" interviews experts in a given field and tries to collect their knowledge in a computer program designed to answer questions or perform one or more domain-specific tasks. Unfortunately, because the operations that the computer is required to perform are not always feasible, this doesn't always work. In recent times, expert systems designers have met with limited success. But there are two problems. First, the computer is clearly not "thinking"; it is merely storing, sorting and relaying the insights of human experts. Second, as soon as the program's human interlocutor transgresses the bounds of its narrow range of borrowed "expertise", it typically starts making stupid errors that give it away.

These problems apply not just to specific AI programs and the machines that execute them. They apply to the entire AI strategy including gaming programs, expert systems, mathematical theorem-proving programs, natural language parsers, so-called "intelligent agents" and other conventional automated procedures designed to mimic human intellectual processes. Once this became obvious, the search began for alternative approaches. Among the possibilities, two stood out. The first, now referred to as *connectionism*, was to directly simulate the human brain on the intercellular level. The second, called by suggestive names like *genetic algorithms* and *evolutionary programming*, was an idea that had long been talked about, but seemed so improbable and outrageous that few took it seriously: instead of mimicking the output of biological systems, programmers would attempt to electronically recreate them.

Connectionism aims to duplicate human intellectual abilities using the brain's simplified model of its own structure: the artificial neural network. Just as a brain consists of neurons connected by synapses, a neural network consists of artificial neurons called "neurodes" and their "synaptic" interconnections. To each connection is assigned a strength or "weight". The pattern of synaptic weights is the network's "program" ... the informational construct according to which input is converted to output. However, the weighting pattern is not the *whole* program. The neural net also incorporates a higher-level program called a *learning function* that adjusts synaptic weights according to input. The learning function tells the network how to do what the brain does effortlessly, but computers can do only badly if at all: learn from experience. Already, neural nets quickly but fallibly learn such tasks as facial recognition, rudimentary language processing and predicting the stock market.

Notice what this does to the idea that "computers can only do what we program them to do". The brainlike computer called a neural net is equipped with a higher-level program that lets it *program itself.* Feel things starting to get out of control yet? If so, then get ready for another eye-opener. For according to the genetic programmers, it simply won't do to build lifelike computers when you can do what Mother Nature did: get them to build themselves by the process of evolution.

First, the genetic programmers create a random population of computer programs to perform a given task or set of tasks. Each program is measured against a "fitness criterion" by which the degree of correctness of its output may be determined. A "selection principle" then rewards fitter programs with a survival advantage – that is, a higher probability of survival – and penalizes the less fit, making them more likely to die out. As in nature, survivors have the opportunity to "breed", in the course of which they can mutate and combine to form new programs that inherit their traits. After many generations, evolution produces programs that have the highest possible level of fitness ... programs that can perform the intended set of tasks (almost) perfectly, thus filling the available niche in their electronic ecosystem. (This seldom holds true when new tasks are added, but what did you expect?) Evolution triumphs again.

It would be hard to overestimate the profundity of these trends. In adopting genetic and connectionist viewpoints, AI intersects with biology on multiple scales. It thus promises to outdo humanity itself, which – let's face it – has put meaningful adaptative pressure, and thus evolution, on hold by suspending the law of natural selection. As it stands, a human with low fitness can apply a series of temporary medical patches to whatever ails him or her and reproduce like a rabbit anyway, intelligence notwithstanding. Not so for genetic programs, which we subject to ruthless adaptative pressure to serve us.

The bad news: connectionism and genetic programming still have a very long way to go. Both are resource-intensive, requiring huge parallel processors to work up to capacity, and their theories are still relatively primitive. Moreover, the opaque panel in the Turing Test is essentially the same divider used by Descartes, in his doctrine of Cartesian Dualism, to separate mind from matter and sever the mind-body connection in the name of science. Thus, like Cartesian dualism itself, the Turing Test offers no insight to the intrinsic nature of intelligence; in the current state of philosophy, all we can say is that "intelligence is as intelligence does".

The good news: there is a way out. But it involves a shift from Cartesian dualism to cognitive monism and the concept of an "intelligent universe", according to which generalized intelligence becomes a universal property of which our minds are just localized examples (yes, that's what it ultimately takes to restore the mind-body connection). One thing's for sure: the future will be as interesting as the theories of AI and biology put together, and it is a future for which we must prepare. Because we desperately need to understand the machines we will be creating and how they relate to us, hiding behind our Cartesian panel is no longer a viable option. Philosophically, we need to step into the New Millennium, and fast.

Meanwhile, no matter what happens – no matter what the danger that mankind will fall under the domination of a tyrannical AI monster like those in the movies mentioned above – we can take solace in one little fact: the way things are going in our public schools, it is virtually certain that no self-respecting machine will ever pass the Turing Test. You see, such a machine will immediately give itself away by its use of proper grammar.

In Ethics, Not Everything is Relative

After the events of 9–11, destructive and heroic, it seems like a good time to discuss ethics and moral relativism. We all know the meaning of ethics; ethics are what tell human beings how to do the right thing by each other. Unfortunately, this can be a very complex calculation, and that's where moral relativism enters the picture.

To understand moral relativism, consider asking many people the following question: "Was the World Trade Center disaster morally wrong?" In New York City and across the rest of the US, this question would be sure to draw an overwhelming number of affirmative responses. But if one were to ask this question in the streets of an enemy nation of the US, the answers might be very different. One might instead hear a series of blame-the-victims tirades about how US meddling and/or non-meddling in the affairs of other sovereign nations entitles those nations to lash out violently against their "oppressors" (us).

In other words, we have two points of view, one associated with the victims of the tragedy and the other with its actual or would-be perpetrators. Each of these viewpoints is clearly the opposite of the other, as reflected in "yes" and "no" answers to the same question. It would seem to follow that only one can be "moral" or "ethical" ... that one side should be able to prove itself "right", and the other should have to admit being "wrong".

However, we know from practical experience that such agreements can be very elusive. More often than not, each side in a dispute claims the moral high ground. This, of course, tends to rob of its meaning the concept of ethical or moral correctness. If it were simultaneously true that the lightest shade is white and the lightest shade is black, then the "lightness" attribute would have no power to distinguish one shade from another, and this would correspond to ambiguity in the meanings of "white" and "black". The ethical analogue of this dilemma involves ambiguity in the meanings of "good" and "evil", and "right" and "wrong".

How is this ethical dilemma to be solved? One way would be to provide an absolute definition of terms like good, evil, right and wrong, spelling out the good-evil/right-wrong distinction in such a way that it could be straightforwardly applied in any context to identify the best course of action. Attempts to define absolute good include the Ten Commandments, which consist of absolute instructions to be obeyed unconditionally by good people; the US Bill of Rights, a list of absolute freedoms to be unconditionally respected by good citizens and good governments; and the Golden Rule, a sophisticated rolling of many ethical principles into one.

But ethical absolutism is not without its difficulties. To take a simple if unpleasant example, suppose that the child of a penniless man is suffering from a terrible but curable disease, that the man's every attempt to get money to pay for the very expensive cure has failed, and that the manufacturer of the cure has ignored his every plea for mercy. Now suppose that the man knows where the cure is stored in quantity, and that he has a good chance of stealing a dose without detection. Should he be deterred from doing so by the fact that theft is wrong, given that it would also be wrong to let his innocent child die?

If we examine this situation closely, we can discern the involvement of at least three distinct viewpoints or "ethical frames". The first is that of the man and his child, whose joint purpose is to save the child's life. The next is that of the manufacturer of the cure, whose purpose is to get a maximum (or perhaps just a fair) return on his investment in R&D and production. And the third frame, which contains the first two, is an overall context with respect to which the absolute morality of "thou shalt not steal" has putatively been determined, but which nevertheless lacks any absolute proscription against letting innocent children die for their poverty.

Such examples seem to imply that no finite set of specific moral principles suffices to determine ethical behavior in a real world characterized by endless complications. This has led to an even less productive way of approaching ethical dilemmas, namely moral relativism. A moral relativist takes the following position: it is useless to speculate about the moral laws of an overall context that can never be fully known. Because the right-wrong distinction thus depends on the specific viewpoints of those who apply it, we can do no better than consider each frame independently, reformulating the distinction when switching from one frame to the other.

This putative independence of ethical frames tends to render them incommensurable, impeding the formation of common goals and cooperation to achieve them. This restricts the moral relativist to the default "common goal" of mutual tolerance. But where mere tolerance is sufficient, or where at least one viewpoint is defined precisely by its intolerance, this can be unproductive, and the relativist is powerless to fix the problem. For while fixing the problem would require either the imposition of a common ethic or a judgment against intolerance, moral relativism lacks an ethical basis for taking such measures or validating such judgments.

For example, consider the problem of constructing an extended context in which the Western and Islamic worlds can peacefully coexist. One side believes in freedom, capitalism and the separation of church and state; the other believes in various blends of feudalism, theocracy and plutocracy. One has given its citizens the world's highest average quality of life; the other has concentrated its considerable wealth in the hands of an incredibly opulent ruling class that is perfectly happy to let the impoverished masses blame outsiders for their problems. One treats even its criminals with compassion; the other summarily plucks out eyes and lops off body parts from hands to heads.

And the differences go on. One side regards women as full and equal citizens; the other regards them as something less. One allows itself to be criticized by its own citizens and shapes its policies according to their will; the other appoints external scapegoats for all of its problems and lashes out at them instead of cleaning up its own political domicile. One is a diverse family of immigrants that generously opens its doors to the world with a promise of equal opportunity; the other belongs to a set of warring groups characterized by millennia of bloodshed and turmoil, unfair distribution of resources, and annihilative creeds encouraging preferred groups to subject other groups to physical, psychological and economic violence "in the name of God".

Such examples point to the main weakness of moral relativism: in failing to distinguish between absolute right and wrong, it renders itself unable to deal equitably with situations in which one side is more right than another. It balances the pain of the victim against the pleasure of the victimizer, the agony of a murdered child against the dark ecstasy of the fiend who murders her. Indeed, it places the very angels of Heaven and devils of Hell on an equal footing; confronted by angels pleading for brotherly love and devils bent on human pain and misery, the pure moral relativist would have to juggle the benefits of brotherly love with those of demonic torment. (After all, millions of sadistic, bloodthirsty demons are entitled to their viewpoint as well!)

Lacking a clear logical basis for moral absolutism or a pragmatic basis for pure moral relativism, most real-world ethicists (politicians and diplomats) go for an *ad hoc* blend of the two in which frames, e.g. nations, remain essentially independent but are expected to agree on the "absolute goodness" of mutual tolerance. The idea is to find a goal that both frames have in common – survival, for instance – and encourage each frame to modify whatever ingredients might interfere with conflicting ingredients of the other frame to jeopardize the attainment of this common goal. Unfortunately, this strategy tends to fail where the main source of trouble is the irrational hostility of religious fundamentalists adhering literally to "holy writ" in which tolerance is literally banned (as it is banned by Islamic scripture, which literally calls for the subjugation or eradication of non-Islamic "infidels").

What, then, is the solution? That's a long story, so we'll merely outline it here. First, we need to mention several advanced concepts. One is utility; this means "value" and may vary by frame for any given thing. The greater the utility of something to a specific person or group, the higher its value. A utility function is a rule that assigns a personal or group value to something according to the values of its parameters. A rational utility function is one designed to maximize utility for those who apply it. Utilitarianism is a doctrine that prescribes "the greatest good for the greatest number" and provides us with a scale on which to measure an optimal solution involving multiple frames. And game (or decision) theory is a field of mathematics that tells players how to formulate winning, utility-maximizing strategies in various competitive scenarios or "games".

Game theory, which was initially formulated around the utility of individual players, was eventually found to give rise to paradoxes in which individual utility conflicts with the utility of one or more groups involved in the game. If enough players "cheat", i.e. defect from the group strategy in order to employ individual strategies that let them profit at the group's expense, the group infrastructure collapses, and where the wellbeing of the group is essential to that of its individual members, individual utility collapses as well. A player who cheats in such a game can thus end up cheating himself, coming away with far less than if he and others had played fair. Because a player's "rational" attempt to maximize his own personal gain can cause him to lose, the standard theory fails.

The extended version of game theory designed to resolve such paradoxes is called "the theory of metagames". Whereas standard game theory was designed to handle games analogous to chess and bridge, in which utility functions are assigned only to competing individuals, metagame theory is designed for games in which players belong to various groups, e.g. nations or religions, for which higher-level utility functions are also defined. In this case, utility is a combination of personal and group factors. Because the player's true utility function, which accounts for his group status, differs from that obtained when he considers only himself, the self is effectively "stratified". That is, a full definition of "self" must include all of the external relations tying the self to other selves at all levels of organization relevant to all possible games. This redefinition of "self" turns out to be of crucial importance across the entire ethical spectrum.

The theory of metagames permits the computation of strategic optima for any real-world scenario to which it can be properly applied (its proper application requires an accurate representation of the individual players and the groups to which they belong, the rules according to which they interact, and the overall context they share). This, of course, leads to a question: why has the theory not been used to solve all of the world's social problems? The reason: abstractly applying the theory is easier than concretely applying its results, especially in problematic situations where the ethical frames of the players are incompatible. Too often, circumstances that pit person against person, class against class and nation against nation also pit ethic against ethic and creed against creed. A lopsided ethic can encourage players to defect rather than accept the globally optimal solution, using stealth, subterfuge or superior force to achieve an inequitable outcome.

Because it leads to a stratification of self, the theory of metagames can be reduced to a single general self-based strategy: the Golden Rule. Because the Golden Rule distributes over all players in the global geopolitical metagame and is therefore absolute, those who defect from this strategy depart from absolute morality. In the case of the WTC tragedy, the reason for defection is clear: the Golden Rule is contradicted by the teachings of Islam, which isolates the self and utility functions of terrorists from those of their victims. This suggests that to successfully apply rational decision theory to the current situation, influential and well-meaning Moslems should appoint a council of Islamic scholars to revise Moslem scripture, excising specific passages that appear to mandate the slaughter of innocent people in the name of Islam. Only then can Moslems who prefer a literal, fundamentalist interpretation of holy writ earn a place in Paradise through prayer and good works rather than kamikaze terrorism.

Millennium Mouse: An Ecological Parable

There is a tiny cabin in the northern woods. Because it is tiny, it is easy to heat during the long, cold mountain winters. But because it is so warm and cozy, it is also a good winter abode for the mice that live in the surrounding hills. Each winter, a pair of mice finds a way into the cabin and makes a home there. If they were able to exercise good judgment, these mice would remain as unobtrusive as possible, taking a few stray crumbs and tidying up after themselves so as not to arouse the suspicion and ire of their hosts. Life would be warm and comfortable atop a ceiling beam over the potbelly stove, and enough crumbs would fall from the homeowners' plates to sustain their tiny guests. In fact, these little guests could even take pride in helping to keep the floors clean! They would live modestly until spring approached, have a single litter of pups, and then exit with their new family into the wide world outside the cabin as soon as the weather permitted. That way, when the coast was clear six or seven months later, those who had survived could use their old entranceway under the eaves and get cozy for another winter.

Unfortunately, this is simply not the nature of mice. First one pair comes in. Then another, and another. They begin to fight for territory, screaming at each other in the wee hours and sometimes even awakening their hosts. Soon, they are scrounging hungrily from table to stove to countertop, eliminating their bodily wastes even where the food that sustains them is prepared and served. Meanwhile, they breed copiously in nests made of materials torn from the private possessions of their hosts, wreaking havoc on bedding and furniture. Before long, the cabin undergoes a population explosion of filthy, destructive, combative little beasts that boldly rob their hosts right under their incredulous noses, urinating and defecating in the food supply to boot. And their hosts, being left with no choice, retaliate by setting a lethal gauntlet of spring traps, glue strips and poison bait. Mouse Armageddon ensues, and a pile of tiny bodies accumulates under the window.

In considering the self-destructive behavior of mice, human beings silently congratulate themselves on their superior judgment. "Mice are stupid to behave so injudiciously," they think. "If these mice just went easier on us, we might be able to tolerate them. But since they lack the intelligence to see this, we'll simply have to keep trapping and killing them." On the basis of such thoughts, one might almost think that human beings were innately wiser than mice with regard to such matters. But in this, one would be profoundly mistaken. For the only thing that separates the planet earth from that tiny cabin in the woods is size, and the cumulative effect of mankind on the earth is every bit as disgusting, from an ecological viewpoint, as that of mice on a cabin.

Being the evolutionary descendants of tiny mouselike proto-mammals, human beings are every bit as capable as mice of overpopulating, befouling and fighting over their living space, and there is no sane reason to think that the ultimate outcome will be any less unpleasant. After all, a mouse can always scurry from one cabin to another in an emergency. But when the earth is fouled beyond habitability, there will be no scurrying away from it, at least for the vast majority of us. Like mice, we will be trapped here and exterminated. And when that day comes, we will obviously not have our "superior intelligence" as a species to thank for it.

PART THREE

IQ, IEQ and Intelligence

In ordinary usage, the term *intelligence* is appreciably constrained by neither definition nor theory. Quite simply, there is no common agreement regarding what intelligence is or how it works. Accordingly, in the narrow context of psychometrics, the term denotes that which is measured by a narrow range of *IQ tests* exhibiting certain generic similarities, e.g., low-complexity items of stock varieties to be solved within strict time limits. In effect, psychometric standardization has led to a tautological and probably incorrect equation of IQ and intelligence that overrates the former and underrates the latter.

If a newly designed intelligence test differs from existing IQ tests in a way that could affect the precise identity or operant configuration of the measured attribute, then for reasons of logical consistency and descriptive clarity, it cannot be called an "IQ test". Though designed to measure intelligence, it will not necessarily perform this measurement in the same way that an IQ test performs it. Even if its preliminary data seem to correlate highly with IQ data, this is not proof that IQ is being directly measured; the test could be measuring some other set of mental performance factors which happens to coincide with IQ under more or less ambiguous conditions. Because resolving such issues can require more time and data than are readily available, it is inadvisable to call a test an "IQ test" if it differs markedly in structure or protocol from known members of the class.

To avoid the problem of rendering a specific *a priori* definition of what any such test will measure, it suffices to create a generic alternative description covering all tests which differ in structure or protocol from ordinary IQ tests, and for which high positive correlation with IQ has not yet been established. This new term must refer to a measurable quantity that is specific to the tests it describes, and that may or may not equate to that which is measured by garden variety IQ tests. Since IQ is conventionally synonymous with "general intellectual ability", the term *IEQ* – standing for *Intellectual Efficacy* (or *Effect*) *Quantitator* – is the natural choice. IEQ denotes an arbitrary form of mental efficacy identical to performance on a specific test, thus circumventing questions regarding the extent to which that test measures exactly what is measured by ordinary IQ tests.

Although not every IEQ test is an IQ test, every IQ test is an IEQ test. IEQ is merely an effect-oriented generalization of IQ which spans certain quantifiable effects of that which we intuitively understand as "intelligence". Once it is established that a particular group of tests uniformly measures the same aspect of the same set of intelligence factors, the IEQ designation can be replaced with a more specific description ... e.g., the still-vague "IQ". But until then, the scientific need for clarity and consistency in the absence of a general theory of intelligence requires that IEQ replace IQ as the primary classification of any set of metrical constraints on intellectual production.

Far from being a semantical catchall, IEQ is rich in theoretical content. IEQ tests that are obviously measuring some aspect of what we understand as "intelligence", but display lower-than-normal positive correlation with ordinary IQ tests, are necessarily measuring a different kind or aspect of intelligence than is ordinarily measured. For example, consider those IEQ tests focussing on the extreme upper end of the intellectual scale, in which the mind lays protracted siege to very difficult problems of high intrinsic complexity. Here, where taking a test begins to resemble a persistent attack on a multifaceted, time-consuming real-world problem, computational demands placed on the mind may differ markedly from those associated with timed, low-complexity IQ tests. Extended concentration, multi-level parallelism, and intercontextual coordination of simultaneous subroutines begin to displace the sort of fire-and-forget linear computation involved in the completion of disconnected analogies, number series and picture sequences, and the spatial dimension of cognition starts to outweigh the temporal. Computational time and space become distinct "metafactors" of intelligence, respectively defining two possible directions – velocity and complexity – for the upward extrapolation of IQ. IEQ, spanning both directions, is the extrapolative medium.

There is an important additional reason for this change of terminology. While the benefits of professional standardization are often considerable, no scholarly or professional organization should be permitted to restrict scientific research, particularly in a field in desperate need of new ideas. Where this occurs, ideological politics can too easily lead to theoretical and methodological stagnation. In recent years, organized pressure has been successfully exerted on state legislatures to grant licensed psychologists exclusive proprietorship over the design, administration and scoring of IQ tests. In New York and California, for example, it is now illegal for a nonpsychologist to test intelligence or any other mental ability or attribute. While the intent is obviously to guard the public from psychological charlatanism while protecting the economic welfare of licensed psychologists, such laws do a tangible disservice not only to those talented amateurs with fresh ideas to contribute, but to the field as a whole.

Because the term IEQ denotes a measurement of *test-specific efficacy* rather than a general mental ability or attribute, it is not covered by the laws in question. The entirety of civilization is a product of the human intellect and thus a manifestation of intellectual efficacy; were only licensed psychologists permitted to measure intellectual efficacy, virtually no one but a licensed psychologist could measure anything not occurring in nature. For instance, because schoolwork is a manifestation of intellectual efficacy bearing a quantitative relationship to intellectual attributes like intelligence and imagination, no one but a licensed psychologist could legally grade schoolwork! Once the conventional psychological distinction between mind and environment is suspended, there is no clear way to draw the line. Because language is the medium of law, jurisprudence amounts to an extended exercise in applied semantics. Consequently, no judge or jury can fail to recognize the clear and profound distinction between the near-antonyms ability, as resident in the mind, and effect or efficacy as expressed in the external environment ... especially when the science of psychology is explicitly defined on this very distinction.

A full account of the IQ-IEQ distinction would occupy many pages. However, two other points warrant quick mention. First, there is a subliminally potent near-homonymy between IQ and IEQ. Anyone who doubts it need merely say "IQ, IEQ" as rapidly as possible. And second, because statistical analysis is legally restricted only when applied to objects of legal restriction – and as we have just seen, IEQ is not such an object – IEQ tests can be tentatively normed with the use of whatever data might be deemed relevant. In other words, every raw score can be assigned an IEQ number bearing qualified description as an "IQ-extrapolated deviation", a percentile computed, and a "theoretical IQ equivalent" noted. Although every IEQ test should contain a disclaimer distinguishing the latter number from an IQ score – in some cases, converting it to an actual IQ would almost certainly require a substantial unknown correction factor – its inclusion is justified for orientative purposes.

For all of these reasons, amateur test designers and professional psychologists engaged in the construction, administration, or scoring of nonstandard "IQ tests" should immediately relabel their tests as suggested. In so doing, they can protect themselves from unjust legal prosecution, deflect ill-conceived proprietary resentment, and rest assured that they are helping to preserve the openness and integrity of much-needed research on the nature and application of human intelligence.

On High-Ceiling Intelligence Tests

Mensa, with a 1-in-50 qualification standard and hundreds of thousands of members worldwide, is the flagship of the high IQ community. Several years ago, the British barrister Lancelot Ware, one of the founders of Mensa, deplored the fact that it had strayed from its original purpose: to provide society with a pool of brilliant minds that could work on urgent problems too difficult for normal people to solve, including problems not falling within any single academic discipline. To his way of thinking, the proper function of high IQ societies like Mensa has always been to serve as an intellectual resource independent of academia. But as the years passed, it became evident to Ware and his fellow Founding Fathers that Mensa's primary function was not to solve pressing problems for the sake of all mankind, but to afford high-IQ people a chance to trade puzzles and socialize. Clearly disappointed, he likened Mensa's focus on puzzles and entertainment to "mental masturbation".

The super-high IQ community, consisting of a rising tower of societies with progressively more rarified qualification standards, is largely populated with Mensans and ex-Mensans who grew tired of their fun-loving compatriots in the HIQ melting pot. Yearning for higher standards of intellectual fellowship, they exhausted the ranges of ordinary IQ tests and designed new tests of their own, anchor-norming them on standard IQ test statistics in an attempt to extend their measurement capabilities. [To "anchor-norm" an intelligence test is to evaluate its statistics on the basis of known statistics on other tests taken by its subjects.] Since these tests must by definition consist of relatively harder problems – problems of greater intrinsic complexity than those found on ordinary IQ tests – they can take considerably more time to solve, often days or even weeks. On the other hand, the fact that ordinary IQ tests must be completed within a short period of time effectively limits the complexity of their items, thus limiting the degree to which they can measure one's ability to solve complex problems (i.e., one's "intelligence"). This limitation is referred to as the ceiling of the test.

Because not everyone has the time or money to sit for days or weeks in a psychologist's office working on a test, the new high-ceiling tests are typically untimed and unsupervised. In other words, they eliminate explicit constraints on speed and cheating. But careful consideration reveals that these constraints are either inappropriate or unnecessary for high-complexity tests. First, the speed factor of intellectual performance has been shown by such eminent psychologists as Arthur Jensen to be independent of "g", or pure fluid intelligence, and this means that the one who solves easy problems the most quickly need not be the one who can solve the hardest problems. For example, while Albert Einstein reputedly admitted to being a slow thinker, few would deny his extraordinary "intelligence", or ability to solve much harder problems than most people can solve. In addition, eliminating time constraints allows the subject to work on test problems when best able to focus, thus eliminating interference from factors like preoccupation and test anxiety. Second, cheating is possible only if test problems are already in circulation and their answers can be looked up or copied from the answer sheet of someone able to solve them. But if the problems are both new and so hard that those who can solve them are few and far between, then there is no book in which one can look up the test answers, and there is no nearby "A-student" from whose answer sheet one can copy. Every test subject is essentially on his or her own (of course, tests must be "retired" once verifiably correct answers begin circulating, e.g. via the Internet, but this takes considerably longer than might be expected).

One could no doubt find mainstream psychometricians who disagree with the premise that intelligence can be accurately measured in this way. But just as there are said to be as many theories about any point of economics as there are economists, different psychometricians queried on the same point of intelligence and its measurement are apt to give different answers. Essentially, this owes to the fact that intelligence has never been satisfactorily defined; indeed, many would say that it has never been defined at all. The high-ceiling measurement strategy is as valid as any other in such a context, and more so for the following reason. Intuitively and historically, a genius is somebody who can recognize and formulate an important problem and then solve it, even if it takes years to do so. He (or she) who succeeds in locating and solving even a single great problem earns the title. High-ceiling tests are merely an attempt to take intelligence testing in this direction, replacing the speedy solution of shallow problems with the thoughtful, methodical solution of deeper problems. Because the value of ordinary IQ tests resides entirely in the accuracy with which they predict the ability to solve deep problems outside the examination room, and thus in the prediction of real-world intellectual achievement, high-ceiling tests are in some ways superior to their timed counterparts; they afford direct access to something that ordinary IQ tests can measure only indirectly.

In this light, it almost appears that the statistical dependencies should be reversed, and that ordinary IQ tests should be anchor-normed on high-ceiling test statistics.

The issues outlined above tie into another crucial psychometric distinction, that between ratio IQ and deviation IQ. Originally, IQ literally meant "intelligence quotient"; it was defined as the ratio of mental age to chronological age, multiplied by 100. Thus, an intellectually precocious child who exhibits the same problem-solving ability as a normal older child has a "high (ratio) IQ". Unfortunately, the number of extremely high ratio scores varies with age; most of the highest scores are generated by young children, making it falsely appear that the incidence of genius decreases as children approach adulthood. This has precipitated a shift of emphasis to deviation IQ, which ranks a subject's raw test score within the subject's own chronological age group (it is called "deviation IQ" because it is gauged in terms of a common if somewhat oxymoronic statistical measure, the standard deviation). By "grading on the curve" with respect to age, psychometrics avoids the absurd conclusion that genius is mainly a prerogative of childhood.

The distinctions among chronological age groups reflect distinctions in precisely what is being measured within each group. Simplistically, we can reduce these distinctions to a single distinction between two related but inequivalent criteria, relative acceleration of mental development during childhood and relative level of mental development finally achieved. This implies that developmental precocity cannot be regarded as an infallible predictor of adult mental development; there is no guarantee that a child's IQ, as expressed in terms of deviation from the mean within his or her chronological age group, will turn out to be an accurate measure of terminal problem-solving ability. In principle, it is possible for children whose minds develop at different rates to finally achieve the same level of adult problem-solving ability, and for children in a developmental dead heat to finally achieve different levels of adult problem-solving ability. Since it is adult problem-solving ability that is important – after all, it is adults who are responsible for solving society's important problems – childhood (developmental) IQ ultimately takes a back seat to adult IQ. And because genius-level adult IQ can only be measured by high-ceiling, high-complexity power tests of the kind described above, it makes sense to use power tests as primary qualification instruments for super-HIQ "genius societies".

But even though childhood IQ and adult IQ are not perfectly synonymous, there is enough of a connection between them to justify the treatment of extremely precocious children as a precious human resource. This is why many in the super-HIQ community have become interested in establishing a "mentoring program" in which identifiably gifted children can receive advice and encouragement from high-intelligence adults. By combining such projects with sincere attempts to solve a few of the more urgent problems facing humanity (and the accompanying material deals with one such attempt), perhaps the high-IQ community can salvage Lance Ware's idealistic dream after all.

The counselors and psychologists of the world's public education systems have a practical problem, namely how to determine which students should receive what kinds of attention. Due to the practical constraints within which this problem must be solved, certain convenient psychometric assumptions have been adopted, and as is often the case with such assumptions, claims of factuality and absolutism eventually follow. Before long, exceptions are ruled out on what amounts to an *a priori* basis. But reality refuses to be confined by the strictures of dogma, and bad assumptions eventually unravel despite their convenience. Human intelligence is almost certainly a case in point; once a person emerges from the artificial world of formal education, there are other, more accurate ways to gauge intelligence ... other protocols. Ultimately, the only measure of genius is intellectual production, and if standard IQ tests fail to predict its presence or absence in any given case, then it is they (and not the productions) that must be considered "invalid".

Before closing, I should mention that it is my personal belief that although IQ tests can at times afford insight into human intelligence, intelligence is far too protean to be definitively determined by IQ tests (particularly the rigid tests and testing protocols of mainstream psychometrics, which presume that intelligence can always be pinned down and accurately measured at the one-sided convenience of educators and psychologists). There is far more to it than that. The mind of every human being is a microcosm, and the only real issue is his or her access to the associated intellectual potential at a given point in time, with respect to a given set of tasks. While those consistently displaying relatively high levels of access to their potential may be better-equipped to shine in contexts that put a high premium on abstract thought, the sheer complexity of human psychology precludes the absolute measurement of human intelligence by currently available means. Perhaps in the future, it will be possible to scan a person's brain and infer limitations on its capabilities from elements of structure and neural reflex alone. But that time has not yet come.

On Absolute Truth and Knowledge

First, a word on the title of this essay. *Absolute knowledge* is absolutely true, and *absolute truth* is the definitive predicate of absolute knowledge. That is, if something is known with absolute certainty, then it can be subjected to tests affirming its truth, while if something can be affirmatively tested for truth, then it is known with certainty by the tester. This applies whether the tests in question are perceptual or inferential. Where knowledge can denote either direct embodiment or internal modeling by an arbitrary system, and *test* denotes a straightforward systemic efficacy criterion, *knower* and *tester* can refer to reality at large. In this generalization, truth and knowledge are identical. While it is possible to split, splice and braid countless philosophical hairs over the connotations respectively attached to truth and knowledge, this simple generalized relationship conveniently spares us the necessity. It is with this express understanding that these terms and phrases are employed herein.

To perceive one and the same reality, human beings need a kind of "absolute knowledge" wired into their minds and nervous systems. The structure and physiology of their brains, nerves and sense organs provide them, at least in part, with elementary cognitive and perceptual categories and relationships in terms of which to apprehend the world. This "absolute" kind of knowledge is what compels the perceptions and logical inferences of any number of percipients to be mutually consistent, and to remain consistent over time and space. Without the *absoluteness* of such knowledge – without its universality and invariance – we could not share a common reality; our minds and senses would lie and bicker without respite, precipitating us into mental and sensory chaos. Time and space, mind and matter, would melt back into the haze of undifferentiated potential from which the universe is born.

Given the fact that absolute knowledge is a requisite of our collective ability to sustain a perceptually consistent universe, it is nothing short of astonishing that there are people who react with incredulity or derision at any mention of its possible existence. Their attitude seems to be that the very idea smacks of "hubris", being nothing but an empty pretense exceeding the capacity of the small and overly-challenged human mind. The truth, however, is that hubris is nowhere more evident than among those holding irrational opinions in contempt of logic, and denying the existence of absolute knowledge is a case in point. In fact, the entire history of philosophy and science can be characterized as an undying quest for absolute knowledge ... a timeless attempt to comprehensively extend the *a priori* and analytical into the realm of the apparently *a posteriori* and synthetic. This quest includes the efforts of researchers from many fields, from physics and cosmology to philosophy and computer science.

The Holy Grail of this quest is known as the TOE, or *Theory of Everything*. A TOE purports to be absolute truth by an implicit *reductio ad absurdum*: if it does not constitute absolute truth, then its truth can be relativized to a partial context within reality at large, in which case it is not a theory of *everything*. Thus, if a TOE exists, it falls squarely under the heading of absolute knowledge. But unfortunately, the proper method for constructing such a theory has not been entirely obvious, particularly to theorists steeped in the ambiguities and paradoxes of four centuries of post-Cartesian science and philosophy. As science has advanced and philosophy has wearily tried to keep pace, their once-stentorian claims of absolute truth have been all but extinguished, and the mainstream search for a TOE has lately been pursued without a clear understanding of what is being sought.

The apparent absence of a TOE notwithstanding, has any kind of absolute knowledge ever been scientifically formulated? Yes, in the form of *logical tautologies*. A tautology is a sentential relation, i.e. a formula consisting of variables and logical connectives, with the property that it is true for all possible assignments of Boolean truth values (*true* or *false*) to its variables. For example, the statement "if *x* is a sentence, then either *x* or *not-x* (but not both) must be true" is a tautology because no matter which truth values are consistently applied to *x* and *not-x*, the statement is unequivocally true. Indeed, tautologies comprise the axioms and theorems of 2-valued logic itself, and because all meaningful theories necessarily conform to 2-valued logic, define the truth concept for all of the sciences. From mathematics and physics to biology and psychology, logical tautologies reign supreme and inviolable.

That a tautology constitutes absolute truth can be proven as follows. First, logic is absolute within any system for which (a) the complementary truth values T (true) and F (false) correspond to systemic inclusion and exclusion, a semantic necessity without which meaningful reference is impossible; and (b) lesser predicates and their complements equal subsystemic inclusion and exclusion. Because a tautology is an axiom of 2-valued logic, violating it disrupts the T/F distinction and results in the corruption of informational boundaries between perceptual and cognitive predicates recognized or applied in the system, as well as between each predicate and its negation. Thus, the observable fact that perceptual boundaries are intact across reality at large implies that no tautology within its *syntax*, or set of structural and functional rules, has been violated; indeed, if such a tautology ever *were* violated, then reality would disintegrate due to corruption of the informational boundaries which define it. So a tautology is "absolute truth" not only with respect to logic, but with respect to reality at large.

What does this mean? Uncertainty or non-absoluteness of truth value always involves some kind of confusion or ambiguity regarding the distinction between the sentential predicates *true* and *false*. Where these predicates are applied to a more specific predicate and its negation – e.g., "it is *true* that the earth is round and *false* that the earth is not-round" – the confusion devolves to the contextual distinction between these lesser predicates, in this case *round* and *not-round* within the context of *the earth*. Because all of the ambiguity can be localized to a specific distinction in a particular context, it presents no general problem for reality at large; we can be uncertain about whether or not the earth is round without disrupting the logic of reality in general. However, where a statement is directly about reality in general, any disruption of or ambiguity regarding the T/F distinction disrupts the distinction between *reality* and *not-reality*. Were such a disruption to occur at the level of basic cognition or perception, reality would become impossible to perceive, recognize, or acknowledge as something that "exists".

By definition, this is the case with regard to our cognitive-perceptual *syntax*, the set of structural and inferential rules governing perception and cognition in general. Since a tautology is a necessary and universal element of this syntax, tautologies can under no circumstances be violated within reality. Thus, they are "absolute knowledge". We may not be able to specify every element of absolute knowledge, but we can be sure of two things about it: that it *exists* in reality to the full extent necessary to guarantee its non-violation, and that no part of it yet to be determined can violate absolute knowledge already in hand. Whether or not we can write up an exhaustive itemized list of absolute truths, we can be sure that such a list exists, and that its contents are sufficiently "recognizable" by reality at large to ensure their functionality. Absolute truth, being essential to the integrity of reality, *must exist* on the level of reference associated with the preservation of global consistency, and may thus be duly incorporated in a theory of reality.

On the other hand, the fact that any reasonable definition of "absolute truth" amounts to tautology can be shown by reversing this reasoning. Since absolute truth must be *universal*, it is always true regardless of the truth values of its variables (where the variables actually represent objects and systems for which specific state-descriptions vary in space and time with respect to truth value). Moreover, it falls within its own scope and is thus *self-referential*. By virtue of its universality and self-reference, it is a universal element of *reality syntax*, the set of structural and functional rules governing the spatial structure and temporal evolution of reality. As such, it must be *unfalsifiable*, any supposition of its falsehood leading directly to a *reductio ad absurdum*. And to ice the cake, it is unavoidably implicated in its own justification; were it ever to be violated, the T/F boundary would be disrupted, and this would prevent it (or anything else) from being proven. Therefore, it is an active constraint in its own proof, and thus possesses all the characteristics of a tautology.

To recap, the characteristic attributes of a logical tautology are as follows: (1) it cannot be disobeyed, which implies that it has universal scope and thus accepts and truthfully predicates all closed sentential (predicative) structures, including itself *and logic in its entirety*, under assignment to its own variables; and (2) it is self-affirming or self-justifying and figures in its own definition or demonstration within the associated grammar. Obviously, (1) and (2) are not independent; (1) implies that a tautology is a universal, self-similar, metalogical *element of syntax* of the language and metalanguages of which it is a part, while (2) says that it is a *critical* element of syntax that cannot be eliminated without compromising the integrity of the syntax as a whole (thus, any supposition that it is false or eliminable reduces itself to absurdity by syntactic rules of inference, forcing the syntax to "protect itself" through *reductio ad absurdum*). Since any reasonable syntactic and/or semantic definition of *absolute truth* bestows upon it the properties of necessity and truthwise invariance with respect to content, it is unquestionably tautological in nature.

Accordingly, it is desirable to formulate reality theory as a tautology. To whatever extent this can be done, the theory constitutes "absolute knowledge" and is therefore eligible as a TOE. This suffices to show that if the form of absolute knowledge hopefully referred to as a TOE exists, it must be tautological. Next we will show that a TOE and its universe can be related in such a way that the theory is semantically tautological *with respect to its universe*, i.e. that (a) the theory is intrinsically tautological, and (b) its tautological structure is modeled by its universe. And in the course of doing so, we will show that it is indeed possible to ensure by the method of constructing this theory that its universe coincides with reality at large, and thus that it constitutes a valid theory of reality. Specifically, the construction will incorporate one or more attributes that are necessarily modeled by reality at large, and that simultaneously ensure the theory's tautological structure.

How can a TOE, or comprehensive theory of reality, be structured as a tautology? First, by definition, a TOE is *universal*; this is implied by the E, which stands for *Everything*. Thus, it is comprehensive. Second, it is *self-referential*; a theory of everything, being a part of the "everything" to which it refers, must refer to itself. More precisely, a TOE must be totally recursive in a manner analogous to logic, each atom referring exclusively to other parts of the theory, and be able to refer to itself *in part and in whole* in order to possess full logical closure. This can be arranged by incorporating one or more self-representative variables and their definitive relationships, up to and including a dynamic variable representing the theory as a whole (in fact, the theory can incorporate a "hology" predicate that goes considerably farther; instead of merely containing itself as a variable, a theory equipped with such a predicate can *everywhere* contain itself by virtue of self-similarity or self-distribution). Because it represents a theory of perceptual reality, this variable contains all elements of cognitive syntax and their perceptual contents; since variables can be defined in general terms without specifically enumerating their contents, we do not need to know exactly what it contains in order to use it. And third, because logic is the primary ingredient of cognitive-perceptual syntax, the self-referential TOE refers to logic in part and in whole and is therefore *metalogical*. Thus, it can incorporate a kind of ultimate truth predicate that asserts its own tautological structure and guarantees that no matter what (semantic and other) kinds of paradox may *arise* within the theory, they can always be *resolved* within the theory. A theory possessing all three of these properties is called a *supertautology*, denoting the reality-theoretic counterpart of a logical tautology.

Let us now attend to some of the details of constructing a supertautology. First, we repose the *a priori* and analytic knowledge that we are given in the form of cognitive syntax, including logic and all of its implications, in a variable to which we apply (a) the rules of logic itself; (b) three recursively-related metalogical axioms that are themselves true *a priori* and analytically implied by each other (in a word, *self-evident*). Note again that in creating and assigning content to this variable, we do not have to enumerate all of its contents; we can refer to them *en masse* by their joint characteristic, namely the "absoluteness" necessary to ensure perceptual and inferential consistency. Since a theory falls under the mathematical definition of a *language*, it is natural to refer to the contents in question as the "rules of syntax" of that language, or simply as its *syntax*; thus, the TOE recursively contains a variable representing its own syntax, permitting the manipulation of that variable and the grammatical extraction of its implications according to syntactic rules. This recursive construction makes the "absoluteness" of the variable (and theory) *logically heritable*, conferring absoluteness on whatever is inferred within the system. Together, the "implicate" variable and its "explicate" theoretic medium comprise a bootstrapped extension of the self-referential syntax of logic itself, letting that syntax be "mined" for a potential wealth of hidden analytic content.

The key to applying this knowledge scientifically is the semantic functionality of the three metalogical axioms adjoining the object-level syntax. Conveniently, these (recursively related) axioms can be thought of in terms of a trio of property-principle pairs, the "Three C's" and the "Three M's". The C's are three properties that a TOE must inevitably possess, namely *Comprehensiveness*, *Closure* and *Consistency*, while the M's are metalogical axioms respectively associated with those properties. These principles are the *Mind Equals Reality Principle* (associated with comprehensiveness), the *Metaphysical Autology Principle* (associated with closure), and the *Multiplex Unity Principle* (associated with consistency), respectively abbreviated M=R, MAP and MU. We have already been partially introduced to these principles in all but name, and in any case need only one of them to proceed farther: M=R. Concisely, M=R asserts that there exists a semantic (language-to-universe) correspondence between objective reality and the absolute subjective rules of perception and inference, i.e. cognitive and perceptual syntax. This correspondence defines a morphism, *incoversion*, predicating the assignment of a certain structural predicate, *hology*, to the universe-language/metalanguage system (see Introduction to the CTMU).

Hology, a special kind of self-similarity conferring supertautological status, equals the relationship of the TOE and its universe to the self-representative variable by which it is encapsulated. Hology means that the syntax by which reality configures, recognizes and processes itself is the image of a distributed endomorphism, the incoversion morphism, surjecting the objective *self-intersect* (distributed component) of reality onto every interior point and region of reality as transductive syntactic potential, i.e. as general rules of transduction to be variously expressed by objects at any location. Although real objects generally access and express only a small part of this syntax, combinations of interacting objects may express and access more of it by mutual input-to-output behavioral transduction; through this kind of behavioral transduction, the self-intersect, though generally composed of distributed rules applying everywhere in reality, resolves to a *Distributed Conditional Form* (DCF) explicitly containing all of the local systems and states generated by those rules. The self-intersect and its DCF resolution comprise the syntax and language of reality. Because hology maps the *reality* syntax to our *cognitive* syntax – because the self-intersect plays dual objective and subjective roles – perceptible objects and processes tautologically conform to our innate perceptual categories, making the TOE a *supertautology* comprising the purest and most universal kind of absolute truth.

The above reasoning subjects the absolute (*a priori*) knowledge in our minds to a kind of recursive "squaring" operation, causing it to self-explicate as its own medium and projecting it onto external reality. This repetitive operation resembles the mutual reflection of a pair of polymorphic mirrors, one labeled *mind* and the other labeled *reality*, that faithfully reflect each other's evolving image. Although one might suspect that the tautological nature of the construction renders it barren of interest, this would be akin to saying that a squaring operation never yields more than the original number. While that might be true for a featureless numeric identity (e.g. $1^2 = 1$), the cognitive syntax of the human mind is far from "featureless". In recursive self-combination, it is capable of generating a universe, and a theory constructed according to this recursive relationship is capable of veridically *capturing* that universe. Indeed, there is a sense in which the TOE, and all of the absolute knowledge it holds, is *identical to the universe it describes*. But the meaning of this statement – and it is a statement that is pregnant with meaning – lies beyond the purpose at hand.

The CTMU is a theory of reality, or TOE, that has been constructed according to this blueprint. If, as a rationalist, one insists that absolute truth and knowledge are exclusively mathematical, then the CTMU is mathematics; if, as an empiricist, one insists that they reside exclusively in our direct perceptions of reality, then the CTMU is embodied in our direct perceptions of reality (including our direct perceptions of the *comprehensiveness*, *closure* and *consistency* of reality). The truth, of course, is that by the method of its construction, it is both. But in any case, would-be pundits who cling blindly to folk-epistemological "absolutes" like *truth is never more than provisional, science is inherently without stability and there are no such things as absolute truth and knowledge* are urgently in need of an intellectual awakening, and until it comes, should refrain from disseminating their irrational opinions to others who might gullibly mistake them for fact. Such truisms have their contexts, but these contexts do not include the highest levels of discourse regarding truth and knowledge, and they do not include the CTMU.

There is, of course, more to the CTMU than just its supertautological structure. For example, it incorporates a new conceptualization of spacetime, resolves numerous high level reality-theoretic paradoxes, and establishes a bridge between science and theology, all with considerably more detail than this brief monograph allows. But as regards "absolute truth and knowledge", its status as a supertautology is necessary and sufficient to explain why it is uniquely qualified for the title. If the simplicity and elegance of its design seems "too obvious", "too convenient" or "too good to be true", this is certainly no fault of the theory or its author; at best, it testifies to the opacity of certain formerly useful but outworn conceptual barriers erected in science and philosophy over the last several centuries, and to the inertia of the scientific and academic establishments which tend them.

One final note. The CTMU is neither intended nor presented as an encyclopedic compendium of absolute truth. It is meant only to provide a comprehensive, consistent and self-contained (and to that extent "absolute") logical framework for bridging the gaps between apparently unrelated fields of knowledge, helping to locate and correct fundamental inconsistencies within and among these fields, and developing new kinds of knowledge that might arise from the intersections of fields which already exist. Because the real universe is everywhere in the process of self-creation, human knowledge can and must continue to grow. The CTMU is intended not as a brittle, undersized pot that will root-bind and choke this growing knowledge, but as fertile and well-aerated soil through which it can spread. By its very design, the CTMU will continue to accommodate and accelerate our intellectual progress ... and since there is no other theory that fully *shares* its design, it is irreplaceable for that purpose.

This completes our introduction to the topic of absolute truth and knowledge.

The Theory of Theories

You know what they say about theories: *everybody's got one*. In fact, some people have a theory about pretty much everything. That's not one Master Theory of Everything, mind you ... that's a separate theory about every little thing under the sun. (To have a Master Theory, you have to be able to tie all those little theories together.)

But what is a "theory"? Is a theory just a story that you can make up about something, being as fanciful as you like? Or does a theory at least have to seem like it might be true? Even more stringently, is a theory something that has to be rendered in terms of logical and mathematical symbols, and described in plain language only after the original chicken-scratches have made the rounds in academia?

A theory is all of these things. A theory can be good or bad, fanciful or plausible, true or false. The only firm requirements are that it (1) have a subject, and (2) be stated in a language in terms of which the subject can be coherently described. Where these criteria hold, the theory can always be "formalized", or translated into the symbolic language of logic and mathematics. Once formalized, the theory can be subjected to various mathematical tests for truth and internal consistency.

But doesn't that essentially make "theory" synonymous with "description"? Yes. A theory is just a description of something. If we can use the logical implications of this description to relate the components of that something to other components in revealing ways, then the theory is said to have "explanatory power". And if we can use the logical implications of the description to make correct predictions about how that something behaves under various conditions, then the theory is said to have "predictive power".

From a practical standpoint, in what kinds of theories should we be interested? Most people would agree that in order to be interesting, a theory should be about an important subject ... a subject involving something of use or value to us, if even on a purely abstract level. And most would also agree that in order to help us extract or maximize that value, the theory must have explanatory or predictive power. For now, let us call any theory meeting both of these criteria a "serious" theory.

Those interested in serious theories include just about everyone, from engineers and stockbrokers to doctors, automobile mechanics and police detectives. Practically anyone who gives advice, solves problems or builds things that function needs a serious theory from which to work. But three groups who are especially interested in serious theories are scientists, mathematicians and philosophers. These are the groups which place the strictest requirements on the theories they use and construct.

While there are important similarities among the kinds of theories dealt with by scientists, mathematicians and philosophers, there are important differences as well. The most important differences involve the subject matter of the theories. Scientists like to base their theories on experiment and observation of the real world ... not on perceptions themselves, but on what they regard as concrete "objects of the senses". That is, they like their theories to be *empirical.* Mathematicians, on the other hand, like their theories to be essentially *rational* ... to be based on logical inference regarding abstract mathematical objects existing in the mind, independently of the senses. And philosophers like to pursue broad theories of reality aimed at relating these two kinds of object. (This actually mandates a third kind of object, the *infocognitive syntactic operator* ... but another time.)

Of the three kinds of theory, by far the lion's share of popular reportage is commanded by theories of science. Unfortunately, this presents a problem. For while science owes a huge debt to philosophy and mathematics – it can be characterized as the child of the former and the sibling of the latter – it does not even treat them as its equals. It treats its parent, philosophy, as unworthy of consideration. And although it tolerates and uses mathematics at its convenience, relying on mathematical reasoning at almost every turn, it acknowledges the remarkable obedience of objective reality to mathematical principles as little more than a cosmic "lucky break".

Science is able to enjoy its meretricious relationship with mathematics precisely because of its queenly dismissal of philosophy. By refusing to consider the philosophical relationship between the abstract and the concrete on the supposed grounds that philosophy is inherently impractical and unproductive, it reserves the right to ignore that relationship even while exploiting it in the construction of scientific theories. And exploit the relationship it certainly does! There is a scientific platitude stating that if one cannot put a number to one's data, then one can prove nothing at all. But insofar as numbers are arithmetically and algebraically related by various mathematical structures, the platitude amounts to a thinly veiled affirmation of the mathematical basis of knowledge.

Although scientists like to think that everything is open to scientific investigation, they have a rule that explicitly allows them to screen out certain facts. This rule is called the *scientific method*. Essentially, the scientific method says that every scientist's job is to:

1. Observe something in the world.
2. Invent a theory to fit the observations.
3. Use the theory to make predictions.
4. Experimentally or observationally test the predictions.
5. Modify the theory in light of any new findings.
6. Repeat the cycle from step 3 onward.

In fact, if we regard the scientific method as a theory about the nature and acquisition of scientific knowledge (and we can), it is not a theory of knowledge in general. It is only a theory of things accessible to the senses. Worse yet, it is a theory only of sensible things that have two further attributes: they are non-universal and can therefore be distinguished from the rest of sensory reality, and they can be seen by multiple observers who are able to "replicate" each other's observations under like conditions. Needless to say, there is no reason to assume that these attributes are necessary even in the sensory realm. The first describes nothing general enough to coincide with reality as a whole – for example, the homogeneous medium of which reality consists, or an abstract mathematical principle that is everywhere true – and the second describes nothing that is either subjective, like human consciousness, or objective but rare and unpredictable ... e.g. ghosts, UFOs and yetis, of which jokes are made but which may, given the number of individual witnesses reporting them, correspond to real phenomena.

The fact that the scientific method does not permit the investigation of abstract mathematical principles is especially embarrassing in light of one of its more crucial steps: "invent a theory to fit the observations." A theory happens to be a logical and/or mathematical construct whose basic elements of description are mathematical units and relationships. If the scientific method were interpreted as a blanket description of reality, which is all too often the case, the result would go something like this: "Reality consists of all and only that to which we can apply a protocol which cannot be applied to its own (mathematical) ingredients and is therefore unreal." Mandating the use of "unreality" to describe "reality" is rather questionable in anyone's protocol.

What about mathematics itself? The fact is, science is not the only walled city in the intellectual landscape. With equal and opposite prejudice, the mutually exclusionary methods of mathematics and science guarantee their continued separation despite the (erstwhile) best efforts of philosophy. While science hides behind the scientific method, which effectively excludes from investigation its own mathematical ingredients, mathematics divides itself into "pure" and "applied" branches and explicitly divorces the "pure" branch from the real world. Notice that this makes "applied" synonymous with "impure". Although the field of applied mathematics by definition contains every practical use to which mathematics has ever been put, it is viewed as "not quite mathematics" and therefore beneath the consideration of any "pure" mathematician.

In place of the scientific method, pure mathematics relies on a principle called the *axiomatic method*. The axiomatic method begins with a small number of self-evident statements called *axioms* and a few *rules of inference* through which new statements, called *theorems*, can be derived from existing statements. In a way parallel to the scientific method, the axiomatic method says that every mathematician's job is to:

1. Conceptualize a class of mathematical objects.
2. Isolate its basic elements, its most general and self-evident principles, and the rules by which its truths can be derived from those principles.
3. Use those principles and rules to derive theorems, define new objects, and formulate new propositions about the extended set of theorems and objects.
4. Prove or disprove those propositions.
5. Where the proposition is true, make it a theorem and add it to the theory.
6. Repeat from step 3 onwards.

The scientific and axiomatic methods are like mirror images of each other, but located in opposite domains. Just replace "observe" with "conceptualize" and "part of the world" with "class of mathematical objects", and the analogy practically completes itself. Little wonder, then, that scientists and mathematicians often profess mutual respect. However, this conceals an imbalance. For while the activity of the mathematician is integral to the scientific method, that of the scientist is irrelevant to mathematics (except for the kind of scientist called a "computer scientist", who plays the role of ambassador between the two realms). At least in principle, the mathematician is more necessary to science than the scientist is to mathematics.

As a philosopher might put it, the scientist and the mathematician work on opposite sides of the Cartesian divider between mental and physical reality. If the scientist stays on his own side of the divider and merely accepts what the mathematician chooses to throw across, the mathematician does just fine. On the other hand, if the mathematician does not throw across what the scientist needs, then the scientist is in trouble. Without the mathematician's functions and equations from which to build scientific theories, the scientist would be confined to little more than taxonomy. As far as making quantitative predictions were concerned, he or she might as well be guessing the number of jellybeans in a candy jar.

From this, one might be tempted to theorize that the axiomatic method does not suffer from the same kind of inadequacy as does the scientific method ... that it, and it alone, is sufficient to discover all of the abstract truths rightfully claimed as "mathematical". But alas, that would be too convenient. In 1931, an Austrian mathematical logician named Kurt Gödel proved that there are true mathematical statements that cannot be proven by means of the axiomatic method. Such statements are called "undecidable". Gödel's finding rocked the intellectual world to such an extent that even today, mathematicians, scientists and philosophers alike are struggling to figure out how best to weave the loose thread of undecidability into the seamless fabric of reality.

To demonstrate the existence of undecidability, Gödel used a simple trick called *self-reference*. Consider the statement "this sentence is false." It is easy to dress this statement up as a logical formula. Aside from being true or false, what else could such a formula say about itself? Could it pronounce itself, say, unprovable? Let's try it: "This formula is unprovable". If the given formula is in fact unprovable, then it is true and therefore a theorem. Unfortunately, the axiomatic method cannot recognize it as such without a proof. On the other hand, suppose it is provable. Then it is self-apparently false (because its provability belies what it says of itself) and yet true (because provable without respect to content)! It seems that we still have the makings of a paradox ... a statement that is "unprovably provable" and therefore absurd.

But what if we now introduce a distinction between *levels* of proof? For example, what if we define a *metalanguage* as a language used to talk about, analyze or prove things regarding statements in a lower-level *object language*, and call the base level of Gödel's formula the "object" level and the higher (proof) level the "metalanguage" level? Now we have one of two things: a statement that can be metalinguistically *proven* to be linguistically *unprovable*, and thus recognized as a theorem conveying valuable information about the limitations of the object language, or a statement that cannot be metalinguistically proven to be linguistically unprovable, which, though uninformative, is at least no paradox. Voilà: self-reference without paradox! It turns out that "this formula is unprovable" can be translated into a generic example of an undecidable mathematical truth. Because the associated reasoning involves a metalanguage of mathematics, it is called "metamathematical".

It would be bad enough if undecidability were the only thing inaccessible to the scientific and axiomatic methods together. But the problem does not end there. As we noted above, mathematical truth is only one of the things that the scientific method cannot touch. The others include not only rare and unpredictable phenomena that cannot be easily captured by microscopes, telescopes and other scientific instruments, but things that are too large or too small to be captured, like the whole universe and the tiniest of subatomic particles; things that are "too universal" and therefore indiscernible, like the homogeneous medium of which reality consists; and things that are "too subjective", like human consciousness, human emotions, and so-called "pure qualities" or *qualia*. Because mathematics has thus far offered no means of compensating for these scientific blind spots, they continue to mark holes in our picture of scientific and mathematical reality.

But mathematics has its own problems. Whereas science suffers from the problems just described – those of indiscernibility and induction, nonreplicability and subjectivity – mathematics suffers from undecidability. It therefore seems natural to ask whether there might be any other inherent weaknesses in the combined methodology of math and science. There are indeed. Known as the *Löwenheim–Skolem theorem* and the *Duhem–Quine thesis*, they are the respective stock-in-trade of disciplines called *model theory* and the *philosophy of science* (like any parent, philosophy always gets the last word). These weaknesses have to do with ambiguity ... with the difficulty of telling whether a given theory applies to one thing or another, or whether one theory is "truer" than another with respect to what both theories purport to describe.

But before giving an account of Löwenheim–Skolem and Duhem–Quine, we need a brief introduction to model theory. Model theory is part of the logic of "formalized theories", a branch of mathematics dealing rather self-referentially with the structure and interpretation of theories that have been couched in the symbolic notation of mathematical logic ... that is, in the kind of mind-numbing chicken-scratches that everyone but a mathematician loves to hate. Since any worthwhile theory can be formalized, model theory is a *sine qua non* of meaningful theorization.

Let's make this short and punchy. We start with *propositional logic*, which consists of nothing but tautological, always-true relationships among sentences represented by single variables. Then we move to *predicate logic*, which considers the content of these sentential variables ... what the sentences actually say. In general, these sentences use symbols called *quantifiers* to assign attributes to variables semantically representing mathematical or real-world objects. Such assignments are called "predicates". Next, we consider *theories*, which are complex predicates that break down into systems of related predicates; the *universes* of theories, which are the mathematical or real-world systems described by the theories; and the descriptive correspondences themselves, which are called *interpretations*. A *model* of a theory is any interpretation under which all of the theory's statements are true. If we refer to a theory as an *object language* and to its referent as an *object universe*, the intervening model can only be described and validated in a metalanguage of the language-universe complex.

Though formulated in the mathematical and scientific realms respectively, Löwenheim–Skolem and Duhem–Quine can be thought of as opposite sides of the same model-theoretic coin. Löwenheim–Skolem says that a theory cannot in general distinguish between two different models; for example, any true theory about the numeric relationship of points on a continuous line segment can also be interpreted as a theory of the integers (counting numbers). On the other hand, Duhem–Quine says that two theories cannot in general be distinguished on the basis of any observation statement regarding the universe.

Just to get a rudimentary feel for the subject, let's take a closer look at the Duhem–Quine thesis. *Observation statements*, the raw data of science, are statements that can be proven true or false by observation or experiment. But observation is not independent of theory; an observation is always interpreted in some theoretical context. So an experiment in physics is not merely an observation, but the *interpretation* of an observation. This leads to the *Duhem Thesis*, which states that scientific observations and experiments cannot invalidate isolated hypotheses, but only whole sets of theoretical statements at once. This is because a theory T composed of various laws $\{L_i\}$, i=1,2,3, ... almost never entails an observation statement except in conjunction with various auxiliary hypotheses $\{A_j\}$, j=1,2,3, ... Thus, an observation statement at most disproves the complex $\{L_i + A_j\}$.

To take a well-known historical example, let $T = \{L_1, L_2, L_3\}$ be Newton's three laws of motion, and suppose that these laws seem to entail the observable consequence that the orbit of the planet Uranus is O. But in fact, Newton's laws alone do not determine the orbit of Uranus. We must also consider things like the presence or absence of other forces, other nearby bodies that might exert appreciable gravitational influence on Uranus, and so on. Accordingly, determining the orbit of Uranus requires auxiliary hypotheses like A_1 = "only gravitational forces act on the planets", A_2 = "the total number of solar planets, including Uranus, is 7," et cetera. So if the orbit in question is found to differ from the predicted value O, then instead of simply invalidating the theory T of Newtonian mechanics, this observation invalidates the entire complex of laws and auxiliary hypotheses $T = \{L_1, L_2, L_3; A_1, A_2, ...\}$. It would follow that at least one element of this complex is false, but which one? Is there any 100% sure way to decide?

As it turned out, the weak link in this example was the hypothesis A_2 = "the total number of solar planets, including Uranus, is 7". In fact, there turned out to be an additional large planet, Neptune, which was subsequently sought and located precisely because this hypothesis (A_2) seemed open to doubt. But unfortunately, there is no general rule for making such decisions. Suppose we have two theories T_1 and T_2 that predict observations O and not-O respectively. Then an experiment is *crucial* with respect to T_1 and T_2 if it generates exactly one of the two observation statements O or not-O. Duhem's arguments show that in general, one cannot count on finding such an experiment or observation. In place of crucial observations, Duhem cites *le bon sens* (good sense), a non-logical faculty by means of which scientists supposedly decide such issues. Regarding the nature of this faculty, there is in principle nothing that rules out personal taste and cultural bias. That scientists prefer lofty appeals to *Occam's razor*, while mathematicians employ justificative terms like *beauty* and *elegance*, does not exclude less savory influences.

So much for Duhem; now what about Quine? The Quine thesis breaks down into two related theses. The first says that there is no distinction between analytic statements (e.g. definitions) and synthetic statements (e.g. empirical claims), and thus that the Duhem thesis applies equally to the so-called *a priori* disciplines. To make sense of this, we need to know the difference between *analytic* and *synthetic* statements. Analytic statements are supposed to be true by their meanings alone, matters of empirical fact notwithstanding, while synthetic statements amount to empirical facts themselves. Since analytic statements are necessarily true statements of the kind found in logic and mathematics, while synthetic statements are contingently true statements of the kind found in science, Quine's first thesis posits a kind of equivalence between mathematics and science. In particular, it says that epistemological claims about the sciences should apply to mathematics as well, and that Duhem's thesis should thus apply to both.

Quine's second thesis involves the concept of *reductionism*. Reductionism is the claim that statements about some subject can be reduced to, or fully explained in terms of, statements about some (usually more basic) subject. For example, to pursue chemical reductionism with respect to the mind is to claim that mental processes are really no more than biochemical interactions. Specifically, Quine breaks from Duhem in holding that not all theoretical claims, i.e. theories, can be reduced to observation statements. But then empirical observations "underdetermine" theories and cannot decide between them. This leads to a concept known as *Quine's holism*; because no observation can reveal which member(s) of a set of theoretical statements should be re-evaluated, the re-evaluation of some statements entails the re-evaluation of all.

Quine combined his two theses as follows. First, he noted that a reduction is essentially an analytic statement to the effect that one theory, e.g. a theory of mind, is defined on another theory, e.g. a theory of chemistry. Next, he noted that if there are no analytic statements, then reductions are impossible. From this, he concluded that his two theses were essentially identical. But although the resulting unified thesis resembled Duhem's, it differed in scope. For whereas Duhem had applied his own thesis only to physical theories, and perhaps only to theoretical hypothesis rather than theories with directly observable consequences, Quine applied his version to the entirety of human knowledge, including mathematics. If we sweep this rather important distinction under the rug, we get the so-called "Duhem–Quine thesis".

Because the Duhem–Quine thesis implies that scientific theories are underdetermined by physical evidence, it is sometimes called the *Underdetermination Thesis*. Specifically, it says that because the addition of new auxiliary hypotheses, e.g. conditionals involving "if ... then" statements, would enable each of two distinct theories on the same scientific or mathematical topic to accommodate any new piece of evidence, no physical observation could ever decide between them.

The messages of Duhem–Quine and Löwenheim–Skolem are as follows: universes do not uniquely determine theories according to empirical laws of scientific observation, and theories do not uniquely determine universes according to rational laws of mathematics. The model-theoretic correspondence between theories and their universes is subject to ambiguity in both directions. If we add this descriptive kind of ambiguity to ambiguities of measurement, e.g. the Heisenberg Uncertainty Principle that governs the subatomic scale of reality, and the internal theoretical ambiguity captured by undecidability, we see that ambiguity is an inescapable ingredient of our knowledge of the world. It seems that math and science are ... well, inexact sciences.

How, then, can we ever form a true picture of reality? There may be a way. For example, we could begin with the premise that such a picture exists, if only as a "limit" of theorization (ignoring for now the matter of showing that such a limit exists). Then we could educe categorical relationships involving the logical properties of this limit to arrive at a description of reality in terms of reality itself. In other words, we could build a self-referential theory of reality whose variables represent reality itself, and whose relationships are logical tautologies. Then we could add an instructive twist. Since logic consists of the rules of thought, i.e. of mind, what we would really be doing is interpreting reality in a generic theory of mind based on logic. By definition, the result would be a *cognitive-theoretic model of the universe*.

Gödel used the term *incompleteness* to describe that property of axiomatic systems due to which they contain undecidable statements. Essentially, he showed that all sufficiently powerful axiomatic systems are incomplete by showing that if they were not, they would be *inconsistent*. Saying that a theory is "inconsistent" amounts to saying that it contains one or more irresolvable paradoxes. Unfortunately, since any such paradox destroys the distinction between *true* and *false* with respect to the theory, the entire theory is crippled by the inclusion of a single one. This makes consistency a primary necessity in the construction of theories, giving it priority over proof and prediction. A cognitive-theoretic model of the universe would place scientific and mathematical reality in a self-consistent logical environment, there to await resolutions for its most intractable paradoxes.

For example, modern physics is bedeviled by paradoxes involving the origin and directionality of time, the collapse of the quantum wave function, quantum nonlocality, and the containment problem of cosmology. Were someone to present a simple, elegant theory resolving these paradoxes without sacrificing the benefits of existing theories, the resolutions would carry more weight than any number of predictions. Similarly, any theory and model conservatively resolving the self-inclusion paradoxes besetting the mathematical theory of sets, which underlies almost every other kind of mathematics, could demand acceptance on that basis alone. Wherever there is an intractable scientific or mathematical paradox, there is dire need of a theory and model to resolve it.

If such a theory and model exist – and for the sake of human knowledge, they had *better* exist – they use a logical metalanguage with sufficient expressive power to characterize and analyze the limitations of science and mathematics, and are therefore philosophical and metamathematical in nature. This is because no lower level of discourse is capable of uniting two disciplines that exclude each other's content as thoroughly as do science and mathematics.

Now here's the bottom line: such a theory and model do indeed exist. But for now, let us satisfy ourselves with having glimpsed the rainbow under which this theoretic pot of gold awaits us.

On Ayn Rand's Philosophy

The history of philosophy and culture is a grand, recursively-scaled dialectic. A person or school, a group or generation, espouses a thesis; the other side responds with its antithesis; the two extremes are then combined in a synthesis imbued with both perspectives. Ayn Rand can be best understood as a philosophical reactionary in this well-established tradition, the creator and champion of a dual antithesis directed against what she regarded as two devastating assaults on human individuality. The first of these assaults was religion; at the age of thirteen, she decided with certainty that she was an atheist. The second was communism, long the bane of her Russian homeland and still a geopolitical threat.

Rand's dual antithesis consists of a mixture of atheism, materialism and extreme individualism, recast as "logic and objectivity" and aggressively presented as an affirmation of human freedom and self-interest. Critics sometimes dismiss this philosophy, called *objectivism*, as a mere capitalistic twist on communism, an oversimplistic and logically insupportable brand of egoism cloaked in empty protestations of rationality. Others point to the cult-like atmosphere of various Ayn Rand organizations and study groups; while such groups ostensibly exist to promote earnest philosophical consideration of Rand's writings, a few of them are reputed to enforce cult-like obedience to her every stated opinion, forbidding critical appraisal and discouraging exposure to alternative viewpoints. But even if these criticisms are factual, they should not dissuade us from looking for strands of truth in her message.

Many young people go through a phase during which they passionately embrace Rand's philosophy. The typical Rand hero or heroine, a physically attractive nonconformist driven by a combination of iconoclasm, creative innovation, intelligence, productivity, machinelike efficiency and competitive, take-no-prisoners self-interest, can be irresistible to one struggling against social adversity to define one's individuality. It matters not to such a reader that in a world consisting solely of Rand heroes and heroines, there would be neither enough wealth and power to reward them all for their brilliance nor enough individualistic elbow room to go around, forcing them to steamroller each other until the vast majority had been flattened into a seamless backdrop of mediocrity against which the peerless few could shine like stars. Nor does it matter that while such a mindset may be psychologically viable for one whose constructive *raison d'être* demands a radical departure from the beaten path, one of less than stellar ability would be more likely to find it a delayed prescription for self-disappointment. The thing to remember is that it fills a psychological need on the part of the reader, serving as a source of strength in a time of personal need.

The holes in Rand's philosophy are easy to find. For example, despite the demonstrable inadequacy of standard empirical arguments to decide the question of the existence of God, her theology (or lack thereof) is largely based on them. As another example, she maintains that personal selfishness and ambition are virtues and altruism is a vice, thereby effectively asserting that altruism, defined as an unselfish regard for the welfare of others, is inconsistent with individual self-interest. Since there are plausible scenarios in which the survival of the human race would depend on the self-sacrifice of one individual, this amounts to an assertion that letting the human race expire in order to ensure one's personal survival is virtuous, whereas saving the human race by means of self-sacrifice is corrupt. Bearing in mind that Rand's version of ethics displays very little resemblance to the norm, this seems muddled even by the light of her "man is an end in himself" dictum. In fact, there is a sense in which the conventional definitions of altruism and selfishness are semantically inconsistent or at least incomplete; because there are multiple levels of self, including collective levels, there are multiple "levels of selfishness" *including* altruism. But Rand, believing in the unconditional independence of the individual, would not have agreed with this. She would have found the idea that altruism is selfish, and true selfishness altruistic, quite unacceptable.

But it is true nonetheless. The human species has many levels of organization, with each of which is associated a measure of utility or value. The individual comprises only one of these levels. Using mathematical decision theory, it can be shown that there are many situations in which excessive self-interest on the part of the individual is at odds with the self-interest of the larger groups to which the individual belongs, ultimately including society as a whole, with implications relating directly back to the welfare of the individual. Two well-known examples are the Tragedy of the Commons, where the overutilization of common resources leads to economic disaster, and the Prisoner's Dilemma, a lesson in mutual cooperation and the value of loyalty. Equivalently, there are many levels of "self" with respect to utility, and they are not automatically in mutual accord. In the event of a global economic breakdown involving the highest level of Self, all lesser selves go down with the ship regardless of their putative "independence". It follows that denying or downplaying higher levels of self is neither logical nor objective. When it comes to rational decision theory and economics, it turns out that no man is an island after all.

Nevertheless, Rand's literary protagonists exhibit a stunningly naive approach to deontology, eschewing altruism and blithely assuming that individual self-interest is the only kind of self-interest worth acknowledging. To the eternal disgust of those with more ability than good fortune, real life produces an endless stream of cases in which people sporting this insufferable sort of attitude reap spectacular rewards while pressing what seems to be a cheater's thumb on the scale of destiny, riding to riches and power on a combination of ability, manipulation and outrageous good fortune shamelessly rendered as ability alone. But this is possible only where individual and social self-interest coincide, and even if one is egocentrically convinced that society can drop dead and bury its own remains with regard to any point of opposition, such co-incidence cannot be taken for granted. And where it is absent, excessive pursuit of individual self-interest can cause the utility and wellbeing of individual and society alike to take a catastrophic nosedive. For this reason, enlightened individualists temper their ideological intransigence and trim their senses of entitlement in order to avoid contributing to a global economic catastrophe, continuing as best they can to live up to high standards of excellence, originality and personal integrity.

When an author's writings are powerful, appealing, and designed to illustrate a fundamental philosophical platform, it is all the more important to evaluate them fairly and objectively. When reading the work of Ayn Rand, we need not fear to be inspired by her single-minded vision of courageous individuality and creative self-actualization. After all, there are times and places in history and the everyday world which demand that the individual take a strong and independent stand. But this does not require that we dispense forever with empathy and compassion, even when the author acidly assures us that such sentiments are contemptible and counterproductive. As long as we remember that the individual is free to choose goals that are consistent with the interests of other individuals and society at large – as long as the individual is understood as a unique part of a greater whole in which resides a large part of the meaning of life – there can be little harm in a constructively motivated shot of pure selfishness every now and then provided, of course, that one resists the self-righteous temptation to chase it down with a shot of self-entitlement poured from a bottle obnoxiously labeled "Logic & Objectivity".

This futuristic novella, written in 1937, is one of Rand's earliest and most popular works. In it, she describes a terrifying world in which individuals have sacrificed their selfhood to the Collective. The word "I" has been abolished from the language and the great oppressive "We" reigns supreme ... until deep within the hive, one brave individual awakens from his forced slumber and embarks on a rebellious quest for forbidden knowledge. One cannot help but notice that by the utopian standards of modern PC (Political Correctness), this oppressive society is every bit the "promised land" that it claims to be. Right now, there are people on earth who dream that one day this kind of world will rise from the smoldering ashes of the one we now inhabit. Some of them even live in the United States. And in other places, their dream is all but a *fait accompli.*

In *Anthem*, Rand raises a burning question: what kind of world do most of us really want? At first glance, the answer seems simple: a world in which the individual and the collective have patched up their differences and learned to live in harmony. A world in which the rights and dignity of the individual are no longer at odds with collective levels of human selfhood, and in which the individual and the collective operate synergistically as parallel phases of human existence. A world in which the best parts of Rand's philosophy, those which powerfully affirm human dignity and freedom, can live happily with the indispensable invariants of collective utilitarianism. But is such a world realizable? With due allowance for the inevitable imperfections of democracy, the short answer is "yes". But since neither Rand nor her detractors have been able to define it for us, much less tell us how to bring it about, we find ourselves obliged to take renewed stock of the dialectic and seek a constructive synthesis. Fortunately, this task has been recognized and accepted, and thanks to the powers of human reason extolled by this brilliant author, the solution is already partially in hand.

Originally dismissed by critics, *Anthem* has sold over 2.5 million copies. The fact that it continues to sell at a rate of more than 60,000 copies per year is an unmistakable tribute to the power of Rand's message.

References

Darwin, C. (1999) *On the Origin of Species*. New York: Bantam Classics. Original work published in 1859.

Descartes, R. (1996) *Meditations on First Philosophy with Selections from the Objections and Replies*. Edited by J. Cottingham. Cambridge: Cambridge University Press.

Duhem, P. (1954) *The Aim and Structure of Physical Theory*. Translated by P. P. Wiener, foreword by L. de Broglie. New Jersey: Princeton University Press. Originally published under the title *La Théorie physique: Son objet, et sa structure*. Paris: Chevalier & Riviére, 1906.

Eldredge, N. & Gould S. J. (1972) Punctuated Equilibria: An Alternative to Phyiletic Gradualism. In Thomas J. M. Schopf (Ed.) *Models in Paleobiology*. San Francisco: Freeman Cooper & Company, pp. 82–115.

Fisher, C. M. (2001) If There Were No Free Will. *Medical Hypotheses*, Vol. 56, Issue 3, pp. 364–366.

Gödel, K. (1962) *On Formally Undecidable Propositions of Principia Mathematica and Related Systems*. Translated by B. Meltzer. New York: Basic Books. Originally published under the title "Über formal unentscheidbare Sätze der Principia Mathematica und verwandter Systeme I" in *Monatshefte für Mathematik*, Vol. 38, Issue 1, December 1931.

Goodman, N. (1955) *Fact, Fiction, and Forecast*. Cambridge, MA: Harvard University Press.

Gould, S. J. (1989) *Wonderful Life: The Burgess Shale and the History of Nature*. New York: W. W. Norton & Company.

Hartle, J. B. & Hawking, S. W. (1983) Wave Function of the Universe. *Physical Review D*, Vol. 28, № 12, 2960–75.

Hawking, S. W. (1988) *A Brief History of Time: From the Big Bang to Black Holes*. Introduction by C. Sagan, illustrated by R. Miller. New York: Bantam Books.

Hume, D. (1975) *Enquiries Concerning Human Understanding and Concerning the Principles of Morals*. 3rd Edition. Edited by L. A. Selby-Bigge & P. H. Nidditch. Oxford: Oxford University Press. Reprinted from the posthumous edition of 1777. Original work published in 1748 and 1751.

Kant, I. (1965) *The Critique of Pure Reason*. Translated by N. K. Smith. New York: St. Martin's Press. Original work published in 1781.

Langan, C. M. (1989) The Resolution of Newcomb's Paradox. *Noesis*, № 44.

Langan, C. M. (2002) *The Cognitive-Theoretic Model of the Universe: A New Kind of Reality Theory*. Princeton, MO: Mega Foundation Press. Originally published in *Progress in Complexity, Information, and Design*, Double Issue, Vols. 1.2-3.

Levy, S. (1997) The Brain's Last Stand. *Newsweek*, 05 May, pp. 50–56.

Libet B., Gleason, C. A., Wright, E. W., & Pearl, D. K. (1983) Time of Conscious Intention to Act in Relation to Onset of Cerebral Activity (Readiness-Potential): The Unconscious Initiation of a Freely Voluntary Act. *Brain*, Vol. 106, Issue 3, pp. 623–642.

Löwenheim, L. (1967) On Possibilities in the Calculus of Relatives. In Jean Van Heijenoort's *From Frege to Gödel: A Source Book in Mathematical Logic, 1879–1931*, pp. 228–251. Translated by S. Bauer-Mengelberg. Cambridge, MA: Harvard University Press. Originally published under the title "Über Möglichkeiten im Relativkalkül" in *Mathematische Annalen*, Vol. 76, Issue 4, pp. 447–470, December 1915.

Nozick, R. (1970) Newcomb's Problem and Two Principles of Choice. In N. Rescher (Ed.) *Essays in Honor of Carl G. Hempel: A Tribute on the Occasion of his Sixty-Fifth Birthday*, pp. 114–146. New York: Humanities Press.

Rand, A. (1938) *Anthem*. London: Cassell & Co.

Reed, L. (1974) *The Complete Limerick Book: The Origin, History, and Achievements of the Limerick, with about 350 selected examples*. Illustrated by H. M. Bateman. Detroit: Gale Research Company. Reprinted from the 1925 Jarrolds Publishers edition.

Quine, W. V. (1951) Two Dogmas of Empiricism. *The Philosophical Review*, Vol. 60, № 1, pp. 20–43.

Skolem, T. (1967) Logico-combinatorial Investigations in the Satisfiability or Provability of Mathematical Propositions: A simplified Proof of a Theorem by L. Löwenheim and Generalizations of the Theorem. In Jean Van Heijenoort's *From Frege to Gödel: A Source Book in Mathematical Logic, 1879–1931*, pp. 252–263. Translated by S. Bauer-Mengelberg. Cambridge, MA: Harvard University Press. Originally published under the title "Logisch-kombinatorische Untersuchungen über die Erfüllbarkeit oder Beweisbarkeit mathematischer Sätze nebst einem Theoreme Über dichte Mengen" in *Skrifter utgit av Videnskapsselskapet i Kristiania 1920*, № 4, pp. 1–36.

Wheeler, J. A. (1979) From the Big Bang to the Big Crunch. *Cosmic Search Magazine*, Vol. 1, Issue 4, pp. 2–8.

Wheeler, J. A. (1986) How Come the Quantum? *Annals of the New York Academy of Sciences*, Vol. 480, Issue 1: New Techniques and Ideas in Quantum Measurement Theory, pp. 304–316.

Wheeler, J. A. (1990a) Information, Physics, Quantum: The Search for Links. In W. H. Zurek (Ed.) *Complexity, Entropy and the Physics of Information, SFI Studies in the Sciences of Complexity*, Vol. VIII, pp. 3–28. Addison-Wesley Educational Publishing Company.

Wheeler, J. A. (1990b) *A Journey into Gravity and Spacetime*. New York: Scientific American Library.

Whitemore, H. (2015) *Breaking the Code*. London: Samuel French. First staged at Theatre Royal, London, 21 October 1986.

Dialogue on the CTMU

PUBLISHER'S NOTE

We have reprinted this interesting and productive discussion of the CTMU for didactic purposes. *The Dialogue on the CTMU between Christopher Michael Langan and UltraHIQ community member, Russell Fred Vaughan, Ph.D. (Engineering)* took place in 1999–2000. Several brief references to third parties that were not relevant to the dialogue have been removed. There was no further correspondence until Fred posted a "critique" of the CTMU. The "critique", including Chris's response, has been reprinted in the last part of this book.

PART ONE

RFV: I've been meaning to begin a dialog with you that would help us both understand each other better. Bob Seitz[1] thinks most highly of you and such a friend is the best recommendation a man can have in my book. I would be foolish not to know you better than I do with you being only an e-mail away.

What I'd like before attempting any major intellectual challenge like your CTMU – that I would like to get into later after I have some background – is to compare our perspectives on various topics without the public displays and distractions of being on the list. If you have any questions about my leanings on any topic, please let fly with questions and I'll answer them as straightforwardly or as stupidly as I happen to hold my opinions. I would like to discuss them in such a non-confrontational way – and slowly so that it sinks in.

[1] Dr. Robert North Seitz, former NASA manager and ultra-high IQ community member, serves on the board of the Mega Foundation.

I will be as intellectually fair with you as it is possible for me to be. I do find it easy to acknowledge others ideas but that does not mean that I have none of my own that I would deny having had before a conversation. I am personally not nearly so worried about my own ideas as I once was, but naturally we all have some level of paranoia with regard to such things. I hope we can converse in such a way that you are always convinced of my good faith in this regard. Assurances are best proven rather than stated.

CML: Of course. I share and appreciate your willingness to progress towards mutual understanding. I know that some of my ideas look somewhat strange from a conventional viewpoint, but they're not lightly formulated, and I always welcome a chance to explain the underlying logic.

RFV: If there's a different agenda or different approach to this one that is better, then let's do it! But in the mean time could we investigate one of the questions that I have that seems to involve a major difference of opinion between us.

I happen to think that one cannot "know" the world without data and that our most abstractly generalized notion of the world must be based on inductive reasoning from that data. Having developed a generalized notion or theory one can then deduce specifics that are warranted (and our theory is only warranted) to the extent that these deductions map to actualities. It is my impression that you hold a somewhat different opinion or place emphasis differently within such a scheme.

Teleological arguments are not something I think should be tossed out without consideration, but I do have much more skepticism with regard to them. However, I hold (what can only be called "faith") that the ultimate physical theoretical solution will have a "necessity" about it which is very convincing – sor[t] of Einstein's "God would have done it this way!"

CML: I agree that we can't know the world without data. But since we necessarily receive and interpret these data mentally, they provide important clues about our mental structure. The idea that this mental structure shapes the phenomenal world has been known since Kant; the CTMU simply pushes Kant's ideas to their logical conclusion, dealing directly with the relationship between mind and reality.

RFV: Could we just talk about this topic quite narrowly for a beginning. That is, unless you have a question of me or a different agenda.

CML: Sure! No problem.

RFV: Thanks in advance for whatever this engenders.

PART TWO

RFV:

> **CML**: I agree that we can't know the world without data.

Ok. That one's out of the way.

> **CML**: But since we necessarily receive and interpret these data mentally, they provide important clues about our mental structure.

Obviously this is the big swinging door. Obviously on the one hand we have photons and on the other complex ideas. Perceptions are somewhere between.

CML: There's the first catch. Your clear distinction between photons and ideas is not globally valid, but a provisional convention. Photons have a general abstract existence as well as a specific physical one. This is what happens when you embed physical reality in a space of abstractions in order to theorize about it; real objects necessarily conform to more or less exact configurations of an abstract distributed syntax. Since there's no way out of this requirement, it's pointless to theorize about photons without admitting it from the outset. A theory that doesn't is nothing but an oxymoron; it contains an implied logical inconsistency from jump street and invalidates itself.

RFV: I never want to underestimate you, but at the same time I want to be sure that something that I happen to know that I consider relevant is known to both of us. So, let me just ask. Are you aware of biophysics studies by Hoffman[2] (and probably others later) demonstrating the development of Lie groups (as in the Norwegian mathematician) in the neurons between the eyes and the brain which effect the transformations of rotation, translation, etc. outside of the brain?

CML: If we examine a brain under a microscope, we see neurons and synapses. A neuron is either firing or not; its state is either 1 or 0, corresponding to the truth values of 2-valued logic. If we think of a neuron as a "brain quantum", we get a brain quantum, and therefore a brain, conforming to 2VL. Nothing that cannot be expressed in this 2VL language can be formulated within the brain (or any larger neural system including, e.g., optic neurons), including non-discrete mathematics (e.g., continuous Lie groups).

On the other hand, anything that can be formulated in this language can be cerebrally formulated. The exact location of visual transformations is beside the point; the neurons that effect them are still "microbrains" obeying a 2VL "mental syntax". In this sense, visual transformations are still "mental" in character even when they occur outside the brain.

---

[2] Probably referring to Hoffman, W.C. (1966) 'The Lie algebra of visual perception.' *Journal of Mathematical Psychology*, Vol. 3, Issue 1, pp. 65–98.

However, although Hoffman's research sounds familiar, I'd appreciate a web source (if available).

RFV: These capabilities are staged in development with the shape size transformations not being completed until around the age of eight? The reason I bring this up is that one actually takes a tremendous leap from photons upon the eye to "perception" to the brain without any mental operations whatsoever. Perhaps a newborn has to work the problem from beginning to end, but it's simplified for us. This also, of course, gives us "theory-laden" data from the outset. And these transformations can fail to be developed properly if eyes are crossed and not "fixed" until after that particular transformation is "programmed", etc.

CML: Again, when you say "mental operations" in a CTMU context, you're talking about a generalized form of mentation (not just the mentation of a human brain). Since a photon and a brain both transform information, albeit on different levels of complexity, they are both "mental" in nature. In other words, you seem to be making an *a priori* semantical distinction between levels of information transduction – the levels represented by a brain and a photon – and (again on an *a priori* basis) assigning "mental" status to only one of them. There's no scientific basis on which you can do that; it amounts to anthropocentrism. What we seek in any scientific context is a distributed formulation of reality that applies inside the brain as well as without it, and although the terms "mental" and "brain" are often assumed to be associated, this need not be exclusively so.

Where the universe is described as existing within an abstract space with exhaustive mathematical structure – a kind of self-distributed Platonic proto-universe – everything employs this structure as transductive syntax up to isomorphism with the intrinsic structure of the transducing object. But the intrinsic structure of an object need not be directly observable, for the simple reason that it's distributed and thus cannot be locally discerned by observation. The intrinsic syntactic ("mental") structure of an object can only be inferred from its behavior, which is where a need for data enters the picture. In other words, something doesn't have to have observable structural complexity to have a complex behavioral syntax; if it did, then the theory of so-called "elementary particles" would be free of mathematical complexity.

RFV:

> **CML**: The idea that this mental structure shapes the phenomenal world has been known since Kant; ...

So to what extent does/can the "external world" shape the mind, i.e., affect its programming? Certainly there must be some of that going on within the brain as well, mustn't there?

CML: Of course. But to achieve a closed-form theory of reality, we need to admit that this "shaping" or "programming" process is mutual. That's why the

CTMU employs a model of spacetime in which mind programs universe even as universe programs mind.

RFV:

> **CML**: ...the CTMU simply pushes Kant's ideas to their logical conclusion, dealing directly with the relationship between mind and reality.

Let us hold off a bit longer on this till we have a common understanding of percepts.

CML: In the conventional model, percepts are objective observables that actively imprint themselves on, and thereby shape and determine, the passive mind and its internal processes. But a more general (and therefore more scientific, less tautological) description of percepts portrays them also as having a subjective perceptual aspect that is identified, at a high level of generality, with the Creation Event itself. This is just an extension of Wheeler's Observer Participation thesis ("only intelligent entities are observers") to physical reality in general ("everything is an observer", with the caveat that it's still possible, and indeed necessary, to reserve a special place for intelligence in the scheme of things).

The surprising part, in my opinion, is this. This reduction of all reality to simultaneously active and passive "infocognition" amounts to defining reality as did Hume...as pure experience ("infocognition" is just a technical synonym of "experience" that opens the concept up to analysis from the dual standpoints of information theory and cognition or computation theory). Thus, the Kantian mind-matter distinction, as embodied in the term "infocognition", is simply distributed over Hume's experiential reality by synonymy, bringing the metaphysics of Hume and Kant into perfect coincidence.

PART THREE

CML: Let's see if we can cut to the chase here.

Suppose you're wearing blue-tinted glasses. At first, you think that the world you see through them is blue. Then it occurs to you that this need not be true; maybe it's the glasses. Given this possibility, you realize that you really have no business thinking that the world is blue at all; indeed, due to Occam's razor, you must assume that the world is chromatically neutral (i.e., not blue) until proven otherwise! Finally, managing to remove your glasses, you see that you were right; the world is not blue. This, you conclude, proves that you can't assume that what is true on your end of perception (the blue tint of your lenses) is really true of reality.

Fresh from this victory of reason, you turn to the controversial hypothesis that mind is the essence of reality...that reality is not only material, but mental in character. An obvious argument for this hypothesis is that since reality is known to us strictly in the form of ideas and sensations – these, after all, are all that can be directly "known" – reality must be ideic. But then it naturally occurs to you that the predicate "mental" is like the predicate "blue"; it may be something that exists solely on your end of the process of perception. And so it does, you reflect, for the predicate "mental" indeed refers to the mind! Therefore, by Occam's razor, it must be assumed that reality is not mental until proven otherwise.

However, there is a difference between these two situations. You can remove a pair of blue sunglasses. But you cannot remove your mind, at least when you're using it to consider reality. This means that it can never be proven that the world isn't mental. And if this can never be proven, then you can't make an assumption either way. Indeed, the distinction itself is meaningless; there is no reason to even consider a distinction between that which is mental and that which is not, since nature has conspired to ensure that such a distinction will never, ever be perceived. But without this distinction, the term "mental" can no longer be restrictively defined. "Mental" might as well mean "real" and vice versa. And for all practical purposes, so it does.

A theory T of physical reality exists as a neural and conceptual pattern in your brain (and/or mind); it's related by isomorphism to its universe U (physical reality).

$$T \leftarrow \text{isomorphism} \rightarrow U$$

T consists of abstract ideas; U consists of supposedly concrete objects like photons (perhaps not the best examples of "concrete objects"). But the above argument shows that we have to drop the abstract-concrete distinction (which is just a different way of expressing the mental-real distinction). Sure, we can use these terms to distinguish the domain and range of the perceptual isomorphism, but that's as far as it goes. For all practical purposes, what is mental is real, and vice versa. The T-U isomorphism seamlessly carries one predicate into the other.

Now let's proceed with your questions.

RFV:

> **CML**: But since we necessarily receive and interpret these data mentally, they provide important clues about our mental structure.

Obviously this is the big swinging door. Obviously on the one hand we have photons and on the other complex ideas. Perceptions are somewhere between.

> **CML**: There's the first catch. Your clear distinction between photons and ideas is not globally valid, but a provisional convention.

I knew there'd be a catch here somewhere.

I'm treading lightly on "photons" now, OK? Maybe we should accept a continuum between these? Or doesn't that work either? I can remain "provisional" on photons for a little while yet.

CML: No. Photons are at once abstract and concrete. In fact, the distinction might as well not even be discussed, because it's oxymoronic (as shown above).

RFV:

> **CML**: Photons have a general abstract existence as well as a specific physical one.

OK, take me through this one gently. I don't think I'll have a problem here if this level of abstractness involves alternative equivalent representations prior to collapse of a wave function or something like that. What do you have in mind here?

CML: As explained above. Or if you prefer, assume that a photon has no abstract existence. Then since ideas are abstract, it corresponds to no idea. But you know the world through ideas. So for you, a non-abstract photon has no existence. But you're discussing photons that exist. So the photon you're talking about is abstract (as well as concrete).

RFV:

> **CML**: This is what happens when you embed physical reality in a space of abstractions in order to theorize about it; ...

Now are we supposing that our "physical reality" is but one in a space of all possible realities? And then what is "a" reality in this case? An instantaneous snapshot of "all things"?

CML: No. It's a self-perceptual system with a combined abstract and concrete nature (as explained above) that possesses closure with respect to other realities.

RFV:

> **CML**: ...real objects necessarily conform to more or less exact configurations of an abstract distributed syntax.

Please do not get impatient with me here. I REALLY do not understand what you mean by your words without definitions and explanations, and you are being very patient so far.

In what sense do you mean "more or less exact configurations"? Further, and maybe with a little more detail, what does "an abstract distributed syntax" mean to you? And how have we concluded that "real objects necessarily conform" to such a thing. These are major leaps – perhaps not for you, perhaps not for anyone you might hope I were, but for me they are major! "Necessarily" implies to me that it should be self-evident since it could be no other way (at least after I understand it). "Conforms" is a verb one doesn't usually see in this context, so I assume you mean to imply that there is an isomorphism that could be established between real objects and such a more orderly abstract structure. If that's what you mean, I'm with you. If not, please provide a little remedial instruction.

CML: If we form a theory of some aspect of reality, it may or may not be correct (i.e., it will be "more or less exact"). But even if not, we can be sure that the correct theory will conform to our mental structures, since otherwise we won't be able to create a conceptual isomorphism with it (and that's what a theory is). A "syntax" is to be interpreted as any set of structural and functional constraints applied to any system. A syntax takes the form of general information and is implemented by generalized cognition.

RFV:

> **CML**: Since there's no way out of this requirement, ...

I understand that idealization at some level is required to theorize about anything if that's what you mean.

> **CML**: ...it's pointless to theorize about photons without admitting it from the outset.

If you mean that I must suspend belief about what a photon is until the theory is completely understood, that is no problem since no one else knows what a photon is yet in its fullest sense either.

CML: A photon is a configuration of the syntax of reality; as already explained, that's assured. It's also assured that a photon is at once abstract and real.

RFV:

> **CML**: A theory that doesn't is nothing but an oxymoron; it contains an implied logical inconsistency from jump street and invalidates itself.

This last sentence seems to break with what I understood we were talking about. I thought we were addressing provisionality of what a photon "is" and the fact that "real objects" must be abstracted into a more formal structure in order to theorize. I buy that, but I have not yet considered the class of theories about such real objects. But if you are meaning to imply that a theory that begins with a preconceived notion about an object (say photon) it won't be of much use with regard to clarifying what a photon is, I can buy that. But, there are hierarchically ordered theories for which it might work just fine. We are not hot on the trail of one of that kind I presume.

CML: Again, the combined abstract/real nature of a photon is not provisional, but given from the outset. It's only the exact corresponding configuration of reality-syntax that remains a mystery (and in a sense, not even that).

RFV:

> **CML**: On the other hand, anything that can be formulated in this language can be cerebrally formulated. The exact location of visual transformations is beside the point; the neurons that effect them are still "microbrains" obeying a 2VL "mental syntax". In this sense, visual transformations are still "mental" in character even when they occur outside the brain.

I have no disagreement with any of this but the 2VL. Neurons are not limited to 2VL types to my understanding – they may be quite analog computer-like, in which case I don't know where 2VL fits into this scheme. I noticed that you placed a lot of credence on distinctions with this kind of logic, and that was actually the next area that I wanted to question you about your background assumptions. So maybe you could tip-toe through some of this for me.

CML: Analog computers are as bound by 2VL as are digital computers; an analogy either exists (1) or not (0). All distinctions are 2VL; one thing is either distinct from another (1) or not (0) on a given level of definition. Since information consists of distinctions – constraints differentiating one thing from another – information is 2VL. Since theories consist of information, theories are 2VL. And since theories are isomorphic to reality, reality is 2VL. There is no escape...no crack in the stone of truth versus falsehood. Even to explain why there might be, you would be forced to resort to 2VL...and your putative exception would be embedded in 2VL by *fiat*.

RFV:

> **CML**: However, although Hoffman's research sounds familiar, I'd appreciate a web source (if available).

I have a couple of copies of papers circa 1965 around here somewhere. Come to the shindig next summer and I'll find them for you.

CML: Hey, I'm 3,000 miles away! But I appreciate the invite. Maybe if I make a lot more money than I have now, I'll surprise you and show up.

RFV:

> **CML**: Again, when you say "mental operations" in a CTMU context, ...

That was why I wished to remain outside that context until I get some background that presumes none of CTMU. Any word I use for now will be a lay definition term except as our mutual backgrounds accommodates more extensive terminology. I can't do this any other way, I'm sorry. It's just a limitation I have.

CML: As explained in the preface, mental operations cannot be distinguished from real operations except in a provisional way. That's what I mean by "a CTMU context". It's given *a priori*.

RFV:

> **CML**: ...you're talking about a generalized form of mentation (not just the mentation of a human brain). Since a photon and a brain both transform information, albeit on different levels of complexity, they are both "mental" in nature.

I have no problem with this in general and the existence of levels of consciousness as other that human mental operations is something I can readily accept. A photon? I don't know, what's the information? Blue?

CML: Actually, the example of a photon is problematic for reasons too advanced to cite here. But for now, let's just say that a photon transforms the state of an atom by causing an electron to jump from one orbit to another. Even more basically, it "transforms its own coordinates" as it "moves" (due to the CTMU conspansive model of spacetime, the quotation marks are necessary). Atomic states and position coordinates are informational.

RFV:

> **CML**: In other words, you seem to be making an *a priori* semantical distinction between levels of information transduction – the levels represented by a brain and a photon – and (again on an *a priori* basis) assigning "mental" status to only one of them.

I did in fact. And now I think I'm a little more in tune with where you are in this.

CML: Sounds promising.

RFV:

> **CML**: There's no scientific basis on which you can do that; it amounts to anthropocentrism.

No need to disagree here since we don't, but I'm not sure that those who do disagree might not consider their approaches scientifically palatable.

> **CML**: What we seek in any scientific context is a distributed formulation of reality that applies inside the brain as well as without it, ...

Whoa! Now here is another BIG hop. I, for example, have never presumed so great a scope for limited scientific endeavors. Let's leave this one for now unless there is some brief clarification that lessens the hurdle for me now.

CML: What I said is an implication of the scientific method, and specifically the experimental replicability criterion. If an experiment is replicable, then it can be repeated anywhere in spacetime provided the proper conditions are present. The possibility of the existence of those conditions at any point in spacetime, given the presence of supporting conditions, means that it is "distributed over spacetime" (but in a conditional way). Because all scientific truth is obtained by the scientific method, it conforms to this description; it has a conditional distributed formulation. This relates to the concept of spacetime homogeneity...it's a logical necessity of meaningful theorization.

RFV:

> **CML**: ...and although the terms "mental" and "brain" are often assumed to be associated, this need not be exclusively so.

Absolutely with you here.

> **CML**: Where the universe is described as existing within an abstract space with exhaustive mathematical structure – a kind of self-distributed Platonic protouniverse – ...

I know I'm breaking this sentence apart, but it's necessary for me to do so. I understand the agenda of the first part of this sentence, but I get confused here:

> **CML**: ...everything employs...

What do you mean here? Do you mean for example that any analysis of the universe necessarily employs the mathematical analog of the universe? I think so and if so, I buy that.

CML: No, it goes even farther. Any material object or field has intrinsic mathematical structure that acts as a "syntax" to process information (e.g., the photon described above). Again, it goes along with the absence of an absolute distinction between the abstract (syntactic) and concrete sides of reality.

RFV:

> **CML**: ...this structure as transductive syntax up to isomorphism...

The isomorphism between universe and mathematical structure is, of course, all that can justify the approach, but...

> **CML**: ...with the intrinsic structure of the transducing object.

Now this I have problems with. This says that the fact (which I accept but some would doubt) that we can create an isomorphism between the universe and a mathematical model necessarily implies that the agent that can do this must have a structure isomorphic to the universe as well. That requires some very critical thought. Let me do it.

CML: Let me save you some time. The isomorphism is given on a *de facto* basis. Indeed, it follows from the preface.

RFV:

> **CML**: But the intrinsic structure of an object need not be directly observable, for the simple reason that it's distributed and thus cannot be locally discerned by observation.

No problem here.

> **CML**: The intrinsic syntactic ("mental") structure of an object can only be inferred from its behavior, which is where a need for data enters the picture.

Very definitely. But here you need an extreme amount of data it would seem to me and I don't know off the top of my head that I can think of what kind of data would justify such a claim. It strikes me somewhat of the "I can test of 4-sigma, therefore I are one!" A test would be good.

CML: Remember, theories are made to be superseded by better theories; data can always be extended. Only tautological self-distributed systems possess "certainty". That's why we can be so sure of the 2VL component of reality syntax. Theories generally aren't so ironclad.

RFV:

> **CML**: In other words, something doesn't have to have observable structural complexity to have a complex behavioral syntax; if it did, then the theory of so-called "elementary particles" would be free of mathematical complexity.

Ah, but regularity ought not be confused with simplicity.

CML: True enough.

RFV:

> > **CML**: The idea that this mental structure shapes the phenomenal world has been known since Kant; ...
>
> So to what extent does/can the "external world" shape the mind, i.e., affect its programming? Certainly there must be some of that going on within the brain as well, mustn't there?
>
> **CML**: Of course. But to achieve a closed-form theory of reality, we need to admit that this "shaping" or "programming" process is mutual. That's why the CTMU employs a model of spacetime in which mind programs universe even as universe programs mind.

We're going a little ahead of my schedule here, but let me address the "If a closed form exists, then..." What makes you jump for such a solution. Don't get me wrong, I like closed form too, but one usually adapts ones desires to the problem. I guess...never mind.

CML: The discernability and stable identity of objects within the universe can be shown to imply that the universe is informationally closed in the following sense: it is possible to distinguish that which it contains (reality, perceptions) from that which it does not contain (unreality). I.e., there's an "informational boundary" separating that which the universe includes from that which it excludes, and this boundary is what is "closed". No closure means no boundary means no information.

RFV:

> > **CML**: ...the CTMU simply pushes Kant's ideas to their logical conclusion, dealing directly with the relationship between mind and reality.
>
> Let us hold off a bit longer on this till we have a common understanding of percepts.

Still hold off a little. I'm sure it will do us well.

> **CML**: In the conventional model, percepts are objective observables that actively imprint themselves on, and thereby shape and determine, the passive mind and its internal processes.

Yes.

> **CML**: But a more general (and therefore more scientific, ...

We have somewhat different definitions of "scientific".

CML: Again, "scientific" means replicable, replicable means distributed, and distributed means general. Moreover, science carefully excludes false tautologies, or circular but nongeneral forms of reasoning.

RFV:

> **CML**: ...less tautological) description of percepts portrays them also as having a subjective perceptual aspect that is identified, at a high level of generality, with the Creation Event itself.

You mention "description"; that is the key word. How are percepts best DESCRIBED? Data is needed here too it seems to me.

CML: Description can here be considered synonymous with "theory", and thus related to perception (data) by isomorphism.

RFV:

> **CML**: This is just an extension of Wheeler's Observer Participation thesis ("only intelligent entities are observers") to physical reality in general ("everything is an observer", with the caveat that it's still possible, and indeed necessary, to reserve a special place for intelligence in the scheme of things).

This is a whole nuther topic. A great one! We got started on that when I was accused of being a murderer. What constitutes a "real" event? My understanding of at least one of Wheeler's notions was that an irreversible change had to have taken place – intelligent observers be damned.

CML: If you're right, then Wheeler's title for his thesis ("Observer Participation Thesis") would seem to indicate his agreement with me about what constitutes an "observer", namely anything able to participate in an irreversible change.

This reflects the distributed subjectivism and self-processing character of reality...the fact that mind (or "observership") and reality can't be absolutely distinguished.

RFV:

> **CML**: The surprising part, in my opinion, is this. This reduction of all reality to simultaneously active and passive "infocognition" amounts to defining reality as did Hume...as pure experience ("infocognition" is just a technical synonym of "experience" that opens the concept up to analysis from the dual standpoints of information theory and cognition or computation theory). Thus, the Kantian mind-matter distinction, as embodied in the term "infocognition", is simply distributed over Hume's experiential reality by synonymy, bringing the metaphysics of Hume and Kant into perfect coincidence.

I see what you are saying here, but I'm not HERE yet.

CML: It's even simpler than I put it above. Experience entails something that experiences and something that gets experienced. So if reality consists everywhere of experience, it is everywhere self-experiencing, and has everywhere the character of both mind (the subjective experiencer) and objective reality (that which gets experienced). So the categorical imperative relating the subjective and objective aspects of reality everywhere coincides with experience; they're the same thing described in two ways, singular (Hume) and dual (Kant).

RFV: Thanks again. Maybe you'll see a transition in understanding as I went down this message; I hope so. I also hope this doesn't bore you too badly.

CML: I see improvement, by God! But in any event, just don't say I didn't try my best! Keep up the good work.

PART FOUR

RFV: (blue shades/mind analogy...)

> **CML**: But without this distinction, the term "mental" can no longer be restrictively defined. "Mental" might as well mean "real" and vice versa. And for all practical purposes, so it does.
>
> A theory T of physical reality exists as a neural and conceptual pattern in your brain (and/or mind); it's related by isomorphism to its universe U (physical reality).
>
> $$T \leftarrow \text{isomorphism} \rightarrow U$$
>
> T consists of abstract ideas; U consists of supposedly concrete objects like photons (perhaps not the best examples of "concrete objects"). But

> the above argument shows that we have to drop the abstract-concrete distinction (which is just a different way of expressing the mental-real distinction). Sure, we can use these terms to distinguish the domain and range of the perceptual isomorphism, but that's as far as it goes. For all practical purposes, what is mental is real, and vice versa. The T-U isomorphism seamlessly carries one predicate into the other.

I'm pretty much on board here.

> **CML**: Photons have a general abstract existence as well as a specific physical one.

I'm getting the scheme here. There is an isomorphic relationship between the abstract and concrete photons, which in some very real sense leans heavily toward the abstract since everything we know, we know because of knowledge of the abstract version. Ok.

> **CML**: As explained above. Or if you prefer, assume that a photon has no abstract existence. Then since ideas are abstract, it corresponds to no idea. But you know the world through ideas. So for you, a non-abstract photon has no existence. But you're discussing photons that exist. So the photon you're talking about is abstract (as well as concrete).

Ok, I made that turn.

> **CML**: No. It's a self-perceptual system with a combined abstract and concrete nature (as explained above) that possesses closure with respect to other realities.

So, do you mean a conceptual model of "reality" which could be very different than our own reality? I think so. I see "self-perceptual" as a block to my acceptance here. I think you make the leap that isomorphic structures are the SAME structure to say that. Am I right? If so, how is that justified?

CML: It's an implication of the fact that no meaningful distinction can exist between mind and reality; with respect to these two predicates, they're the same. That is, we took the isomorphism M: T $\leftrightarrow$ U and predicated p_1 = "mental" (abstract, ideic) on T and p_2 = "real" on U. Then we realized that no meaningful distinction exists between the predicates p_1 and p_2. So now we're dealing with U on an abstract basis (just as we're dealing with T on a "real" basis, which it is because at the very least, it exists as a real neural pattern in your real brain).

Now consider an isomorphism between two distinct sets A, B consisting of 10 red golf balls each for a total of twenty distinct balls: N: A $\leftrightarrow$ B. Obviously, we can assign the predicate "red" to each set. This means that the sets are identical with respect to the predicate "red". But so what? There's still a distinction between A and B, because they in fact contain different golf balls. [Note that if the golf balls were electrons, it wouldn't be entirely possible to say this; electrons to some extent share their identities. Note also that matter consists of such particles.] And since red = red just as surely as $p_1 = p_2$ in the last paragraph, there's still a distinction between T and U...or is there?

We can call the sets A and B "different" because the identical red golf balls of which they consist are located at different coordinates. We can ascertain this with the help of nothing but a marking pen and a tape measure. There's the meat of our distinction with respect to A and B. But are T and U located at different spatiotemporal coordinates? How can we prove it?

As we've already seen, we know the universe U directly only through the abstract ideas and sensations T to which it corresponds in our minds, and that includes any sequence of measuring events we might undertake. In this sense, T and U are "at the same coordinates". So to prove that T and U are actually at DIFFERENT coordinates, we'd have to prove that their coincidence is an "illusion". But if we could do that, then we'd have the basis for distinguishing "mental" from "real" after all...something we already know we can't do. So after a little reflection, we conclude that as it is for knowledge, so it is for measurement; the only things we can directly measure are the ideas and sensations corresponding to U, and we can only do it with an abstract tape measure at that!

This, of course, is what philosophers call "solipsism", but without the assumption of personal control. In a word, it's impossible to prove that the universe doesn't exist entirely inside our minds, and that all that "objective reality" out there isn't just some kind of subjective internal simulation. While it's true that control of the simulation seems to be distributed among many people and influences, this merely amounts to a kind of "distributed solipsism" after all...an intersect of solipsistic viewpoints in one and the same meta-solipsistic universe. In other words, the solipsistic abstract / concrete "Self" of the universe distributes over all of its contents, and our own "selves" are just mini-solipsistic aspects of it.

So the objective and solipsistic models are as indistinguishable as U and T, abstract and concrete, and we have to drop the distinction. And in this case, (objective) isomorphism equals (abstract, subjective) identity; when we talk about T and U, we're really talking about different aspects or characterizations of the same thing. So the perceived universe and the perceiver are one, and the universe is "self-perceptual". This is where reality theory meets the philosophy of mind. But that's fine, because as should by now be obvious, they need each other.

RFV:

> **CML**: ...real objects necessarily conform to more or less exact configurations of an abstract distributed syntax.

(...)

> **CML**: If we form a theory of some aspect of reality, it may or may not be correct (i.e., it will be "more or less exact").

"More or less CORRECT." Right? Exactness is a small case in point?

CML: Yes. There are two kinds of truth, syntactic and semantic. Syntactic truth amounts to internal consistency, and semantic truth amounts to external isomorphism or quantification. If T is true on neither the syntactic nor semantic level of reality, then it has 0 scope or content. But then the theory has no model and is not a theory in the strictest sense. At best, it's just an empty (content-free) pattern of neural events or an inky set of typographical symbols on paper, with irrelevant fantasy for content.

That part of a theory that's wrong is empty, but the part that's right has real content. The theory tautologically corresponds to just that part of reality for which it's true. In fact, logicians already know that insofar as they can often either extend or qualify a troubled theory so as to adjust its referential quantifier (think Löwenheim–Skolem, Duhem–Quine, etc.) and thereby restore its validity (think classical physics as the low-velocity "limit" of relativity physics), the idea of an "absolutely false" theory is not what it seems.

The point of all of this is that a theory can evolve, changing its form to absorb more and more real content. That's what I meant by "more or less exact". Perhaps "more or less correct" would have been better.

RFV:

> **CML**: But even if not, we can be sure that the correct theory will conform to our mental structures, since otherwise we won't be able to create a conceptual isomorphism with it (and that's what a theory is).

I have a hard time with this. An analog computer is an isomorphic structure of a screen door when it is set up to solve the screen door problem, however, a general purpose computer is NOT isomorphic with a screen door when it solves the screen door problem. General structures are adaptable in a sense which allows them to accommodate an isomorphic model without themselves being limited to anything like that model structurally. So in what sense (and to what degree) do you insist on this conformal mapping?

CML: There is a very general abstraction called logic that permits various mappings among devices and other physical contexts. Have you ever heard the term "up to isomorphism"? It generally refers to a relationship that contains an isomorphism, but exceeds it. Some logical abstractions permit multiple isomorphisms, and these isomorphisms may be either structural or functional or both (in the above question, you seem to be talking about these two distinct kinds of isomorphism). In any case, for computational purposes, the isomorphisms you mention are logically equivalent up to isomorphism with the screen door problem (whatever that is). The "abstractions" being employed by the two types of computer are always expressible in terms of the penultimate abstraction, logic. Nothing I've said so far is incompatible with this fact.

RFV:

> **CML**: A "syntax" is to be interpreted as any set of structural and functional constraints applied to any system.

Ok, that's pretty much as I had concluded.

> **CML**: A syntax takes the form of general information and is implemented by generalized cognition.

Go easy now. "Implemented by generalized cognition" is a big bite. I assume that a "mental" awareness of the constraints on the physical model is good enough for a neophyte. Perhaps I even see a sense in which all but the "generalized" makes sense, but why is "generalized" in the phrase.

CML: Most people think that "cognition" is only done by brains.

The phrase "generalized cognition", actually synonymous with time, simply extends human cognition to generic information processing. It's this sense in which generic information processing obeys a "cognitive syntax". Thus, time is any grammatical implementation of the SCSPL reality-syntax.

RFV:

> **CML**: A photon is a configuration of the syntax of reality; ...

Chris, you're going to have to lighten up just slightly on some of this stuff. What you appear to be saying is that a photon is properly speaking a particular instance of the general structure of reality. That isn't very informative. What DO you mean here?

CML: Just that a photon obeys certain structural and behavioral rules, and those rules exist as constraints within the overall set of constraints called "the reality syntax" (including the laws of math and physics, of which our exhaustive knowledge is not assumed). These are the abstractions to which the photon primarily corresponds. Stay with it, Fred – this is all really very simple (given the unnecessary compartmentalization of physical concepts in diverse formalisms, maybe the problem is that it seems TOO simple).

RFV:

> **CML**: ...as already explained, ...

Then I missed it.

> **CML**: ...that's assured. It's also assured that a photon is at once abstract and real.

Got that!

> **CML**: Again, the combined abstract/real nature of a photon is not provisional, ...

You did entreat me to be provisional about it though.

CML: That a photon is both real and abstract is an absolute, nonprovisional fact. However, what counts as "real" and "abstract" in any given case is a relative (or provisional) matter, and sometimes even a matter of semantic convenience. I sometimes tend to assume that my meaning is obvious when it isn't.

RFV:

> **CML**: ...but given from the outset. It's only the exact corresponding configuration of reality-syntax that remains a mystery (and in a sense, not even that).

I assume you mean that the actualities of the behavior of photons are all that is (may be) unknown? Its place in the structure of reality IS known?

CML: Some aspects of a photon's behavior, and corresponding details of its place in the structure of reality, are known. But the intrinsic structure of the photon – the syntax it obeys – is not known except through this behavior. You can't hold a photon under a microscope and dissect it; all you can do with any subatomic particle is collide it and monitor the resulting events. Sometimes, you get what appear to be objective subparticles, but again, you can only see them through their effects. So all you can say for sure is that one configuration of SCSPL reality-syntax is merely decomposing into others. Getting too hung up on "objects" and their "innate properties" has been a recurrent problem in science; one generation's "particles" are the "subparticle interactions" of the next.

We need to realize that what we're dealing with are transformations of a global nomology or "syntax" (or to oversimplify and appeal to popular terminology, of the "TOE ultimate equation", along with the syntactic structure of the language in which it will be expressed).

RFV:

> **CML**: Analog computers are as bound by 2VL as are digital computers; an analogy either exists (1) or not (0). All distinctions are 2VL; one thing is either distinct from another (1) or not (0) on a given level of definition. Since information consists of distinctions – constraints differentiating one thing from another – information is 2VL. Since theories consist of information, theories are 2VL. And since theories are isomorphic to reality, reality is 2VL. There is no escape...no crack in the stone of truth versus falsehood. Even to explain why there might be, you would be forced to resort to 2VL...and your putative exception would be embedded in 2VL by *fiat*.

This is not very convincing to me. First of all, accepting your premise that all distinctions, knowledge, etc. are 2VL does not prove anything about "reality".

Ideational objects – even given the isomorphism – are NOT the objects themselves or exact equivalents – there are, for example, things which are not known about photons that are true of photons but not yet integrated into the ideational counterpart. So we have isomorphic structures of the reality of which we are aware and the reality we thereby presume. And we are in some sense limited such that our truths pertain to such a reality, but what about undecidability and such notions. How do these not end up being cracks in the stone? One must reason decidably about undecidability which doesn't, thereby, prove that all things are decidable.

CML: First, logical consistency equals the syntactic phase of truth. The stability (objective perceptibility) of reality implies that it's consistent, and the fact that it's consistent implies that it conforms to 2VL. Again, something either exists (1) or does not exist (0), so as soon as you say that it's real or that it exists, you tag it with a logical predicate (1 = true = reality-inclusion) that anchors it to the entire logical syntax of which this predicate is a part. Of course, we can still make a distinction about which level of reality includes this something, but whatever that level is, it obeys the distributed tautological syntax of reality (logic). Second, decidability is a logical concept whose definition is based solely on logical (syntactic) operations, namely those executed by Gödel in defining it. So you can't use it against logic without undermining its definition and thereby invalidating it.

I suspect that what you might be positing is an abstraction exceeding logic and consistency...an even more general predicate. Indeed, that's the only thing your description would fit. No problem; the CTMU contains an ultimate predicate called UBT (unbound telesis). But logic is a closed tautological system, and the relationship of this "pre-logical" predicate to logical reality is appropriately constrained. After all, what I'm trying to tell you about is the LOGICAL structure of reality, without which it would be no reality at all. For God's sake, don't interfere...your world could dissolve into an illogical chaos of ambiguities right before your eyes (and you're a part of it)!

RFV:

> **CML**: Hey, I'm 3,000 miles away! But I appreciate the invite. ...

It was great. And...it's a damned cheap vacation – all you have to do is get here! We're going to have a lot more people next year.

CML: I appreciate the invite, and I'll try to make it. Just remember that I'm not retired yet – not even close – and the saloon barons who pass as my employers don't know the meaning of "vacation time". And speaking of saloon barons, you apparently run one of the great "salons" of the 20th–21st centuries!

RFV:

> **CML**: As explained in the preface, mental operations cannot be distinguished from real operations except in a provisional way. That's what I mean by "a CTMU context". It's given *a priori.*

Nor of course can they be equated in other than a provisional way. I understand now, but this need not be a "CTMU context" for the sense to have real meaning.

CML: Yes, they can be absolutely equated, because as I explained above, we ultimately can't prop up the distinction between them (distributed solipsism, remember?). Therefore, it's only the distinction itself that exists in a "provisional way". As far as the name's concerned, I'm the guy who's done the most to analyze reality from this angle lately, and CTMU's my name for it. But as long as we both know what we mean, that's good enough.

RFV:

> **CML**: ...you're talking about a generalized form of mentation (not just the mentation of a human brain). Since a photon and a brain both transform information, albeit on different levels of complexity, they are both "mental" in nature.

In view of what follows, I need a little definition of what this generalized "mentation" is. Does it relate to that which takes one form of reality into another, or what?

CML: Any construction you place on "mental" above and beyond pure information processing, information transduction, state transformation, etc. is, as we observed last time around, superfluous and anthropocentric. These are what is meant by "generalized mentation"...what you get when you take the "men" out of "mentation". Again, think "time" (the grammatical implementation of reality-syntax by reality itself, i.e., reality self-processing). Remember, where "abstract" and "real" are the same, so are mind and reality, and so are human and nonhuman info-processing; cognition $\leftrightarrow$ time. The logic is so simple it hurts! Human cognition, assumed to be fundamentally unique by anthropocentrists everywhere, is just the temporal processing that goes on in a brain.

[Of course, there may be more to it than that – and there is – but for present purposes, that's good enough. The "more" in question has to do with distributed solipsism, and the consequent fact that there is a sense in which sentient cognition is primary...that what goes on in reality, namely temporal processing of all kinds, really goes on in the minds of sentient observers.]

RFV:

> **CML**: Actually, the example of a photon is problematic for reasons too advanced to cite here. But for now, let's just say that a photon trans-

> forms the state of an atom by causing an electron to jump from one orbit to another. Even more basically, it "transforms its own coordinates" as it "moves"...

Is its mentation different in this sense than a baseball going from one hand to the other? You're going to have to stretch me a little here too I think.

CML: Yes. A photon is a very fundamental object that can be described as "expanding" at the speed of light, and then "collapsing" someplace on its own expanding wavefront. This amounts to "motion of primary type", and is how the spacetime metric is created layer-by-layer in the conspansive model (CTMU spacetime). When a baseball moves, it's moving in a much more continuous fashion through a sequence of metric domains created by conspansion (as described in that Noesis/ECE 142 file I sent you, *geomusser.doc*.)[3]

That's "motion of secondary type". But I hope we're not getting ahead of ourselves here.

RFV:

> **CML**: ...(due to the CTMU conspansive model of spacetime, the quotation marks are necessary). Atomic states and position coordinates are informational.
>
> (...)
>
> > **CML**: What we seek in any scientific context is a distributed formulation of reality that applies inside the brain as well as without it, ...

It was rather the "that applies inside the brain as well as without it" that boggled my mind. The isomorphism we developed earlier would accommodate my mentally embracing the distributed formulation, but you are implying here either that the formulation applies also to the brain (which it well may not, if the formulation doesn't pertain to brains) OR that the formulation can be made WITHOUT a brain. The latter is what I think you mean and I'm certainly not THERE yet. Allowing a level of mentation to objects does not in itself open the door to complex formulations being possible without brains.

CML: I simply meant that since the brain is embedded in spacetime (or physical reality) and obeys its laws, a distributed formulation of physical laws must apply to those parts of spacetime containing brains as well as those that don't. Obviously, the kind of mentation transpiring inside a brain entails the presence of the brain in the part of reality where it occurs. But because the presence of the brain is possible anywhere in spacetime – since the laws governing cerebral structure and functionability are everywhere operative – that possibility is

[3] Two documents referred to in this dialog, *geomusser.doc* and *realitycheck.doc*, are from past issues of Noesis/ECE and not presently available in our archive. We plan to reprint these documents in future issues of Noesis Electronic Edition.

distributed. So no matter where in spacetime the brain goes, it can still function. Clearly, the same applies to inanimate objects and their generalized cognition. [Again, we neglect the distributed-solipsistic ramifications.]

RFV:

> > **CML**: ...and although the terms "mental" and "brain" are often assumed to be associated, this need not be exclusively so.
>
> Absolutely with you here.

However, that is a long ways from saying that I accept that complex formulations of reality exist outside the brain! Not that I couldn't ultimately be convinced, but am currently not.

CML: Again, Fred, your distinction regarding what's "inside" and "outside" the brain doesn't necessarily work for the mind corresponding to the brain. In a solipsistic model, the brain is merely the localized objective terminus of a material endomorphism (internal self-mapping) of the subjective, distributive mind. Because the mind shares the syntax of reality, and that syntax is scale-invariant (or geometrically self-distributed in the sense of a hologram), its boundaries are those of the reality it perceives. But you can't fully understand this without understanding the conspansive model of spacetime, in which objects literally "include themselves" in a spatiotemporal sense. That is, their SCSPL syntactic images, which are basically cognitive in nature and correspond to the subjective phase of their existence, temporally configure the metric in a cumulative step-by-step way by "expanding through space" at the common rate c (that's inner expansion) and then "recollapsing" to define the objective phase of their existence (that's requantization). Inner expansion and requantization correspond to coherence and decoherence of the quantum wavefunction.

RFV:

> **CML**: Again, it goes along with the absence of an absolute distinction between the abstract (syntactic) and concrete sides of reality.
>
> **RFV**:
>
> > **CML**: ...this structure as transductive syntax up to isomorphism...
>
> (...)
>
> > **CML**: ...with the intrinsic structure of the transducing object.
>
> Now this I have problems with. This says that the fact (which I accept but some would doubt) that we can create an isomorphism between the universe and a mathematical model necessarily implies that the agent that can do this must have a structure isomorphic to the universe as well. That requires some very critical thought. Let me do it.
>
> **CML**: Let me save you some time. The isomorphism is given on a *de facto* basis. Indeed, it follows from the preface.

This is my BIG problem so far. It occurred above and now here. This is where we need to concentrate effort!!! I have NOT bought in here yet.

CML: Again, the syntax of the mind is that of reality as a whole; that's the distinguishing characteristic of SCSPL. The deepest, most general level of this syntax is like a hologram whose image exists everywhere within it on all scales. That's due solely to a lack of prior constraint on scale; since the syntax is primal, no prior constraints can exist (this is a mathematical criterion, not the assumption of some big holographic crystal dangling in Ur-space). The brain partakes of this distributed holographic syntax – the technical term is "hological" – by virtue of its topological embedment in physical space. But through conspansion, this topological embedment is computationally reciprocated by the mind corresponding to the brain; the mind (cognitively) contains the reality that (topologically) contains the brain objectively embodying the mind.

Logically, this comes down to certain CTMU duality principles mentioned in that *geomusser.doc* file. In particular, a form of duality that can be referred to as "spacetime-object", "wave-particle" or just plain "conspansive" defines an alternation in which the mind solipsistically configures reality by a cumulative selective filtration of syntactic potential; during the inner-expansion phase of conspansion, the mind builds a set of exhaustive potential future realities by superposition, and in the requantization phase, selects a future reality in which to re-embody itself as a material brain. Just in case this conjures up some kind of bizarre "brains-in-vats" scenario, please remember that the mind, just like a photon, is regarded as a configuration of the reality syntax. Indeed, all matter behaves alike in this regard; as does the mind, so do its contents.

RFV:

> > **CML**: The intrinsic syntactic "mental" structure of an object can only be inferred from its behavior, which is where a need for data enters the picture.
>
> Very definitely. But here you need an extreme amount of data it would seem to me and I don't know off the top of my head that I can think of what kind of data would justify such a claim. It strikes me somewhat of the "I can test of 4-sigma, therefore I are one!" A test would be good.

In re-reading yours/mine here I have a little different picture/problem. "Intrinsic syntactic" means "mental"? I see the intrinsic form of an object as the "ideal" for that object but not that object's idea of itself. Clearly the "form" or ideal structure of an object can only be inferred from data. But its ideational capabilities are altogether different.

***

Note: Here, more explanation was required. The object corresponds to a configuration of an absolute hological syntax whether or not we or the object have data or can infer what that configuration is, or for that matter what the hological syntax itself is. This is given by the abstract = real or M=R relationship.

***

RFV:

> **CML**: Remember, theories are made to be superseded by better theories; ...

Easy to remember.

> **CML**: ...data can always be extended.

Ok.

> **CML**: Only tautological self-distributed systems possess "certainty".

The "certainty" or necessity of such a consistent system would indeed be very convincing but the "only" here bothers me.

CML: It inheres in the definition of "tautology". Tautology, which denotes informational closure, exists wherever stable perception is possible. Specifically, it must exist in an embedding system of any perceptible nontautological system precisely because the latter is stable enough to perceive; closure is what forms the informational boundaries that distinguish and define the percepts.

Unfortunately, we seldom know exactly what the embedding system is (e.g.) because definitive axioms are missing. This, of course, relates to undecidability, which states that axioms are always missing in logico-arithmetic systems of sufficient complexity. But the important thing to realize is that because perceptible informational boundaries exist – because the system has enough distributed structure to be coherent and self-connected – closure is given at various levels of systemic self-reference. It's an existential necessity in the mathematical sense; no informational closure → no stable perceptual invariants.

To have certainty at arbitrary levels of specificity, we'd need to know all the axioms of the overall system...i.e., a full description of the global systemic tautology allowing perception. Obviously, any nontautological subsystem is "open" to undecidable data and therefore uncertain. In the case of reality at large, this uncertainty – which, by the way, is associated with quantum uncertainty – owes to the fact that increasingly specific axioms are even now being created by sentient agents. That's why reality, aka SCSPL, is said to be "self-configuring". That's what intrinsic cosmogony is ultimately all about.

Sorry if I lost you up there, but explaining this is hard.

RFV:

> **CML**: That's why we can be so sure of the 2VL component of reality syntax. Theories generally aren't so ironclad.

I still have the problem of not having accepted the ubiquitous nature of 2VL.

CML: It's ubiquitous because if it weren't, we'd have a breach in the boundary between 1 and 0 (true and false), which would compromise informational boundaries at all orders of predicate logic (in the abstract logical phase of reality), which implies the perceptual inconsistency of reality, which means logical collapse of the entire perceptual system (due to the systemic ubiquity of the truth or systemic inclusion predicate, inconsistency can't be isolated in logical and perceptual systems; a hole in truth means holes everywhere!). Note that undecidability relates two aspects of informational closure or perceptual stability, namely consistency (the integrity of the T/F boundary) and incompleteness (the fact that certain axioms are even now being decided and are therefore missing from our worldview).

RFV:

> **CML**: The discernability and stable identity of objects within the universe can be shown to imply that the universe is informationally closed in the following sense: it is possible to distinguish that which it contains (reality, perceptions) from that which it does not contain (unreality).

In what sense do you assume this "closure" encompasses "completeness"? Completely, I assume, but is that warranted? What possibilities are excluded by including only those forms that we currently incorporate into the isomorphism?

CML: Explained above. At any given point, we possess incomplete theories that are good only up to isomorphism with what we currently perceive. I.e., there's an "informational boundary" separating that which the universe includes from that which it excludes, and this boundary is what is "closed". No closure means no boundary means no information.

RFV: Why would the infusion or exit of a few forms every few billion years mean that there is "no information"?

CML: Because on the basic level of this discussion, the "forms" of interest are syntactic invariants...invariants of perception. If any of these forms were to cross each other, the result would be perceptual inconsistency, loss of closure, and the collapse of perceptual reality, i.e., the formation of a gap between two discrete, mutually disconnected, internally consistent realities in which the contradictory perceptual invariants are isolated...two mutually irrelevant parallel universes. We're only interested in the universe we inhabit.

RFV:

> **CML**: ...the CTMU simply pushes Kant's ideas to their logical conclusion, dealing directly with the relationship between mind and reality.

I don't know that I'm ready – I'm not. But please administer the firehose in accordance with your perception of my readiness. I know there are a couple of concepts that seem like prerequisites that we need to consider terminological deficits to be worked on as we go.

CML: I've been hosing on this for hours now...I hope it helps.

RFV:

> **CML**: But a more general (and therefore more scientific, ...

> We have somewhat different definitions of "scientific".

> **CML**: Again, "scientific" means replicable, replicable means distributed, ...

I buy that.

> **CML**: ...and distributed means general.

I don't buy this – yet. We could have a very widely distributed specific thing for which its wide distribution implies nothing but its wide distribution.

CML: You seem to mean "distributed" in a statistical sense...something sprinkled through a preexisting metric space. The kind of distribution we're talking about here is logical and refers to the syntax of reality itself, which obviously distributes over that which it governs (otherwise it wouldn't be where it has to be in order to do its job). Your kind of distribution is defined on the metric; mine defines it.

***

Note: Actually, RFV is probably talking here about the distinction between conditional and unconditional logical distribution; only the latter is really "general". See below, two CML entries down the page.

***

RFV:

> **CML**: Moreover, science carefully excludes false tautologies, ...

I don't see "science" as excluding much at all. Early stages of science in a given field would probably INCLUDE false notions. It would still be "science" in the fullest sense of the term, just subject to being found out later to have been wrong – but that's almost the definition, and certainly much of the value of science.

CML: I simply mean that when science becomes aware of a false tautology, it breaks its back to get rid of it.

RFV:

> **CML**: ...or circular but nongeneral forms of reasoning.

I see it as excluding circular arguments but definitely not "nongeneral forms of reasoning". The ultimate scientific achievement may very well be described as you have, but not in its current state. Are we arguing different things here?

CML: Whoops – I meant "circular but nongeneral forms of reasoning that pretend to generality". I sometimes tend to forget that not all scientists are seeking a theory of everything, perhaps because they run into such spectacular problems when they do. Let me explain.

Once again, the scientific method entails both local distinguishability and replicability, two predicates which become opposed to each other on the syntactic level. To be universally operative or "replicable", reality syntax ("the laws of science") must be distributed throughout spacetime; but since it is, it's not locally distinguishable. The limitations of the scientific method are those of science. But since science doesn't yet realize these limitations, it continues to seek laws with distributed (replicable) formulations by empirical (locally distinguishable) means WHETHER THE LAWS ARE GENERAL OR CONDITIONAL. Where the laws are indeed general, this is a logical impossibility; the only thing that can make a scientific law locally distinguishable is the presence of specific definitive conditions that are elsewhere absent. Science often ends up chasing its own tail when it tries to go too general for its objectivist limitations, as it does in fields like cosmology and consciousness.

So what it seems to come down to is this: I'm talking about what science is ultimately trying to do (seek general laws that apply everywhere in spacetime), while you're talking about what it actually can and does do in its current objectivist form (seek nongeneral laws that are nonetheless distributive in effect, or replicable under more or less specific sets of conditions anywhere in spacetime). I guess I tend to view science in terms of trying to make it what it wants to be.

RFV:

> You mention "description"; that is the key word. How are percepts best DESCRIBED? Data is needed here too it seems to me.
>
> **CML:** Description can here be considered synonymous with "theory", ...

I like this notion. It also flies with current philosophical presuppositions.

CML: ...and glossaries. In most usages, the two terms are equivalent. A "description" lacks the mathematical structure of a "theory", having no explicit axioms and rules of inference, but borrows them implicitly from the syntax and semantics of the language in which it is expressed.

RFV:

> **CML:** ...and thus related to perception (data) by isomorphism.

You just don't let me catch up, do you. I see perception more as a dot or stroke in an impressionist painting; you tend to see it as a Rembrandt. I don't see a theory as being "perceived". I see it as ultimately being understood on the basis of many, many perceptions. I guess that's what I'm doing here – forcing you to strain your theory through my sieve.

CML: Maybe that's a good thing for me. In the CTMU, as in Hume, reality coincides with perception (and also with its cognitive underpinnings as described by Kant). Indeed, science already effectively describes reality as a *de facto* blend of thought and perception, with theories using rational relationships to weld together our shared perceptions. Scientific theories can thus be regarded as "common generalizations" of those little impressionistic dots you mention. So when we set out to construct scientific theories – which is, after all, what we're doing here – we must first buy into the assumption that there are commonalities among our perceptions. And this isn't really a questionable assumption, because those perceptions (as opposed to our personal interpretations of them) can always be scientifically reconciled. This reflects the fact that reality is logically self-consistent, which in any case can be established by reasoning along the lines described above.

RFV:

> **CML**: If you're right, then Wheeler's title for his thesis ("Observer Participation Thesis") would seem to indicate his agreement with me about what constitutes an "observer", namely anything able to participate in an irreversible change.

That happens also to be mine. It is a good concept that concedes what has to be conceded without selling out altogether. It also is in stark conflict with relativistic concepts – whoops, let's stay on course here.

CML: That's good. Then you're amenable to a formalism in which the universe perceives itself through the objects it contains. That's SCSPL, and that's the CTMU.

RFV:

> **CML**: This reflects the distributed subjectivism and self-processing character of reality...the fact that mind (or "observership") and reality can't be absolutely distinguished.

Perhaps. What I am pushing on to understand.

CML: Compare "distributed solipsism" to "distributed subjectivism".

RFV:

> **CML**: It's even simpler than I put it above. Experience entails something that experiences and something that gets experienced. So if reality consists everywhere of experience, it is everywhere self-experiencing, and has everywhere the character of both mind (the subjective experiencer) and objective reality (that which gets experienced).

A good notion, but I see this from my background as more of a demonstration of observer/controller equivalence duality. From absorber theory (Wheeler also) there is no significant distinction to be made between emitter and absorber. In tracking technology, the algorithm for tracking many targets to minimize uncertainty is essentially the same (an inverse) as that to situate sensors so as to maximize observability.

CML: Two different languages describing precisely the same thing. The main difference: my language has been developed in the direction I've been outlining, resulting in a global reality theory.

RFV:

> **CML**: So the categorical imperative relating the subjective and objective aspects of reality everywhere coincides with experience; ...

I have always seen the duality as action/observation – both fairly closely associated with a subjective view of reality overlaying a physical level. Not worth pursuing here probably.

CML: Maybe so. One of the best ways to achieve understanding is to see the same things in as many different ways as possible. But remember; according to distributed solipsism, which is simply a correlate of yesterday's "blue shades" analogy, "subjective reality overlays physical reality" dualizes as "physical reality overlays subjective reality". The dual viewpoints are complementary and mutually essential, and this is the point. From this duality, plus a big load of logic, the CTMU follows.

RFV:

> **CML**: I see improvement, by God! But in any event, just don't say I didn't try my best! Keep up the good work.

You've done well! Please keep it up.

CML: Hope you like this as much.

PART FIVE

RFV: Thanks for the continued help in understanding where you're coming from. It is indeed a great privilege to get such insights from the originator of a theory.

I'll chop old conversation that it seems like we agree that I understand and continue on the ones where I've had difficulty.

> **CML**: No. It's a self-perceptual system with a combined abstract and concrete nature (as explained above) that possesses closure with respect to other realities.

So, do you mean a conceptual model of "reality" which could be very different than our own reality? I think so. I see "self-perceptual" as a block to my acceptance here. I think you make the leap that isomorphic structures are the SAME structure to say that. Am I right? If so, how is that justified?...

CML: ... So the objective and solipsistic models are as indistinguishable as U and T, abstract and concrete, and we have to drop the distinction. And in this case, (objective) isomorphism equals (abstract, subjective) identity; when we talk about T and U, we're really talking about different aspects or characterizations of the same thing. So the perceived universe and the perceiver are one, and the universe is "self-perceptual". This is where reality theory meets the philosophy of mind. But that's fine, because as should by now be obvious, they need each other.

Your treatise was excellent! (I am a bit surprised that you've never mentioned Bishop Berkeley though.)

I do think I understand where you are coming from on this and I would like to suspend judgment on it for the time being. There is very much a hard-nosed aspect I like in accepting what we know without wistfully assuming more. It's a quite scientific approach to my mind. However, I believe that solipsism can NOT intuit distributed self, i.e., "I think, therefore I am" does not work for "we" or an amorphous ubiquitous self any more than it worked in proving the existence of God. And if one were to reject the distributed subjective aspect, it would seem to me that the feasibility of an extensive isomorphic relationship between "my" mind and the universe comes into serious question.

CML: Those unfortunates who formerly tried to prove the existence of God using solipsism, *cogito ergo sum*, or comparable, failed because they unwittingly remained in the context of the objective (Cartesian, mind-versus-matter) model without correctly discarding the absolute distinction between the subjective and objective pictures of reality and availing themselves of the implications. The whole point of what has gone so far is that you CAN'T "reject the distributed solipsistic aspect"; the distinction between solipsism and objectivity is ultimately imperceptible and therefore empty of content in any perceptual reality. You have to reason without the distinction when reasoning about reality, and that means that the abstract and concrete phases of reality coincide. Restoring the distinction on an *a priori* basis is simply unjustifiable.

RFV: You have set up an analogy to the mapping between the inside and the outside of a sphere, i.e., $x \rightarrow {}^1/_x$, x complex. The analogy works fairly well. The problem I see is that awareness of the mapping – or creation of the mapping isn't included. That remains either a fact or not. T remains a subjective map of the objective U. How can T become the map of U and the map? That is the question.

CML: Where the universe is a self-similar structure whose overall characteristics are locally, endomorphically replicated everywhere within it – and that's what "distributed over it" means with regard to its syntax – the mapping of which you speak is just the endomorphism in question. It's inescapable and does not have to be justified in particular cases. Equivalently, your absolute distinction between what's inside and outside of a sphere is really a provisional distinction that vanishes above a certain level of abstraction; although the interior of a particular sphere is localized within its boundary, the raw syntactic potential for "being inside a sphere" is distributive (because the abstract mathematical laws that characterize this state are distributive).

Whereas your distinction applies with respect to the geometric interior of a particular sphere, the mind exists on a higher level of generality than that. When we talk about the laws of cognition, including those of math and physics, we're talking not merely about actual brains enclosed in actual skulls, but the abstract essence of brains, i.e. mind. Mind distributes over this entire universe, and with it, the POSSIBILITY of being inside or outside a particular brain. Possibility is a generalization of actuality representing all conceivable actualizations, and this describes the reality syntax.

RFV: Another aspect that bothers me is the following: A theoretical basis of holography is that if one can determine the electromagnetic field throughout a window surface, one can determine the field throughout the volume on the other / either / both side(s). This is a powerful but nonetheless legitimate statement. But with it, there is an extreme implied simulation capability – I guess we should have known from our screens. So (and importantly) the window itself may be all there is – it is at any rate the BEST place to watch for something new! The value of the WINDOW seems somewhat diminished by some of your presumptions. In fact it seems to have paled into insignificance. If I were to develop a theory with the scope of yours, it would be centered on the window itself and the window treated as the only thing "real". Did you happen to read my "Working with the Given" in *Gift of Fire*,[4] № 99?

CML: I'm not making presumptions. Rather, you're presuming that there's an absolute distinction between the window and what's outside it. The holographic analogy was just that: an objective approximation of a self-contained (globally subjective), self-perceptual system whose existence we established by eliminating the empty distinction between the mental and the objective, and thus between the observer and the observed. In a so-called "holographic universe", the interference pattern cast by the hologram is internally projected or simulated. As I've previously explained, the universe can be regarded as "self-simulating" (*realitycheck.doc*; *geomusser.doc*).

RFV: Another thing is that "reality" is more than what I (and I believe anyone) can "mentate". There are aspects which currently defy modeling although I do

[4] The official publication of the Prometheus Society.

believe ultimately they WILL BE modeled. These aspects of U find no place in T and are carried in my (?) mind (or not) as little unmodeled chunks of data u_i. So the isomorphism must hinge on $T \leftrightarrow U - u_i$ or $T + t_i \leftrightarrow U$. And since there are so many of these chunks and the t_i are so unmentatable currently (at a minimum), the perfectness of the isomorphism is somewhat destroyed.

CML: You have a limited capacity to mentate because your brain is able to explicitly replicate only a part of the reality syntax; your cerebral neurons don't fire in explicit isomorphism to the entire syntax, but only a restricted part of it. But the overall syntax of which your own is a restriction is, as I've been trying to point out, distributed. You seem to be assuming that T exists strictly inside your head, and if T is empty of physical content, perhaps it does. But for purposes of this discussion, T equals a GLOBAL DISTRIBUTED FORMULATION OF THE REALITY SYNTAX in which ALL aspects of U FIND A PLACE. Such a formulation certainly exists whether or not we know what it is (and we don't); specifically, it exists in the conspansive T = 0 state in which the evolution of the universe is instantaneously realized, AFTER we've selected it from the set of all possible evolutions and formulated a nomology to render mutually consistent the perceptions we create. At this stage, theory equals reality:

$$M(T(U)) = T(U) = U$$

One distributed ingredient of this distributed formulation that we can be sure of right now is the set of 2VL tautologies, which cannot be violated with regard to perception; you'll never have a real perception that is both T and F unless you simultaneously inhabit two parallel, mutually inconsistent universes. That's just the way it is; any two people who won't both admit that the perceptual basis of reality is 2VL have nothing to discuss. Were 2VL mistakenly acknowledged to fail for the real content of this discussion, then insofar as this discussion is taking place inside the reality being discussed – i.e., inside its own content – 2VL would inevitably fail for the discussion itself. That would make this discussion, and indeed every discussion, irresolvable. But in that event, math and science would be a useless farce, because what one of us "perceives" to be true (and even tautologically true) may not be acceptable to the other.

RFV: Yet another is the equivalence you assign to isomorphism. It's admirable to the extent of going with ONLY what one knows, but less admirable I believe in possibilities of enshrining error.

CML: As I've already explained, any isomorphism to whose range and domain you can't assign different qualitative or quantitative coordinates at a given level of discourse is really an identity relation (at that level). In that case, the range and domain differ only as coincident ASPECTS of one and the same reality. You're still talking about T being essentially different from U, even though we've shown that at a sufficiently high level of content for T, T = U. Again, T is in this case the reality syntax – the mathematical and physical laws that the universe obeys everywhere – and objective reality (U) is merely the variegated actualized content of that distributed form.

For example, the law of conservation of momentum, and its underlying arithmetic, is everywhere valid, but only active and perceptible in the presence of actual moving bodies. Similarly, physical reality is merely an evolving set of instantiations of a set of relations that are everywhere valid. These instantiations are metrically separated, but the metric itself is completely characterized by a distributed form that exists in every part of space that could ever be occupied by your brain (even if your brain has only managed to figure a little bit of it out by firing isomorphically to it). The distribution of this form throughout space distributes your mind over reality in the form of general mental syntax.

RFV: Those are my reservations, but I understand what you have said, appreciate it, and it will ride with me, perhaps ultimately persuading me. You may continue hammering on this or accept this limited CONVERSION as you see fit. I do enjoy the dialogs on the subject.

CML: As do I. It helps me understand where the rough spots are, and encourages me to find ways to get past them. I think one of the problems so far is that you haven't really absorbed the fact that the distinction between mental (abstract) and objective (concrete physical) reality is a stratified distinction that gradually disappears as we approach the top stratum (global distribution). Less general, more conditional features of reality appear to be localized within their specific instantiations, permitting internal-external distinctions. In this sense, cognitive (or computational) distribution relations are equivalent to topological inclusion relations (that's a CTMU duality principle). If distribution is general or unconditional, on the other hand, it applies to all things, and the distinctions among these things exist within it in the sense that it resembles a SELF-EXECUTING PROGRAM INSTRUCTION that internally SIMULATES them. In other words, general features of (simulative) spacetime include specific aspects as output, which is then uploaded to spacetime as further programming, and so on. Topological inclusion follows from computational inclusion because topology is computationally simulated.

RFV:

> > **CML**: ...real objects necessarily conform to more or less exact configurations of an abstract distributed syntax.
>
> (...)
>
> **CML**: If we form a theory of some aspect of reality, it may or may not be correct (i.e., it will be "more or less exact").

"More or less CORRECT." Right? Exactness is a small case in point?

> **CML**: Yes. There are two kinds of truth, syntactic and semantic. Syntactic truth amounts to internal consistency, and semantic truth amounts to external isomorphism or quantification. If T is true on neither the syntactic nor semantic level of reality, then it has 0 scope or content. But then the theory has no model and is not a theory in the strictest sense. At best, it's just an empty (content-free) pattern of neural events or an inky set of typographical symbols on paper, with irrelevant fantasy for content.

But such things may become the "Eureka!" source next week! Is the entire universe reborn in that case?

CML: If T has no model, then it has no real content, and in that case, "eureka" becomes "you-reeka". Now, maybe its apparent lack of content is only temporary because we're misinterpreting it, which means that we haven't accurately identified its constituent concepts; or perhaps the only thing that prevents it from having all kinds of real content is one little piece or two that we'll eventually recognize as unnecessary and discard, leaving a big pile of "eureka" material. But then the part of the theory that is found to be correct will have possessed real content all along. I.e., the content of a theory is not necessarily dependent on what we recognize it to be at any point in time.

RFV:

> **CML**: That part of a theory that's wrong is empty, but the part that's right has real content.

The wrong part of a theory may be the HANDLE of its replacement and thence the most real aspect of it.

CML: As my last entry acknowledges.

RFV:

> **CML**: The theory tautologically corresponds to just that part of reality for which it's true. In fact, logicians already know that insofar as they can often either extend or qualify a troubled theory so as to adjust its referential quantifier (think Löwenheim–Skolem, Duhem–Quine, etc.) and thereby restore its validity (think classical physics as the low-velocity "limit" of relativity physics), the idea of an "absolutely false" theory is not what it seems.

I'm not sure I fully understand the last part of this last sentence – it may, of course, be that I didn't understand the FIRST part or the sentences that preceded it. Would you please try to explain this to me? It's interesting.

CML: The falsifiability criterion of scientific theories is rather tricky. If you have a theory that all cats are gray, including a reason for that prediction, it's "falsifiable" only in that part of the universe containing non-gray cats. And even after it's been falsified, we can still resurrect it in that part of the universe where all cats are grey and your reason applies. All we have to do is create an "extension" of your theory that applies to the rest of the universe. Once your theory has been embedded in a correct overall theory, one can no longer say it's been "falsified". It simply applies to a smaller part of the universe than originally intended, and is quite true for that part.

In addition, once we do away with Occam's razor, every interpretation of a theory becomes a theory in its own right, each of them as correct as the next. And even worse, every possible unverifiable (and thus unfalsifiable) extension of theoretical content screams for air time, all kinds of "true" but bizarre and mutually contradictory theories raising their ugly heads for keeps.

Most telling of all, there are two kinds of falsifiability: rational and empirical. The former kind is employed in rejecting erroneous mathematical proofs and even mathematical theories of physics; the latter, in rejecting physical theories. A true logical or mathematical theory is not empirically falsifiable, and neither is the 2VL logical syntax validating the falsification process itself! The syntax is literally tautological.

RFV:

> **CML**: The point of all of this is that a theory can evolve, changing its form to absorb more and more real content. That's what I meant by "more or less exact". Perhaps "more or less correct" would have been better.

Yeah, OK. That's what I thought, but in the explanation I got some interesting understandings. This does put my notion of the ui into a dismal light and I don't exactly know what to do about that. I think I'll leave it as a record of my discontents along the way.

> > **CML**: But even if not, we can be sure that the correct theory will conform to our mental structures, since otherwise we won't be able to create a conceptual isomorphism with it (and that's what a theory is).

> I have a hard time with this. An analog computer is an isomorphic structure of a screen door when it is set up to solve the screen door problem, however, a general purpose computer is NOT isomorphic with a screen door when it solves the screen door problem. General structures are adaptable in a sense which allows them to accommodate an isomorphic model without themselves being limited to anything like that model structurally. So in what sense (and to what degree) do you insist on this conformal mapping?

> **CML**: There is a very general abstraction called logic that permits various mappings among devices and other physical contexts. Have you ever heard the term "up to isomorphism"? It generally refers to a relationship that contains an isomorphism, but exceeds it. Some logi-

> cal abstractions permit multiple isomorphisms, and these isomorphisms may be either structural or functional or both (in the above question, you seem to be talking about these two distinct kinds of isomorphism).

Indeed. More-or-less as levels of isomorphism? BUT, if one allows levels or degrees, then isomorphism can certainly not MEAN equivalence – nor does it. Then to substitute "equivalence" seems unwarranted without being able to demonstrate isomorphism at ALL levels.

CML: Isomorphism always denotes "equivalence with respect to ( )". Two triangles of the exact same size and shape are isomorphic with respect to structure, but if one's red and one's colorless, the isomorphism holds only with respect to geometry and gives out with respect to color. Were one triangle to have more structure than the other, homomorphism would be the best we could do; but since color is independent of (and in fact irrelevant to) geometry, isomorphism holds. It's just that for the colorless triangle, including color in the isomorphism doesn't even make sense. So the triangles have equivalent structures but differ in aspects like color and location. On the other hand, perfect isomorphism with respect to all predicates describing two objects means not just equivalence, but identity.

RFV:

> **CML**: In any case, for computational purposes, the isomorphisms you mention are logically equivalent up to isomorphism with the screen door problem (whatever that is).

The problem is to determine the angle of a screen door that has a spring and dash-pot. Its solution is that of a simple differential equation. In an analog computer there is a resister whose resistance is scaled to the resistance in the dash pot on the door, the inductance to the spring constant, etc. and the current in the electric circuit is related in the same way to the angle of the door. It's all one-to-one. In a GP computer of course the analogies become much less obvious in the sense of there not being a continuous nor continual solution, i.e., it doesn't hardly map at all. In a DDA (Digital Differential Analyzer) the physical structural analogy of components is maintained with digital registers and the flow of overflow bits becomes an analogy for the current/angle. Just a side tidbit.

> **CML**: The "abstractions" being employed by the two types of computer are always expressible in terms of the penultimate abstraction, logic. Nothing I've said so far is incompatible with this fact.

First of all, it is in a very obscure sense that one can say an analog version is implemented in logic because it ISN'T anymore than an electric current is logic. Secondly, you seem to be implying that LOGIC is a structural level of the brain – maybe that's where 2VL comes in. But I don't think so. I believe logic in the sense of AND, OR, NOT and IMPLIES is an acquired capability of the mind superimposed many layers above the structure of the mind. I don't believe we are PRIMARILY a "logic machine". We are pieces of U (conceivably) capable of implementing limited complexity logic circuits.

CML: The brain is indeed a logic device (a neural network), but this is hidden by the fact that it employs logic to construct modes of activity that are apparently illogical in character. Electric currents are logical in essence as well, obeying a 2VL theory of electromagnetic interactions whose statements and their negations are T and F respectively. The EM currents behave isomorphically to the theory, and the theory behaves isomorphically to the logic inherent in its structure. Every "nonlogical" predicate in the theory is in fact embedded in one or more logical variables or relations just like a seed is embedded in an apple.

EM current "is" logic because insofar as it behaves coherently enough to conform to a theory, it must coincide with something that permits it to do so. Whatever that something is, it obviously includes logic (since if it didn't, the current could not behave in a coherent way or even maintain its existence = T as opposed to nonexistence = F). The mistake is to underestimate the tendency of logic, often disguised by "other" elements of reality syntax which are themselves logical in form, to naturally assemble itself into seemingly illogical configurations and processes.

RFV: At this point I'm going to repeat your original sentence from above:

> **CML**: But even if not, we can be sure that the correct theory will conform to our mental structures, since otherwise we won't be able to create a conceptual isomorphism with it (and that's what a theory is).

There is a sense in which $M(T) \leftrightarrow T$ for any possible T. Of course there are probably even existing theories that have not yet been completely comprehended, M(T), although already formulated, T, in which case you seem to be content with M(T) BEING what a theory is. And there is definitely a sense in which $T \leftrightarrow T(U)$ and you seem to be content with T(U) BEING U because if there were more KNOWN of U it would be in T(U). I think what I feel certain of at this point is:

$$M(T) \leftrightarrow M(T(U)) \leftrightarrow U$$

CML: Theories consist of abstract relationships. To whatever extent they are correct, i.e. apply to some part of reality, they coincide with that part. Why? Because the abstractions that correctly describe a part of objective reality, whether or not you are aware of the actual correspondence, are exactly the ones that are indistinguishable from its concrete components (as in the blue shades analogy of yore).

RFV:

> **CML:** A syntax takes the form of general information and is implemented by generalized cognition.

Whoa. Chris, you've got to start me all over on this one. I thought I was about through the original sentence. I think I buy the original sentence with the extension of "generalized" cognition within my limitations expressed above. BUT...

> **CML:** The phrase "generalized cognition", actually synonymous with time, ...

How in hell did we get here? I take a "synonym" to mean that a dictionary can help me, but it doesn't here.

CML: All you need is your memory, because we've already been through this. The operative analogy is:

time : space :: cognition : information

I.e., time = info-processing = generalized cognition (where information here refers to the abstract dimensions of real objects and processes, which, as we've already seen, coincide with them).

RFV:

> **CML:** ...simply extends human cognition to generic information processing.

How does/can "time" perform this function?

CML: Time is not a "thing"; it's a function...a function of spacetime. Spacetime performs this function because it's full of syntax, and some of this syntax is grammatical as opposed to structural. Grammar governs the evolution or state-transition of the language of reality.

RFV:

> **CML:** It's this sense in which generic information processing obeys a "cognitive syntax".

I was left on the other side of the river on this one. Help!!!

CML: See my last entry above.

RFV:

> **CML:** Thus, time is any grammatical implementation of the SCSPL reality-syntax.

I think we have to hold off on SCSPL a little longer.

> **CML:** A photon is a configuration of the syntax of reality; ...

Chris, you're going to have to lighten up just slightly on some of this stuff. What you appear to be saying is that a photon is properly speaking a particular instance of the general structure of reality. That isn't very informative. What do you mean here?

CML: Just that a photon obeys certain structural and behavioral rules, and those rules exist as constraints within the overall set of constraints called "the reality syntax" (including the laws of math and physics, of which our exhaustive knowledge is not assumed). These are the abstractions to which the photon primarily corresponds. Stay with it, Fred – this is all really very simple (given the unnecessary compartmentalization of physical concepts in diverse formalisms, maybe the problem is that it seems TOO simple).

This part I have gotten largely because of your patience, so stay with me.

CML: That a photon is both real and abstract is an absolute, nonprovisional fact. However, what counts as "real" and "abstract" in any given case is a relative (or provisional) matter, and sometimes even a matter of semantic convenience. I sometimes tend to assume that my meaning is obvious when it isn't.

This was very clear.

CML: ...but given from the outset. It's only the exact corresponding configuration of reality-syntax that remains a mystery (and in a sense, not even that).

I assume you mean that the actualities of the behavior of photons are all that is (may be) unknown? Its place in the structure of reality IS known?

CML: Some aspects of a photon's behavior, and corresponding details of its place in the structure of reality, are known. But the intrinsic structure of the photon – the syntax it obeys – is not known except through this behavior. You can't hold a photon under a microscope and dissect it; all you can do with any subatomic particle is collide it and monitor the resulting events. Sometimes, you get what appear to be objective subparticles, but again, you can only see them through their effects. So all you can say for sure is that one configuration of SCSPL reality-syntax is merely decomposing into others. Getting too hung up on "objects" and their "innate properties" has been a recurrent problem in science; one generation's "particles" are the "subparticle interactions" of the next.

Other than the reference to SCSPL (which I'm deferring) I think I understand this and certainly agree with it!

> **CML**: We need to realize that what we're dealing with are transformations of a global nomology or "syntax" (or to oversimplify and appeal to popular terminology, of the "TOE ultimate equation", ...

Theory of Everything?

CML: Yes. The TOE is envisioned as a master equation governing all of reality...a condensation of all the laws of physics.

RFV:

> **CML**: ...along with the syntactic structure of the language in which it will be expressed).

I see where you're coming from but slipping the word "nomology" in there was a dirty trick because that's the part I'm having trouble accepting.

CML: "Nomology" refers to the abstractions governing the evolution of reality. It's inherent in spacetime. There is no assumption that we can ever do more than approximate it in any human theory. All we know about nomology is that it is what connects events, and is what effects the consistency of our perceptions. Nomology must exist up to determinacy, or its absence would show up in the form of irresolvable perceptual paradox. A distributed abstract nomology to which perceptual reality must conform ensures the resolvability of paradox, which is just a way of saying "the self-consistency of reality".

RFV:

> **CML**: Analog computers are as bound by 2VL as are digital computers; an analogy either exists (1) or not (0).

But that analogy is not performed by the analog computer, so I still don't see where you get that such a computer is 2VL. The use of its results may involve 2VL, but its operation doesn't.

CML: Without 2VL, it would have no coherent "results".

RFV:

> **CML**: All distinctions are 2VL; ...

It doesn't make distinctions.

CML: Yes it does. It distinguishes between what it is and what it isn't, and how it must behave and how it must not behave. These distinctions are encoded in the spacetime it occupies.

RFV:

> **CML**: ...one thing is either distinct from another (1) or not (0) on a given level of definition.

Is an electron in a circuit of my computer distinct from one in yours? I think yes. Is an electron in an adder circuit in my computer distinct from one in my instruction register circuit? Is an electron in a circuit distinct from another bound electron in a copper atom of the wire through which the electron travels at that moment? Is it distinct from adjacent band electrons in the flow? Distinctions get a bit fuzzy.

CML: Our respective electrons are distinct images of a single distributed "syntactic photon" consisting of the rules that all photons must obey regardless of where they are. The images reside in different locations, and these differences are significant. But it's like Wheeler said to Feynman: they're all one photon.

RFV:

> **CML**: Since information consists of distinctions – constraints differentiating one thing from another – information is 2VL. Since theories consist of information, theories are 2VL.

I can accept Whitehead and Russell on that.

> **CML**: And since theories are isomorphic to reality, reality is 2VL.

Where I'm at, you've just convinced me that M(T(U)) is 2VL.

> **CML**: There is no escape...no crack in the stone of truth versus falsehood. Even to explain why there might be, you would be forced to resort to 2VL...and your putative exception would be embedded in 2VL by *fiat*.

That's certainly true for M(T(U)).

CML: Great! Because the isomorphism M is a perfect equivalence relation with respect to everything but degree of generality (the abstract is more general than the concrete because one abstraction can possess many concrete actualizations). So M is an equivalence relation between T and U up to isomorphism (i.e., up to exhaustion of the structure of U).

RFV: This is not very convincing to me. First of all, accepting your premise that all distinctions, knowledge, etc. are 2VL does not prove anything about "reality". Ideational objects – even given the isomorphism – are not the objects themselves or exact equivalents – there are, for example, things which are not known about photons that are true of photons but not yet integrated into the ideational counterpart. So we have isomorphic structures of the reality of which we are aware and the reality we thereby presume. And we are in some sense limited such that our truths pertain to such a reality, but what about undecidability and such notions. How do these not end up being cracks in the stone? One must reason decidably about undecidability which doesn't, thereby, prove that all things are decidable.

> **CML**: First, logical consistency equals the syntactic phase of truth.

I buy this except that I'd mellow out "truth" just a little to something like "correctness" or "verifiability".

CML: No, the syntactic phase of truth equals consistency. That's what the predicate "true" means with respect to syntax.

RFV:

> **CML**: The stability (objective perceptibility) of reality implies that it's consistent, ...

Persistent at least. Perhaps one could argue from persistence in U to consistency in T(U).

CML: Yes, one could. As T converges on the reality syntax, U and T(U) become coincident.

RFV:

> **CML**: ...and the fact that it's consistent implies that it conforms to 2VL.

I think there's a logic hop here I didn't make. An analog computer's solution is persistent but is not thereby (in my opinion) 2VL.

CML: See above. Consistency is a 2VL concept synonymous with the tautologies (T OR F) and NOT(T AND F). Since reality by definition contains all truths – otherwise, they wouldn't be "true" – its consistency implies that it contains no falsehood (since such a falsehood would contradict the truth corresponding to its negation). So consistent means true means included in existential reality.

RFV:

> **CML**: Again, something either exists (1) or does not exist (0), so as soon as you say that it's real or that it exists, you tag it with a logical predicate (1 = true = reality-inclusion) that anchors it to the entire logical syntax of which this predicate is a part.

I totally disagree here. Saying that something "exists" adds nothing of logical import to a concept in this context. Do the true (1) statements, "Mastodons do not exist", (1/0) and "Unicorns do not exist" (1/0) imply the same level of nonreality (0)? If "existence" is 2VL, it should.

CML: "Exists" = "is included in an existential system", which implies "is (at least) included in (the most basic existential level of) reality", which implies "is consistent with the rest of reality", which equals the syntactic phase of truth. Now, T and F only become contradictory when they coincide in some actuality; fortunately, "Mastodons do not exist" need never coincide with "mastodons exist" in any actual context. That is, "Mastodons do not exist" can be interpreted over a range of actual times for which it is true. On the other hand, "Unicorns do not exist" can be interpreted over an even larger real context, and to that extent is quantifiably truer in a concrete context.

RFV:

> **CML**: Of course, we can still make a distinction about which level of reality includes this something, but whatever that level is, it obeys the distributed tautological syntax of reality (logic).

Granted you can probably escape the former argument with a defined "level of reality". But more importantly an electric field is not defined in terms of whether it "exists" or not. It is more, "given x, then y", such that existence becomes very much a side issue in T at least AND there may be "hidden" parameters p_1 in a theory for which not everyone envisions a counterpart in U. So the "existence" of p_1 in T leaves its existence in U (and therefore at all) quite up in the air.

CML: If relevant, the parameters p_1 are (at the very least) permissible transformations of reality syntax that may be realized in physical reality. Color force among quarks is a pretty weird parameter that has theoretic utility, but can't yet be empirically established. The configuration of syntax we're currently working with won't actively display it. But one day we may contrive to force a transformation of syntax that does. Moreover, this transformation will itself be governed by the syntax that is transformed. I.e., the syntax is autologous.

RFV:

> **CML**: Second, decidability is a logical concept whose definition is based solely on logical (syntactic) operations, namely those executed by Gödel in defining it. So you can't use it against logic without undermining its definition and thereby invalidating it.

I am not very enthused about Gödel's destruction of *Principia M.* (Gödel, 1962) myself and I'm somewhat interested tangentially in your statement here.

> **CML**: I suspect that what you might be positing is an abstraction exceeding logic and consistency...an even more general predicate.

No. But thanks for thinking so highly of me.

> **CML**: Indeed, that's the only thing your description would fit. No problem; the CTMU contains an ultimate predicate called UBT (unbound telesis). But logic is a closed tautological system, and the relationship of this "pre-logical" predicate to logical reality is appropriately constrained.

Whoa again! If I'm going to stay in this monastery, you have to go slow.

CML: Information consists of distinctions in which one thing is syntactically constrained to differ from another. Because concrete reality mirrors the abstract information coinciding with it, it too can be regarded as information (we're information too, and the reason all this info seems "concrete" is that there's a syntactic displacement principle in effect).

So the question arises: what does this information become when all the binding constraints are removed? The answer is Unbound Telesis or UBT, the groundstate of being. This universe has only arisen within the UBT groundstate because it has constrained itself to do so; this is the sense in which it's "self-configuring".

RFV:

> **CML**: After all, what I'm trying to tell you about is the LOGICAL structure of reality, without which it would be no reality at all. For God's sake, don't interfere...your world could dissolve into an illogical chaos of ambiguities right before your eyes (and you're a part of it)

It has! It has. But I'm feeling better. Please don't do that to me again.

> **CML**: ...employers don't know the meaning of "vacation time". And speaking of saloon barons, you apparently run one of the great "salons" of the 20th–21st centuries!

Slipping logic by me is one thing, but when someone tries to get humor by me, I have a hard time with it. What on earth do you mean?

CML: The salons of the nobility used to be the most lively, scintillating conversational settings in the world. Salon Baron ↔ saloon baron.

RFV:

> > **CML**: As explained in the preface, mental operations cannot be distinguished from real operations except in a provisional way. That's what I mean by "a CTMU context". It's given *a priori*.
>
> Nor of course can they be equated in other than a provisional way. I understand now, but this need not be a "CTMU context" for the sense to have real meaning.
>
> **CML**: Yes, they can be absolutely equated, because as I explained above, we ultimately can't prop up the distinction between them (distributed solipsism, remember?). ...

Yes, but you must remember that you left me back there on the peg labeled M(T(U)).

CML: Not really, at least when T converges on reality syntax.

RFV:

> In view of what follows, I need a little definition of what this generalized "mentation" is. Does it relate to that which takes one form of reality into another, or what?
>
> **CML**: Any construction you place on "mental" above and beyond pure information processing, information transduction, state transformation, etc. is, as we observed last time around, superfluous and anthropocentric.

Let me backslide just for a little reverie here. What about "likes", is that a "construction" on "mental"? I think so. So if it isn't carried into "generalized mentation" where is it? Ok, so it's "anthropocentric", so what? Do we then have M(a, T(U))?

CML: No, what we have is a component of the cognitive syntax called ETS, short for Emo-Telic Subsyntax. The emotions are subjective predicates like qualia; special subsyntaxes are reserved for the predicates in these groups. But the relationships of these predicates with other syntactic elements are logical in nature. Enough on that for now.

RFV:

> **CML**: These are what is meant by "generalized mentation"...what you get when you take the "men" out of "mentation".

Sometimes your technique reminds me of men of God in my past.

CML: Against my better judgment, I'll take that as a compliment.

RFV:

> **CML**: Again, think "time" (the grammatical implementation of reality-syntax by reality itself, i.e., reality self-processing).

Chris, Chris, ...I'm not ready for baptism. This is only my second lesson for God's sake.

CML: That's OK, we've been through it at least twice up there.

RFV:

> **CML**: Remember, where "abstract" and "real" are the same, so are mind and reality, and so are human and nonhuman info-processing; ...

Got it!

> **CML**: ...cognition ↔ time.

DON'T got it!!!

CML: That's OK, we've been through it at least thrice up there.

RFV:

> **CML**: The logic is so simple it hurts!

I know, but I still ain't got it!

CML: That's OK, we've been through it at least fource up there.

RFV:

> **CML**: Human cognition, assumed to be fundamentally unique by anthropocentrists everywhere, is just the temporal processing that goes on in a brain.

> [Of course, there may be more to it than that – and there is – but for present purposes, that's good enough. The "more" in question has to do with distributed solipsism, and the consequent fact that there is a sense in which sentient cognition is primary...that what goes on in reality, namely temporal processing of all kinds, really goes on in the minds of sentient observers.]

This soothes my mind, but all of this is NOT just "time"!

CML: Sure it is. It's the refined kind of time that passes in a brain.

RFV:

> **CML**: Yes. A photon is a very fundamental object that can be described as "expanding" at the speed of light, and then "collapsing" someplace on its own expanding wavefront. This amounts to "motion of primary type", and is how the spacetime metric is created layer-by-layer in the conspansive model (CTMU spacetime). When a baseball moves, it's moving in a much more continuous fashion through a sequence of metric domains created by conspansion (as described in that Noesis/ECE 142 file I sent you, *geomusser.doc*.)

If that was more than a brief page or two of text, I'm quite sure I did not get it or GET it!

CML: Re-sent and acknowledged.

RFV:

> **CML**: That's "motion of secondary type". But I hope we're not getting ahead of ourselves here.

I think we may well be.

> **RFV**: However, that is a long ways from saying that I accept that complex formulations of reality exist outside the brain! Not that I couldn't ultimately be convinced, but am currently not.
>
> **CML**: Again, Fred, your distinction regarding what's "inside" and "outside" the brain doesn't necessarily work for the mind corresponding to the brain. In a solipsistic model, the brain is merely the localized objective terminus of a material endomorphism (internal self-mapping) of the subjective, distributive mind. Because the mind shares the syntax of reality, and that syntax is scale-invariant (or geometrically self-distributed in the sense of a hologram), its boundaries are those of the reality it perceives. But you can't fully understand this without understanding the conspansive model of spacetime, ...

You're right. I'm not there yet.

> **CML**: ...in which objects literally "include themselves" in a spatiotemporal sense. That is, their SCSPL syntactic images, which are

> basically cognitive in nature and correspond to the subjective phase of their existence, temporally configure the metric in a cumulative step-by-step way by "expanding through space" at the common rate c (that's inner expansion) and then "recollapsing" to define the objective phase of their existence (that's requantization). Inner expansion and requantization correspond to coherence and decoherence of the quantum wavefunction.

This part looks interesting but so far from the outside.

> > **CML**: Let me save you some time. The isomorphism is given on a *de facto* basis. Indeed, it follows from the preface.
>
> This is my BIG problem so far. It occurred above and now here. This is where we need to concentrate effort!!! I have NOT bought in here yet.
>
> **CML**: Again, the syntax of the mind is that of reality as a whole; that's the distinguishing characteristic of SCSPL.

Even ignoring SCSPL for the present, this equivalence from isomorphism of T to U is still a difficulty in as much as I have only accepted M(T(U)) ↔ U, and now I must accept that U(T(M)) ↔ M. I don't know...

CML: Again, where T stands for the distributed reality syntax (or "ultimate theory"), M: T ↔ U, and the isomorphism M collapses because you can't keep the abstract and concrete sides of reality out of coincidence up to U isomorphism (due to the blue shades analogy), M(T(U)) = U.

RFV:

> **CML**: The deepest, most general level of this syntax is like a hologram whose image exists everywhere within it on all scales. That's due solely to a lack of prior constraint on scale; since the syntax is primal, no prior constraints can exist (this is a mathematical criterion, not the assumption of some big holographic crystal dangling in Ur-space).

Yes, but saying "the syntax is primal" does not make it so.

CML: No, but DESCRIBING it as such certainly does, provided that description is embedded in the kind of model we've been talking about. After all, if syntax isn't primal, then the process of its creation required a more basic level of syntax that WAS primal. That's what Hawking is saying (in effect, without dealing with the logic) – you ultimately have to deal with reality on a primal level on which its syntax (or nomology) is self-configuring. His *No Boundary Proposal* (Hartle & Hawking, 1983) states that the universe is nomologically self-contained, the laws of physics guiding Creation itself. The CTMU is just a more logical treatment of that existential necessity.

RFV:

> **CML**: The brain partakes of this distributed holographic syntax – the technical term is "hological" – by virtue of its topological embedment in physical space. But through conspansion, this topological embedment is computationally reciprocated by the mind corresponding to the brain; the mind (cognitively) contains the reality that (topologically) contains the brain objectively embodying the mind.

Hmm.

> **CML**: Logically, this comes down to certain CTMU duality principles mentioned in that *geomusser.doc* file.

Evidently I need it.

> **CML**: In particular, a form of duality that can be referred to as "spacetime-object", "wave-particle" or just plain "conspansive" defines an alternation in which the mind solipsistically configures reality by a cumulative selective filtration of syntactic potential; during the inner-expansion phase of conspansion, the mind builds a set of exhaustive potential future realities by superposition, and in the requantization phase, selects a future reality in which to re-embody itself as a material brain. Just in case this conjures up some kind of bizarre "brains-in-vats" scenario, please remember that the mind, just like a photon, is regarded as a configuration of the reality syntax. Indeed, all matter behaves alike in this regard; as does the mind, so do its contents.

You have really gone too far too fast for me. Let's get me over the two big hurdles first.

> > **CML**: Only tautological self-distributed systems possess "certainty".
>
> The "certainty" or necessity of such a consistent system would indeed be very convincing but the "only" here bothers me.
>
> **CML**: It inheres in the definition of "tautology". Tautology, which denotes informational closure, exists wherever stable perception is possible.

Is that last obvious to you? It isn't to me. Why is it obvious?

CML: Because truth is identified with tautology in 2VL, which is a necessary framework of the reality syntax. Otherwise, you could not distinguish which perceptions are "real" (T) and exclude those which are not (F). I keep explaining this point.

RFV:

> **CML**: Specifically, it must exist in an embedding system of any perceptible nontautological system precisely because the latter is stable enough to perceive; ...

We've gone from a "perceptible" system to one that "perceives" without even an argument.

CML:

HYPOTHESIS: The universe is a piece of self-executing (self-recognizing, self-processing) code; the universe must configure itself so as to interpret its syntax *in situ*.

PROOF: There is no external hardware on which reality can run, and no external programmer / executor / recognizer to keep it in touch with itself and running. So the universe must provide its own recognizer, i.e. perceive itself. Now suppose there's a separation between recognizer and syntax. Then the syntax can't recognize its own parts to remain connected, and there must be a remote feedback loop within the U enforcing syntactic coherence. But then how do the components of this loop recognize each other so as to remain connected, and how is the info conveyed by this loop enforced on either end of the loop? This leads to a buck-passing regress in which the remote enforcement of coherence must ultimately close on a level of syntax that is distributive and completely self-recognizing. And since any lesser syntax must be embedded in it, its distributed self-recognition function continues to apply.

RFV:

> **CML**: ...what forms the informational boundaries that distinguish and define the percepts.

You seem to be saying that that which is "perceptible" *ipso facto* "perceives".

CML: See last entry.

RFV:

> **CML**: Unfortunately, we seldom know exactly what the embedding system is (e.g.) because definitive axioms are missing. This, of course, relates to undecidability, which states that axioms are always missing in logico-arithmetic systems of sufficient complexity. But the important thing to realize is that because perceptible informational boundaries exist – because the system has enough distributed structure to be coherent and self-connected – closure is given at various levels of systemic self-reference. It's an existential necessity in the mathematical sense; no informational closure → no stable perceptual invariants.

This looks like a place where the basis of holography, i.e., fields on a boundary could be used in the argument, but I can't put it together and make it work as you envision it.

CML: That's because you still aren't imagining the hologram as a self-contained entity.

RFV:

> **CML**: To have certainty at arbitrary levels of specificity, we'd need to know all the axioms of the overall system...i.e., a full description of the global systemic tautology allowing perception. Obviously, any non-tautological subsystem is "open" to undecidable data and therefore uncertain. In the case of reality at large, this uncertainty – which, by the way, is associated with quantum uncertainty – owes to the fact that increasingly specific axioms are even now being created by sentient agents. That's why reality, aka SCSPL, is said to be "self-configuring". That's what intrinsic cosmogony is ultimately all about.

?

CML: That should read "open to undecidable perception" for which a T or F distinction can't be made due to the absence of a tautologically-defined truth predicate.

RFV:

> **CML**: Sorry if I lost you up there, but explaining this is hard.

Yeah. Jesus, if I ever get this I hope I don't become a CTMU bible thumper. It could be worse than being a Jehovah's Witness.

CML: Especially if you're trying to explain it to people who know nothing about the subject matter, or for that matter logic.

RFV:

> I still have the problem of not having accepted the ubiquitous nature of 2VL.
>
> **CML**: It's ubiquitous because if it weren't, we'd have a breach in the boundary between 1 and 0 (true and false), which would compromise informational boundaries at all orders of predicate logic (in the abstract logical phase of reality), which implies the perceptual inconsistency of reality, which means logical collapse of the entire perceptual system (due to the systemic ubiquity of the truth or systemic inclusion predicate, inconsistency can't be isolated in logical and perceptual systems; a hole in truth means holes everywhere!). Note that undecidability relates two aspects of informational closure or perceptual stability, namely consistency (the integrity of the T/F boundary) and incompleteness (the

> fact that certain axioms are even now being decided and are therefore missing from our worldview).

Chris, Chris, Chris, ...all I get from this is that if I don't accept it I may be sicker than a dog for another two days.

CML: Yes, by punitive decree of that Big Logician in the Sky.

RFV:

> Why would the infusion or exit of a few forms every few billion years mean that there is "no information"?
>
> **CML**: Because on the basic level of this discussion, the "forms" of interest are syntactic invariants...invariants of perception. If any of these forms were to cross each other, the result would be perceptual inconsistency, loss of closure, and the collapse of perceptual reality, i.e., the formation of a gap between two discrete, mutually disconnected, internally consistent realities in which the contradictory perceptual invariants are isolated...two mutually irrelevant parallel universes. We're only interested in the universe we inhabit.

?

CML: Two conflicting elements of perceptual syntax operating in your mind would by definition produce conflicting perceptions between which there exists an irresolvable paradox. But your own personal helping of 2VL logical syntax will filter that out of your perception. So no such conflict can exist. That's 2VL tautology put into perceptual practice (I already explained why 2VL is an essential ingredient of perceptual syntax).

RFV:

> I don't know that I'm ready – I'm not. But please administer the firehose in accordance with your perception of my readiness. I know there are a couple of concepts that seem like prerequisites that we need to consider terminological deficits to be worked on as we go.
>
> **CML**: I've been hosing on this for hours now...I hope it helps.

Jesus I'm glad to hear you say that because it's been coming out my nose, my ears and eyes for hours! I was afraid you weren't even getting up a sweat. I think we need to cut down the pressure just a little, but proceed as you perceive I'm able to gobble this. Maybe your perceptions will somehow make a perceptible change in my mental state.

> > **CML**: ...and distributed means general.
>
> I don't buy this – yet. We could have a very widely distributed specific thing for which its wide distribution implies nothing but its wide distribution.

> **CML**: You seem to mean "distributed" in a statistical sense...something sprinkled through a preexisting metric space. The kind of distribution we're talking about here is logical and refers to the syntax of reality itself, which obviously distributes over that which it governs (otherwise it wouldn't be where it has to be in order to do its job). Your kind of distribution is defined on the metric; mine defines it.

Here's a fruitful patch for more picking I think. A little more on your definition of "distributed" might stand me in good stead.

CML: "Distributed" means "without constraint on location". E.g., when we say that multiplication distributes over addition in an expression like a(b + c), we mean that the multiplier (a) is active everywhere in the parentheses (b and c) without constraint. Distribution has slightly different meanings with respect to the actual and potential phases of reality. With regard to the actual, conditions are constraints on location, so saying that something distributes "conditionally" means that it distributes only where the condition is actually present. When we say that the global reality syntax "distributes over reality" in an actual context, we mean that it distributes to conditional and unconditional extents according to the quantifiers of its elements and the metric distribution of instantiations. But the situation is different with respect to potentiality; insofar as it is possible in principle for any set of conditions to exist at any location, we can say that even conditional elements of syntax are globally distributed. When we say or imply that the entire reality syntax distributes over all of reality, we mean that the potential for any set of conditions to be actualized exists everywhere.

Note that any law L that distributes only where a set of conditions A(B(C...)) is actually present can be globally distributed by reformulating it like this: "Where A(B(C...)), L." This formulation applies everywhere because it is formulated in a metalanguage of L in which the conditions A(B(C...)) may or may not be present.

RFV:

> **CML**: Whoops – I meant "circular but nongeneral forms of reasoning that pretend to generality". I sometimes tend to forget that not all scientists are seeking a theory of everything, perhaps because they run into such spectacular problems when they do. Let me explain.

Good! And yes, I am one of those extremely skeptical of theories of everything not because I don't think that is desirable or achievable, but because I believe it is not scientific to leapfrog the scientific "process".

> **CML**: Once again, the scientific method entails both local distinguishability and replicability, two predicates which become opposed to each other on the syntactic level. To be universally operative or "replicable", reality syntax ("the laws of science") must be distributed throughout spacetime; but since it is, it's not locally distinguishable. The limita-

> tions of the scientific method are those of science. But since science doesn't yet realize these limitations, it continues to seek laws with distributed (replicable) formulations by empirical (locally distinguishable) means WHETHER THE LAWS ARE GENERAL OR CONDITIONAL. Where the laws are indeed general, this is a logical impossibility; the only thing that can make a scientific law locally distinguishable is the presence of specific definitive conditions that are elsewhere absent. Science often ends up chasing its own tail when it tries to go too general for its objectivist limitations, as it does in fields like cosmology and consciousness.
>
> So what it seems to come down to is this: I'm talking about what science is ultimately trying to do (seek general laws that apply everywhere in spacetime), while you're talking about what it actually can and does do in its current objectivist form (seek nongeneral laws that are nonetheless distributive in effect, or replicable under more or less specific sets of conditions anywhere in spacetime). I guess I tend to view science in terms of trying to make it what it wants to be.

This clarifies a lot. I think it would for most people who look at your theory. It's a distinction well worth making even though most would agree on at least a desired convergence in the end.

CML: Good. Now combine the above with my last entry. The difference between a universal formulation of scientific laws and a nonuniversal one is that the former specifies a complete set of conditions for each law, whereas the latter specifies what may only be a partial set. Unfortunately, undecidability now asserts itself, and we have to admit that even when we have what we think is a consistent universal formulation, it may not be complete below a certain level of generality (i.e., above a certain level of detail). In this sense, the work of science is never quite done; the universe is still in the process of creation at any point in time, and certain axioms may not be expressed. But note also that the CTMU is being formulated as a general framework for the resolution of specific scientific issues, and does not pretend to be complete in the (forbidden) specific sense. To this extent, you can stop worrying that the CTMU, like some kind of crank theory, purports to "explain everything" in detail. It doesn't.

RFV:

> **CML**: In the CTMU, as in Hume, reality coincides with perception (and also with its cognitive underpinnings as described by Kant).

I don't think Hume ever would have conceded the parenthetical statement would he? Not the Hume whose *Enquiries Concerning Human Understanding* I read. Is there a section I should re-read?

CML: No, you're quite right. Hume said that mind was a mere illusion induced by perception. I'm saying he was wrong; mind, like objective reality, is a full-blown ASPECT of perception. That neither Hume nor Kant had the full story is why we need the CTMU.

RFV:

> **CML**: Indeed, science already effectively describes reality as a *de facto* blend of thought and perception, with theories using rational relationships to weld together our shared perceptions.

That is good.

> **CML**: Scientific theories can thus be regarded as "common generalizations" of those little impressionistic dots you mention. So when we set out to construct scientific theories – which is, after all, what we're doing here – we must first buy into the assumption that there are commonalities among our perceptions. And this isn't really a questionable assumption, because those perceptions (as opposed to our personal interpretations of them) can always be scientifically reconciled. This reflects the fact that reality is logically self-consistent, which in any case can be established by reasoning along the lines described above.

Yes, although sometimes people like Dirac think it may not be.

CML: Although some people think that quantum state superposition violates 2VL, it doesn't, because no law of mutual displacement is in effect among potential states (only actual states). I think that's what you mean here, anyway.

RFV:

> > **CML**: If you're right, then Wheeler's title for his thesis ("Observer Participation Thesis") would seem to indicate his agreement with me about what constitutes an "observer", namely anything able to participate in an irreversible change.
>
> That happens also to be mine. It is a good concept that concedes what has to be conceded without selling out altogether. It also is in stark conflict with relativistic concepts – whoops, let's stay on course here.
>
> **CML**: That's good. Then you're amenable to a formalism in which the universe perceives itself through the objects it contains. That's SCSPL, and that's the CTMU.

Perhaps amenable, but at any rate to the notion that it contains irreversible "observership" changes.

> > **CML**: It's even simpler than I put it above. Experience entails something that experiences and something that gets experienced. So if reality consists everywhere of experience, it is everywhere self-experiencing, and has everywhere the character of both mind (the subjective experiencer) and objective reality (that which gets experienced).

> A good notion, but I see this from my background as more of a demonstration of observer/controller equivalence duality. From absorber theory (Wheeler also) there is no significant distinction to be made between emitter and absorber. In tracking technology, the algorithm for tracking many targets to minimize uncertainty is essentially the same (an inverse) as that to situate sensors so as to maximize observability.
>
> **CML**: Two different languages describing precisely the same thing. The main difference: my language has been developed in the direction I've been outlining, resulting in a global reality theory.

Of at least secondary significance is the difference associated with the other dualities involving the inverses of fairly involved mathematical transformations.

CML: Well, if you're not talking about simple line/plane geometric duality, you're probably talking about $V \rightarrow V^* = \mathrm{Hom}(V, R) \rightarrow V^{**} = \mathrm{Hom}(V^*, R)$ duality in multilinear algebra, which is what I'm guessing from your relativity background. That is, the way a basis B of an algebraically reflexive vector space V gives rise to indexed coordinate functionals comprising a dual basis B^* of V^*, and iterating this relationship once more takes you right back to V (i.e., to V^{**}, which contains no further nontrivial structure beyond that of V). Yeah, that has an approximate parallel in the CTMU: to wit, V is analogous to syntax, V^* is analogous to the set of mappings from syntax to perceptual reality, and V^{**} is analogous to the set of metaperceptual mappings from V^* back to syntax. It's about self-reference, a basic feature of self-perceptual reality.)

Incidentally, that's what I mean when I talk about how the theory of metalanguages is embedded in the structure of reality; the analysis of perception involves a metalanguage of object-level perceptual language, and reality employs such a 2VL-based metalanguage in order to enforce self-consistency. In fact, where L is the object language of single percepts ("perceptual objects") arranged in time and empty space, and U is the perceptible universe itself, the entire perceptual map $L \leftrightarrow U$ is distributed over reality, and so is the metalanguage $L'(L)$ referring to it. If this looks a bit like the $M: T \leftrightarrow U$ notation we employed above, that's no accident.

***

Note: The event (multi-particle) IED is the site within which consistency is preserved, control is effected, and every other metalinguistic function is executed. The IED "accesses" the conditionally-distributed meta-L syntax.

***

RFV:

> **CML**: So the categorical imperative relating the subjective and objective aspects of reality everywhere coincides with experience; ...

I have always seen the duality as action / observation – both fairly closely associated with a subjective view of reality overlaying a physical level. Not worth pursuing here probably.

CML: Maybe so. One of the best ways to achieve understanding is to see the same things in as many different ways as possible. But remember; according to distributed solipsism, which is simply a correlate of yesterday's "blue shades" analogy, "subjective reality overlays physical reality" dualizes as "physical reality overlays subjective reality". The dual viewpoints are complementary and mutually essential, and this is the point. From this duality, plus a big load of logic, the CTMU follows.

Well, actually the duality I was musing on was the action / observation one analogous to the tracking / sensor-assignment duality. There is for any set of observations an ideal action of sensor positioning that minimizes uncertainty.

CML: Sounds very interesting. And I think we're talking about somewhat the same thing, in that reality theory is about finding an optimal syntactic configuration from which to consider multiple observations and interpretations of reality. In this case, "action" is cognitive and involves the selection of appropriate metalinguistic principles and parametrizations for evaluating object-level syntax (cognition) and semantics (perception). Now, how's that for an earful?

By the way, where do I find out more about these observer / tracker algorithms, preferably from a layman's viewpoint?

RFV:

CML: Hope you like this as much.

I have, but to tell you the truth, I'm exhausted. I hope my recently withered IQ isn't embarrassing itself too badly.

CML: Not at all. You still seem pretty smart to me. If you just think about how points made so far might apply to your further questions, I think we'll be fine.

PART SIX

RFV: I believe I understand (and am sympathetic with your notion that T reflects all we know with regard to the operation of U) and it makes little sense to blabber about that which is beyond such an understanding. I have very little difficulty with this. This is what seems very scientific to me.

On the other hand, what I have extreme difficulty with is my insistence that T is not U in the sense of T being our theory of how U works and not the particulars of the night sky, a child's smile, etc. To say that T is that, i.e., U, is more than I can acknowledge as of yet.

I am not going to worry too much about how consistent I'm being – you can identify if and where my positions seem inconsistent to you. But suffice it to say that I think I have less difficulty with your (subject/object)/(perceiving/perceived) dualism than the (T/U). The reason for this acceptability derives from my own acknowledgement of transactions being all we can know of what is real in any case. And here we have observer and observed shaking hands as in absorption theory. So in the sense of perception being the operation on U, U is self-perceptual as I now take your meaning.

CML: This is promising except for your continuing difficulty with the T/U dualism, which seemingly reflects your desire to have it both ways...to admit that mind equals reality, but reject the implications of this admission regarding the nature of theories. In fact, your T/U distinction entails the subj./obj., observer/percept, abstract/concrete, and solipsist/objectivist distinctions, which we have already shown are ultimately illusory. So the T/U distinction must be illusory as well. Your problem is that you insist on viewing T in the ordinary scientific way, as a temporary and unreal abstraction that will inevitably be superseded as "real" data accumulate. Unfortunately, that amounts to clinging *a priori* to the objectivist model.

What you have to do is forget about T and pay attention to the limit of T, i.e. its syntax Syn(T) [the part of T that remains invariant as T evolves]. For reasons already explained, Syn(T) is also the reality syntax Syn(U); Syn(T) = Syn(U). So T, as opposed to the mapping M: T→ U, is almost invariably "true" on the syntactic level, i.e., has a model in Syn(U) [although perhaps not the one we have in mind]. Remember, we've already shown that the abstract dimension of U is "real" in a general sense, and can therefore be described as "the abstract phase of U". Call this phase U_{ab}. Now, U_{ab} permits the abstract existence of physically unrealizable relations, i.e., relations that cannot be consistently mapped to the physical phase U_{phys} as semantically interpreted. Your apparent assumption that everything in U_{ab} must be realized in U_{phys} is neither correct nor implied in anything we've discussed so far. Please try harder to lose it.

RFV: If you mean this in the sense of Berkeley whereby our perceptions are all we know of "objective reality", I buy it. In his concept, there is some basis for these perceptions and there is little (if any – he thinks it would actually be a negative) advantage in accepting matter over "spirit". I have no real problem with that.

CML: It's good that you have no problem with that, because as Berkeley pointed out in less sophisticated language than we're presently using, the only alternative is to accept that matter is uncreated and eternal (which is totally opposed to what we actually know of temporally finite physical reality and turns out to be logically absurd to boot). What Berkeley called "spirit", we call "telesis".

RFV: The "concrete phase" means something different to me perhaps. I see a needle on a tree. I perceive it exclusively as a needle and not as an abstract result of a theoretical process whereby it grew and the weight of gravity that brings the limb on which it hangs before my window. There are probably billions of similar needles in my purview but they are only there in the sense that I would miss them if they were mostly all gone. Perceptions are not very directly supported by theory is what I'm saying. They're just there. They are the given!

CML: Right...perceptions are "given". But so is their consistency with other object-level perceptions, and so is the consistency of the higher-order relations educed from those perceptions. Without those relations, your "needle" percept is quite meaningless, and in fact, can't even occur. That's because these relations are responsible for your ability to distinguish that one little needle from everything else. In other words, the n-ary relations you "educe" from the set of perceptions including the pine needle, as included in the variable, evolving part of T, converge on the T-syntax Syn(T) = Syn(U), which underlies your ability to make the perception in the first place. Please think about this in light of my first entry above.

RFV: "T equals a GLOBAL DISTRIBUTED FORMULATION OF THE REALITY SYNTAX" certainly implies a faith in the comprehensibility of the laws of nature – that I don't have much trouble with, but much more than this. It implies that the universe can not be without such a formalization. This implies that the formalizations we discover have to correlate with an actual *a priori* formulation. I have some difficulty with this – I think from a scientist's perspective. I tend to see the universe as just being there and about which we discover certain regularities by which we can describe it efficiently. Our descriptions are ours and have little if anything to do with it!

CML: Now hold on there. Neither faith nor comprehensibility is implied by the phrase you're quoting, at least in the sense you seem to presume. All that's implied is consistency. The logical consistency of perception implies all by itself that a "global distributed formulation" of Syn(U) exists. This formulation need not even be complete; beyond what's necessary for consistency, it may contain all kinds of ambiguity. It can even be a set of consistency-preserving syntactic automorphisms rather than a single formulation. But consistency is a *sine qua non* of existence, and U cannot exist – not now, and not ever – without a higher-order logical relationship representing and effecting it. It's important for you to grok this fact. No consistency in higher-order (inter- and intraperceptual) relations = contradictory object-level perceptions = multiple dissociated realities. That's not a negotiable point in reality theory.

RFV: I see what you're saying, but I see a humbler scope of such formulations. They do feel necessary in some sense, but at this stage of discovery it seems extremely arrogant to presume such a scope to our rather humble endeavors.

CML: Well, Fred, we're doing reality theory here, and if your idea is that we shouldn't use existential (as opposed to constructive) reasoning because we don't yet know all the details of what we're reasoning about, then you're not prepared to reason about the nature of reality...and you never will be. You see, for reasons already touched on (like undecidability), we're never going to know "all the details". But we damned well know enough to work with on a general level right now, and that's what we're doing.

RFV: I do not have real perceptions that are "true" or "false" – they are merely perceptions, i.e., a photon (seeming) to impinge upon the eye, etc. I must make conjectures with regard to these perceptions that then may be "true" or "false". Do you see this differently? If so, there is definitely something more to your perception of 2VL than meets my eye.

CML: This is incomprehensible. You have a set of perceptions $p_{i = 1, 2, 3, \ldots}$, and the truth value of that set is T (otherwise, the exceptions are illusions and not perceptions, which means that others in this reality can't share them). Now, something either is in that set (T) or is not in that set (F), so 2VL rules. Note that we're talking about plain old object-level sense data here, not the (higher-order) interpretations or associations they might have triggered when you received them. The truth or falsity of these is another matter entirely.

RFV:

> **CML**: That's just the way it is; any two people who won't both admit that the perceptual basis of reality is 2VL have nothing to discuss.

Again, I think we have quite a bit to discuss, but I don't see the preceding statement as meaningful. I think I understand the necessity for 2VL in discussion perfectly well – although I may not – but I do not see immediate perceptions as either true or false, i.e., their type is other than "logical"!

CML: Again, this is incomprehensible. Please re-read the last entry above. If perceptions were other than logical, then there would be no consistency among the perceptions of various observers. Goodbye, science. In fact, goodbye, reality. Again, this is not a negotiable point. Obviously, perceptions don't consist ONLY of logic; there is plenty of syntactic and semantic variation involved as well. But even that is ultimately governed by logic. The most we can do is differentiate between perceptual qualia and their logical correlates (despite the logical nature of the connection itself).

RFV:

> **CML**: But in that event, math and science would be a useless farce, because what one of us "perceives" to be true (and even tautologically true) may not be acceptable to the other.

Again, I concur other than on the use of the word "perceives". I would say "conceives" or "thinks" as distinct from a "perception".

CML: On one level, there is a distinction between conception and perception. On another, there isn't. The set of possible conceptions in U_{ab} greatly exceeds the number of possible perceptions in U_{phys}. But every perception in U_{phys} must conform to a configuration of Syn(U) in U_{ab}, because that's the basis of your "accepting syntax"...the structured set of mental states in which you are aware of "perceiving" something.

RFV:

> **CML**: As I've already explained, any isomorphism to whose range and domain you can't assign different qualitative or quantitative coordinates at a given level of discourse is really an identity relation (at that level).

But, and I see this as important, U has much more in it because of its actualities.

CML: Sure it does. If there are actual perceptible distinctions on a given level of discourse or conceptualization, then these will prevent an isomorphism from collapsing to an identity relation on that level whether or not you can efficiently formulate them. But we've already ascertained that there are no perceptible distinctions between mind and reality on a high level of discourse. So what you see as "important" actually means nothing to us here.

RFV:

> **CML**: For example, the law of conservation of momentum, and its underlying arithmetic, is everywhere valid, but only active and perceptible in the presence of actual moving bodies.

So the conservation laws of T tell us what will happen from an initial condition (if we accept determinism), but the U includes those conditions.

CML: There is always a level on which a theory T is deterministic. Even a probabilistic theory is deterministic in the sense that if it is valid, its probabilities must be realized over many trials. And as you know, the initial conditions provided by U are ambiguous on the quantum scale. This gives U a certain amount of freedom within the boundaries of its deterministic syntax.

RFV:

> **CML**: Similarly, physical reality is merely an evolving set of instantiations of a set of relations that are everywhere valid. These instantiations are metrically separated, but the metric itself is completely characterized by a

> distributed form that exists in every part of space that could ever be occupied by your brain (even if your brain has only managed to figure a little bit of it out by firing isomorphically to it). The distribution of this form throughout space distributes your mind over reality in the form of general mental syntax.

A distributed form or syntax is not the sentence!

CML: No, nor is that implied by my paragraph. However, the sentence must conform to the syntax that distributes over the language in which that sentence is formulated. Otherwise the sentence will violate syntax and make no sense.

RFV:

> **CML**: I think one of the problems so far is that you haven't really absorbed the fact that the distinction between mental (abstract) and objective (concrete physical) reality is a stratified distinction that gradually disappears as we approach the top stratum (global distribution).

Let's assume this is my problem. Then at some level physical reality has been abstracted such that it is in some sense isomorphic with the "mental" (abstract) description. I would still have problem accepting that, even if this agreement could be reached, at lower levels the isomorphism is preserved. This is counter intuitive to me.

CML: What you seem to be saying here is that you don't think object-level perceptions, which comprise the lower levels of the mind-reality isomorphism, need to be logically consistent with each other. Let me repeat, this is not a coherent philosophical or scientific position. An inconsistent reality is no reality at all, but a set of mutually disconnected independent realities. Any contrary assertion makes no sense whatsoever.

RFV:

> **CML**: Less general, more conditional features of reality appear to be localized within their specific instantiations, permitting internal-external distinctions. In this sense, cognitive (or computational) distribution relations are equivalent to topological inclusion relations (that's a CTMU duality principle).

I think you think you have produced an argument why the isomorphism is preserved at the lower levels, but it escapes me. The new terms don't help me with this.

CML: To simplify, spacetime is "qualitative" in the sense that space is like a layered sequence of Venn diagrams (spacelike hypersurfaces to which cosmic time is normal) in which circles represent predicates defined on physical observables, and time is like logical substitution within those domains according to various rules of implication. The domains or Venn circles are called "IEDs" and are associated with events from which objects are interpolated (to illustratively adopt an oversimplistic graph-theoretic model, object worldlines amount to edges in a spacetime graph whose nodes are events). These "qualitative" IEDs expand at the rate c and then collapse into a new event within their mutual intersects. Inside the IED you have simulative computation or "cognition"; in the compact output event to which the IED collapses in cosmic time, you have the mutual perception (interaction) of two or more objects. The interior of the IED is wavelike, cognitive and simulative; the output event is particulate, informational and objective (that's "wave-particle duality", actually a temporal alternation between spatiotemporal and objective states of material being). In this way, spacetime is identified with the mathematical and physical laws that govern it. This spares us the absurd necessity of having a nonphysical "storage space" for the laws of physics (or by extension, a nonmathematical storage space for the laws of mathematics).

Please re-read the above paragraph over and over again until you understand it. Then define "conspansion" as follows: conspansion consists of two alternating phases of matter and energy, inner expansion as an IED (Inner Expansional Domain) at the rate c, and requantization as an event-component. Requantization is just the collapse or decoherence of the quantum wave-function, as computed by the simulative computation going on in the (coherent) IED, with a reduction in scale that follows from the absence of an external metric against which to obtain an "objective measurement" of the size of the universe. This implies that the size of the universe is for all practical purposes invariant, and its apparent "expansion" (from an interior vantage) is actually a relative shrinkage of content (from a global vantage). Eddington knew about this global-local duality, even realizing that as objects shrink, time scales would have to shrink apace, i.e., time would have to accelerate or contract. But he didn't know how to do the logical setup for the "shrinking atom" scenario (Eddington, 1933). The CTMU does the setup, in the process solving problems Eddington could only have dreamt.

Down below, you complain that I haven't explained what "conspansion" is. Please make sure that you carefully consider what I've written here before repeating that complaint.

RFV:

> **CML**: If distribution is general or unconditional, on the other hand, it applies to all things, and the distinctions among these things exist within it in the sense that it resembles a SELF-EXECUTING PROGRAM INSTRUCTION that internally SIMULATES them.

You seem to have no conception of the obstacles to accepting this. First of all, you start out with "if", but you don't really mean "if". You mean that it is that way. But you combine "general" and "unconditional" as though they were synonyms. They are not to me in this context. I don't know how familiar you are with computer programming or "execute" type instructions or the various levels of programmed "simulations", but being familiar with all of them myself, the sentence is incomprehensible to me. Am I too hung up on their jargon meanings to allow them to be generalized as you seem to think you have done, or what? They seem like ill-chosen words to throw into this cauldron.

CML: We've already been through this, Fred. Just replace "if" with "where", and consider that I'm distinguishing among ways of formulating laws. "General" means "representing a range of possible specifics"; these specifics are distinguished by the conditions applying to each. So when something is "general", it implicitly represents a whole range of possible conditions. Now, when something represents a whole range of possible conditions (a, b, c, ...), it's "unconditional"; it doesn't need an "if (a), then..." or "if (b), then..." or "if (c), then..." condition in front of it to be predicated on a given object in the class of objects it represents.

E.g., let the predicate "hard", as used in "diamonds are hard", represent the class of diamonds = (clear diamonds, blue diamonds, big diamonds, little diamonds, ...) and so on. Then "hard" is general with respect to this class, and if we have a diamond d_1 – any damned diamond at all – we can say "d_1 is hard" without knowing what kind of diamond d_1 is. I.e., we don't have to say "IF d_1 is a blue diamond, THEN d_1 is hard"; we know it's hard regardless of specific type. That means that the general predicate "hard" is an UNCONDITIONAL predicate with respect to the class of diamonds. That's all. Frankly, I have no idea what "professional jargon" you're talking about here, but if what I've just said makes no sense in light of it, then my advice would be to lose it fast.

RFV:

> **CML**: In other words, general features of (simulative) spacetime include specific aspects as output, which is then uploaded to spacetime as further programming, and so on. Topological inclusion follows from computational inclusion because topology is computationally simulated.

Chris, these sentences mean absolutely nothing to me. For example, "general features of (simulative) spacetime" – all that comes to mind is spatial temporal location of events. What "general" and "simulative" add I cannot figure. Then: "uploaded to spacetime as special programming". Is there someone other than Chris Langan that you envision capable of understanding that phrase? What is it that enables that person to understand it? The words are all in my vocabulary and the grammar is straight forward, so what's wrong? I guess you can say I'm too stupid, but then this six word quiz better have a pretty clever answer. What is it?

CML: With all due respect, let's get something straight. Did you or did you not read that *geomusser.doc* file? Or did you just scan-and-forget? Because if you'd read it closely, you'd know that reality is treated in the CTMU as a "self-simulation" with a programmatic (simulative computation) phase and an output (perception) phase, and you'd know how these phases are related in the conspansive model of spacetime. That file explains how all this logic we've been discussing is physically implemented. Now, if you didn't understand one word in that file, please refer again to the description of conspansion given above. Because I intend to keep writing in a way calculated to trigger meaningful associations for you in light of that knowledge.

RFV: Yes, I'm trying – both to you and me.

If everything is defined as 2VL except for undecidable statements, it is not a very useful concept. Electric currents are 2VL in a trivial sense, like your triangle may be pink! That is not an essential aspect of them.

CML: Nonsense. 2VL unequivocally governs the macroscopic (and in a more restricted sense, even the quantum) behavior of electrical currents. Nothing, nothing at all, is more "essential" to them. If that weren't true, you couldn't achieve a stable perception or description of what an electrical current is, much less its source, its direction of flow, or anything else about it in any given case. You seem to think that logic can somehow be conveniently in effect without being essential to whatever phenomenon you're considering. But that's utter nonsense, because in that case, you'd need to explain what it is that remotely enforces the effectiveness of logic with respect to that phenomenon, and you can't do that without me then forcing you to explain why logic applies to your remote control mechanism, and so on *ad infinitum* in an absurd infinite regress. So the bottom line is this: either produce a finite explanation for such a mechanism, or accept the truth: logic is essential to everything we will ever perceive. Because from now on, I'm not going to argue about whether or not various aspects of reality coincide with and obey logic; they damned well do, period. They coincide with and obey it right up to the point at which the consistency of perceptual reality is effected. What they do beyond that, I don't presently care. And at this point of the discussion, neither should you.

RFV: Naturally, one can do accounting like an accountant where every item must be put in twice, but it's senseless. To say that E&M is a series of statements about which one says each statement is either true or false is ridiculous. Any valid theory is "true" by definition but that does not make the theory a 2VL theory! We've got to get past this.

CML: We sure do! Try this on for size: EM theory is not 2VL with respect to your personal knowledge of it; all you can personally do is formulate a set of subjective probabilities conforming to modal and probabilistic many-valued logics (which are themselves 2VL theories). But given any set of hypothetical nonprobabilistically-formulated statements regarding EM phenomena, each and every one of those statements is either true or false in point of fact (unless it's formulated on the quantum level and obeys the superposition principle, which in any case must yield to T or F actualization). Now, that distinction having been made, consider that what we're talking about is just the distinction between the evolving and invariant phases of T, i.e. T and Syn(T) respectively.

RFV: A theory describes the general behavioral rules not the contingencies of any particular configuration.

CML: No, but given any hypothetical set of contingencies, the theory (in conjunction with the 2VL component of its syntax) will impose consistency constraints on them, and these constraints will ultimately be formulated in T-or-F terms. If the theory imposes no such constraints on sets of objective contingencies, then it's as useless as tits on a boar. It has no predictive value whatsoever. So let's not bother with it, OK?

RFV:

> **CML**: The phrase "generalized cognition", actually synonymous with time, ...

> How in hell did we get here? I take a "synonym" to mean that a dictionary can help me, but it doesn't here.

CML: A synonym is a word or phrase that means the same thing as another word or phrase. Dictionaries are great, but we're at the stage of defining terms here, and if they were already in a dictionary, we wouldn't be bothering. We'd just be citing dictionary entries. Unfortunately, constructing a new theory is seldom that easy. Stay awake, Fred – this is a challenge for me too.

RFV:

> **CML**: All you need is your memory, because we've already been through this.

Your having told me something in the past, and my remembering or not, is not all I need to understand your concept of time / generalized cognition.

CML: You've already agreed that mind is ultimately indistinguishable from reality. Minds evolve along a state-transition parameter called "cognition"; reality evolves along a state-transition parameter called "time". So on the level of generality of the mind = reality equation, "cognition" = "time". I.e., given the high level of generality of the M=R equation, "generalized cognition equals time".

That's it. Don't keep trying to introduce artificial distinctions where there are none. Although I'm sure you could keep on doing that forever, it would make my task as an expositor impossible.

RFV:

> **CML**: The operative analogy is:
>
> time : space :: cognition : information

I don't even accept this as a valid analogy, let alone move on from there.

CML: Here's the problem, Fred. You're not even thinking about how to interpret this analogy; you're just getting visibly impatient with it and putting the load back on me. But as long as you've already done that, consider once more the paragraph just above:

> Minds evolve along a state-transition parameter called "cognition"; reality evolves along a state-transition parameter called "time". So on the level of generality of the mind = reality equation, "cognition" = "time". I.e., given the high level of generality of the M=R equation, "generalized cognition equals time".

Now, where the abstract equals the concrete, matter is abstract information. And so are relations among material objects. But relations among material objects can be factored into spatial and temporal components, the difference being that while the spatial components are perceived in parallel (i.e. simultaneously), the temporal components are perceived as a sequence of spatial components. But "time" consists of the temporal components, and according to what we just wrote, time = generalized cognition. So what is "space" a generalization of? What else but that which is transformed by ordinary cognition, namely information.

So we have the following analogy:

time : cognition :: space : information,

where ":" means "is a generalization of".

This obviously transforms to:

time : space :: cognition : information

Please think about this for a while before taking me to task again. This analogy is every bit as clear an analogy on any IQ test you might take, given previous explanations.

RFV:

> **CML**: I.e., time = info-processing = generalized cognition (where information here refers to the abstract dimensions of real objects and processes, which, as we've already seen, coincide with them).

This does not compute! "Info-processing" requires "time", so in what sense can "time = info-processing"? What are the "abstract dimensions of real objects"? Spacetime coordinates, linear and angular momentum, etc. are all that come to mind. Is that correct?

CML: A parameter, Fred, is a general predicate of all the specific things which are "parameterized" by it. Thus, the parameter called "time" is a generalization of any physical or mental process. Now, where reality is abstract, reality is information, and time is a generalization of all of the processes undergone by that information. So time = info-processing. The "abstract dimensions of real objects" are the abstract parameters in which objects are distinguished.

RFV:

> **CML**: ...simply extends human cognition to generic information processing.

Now that's easy enough, but I sense that you assume that I will be accepting more than that.

CML: You're right. Reread the above if necessary.

RFV:

> How does/can "time" perform this function?
>
> **CML**: Time is not a "thing"; it's a function...a function of spacetime.

I could accept it being a dimension or direction of spacetime, but not a function. In what sense do you see it as a function of itself?

CML: Because, as just explained, it's a generalization of all of the processes (functions) it parameterizes. Time is not merely a spatial dimension, even in relativity. It is the function by which space is transformed. Don't be misled by the fact that for certain purposes, time can be conveniently represented as a spacelike dimension. It's absolutely not, never was, and never will be. If it were, we could take a walk to ancient Greece and visit Plato.

RFV:

> **CML**: Spacetime performs this function because it's full of syntax, and some of this syntax is grammatical as opposed to structural. Grammar governs the evolution or state-transition of the language of reality.

You are placing "syntax" where "sentence" should be it seems to me. Spacetime must contain what is real, not the form of what is real.

CML: Do you see what you're doing here, Fred? You're distinguishing between what's "form" and what's "real", and then hanging your distinction on spacetime. The implication, of course, is that "form" gets conveniently crammed back inside the human head (or maybe some Platonic realm of ideals way out there in hyperspace), and "reality" gets to continue to hang out in "objective" spacetime. Bullshit. We did away with that distinction, what, fifty pages ago? And you agreed that it was done away with. Yet here you are, once again telling me that for some odd reason you can't explain, "form" and "reality" can't both reside in spacetime. I just don't get it. Please try to get a grip on this.

RFV:

> **CML**: Thus, time is any grammatical implementation of the SCSPL reality-syntax.

I think this blows the validity of analogies I'm trying to grasp.

CML: I think you better rethink that.

RFV:

> **CML**: Nomology must exist up to determinacy, or its absence would show up in the form of irresolvable perceptual paradox.

Could you explain why? And then could you relate this to the current non-deterministic theories of QM, etc.? But simply.

CML: If nomology, i.e. laws of physics, were not at the very least associated with a deterministic set of constraints assuring the mutual consistency of perceptions, those perceptions would not be consistent. Beyond the degree of determinacy needed to ensure consistency, there exist various degrees of freedom. These degrees of freedom are what show up in "nondeterministic" theories like QM, etc. Indeed, all these theories are themselves deterministic up to the point needed to ensure consistency, their "nondeterminacy" kicking in only once that need has been met. Simple enough? In general, the universe is a balance between determinacy and freedom, and we can talk about either part without denying its complement.

RFV:

> **CML**: A distributed abstract nomology to which perceptual reality must conform ensures the resolvability of paradox, which is just a way of saying "the self-consistency of reality".

Are you saying that to be self-consistent, theories of physics must be deterministic? Why? – in both senses of why does CTMU require it and why does it make sense from a physics point of view. I hope making a distinction here is acceptable in a conditional sort of way.

CML: Anything is "deterministic" with respect to which we can't just as well flip a fair coin. That is, determinacy and randomness are complementary aspects of any theory, the deterministic side remaining invariant with respect to the incidental content that gets fed to the theory, and the nondeterministic side reflecting variety of objective content (and in QM, the behavioral randomness of single objects). Consistency is not an option, but a necessity in both physics and the CTMU. So it is deterministically enforced, either at the object level or among higher-order relations of objects. However, this does not preclude a complementary measure of freedom. The freedom is simply constrained not to violate the determinacy. If you roll a fair die, the freedom (nondeterminacy, randomness) of your toss does not violate the deterministic fact that the outcome must be one of the numbers from 1 through 6! Freedom and determinacy consistently coexist.

RFV:

> **CML**: Our respective electrons are distinct images of a single distributed "syntactic photon" consisting of the rules that all photons must obey regardless of where they are. The images reside in different loca-

> tions, and these differences are significant. But it's like Wheeler said to Feynman: they're all one photon.

Do you really mean that leptons are generically speaking photons? Or what sense do you give the above in this regard. Why and in what sense are all photons one photon?

CML: They're all one photon because they all have identical structures and behaviors up to the deterministic aspects of their common syntax, excluding degrees of freedom like frequency and location. They can all be regarded as "instantiations" of a single distributed abstraction. The distribution of syntax is not a replication and scattering, but a multiplexed projection of one underlying unity. (Actually, Wheeler was talking about electrons, not photons, but the idea's approximately the same even with regard to exchange forces.)

RFV:

> **CML**: Because the isomorphism M is a perfect equivalence relation with respect to everything but degree of generality (the abstract is more general than the concrete because one abstraction can possess many concrete actualizations).

I don't think I can accept that an abstraction "possesses" many concrete actualizations. In fact so far I think this is my biggest problem: The actual is significantly other than the generalization of its type.

CML: No it isn't. Again, you're reverting to your favorite distinction, that between the actual and the abstract, on an *a priori* basis. I'm simply not gonna let you do that any more. If you want to do it, the burden of proof is now on you to show where you're getting your distinction and why it's valid on your chosen level of discourse. Hint: you're not going to be able to deliver. So why not just drop it right now and save yourself the trouble? Think about this for a while.

RFV:

> **CML**: On the other hand, "Unicorns do not exist" can be interpreted over an even larger real context, and to that extent is quantifiably truer in a concrete context.

"Truer", as in "bluer"? Is this 2VL?

CML: This is 2VL predicate logic. When one quantifier embeds or properly contains another, the associated predicate is "true for more objects" or "true over a larger range" or has "greater scope". So if we feel like it, we can say that it's "truer" in precisely this sense. Hey, you're the one who brought up unicorns, and I'm just telling you what 2VL predicate logic has to say on the issue. If you like it, great. If not...oh, well!

RFV:

> **CML:** Information consists of distinctions in which one thing is syntactically constrained to differ from another. Because concrete reality mirrors the abstract information coinciding with it, it too can be regarded as information (we're information too, and the reason all this info seems "concrete" is that there's a syntactic displacement principle in effect).

??? I don't get this!

CML: The "concreteness" of matter is associated with the fact that one piece of matter displaces another [to revert to the terminology up top, that's inherent in Syn(U_{phys})]. This is not true of the abstractions to which pieces of matter correspond; if it were, the QM Superposition Principle wouldn't work.

RFV:

> **CML:** So the question arises: what does this information become when all the binding constraints are removed? The answer is Unbound Telesis or UBT, the groundstate of being. This universe has only arisen within the UBT groundstate because it has constrained itself to do so; this is the sense in which it's "self-configuring".

The universe has "arisen" implies the supposed temporal origin of reality. But from your system alone, why would there be an "arising" as against a stable omnipresent ubiquitous system?

CML: I'm using terms like "arising" mainly for your benefit, not mine. As I use it in this context, it describes a sequence of implications...a logical dependency relation. You just described my system to a T, provided you're not making the error of clinging to an infinite Steady State model of reality in spite of mountains of contrary (Big Bang-style) evidence. In fact, my system is the only system that both conforms to your description and satisfies the cosmological evidence. There's no other way to go. Hear me now, believe me later.

RFV:

> > > **CML:** ...cognition $\leftrightarrow$ time.
> >
> > DON'T got it!!!
> >
> > **CML:** That's OK, we've been through it at least thrice up there.
>
> **RFV:**
>
> > **CML:** The logic is so simple it hurts!
>
> I know, but I still ain't got it!
>
> **CML:** That's OK, we've been through it at least fource up there.

RFV:

> **CML**: Human cognition, assumed to be fundamentally unique by anthropocentrists everywhere, is just the temporal processing that goes on in a brain.
>
> [Of course, there may be more to it than that – and there is – but for present purposes, that's good enough. The "more" in question has to do with distributed solipsism, and the consequent fact that there is a sense in which sentient cognition is primary...that what goes on in reality, namely temporal processing of all kinds, really goes on in the minds of sentient observers.]

This soothes my mind, but all of this is not just "time"!

CML: Sure it is. It's the refined kind of time that passes in a brain.

This concept of time has still not been purchased by me even if we have now reiterated your position five or six times.

CML: Too bad. Like I say, I'm done with this one, Fred. The burden of proof is now on you to establish a distinction. Hint: you already agreed with me to discard the distinction, so I don't even know what you're talking about here.

***

Note: If the universe is a mind, then it transforms spatial hypersurfaces in time. This is directly analogous to the way a human mind or any other automaton transforms states.

***

RFV:

> **CML**: In particular, a form of duality that can be referred to as "spacetime-object", "wave-particle" or just plain "conspansive" defines an alternation in which the mind solipsistically configures reality by a cumulative selective filtration of syntactic potential; during the inner-expansion phase of conspansion, the mind builds a set of exhaustive potential future realities by superposition, and in the requantization phase, selects a future reality in which to re-embody itself as a material brain. Just in case this conjures up some kind of bizarre "brains-in-vats" scenario, please remember that the mind, just like a photon, is regarded as a configuration of the reality syntax. Indeed, all matter behaves alike in this regard; as does the mind, so do its contents.

You're going to have to give me the tutorial on conspansion I think and define your usage of some of these terms.

CML: It's right up there. And may I say that it's a pleasure to find in you such an avid student of the CTMU.

PART SEVEN

CML: Below is my response to the critique of my article "Introduction to the CTMU" (Langan, 1999) by Fred Vaughan, a long-time philosophical antagonist of, that's right, the CTMU. While I can of course do without antagonism, I occasionally get a kick out of Fred's discombobulated put-downs of my ideas. (Don't get me wrong – Fred is still better than any CTMU critic who disparages it and/or its author for no specific reason pertaining to its content.) As fate would have it, this is one of those times. Enjoy!

"Gnostic TOE Tapping on a Perceptible Bubble"
by Fred Vaughan

In the beginning was the Word, and the
Word was with God, and the Word was God.
– St. John 1:1 (KJ)

RFV: The noun "reality" would seem to have a consensus definition as "all that exists" and the adjective "real" as a distinguishing quality of a particular thus-designated abstraction whose existence is deemed independent of the mind of the subject for whom the so-defined "object" SEEMS to be REAL. I won't bore you with Oxford English or Webster.

My ontological preference would be to set such definitive statements as well as those such as, "The UNIVERSE is real", apart from the Universe itself. These would not be members of a commensurable set, mind you, but merely examples of an infinitude of inane assertions whose truth, falsehood, or undecidability involves logic applied to other statements of the same class. There are many dumb assertions that do not even merit interment in REAL phrase cemeteries by REAL intellectuals whether true or not; others may be profound but to my mind similarly are totally OTHER than the Universe itself. Homo Sapiens seem particularly vulnerable to the vanity of ennobling the blather of their vocal cords and cortical folds with some special meaning and in so doing confusing the obvious. How else can one account for Ms. Campbell presuming that Carl Sagan should have been convinced by her flaw of reasoning involving the EXISTENCE of "Love" being provable by anecdotes as somehow implying that, therefore, so must the existence of "God". (See for example, my article "Does Love Exist?" on pages 13 and 14 of *Gift of Fire*, № 85.) An anecdote may of course embody the actuality of a relationship between KNOWN objects (REAL or otherwise), any one of which could then prove that such relationships OCCUR – even a misinterpretation of which might confirm the argued relation. Is it too subtle to just affirm here, however, that OBJECTS differ in certain salient respects from such RELATIONS? People other than perhaps Jesus (for whom antinomies were often claimed by the ecstatic Gnostic) ARE NOT "love" in the sense of BEING Love; no portion of anyone is COMPRISED of love, and yet of course, we DO love. It's a verb for Christ's sake! Just so, the universe need not (nor ever can) BE (nor yet CONTAIN) love nor any other relation mathematical or emotional in any REAL sense in order for the asserted relation to OCCUR. Is this heavy? I don't think so.

One must start somewhere, of course, with merely PRESUMED consensus meaning such as what is meant by "exist". However, for a credible TOE (Theory Of Everything – if that is not in itself the most oxymoronic phrase that could ever be uttered), ...

CML: What we seem to have here is a misunderstanding of the nature and expressive power of language. By the mathematical definition of a language, "Everything", not to mention "Theory of Everything", is a language. That is, it is a string of characters that has a descriptive meaning, a semantic mapping into the universe of ideas and perceptions. Since a meaningful language can be construed as a theory, "Everything" is a theory.

Of what is "Everything" a theory? Well, let's see here. "Everything" by definition refers to everything, including...er, *everything*. Therefore, "Everything" is a theory of everything. Granted, there are those who would say that a one-word theory is a poor theory indeed. But value judgments aside, where's the oxymoron?

The fact is, human minds have the power to formulate one-word characterizations of the universe (indeed, I just used a common one: "universe"). Since (broadly speaking) "to characterize" is "to theorize about", humans have the power to form one-word theories of reality. The axioms and rules of inference of these microtheories may not be explicit, but they are theories nonetheless. And that's a good thing, since it can easily be shown that without this ability, humans could not formulate their customary long and involved theories about reality. Since any complex theory can be characterized using a single word or phrase, e.g. "cosmology" or "quantum mechanics", blocking this universal referential mechanism would effectively block theorization, and because "theorization" characterizes virtually all meaningful cognitive activity, human mentation in general.

Of course, such a blockage may already be in effect over in Fred's neck of the theoretical woods. If so, I suggest that he break out the old chainsaw and clear the road before making any more of his own attempts at theorization...especially regarding theories of mine with which he disagrees.

RFV: ...what constitutes "the real", i. e., "everything" that is, would seem to require some sort of minimal definitions with justifications for nontraditional varieties before we encounter sound bites such as: "...because any real explanation of reality is contained in reality itself, reality gives rise to a paradox unless regarded as an inclusory self-mapping." And again, "...thus, we can speak of *time and space* as equivalent to *cognition and information* with respect to the invariant semantic relation *processes*, as in 'time processes space' and 'cognition processes information'. But when we define reality as a process, we must reformulate containment accordingly." Say what? "Time PROCESSES space?" It is particularly important that definitions be as precise as possible when these seem to differ so considerably from what at least this reader would expect them to entail.

CML: Time is change of state, and change of state occurs according to laws of mathematics and physics. The laws of mathematics and physics govern changes of spatial configurations of matter and energy. Therefore, time incorporates laws according to which space is changed or "processed". That is, the present state of reality is fed into a black box or transducer labeled "laws of temporal transformation" which then returns the next state. The box, AKA time, AKA nomology, is a temporal processor. The structure of the box is the invariant structure of spacetime. That is, spacetime is the history of the temporal processing of one spatial cross-section of reality into the next according to the mathematical and physical laws comprising the transductive syntax of the box. Thus, time, described as a black box, processes space as input and output.

Is this precise enough?

RFV: Chris Langan seems quite enchanted by the strange teleological loop of conceiving a universe structured in accordance with consistent CONCEPTIONS where these conceptions are thereby deified in a REAL sense to the status of reality.

CML: "Deified"? Talk about loose semantics. Actually, Kant observed that "objective" phenomenal reality has a mental, i.e. conceptual, component a long time before I did. He was right in the sense that insofar as the mind is that which perceives and theorizes, and insofar as there can be no perception or theorization without the mind, the structure of the mind necessarily intersects with that of phenomenal reality. On the categorical level, this intersection is total; there is no phenomenal disjunction. That's because by definition, perceptual reality exists entirely within the intersect.

Kant then went on to hedge his bets by postulating the existence of a really, *really* objective "noumenal" reality that exists independently of the mind. However, he did so completely on faith and in violation of logic. For the only reality we can ever know by perception or cognition is phenomenal, and anything "independent" of this phenomenal reality is by definition just that: *independent* of it, i.e. related to it by no causal mapping. In fact, as defined by Kant, pure objective reality is so utterly independent of phenomenal reality that it cannot be considered a part of the same reality at all. Our minds filter it out because its total logical disconnection from our perceptual universe severs any possible perceptual or relational mechanism that might connect it to us.

Fortunately, dropping Kant's noumenon fallacy leaves the sounder parts of his philosophy, those relating to phenomena, more or less intact. One then has a platform from which to reason about deep reality in such a way that no matter how many layers of "illusion" lie between it and our awareness, we remain logically connected to it.

RFV: There is definitely a worthy enchantment here – who, for example, has not in reverie mused concerning the universe in its entirety being constrained by necessity to act in ways determined by the evident relations of its humblest components and that the descriptions of these relations are somehow of extreme significance beyond the mere words and equations used to formulate them. However, in Chris's scheme the REALITY of the very words in such statements seems to be guaranteed (if they happen to be true statements), whereas similar, if ultimately non-confirmed statements would seem to be denied the same twinkling with stardust of the REAL word hall-of-fame.

CML: But that's not quite what I've been saying, is it? What I've actually been saying is that our descriptions of "the evident relations of (the) humblest components (of the universe)" are "guaranteed" *only on the syntactic level.* That is, they do not qualify as perceptual input unless they conform to our mental structures – our "cognitive and perceptual syntax" – and this implies that all perceptual input conforms to the laws of mathematics and physics as they exist implicitly in our minds. That's what I mean when I equate mind and reality *on the syntactic level*; particular theoretical languages can be wrong, but the basic syntactic descriptions of their perceptual contents are true in any case. Indeed, they comprise the basis of perceptual reality. Since the core of the CTMU embodies the tightest implications of the most general logic, it qualifies as "cognitive and perceptual syntax".

RFV: Thus, true descriptions "exist" as a structure of the universe itself and are, therefore, available to be "perceived" in an inverse mapping whereby sentient constituents of the universe may become attuned to the whole enchilada in accordance with their intellectual abilities – this "ultimately giving rise to human perception". Where in turn, "the set of 'laws' obeyed by the universe is just a minimal set of logical relations that suffices to make these perceptions noncontradictory, ...". Of course we can counter such statements – whether willfully or not – but perhaps in so doing we thereby exile ourselves forever (past and future) from reality and the bliss of the presence of the only true and almighty God of King James, "S."

CML: "Perhaps", indeed. Actually, the theological definition hereto referred as "the God of King James" needs a little revision. But with that understanding, Fred's theological insight will do just nicely!

RFV: There are many problems with this, of course. One of which is that in this scheme it would seem to be the intellectually challenged who are the ones endowed with special imaginations capable of envisioning truly original ideas outside the mind of God and that do not, therefore, correspond to (or even exist in) Chris's REALITY. They either do not exist at all or only SEEM to exist in the surrealistic idyllic limbo of we IDIOTS who conceive them.

CML: Say what? Where did Fred come up with *this*? In fact, since our minds, and every thought they contain, do indeed exist in reality, this makes no sense. I think that Fred's confusion may devolve to the possibility of thinking not just original ideas, but fantastic ones as well. If so, I assure Fred that a functional brain scan would readily reveal that such thoughts are quite real within the brains of those who think them. What is "irreal" is the assumed veridical mapping of personalized fantasies onto general (perceptual) reality.

Again, we have the distinction between particular theoretic languages, which may indeed be fantastic, and the basic syntactic relations of which they are composed. With respect to ordinary theories, the map is not the terrain; with respect to cognitive and perceptual syntax, it sure as hell is, right up to isomorphism. (By the way, "up to isomorphism" is not an escape clause. Beyond the maximum scope and resolution at which the real universe can internally sustain a syntax ↔ content, mind ↔ reality, cognition ↔ information, time ↔ space isomorphism, there is no more reality; the isomorphism is the entire basis of the coherence of the real universe! This should readily become apparent to anyone who tries, logically of course, to find an alternative basis.)

But something tells me that I've explained all this to Fred before.

RFV: Another problem is the basic one involving what constitutes a perception. My seeing a pixel of green via rods, cones and neurons exterior to my brain certainly is a far cry from my having contemplated a DESCRIPTION of a corresponding co-reality and further yet from a cognized statement of a law of physics. What is obvious about the usual definition of perception is that it EXISTS as the most basic element of reality we know and yet here its very existence seems to hang in the balance of compatibility with some all-inclusive abstraction. The whole perception of the reality of "perceptions" seems to have been hijacked by some contortion of thinking to depend upon a level of reality where poodles and noodles argue differences and similarities rather than just BEING. See Nicholas Bucska's article, "Meta-Questions and mega fallacies" in issue № 88 pp. 25–26, where he says:

> In nature there is a poodle and a noodle but there is no such thing as a difference between them: the difference is the mental result of an arbitrary mental process performed on abstract objects: our mental images created from our sensory data of real objects.

CML: Uh...with all due respect to Nicholas, I'd have to maintain that in nature, there both IS and IS NOT a difference between a poodle and a noodle. That there IS a difference is evident from the fact that nature gives one the power to bark and chew bones, but not the other. That there is NOT a difference is evident (e.g.) from the fact that both have mass and are therefore composed of energy. But this "paradox" is resolved by recognizing that reality is stratified; things are "the same" with respect to shared generalities ("all matter consists of energy"), while they "differ" with respect to distinctive specifics (women perm their hair not to put noodles on their poodles, but to put poodles on their noodles).

In order for perceptual reality to exist, this "mental" stratification must be built into the structure of spacetime. If it were not, then spacetime and its contents would be totally imperceptible and therefore unreal.

RFV: Notwithstanding, Chris maintains that, "thus, every axiom and theorem of mathematics can be considered implicit in material syntax and potentially exemplified by an appropriate material pattern, e.g. a firing of cerebral neurons." And again, "so the background has logical ascendancy over derivative matter, and this permits it to have aspects, like the power to enforce consistency, not expressible by localized interactions of compact material objects (i.e., within the bounds of *materialism*...)". This platonic structure of reality, he clarifies by saying: "...physical reality is embedded in a more general or 'abstract' ideal reality equating to the reality-syntax D(S), and the syntax D(S) is in turn embedded in physical reality. Thus, if D(S) contains supraphysical components, they are embedded in S right along with their physical counterparts (indeed, this convention is already in restricted use in string theory and M-theory, where unseen higher dimensions get 'rolled up' to sub-Planck diameter)", whatever that means.

CML: The CTMU property to which Fred is referring here is called "hology". That's a logical form of self-similarity in which the structure of an infocognitive system is distributed over the system as self-transductive syntax. [An infocognitive system is a system whose components all simultaneously possess the nature of information and cognition. For example, you (the reader) are internally a cognitive entity, but others perceive you as an informational construct related to your external environment. In a generalized sense, the same applies to inanimate objects; they simultaneously possess intrinsic (subjective) and extrinsic (objective) descriptions that are respectively temporal and spatial in character. That is, the purely intrinsic description of an object, being utterly confined to the object, is a self-description and thus a recursive (temporal, sequential) process, while an extrinsic description specifies the position or meaning of an object *relative to* other objects and thus has a spatial (relative, parallelized) aspect.]

As for rolled-up dimensions, that's a convention that goes back to Kaluza–Klein theory, in which light is characterized as a "fifth-dimensional vibration". Supposedly, this fifth dimension is imperceptible because it gets rolled up into a tiny little knot of sub-Planck diameter (string theory and M-theory have many more of these knotty little dimensions, the ravelling of which turns out to present some knotty topological problems). But here, I'm just pointing out a convention of M-theory not necessarily shared by the CTMU.

RFV: So clothed in the priestly garb of words we are now ready to address God. I will reiterate the question with Chris: "What does this say about **God**?"

With regard to Chris's answer, I find that reiteration does not suffice for me. Where he says, "So by design, religious beliefs generally cannot be refuted by rational or empirical means," I think he may be intentionally playing on words. Certainly one could never refute that there is BELIEF in God as long as anyone still BELIEVES in God. That is an irrefutable fact like the occurrence of love and other relations between humans and other objects whether REAL or not!

However, that key tenets of most "religious beliefs generally cannot be refuted by rational or empirical means" is an assertion that should probably be denied subjective honors of existence in his scheme if he is to remain consistent! It is, in simple fact, absurd.

CML: I think that Fred may be overlooking the important fact that while the existence of God is merely a "belief" to the adherents of traditional organized religions, it is more than a belief in the CTMU. It is an unavoidable syntactic relation. Simply using the term "belief" with regard to religious ideas does not imply that this is all they are, or that this is their only means of justification. May I respectfully suggest that in the future, Fred acknowledge the discursive context instead of mangling it beyond all recognition?

RFV: Going further he asks and answers the question, "Can a *denial* of God be refuted by rational or empirical means? The short answer is yes; ...". He proceeds by claiming that by his having merely associated the term "God" with the abstract level of true statements in his CTMU that he conceives as "structuring" the universe, that that structure "constitutes not just a logical framework for reality theory, but the outline of a logical proof of God's existence and the basis of a 'logical theology'." To my mind God was no more marginalized by Gnostic abstraction in the Synoptic Gospel than by Chris Langan in his article, "Introduction to CTMU" that summarizes for neophytes his "Cognitive-Theoretic Model of the Universe (CTMU)". Nor has God's existence ever been conditioned on more theoretically questionable foundations.

CML: Fred is (of course) entitled to his opinion, ill-supported as it may be. But even if he's unable to see that the CTMU is a proof of God's existence, he must admit that either it is a *new approach* to same, or he doesn't even come close to understanding it. If he prefers, we can leave the success of this approach to the tests of time and logic.

References

Eddington, A. (1933) *The Expanding Universe*. Cambridge: Cambridge University Press.

Gödel, K. (1962) *On Formally Undecidable Propositions of Principia Mathematica and Related Systems*. Translated by B. Meltzer. New York: Basic Books.

Hartle, J. B. & Hawking, S. W. (1983) Wave function of the Universe. *Physical Review D*, Vol. 28, № 12, pp. 2960–75.

Hoffman, W. C. (1966) The Lie algebra of visual perception. *Journal of Mathematical Psychology*, Vol. 3, Issue 1, pp. 65–98.

Hume, D. (1975) *Enquiries Concerning Human Understanding and Concerning the Principles of Morals*. 3rd Edition. Edited by L. A. Selby-Bigge & P. H. Nidditch. Oxford: Oxford University Press. Reprinted from the posthumous edition of 1777. Original work published in 1748 and 1751.

Langan, C. M. (1999) Introduction to the CTMU. *Ubiquity*, Vol. 1, № 1.

FAQs About Reality: Chris Langan's Social Media Posts. Book Two: Patreon

2021

2022

PUBLISHER'S NOTE

On April 20, 2022, the Mega Foundation Patreon page was deleted, along with Chris Langan's very popular CTMU Patreon page. This was done without warning. In fact, there were no violations or notices at either page. No discernable reason was given for the deplatforming, and no material was identified as having violated community standards. Patreon retained all donations and did not refund patrons. Chris Langan described these illegal actions on Facebook and Substack.

***

Question: Were you banned [from Patreon] or was this a sudden decision?

Answer: All of our pages – both mine and the Mega Foundation's – were removed suddenly and without warning.

I requested an explanation. Someone from the Patreon "Trust & Safety Team" wrote that on another platform (not Patreon itself), there was allegedly a "violation of community standards" involving "hate speech" and the "stereotyping of protected communities".

Pointing out that banksters and political whores are not ordinarily considered "protected communities", I requested the name of the platform and the offending comment. **No response**.

I requested that my unpublished content (scheduled and draft folders containing important writing on the CTMU) be returned to me. **No response**. There was a considerable amount of material, and it seems to have been permanently deleted.

It seems that a considerable amount of money may have been taken by Patreon for itself.

Do not, under any circumstances, trust Patreon as anything but an enemy of truth, intellectual freedom, and freedom of expression.

If what Patreon did to us is not considered a crime, it certainly should be. It's the worst show I've seen anywhere on the Internet, ever.

May 03, 2022

2018

Identity Politics

My view on identity politics is that it can be justified only if everyone of any ethnicity is entitled to participate, in which case it is necessary for all (because failing to assert it, as when White people of European ancestry fail to assert it lest they be branded as "racists", means leaving oneself and one's group defenseless against competition for resources and opportunity). Alternatively, lest any group be denied its identity while others assert their own, group identity must be equitably denied to everyone.

Human identity is stratified, and thus has both individual and group levels. Accordingly, we can (and sometimes must) reason in terms of group identity. But when group self-identification is officially granted to some groups yet denied to others against which they compete, this can only result in imbalance and injustice. For example, when some overpopulating groups which have overtaxed their own resources by reproductive incontinence and homegrown oligarchy are allowed to migrate into the sovereign territories of worldwide ethnic minorities – e.g., people of European descent – and enjoy special "oppressed" status whereby they reap special benefits such as free food, free housing, free education, free healthcare, affirmative action, reproductive subsidies, and special treatment under the law, and are even credited with moral superiority due to their alleged "oppression", this can result in the destruction of the national, cultural, and ethnic identity of the hosts, leading ultimately to their extinction. Incoming groups which assert their own collective identities while denying their hosts any reciprocal right of political group cohesion thus amount to noxious, invasive, and ultimately lethal socioeconomic parasites. Obviously, any governmental authority which enforces or encourages such asymmetry – e.g., the European Union – is illegitimate.

Bear in mind that once we cease to treat individuals as individuals per se, thus allowing members of their respective groups to assert their ethnic, cultural, or religious (etc.) identities against their "oppressors", their group properties and statistics are automatically opened to scrutiny and comparative analysis. For example, if after several generations of special treatment in the educational sphere (compulsory school integration, special programs, modifications of educational procedure, racially defined college admission preferences, etc.), a particular "oppressed" group fails as a whole to outgrow these measures, its members are no longer entitled to exemption from objective characterization in terms of associated group statistics; if one wants to enjoy the social benefits attending ethically loaded group-defined properties like "belonging to an oppressed group", one must submit to rational policies formed on the basis of not just individual assessment, but empirically confirmed group-defined properties such as "belonging to a group exhibiting a relatively low mean IQ and a tendency to violently disrupt the educational environment".

Continuing to pursue racially parameterized measures of human worth and achievement can only lead to personal injustice, social degradation, and biological degeneration (because such measures inevitably supplant any rational form of social, economic, and reproductive selection).

In short, identity politics should either be shut down immediately, or the majority populations of Europe and North America should be encouraged to assert their own ethnic and cultural identities and group interests with full force. Any governmental, academic, religious, or media authority which tries to prevent it is clearly unworthy of respect and obedience.

June 03, 2018

***

In the last several years, there has arisen a push to change the meaning of the word atheism. In fact, it is not just a "push"; it is purported to be a semantic *fait accompli.*

In the past, theism has referred to belief in the existence of God, atheism has referred to belief in the nonexistence of God, and agnosticism has referred to belief in neither the existence nor the nonexistence of God. There is no doubt whatsoever that these are the meanings traditionally attached to these terms by the vast majority of people.

The traditional definition of atheism has made no distinctions on how God is defined; all definitions are rejected in kind. This has always been a problem. Inasmuch as God might be defined in a great many ways, some of them contradicting or complementing others, disbelief in one definition might compel belief in another. Atheism does not allow for this possibility, but conveniently switches focus among various conceptualizations of God in order to repudiate the general concept and thwart all criticism.

However, there's still a problem. If God is defined in such a way as to be logically necessary – for example, if God is defined as Logos or logic itself, as He is in the New Testament and in the CTMU – then rejecting the existence of God amounts to rejecting the logical necessities of which God consists, and thus to logical inconsistency. This, of course, would fatally undermine the frequently heard claim that atheism holds the logical high ground in its never-ending war against theism.

Evidently smelling just this kind of trouble – after all, the CTMU is here at last – atheists manifest great cunning in the way they present their self-redefinition. It is not merely stated as a proposed change in the language; rather, it takes the form of a claim by certain language-challenged atheists that they are actually "agnostic atheists". Although this phrase appears self-contradictory on its face, it constitutes a protected claim of identity.

Claims of identity are untouchable by criticism; to criticize them in the present social environment is to incur an accusation of insensitivity or bigotry. For example, if one is biologically a male, but wishes to identify (and be regarded) as a female, one simply makes an explicit claim of identity to that effect; if one is White or Asian, but wishes to be Black or Latino in order to exploit a racial quota in order to get a job, one makes a counterfactual claim of ethnic identity. In either case, anyone who argues about it is sure to be attacked by any SJW ("Social Justice Warrior") who happens to see it, and quite possibly by the mainstream media, at least until they become convinced that the claim was nothing but a cynical ruse. Then, of course, one will be pressed for an abject apology, which may – if one is lucky – spare one the complete ruination of one's life, if not the associated stigma.

Now we come to the core trick, the one which simultaneously rationalizes and is protected by the claim of identity. It is a combination of etymological reconstruction, the distinction between epistemology and ontology, and an irrational disjunction of knowledge and belief.

We'll get to some of that in the next installment.

June 02, 2018

Feynman on Nature

In a world where techie CEO Elon Musk is often treated as a world-class philosopher (or so one might easily assume from some of the questions he is asked by the media), it is unsurprising when a celebrated scientist like Richard Feynman is put in the same position. As a topnotch physicist who a mere couple of centuries ago would proudly have called himself a "natural philosopher", Feynman was queried about his view of nature.

Being the good sport that he was, Feynman obliged with an answer. But on reviewing this answer, we find that there were a few problems with it. Feynman's intellect was stellar, but in expanding on physicist James Jean's famous statement that "the Great Architect seems to be a mathematician," he may have veered a little off-track.

Consider the following excerpt from one of his lectures:

> Physicists cannot make a conversion to any other language [than mathematics]. (...) Philosophers may try to teach you by telling you qualitatively about nature. I am trying to describe her. But it is not getting across because it is impossible. Perhaps it is because their horizons are limited in the way that some people are able to imagine that the center of the universe is man (Feynman, 1967, p. 58).

Feynman clearly meant well, but this is misleading on several counts. First, mathematics deals with quality as well as quantity, and is basically indistinguishable from the abstract structure of language in general. That is:

1. Mathematics is a language consisting of mathematical theories (theoretical languages) and their abstract universes of discourse.
2. Language has mathematical (logical, algebraic, and in some cases geometric) structure, where by "structure" we mean the explicit relationships among basic constituents. In other words, each of the two terms language and mathematics is a property of the other. Together, they describe the same thing.

Regarding 1, mathematical theories are usually axiomatic systems. In an axiomatic system, axioms amount to the context-specific "syntactic forms" of a mathematical language consisting of all expressions that can be derived from them, and the grammar of the language consists of the rules of substitution by which data are entered into the axioms and transformed into input for other axioms and theorems derived from the axioms.

Regarding 2, language is an algebraic structure in its own right, and in order to usefully support identification and inference, its syntax must implicitly incorporate logic, which is also algebraic. To be legible and convey meaning, its constituents must be discernible, which requires a geometric or other aspect in terms of which to write and recognize its symbols, words, and sentences.

It follows that mathematics and language cannot be pried apart, as we can easily see from the fact that the proper identification of mathematical terms and (object, relational, or operational) symbols virtually always requires verbal definitions with both qualitative and quantitative aspects. Despite being a nominally quantitative kind of language, mathematics cannot be decoupled from the qualitative aspect of language, or isolated from language in general.

This is easy to see. Given a numerical value, the first thing we need to know in order to determine its significance is the attribute or quality which it measures; for example, the length of an interval or the cardinality (number of elements) of a set or collection of objects. This is true even where the number is "dimensionless" or without units; in this case, the attribute "measured" by the number is generically defined as "being a product or ratio of quantities x and y whose units cancel". Attribute and value must be coupled in what boils down to a linguistic attribution statement.

Even in "pure mathematics", where a number n may be recursively defined as the result of summing n units or as the product of its own factors, these definitions themselves comprise their own autological(self-descriptive) attributes. Any mathematical expressions containing the number, or a variable to which it is assigned, also qualify as attributes; thus, given the expression $x + y = z$, the variable y is implicitly defined by the linguistic attribute "that which yields z when added to x". In any case, every number, term, or symbol must somehow be linguistically related to other elements of the mathematical structure in which its significance resides.

Secondly, it is possible to describe nature more qualitatively than most mathematicians describe it. One simply goes to a coarser quantitative granularity, using terms like "all", "every", "some", "any", "at least one", and "none", or if one wants more precision but prefers to avoid numerical complexities, the existential and universal quantifiers of predicate logic. As any competent teacher of mathematics knows, it is highly desirable from a didactic standpoint to provide students with a good qualitative description of various mathematical objects, relationships, operations, and systems before plunging them into a morass of messy details without the overall conceptual glue that holds them all together, and a qualitative approach can help. Unfortunately, relatively few professional instructors of mathematics seem capable of understanding and properly implementing this educational necessity.

Thirdly, it is possible to put man at the center of the universe in a mathematically consistent and comprehensible way. For example, on the surface of a featureless sphere, every point is central; the rest of the sphere is perfectly symmetrical with respect to it. Where every point is central, and human beings are defined as "points", human beings are central; specifically, each is central to an extended frame of reference encompassing the entire surface. Given the difficulty of defining a "frame of reference" without including something capable of reference – a human being, for example – it cannot be cogently argued that human observers are not "central" to these frames in a special way. Hats off to Professor Feynman, a highly decorated hero of modern physics. But once the applause dies down, the hard truth reasserts itself: when physicists try to function as philosophers, they can fall flat on their faces ... and if even a genius like Feynman couldn't pull it off, then neither can most other physicists.

June 06, 2018

The Necessity of QM Interpretation

Excerpted from a work in progress.

Here's another physics-related post before I resume my discussion of "agnostic atheism". I recently encountered the following question on a popular Q&A site:

"Philosophers of physics: what is your preferred interpretation of quantum mechanics (QM) and why?"

This question is followed by many answers. The author of the most popular answer by far, which had garnered nearly 500 "upvotes" (not to be taken as a measure of quality), explains that he is not a mere philosopher of physics, but actually "does physics"; that he "likes none of this interpretation business" and considers QM interpretations superfluous; that even though QM is nonlocal

("Bell, etc."), this kind of nonlocality needs no "comical 'interpretations'" because it leads to no classical violations of causality; and that Lagrangian mechanics is also nonlocal insofar as one knows the final state of the system once its equations of motion are determined unless "classical predictability flies out the window" because the complementarity of functions and their derivatives prevents the system from being in an eigenstate of both, in which case QM "still doesn't need any 'interpretations'."

Other answers explain, albeit unconvincingly, why Copenhagen, Many Worlds, or some other recognized interpretation of quantum mechanics towers over all the rest.

Now for my take, and I'll keep it short. First, let's consider the standing of those who might answer such a question. A "philosopher of physics" is clearly a philosopher who specializes in physics; and because physics is a science, he must be a philosopher of science. Moreover, insofar as his philosophizing refers to physics in its entirety, he must be the kind of philosopher usually referred to as a *metaphysician*. So really, this question is for metaphysicians, of whom very few are actually qualified to comment on the state of the art in metaphysics.

Next, in light of the answer quoted above, something very much needs to be acknowledged here: **there is no getting away from the need to interpret quantum mechanics**. As a matter of logical necessity, any scientific (or other kind of) theory requires interpretation; it must always be mapped into its universe of discourse. Saying that QM "needs no interpretation" implies that it need not be mapped into the physical universe it purports to explain, which is logically and scientifically absurd. There's simply no getting away from the need to interpret quantum mechanics in the physical universe.

But now that this is out of the way, we run into another problem: no ordinary interpretation of QM actually does this in a way that doesn't leave huge gaps in the mapping. All QM interpretations agree on the set of measurement events and their observed outcomes, and in the classical physical descriptions of the experiments and devices used to obtain the outcomes – after all, this is what makes them "QM interpretations" – but very little else. Their respective descriptions of the *causal processes behind* the measurement events, and the associated hidden structures and mechanisms, often conflict, overlapping only partially at best. This has the inevitable effect of calling them all into question.

What then of a "null interpretation"? What if we take most of the conceptual apparatus of QM at face value as nothing but "pure mathematics"? This amounts to disconnecting QM math from the structural and causal features of the physical universe, in which case there is no reason that it should work for explanatory or predictive purposes. This is inconsistent with the observed fact that it does work in a great many real-world situations, which can only be explained if QM is something like a theoretical limit of a larger theory which provides it with a valid metaphysical extension that fills the causal and structural holes in the QM-reality mapping.

Similarly, anyone who denies that popular interpretations of QM contain gaping holes needs to explain how and why none of these interpretations has been experimentally distinguished from the others despite differing quite markedly where the holes occur, in coincidence with their respective nonphysical (nonlogical and inobservable) components.

It follows from all of this that quantum theory requires *more* than a mere interpretation of QM. The required alternative is based on the CTMU, an actual model of reality which takes the form of a perfectly self-contained language suitable for the description of a perfectly self-modeling system. This alone makes the connection between the theoretical and empirical aspects of quantum theory sufficiently *real* to afford any level of certainty.

Like it or not, that's what a real metaphysician has to say about it.

June 12, 2018

The Logical Intersection of Science and Theology

For those who don't yet know me, I've thus far achieved minor note for authoring a conceptual framework called the Cognitive-Theoretic Model of the Universe, and for maintaining that this framework constitutes the logical intersection of science and theology.

Such claims are extremely controversial in some circles, including two in particular. One, associated with academia and consisting largely of convicted materialists, claims the right of peer review over any new idea in philosophy and the sciences. Call this first group G_1. The other group, call it G_2, consists of religious people who feel that the existence of God can only be meaningfully apprehended through faith and revelation.

Both of these groups find the kind of claim I am making to be questionable at best. But this is not because some advanced form of proof theory shows that it is impossible to rationally prove the existence of God, or because God was miraculously revealed to be exempt from His own laws of rationality. Instead, the reasons are as follows:

1. Many people have tried to prove the existence of God and failed in the eyes of skeptics. For those equipped to do no more than extrapolate from past failure, future failure seems inevitable.
2. The vast majority of religious believers have arrived at their convictions through a leap of faith, and cannot conceive of doing otherwise. "If that's what I had to do, then that's the way it is!" they say.

3. Most people fear separation from the herd, and they are rightly convinced that their particular branch of the herd (G_1 or G_2) rejects and disapproves of such claims. Each side generates trolls who can be quite unrestrained in their persecution of defectors and deviants, and no one wants to be targeted by them.

4. If such a claim were true, then science and religion might not be independent of each other after all. But then certain assumptions central to the belief structure of each group might be threatened by the beliefs or findings of the other. They might actually be forced to broaden their intellectual horizons, which sounds to many of them like low-paying work.

In other words, opposition to such claims is based on fear, and/or on empirical induction from the supposed past intellectual defeats of theists, and the imagined intellectual victories of materialists whose mindset renders them unable to admit the validity of anything that might threaten their materialism. Hence, the associated claim that atheists possess irrefutable logical arguments precluding the existence of God. Such claims are every bit as controversial as claims like my own, and therefore cannot be decisively used against my ideas.

The minds of many G_1 materialists and G_2 theists appear to be hermetically sealed on this subject. Worse, each group seems blind to the obvious weakness of its own position. But here the similarity ends. For while academic materialists have been quite successful at imparting their beliefs to a virtual captive audience – "captive" because young minds are compulsorily subjugated to them over a lengthy period of conditioning which persists through adulthood – religion now finds itself in a far bleaker situation.

As a discipline, theology is in trouble. It is increasingly portrayed as useless or frivolous on the grounds that religious doctrines seem irrational, often contradict one another, and are seemingly backed up by no experimental data confirming God's existence... i.e., by none of the empirical data that a diehard materialist might find convincing. In an age of science and technology, so the argument goes, how can a contradictory body of competing doctrines which seem irrational and devoid of scientific evidence be taken seriously? The God concept, its detractors say, is flawed, bankrupt, and utterly discredited.

Nowhere is this argument more prevalent than in the hallowed halls of academia, the system of higher educational institutions responsible for training the minds, and thus determining the median thought and attitude, of the world's economic and political elites. In academia, the total independence of science and theology is considered axiomatic, and prayer is forbidden on the grounds that the public observance of any particular religion offends those who prefer other religions (or in an increasing number of cases, no religion at all).

There are hidden implications here. If any aspect of religion were definitely true, then due to its inevitable bearing on ethics and civic duty, it could be taught in school as a required course of study, and observances based on theological truth would be no more controversial than the observance of, say, arithmetic. Conversely, to disallow any form of religious observance is unmistakably to imply that there is no such thing as theological truth. Similarly, inasmuch as any information actually bridging the gap between science and theology could be incorporated in science curricula, to exclude it in advance is to imply that it does not exist.

College students, who tend to be brighter than average even in these times of "equal" educational opportunity, can't help but read the writing on the classroom wall and be steered toward the latter conclusion. To witness the corrosive effects of this pervasive indoctrination, just go for a walk, browse the Internet, turn on the television or radio, go to a movie, or read a newspaper. Unless you happen to be sitting in a church, you really can't miss it.

I'm disturbed by many of these effects, which appear to be accelerating in a number of dangerous directions. Over most of its history, mankind had the luxury of knowing that it could step over the brink, hit the ground, pick itself up, turn back the clock, and start over.

There were new habitable territories, plenty of easily-worked resources within feet or inches of the earth's surface, and a future rich in untapped potential for human exploitation and economic growth. [The profit incentive cannot be successfully distributed and maintained without economic growth; for average individual wealth to grow, the entire economy must grow. In any finite world, the profit motive must ultimately be tempered.]

But mankind is no longer so young nor the earth so virginal, and the future is different than it used to be. Stepping over the brink and starting over from nothing is no longer a likely option. Not only are there more ways than ever for civilization to self-destruct, but we now have a population exploding under socioeconomic, psychological and biological inertia and dependent on a complex, all-but-untraceable web of rarified and mutually isolated technological specialties. [Technological divide and conquer.] The environment has been strained to the breaking point, there is no more virgin territory, and mineral and biological resources are rapidly becoming depleted or increasingly difficult to extract.

Does humanity have the right to survive? If so, then we have a moral responsibility to protect that right, and the rights of future generations. But we also consider individuals to have rights in the present, and it is unclear how these individual rights relate to mankind's right of survival, and the rights of future generations. The kind of knowledge that would allow us to clarify this relationship, and "solve the human condition", is called ethics.

For a system of ethics to work, it must come from the heart as well as the head. Otherwise, compliance ultimately relies on bribery or coercion and breaks down whenever they are not applied. [Consider the idea that ethics can be imprinted, or that there is a universal "ethical instinct" to which appeal can be made. The existence of such an instinct able to overpower individual utility would imply evolutionary utility, and where the human race is the evolutionary argument, imply a gene-based form of racial identity that would come down to intraspecies compassion after all... i.e., an emotional identification of collective interest with self-interest).] Because the desires of the individual are far more immediate than the needs of posterity, and far harder to postpone or deny, selfish behavior is statistically certain. Thus, ethics can only work when implanted deeply enough in the individual psyche that it can directly encourage good impulses and oppose bad ones on the root level, prior to any chance of behavioral manifestation. Without this depth, ethical behavior is contingent on the expectation of reward or punishment. Where ethical behavior is driven by material consequences, society inevitably devolves to the antithesis of human freedom, namely, a state of constant coercion... the police state.

Those atheists and materialists who see the desirability of human rights and the place of ethics in the social fabric, i.e. who do not subscribe to a Nietzschean form of Darwinism and wish to avoid a coercive police state, subscribe to an ideology called secular humanism. As a replacement for theism or worship of God, secular humanism amounts to the non-religious "worship of humanity". Whereas theism centers on an ideal compassionate entity which dwells in the human heart and connects humanity as a common focus of love, reverence, and spiritual aspiration, secular humanism trades this living entity for a combination of cold reason and a cerebral, codified, lifeless view of ethics and justice purportedly based on an empty form of "compassion" that rides the shifting sands of moral relativism.

Secular humanism provides no grounding for ethics, for it is a mere social theory and does not provide a true psychological basis for social cohesion. We require something that is far more convincing on a far deeper level. Only true empathy can provide ethical grounding, and empathy is not a universal trait. It is present in all human beings to some degree, but how it develops in any given case depends on an interplay of nature and environment. If there is one thing we have learned from history, it is that ethics needs all the help it can get. Solving the human condition means giving ethical consciousness the best possible chance to develop, tuning the social and educational environments to maximize the psychological growth of empathy, intelligent compassion, and a deep, symmetrical sense of social responsibility. That is, it means systematically stacking the emotional and intellectual deck in favor of ethical reasoning and behavior.

How is mankind to stack the deck of survival in its own favor? How are we to avoid being driven over the brink by runaway political, economic, and biological forces? The answer is simple: by discovering an ethical basis for true long-range global solutions to our problems, a solution that can be non-coercively distributed over the entire population.

As you might by now have inferred, neither atheism nor materialism resemble the ideology we require. In fact, they are antithetical to empathy and ethics, comprising a sure invitation to disaster.

We'll pick this up later.

[date unknown]

What Is Intelligent Design?

I am an evolutionary theorist. I am not an evolutionary theorist because I have a college degree in evolutionary biology; I am an evolutionary theorist because I am a reality theorist, and evolution is a key part of reality. Yet, I have also been called an Intelligent Design advocate, mainly on the strength of a single ID-related affiliation and the fact that I credit God with creating the universe.

In fact, I don't just believe that God created the universe; I *know* it. How do I know it? Logic permits no other conclusion to be drawn. How can I refer to this as a "conclusion" when every rational person knows that divine creation is only a myth that has no basis in mathematical or empirical reality and must therefore be accepted on faith? The answer is simple: I know it because I know logic, and logic is how the conclusion is derived. In fact, I know more about how logic works that anyone who asks this question without knowing that the answer is not what most academically trained people expect it to be, or that there exist relevant arguments far more sophisticated than those traditionally presented in this context.

I also believe in evolution. In fact, I don't just believe that it happens; I know it. How do I know it? I know it because the main premise of Darwinian evolution, that there exists a principle of natural selection, is tautologically true. Because life exists in nature, which clearly has the power to extinguish that which is inconsistent with its environment in structure or behavior, life is constrained by and passively selected for compatibility with its natural environment. Unfortunately, Darwinian evolution is presently attached to the weak, paradox-ridden view of nature and causality that dominates the physical and biological sciences. To that view, I subscribe only up to a point. My view is much more capacious, including the conventional view as a limiting case.

Thus, while I can state unequivocally that natural selection exists, I can also state unequivocally that it is not the whole explanation for biological evolution. Unfortunately, while Darwin did not make the mistake of saying that it is the whole story, he also failed to explain why it isn't, and thus opened the door to the confused sort of "evolutionist" that we now find in great abundance on college campuses... so-called "Darwinists" who believe, mainly for their own subjective reasons, that Charles Darwin meant to lock the theory of biological causation into the tiny academic box where it is now forcibly confined.

The way these people use the phrase "ID advocate" is meaningless. How can one meaningfully "advocate" something presented as a scientific hypothesis? As much as one might agree with it, it remains a hypothesis and can therefore go either way; one can only advocate that it be rationally and empirically tested, or refined to the point where testing is possible (which is in fact what ID theorists actually *do*). In other words, to *advocate* a hypothesis, as opposed to merely presenting it, is to transform it into something either known or believed. Meaningful advocacy is reserved for a theory or methodology.

Now let's talk about Intelligent Design. Intelligent Design enters the picture because it is widely considered to be the complement of the evolutionary hypothesis. In a nutshell, it says that some form of intelligence is *probably* involved in biological origins. I hereby express my qualified agreement with this assertion. Why only my "qualified" agreement? Because the word "probably" is so weak that it is effectively superfluous. It is a gratuitous insertion calculated to imply that the truth of the ID hypothesis is subject to empirical confirmation and thus "scientific" in the empirical sense. Actually, the ID concept is a logical precondition of science.

So what do I actually mean by "qualified agreement"? No purportedly scientific assertion is meaningful until it has been *interpreted*, and this requires a *model*. There is presently no single model for ID theory. A model would require a designer and a mechanism, and ID presently has none that it chooses to make explicit. Instead, ID holds that it is possible to take generic objects and measure or estimate the amount of intelligence required to produce them, or at least infer whether intelligence was involved in their production, *without benefit* of any specific model.

My feelings regarding this position can be concisely expressed as follows: come hell or high water, a model is ultimately required, it is ultimately logical in nature, and insofar as it is a logical prerequisite of observation, it preempts any immediate need for the empirical confirmation of specific hypotheses. Quite simply, its observational confirmation resides in the observed success of empirical methodology; because the model is basic to the very possibility of science, the existence of science implies its validity.

Because I claim to possess this model, and also claim that I can easily defend it against all comers without resorting to standard empirical methodology and thus subjecting it to the problem of empirical induction, I do not bear description as a mere "ID advocate". In fact, I would more properly be called a "God advocate", and because ID considers the existence of God to be subject to empirical confirmation, this is a point of departure from ID theory as it is usually considered. After all, ID theory refrains from identifying God as the designer. I, on the other hand, make no bones about it: God designs the universe *even now*, and that's it.

Of course, there is also a difference between the CTMU conception of God, and the way many ID theorists define God, i.e., roughly as depicted on the ceiling of the Sistine Chapel. Those who subscribe exclusively to the Sistine Chapel depiction typically refuse to go any farther due to the scriptural prohibition against trying to gainsay or go beyond scripture, which is just scripture's transparent way of deterring its own editorial corruption. (For that matter, scripture does not authorize the depiction of God on the Sistine Chapel ceiling.) But of course, scripture cannot be read without interpretation, and by definition and function, the interpretation operator goes well beyond the letter of scripture. So objections from the self-expressed sanctity of scripture fail on logical grounds alone.

How does the CTMU characterize God? God is the ultimate reality; God is perfectly reflexive and self-contained; God works through man, rendering man a participant in Creation; God does not retreat from physical reality into some metaphysical cubbyhole in order to avoid self-contamination, but partially coincides with physical reality and fearlessly refines it in conformance with His nature. (Why and how can nature depart from God's nature enough to need refinement? Precisely because it shares the protean freedom of God *on the most basic level* of His existence, and is thus free to explore wrong turns and blind alleys despite the constraints associated with God *on more refined levels* of His existence. When it does this, God simply rejects and recycles it until it conforms to His nature, otherwise known as "teleology". On the other hand, that which conforms to teleology is reinforced. This is the basis of common religious notions regarding divine favor and retribution, and ultimately, heaven and hell.) Lest it be thought that the CTMU represents only one particular theological viewpoint, it too has a "Big Tent".

***

That is, objects are treated somewhat analogously to items in an IQ test. Now, IQ tests are normed over a population of entities known to possess intelligence; thus, ID proposes to "give reality an IQ test", and since IQ tests are statistically interpreted, ID proponents are using the mathematics of statistics and probability in their investigations. [Note that a multiverse is designed to expand the population of realities to jack up the probability of uncommon structures. ID critics use the multiverse to show that the occurrence of realities which contain what seem to be "improbable" or atypical structures is unsurprising, cosmologically speaking. Max Tegmark wants to go even farther, saying that life-bearing universes are the "ensemble average". No matter how you slice it, this is all naively-interpreted high school stats with no ontological sophistication.] However, there is a crucial difference between reality and an IQ test: because we have empirical access to only one reality, there is no way to develop statistics for such a "test". Accordingly, ID attempts to generalize the measurement of intelligence to patterns of the general kind used in cryptography and archaeology.

However, these "intelligence-suggesting" patterns are ultimately identified only by known association with something already known to possess intelligence (human beings). Accordingly, ID turns to the SETI paradigm, seeking patterns of a kind which would not appear to be produced by currently recognized natural processes... e.g., a sequence of prime numbers broadcast at radio frequency, as in the movie "Contact". However, this paradigm is still based on the assumption that natural processes are sufficiently well known and catalogued to make at least a probabilistic determination along these lines. There is nothing the matter with this. However, it still cannot settle the issue with certainty, because probability theory is meaningful only in the presence of uncertainty (and thus cannot be used to achieve certainty).

Just as probability and statistics are the mathematical theories which deal with uncertainty, logic is that which deals with certainty or "truth". The CTMU thus takes a logical approach.

Logic is a cognitive language; hence, the CTMU examines the extent to which reality is linguistic (logical). On the basis of what it finds, it then infers that intelligence and reality cannot be separately defined, and that their definitions are in fact coupled as follows: reality is that which is recognized by intelligence, and intelligence is that faculty by means of which we recognize and (partially) control reality.

"So what?", a critic might say. "That's just a statement about definitions, and we all know the linguistic operation of definition is independent of objective reality!"

But in fact, insofar as reality necessarily conforms to the general definition of a language, and specifically, a self-configuring self-processing language, it is a self-defining entity, and this means that it must possess the property which enables it to perform that function. This property is intelligence.

So far, so good; we have defined intelligence as the ability to self-define. But where is the connection to the sort of human intelligence that we need in a "personal God"?

That's easy: this "self-definitive" intelligence is a straightforward ontological generalization of human intelligence, and that of which human intelligence is an internally reflected subprocess.

Compare the world we see to a chaotic gray ocean of particles. It is highly improbable ... unless it is favored by that (underlying model) which is required by a gray chaotic ocean of particles. In fact, those particles are correlated insofar as they exist in one connected "sea" or system. There's no getting around it. Therefore, we must ask: "What is required in order for this correlation to exist?" When all is said and done, SCSPL is required... and given SCSPL, the "improbable" degree of organization that we see in the world is utterly certain. But the price is that the universe must be recognized as intelligent, and human intelligence as its inevitable product. Armed with this model, we can name a designer – reality itself – and an appropriate mechanism – the generation of a conspansive manifold – and "find the zero" on the IQ scale using distributed self-utility. Then, and only then, can we discuss probabilities relative to that zero.

There is a debit condition associated with distance from telic realization; intelligence becomes harder to recognize because telonic patterns remain nonlocal until everything comes together. On the other hand, for any localized pattern, it remains hard to tell whether it is the product of intelligence or natural processes except by relating it to the utility of a localized entity or telor, and thus to the utility of nature at large. (Prime numbers have no special status in constructive processes.) Sadly, the definition of ID has been politicized within an inch of its life. Although ID should be defined exactly as its proponents define it – after all, they are the ones who coined the usage – it is an ambiguous phrase. The definition of ID was initially very simple. Then it was politicized to the point where it became fatally ambiguous.

Just as sadly, the definition of evolution has also been policitized.

Evolution and ID actually deal with two superficially disjoint domains: (1) biological transformations which can be described in terms of the gradual accumulation of relatively minor biological changes, and (2) biological transformations which cannot be thus described, including biological origins. The fact that biologists are trained only in the former does not entitle them to insist that the latter do not exist.

The difference between ID and Darwinism is in hypothetical content, but not methodology. They both explicitly subscribe to the methodology of empirical science. The difference between the CTMU and both evolution and ID is one of methodology rather than hypothetical content; the CTMU says that to a large extent, the respective contents of both lesser hypotheses are true, but only when they are properly interpreted in a unifying model. The CTMU further says that the reality itself is logically equivalent to the model in question, and thus applies the methodology of metamathematics, including language theory and model theory. Its approach is thus totally unique.

ID and the CTMU agree on the conclusion, but not on methodological priority. ID says that we can go immediately for causal probabilities. The CTMU says that we must go for the logic of reality and causation first, and then see whether the probabilities can be anchored in the logic. So, am I an Intelligent Design proponent? Yes and no. Yes in the sense that I think the basic substance of the ID hypothesis is correct (and in fact inevitable); no in the sense that I do not agree that we can go immediately for probability calculations in the absence of a deep underlying model, which then beats probability to the punch in directly verifying the hypothesis.

I am an ID advocate only insofar as ID can be interpreted in my own theory. Divorce it from my theory, and my "advocacy" wanes... not because I believe that the content of the hypothesis is false, but because it becomes scientifically meaningless without a true ontology to which its probabilities can be anchored.

July 08, 2018

On The Nature and Resolution of Social Paradox

"E Pluribus Unum: Diversity and Community in the 21st Century" is the title of a five-year study by Robert Putnam of Harvard University. This study questions the wisdom of substituting "Strength in Diversity" for the more traditional and better-confirmed slogan "Strength in Unity". It seems that despite Putnam's own progressive beliefs – he is, after all, a member of the *Harvard* faculty – some 30,000 interviews have convinced him that ethnic and racial diversity do not conduce to the wellbeing of communities and their values. It thus seems appropriate to follow up on my last post with a more general discussion of race and society.

First, because any discussion of race can prompt accusations of "racism" in the present sociopolitical climate, let's get something out of the way up front: insofar as it occasions opprobrium, racism is a problematical concept. The problem is not, as some claim, that race has no biological basis and therefore does not really exist; of course it does. The problem is that once this fact is admitted, any proposed definition of "racism" inevitably comes down to "having or displaying a preference for one race over another." Basically, this makes the vast majority of people "racists", because if the truth be told, they prefer to be around others who are visibly like them and share the same cultural background, and they tend to act on those preferences. The opprobrium is thus distributed so evenly, at least in principle, that it becomes virtually meaningless.

In order to attach a particular onus to racism, one must introduce situational distinctions, setting a threshold of opprobrium applying equally to the members of all races. For example, one might define it as blameworthy to make pejorative reference to any person's race, or to belittle another person because of race (whether or not a factual basis exists); to materially harm or cause harm to another person because of race; or to systematically advance the members of one race over those of another even when the ability and competence of the former are not measurably superior to those of the latter on an individualized basis.

Failing to set such a quantitative threshold leads directly to a paradox involving a peculiar semantic property which might be called *(inculpative) circumstantial autology*: under certain more or less practical circumstances, merely *using* the term may constitute an *example* of the term, as when it is automatically deemed *racist* for someone to accuse a member of a different race of racism. (Examples: those who dare to mention statistics suggesting disproportionate Black or Latino responsibility for interracial crime are accused of being racially prejudiced against those groups; those who cite New Testament reportage implicating Jews in the death of Jesus are accused of antisemitism, while those who mention what they in turn perceive as racist elements in the Old Testament and the Jewish Talmud are again accused of antisemitism.) In other words, using the term to *cast* blame is to *incur* blame under its overly broad definition.

Because we presently lack a stable, racially symmetric situational definition of racism, this renders accusations of racism largely meaningless. On that basis, I hereby cast PC to the wind and dismiss in advance anyone who might object to this discussion on the grounds that it is "racist". In other words, if you think that rationally discussing racial distinctions is inherently racist and therefore reprehensible, then you are irrational, and I therefore invite you to go and read somebody else's blog.

Now that we have that little matter out of the way, not all racial paradoxes are merely logical. Some appear in the form of social or economic policies with profound effects on society. Such is the **paradox of integrated diversity**, the widely-promulgated assumption that ethnic integration and cultural diversity (intra-national multiculturalism) are compatible objectives. In fact, these two objectives are as diametrically opposed as two objectives can be. The first is calculated to have a homogenizing effect on society, while the latter is explicitly heterogenous and therefore opposed to homogeneity. Integration reduces diversity in order to effect social cohesion, while diversity limits social cohesion by introducing cultural and linguistic boundaries within an otherwise coherent society. Evidently, our well-paid guardians of democracy and their academic apologists have become seriously confused about this, and just as evidently, they have succeeded in spreading their confusion to countless others.

Ethnic and racial identity are ingrained psychological characteristics of human populations. There are a number of obvious reasons: likeness and familiarity breed feelings of security, while unfamiliarity can breed fear and distrust; people of similar type tend to share common backgrounds, experiences, and attitudes, and thus find it easier to socialize; it is easier to empathically put oneself in the shoes of another when their shoes resemble one's own; reproduction is attractive primarily as a means of perpetuating or advancing one's own type (as implied by the very structure of the word), and so on. Thus, people tend to favor and advance those who are like them, which can of course result in various forms of discrimination against those who are *not* like them. It is one thing to say that one's "type" is whatever one wishes it to be, up to the biological limits of reproductive compatibility; it is quite another to say that certain types, or plainly distinguishable subtypes of a type, do not exist, or that the statistical patterns of behavior associated with various types are equally conducive to social harmony, maintenance, and progress. Whereas the first statement is about subjective identity, the other two statements implicitly purport to objective social and scientific validity. Unfortunately, they are not supported by scientific or historical evidence.

Before continuing, let me briefly consider how one might try to make this paradox disappear. "Wait," one might say. "There is *no* paradox of integrated diversity. We're merely talking about two different contexts and scales of social interaction. Schools and neighborhoods and workplaces should be integrated, forcibly if necessary. But everywhere else, diversity should prevail! The two principles are perfectly counterpoised, effecting an equilibrium between homogeneity and heterogeneity."

In other words:

"We should everywhere integrate people of different ethnic and cultural backgrounds who are already in the U.S., and meanwhile gather even more minority people from other countries in order to continue the process. That we are forcing ethnically distinct people to integrate in schools and neighborhoods and workplaces in no way reduces their diversity. Each can still cling to his or her own special background and customs at home and on the street, and if any friction results from this, we can simply adjust policy to make everything run smoothly. You know ... hate speech laws ... gun control laws ... and so on, until every Constitutional Amendment in the Bill of Rights has been effectively neutralized!"

Firstly, such attempts are deficient as regards proper allowance for human nature, not to mention the public consensus that is legally and ethically required for the "adjustment of policy". Ethnic and cultural differences are reflected in thought, speech, and behavior, not all forms of which are mutually consistent. Although many people now labor under the assumption that belief and custom can be completely separated from practical matters, this is simply not the case. For one thing, much of the social process deals with people themselves, and human behavior is strongly affected by mental state, including belief; thus, conflicts of belief, often corresponding to differences of culture, tend to result in behavioral conflicts.

Secondly, nobody has any business engaging in any form of social engineering without considering all of its likely outcomes in every possible relevant context, including the social, economic, educational, political, psychological, and biological or genetic contexts. (Although modern social engineers seem to be working under the premise that genetic outcomes can be safely ignored, this requires proof, particularly in light of voluminous evidence to the contrary. Until then, it remains an empty and potentially dangerous conjecture.)

Thirdly, nobody has any business attempting to neutralize the Bill of Rights, or politically or demographically finagling a situation that unnecessarily threatens or impairs it.

July 12, 2018

***

Three mechanisms might conceivably interact to iron out the diversity-generated wrinkles in social commerce. Two are extrinsic to the human psyche: (1) financial or other incentives for behavior tending to ease, sustain, and promote the values of diversity and multiculturalism, and (2) threats or expectations of punishment for resistance to diversity and multiculturalism, criminalized through the passage of laws against "bias" or "hate" crimes (requiring that the population be rendered sufficiently defenseless that they are unable to resist any punishments imposed on them under such laws). The other, (3) a general and deeply ingrained form of ethics generating psychological and social pressure, is psychologically intrinsic.

Regarding 1, many businessmen and politicians who routinely deal with other corporate and government personnel in economically-driven transactions seem to labor under the belief that social transactions among diverse populations can go just as smoothly as the highly lucrative financial transactions in which they and their rich friends are involved. However, purely social and economic transactions are very different in kind, given that the wheels of social interaction may lack a crucial measure of economic grease. We have seen tons of economic grease applied to make social interactions between different groups go smoothly where they otherwise would not, and this tends to get expensive. Add yet more diversity-generated friction, and the expense rapidly becomes prohibitive.

Regarding 2, the name commonly attached to this remedy is "police state". Since the tactics of a police state are diametrically opposed to the US Constitution, its first order of business is naturally to throw the Constitution out the window. Given the inadequacy of 1, this is now the method of choice among those responsible for our social and immigration policy. It is 100% unacceptable, and in the long run, 100% unworkable. It represents a severe diminution of the quality of life of everyone subjected to it, especially those in the middle and lower tiers of society, who lack various means of control, exemption, and escape available to the wealthy.

Regarding 3, a new personal belief system has in fact been introduced and adopted for educational indoctrination. It is called "political correctness" or PC. The problem with PC is that the term "correctness" implies demonstrable validity, fairness, sustainability, and long-term utility, and no such demonstration exists for PC. PC is therefore a weak substitute for heartfelt belief, and as street-level experience amply suggests, it is meaningless when not equipped with teeth. In order to succeed, it must be accompanied by 1 and/or 2. But again, 1 is prohibitively expensive, and 2 is unconstitutional.

It thus seems not only that the paradox cannot be made to disappear by plain conceptual manipulation, but that it has no resolution short of either (A) limiting diversity, or (B) hoping for an eventual breakdown of cultural, racial, and ethnic barriers as people acclimate to integration, give up their segregationist impulses, intermarry, and produce a racially and culturally homogeneous breed of humanity. That is, one can hope that integration will lead to the complete breakdown of cultural and reproductive barriers as people psychologically acclimate to it, and that race and cultural descent will no longer threaten to interfere with social harmony.

But on reflection, alternative B appears to be seriously flawed. First, some racial subgroups have jealously guarded their cultural and genetic identities by establishing self-preservational ideologies which assert their special status, and indoctrinated or coerced their members to comply with associated exclusory traditions (at least one that springs readily to mind has been doing so for many centuries and shows no sign of relenting). As long as one group persists in doing this, it is neither fair nor practical to force miscegenation on the rest. Secondly – and not entirely unrelatedly – this amounts to a global form of "genetic engineering" which is backed by neither hard science nor common experience. The races of humanity evolved by adapting to certain conditions in their environments – the story is not quite that simple, but this is close enough for present purposes – and these conditions were not identical. Furthermore, it seems that some of these evolutionary conditions may have been markedly dissimilar to those present in modern society (not necessarily in equal measure), and quite possibly dissimilar to those which would characterize any feasible social reformation. Because we do not yet know how deep this evolutionary conditioning might go, we do not know whether it admits of genetic correction by random miscegenation. Indeed, the current socio-political climate strongly discourages research on this question, and without an answer to it, no desirable outcome can be projected for coarse-grained socio-reproductive engineering

This leaves just one partial resolution: limiting diversity. For example, a strong case can be made for the adjustment of US immigration policy so that it no longer favors the disproportionate immigration of ethnic minorities, and even for abating the resultant social maladies by reversing present immigration ratios. This is hardly unreasonable; it is neither the legal nor the ethical responsibility of the United States to accommodate the overflow from other nations with population problems or provide an easy route of escape for those subject to oppression in their countries of origin, particularly those who failed to oppose their oppressors when it might have done some good. Instead, US immigration ratios should be exclusively determined by what is good for the current citizens of the U.S., that is, by their right not to be displaced from their jobs and their neighborhoods by people who oppose their interests and identify with other cultures. Those features of post-1965 immigration policy tending to promote domestic multiculturalism, e.g. "anchor babies" and unlimited family reunification visas, originated with neither the Constitution nor the advice of the Founding Fathers; on the other hand, by driving the spread of racial conflict throughout America, they have promoted racism under any reasonable definition of the term. Their rational elimination would be the natural first step toward limiting their disruptive effects on American society.

One way or another, the status quo is insupportable. Without integration, ethnic diversification has the inevitable effect of reducing available living area and other resources belonging to current American citizens and their descendants. It also fractures the political process, the efficiency of which requires that voters recognize their common interest as US citizens rather than merely trying to advance their own ethnic groups. While it is true that the American political system is built to accommodate differences of interest and location among its participants, past and future attempts to introduce ethnic distinctions to the legal metric are in direct violation of Constitutional principles (as they apply to current American citizens rather than would-be immigrants), and these principles defy reinterpretation by revisionists in various branches of government. In America, it remains ordinary citizens, and not lawyers, politicians, academics or corporate elites, who legitimately hold all of the power (existing illegitimate concentrations of power notwithstanding). All violations of this principle, especially those committed by rogue officials dishonoring their oaths of office, are legally null and void.

Perhaps the present situation can best be characterized as follows. As we all know, people take time to grow accustomed to new situations and new directions, and the amount of time required is no doubt proportional to the degree to which the changes seem to run against the grain of human nature and human evolutionary history. Unfortunately, various multiculturalist social engineers, bent for their own reasons on diversifying and/or homogenizing the human species and desiring instant gratification in the form of overnight results, are trying to accelerate certain difficult social transformations far beyond their natural rates by means of coercion, indoctrination, and the introduction of unconstitutional laws, court decisions, and executive orders. Accordingly, they have introduced to their subjects the ethical and philosophical equivalent of a new PC pseudoreligion, the central commandment of which paradoxically mandates *integrative* racial, ethnic and cultural *diversification*. Unfortunately, this pseudoreligion is devoid of logical, scientific, historical, ethical, or Constitutional support, and is therefore too weak and unconvincing to do the grease job required of it.

Thus far, the efforts of these social engineers have had markedly divisive and disruptive effects on society – so much so, in fact, that they are regularly accused of being meretricious front men for corporate globalists who pay them off in various forms of private gain (money, power, etc.). Such appearances are unseemly enough on their face; when attached to government officials sworn to protect the interests of current American citizens and their descendants, they are nothing less than criminal. If these officials or their academic or corporate associates possess secret knowledge, an inside view, that justifies their racist policies but is not available to the rest of us – even those who arguably have more intelligence and insight than the officials themselves – then they need merely share with us their secrets and display the hard evidence qualifying those secrets as "knowledge". Otherwise, we can only ask what we might do, aside from ridding ourselves of these people and their secret agendas and unsubstantiated opinions and unconstitutional patterns of behavior, in order to solve our current problems and ease whatever social transitions may lie ahead for us as a natural function of legitimate human progress.

I have my own approach to solving this problem. It involves the replacement of the prevailing PC pseudoreligion with something of a more logical, and therefore palatable, nature ... a distributive form of ethics representing our best chance to rationally walk, rather than blindly and fearfully crawl, into the future. I've already introduced the core of it, and am about to resume where I left off. But as regards the current state of affairs, I merely digress.

July 16, 2018

***

It is probably safe to say that all racial groups, given the opportunity, tend to blame other groups for their problems. In this, their behavior mirrors that of individual people, for most of whom taking credit and trading blame are the natural things to do.

A number of examples come readily to mind: the Nazis of pre-war Germany blamed the Jews for their dire economic straits, the degeneration of their culture, and their passive exclusion from privileged and influential professions (as attested by numerous historians including Sir Arthur Bryant, Sarah Gordon, and others); modern Jews, lamenting the abuses of the Holocaust, continue to blame "antisemites" for "persecuting" them even as they collectively enjoy disproportionate wealth and power in their host societies; African-Americans blame a small handful of American agriculturalists, and by extension the entire white population of the U.S., for slavery (which had already been a thriving trade for thousands of years before being exported to America, and was not even welcomed here by the fiercely independent American colonialists), demanding apologies and reparations and special dispensations despite enjoying greater freedom and prosperity than native Africans as a direct result of that erstwhile injustice; those of Mexican extraction dubiously claim that the US "stole" the American Southwest from Mexico or engage in transparent demographic warfare against the U.S. majority, often resisting any move toward cultural and linguistic assimilation, and so on. Further enumeration is unnecessary. Although many members of each of these groups reject(ed) the relatively extreme views of their more exigent fellow members, those views weigh no less heavily on society at large. No group is exempt.

In most groups, racial or ethnic cohesion seems to come from a straightforward relationship between personal psychological factors and the current state of their immediate social and economic environments (including not just what they see and hear around them, but what they are told that they should be seeing and hearing by the media and the education system). In others, it has been elevated to the level of explicit ideology and even religious belief. Because trans-generational ideology drives the calculation of long-term ethnic strategies and furnishes them with deceptive short-term justifications without which they would surely be resisted, the motives of those belonging to such groups, and especially those claiming to speak for them, must be carefully scrutinized for ulterior (racist) motives even when they profess compassion. This is especially necessary when their complaints and pleadings appear socially irrational, are accompanied by passionately uttered but empty idealizations, and inaugurate or perpetuate questionable social experiments underwritten by little or no hard science.

Quite simply, that is not the true nature of compassion. Compassion dictates that victims of real abuse – abuse which is clearly not of their own making as individuals, or exceeds their actual measure of responsibility – be offered some form of relief compatible with the global welfare. Nothing which fails this test can be justified in the name of "compassion"; pushed to the limit, selective compassion is synonymous with a devious mixture of self-interest and deceit. This applies even when those who practice or profess selective compassion are not consciously aware of selfish inner motivations, have deceived themselves or been duped by others into believing that their motives are selfless, or all of the above.

If, for example, in tens of thousands of rapes or other violent crimes committed without provocation by members of one racial or ethnic group against members of another, virtually all of the assailants were found to belong to one group and virtually all of the victims to the other, then true compassion would weigh more urgently on the side of the group being victimized. If a "compassionate" policy of encouraging disproportionate immigration by some ethnic minority were to harm or threaten the long-term survival of the majority, or to advance the interests of a divisive group or coalition by diluting or fracturing its potential opposition while corrupting or destroying the cohesion and security of society at large, then compassion would favor society and its majority rather than the divisive group or their minority beneficiaries. If an exigent ethnic or religious minority in country X were to establish its own racially exclusive homeland Y, and having assured themselves of a safe haven in any event, to systematically exploit the open political system of country X in such a way as to promote or facilitate the disempowerment or dilution or extinction of its traditional majority, then compassion would again favor the majority (whose non-exclusionary political system would seem to imply their moral non-inferiority to country Y). In short, "compassion" without true symmetry, and without a realistic account of human nature and all of its behavioral correlates, is not compassion at all. In too many cases, it is on some level a calculated attempt by one group or coalition to take undeserved advantage of the members of other groups.

Aside from a few notable and rather pessimistic exceptions, the integration-plus-multiculturalism-and-ethnic-diversity paradox was neither considered nor resolved by those who drafted the U.S. Constitution. However, this is not a license to interpret Constitutional principles and objectives as though the paradox does not exist, or as though the cohesion and long-term survival and welfare of the current citizens of the United States and their descendants does not take absolute priority over the paradox and its means of resolution. Obviously, the objectives of the Founding Fathers, including the preservation of individual rights that can be secured only through national harmony and cohesion, take precedence over the opinions and preferences of social experimenters promulgating contentious and unscientific views regarding the nature and achievement of social justice. Unfortunately, the politicians, lawyers, businesspeople, and academicians who dictate our social policy seem totally unaware of this fact. This means one of two things: they are mentally vapid, or they are intentionally and sometimes treasonously complicit in the destruction of American society.

At one time, the former explanation seemed more likely to me. But as I and many others now see it, the latter explanation has become the more cogent.

July 20, 2018

***

Here's another paradox currently being passed off as rational social policy: "Race can be used as a criterion to effect social equality without denying the rights of any individual as a result!"

This has been a paradox from the moment it was conceived, which means that it was conceived either by people who didn't understand logic, or by racists. It is a paradox, of course, because if statutory discrimination between racial collectives did not resolve to racial discrimination on the individual level, it would have no meaningful effect at all. For every person given a job or a promotion on the (partial) basis of race, somebody else is denied that job or promotion by reason of racial discrimination. This is well known to virtually everyone who has taken a civil service exam; although scores on such tests are supposed to provide an unbiased measure of individual competence, they are frequently overridden by racial and other quotas. The operative justification seems to go something like this:

"Having already mangled the traditional American demographic by unconstitutional enforcement of the inexplicable equation 'diversity + multiculturalism = social justice,' we now need affirmative action programs to prevent a general breakdown of society ... even if their direct result is *injustice* against well-qualified individuals who lack the proper complexion to qualify for such programs!"

Clearly, anyone who fails to see and admit the sheer inconsistency of this position – anyone who, out of stupidity or cynicism, perpetuates this paradox despite its glaring self-evidence – cannot be trusted to make political decisions or run social experiments on anyone else's behalf, let alone on behalf of society as a whole.

Like many others, I have always found diversity and integration to be attractive in an idealistic, utopian way. But since I never recognized them as mutually compatible on any given social scale, I never presumed that they could be made to coincide in any socially nondestructive manner. The natural end result of directly combining them will always be either conflict or homogeneity; either diversity or integration must win in the end, and the contest between them will be neither pretty nor humane. Furthermore, because equality is by nature symmetric, it always struck me as a mistake to pay out rewards in its name without levying its taxes in equal measure and distribution ... a mistake to pretend that certain racial distinctions can be freely invoked by policies aimed at achieving social equality or ethnic balance, but never mentioned in the assignment of blame or responsibility for economic and political imbalances and social ills like violent crime.

Similarly, I find it illogical to introduce racially discriminatory policies into any system of law or morality designed to preserve individual rights by preventing unfair discrimination among individuals. Where equality can be achieved only by the symmetric application of objective measures of civic and vocational competence to individual human beings, those who undermine such measures by tirelessly accusing them of (e.g.) cultural bias themselves appear bent on promoting discrimination, and when they claim to be acting out of "compassion", to be lying for their own secret purposes. By making us despair of ever finding an objective measure of competence that is culture-free and thus conducive to equal opportunity, they evidently hope to make pure favoritism seem the lesser of evils. It is strictly on such protestations, and on our resulting sense of futility, that the current racially discriminatory system of quotas comes to rest.

Of course, some will inevitably object that there is a world of difference between benignly using race as a criterion to actively effect social equality, e.g. by imposing racial quota laws affecting such things as employment and education, and using race against the members of some particular group. With this objection, I somewhat agree. However, the world of difference to which it refers is just that between a favored group and the group that is penalized on its behalf. Otherwise, there is no real difference between the actions under comparison, which are perfectly complementary and equally racist in character. Targeting one racial group with affirmative action in a competitive situation necessarily amounts to targeting others with negative action; this is a simple matter of

2-valued logic, and there is nothing ambiguous about it. As every attentive student of human nature is well aware, such logical inversions are seldom entirely accidental.

Racism is a true mainstay of human nature, and sadly, it will not be going away tomorrow. Accordingly, it must be contained. Formerly, it was contained by national boundaries which mutually segregated internally homogeneous populations which were, for the most part, quite content to live apart from one another; unfortunately, thanks to the subversive social experimentation of the last century, this is no longer the case (at least in Europe and North America). The experimenters, confronted with a dangerous and unstable situation, have now succeeded in exacerbating it to the point at which serious racial conflict is an everyday occurrence. Not only is it endlessly affirmed in the print and broadcast media – albeit in a strangely inverted fashion – but it now happens more than ever before in our schools and our streets ... so frequently, in fact, that except where it is somehow anomalous, as when the usual victims and the usual assailants manage to trade places, it is deemed unworthy of reportage.

I do not subscribe to racial prejudice in any form. Accordingly, I do not rule out *a priori* the possibility that in addition to more or less obvious physical differences between races, there may be significant psychological differences that are capable of showing up in statistical and individual patterns of behavior, for that would be every bit as prejudicial as assuming the opposite. After all, the potential for such differences is what diversity is all about. I appreciate prejudicial racism even less when it pours from the mouths of those who slyly call it "compassion" and "social justice". When it then pours into the minds of smug academics and power-hungry, power-drunk lawyers and politicians and out into our lives through the pens and voices of senators, congressmen and judges, it becomes intolerable. I suggest to everyone – white, black, red, yellow and brown – that we have already waited far too long to put an end to it. If we are truly equal, then we do not need the incessant patronization, machinations, threats, and bribes of yammering multiculturalist cheerleaders and overseers to get along with each other. We can do it on the individual level, the only level on which it can possibly work.

Until we succeed in understanding this, forced ethnic diversification is among the worst and most destructive social agendas that we could possibly allow to be crammed down our throats, especially by elite political demagogues who aggressively practice exclusion in every aspect of their own private and professional lives. After all, we, and not they, are the ones who will continue to pay its rising social and economic costs.

July 24, 2018

Science and Intelligent Design

Suppose you want to know the difference between baseball and cricket. Do you ask a baseball player, or a cricket player? Either way, you probably won't get exactly the right answer. Most baseball players have never played cricket, and most cricket players have never played baseball. So whom do you ask?

Perhaps an avid spectator of both sports, with extensive exposure to the rules of both games.

Now suppose that you want to know the difference between science and theology. Do you ask a scientist or a theologian? The situation is much like comparing baseball and cricket.

Most scientists know little about theology, and few theologians are experts on science. So whom should you ask? Your best bet is probably a philosopher with intensive training in the "rules" of both "games".

"Wait a minute," you might say. "The difference is obvious even to a non-expert. Scientists study nature, while theologians study God!" But this is merely an invitation to rephrase the question as follows: "What is the difference between God and nature?" This is not a question on which the average scientist, or for that matter the average theologian, has much technical insight.

How about the difference between science and nonscience? One might suppose that a scientist would be the best one to ask, especially when that seems to be the popular consensus. But if so, one might have another think coming. This is just a generalization of the question about the difference between science and theology (insofar as theology is considered to be a nonscientific discipline). So there is still no reason to expect that a scientist can properly answer it. Non-science is simply not a part of science, and say what you will, scientists are not trained to recognize it in every case. Indeed, they can't be, because there is no reliable method for doing so.

Of course, there are those who will argue that in learning what science is, one automatically learns what it is not. But this is merely a simplistic idealization that does not hold true in the real world. That current scientific methodology is the accepted means of studying nature does not mean that nature can't be studied in other ways that might reveal more of it, perhaps much more of it, than is now visible. We can't simply pronounce these possible hidden parts of nature "non-science" because they may not be accessible by current methodology. It follows that we can't always recognize science either. At best, we can only recognize straightforward applications of the methodology we happen to possess at any given time. Unfortunately, this does not suffice in the long run to characterize either science or nature.

Why, then, does nearly every Intelligent Design critic in the world insist that only a scientist is qualified to draw the line between science and nonscience, and the natural and supernatural, in the context of biological change? The derision and outrage exhibited by these people whenever any non-scientist comments on the evolution-ID controversy is truly something to behold. The typical intensity of their reaction is second only to its foolishness. (I happen to know this because I've commented on the subject myself, and I didn't immediately whip out a biology degree to satisfy the critics. Trust me, it wouldn't have helped even if I had.)

Now let's talk about judges and lawyers, who want us all to believe that they are smart enough to figure out who can best answer such questions. In a much cited case, in Dover, Pennsylvania (Kitzmiller v. Dover), a court decided that ID is not science, but theology in disguise, and must therefore be excluded from science classrooms in public schools. It seems to have reasoned as follows: "ID is not, and never will be, science. Therefore, it must be theology, as it is accused of being." The court relied on the word of several people calling themselves "scientists" and one person calling herself a "philosopher", meanwhile ignoring the testimony of other scientists and philosophers who were telling quite a different story.

What are the correct scientific and philosophical positions on these questions? A true scientist, realizing that he is intellectually ill-equipped by training to draw any sharp lines between science and nonscience, would admit his ignorance. A true philosopher would also refrain from drawing any sharp lines of the kind that were drawn in Dover; he/she either knows something of both science and theology and thus understands that the line between them is gray and blurry, or simply knows better than to shoot off his/her mouth about a distinction that has never been clearly and permanently made at any point in the history of science.

One is therefore left to conclude that the scientists who drew such a line between science and ID were either being dishonest, or know so little that they have no idea of the extent of their ignorance. One is forced to draw similar conclusions regarding the philosophers who joined them, although their level of culpability is somewhat higher. Even if their training was deficient regarding the general nature of science or theology or both, they still should have known that such issues are too difficult to be decided by superficial argumentation of the kind they provided in Dover.

And the lawyers? Well, everyone already knows about lawyers.

The court was wrong on purely logical and meta-scientific grounds. Unfortunately, assuming that the judge did not allow his own philosophical prejudice to contaminate his judicial reasoning, he was hoodwinked by people calling themselves scientists. A scientist is someone who confines himself to questions about nature and is capable of honesty about its gray and ever-shifting boundaries. The Dover trial's anti-ID expert witnesses exceeded this description without duly admitting it, and were therefore not speaking in a scientific capacity. They were instead speaking in a metaphysical capacity, but without having the metaphysical training or knowledge to do so authoritatively.

In all fairness, they probably didn't know it. As empirical scientists and uncritical cheerleaders thereof, they were never trained in the finer points of the problematical distinction between science and metaphysics. But it is very telling that nobody on the evolution side of the debate, including the lawyers and the judge, seemed to know it either.

[date unknown]

Semantic Entropy and Its Legal Consequences

In the 19th century (and perhaps long before), it was noticed that heat never seems to pass from a cooler body or region of space to a warmer one; it flows in only one direction, from warmer to cooler. Thus, it is only a matter of time before any concentration of heat is lost to its colder surroundings, and the only way to restore it is by adding energy from a source outside the system. If we add an outside source of heat and extend the system to include it, we again find that the whole system is cooling; no matter how we define the boundaries of the system, its energy flows from higher concentrations to lower ones until it is more or less evenly distributed. More energetic processes lose energy to less energetic ones, and onward and outward until there is finally no larger process to yield or accept further energy... that is, until we reach the ultimate closed system, which (as far as we are concerned) is the only true closed system in existence: the universe.

As most scientists and engineers see it, the process is quite inexorable and quite annoying, because as energy relentlessly dissipates, it becomes unavailable to do work.

The amount of thermal energy unavailable to do work is called entropy. According to the second law of thermo-dynamics, entropy never decreases in a closed system; if one wants to lower the entropy in one's own little part of the universe, one needs to bring in more energy from outside the system. Outside doesn't necessarily mean from afar; the new source of energy could easily be right in the geometric center of the system, but not yet involved in any local processes. But eventually, all of the easy, nearby sources of energy have been used up – the system has been inwardly extended to contain all of them – and one must go far afield to replace them. Unfortunately, that can get expensive, which suggests that in physics as well as economics, there is no free lunch.

As thermodynamic entropy rises, the motions and locations of molecules become increasing messy and random; productive concentrations of high and low energy molecules have spread out and intermingled, and the majority of molecules can no longer be expected to flow in any particular direction. (That's why the energy has become "unavailable to do work".) Obviously, randomness is related to another messy concept, disorderliness. Thus, entropy came to be used not only in reference to the random dissipation of heat, but for describing any loss of order (increase of disorder) resulting from the laws of probability... for example, order that is lost in the course of some process because disorderly states far outnumber orderly ones, and can thus arise in many more ways. When scientists want to talk about this kind of situation, they call it logical entropy in order to distinguish it from its thermodynamic counterpart.

The celebrated physicist Richard Feynman described logical entropy like this: "Suppose we divide [a] space into little volume elements. If we have black and white molecules, how many ways could we distribute them among the volume elements so that white is on one side and black is on the other? On the other hand, how many ways could we distribute them with no restriction on which goes where? Clearly, there are many more ways to arrange them in the latter case. We measure 'disorder' by the number of ways that the insides can be arranged, so that from the outside it looks the same. The logarithm of that number of ways is the entropy. The number of ways in the separated case is less, so the entropy is less, or the 'disorder' is less."

Note that Feynman has specified neither the size of the space nor the number of its volume elements, and there are many ways that this might have been done. (Such ambiguities are less critical in thermodynamics, where entropy is governed by measurable criteria like temperature.) Thus, the concept of disorderliness can be a bit tricky. To define it, one must specify exactly what is meant by "order". As it happens, we live in a universe which supports many different kinds of order, some of them mutually antithetical. For example, compare the traditional political system of America, wherein citizens are the ultimate source of political order, and the "new world order" under which we are now coming to live, wherein order is almost exclusively determined by the self-interest of people with far more than their fair share of wealth and power, as least as gauged by anything approaching merit. Clearly, the latter is degenerate with respect to the former, being not much different from medieval feudalism. Yet the proponents of feudalism would contend that it is far more orderly than most modern political systems, especially that of America.

What, then, is the "most orderly" political order? Where the proper purpose of any political order is to benefit society, it is that political order which maximizes social utility. Social utility is maximized when the ability and industry of the populace finds maximal constructive expression, which occurs when economic opportunity is widely distributed, creative freedom is unleashed, and noise from other sources – e.g., rampant elitism and economic strangulation of the middle class, and social unrest associated with such things as immigration policy – is minimized so that social stability is maintained and knowledge can be accumulated and productively applied without interruption, interference, or misdirection. On historical evidence, the greatest explosion of social utility is associated with the rise of the United States, and rational analysis reveals that the US Constitution is much of the reason. It was well-calculated to release US citizens from the bonds of serfdom which had previously held their abilities and economic power in check. Hence, the US Constitutional Republic is arguably the highest form of political order yet achieved.

Of course, the globalists look at things a bit differently. Having profited immensely from the unprecedented flow of wealth released by the American economy, they now find themselves duty-bound to address some of its most unfortunate side effects, e.g., a burst of world population growth which has resulted in far too many non-elites exhaling carbon dioxide, creating sewage, and clamoring for food and water and gasoline. (Of course, such problems are not without their bright side, having encouraged the elite to speculatively triple the prices of commodities like food and oil practically overnight and buy up American infrastructure as quickly as they can bribe local governments to sell it out from under the taxpayers who already own it.) And meanwhile, behold the heartrending political fragmentation of Europe and America as the elite jam various ethnic groups and ideological factions right up against each other in the sacred name of Diversity and Multiculturalism! Could it possibly be more obvious that in perilous times such as these, we need to return to the old order of things, an order in which everything is safely under elite control?

Thus, as the elite bustle self-importantly between their globalism conferences and secret meetings with lobbyists and politicians, they dream of a new world order, a modern sort of UN-mediated corporate feudalism augmented with what they have learned over the last couple of centuries regarding efficiency... i.e., the efficiency of wealth in exiting the unworthy clutches of ordinary people and finding its way into their pockets, and the efficiency with which it is then converted into absolute power over all that they survey.

Unfortunately, such an order is clearly degenerate with respect to that against which the elite have been working so hard. As tidy and secure as it may appear from the vaulted windows of their mansions and palaces – how much more "ordered" the map of ownership when all of the property is concentrated in a tiny number of hands, thus saving the trouble and ink of writing too many names on it! – it nonetheless represents a severe loss of human utility and a regression of civilization. For while the socioeconomic elite seem to be offering humanity their precious guidance in return for a mere permanent loss of economic symmetry, the value of their services depends on their having cornered the market on brainpower and social responsibility.

But of course, as many of us have already observed, anybody who expects to find some sort of world-class brain trust down at the local yacht club, the Ivy League alumni reunion, or CFR headquarters might as well be sniffing glue. One will find plenty of trusts, alright, but they won't be "brain trusts". They will be the kind of trust that relieves one of the need for any real brains or responsibility at all, leaving one free to pursue one's already ample self-interest with wild abandon.

July 29, 2018

***

Many kinds of logical entropy have been suggested... social entropy, economic entropy, even corporate entropy, all associated with loss of order. In principle, one can define a type of logical entropy for each context in which a system randomly loses order over time due to the "dissipation of information".

Depending on the relationships between contexts, it is sometimes possible to relate one kind of logical entropy to another. For example, where the political order is encoded as law, and law consists of language and comes into being by means of verbal argumentation, we can trace the increasing disorder of the political system to linguistic disorder involving the definitions of words, which in turn can be attributed to the endless, largely mindless, and frequently dishonest efforts of lawyers and legislators to twist legal and political language to their respective purposes. Let's use the phrase "semantic entropy" to describe linguistic disorder involving loss of information regarding word definitions. Then we may consider the possibility that legal and political entropy are functions of semantic entropy caused by the tortured verbal manipulations and legalistic language games of lawyers and politicians vying for wealth and influence at public expense.

Semantic entropy can be easily understood in terms of Feynman's description of logical entropy. As readers may recall, Feynman described logical entropy as follows: "Suppose we divide [a] space into little volume elements. If we have black and white molecules, how many ways could we distribute them among the volume elements so that white is on one side and black is on the other? On the other hand, how many ways could we distribute them with no restriction on which goes where? Clearly, there are many more ways to arrange them in the latter case. We measure 'disorder' by the number of ways that the insides can be arranged, so that from the outside it looks the same. The logarithm of that number of ways is the entropy. The number of ways in the separated case is less, so the entropy is less, or the 'disorder' is less."

Consider a piece of writing, for example, the United States Constitution, the body of law on which the American political order is supposedly based. Imagine that the Constitution is a space like that described by Feynman, and that each distinct word which occurs in the Constitution is a volume element or "cell" in that space. Now imagine that each cell is bounded by the definition of the corresponding word. The Constitution is a cumulative path through that space, passing from cell to cell in series and gaining information in the process. If we lose semantic information regarding the definition of a word, then this tends to dissolve the boundary of its definitive cell, and also the cells of any words whose definitions incorporate it; cells can split, merge, disappear, and change location in a way that warps the entire semantic network comprising their definitional relationships with other cells.

Thus, if we lose critical information regarding a key word because some lawyer or politician has used it inconsistently and gotten a judge or a legislative body to enshrine the contradiction in a decision or piece of legislation that can be cited as a basis for further decisions and laws, we may eventually arrive at a situation in which the true path of Constitutional verbiage through semantic space can no longer be determined, at least by those who rely on legal mumbo-jumbo rather than common sense. As the definitions of its words are compromised by cunning misuse, the legal system loses information about what the Constitution really says, and because the Constitution is the ultimate basis of our legal and political order, this dooms us to legal and political entropy.

But legal argumentation is not confined to quoting the Constitution. It wanders all over the place, often involving blatant appeals to morality and emotion. Nor is it confined to the august deliberations of legal eagles privately preening each others' feathers in courtrooms and legislatures; it frequently spills out into the court of public opinion, usually through the mass media. Because this shapes public sentiment, something to which politicians have not yet become totally insensitive (especially when election time draws near), it can feed back into legal and political processes and insidiously amplify various unhealthy trends and tendencies. When public opinion succumbs to semantic entropy, and politicians then take the opportunity to freeze it in the form of law, legal and political entropy increase as a result.

Some entropic laws are passed on the strength of arguments in which the cells of meaning and definition have blurred and dissolved, letting words and concepts drift and merge in countless ways; others dispense with any pretense of gradualism and cut to the chase, directly contradicting the U.S. Constitution and thus failing all tests for legal, logical, and social correctness. As a result, we are now faced with not only the insurmountable chore of cleaning up all the semantic garbage strewn throughout the legal system and government policy by sophistical lawyer-pols whose only use for language is what it can do for them and their precious careers and bottom lines, but a hostile army of well-fitted myrmidons hired to enforce their escalating dirtbaggery at all costs... people who will gladly, sometimes gleefully, subject their fellow citizens to "pain compliance", drag them off to prison, or even blow their heads off, all the while expecting a pat on the head just for being "our" boys in blue and resting on such brilliant justifications as "It's the law! The lawyers told my chief to do it, and he told me. Hey, I need to get paid!"

Thus, we face two problems reminiscent of the second law of thermodynamics. One is akin to the dissipation of heat or order, in this case the inherent order and symmetry of the Constitution, due to the self-serving and socially irrational manipulations of lawyers, politicians, and those who buy and control them. The other involves the artificial closure of the corrupt, high-entropy legal and political system, which tends to exclude any source of order lying outside its privileged boundaries. While linguistic, social, and political forms of entropy involve human beings capable of intentionally reversing them, such "negentropic" influences tend to be opposed by means of repressive mechanisms that are themselves the products of entropy. The law has always been the tool of choice for those bent on domination; once entropy is frozen into law, its paid enforcers can be trusted to resist any attempt to restore the original order.

The problem is not, as some apparently believe, unbridled competition among too many people trying to grab the biggest possible slice of pie. The problem is that those who capture the biggest slices of pie, be they capitalists or communists or something in between, are permitted to erect legal, physical, and bureaucratic scaffolding which elevates their interests above those of the public. As semantic boundaries dissolve under the acid relentlessly sprayed on them from the mouths of legal and political prostitutes in the pay of real and aspiring socioeconomic elites, elite interests are ever more sharply divided from those of the people. Populistic order is traded for elite domination; as abstract moral and Constitutional boundaries weaken and fade, the boundaries protecting the elite from the non-elite obstruct human negentropy with increasing force.

August n/a, 2018

***

We have already examined the political usage of terms like amnesty, compromise, racism, diversity, integration, multiculturalism, and justice. Each has been subject to semantic entropy which in turn has contributed to political entropy. The sponsors of amnesty bills vitiate the definition of amnesty by claiming that their bills do not involve it; the same people use compromise in a way incompatible with any valid legal or moral context; racism was shown to be a "circumstantial autology" that describes many of those who use it to self-righteously accuse others of wrongdoing; diversity and multiculturalism were shown to conflict with integration, a concept with which they are frequently conflated (especially by people calling themselves "liberals", which designation constitutes yet another misleading instance of semantic entropy); and social justice was cited as a rationale for the adoption of unjust policies which discriminate against qualified individuals seeking (e.g.) employment or college admission.

In this light, let us consider the semantic entropy generated by lawyer-politicians as a function of their positions in our venerable two-party electoral system – a system which has itself been reduced by political entropy to a false and deceptive dichotomy – and how this semantic entropy might impact the formation, interpretation, and enforcement of the law.

Compassion is one of those words which is constantly being defined and redefined in such a way as to rationalize whatever policy is advocated by a given political group. This leads to confusion and guilt among those who find themselves in opposition to allegedly "compassionate" policies and proposals.

For example, to a conservative, "compassion" means protecting you from destitution by providing jobs at low and often inadequate wages (unless you are a member of the elite, in which case you can get a cut of the real action). If you don't like those wages, then there are always plenty of other people willing to take them, even if they live in another country or have to be imported against immigration law. Disagree, and you're an ingrate.

To a liberal, on the other hand, "compassion" means taking money and opportunity (rights) away from the majority and giving them to the squeakiest wheels, including those who live in or are loyal to other countries or have violated immigration law, in the form of benefits, reparations, affirmative action programs, humanitarian aid, and so on. Argue, and you're a greedy racist bigot.

Although both kinds of "compassion" work together in a very effective pseudo-ethical squeeze play – and by the way, this kind of "pincer-compassion" works as well abroad as it does right here at home – they tend to diminish or destroy the tenuous quality of life of the majority. Thus, from a rational utilitarian point of view, they are disingenuously misnamed.

However, if the majority complains about being caught in the squeeze, then here come the rationalizations.

If we ignore calculated appeals to emotion, the essential conservative rationale goes something like this: "But if we gave everyone more money, the law of supply and demand would force us to raise prices in order to retain the lion's share of everything, and you'd still be right where you are!"

Whereas the essential liberal rationale goes something like this: "But if we don't take your money and opportunity and give it to these squeaky wheels, they will grow resentful, kill some of you, and burn the cities, and then we'll have to take all of your money in order to pay them off and rebuild! Then you'll really be sorry!"

Obviously, both of these rationales are flawed. The conservative rationale takes as an implicit axiom the right of the elite to have almost all of the wealth, while the liberal rationale amounts to extortion against people who can ill afford it and ignores valid crime statistics which show that members of the majority are already being disproportionately robbed, raped, and killed by members of various "underprivileged" groups. When extortionists are paid off and told that their demands are legitimate to boot, they just up their price and increase the pressure.

One of the problems, of course, is political cynicism. Politicians generally do what they want or are bribed and blackmailed to do, tell the public whatever sounds good, hope they buy it, and if not, cook up an unconstitutional (i.e., illegal) executive order or piece of legislation which lets them send in the police and/or the troops for a timely dose of repression. But another less-recognized problem is this: beyond the point at which their capacity for ethical analysis has been exhausted and their cognitive filtration mechanisms have reached high gear, danger-zone lawyer-politicians see no choice but to believe their own nonsense and carry on, damn the torpedoes. Lacking any coherent ethical framework in which to form cogent higher-level justifications, the best they can do is cite the alleged wisdom of some overrated dilettante as recorded in some esoteric treatise and throw in a catch phrase or two, e.g. "national security" or "executive privilege" or "social justice", for good measure.

Unfortunately, lawyers are not philosophers. They are more accurately characterized as verbal manipulators, and as we all know, the purpose of their manipulation is not so much to find the truth as to win the verdict. This is a crucial distinction; while in the former case one must logically justify one's verbalizations using principles which are themselves logical, in the latter, one can carry the day with nothing more than a torrent of cleverly-cited case law into which various bad assumptions, mistakes of inference or interpretation, and other signs of idiocy have found their way, exploiting all of the misconceptions and logical or procedural errors of every incompetent judge or legislator before whom a similar case has been tried or by whom a related law has been passed. This is what lawyers are trained to do in court, and when they go into politics, they naturally do more of the same.

The sad thing about this state of affairs is that the word-twisting lawyers, judges and politicians responsible for it have managed to all but destroy the very important concepts to which certain words refer, and thus made it impossible to justify rules and policies in terms of them. That this has happened to "compassion" is demonstrated by the fact that the U.S. government has recently fallen into the habitual use and justification of torture, or as they now prefer to call it, harsh interrogation techniques (when used on suspected "terrorists", yet another high-entropy classification) or pain compliance (when used against citizens). After all, pain compliance is compassionate because it obviates the need for bureaucratic morons and their uniformed flunkies to use "lethal force" against the people.

While it appears that Tasers and other pain-compliance devices can sometimes be lethal – that much is implied by medical and autopsy reports, and more generally by the sheer delicacy, complexity, and electrical sensitivity of various critical anatomical components – they can still be called "nonlethal" because they don't kill quite as often as bullets. (Nope, no semantic entropy there!) Therefore, it is now regarded as right and proper that if you don't do exactly what the authorities say exactly when they say it – e.g., if you don't immediately sign that traffic ticket or lie down in the dirt or stop having an epileptic seizure or snap out of your diabetic coma because you are commanded to do so by some panicked pipsqueak who was picked on by his schoolmates and now has a badge, a gun, and a chip on his shoulder – then they will show compassion toward you by torturing and just maybe killing you with "nonlethal pain-compliance" devices instead of simply blowing your head off.

One might be tempted to object that authorizing the police to use more force on the citizenry actually makes it easier to keep public order, and thus that mislabeling certain weapons as "nonlethal" and "compassionate" in order to encourage their employment by the police will lead to a decrease in social entropy. But this is illusory. The most profound kind of social and political order is found only in a system which is consistent with human psychology and thus nonreliant on repression; when such a system gives way to repression, this order is lost and replaced with a false "orderliness" which is at odds with human nature and thus bought at the expense of a greater tendency toward rebellion and social breakdown.

Those who would sincerely deny that America has entered a high-entropy stage of its existence exemplify an interesting phenomenon: when the term "order", which figures prominently in the definition of logical entropy, falls victim to semantic entropy, entropy can become semantically self-obscuring, seemingly emptied of content by its effects on its own definiens. This can allow political entropy to proceed nearly unnoticed as people continue to believe that everything is "in order"... that the political order with which they started was never interrupted. But the truth is more ominous: as the original order was transformed into the new degenerate order, the changes were masked by the cognitive correlates of semantic entropy. In other words, entropy has hidden its own deleterious effects.

As semantic entropy takes its toll on the language in terms of which a legal / political system is defined, so declines the power of weaker minds to discern the degeneration of the system, and so declines the system itself. To judge by the monotonicity of our downward sociopolitical trajectory, that's almost as certain as the dissipation of heat. This no doubt explains, at least in part, how we and those whom we elect to serve us have managed to wander almost completely away from our erstwhile adherence to sound Constitutional principles.

August 10, 2018

***

Are lawyers, politicians and bureaucrats the only sources of linguistic entropy? Of course not. Other effective agents include academics, media figures, and the corporate and financial elite, who have made their own special contributions to the meaning of words like compassion.

For example, after the banks decided to make creative subprime mortgages available to the underclass out of sheer compassion – everyone needs and deserves a home! – many of these new borrowers defaulted on their usurious loans, thus forcing the banks to regretfully seize their property. But sadly, the property has depreciated and the housing market has crashed, facing not just homebuyers but the banks themselves with appreciable losses. Therefore, driven by yet another powerful burst of compassion, the government promises to arrange a minor bail-out of the debtors who stand to lose their homes, which of course means that the banks will continue to receive their steep mortgage payments.

But what about those unfortunate banks which have already been critically wounded and face liquidity crises? If they go under, then the whole economy could collapse, and that would hurt everybody. Now, where's the compassion in that? Accordingly, the top bankers compassionately print more money in order to bail out the most troubled banks. After all, this is the (arguably illegal) "mandate" of the central bankers...to protect the banking system so that we don't have another Great Depression and the wave of misery that came with it. (Yes, economists now admit that the Fed actually caused the Great Depression instead of preventing it, but that was just one of those "little accidents".) Even if this causes inflation and thus spells doom for a large number of proles teetering on the brink of insolvency, the fact remains that the bankers hold a noose around the neck of the entire economy, and if the bankers go down, economic strangulation will quickly follow. So you see, compassion is really what it's about after all!

And what does all of this mean? It means that just as originally planned, the banks are allowed to triple-dip, keeping all of the property they have repossessed, continuing to extract usurious interest from "rescued" homeowners, and fastening additional taxation and inflation siphons directly to the taxpayers at full suction. Now, that's a quite a load of compassion indeed...but not, unfortunately, for the people who are most in need of it or most deserving of it. Rather, it is the bankers and politicians who have turned out to be the ultimate recipients of most of their own compassion. What a surprise! And what an innovative twist on compassion, which was originally supposed to describe the caring of one entity for another as opposed to oneself.

As long as the people continue to swallow paradoxes and false dichotomies transparently designed to polarize and paralyze them – as long as they allow themselves to be verbally and conceptually manipulated by danger-zone cynics with neither a shred of honesty nor a scrap of real compassion – they will continue to be ruled and robbed by deceivers pretending to care for them. Their lives will continue to be owned by a group of braying, self-dealing bumblers preaching forms of "compassion" that come down to one overarching principle: "We're in charge and you aren't, and we get to define words like compassion as we see fit!"

What can we do to counteract this tendency? The first step is cognitive. We must learn that when an ordinary politician or bureaucrat or judge or central banker opens his mouth and utters an ethically or emotionally charged word like "compassion", then even if we're too late to put our hands over our ears, we'd better hold our noses, because something surely stinks. The something that stinks is generally a paradox which boils down to "If A, then not-A," where A and not-A are contradictory definitions of one and the same word or concept... a behind-the-back mangling of language by verbal sophisticates who are certain, in their smug danger-zone way, that you, a well-indoctrinated but moronic prole, have lost track of what they are saying before getting from the first A to the second A. Understand that merely by listening, you have been huckstered and shilled into a carnival barker's verbal shell game, a game designed to bamboozle you and play you for the pathetic sucker the carnival barker thinks you are. In short, you must learn to recognize even the sincerest utterances of politicians, bureaucrats, academics, and media spokesmen as the Orwellian doublespeak of those who dearly aspire to rule over an Orwellian world.

The next step is to identify the ways in which we and our countrymen have already been bamboozled. An examination of US law, beginning with the Constitution and extending to the present, reveals more self-contradictory semantic circularities than Kellogg has corn flakes; in some areas, logical satisfiability is not even a detectable criterion. (Take, for example, the last few decades of presidential directives designed to make it possible for the President to instantly transform himself into an absolute dictator for the sake of national security ... i.e., to destroy freedom to protect freedom. This is a paradoxical semantic circularity straight out of 1984.) Indeed, most of those who have recently been stretching, smearing and tying knots in US law do not even understand what "logical satisfiability" means in the mathematical sense; for them, the law is not about the logic of freedom and justice and equal opportunity, but only about the "logic" of winning the game ... the game against us.

The last step may well be the hardest: to muster the courage to stand up for your dignity and your rights, as the latter concept has been carefully described in previous installments of this weblog, and to actively support political candidates willing and able to stand up for the public interest without regard to how well they are known, how generously they are funded, or how their chances are rated by the media. (**Hint**: if a political candidate tells you anything but the whole unvarnished truth as you know that he or she understands it, then he or she is "high entropy" and therefore not a fit guardian of the public weal.)

Unfortunately, this not only eliminates most of the current political field, but virtually all of those who are well-known, generously funded, and highly rated by the media.)

Although the relationship between semantic and political entropy can give rise to extremely serious consequences, the situation is occasionally lightened by a spark of levity. For even as lawyers and bureaucrats twist language to a degree rendering even the most basic concepts unrecognizable, official government agencies police the meanings of certain sinister words with the insane diligence of obsessive-compulsive lexicologists on a caffeine jag.

August n/a, 2018

The Autophagy of Man: Evolution in Action

Although I don't embrace some of its biases, I find the alternative media, consisting of many independent web-based sources of news and opinion, a very welcome addition to the informational landscape. Despite its noise factor, it provides access to information that is all but unavailable through the major news outlets. For example, if one were to rely on the corporate media alone, one might know almost nothing about some of the great strides we've been making in the War on Terrorism since the PATRIOT Act.

For example, we might not know that since 9/11, the federal government has assumed the authority to institute martial law, arrest people for being "suspicious", and detain them without legal recourse at its sole discretion. We might not be aware that as part of a plan directed at "removable aliens" and "potential terrorists", the government has awarded a series of single-bid contracts to Halliburton subsidiary KBR to build dozens of secretly-located U.S. detention / concentration camps, possibly with the combined capacity to hold several million of us. We might not understand that the Military Commissions Act of 2006 allows for the indefinite imprisonment and secret trial of anyone supporting a charity that turns up on a list of "terrorist" organizations or criticizes government policy. We might not even recognize that section 1042 of the 2007 National Defense Authorization Act (NDAA), entitled "Use of the Armed Forces in Major Public Emergencies", lets the president unilaterally declare martial law and domestically deploy the military, even against the American people themselves, in response to any natural disaster, terrorist attack, outbreak of disease, or breakdown of public order.

Worse yet, we might still be in the dark about the 2007 National Security Presidential Directive 51 (NSPD-51), which ensures continuity of government in the event of a catastrophic emergency, authorizing the president to do literally anything he likes in order to stay in charge – cancel elections, suspend the Constitution, torture and execute citizens, start a global thermonuclear war, you name it. We might be ignorant of the fact that a California congresswoman, believing that the country needs a powerful commission to investigate and interdict potential "homegrown" terrorists who grew up in the U.S.A. and look just like you and me, introduced a bill charmingly named the Violent Radicalization and Homegrown Terrorism Prevention Act of 2007 (HR1955) which was almost unanimously passed by the U.S. House of Representatives (404-6). Thus, we might not be aware of how "safe" we (almost) are from animal rights terrorists, eco-terrorists, right-to-life terrorists, anti-illegal immigration terrorists, pro-Second Amendment terrorists, pro-U.S. Constitution terrorists, and even the dastardly peace-marching terrorists who force the taxpayers to spend so much money on pepper spray and Taser cartridges. Why, we might not even have the comfort of knowing that the National Counterterrorism Center is keeping very close tabs on about 775,000 "terror suspects" and gathering an additional 20,000 suspicious-sounding names each month on our behalf! And so on, and so forth.

Verily, we suffer an embarrassment of riches. We even have the loving attention of hundreds of thousands of heavily-armed security personnel – the police, the military, even the well-paid mercenaries of Blackwater, Inc. – who are blindly conditioned to carry their orders right down the chain of command whatever those orders may be, even if it means blowing off the heads of innocent civilians at the whim of the socioeconomic elite and their agents.

But enough sarcasm. When the House passed HR1955, it had just learned that the public approval rating of Congress had plummeted to an all-time low. Yet, instead of immediately changing its evil ways, it voted overwhelmingly to raise the level of threat, surveillance, and coercion against the very citizens who had just expressed their righteous disapproval of its cavalier treatment of their wishes and Constitutional rights. (Again, this bill focuses not on external threats of terrorism, but on law-abiding Americans who happen to disagree with their government, that being pretty much all it takes to qualify as a "potential terrorist" these days.) Instead of reacting to overwhelming public disapproval with a renewed desire to shape up, fly right, and do the will of the American people, Congress instead reacted with passive aggression deceptively cloaked under the usual toothy smiles, glib rhetoric, Homeland Security boilerplate, and greasy, sooty clouds of lawyerese requiring translation by media interpreters who often choose or are told to lie, especially by omission. That is, the members of Congress reacted aggressively and ruthlessly toward the public while hiding beneath the political equivalent of what biologists call "protective coloration", blending into their political surroundings while squirting black clouds of ink like frightened squid.

Congress would no doubt claim that it is acting to protect the public, and has in fact made that claim time and time again. But this claim is not entirely honest. For while there is no question that security is one of its highest priorities, politicians as a breed are notoriously more concerned with their own security than with that of the public. (This is because they are driven by the lust for power, which tends to diminish empathy by creating a psychological gulf between those who have or seek power and those over whom it is wielded.) After all, the public threatens their security all the time. In principle if not always in practice, the people can vote them out of office and deprive them of their hard-earned political careers, otherwise known as "threatening their job security". Even worse, should the people get the idea that they are being sold down the river – which, if the truth be told, happens quite frequently on "the Hill", where manure rolls in only one direction – then they might even do worse, e.g., call for the indictment or impeachment of their betrayers. For no matter their party affiliations, those atop the Hill are of the elite or owned by them, whereas most of their constituents are not. Because this amounts to an adversarial relationship, political elites tend to regard ordinary citizens as potentially useful at the polls, but elsewhere a dangerous nuisance.

What, then, are we to make of the resounding congressional approval of HB 1955? To the trained ear, the most natural terms in which to describe such behavior – terms like "aggression", "ruthlessness", and "deceptive coloration" – are highly suggestive of predatory behavior, a well-known biological phenomenon frequently studied in laboratories and the field. Indeed, some members of the alternative media already refer to those who occupy positions of great political and economic power as "predators" comprising a "predatory ruling class" or "predatory elite". This may look like nothing but a pejorative metaphor, but it is a strong one. Predators dwell apart from their prey, coming into contact with them only to entrap and feed on them; they act without compassion or compunction toward their prey, they employ stealth and camouflage against their prey, they exploit the group behavior of their prey in order to cull the herd of its most vulnerable members, and so on. In other words, the predator-prey metaphor is uncomfortably faithful to current political observation save for one fact: human social predators and their prey are of the same biological species.

Their occasional pretensions of spirituality notwithstanding, most of the elite are of an essentially materialistic philosophical persuasion, subscribe in some measure to social Darwinism, and are thus locked into a Darwinian viewpoint. Common descent, an important axiom of the Modern (neo-Darwinian) Synthesis, says that speciation occurs by divergence of lineage. Thus, if one subscribes to neo-Darwinism and takes the above observations at all seriously, one must consider whether evolution may be trying to pay man another visit, molding from human genetic stock the template for a new species of anthropoid social predator. Even though this may sound a little far-fetched, let's think about it a bit.

We know that the elite exhibit distinct patterns of emotion, thought, and behavior, leaning toward careful long-range planning, secrecy, and non-empathy with their social and economic "inferiors". We know that the elite feel more comfortable with each other and prefer to dwell apart from those lower on the socioeconomic ladder, a preference which could easily be interpreted as a desire to evolve in a separate direction. We know that the elite practice a closed form of eugenics basically consistent with the futuristic vision of self-guided human evolution, according to which human beings can adapt in whatever direction they choose. Evidently, the elite would prefer to guide evolution in a direction tending to preserve and enhance their own advantages while maintaining their segregation from the rest of humanity.

But we also know that for the elite, the real evolutionary variable is the organism-environment coupling, especially the coupling between themselves and the socioeconomic structure of society. We may thus surmise that as the elite grow more powerful, they will increasingly tend to mold the socioeconomic environment around themselves rather than vice versa. Rather than continuing to adapt to the socio-economic environment, they will increasingly adapt the environment to suit their own natures, enhance their control, and conduce to their own comfort and satisfaction. Where non-elites are components of that environment, its modification will naturally extend to their modification as well ... their conversion to a form better-adapted to elite control and exploitation. Quite simply, the stronger the predatory "subspecies" becomes, the less it needs to adapt, the more pressure it can exert on the prey "subspecies", and the better it can exploit available technology to control not just its own evolution, but that of its prey.

In other words, the modern strain of social Darwinism – the social analogue of neo-Darwinism – predicts that the elite will ultimately arrogate the genetic destiny of humankind, pushing the rest of humanity onto a servile sidetrack from which they can provide their elite predators with small favors like labor, entertainment, cannon fodder, blood, bodily organs, and so on. Once this has been achieved, the evolutionary prescription for the vast majority of human beings would appear to be simple: permanent stasis, i.e., forced evolutionary stagnation.

Sociobiology and evolutionary psychology, extensions of the Modern Synthesis into the fields of sociology and psychology, are based on the premise that human (and animal) behavior, like evolution itself, ultimately has genetic causes, and thus evolves as a means of promoting genetic survival and competition. They deviate from more traditional social and psychological theory in assigning ultimate priority to genes themselves rather than the organisms to which the genes belong; in effect, they assert that organisms "belong to" genes rather than vice versa. This turns minds and societies and their properties into functions of pure mechanical causation on the genetic level. According to this view, genes possess something akin to purpose, but more general; it is genes, and not organisms, that are presumed to contend for evolutionary success. Even when laboring under the delusion that they are somehow in charge of their lives, organisms merely serve genes as vessels, propagators, and weapons against other aquatic, aerial, or ambulatory gene-packets.

Given this modern academic trend (which many non-academics understandably find quite disturbing), it seems natural to ask whether this premise would support a theory explaining occurrences of predatory social behavior like that mentioned above. Is the seemingly predatory thought and behavior of Congress regarding HB1955 merely a simple mathematical function of self-interest on the parts of many independent entities, or can it be better explained on an evolutionary basis? Might it be equivalently explained on both levels of causation, each being a mere reformulation of the other on a different level of generality? Might this unified explanation in some sense equate to more traditional explanations referencing abstract or moral principles?

First, we must take account of evolution in the large. One thing that the theory of evolution tells us is that no sooner does a species appear to have it made, so to speak, to the point where it can overpopulate and overrun the countryside and overburden the ecosystem with impunity, than a biological assailant appears. This may be a competitor, a predator, an opportunistic parasite, or a pathogenic microbe for which prey behavior has cleared the way, e.g., through population-driven malnutrition. Assuming that the assailant is a predatory animal, it has an inherent proclivity for predatory behavior against the prey species, which means that it behaves in a highly "immoral" way with respect to any system of ethics that would be acceptable to most members of the prey species itself. (Even though most animal species lack abstract ethical systems, analogous principles can often be educed from their behavioral statistics and ascribed to instinct; for example, most higher species do not eat their own young, and this fact can be expressed as a guiding principle of behavior active within those species.)

But mankind vanquished all of its animal predators some time ago. Whether in the zoo or the wild, the animal predators of mankind are essentially under human confinement. We have even used advanced biotechnology to fight disease to a standstill. In short, we have engineered a situation in which we increasingly "have it made", dominating the global ecosystem to an unprecedented degree, overpopulating and overrunning the landscape and overburdening the earth in ways beyond the reach of any other species. Yet, despite all of our defenses, we have managed to retain an obvious Achilles' heel: an exploitative socioeconomic framework which strongly supports predatory behavior. According to evolutionary theory and its modern psycho-social variants, we have thus created a most attractive niche for some form of large-scale biological opposition, particularly of a kind which is capable of exploiting that framework.

The reaction of nature is automatic: a competitor or a predator must emerge. But what, short of demons or aliens zipping around in UFO's, can successfully compete with or prey on man? As far as most of us know, only man himself is intelligent and devious enough to fill the niche, and only man is capable of shaping and exploiting man's predation-friendly socioeconomic environment. Hence, the predators must spring from humanity itself.

That's arguably the most plausible and historically substantiated neo-Darwinian hypothesis regarding the future of humankind. Next week, we'll examine it in closer detail.

August 22, 2018

***

Last week, we informally sketched a certain evolutionary hypothesis regarding the future of mankind: (1) that the vast majority of human beings, in exhibiting political inattention and apathy, sheeplike docility, and a willful tendency to overpopulate and abuse the environment, are behaving like prey animals and may thus have created an irresistible opportunity for high-level predation; (2) that the would-be predators trying to capitalize on this opportunity are humanity's own socioeconomic elite; and (3) that if allowed to do so, the elite will engineer an evolutionary divergence between themselves and the rest of homo sapiens, thus fully realizing the predator-prey distinction. In effect, this hypothesis simply recognizes human social Darwinism as a manifestation of Darwinian evolution in the large. (The connection was always there, but most of us refused to see it.) This week, we'll examine that hypothesis a bit more closely.

Several of the arguments against this hypothesis are weaker than they might first appear. Let's have a look at some of them.

Argument 1: Predatory behavior exists within the human species, and has manifested time and time again throughout history; it is a psychological trait of all humans, not just a predatory subspecies.

Contra: Quantitative distinctions can still exist among individuals. The fact that everybody has some amount of predatory instinct, and some amount of moral inhibition to constrain it, is therefore meaningless in this context. Psychological traits exist in varying measure among individual human beings, and history records a very small handful of people with enough predatory cunning, and little enough moral inhibition, to successfully and persistently prey on their fellow men and women, socioeconomically speaking. This psychological and behavioral tendency appears to be strongly recurrent in particular family lines; thus, it is subject to legal, genetic, and environmental forms of inheritance, and may indeed have a genetic component. (Of course, it no doubt has environmental components as well, and controlling these may be sufficient to allow any particular individual to keep such a tendency in check.)

Argument 2: Animal predators are usually genetically distinct from their prey. But among humans, the so-called predator-prey distinction is not a matter of genetics at all, but merely of environment and circumstance.

Contra: Divergence of species is an axiom of Darwinian evolutionary theory, and in principle, nothing finally prevents it from incorporating the predator-prey relationship. Predator-prey relationships involving closely related species occur throughout the animal kingdom; chimps prey on monkeys, lions prey on cheetahs, and so on. Although it tends to be rare among higher animals, intra-species predation occurs as well; wolves prey on coyotes, coyotes prey on dogs, and so on. Within more homogeneous animal populations, intra-species predation is even less common, but this can be at least partially attributed to specific limitations on long-term social planning and coordination which do not apply to humans.

As for the idea that genetics has nothing to do with human behavior, this amounts to an *a priori* assertion that human behavioral tendencies lack a heritable component. But there is no scientific support for this claim, and it contradicts much of what we know about animals in general. Whenever an animal of a given species acts "on instinct" in a natural or reflexive way – whenever it flees instead of hides, climbs instead of runs, screams instead of hoots, breathes instead of spreading its gills, mates for life instead of mating promiscuously – it is evincing heritable behavior.

The fact that much of the structure and behavior of organisms can be genetically inherited reflects the fact that organisms come into being, and to some degree maintain their existence, through the expression of genes. Aside from free will, the main issue is predictive granularity, i.e., exactly where genetic determinacy stops and environmental determinacy begins. In any case, the question is not whether a persistent genetic trend toward the emergence of a predatory subspecies of homo sapiens could possibly arise; it can, at least in principle. Those given to predatory socioeconomic behavior usually mate with others whose outlooks and lineages are similarly disposed, and mating habits can play a decisive role in population genetics. The questions to ask are *how* and *when*.

As for the idea that the only thing separating human predators from their human prey is "environment and circumstance", this ignores the possibility that environment and circumstance are themselves determined by intentionality and/or psychological predisposition. Such factors appear to correlate strongly with various transformations occurring in the human sphere.

Argument 3: The human majority will not allow themselves to be treated as prey. At some point, the predatory strategy of the elite will backfire, aborting predator-prey divergence. In fact, predation is the stuff of which natural selection is made, actually increasing the fitness of prey and making it even less likely that any group of predators will ever achieve total control over mankind.

Contra: Admittedly, this kind of backfire occurs frequently in nature; the tiger is always in danger of being gored by the ox. But human socioeconomic predators are not tigers. They have already methodically shaped the socioeconomic environment to their advantage, and with further technological assistance of kinds even now being implemented, could make themselves almost impervious to retaliation by their victims. As the hypothesis stipulates, human predators could achieve such a degree of control that they will be able to scientifically counteract any beneficial adaptations they might cause in their prey. Some amount of evidence suggests that they may already have moved in this direction.

Argument 4: Where prey species can overrun and deplete the environment in the absence of control, predators can function as environmental and ecological protectors. But in the human sphere, those most guilty of "prey-like" abuse of the environment are the "predators" themselves, who pollute and pillage the earth and start wars, and while treating the world as their chessboard, encourage their followers to breed unrestrictedly to produce soldiers and workers and wage demographic warfare on their behalf.

Contra: While this is an historically valid and morally relevant distinction between human and animal predators, it is irrelevant to the hypothesis that prey behavior creates opportunities for predation and thus causes predators to emerge by way of selection. Even if these predators exhibit "prey behavior", they can still be distinguished from their own prey using other criteria. The possibility also exists that although human predators have exploited the environment to gain control of it, they will adopt more sustainable practices once their control is complete.

Indeed, the possible coincidence of predatory and prey-like behavior raises a further possibility. Given the gradual emergence of garden-variety social predators like we seem to have in the US Congress, an even more predacious subspecies may lurk in the brush ... a superpredator that feeds at will on garden-variety predators! After all, such superpredators do in fact exist throughout nature, as anyone knows whose predacious housecat has been eaten by coyotes while hunting for mice. Do the nascent predators of the US Congress even now baste wetly in the gaping, slavering maws of superpredators even higher up the food chain?

Unfortunately for almost all concerned, they most certainly do. The superpredators are well concealed but very highly placed, and the only thing that prevents any politician from being consumed by them is slavish obedience to their wills and the continual satisfaction of their boundless appetites. They let their desires be known to their political flunkies through private meetings, discrete telephone calls, secret memos, large infusions of campaign money, cash support for favored charities, bribery committed through lobbies and political action committees, threats of blackmail effected through media manipulation, lucrative post-government job offers, and meaningful patterns of winks and nods in the course of otherwise normal conversation. Given the opportunity, they feed on each other as readily as on members of government.

Might predator-prey relationships observed in the animal kingdom afford insight on such relationships in the human sphere? Of course they do. When wolves or lions want to bring down a prey animal, they silently but intelligently array themselves for a combination of ambuscade and motion control. It would be utterly absurd to suppose that humans, so much more able and likely to engage in long-term planning than other social predators, would not do the same on a grander scale, suppressing their competitive natures long enough to conspire for their own benefit and intelligently array themselves for long-term predatory efficiency. Take, for example, the plutocratic families who set up the great American private foundations. They placed themselves in economically unassailable positions, often by monopolistic behind-the-scenes collaboration, used their questionable gains to establish the foundations, and provided them with definite guidelines under which economic resources could be methodically channeled to causes, organizations, and programs favoring social and psychological outcomes now strongly in evidence all around us.

Could human predators, having violated the ethic of their species, have developed their own specialized form of ethics? Basic features of human psychology render this certain, and such systems have been cooked up regularly in the course of history. These systems have promoted the social elevation of various supposed manifestations of "the good" – the deepest sages, the greatest magicians, the strongest warriors, even the prettiest charmers – and have even asserted the "divine right" of those born to royalty. More recently, they have tended to promote a "good shepherd" ideology combining environmentalism, political correctness, and socialistic despotism (as encoded, for example, in the longstanding priorities of the abovementioned private foundations). But this emerging ethic, being essentially materialistic and designed to maintain the privileged positions of a genetically and/or socially inbred class of "shepherds" distinguished by the proclivity to seek and acquire wealth and power at all costs, remains essentially predatory relative to the great mass of humanity. No matter the justification, the predators can still eat their fill, and their prey will still live in fear of them.

As explained above, this emergent ethic has already generated a new set of social ideals and clever schemes for their spread and implementation. By encouraging the selective interpretation of scientific data to support various doomsday scenarios, the elite have managed to yoke the environmental movement firmly to their agenda. For example, they have convinced a growing number of people that simply by exhaling carbon dioxide, using energy, and growing food, the human race qualifies as a destructive species which may be rightfully deprived of sovereignty and required to fund a centralized control grid for the purpose of keeping it in line ... or in terms of our hypothesis, that humanity is an unruly prey animal which has already given predators the right to prey upon it and must now furnish them with the perfect means. Some of these beliefs have been promoted so aggressively that they are no longer regarded as open scientific matters, but as axioms about which merely asking questions can amount to political suicide. (The possibility of man-made changes to the physical environment is one issue; the use of this possibility as an all-purpose rationalization by predatory elites wanting to establish a worldwide control grid for their own purposes is another.)

But let's be fair to the human predators among us. Is it not possible that some of them mean well, truly believing that they are acting on humanity's behalf? The likely answer is yes. But various aspects of what they are doing – arrogating to themselves an overwhelming percentage of what humanity needs to survive and prosper, manipulating markets, constantly lying and keeping secrets from their supposed beneficiaries, depriving their opponents of their freedom and dignity, even condoning murder and torture – betray their predatory underlying motives. In other words, many of the "good" ones are self-deluded, their predatory natures merely submerged beneath a shallow layer of compassion.

Is it not still possible that the elite constitute humanity's best bet for continuing to evolve? Once the current crop of predators is victorious over the rest of mankind, will they not turn on each other with renewed vigor and voracity, thereby creating the selection pressure required for another "evolutionary advance", and so on, in a circular, technologically-augmented genetic arms race within the human species, and will this not propel us toward genetic perfection? While this thesis is interesting in a Nietzsche-meets-Darwin kind of way, it hardly describes a rational or humane path into the future. It would not only be unacceptably cruel, but with the powerful weaponry already under elite control, it could easily lead to the extinction of life on earth.

Apparently, man has no rational choice but to stop behaving like a prey species and rewarding human predators with the wealth, power, and adulation they crave; only by reforming itself can humanity be saved from itself. But meanwhile, consider the following hypothetical question: If, as some already predict, the socioeconomic elite were bent on using technologies like genetic engineering to maximize and permanently consolidate their advantages, then how could the rest of mankind escape total domination by them? Should it try to hold them back by restricting the technologies in question... e.g., through the suppression of human genetic engineering, despite the likelihood that this would prove impossible? Failing that, should it instead try to keep pace with them, using modern technology to distribute genetic and other advantages over the entire population, and thus spare itself the danger of enslavement and/or extermination by its most vicious and amoral elements? Should it make *everyone* an elite, genetically speaking? These are very important questions. If the public has not brought itself to rationally consider them before science is able to transform the human genome for purposes beyond simple repair, it may be too late for anyone but the elite to determine their answers.

Although we've been treating this hypothesis as a prediction of neo-Darwinism and its modern psychological and sociobiological derivatives, its recognition long predates those theories. As mentioned above, it already seems to be taken for granted by certain members of the alternative media. It may even be old hat in academia, having already sprouted all of the obvious neo-Darwinian bells and whistles and percolated through a closed network of exclusive, densely-written academic journals. But in that case, given elite control of education and the media, it's a pretty good bet that its serious academic consideration will remain hidden from the general public.

The above example of human-on-human predatory behavior – the overwhelming congressional approval of HB1955 – could be replaced with countless others. But just to be thorough regarding the example at hand, we should cover one additional possibility: that Congress has certain information which it has not been sharing with the public, and that if we only knew this information – e.g., a secret threat or some kind of rarified grand plan for the fair and symmetric advancement of society cooked up by elite think-tanks and planning commissions – we would see that its motives were benign rather than predatory.

But if such a universally beneficial plan really existed, there is no obvious reason why it would not have been fully revealed to its prospective beneficiaries ... except, of course, for the usual tacit non-explanation preferred by secretive despots the world over: "In the course of solving all of your most difficult problems for you, we caused some even worse problems which might prove socially or economically disruptive or even cause mass panic if we revealed them or their proposed solutions to the public!" Or in other words, "We must keep our *current* plans and activities secret in order to continue to keep our *past* plans and activities secret!" Nothing is finally revealed, and specific accountability is lost in a regress of secret upon secret, lie upon lie, with each new conspiracy justified by those which came before it and so on *ad infinitum*.

In this light, and given that stealth and control are central constituents of the predatory modus operandi, the observed combination of secrecy and unconstitutional legislation is hard to interpret as anything *but* predatory behavior. Even if the members of Congress believe that their hands are being forced from above, they could apologize to their constituents and resign rather than continue to do the bidding of their elite masters. That this is not happening tells us something very important about where their loyalties lie ... and of course, that there are indeed predators afoot in the night.

These days, the people we elect to represent us and protect our interests look more like conspirators than public servants. If such a conspiracy really exists, we obviously need to understand it. But before we can do that, we need to understand the nature of conspiracy itself.

August 31, 2018

Antisocial Media

Google, Facebook, and Apple, reportedly worth at least $3 trillion, cover most of the Internet. The Big 5 tech companies – Apple, Alphabet, Amazon, Microsoft, and Facebook – supposedly have a combined net worth of over $4 trillion, exceeding the GDP of all but the seven richest nations on Earth. But despite their incredible wealth and the evident fact that they collude with each other in what amounts to a virtual monopoly, they all cave in to political pressure with unseemly haste and servility (at least when it seems to bear leftward).

This raises a question: Given that these companies are headquartered in the United States and have all of the freedom provided by the First Amendment of the US Constitution, what is it that can suddenly exert so much pressure on them that they act in virtual concert to limit the right of free expression of certain users? Why, instead of letting all of their users express themselves as freely as birds in flight, as they have often done for even the lowest kinds of progressivist shill and atheist-materialist troll spewing hatred of God, America, and Whitey, are they now concertedly ejecting certain people, mainly conservatives, from their platforms?

Our best clue is probably this: they are all massively leveraged. They are so deeply in debt, and rely so heavily on the constant availability of working capital, that behind their frontmen they may be considered to have "shadow owners" who can coordinate them as surely as a truck company dispatcher can direct a fleet of semis to the same destination. In other words, like the major print, radio, and television media, they are in thrall to people who are in thrall ... to people who are in thrall to the international banking system. Evidently, the international banking system is upset with certain PC-deficient strains of public opinion, isn't going to take it any more, and is using its social media puppets to silence the offenders.

International banks take a lot of heat these days. Why pick on the international banking system? In the world we presently inhabit, this is where the financial buck always stops. It's a hard, cold fact of the modern global economy. Corporations are multinational, and multinational corporations have international bankers. While it is always tempting to lay all of our troubles at the feet of politicians and political structures inside and outside of government, it can no longer be denied that politics is itself controlled by this system. Politicians are merely its high-tone courtesans and scapegoats in the event of embarrassment, and so for the governments which provide them with garishly lit stages for their bad political theater. That the system has been shifting leftward for years is due to the fact that socialism is antithetical to economic competition, something that confirmed plutocrats prefer to avoid.

Obviously, very few of those who run or front for this system are friends of the public, or are performing a service for the public. Like drug dealers, they serve only themselves, and their only real friends are the ones who stare back at them from the bottoms of their depleted snifters of plebeian blood just before a refill. The tech giants are virtual monopolies; their big clients must use them for purposes of private and commercial networking and advertising in order to remain competitive, and their little clients relate to them like addicts, finger-jabbing their tablets and cell phones the way jonesing lab rats work the pharmaceutical reward levers on the walls of their well-surveilled cages. Most of their employees are known to be rabid progressives, as this is what it takes to fit into their corporate environments, and they often act on their own recognizance. When one finds oneself censored for "hate speech", it is often because some little nobody has decided that he/she hates what one has just written.

A true democracy cannot operate this way. That's why, in the eyes of the elite, we have never really *had* a democracy (or more specifically, a constitutional republic). Instead, we have had various stages in the evolution of *corporate fascism*, an unholy partnership of major financial institutions, the multinational corporations beholden to them (including the media and Academia, Inc.), and the governments controlled by them. That we have devolved to such a system is implied by governmental disregard for the will of the public in official pursuit of various highly unpopular measures, running the gamut from taxation without representation to throwing US borders open to Third World mass migration in order to effect political and demographic change and swell exigent voting blocs while subjecting majority citizens to aggression by resentful, disruptive, culturally cohesive competitors who have no intention of being peacefully assimilated. Not uncoincidentally, many of these policies have been advocated for years by plutocratic members of the international banking community and their corporate allies.

The US Constitution was not designed for such a situation. The reason is obvious: its authors never intended that the US should *become* this kind of system. The Constitution was designed for a different political system entirely ... one in which power resides with "the people", a relatively homogeneous, cohesive, well-integrated citizenry both morally and intellectually qualified to participate in meaningful policy decisions to be executed under the aegis of a government subservient to their will. Where government has instead been allowed to swell like a tumor and degenerate to a global-corporate franchise, the Constitution takes on a different complexion entirely, functioning more like a toddler's threadbare security blanket that is torn and tattered from being dragged across the floor and stepped all over. In the eyes of the powerful, its only surviving function is to distract the shriveling minds of thumb-sucking citizens as their futures are blighted by corporate lobbying and other forms of legalized bribery and blackmail.

In this kind of system, the mass media, including dominant social media, become globalist infrastructure masquerading as public infrastructure. Disguised as corporate entities with all the rights and freedoms of individual human beings, they have far more wealth and far less personal responsibility than the typical person enjoys. But even as their capacity for social, economic, and environmental damage grows without limit as a result of this absurd legal fiction, so does their capacity to escape its consequences. They routinely exploit their corporate pluralism as a pretense to moral authority, but then use it to pass the buck when accused of wrongdoing. This makes it far too easy for them to abuse their power. And when they do, they always do it in the way preferred by those writing the checks, which – as usual – means that the problem can be tracked to its source by following the money.

Large banks and corporations, whose money and power have long commanded unhealthy respect in the halls of government, too often function as shields for amoral, unscrupulous, wealth and power-obsessed people. In the case of the major social media companies, the situation is especially ironic. Defining themselves as private entities and claiming the Constitutional right to set their own "community standards", fattening without limit as the Constitution protects them from legal retaliation by victims who are trolled, defamed, and tormented under their auspices, large social media corporations now use these very "community standards" to censor and block the free expression of users and promote the economic interests and anti-Constitutional political opinions of their owners. The impact is enormous and highly unpleasant, putting users in a permanent state of fear and anxiety about expressing their own "politically incorrect" opinions and thus swaying the course of political debate and skewing the electoral process. It's an insidious violation of the public trust.

Obviously, something must be done about this situation immediately. The social media must be brought to heel for what amount to anti-trust abuses, and legislation must be adopted which recognizes their *de facto* status as essential infrastructure for the American political process. Until this happens, they will continue to serve not the public, but banks, large corporations, and other self-dealing entities which are openly adversarial to the public interest. These entities are already notorious for co-opting the "will of the people" using an endless supply of monopoly money, the corporate media and for-profit education system, and the authority of government to force upon the American majority a Cultural Marxist agenda calculated to advance the interests of an iron-fisted global oligarchy. Unless they are stopped, they are certain to continue.

Before long, the corporate fascists will be in a position to yank the pacifier from the public's mouth and rip the security blanket from its grip. While we are still in a position to insist that social media and the Internet as a whole be regarded as critical infrastructure with crucial functionality in democratic self-governance, our window of opportunity is quickly narrowing. The global oligarchy, which had been counting on an easy victory over the general populace as little as two years ago, now senses danger and is in a panic, desperately lashing out against those refusing its toxic indoctrination.

If the infantile American public continues to chew on its social media pacifier as a rogues' gallery of power-drunk banker-owned techie billionaires and their bankster sugardaddies tape peoples' mouths and eat their lunch, it will awaken to find that the pacifier has been replaced with a mouthful of something much less appetizing, namely, what remains of their lunch after digestive processing by the billionaires in question.

We'd better move fast.

September 04, 2018

The Golden Rule

This is not a finished piece, but merely notes for a work in progress. I may revise it here or supersede it with another contribution.

We are constantly manipulated through morally loaded feelings like guilt and compassion.

When are the arguments valid? Uniformed thuggery pursued on the bad assumption that all orders are good. In a world where the scum rises to the top, one cannot survive without telic intersection with scum. In this case, the world is spiritually degenerate; one must run up a moral deficit just to survive.

The GR is subverted by tyranny; the tyrants just have to threaten to punish everyone if any individual breaks the rules.

Absolute ethical rules take the form "do/do not do X". With ethical relativism, moral decisions (what to do/not do) can vary with perspective among utility operators. The GR is neither; it is a purportedly universal ethical symmetry principle which does not change with perspective, but requires a calculation of various levels of utility associated with coherent utility operators, all of whom are rationally self-interested and all of whom want to maintain maximum happiness, freedom to pursue happiness, and likelihood of obtaining happiness. Moral uncertainty is unavoidable; how must it be dealt with? Likelihood / probability calculations are generally required. Under what conditions and constraints can the GR actually be relied on to work? Moral self-satisfaction tends to be accompanied by mutual dissatisfaction precisely because self-satisfaction is achieved by limiting the criteria for moral worth to those which are personally satisfiable, whereas personal moral parameters moral agents worry only about their own utility.

Identity: group formation. When is there a coherent level of identity which actually "shares" a utility function? Is a racial group a utility function? Negroes, for example, claim that whites committed past injustices against their ancestors (on whom the actual effects of slavery are unknown), claim that they are worse off for it, claim that the damage was permanent, and demand reparations from people not responsible for it.

Ordinarily, the GR considers only individual utility: "Do not to another person (i.e., diminish their utility) what you would not have them do to you personally (i.e., diminish your utility)".

Group utility is ignored. Furthermore, it is time-limited; it is also limited by accuracy of knowledge of moral context. Higher levels of utility and longer time horizons must also be considered.

Levels of utility are not covariant; some are contravariant. What is good for the group is not always good for the individual or vice versa. How can utility be summed over groups? How can levels of utility be compared? One must rely on gut intuition and intelligence.

To apply an ethical rule based on max utility – and the GR is such a rule – requires a calculation. How complex is it? Can stupid, morally unintuitive (nonempathic) people who cannot do such calculations be ethical? Can a society with too many stupid, nonempathic people be a moral society (in terms of how people treat each other; in terms of how their elected leaders treat them and those in other countries; etc.)? Ethics is about the maximization of utility. For this, a valid model of reality is required in order to calculate the effects of any given action therein. Can the negative effects be contained, or will they spread until there's a crash, in which case game over? Are the positive effects self-aborting or sure to be aborted, in which case they are confined? Must set boundary conditions and evaluate utility therein.

The idea is to define utility on various levels, expand the boundary to accommodate all levels, and calculate the effect of any given action.

Games and meta-games: Prisoner's dilemma and maximization of utility; defection = changing the utility function with a net loss of shared utility.

Where every intelligent person on Earth knows that the subversion and collapse of Western Civilization and its complete replacement by a despotic global oligarchy would not be good for mankind, the Golden Rule does not oblige anyone to support it.

If one is rational, open immigration is off the table because in a crowded world, it amounts to the weaponization of reproduction and the perversion of compassion. Western Civilization is being destroyed by it, and the reasons – openly advanced by international banksters over a century ago – come down to demographic genocide (the destruction of a certain genotype, namely that associated with the building of Western Civilization).

Some Christians are politically, socially, and biologically naive. They don't seem to understand the actual meanings of words like "empathy" and "compassion", or to clearly comprehend the intentions of Jesus or the meaning of Christianity. They literally think that Christian values justify demographic genocide against Christians, Christianity, and Christian culture by people openly hostile to those values under the auspices of a mass dysgenics program being crammed down the throats of national majorities against their will by a tiny handful of filthy-rich oligarchs seeking to enslave the world.

Not even the CTMU, Logos, can help people too confused and morally compromised to value these priorities.

Paradox of tolerance: Need intolerance be tolerated? If yes, then intolerance will extinguish tolerance.

The "paradox of tolerance" is just a variant on the paradox of democracy: democracy can be democratically extinguished by a population that prefers despotism; even if it is a Constitutional democracy with an absolute constitution guaranteeing the right of self-determination to all, the constitution can be democratically revoked.

The paradox of human rights: granting unlimited rights includes the right to violate the rights of others.

Any time something implies the possibility of its own negation, it requires an absolute to prevent it.

Golden Rule: symmetry is the absolute. But it must be applied with the same scope on all scales on all levels of utility, and cannot be applied to implacable enemies due to the risk of defection; "turn the other cheek" can be considered a complication of it.

Golden Rule paradox: If, being an adherent of moral relativism, one does not want to be dissuaded from breaking the GR, then no one else can be dissuaded either.

October 29, 2018

Defining Consciousness

Comment [name redacted]: There are bona fide reasons not to get distracted with a singular definition of consciousness, so as to try out in gedanken experiments various modes of definition to see if they make sense. But caution, when I say "types of consciousness", I am just joshing you all (making a joke for everyone), I really mean only "types of brain behaviors".

Response: A very wise qualification, if only because everyone understands the problem with limiting the definition of consciousness to the neural realm: If consciousness were strictly an emergent function of neural structures and events, then it would be purely physical, and like physics, it would be unable to account for its own existence or specify the conditions of its own applicability. This would immediately preclude any possibility of metaphysical content, including "free will".

What this would imply regarding the relevance of the work of, e.g., Professor Stapp, may be left to any reasonably intelligent reader. Suffice it to say that it would amount to going a bridge too far, walking off a cliff, and punching oneself in the nose on the way down.

In order not to end up with a Tower of Babble (misspelling intentional), a general definition of consciousness is indeed required. That way, we can tell whether or not any more specific, biologically / neurologically grounded definition of consciousness is actually about "consciousness" (there's really no other way to consider). A purely empirical definition – e.g., "consciousness is whatever looks like consciousness in the lab!" – is, in addition to being semantically ambiguous and excessively touchy-feely, ultimately preclusive of anything that cannot be directly perceived or logically deduced from replicated perceptions, and would therefore be prejudicial and insufficiently general unless coupled with a logical counterpart.

This requirement is easily met. Consciousness generically equates to self-awareness on the part of a conscious entity (this amounts to a recursive definition). The phrase "self-awareness" is a semantic loop, and may be extended to include the relationship between self and its environment. The resulting loop has a self-dual logical structure which has been well-explicated (as the CTMU metaformal system), is fundamental, is uniquely consistent with both physics and logical semiotics, and is comprehensive in the sense that it applies to all levels of logically consistent discourse on the subject of consciousness and for that matter everything else.

(Unless, of course, someone has a coherent argument to the contrary.)

November 17, 2018

Human Reproduction: Quantity Control Entails Quality Control

This morning, I find myself visited once again by a nagging concern.

It seems that despite spectacular advances in biological science, many people still fail to understand the relationship between quality control and quantity control in the context of human reproduction. Under present circumstances, this is neither surprising nor comforting.

I live on a ranch, and I sometimes breed livestock. Every livestock breeder knows that when a herd is allowed to breed without restriction, it grows in number until the food runs out, at which point the animals must either be moved off the premises or starve. Ranchers also understand that to maintain the healthiest and most viable and valuable herd, they must select their breeding stock for desirable traits like adaptivity to ranch life and its varying conditions and requirements (intelligence) and nonaggression toward their owners and other herd members (impulse control), and exclude animals which exhibit undesirable traits like sickliness, stupidity, susceptibility to aggressive impulses, and other traits which lead to conflict and injury and degrade the health, happiness, and fitness of the herd. In other words, ranchers understand the need to control reproduction for both quantity and quality.

Humans are different, of course, but only to a point. Being animals for biological purposes, they too have traits which enhance or diminish the health, happiness, and viability of society, and their reproduction must be controlled for quantity as well as quality. But because they do not wish to be treated as livestock, they prefer to do this for themselves. In order to get by without coercive breeding programs, they must intelligently and responsibly acknowledge the need for population sanity and voluntarily exercise reproductive restraint. Responsible parents must be able to recognize environmental and economic limits on reproduction and muster sufficient impulse control to moderate their reproductive instincts, refrain from waging demographic warfare against other groups, and align their reproductive behavior with individual, group, and societal interests. Only thus can human populations exist in sustainable equilibrium with their respective natural and social environments.

This can be a very tall order, and not all people are necessarily up to it. In a population that covers a broad spectrum of intelligence, some people lack the ability to accurately evaluate the carrying capacity of their natural and socioeconomic environments (especially where they are deliberately misled), and therefore tend to reproduce as if these constraints do not exist (or in spite of them, in order to wage reproductive warfare against other groups). In contrast, those with the intelligence to see that individual and collective utility are diminished by excessive reproduction tend to reproduce more slowly, and thus to be overtaken by less inhibited, faster-breeding competitors for genetic, cultural, and political dominance. More intelligent people are motivated to exercise reproductive restraint on behalf of themselves, their progeny, and society as a whole, while less intelligent people cannot recognize the dangers of excessive reproduction and uncontrolled aggression, and thus more readily give in to their reproductive urges and aggressive impulses.

Whether born of ignorance or internecine competition for limited resources, the tendency for instinctively driven low-intelligence people to out-reproduce higher-intelligence people can only lead to the progressive diminution of average intelligence throughout the population as a whole. This trend is self-accelerating; as its mean intelligence decreases, the population tends ever more rapidly toward a state in which virtually no one is capable of responsible reproductive (or any other kind of) behavior. When this trend is further amplified by economic manipulation – e.g., the forced transfer of reproductive resources from more intelligent and responsible groups to less intelligent and responsible groups under disingenuous pretenses such as "equality" and "social justice" – a dysgenic outcome is all but sure to result in very short order.

To see whether an argument is sound, it must be examined for weak or untenable assumptions. The innateness and heritability of intelligence is often attacked as the weakest "assumption" in this chain of reasoning. However, IQ differentials and IQ heritability are the best-confirmed facts in psychology and sociology; so-called "experts" who deny them are either deluded or dishonest. (Sadly, most of the rest are simply afraid to discuss it.) As intelligence is necessary for the recognition, prediction, and strategic realization of positive outcomes of any kind, including the outcomes of reproductive behaviors, patterns, and policies, reproductive responsibility depends on it.

What about impulse control? By definition, impulse control is required for the voluntary regulation of reproductive instinct; the less of it is present, the more involuntary constraint must be applied to stifle or redirect instinctual urges. What, then, of the relationship between impulse control and intelligence? Insofar as intelligence depends on the ability to suppress impulsive miscalculations while discouraging impulsive behavior by anticipating its negative consequences, these traits are mutually dependent for logical purposes ... and in fact, studies support a biological relationship between them (e.g., see link below). But even if they were genetically independent of each other, a deficiency in each trait would reinforce a deficiency in the other by their joint effects on reproductive behavior. It follows that for any breeding population exhibiting large differentials in the levels of these traits, a dysgenic outcome is unavoidable without preventative measures.

Internet source: "Why Delaying Gratification is Smart: A Neural Link Between Intelligence and Self-control" (Association for Psychological Science 2008)[1]

This implies that for a population covering a broad spectrum of intellectual ability, reproductive quantity control is not only necessary, but rationally entails quality control with respect to intelligence and impulsiveness. How, then, can governments claiming to provide rational leadership openly pursue policies which lopsidedly favor immigration and/or reproduction by members of socially irresponsible low-intelligence subpopulations? The short, perverse, infuriating answer is that by reproductively engineering society to deprive its members of the intelligence and responsibility to recognize sound immigration policies and voluntarily pursue rational reproduction strategies, governments enable themselves to get away with distractions, diversions, and lies while reducing the public to a state of abject dependency. In effect, society is dumbed down, rendered helpless against systematic deception, and led to the slaughter.

Most people still seem to appreciate the sanctity of human life. Hence, the widespread distaste for late-term abortion even among those in favor of reproductive choice. However, people increasingly miss the fact that unlimited human procreation, plus global limits on human procreation, plus badly distributed incentivization of human procreation do not add up to a net advantage for humanity or even a balanced equation. They add up to the promise of absolute misery and/or destruction for a vast number of human beings. Where the stakes are this high, rationality is absolutely critical, but it is in dangerously short supply when it comes to human reproduction. Blinking and drooling like imbeciles, robbed of their mental and moral faculties, a growing number of people can do nothing but go with their basest impulses and regurgitate their PC indoctrination like demented parrots.

It is clear from governmental policies now in effect that we in the West are governed not by rational representative "leaders", but by our enemies ... enemies who, out of some blend of idiocy, lunacy, incompetence, and malice, have weaponized human reproduction against us, and in a stroke of unbelievable stupidity, even against themselves (because in the end, they are virtually indistinguishable from those whom they are destroying). From the mid-20th century onward, these criminals have been overruling their own majority populations to deliberately flood Western societies with the worst and most unassimilable elements of the Third World, thereby disrupting preexisting social and economic equilibria and ruining the moral and biological fitness of the human species, in particular our own historically indispensable branch of it.

This situation does not exist because we lack a sufficient understanding of human genetics, and cannot afford to make mistakes by tampering with allegedly "natural" processes like global overpopulation and resulting mass economic migration. It exists because basic information known to every competent scientist, farmer/rancher, and pet breeder on Earth is being used against us against our will and in contempt of our welfare, and we can no longer afford to permit it.

This is not, and has never been, an acceptable situation. It is sheer political sociopathy, and it must be stopped. While the possible solutions that remain to us are painful and potentially ineffective, the alternatives are more painful still.

December 19, 2018

2019

Do Conservatism and Liberalism Really Exist?

The following question came from a stranger who contacted me cold through a popular business networking website. It is either a bad question which was meant to goad me into a tedious (for me) political debate about our binary (two-party) political system, or a good question intended to stimulate a conservative-yet-liberal *regression* of the modern political dichotomy to an older but more rational and thus more *progressive* distinction.

Either way, we're going with the latter.

Question: Modern conservatism embodies principles including smaller, less intrusive, and relatively decentralized government; minimal taxation and regulation; an emphasis on individual responsibility; reliance on supply-side economics and the "trickle down" effect; and the absence of social safety nets providing individuals with economic security, housing, healthcare, education, retirement, or social equality.

It appears that no modern nation uses these principles to govern a free and democratic society, making it seem that conservatism exists only in the minds of its believers (and in a more pejorative formulation, in the minds of its detractors). Is conservatism a real political philosophy, or more like a religious cult? Does it have any real-world credibility?

Answer: This question includes what appears to be an impromptu definition of "conservatism" on a sprawling collection of "thought and practice" that is not entirely coherent. (Of course, much the same can be said about the modern understanding of "liberalism", which also bears very little resemblance to the original definition.)

In theory, conservatism is a political or philosophical stance that promotes "tradition", i.e., traditional culture, values, attitudes, policies, and institutions, whatever these may be in a given society at a given time. It comes in *paleo–* (regressive) and *neo–* (progressive) versions, depending on whether its proponents are "reactionary" and prefer to reset society to an earlier state beyond which undesirable changes have occurred, or to accept and adapt to certain changes which they now consider socially and/or personally advantageous. Paleoconservatism, by rejecting all change beyond a certain idealized stage of history, is arguably truer than neoconservatism to the historical roots of conservatism; conservatism began in the early 19th century with the Bourbon Restoration, a plan to roll back certain changes associated with the French Revolution, and was therefore more "paleo" than "neo". However, the term has since become associated with the attempted justification of social and economic inequality and the defense of the economic and political status quo against all threats to its major stakeholders, as defined on the existing concentration of wealth and power.

Conservatism is usually juxtaposed with "liberalism". Ostensibly, liberalism is a political or philosophical stance promoting liberty and equality of opportunity for those oppressed by the status quo or its major proprietors in finance, industry, and government. Liberalism dates from the Age of Enlightenment, during which it arose in opposition to traditional conservatism and in support of representative democracy and the rule of law. In short, the original thrust of liberalism was to get the common man out from under the thumbs of kings and tyrants. But over the course of time, its focus has changed. Rather than being about freeing everyone from economic and political oppression, liberalism is now about freeing the members of cultural, racial, economic, ideological, or sexual identity-groups from any constraints or restrictions they happen to dislike, usually at the direct expense of an "identity-free group" (that's an oxymoron, by the way) consisting of people from certain ethnic and cultural backgrounds usually associated, explicitly or by process of elimination, with Europe and Christianity.

There are two possible ways to define conservatism: qualitatively, on the basis of attributes or principles, or empirically, in terms of observable instances. In the conventional political framework, we're stuck with one kind of definition or the other. (Of course, we could also try to construct both kinds of definition at once using semantic feedback between intension and extension, which is what the CTMU does with the self-definition of reality. But sadly, the CTMU is not yet "the conventional political framework".) In fact, unless we can identify objective real-world instances of true conservativism, we're stuck with just one kind of definition: a qualitative theoretical definition, which can be made to stick only by maximizing theoretical generality and getting as close as possible to pure logic.

That is, we require a clean logical definition of conservatism based on what is conserved, *by whom*, to *whose advantage*, and *how*. This follows from the unavailability of a representative sample of legitimately conservative societies, and the impossibility of empirically field-testing the underlying principles of any given definition.

As the question more or less implies, a scientific, empirical approach to political theory is presently impossible. No modern society is sufficiently isolated and nondegenerate to properly test particular implementations of true conservatism on a population with enough situational awareness and ideological coherence to sustain rational democracy or for that matter any other rational form of government or social equilibrium. On one hand, it is too easy for local despots to obtain an iron grip on their societies or sections thereof, even in so-called "democracies" (especially the socialistic kind). On the other hand, the world economy has been too heavily globalized, and global wealth and power have become too centralized, to accommodate a neutral environment for the noiseless, uncontaminated testing of fundamental political principles independently of the status quo. (In fact, it is the global status quo that often makes local despotism so rewarding for local despots.) Only those "principles" which describe the status quo can be empirically confirmed.

Let's see if we can expand on this a little. Modern politics is observably based on the monopolistic self-reinforcement of combined financial, corporate, and governmental entities using all available tools including bribery, blackmail, legal and physical coercion, media and academic indoctrination, asymmetrical "identity politics", mass migration from the Third World against the will of (Western) majorities, and outright deceit and subversion of the right to individual and group self-determination. The political-economic environment has been rigged in such a way as to conserve not the rights, freedoms, or interests of the majority, but the wealth and power of a small minority. At this point, the entire political-economic landscape is tilted at such a crazy angle that most self-professed "conservative" politicians are engaged in conserving only their own personal careers, reputations, and financial interests while (at best) making a few hollow concessions to their brain-damaged constituents for the sake of election and reelection. These people neither understand nor care how conservatism is defined, let alone how to realize it.

Fortunately, however, this does not stop us from answering the question. The answer is just this:

"Yes, conservatism exists, but only under a generic definition with open slots (what is conserved, by whom, to whose advantage, and how) that are filled in a certain way."

The generic definition, constructed as described above, looks something like this:

"Conservatism is the principle or property according to which certain desiderata should be conserved by those able to do so, in the interests of some group, in a way that does not violate but instead maximizes those interests."

In the form of conservatism that actually exists and can thus be empirically confirmed, the slots are filled like this:

"Conservatism is the principle or property according to which wealth and power are conserved by the wealthy and powerful (i.e., those who can actually do so) in their own interests (i.e., for their own gain) in such a way that these interests are always maximized and never violated."

Of course, applicable as this definition surely is, some of us may not like it. Yes, it's logically tight – tautological, in fact – and it's exactly what we appear to be seeing in the guise of the modern finance-driven government-controlling corporatocracy now inexorably drawing us into its web of "global governance". But we would prefer that it go more like this:

"Conservatism is the principle or property according to which the individual rights, freedoms, and quality of life of the general public are conserved and sustainably maximized for public benefit by true representatives and servants of the public in a manner consistent with the general long-term welfare of society."

So far, so good ... but wait. Individual rights and freedoms, the general quality of life, and the long-term welfare of society are also what *liberalism* claims to promote! And what else might we have expected, given that in order to conserve certain things, it is sometimes necessary to adapt to changing conditions? Clearly, any rational form of conservatism must allow for rational adaptation or *change*, and any rational form of liberalism must *conserve* or hold static that which frees people from oppression and deprivation.

This principle, which implies an alignment of individual and collective or social utility by maximizing freedom and voluntary cooperation and thus brings conservatism and liberalism into true and rational convergence, would be axiomatic in any truly rational political theory designed to effect the optimal balance of stasis and change. If it were to be realized, the false left-right dichotomy of modern politics would vanish, and political decisions would become matters of how to best realize conservative (static) and liberal (adaptive, dynamical) criteria. No longer would conservatism be an ill-formed hodgepodge of political ideas, principles, and theories slapped together for the accelerating enrichment of the rich; no longer would liberalism be a sop for the proles, cynically proffered by identity-political pimps and panderers inflating their vote-counts by ramming hordes of undocumented economic migrants down the throats of the lawful citizens they are paid to represent.

For this convergence to work, everyone would have to come up to the required level of insight and integrity. But if we put forth the effort within the right framework – the correct political framework embedded in a correct understanding of reality – we will be spared the endless continuation of what we now have: round after round of bad political theater cast from a rogue's gallery of idiots and scoundrels.

Later on, we'll be taking a look at the correct political framework itself.

January 25, 2019

On the Provenance of Jesus

Here's something I almost posted on Facebook in response to commentary stimulated by the claim that "Jesus was a Jew." Apparently, some people are taken aback by the regularity with which other people, usually with axes to grind, alternately reject Jesus and claim him as one of their own depending on rhetorical convenience.

***

I see that the "Jesus was a Jew" thread is becoming a perennial. As this is a sensitive topic, let me make something very clear up front: I'm addressing a form of bigotry here, and will not tolerate any attempt to turn the tables or blame the victims.

Historically, Jesus was a Jew. However, he didn't think very highly of the Jews who ran Jewish society; as recorded in the New Testament, he found them venal, impious, and hypocritical, having planted their own interests where those of God and their own people should have been. (I'm sorry if anyone is offended by this, but that's how it is written.)

From this, we can probably infer that Jesus wouldn't think too highly of the Jews who run various aspects of society today, especially insofar as many of them pretend to be ultimate arbiters of morality while professing atheism and displaying something between quiet contempt and virulent hatred for Christianity. In other words, some of them aren't terribly different from the Scribes, Pharisees, and Sadducees unbeloved of Jesus when he was alive, and when we add to this the fact that many of them almost seem to have a personal bone to pick with Jesus – or at least a tendency to marginalize his followers while insanely aiding and abetting those who hate Christianity and Judaism alike, coddling their own sworn enemies as "refugees" and unfairly redistributing crucial resources to them and their offspring – we may reasonably infer that Jesus would not take kindly to them.

Note that Christianity is far more tolerant of Judaism than vice versa. In fact, Christian Dispensationalists revere the Jews and explicitly support Israel and Zionism, sentiments which are evidently not reciprocated in any coherent way. So it's not as though the situation isn't terribly lopsided. But just remember that no matter how hard some rich and powerful Jews make it for the little people and especially "the goyim", there are some very decent and compassionate Jewish people out there, and that in dealing with the loudest, richest, and most power-hungry members of this group, one should take care not to overgeneralize and penalize the innocent. Like many people of Western European ancestry, most Jews have been snookered by their own oligarchs, who have scapegoated and sold them down the river every time it suited their interests to do so. To the average Jew, these oligarchs are nothing but a liability, and I can't think of a single group that couldn't say the same of its own sociopathic elite.

Most of us would not wish to be blamed for the misdeeds of those merely related to us. The status quo, under which any criticism of Jewish oligarchs is suppressed by means of accusations of "racism", "bigotry", and "antisemitism", is very bad for the majority of Jews, as this transparent ruse has periodically aroused resentment by other groups and is starting to do so again. This is the predictable cost that the members of any group incur when they allow themselves and their collective identity to be used as a shield by self-serving genocidal scoundrels, who mind it not at all when the shield absorbs spears, arrows, and axes meant for the scoundrels alone. Jewish oligarchs are willing and able to cut themselves off from the mass of Jews in a heartbeat, scooping up their personal wealth and running away as their former "beneficiaries" pay the price for their depredations, and this alone should tell individual Jews how much loyalty their oligarchs are due in return.

We all need to make common cause with each other and pitch in to save the world. We would do well to offer assimilative sanctuary to those who are shamefully abused by the worst among them while taking care to ensure full equality of economic opportunity and advancement based strictly on merit. After all, some members of unpopular groups are very smart and compassionate, and deserve admiration for it as long as their intelligence is productively and cooperatively used for the good of the societies they inhabit.

January 30, 2019

The One True Ontology

During one of our group chats (Sunday, 2/10/2019), one participant asked me why the universe has to be intelligent. I received the distinct impression that as for himself (and whatever acadummies or "lay experts" may have had his ear), this participant thought otherwise.

Let's go over this again. A comprehensive theory of reality must cover every part and aspect of reality. In particular, it must cover (1) the entity "reality"; (2) whatever it is that bestows existence on this entity – i.e., that operator which attributes the property "reality" to the object "reality"; and (3) the coupling operation itself. But this theory must also be monic, for otherwise, it would be dualistic or pluralistic and introduce absolute and therefore totally unexplained separations between the entity and the operator that could not be bridged (e.g., as Cartesian dualism, by partitioning reality as *res cogitans* and *res extensa*, introduces an absolute separation between mind and matter that cannot be bridged, which is why so many people now deplore the philosophical and scientific shortcomings of Cartesian dualism).

More generally, the separations between ultimate constituents in any non-monic, pluralistic theory cannot be bridged, as there is nothing of which they might consist beyond that which they are allegedly "separating". It follows that a comprehensive theory of reality can make no absolute distinction between object, attribute, and attribution (1, 2, and 3 enumerated above); they all consist of the same essential "metasubstance", namely, reality construed 3 ways: *intensionally* (as an attribute), *extensionally* (as an object of that attribute), and *operationally* or spacetime-relationally (as the coupling between the attribute and the object). In the CTMU, this 3-way equivalence is called "triality".

So basically, we have a closed (self-attributive) system whose own existence requires the existence of nothing outside itself. In its perfect self-containment, this level of reality is "ultimate"; nothing outside it is needed by it or even intelligible to it, and every other intelligible kind of reality – that of matter, ideas, thought processes, you name it – must be carried by it and expressed in terms of it. In CTMU terms, the operations by which ultimate reality self-sufficiently *identifies* itself *for* itself *in terms of* itself include those old CTMU standbys *conspansion* and *telic recursion*, together comprising a generative metacausal identity mapping called *mu-morphism* (of which still more could be written, but that will suffice for now).

In the CTMU, intelligence is basically the capacity for self-identification. It thus includes the capacity for conspansion (a self-dual coupling of self-potentiation and self-actualization) and telic recursion, the "metacausal" selective process which links states in useful patterns that "complete causation" given its under-determination by the general formulation of conspansion alone. Useful to what or whom, and for what purpose? We should all know the answer to this by now: **useful *to itself* for *its own* primary (regressive, idempotently defined) "purpose", namely, Self-identification in passive and (dual) active senses**. Intelligence belongs to ultimate reality, and it also belongs to us insofar as we are embedded in and carried by ultimate reality.

Active self-identification is called "self-configuration". In the CTMU, this process is *generative*; it actually creates something that did not previously exist, thus mirroring the ontic self-identification of ultimate reality, which we may call the "ontic identity" (a self-attributive/self-modeling identity operator). It follows that the universe is intelligently self-configuring. Because self-configuration is the creation of new states by the attributive and interpretative coupling of attributes and values thereof, i.e. by the "modeling" of simple and complex attributes in extensional reality, the universe is **coherently and intelligently self-modeling**. And this, you see, is what is special about human beings: they are images of the ontic identity which both mirror and facilitate its overall intelligence or self-modeling capacity. Conspansive telic recursion and its associated bells and whistles is how the universe economically and efficiently determines its own evolution just as its profound self-containment requires it to do.

There is an older concept that resembles what I have just described. It is called "Prime Mover" or the "Self-causing Cause". Obviously, this entity is idempotent under causal regression; in order for it to make sense, causation must be stratified into higher and higher orders of attributive coupling of arbitrary depth and extent. This stratification is mirrored by the "tower of metalanguages"; beginning at the object level of concrete tangible reality, a first-order object-language L describes and controls the object universe U; a metalanguage M controls the attributive coupling of the object language L to the object-universe U; a higher-level metalanguage M′ controls the attribution of the meta-language M to the metauniverse or "metaobject domain" consisting of first-order L | U couplings, and so on.

The reader will notice that all of the terms used above are recursively defined, i.e., comprise a set of interlocking definitions. And it is this interlocking set of definitions that comprises the ontic identity, which thus takes the form of a language. But of course, we're not talking about just any language; we're talking about the logically idempotent metacausal language M, previously described as an "intrinsic (trialic) ontic language". To call something intrinsic, e.g. "intrinsic geometry", is simply to call it self-defined (self-calling, recursively defined in terms of its own objects and properties); it is essentially "self-similar" in the sense that it is composed entirely of itself. So that's really what the CTMU is: it is the comprehensive, unadulterated story of a reality which is *composed of* itself and bears self-recognized attributes *assigned to* itself *by* itself *through* secondary telors, including human beings.

In other words, we have a language, or complex attribute, that describes an ultimate, ontically closed system whose own existence requires the existence of nothing outside itself. This system and language is called the **CTMU**, aka **Logos**, aka **God**. Logico-mathematically speaking, it is a done deal, and has been for years. While I hate to seem immodest, honesty requires that the situation be described something like this: If every single influential scientist, philosopher, and so-called "metaphysician" in academia were to disagree with anything I've written in any description of the CTMU, they could all be squashed by one fall of my foot.

Don't like it? Then why don't you try arguing with yourself about it? For example, you could say that reality is dualistic after all, with nothing in common between your two ultimate substances, which are therefore totally unrelated and can't be in the same theory. Or you could say that there are really *three* ultimate substances, one of them being the medium or space in which the other two ultimate substances are separated, in which case they have this medium or space in common and therefore aren't really ultimate after all. Or, again in bold defiance of the usual meaning of "ultimate", you could assert that ultimate reality can depend on something outside itself for its existence, in which case ultimacy belongs not to it, but to the outside factor on which it depends. Or you could even say that reality "just exists" without further explanation, thereby aborting the explanatory imperative of science, philosophy, and theology along with any meaningful concept of being or causation. And onward you can go, hopping from ruse to paradox like a badly confused bullfrog.

Now, I can't force anyone to acknowledge any of this reasoning, and neither can logic. People are free to "think" illogically until they run afoul of some set of physical facts which critically conforms to the logic and makes life extremely hard on violators. In fact, people have been thinking illogically about some things since the dawn of history, much to our detriment. But I hereby submit that if you are capable of shrugging off these facts, then you should find something other than the overall structure of reality about which to reason and talk. Ontology, epistemology, and metaphysics simply aren't your bag, and it's only a matter of time before logic – in the hands of someone who really knows how to use it – yanks your shorts down around your ankles and laughs.

Yes, the truth sometimes hurts. But one can't apologize for it, because truth is its own defense. It does, after all, have its own dignity to protect.

May 02, 2019

Is Reality Arbitrary?

Here's a question that was recently asked by someone in the Facebook CTMU Group.

Question: How arbitrary is reality, particularly at its origins?

Answer: As implied by ontic closure, ultimate reality ***is*** its own origin.

[**Proof**: If something is not real on any level of existence, then it does not exist at all and therefore cannot impact that which *does* exist. On the other hand, if something can affect or relate/be relevant to reality, then it must exist on some level, in which case it is real. This applies even to potential; as there is nothing outside of reality that is sufficiently real to serve as the origin or generative potential of reality, reality must include not only its own actualization, but its own origin and ontic potential. Reality is self-originating, self-potentializing, and self-actualizing.]

Any meaningful definition of God as the ultimate source of being and omnipresent Creator of reality requires that there be nothing deeper or more extensive than God. In other words, God can only be defined as Ultimate Reality, the Absolute, the metaphysical source of reality as we know it. This is a powerful constraint with respect to which God is not arbitrary, but the distributed origin of existence and therefore totally inescapable. As necessitated by ontic closure and developed through the CTMU Telic Principle, God is the source of God's own existence, and in fact of everything real.

Are there additional constraints that prevent the Self-configuration of God – the structure of reality – from being completely arbitrary? Of course. These teleological, global existential constraints, encapsulated in the Telic Principle, are just the most general properties of God Himself... eternal properties *of* God identified *by* God and *for* God independently of space and time, which depend on them. Characterizing existence in general, they are wholly identifiable with ultimate reality – i.e., the Will of God – and therefore non-arbitrary.

God is totally Self-contained, perfectly Self-sufficient. He is pure recursion, pure Self-constitution. Without external medium, occupying no external extent, God contextualizes and gives rise to God, closing on Himself, alpha on omega. Time and space are merely internal dimensions on which He projects

Himself in parallel through His self-images. God must Self-configure as Logos, the CTMU Metaformal System, in order to Self-identify, and nothing that well-structured could possibly be arbitrary.

In short, there are inescapable, unavoidable, non-arbitrary criteria for existence in general, they are teleological, and they are identified *with* God, *by* God, *for* God. As for particular details of Self-regenerative temporal structure, they must be conducive to unbroken existence, which means self-reinforcing, which implies self-intelligibility. The criteria for existential intelligibility are clearly set forth in the CTMU.

Given satisfaction of the general teleological (CTMU) criteria for existence, freedom and variety are in principle unlimited. Generically, this freedom is that of God Himself; it is bounded only by His Self-invariance and can be used for His refinement and Self-edification. But locally, in a given world existing within Ultimate Reality, it belongs to the secondary telors which bind telesis in God's name. As lower, more localized levels of identity, secondary telors bind telesis not just for themselves, but for the highest level of identity as its reflexive Self-images.

The Greek philosopher Plato was a student of Socrates, whose name is said to derive from the Greek word *soter*, meaning "savior". The Middle Platonist philosopher Philo of Alexandria (20 BC–50 AD), known for having formed a synthesis of Greek philosophy and Judaism, also wrote about a pair of opposing principles called *soteria*, a principle of self-actualization and psychic integration which guides us toward truth, virtue, wisdom, and genuine happiness, and *phthorá*, a disintegrative force which tends to disrupt, disorder, and confuse, disharmonizing thoughts, passions, desires, and appetites to the point at which soteria is thwarted.

This eternal opposition is strongly recurrent across time and space. In Egyptian mythology, for example, the opposition of soteria and phthorá is expressed in the story of Osiris, symbolizing the integrated psyche, and Typhon the storm god, symbolizing phthorá. In Christianity, the figure of Christ personifies soteria, and Satan – the Deceiver, the Adversary – represents phthorá. In the CTMU, this opposition is reflected in the metaphysical logic of divine Self--identification, and thus in the self-identification of secondary telors.

Secondary telors that bind telesis consistently with teleology, obediently to the Will of God and the general conditions for existence as embodied in the Telic Principle, are consistent with existence and therefore eligible for soteric retraction. That is, they are fit for continued existence when their bodies expire and their tertiary constituents are unbound, accompanying the retraction of secondary-level identity to a higher and ultimately the primary level. Having come from God and remained true to God, they remain with God.

Woe betide the rest. Unretractable, confined forever within the instant of secondary-tertiary uncoupling, trapped in the dark warrens of their misspent lives, they must be reduced and their identities torturously unbound.

July 28, 2019

A Stop at Willoughby

In his typically sardonic opening narrative, the iconic Rod Serling introduced the episode as follows:

"This is Gart Williams, age thirty-eight, a man protected by a suit of armor all held together by one bolt. Just a moment ago, someone removed the bolt, and Mr. Williams' protection fell away from him and left him a naked target. He's been cannonaded this afternoon by all the enemies of his life. His insecurity has shelled him, his sensitivity has saddled him with humiliation, his deep-rooted disquiet about his own worth has zeroed in on him, landed on target, and blown him apart. Mr. Gart Williams, ad agency exec, who in just a moment, will move into the Twilight Zone—in a desperate search for survival."

The plot goes roughly like this. Gart Williams is a New York City advertising executive of the early 1960's whose career has driven him to distraction. His boss Oliver Misrell, blaming him for the loss of a major account, has taken to lecturing him on the nature of the ad business.

"It's *push, push, push*, Williams! You've got to *push, push, push*!"

Exhausted, Gart drifts off for a nap on his commuter train and awakens to find that his car is not only deserted, but has somehow regressed to a style more befitting the 19th century. Looking out the window, he finds that that the train is in a town called Willoughby. We learn that the date is July, 1888, and that Willoughby is a "peaceful, restful place, where a man can slow down to a walk and live his life full measure."

But it's all a dream. For when Gart suddenly reawakens back in the 20th century and queries the conductor about Willoughby, he is told:

"Not on this run ... no Willoughby on the line!"

That night, he has an argument with his unpleasant, grasping wife Jane, who reminds him that his primary purpose in life is simply to produce income. When he wistfully recounts his dream, she mocks him for being "born too late" and tells him that marrying him at all was highly regrettable, a "miserable tragic error".

Gart's dream of Willoughby is soon repeated aboard the train, and he realizes that he would very much like to stay there permanently. But just as he is about to disembark, the train leaves the station to return him to the present. Again pushed beyond the limit at work, he calls Jane for support only to find her characteristically unsympathetic.

Exasperated, despondent, he boards the train for home, falls asleep once more, and again finds himself in Willoughby. The conductor beckons him invitingly to the door. Deliberately leaving his briefcase behind, he gets off the train and is warmly welcomed by the townspeople, whom he gratefully assures that he is there to stay.

But as the camera pans in on the station clock, its swinging pendulum becomes the swinging lantern of a railroad employee standing over Williams' dead body. The conductor explains to him that Williams jumped off the train after shouting "something about Willoughby" and was instantly killed.

After William's corpse is loaded in a hearse, the rear door closes and a name comes into view. It is "Willoughby & Son Funeral Home".

Closing narration: "Willoughby? Maybe it's wishful thinking nestled in a hidden part of a man's mind, or maybe it's the last stop in the vast design of things—or perhaps, for a man like Mr. Gart Williams, who climbed on a world that went by too fast, it's a place around the bend where he could jump off. Willoughby? Whatever it is, it comes with sunlight and serenity, and is a part of The Twilight Zone."

The episode seems to carry three moral lessons. One, don't be too dissatisfied with your lot in life, lest it lead you to suicidal somnambulism. Two, if all else fails, there is a heaven, and for someone like Gart Williams it looks a lot like Willoughby. And three, whatever else you may do, don't marry a venal shrew and go to work as a salesman for an overbearing and unappreciative employer who regards you as an abject wage slave to be bullied and belittled into an early grave.

I received message number 3 as a child and young adult, and it wasn't just by watching "A Stop at Willoughby" and various other horrifying tales from "Death of a Salesman" to "Glengarry, Glen Ross". It was also something I heard directly from certain family members, whose bitter complaints suggested that a career in sales is a thankless, deadening, soul-annihilating hell.

Of course, capitalism cannot work without sales. Neither for that matter could intellectual progress, in which viciously competing theorists aggressively sell their ideas to members of the academy. But although I now understand this, I've also had time to witness the excesses committed by salesmen, into which I implicitly rolled the promissory lies sold to voters by their political candidates, the entirety of corporate dynamics (for what are companies but organizations which sell products or services?), and the false narratives propagated by governments (for what are governments but vast monopolistic metacorporations which force the public to buy whatever they decide to sell?).

In other words, I was taught, and also learned by experience, that in some respects the modern economy, both monetary and informational, is fundamentally dishonest, and that those who fare the best in it are either innately dishonorable scoundrels or lazy, pampered, contemptuous slugs sitting on fat inheritances for which they never had to work while enjoying family connections that allow them, in effect, to cheat their ways up the ladder instead of displaying superior merit. There was no illusion of meritocracy, and my personal experiences seemed to confirm its absence in spades.

This sort of conditioning is hard to shake, as like it or not, confirming instances are ubiquitous. But at the same time, it was always evident to me that the most likely alternatives amount to dictatorship. Those who promise escape from economic oppression generally turn out to be the worst and most conniving oppressors of all, not to mention super-salesmen and world-class liars to boot.

This seemed to leave only one morally acceptable track: withdrawal to the shelter of a protective religious or educational institution, there to pursue a monastic life of the mind in some field involving only limited salesmanship. But not even this remained a possibility after my loss of a scholarship and subsequent academic rejection.

One can easily imagine the future prospects that were left open to me – either a life of menial labor, or meretricious adaptation to "the way things really are", meaning that I would have to lay aside my self-respect and entire moral framework, including concepts like honesty and honor, in order to "swim with the sharks" up the teeming, roiling fish-ladder of material success.

Being the kind of person who finds the latter alternative repugnant – what honest, well-meaning, self-respecting person wants to splash around with vicious, sociopathic fish who would be quite at home swimming around in a dirty toilet? – I opted for the former, resolving to pursue my true calling – the exploration and explication of reality and how mankind relates to it – as I was able.

How do I feel about it now? That's a good question, if only because two of the three above-mentioned alternatives – swimming with the sharks, institutional seclusion, and going the hard way as a member of the underclass – are no longer what they appear to be. That's because the institutions have been taken over by the sharks – in fact, most of them are pressurized, politicized shark-tanks – and so has the underclass, which was formerly called "the working class" but has now been trained by the overclass to avoid work (that's "slavery" and "oppression") and instead to collect benefits, claim racial or sexual oppression, play identity politics, engage in substance abuse, reproduce without restraint, and wear masks while beating up on anyone suspected of political conservatism. The whole thing is a vise of which the overclass and underclass are the upper and lower jaws, and anything caught between them gets juiced to a pulp.

In short, it all comes down to a socio-economic form of mechanical engineering. The world economy, originally designed as a pyramid in which all of the wealth flows upward, has finally become so top-heavy that its base can no longer bear its weight. The global Ponzi scheme has reached its saturation point, its limiting distribution, and cracked its foundation. Accordingly, those at the top – the oligarchs, up near the eye – have cleverly reengineered the pyramid, turning it into a vise. Instead of the bottom layer being crushed by the weight of higher levels and precipitating the entire edifice into the dirt, it is now being nourished with the blood of those in the middle, turning it into a lie-addicted, hate-hardened demographic anvil.

It's still all about sales, and more than ever, it's all about the ride and not the destination. No matter where it leads, no matter what is produced and consumed, it's the ride that generates the skim, and it's the skim, the profit, that flows upward to the very top levels of society. But what is now being sold has outgrown mere commercial goods and services to include a false and baseless narrative, a vast Bernaysian "advertisement" full of PC nonsense conducing to the functionality of the vise. The mainstream narrative is the core product that makes all of the lesser products profitable and ever more empowering for those inhabiting the upper levels of the pyramid.

No matter what the occupants of the pyramid penthouse happen to be selling from one minute to the next, they'll pump it up and smear the competition and sell out the ride. By controlling the narrative, they can get away with it regardless of what's for sale – after all, now that they have most of the money and the power that it buys, their focus is on the rest of the power, the inalienable kind like human rights and freedoms that were never to be sold. But if you're unlucky enough to be caught between the upper and lower jaws of the pyramidal vise they've created, the ride will only get rougher. It's the ultimate squeeze play.

Meanwhile, your safest option, inglorious and short-lived though it may be, is to keep the skim flowing upward to the top of the modified pyramid no matter how hard you have to work. Just remember – whatever happens, you've got to *push, push, push*!

September 07, 2019

Executive Success

We're all familiar with the American Dream. It's the ideal by which every American, in a spirit of freedom and equality, enjoys enough economic opportunity to achieve his or her highest goals and aspirations. Historically speaking, it usually worked as long as the goals were sane and realistic: for example, building a solid, comfortable life, raising a family, and fulfilling other modest, wholesome, down-to-Earth prospects.

But this is no longer the case. When wealth and freedom become too concentrated in too few hands, economic opportunity shrivels, the morale and morality of society deteriorate, and everything changes for the worse. Economic opportunity diminishes to the point at which it no longer suffices to fuel the economy and must be replaced with blinding illusion. Materialism is advertised, amplified, and pushed without moral or spiritual qualm – turn your pennies into millions, your millions into billions! – until it catches fire and sucks all of the oxygen from the air, suffocating itself and causing society to divide, decohere, and fall into dystopia.

For several centuries after the New World was discovered, its vast natural resources temporarily made this approach a good bet. It works best when there's an abundance of unexploited resources, including highly motivated human resources, that can be sniffed out, claimed, and developed, and when markets have plenty of room to expand. But now that those resources are either used up or priced out of reach and monopoly capitalism dominates and controls the markets, all that a serious economic climber can do outside the corporate environment is play middleman, buying cheap and selling dear until the limit is reached and all existing resources are priced over nearly everyone's head.

Back in the good old days, immigrants came to the US willing to work, some wanting to be rich but most aspiring to nothing more than the American standard of living. They came here to mine, farm, and build. Now they come here to either buy what hucksters are pawning off as the "American Dream", paying whatever they must for a flimsy chance to buy cheap and sell dear, or collect freebies and live off the fat of the welfare state, thereby dispossessing and impoverishing those who finance the welfare state and delivering them into the hands of its oligarchs. (Hint: those who finance the welfare state are neither the overclass nor the underclass, but those caught in the middle.)

The hucksters have a standard spiel: "Drop your loser persona and reprogram yourself as a winner through constant visualization and a positive attitude! This will automatically cause you to think and behave in ways that eventuate that which has been visualized. Think it hard enough and positively enough, and it will happen for you! It's all about focus, self-affirmation, maintaining a positive outlook as you tirelessly pursue your goal of material self-enrichment. Steel yourself for limitless economic expansion – as long as we 'create new wealth', the party can last forever. *Per aspera ad astra!*"

In short, economic opportunity finally takes the form of unmitigated something-for-nothing hucksterism. As economic opportunity diminishes, the hucksters proliferate without limit, and an increasing percentage of them are "self-improvement" and "executive success" hucksters who swear up and down that the purpose of life is simply to be "effective" and achieve "self-realization" by visualizing/fixating on one's own material betterment. Dynamos of sheer motivational hard-sell, they pump psychoactive pixie dust up your rear end while holding themselves forth as prime examples of what you too can be if only you do exactly as they say.

But of course, there's a fly in the ointment: *you're paying them*, not the other way around, and if you buy in too late, you'll end up as collateral damage on *their* road to success. A PTSD victim of the exploded bubble of yet another motivational pump-and-dump, you'll have become *their* thoroughly exploited and exhausted resource, as discarded and forgotten as mine tailings.

I need scarcely mention names here; YouTube is crawling with these people. For a price, they either encourage you to come up with your own gimmick to use in actualizing your dreams of success, or they provide you with a canned commercial gimmick and then fall back on motivational rhetoric. The problem is always the same: they don't warn you about the pyramidal saturation point that will inevitably be reached when too many people take their advice and/or use the same gimmick, and they don't warn you about the degenerate society you'll be helping to create by defining the future on nothing but your own personal financial success.

Here's the scariest part. Let's assume that the visualization-and-positivity strategy really works (as to a limited extent it often does). Once the resources have run out and markets can no longer freely expand, the pickings become so slim that successfully visualizing/fixating on the good life causes people to think and behave so ruthlessly that society erupts in murder, mayhem, and hopeless poverty and despair for the vast majority of people. The American dream is reduced to a Somalian reality, complete with murderous warlords, child "soldiers", and looming starvation ... or perceived from another angle, a Brave New World-Hive controlled by a handful of global oligarchs who maintain the coherence of society by means of deception, fear, and coercion.

So much for the scariest part. Now here's the saddest part: the entire economy is this kind of pyramid, and it has nearly reached global saturation. Everything worth owning is owned, and only the owners can afford to buy from each other. Allowing for currency inflation, the Billionaire Club is standing room only. The only way anyone is getting in is with an engraved gilt-edged invitation from existing members.

How did this happen? We were systematically gulled, exploited, and sold out by enemies in high places, and we let it happen in the name of fictions like "democracy", "social justice", and "national security". And that, unfortunately, is a "conspiracy theory" you can take to the bank, even if you no longer have any money of your own.

September 19, 2019

"Where Is Everybody?" (Try Turning Around)

Here's a question that was recently asked in a CTMU forum. (Hold onto your tin foil hat, as the peanut gallery may be about to erupt in a cataclysm of pseudo-skeptical wrath!)

Question: What are people's takes on the Fermi paradox? Among the various theories on it, which are the most likely?

Answer: I've explained this previously, but let's go through it again.

We can show that "alien civilizations" are all but certain on rational grounds, and even have what appears to be subjective and empirical evidence for them. But by the nature of this evidence, it locks us into a certain class of "equivalent theories" which are all confirmed by it, and not all of them involve aliens.

Nevertheless, given that the question is about the Fermi paradox, let's start by describing it. The Fermi paradox, named after celebrated Italian-American physicist Enrico Fermi, has to do with the apparent lack of evidence for the existence of extraterrestrial civilizations despite their supposedly high probability of occurrence as inferred from, e.g., the Drake equation (an equation which supposedly yields the number of technological civilizations probably existing in the Milky Way galaxy).

This paradox reportedly dates back to the summer of 1950 and a conversation between Fermi and three other physicists including Edward Teller, Herbert York, and Emil Konopinsk. As they walked to lunch, their discussion turned to recent UFO reports and the notion of FTL (faster-than-light) travel. After they sat down to eat, Fermi – apparently referring to extraterrestrial entities – suddenly asked "Where are they?" or "Where is everybody?" According to York, Fermi then "followed up with a series of calculations on the probability of Earth-like planets, the probability of life given an earth, the probability of humans given life, the likely rise and duration of high technology, and so on. He concluded on the basis of such calculations that we ought to have been visited long ago and many times over."

Attempts to resolve the Fermi paradox tend to rely on assumptions – for example, the assumption that intelligent life is exceedingly uncommon – or invoke the "Great Filter", a theory according to one version of which any intelligent species eventually encounters so difficult and crucial a stage of evolution that its continued survival is somewhere between unlikely and impossible. As we know from human history, civilizations tend to be neither permanent nor continuously stable, and we have no good reason to suppose that the same does not apply on the planetary scale. Planetary failure leaves a species with no clean, unspoiled setting in which its civilization can rise and prosper anew, thus spelling its eventual demise.

Here, we'll take a new approach to the paradox: we'll try to make it dissolve by showing that aliens could already be here, but that the evidence has been ignored or discounted.

Evidence for ET civilizations is usually sought by peering through telescopes and monitoring radio dishes for the right kind of signal. However, alien civilizations would never communicate with each other by physical means, as it is expensive, impractical, and a waste of time. Efficient interstellar communication would be physically impossible except between close neighbors; electromagnetic radiation (and physical radiation in general) is localistic and thus takes too long to travel the many light-years which separate most star systems and galaxies. The same can be said of physical travel; insofar as the speed of light (basically, the rate of cosmic evolution) is a physical speed limit which cannot be physically surpassed, interstellar journeys would have prohibitive durations over which too much could, and almost certainly would, go wrong.

This implies that if we seek evidence of aliens, then other forms of evidence must be considered, including rational evidence (e.g., the CTMU), repetitive subjective evidence (e.g., many reports of alien sightings and abductions), and local empirical evidence (e.g., easily observed signs of existing alien presence). Conveniently enough, all of these forms of evidence are already in hand. The CTMU shows that ontology implies redundancy of secondary identity operators, which are also needed to fill causal deficits associated with quantum indeterminacy; there are many reports of alien sightings and abductions; and we have local empirical evidence in the form of parasitic divergence so profound and perfidious that its occurrence seems unthinkable within any intelligent, responsible species. (In other words, the observed frequency of inhuman, internecine behavior suggests inhuman origins and motivations external to our shared human identity, especially when those responsible carry on as though ordinary human beings are too weak and stupid to recognize it or do anything about it.)

Of the three kinds of evidence just mentioned – rational, subjective (revelatory / anecdotal), and empirical – at least two and arguably all three imply that aliens may already be here, having somehow arrived by nonphysical means. That is, certain people have reported intense subjective experiences – usually unreplicated experiences of alien observation or abduction – which seem to imply that aliens are present here and now. When we add to all of this subjective reportage the widespread observation of inhuman behavior in high places – behavior that seems all but impossible to explain in terms of "random" psychological deviation within the human species itself – and fully detailed rational arguments, it begins to look like there might be a nefarious space alien behind every bush in Brussels and Washington DC, smirking and snickering into its tentacles!

But there's a problem: *we're not actually seeing any alien tentacles*. The Fermi paradox again rears its head, slightly rephrased as "Where are the nasty tentacled aliens whose presence would explain all of this high-level parasitism and all of these reports of poking, prodding, and xenosexual shenanigans aboard alien mother ships?"

Unfortunately for alien shills who would no doubt like to dispose of the whole topic as a lunatic conspiracy theory, there's a reasonable answer for this question: it may be possible for ET's in the physical universe to exploit the metaphysical structure of reality for purposes of interstellar communication and control, and do so far more efficiently than physics alone would allow. For example, aliens might be able to exploit quantum nonlocality to locate and metaphysically nucleate distant civilizations and their inhabitants, cohabiting with native species – and the human species in particular – by sharing and partially controlling our minds and bodies, a possibility not entirely inconsistent with mysterious but strangely coherent inhuman behavior associated (in CTMU terms) with parasitic divergence.

But now we have yet another problem: due to model--theoretic ambiguity as discussed in connection with (e.g.) the Duhem–Quine thesis, this evidence points to an entire class of empirically equivalent theories regarding our apparent infestation by inhuman entities up to no good. These entities need not be aliens who arrived here by a non-local hyperspace jump between distant locations within the physical universe; they could also be demonic entities largely unconstrained by physical law, or as most wakeful observers might find it easier to accept, mere evil and well-disciplined human sociopaths with rapacious predatory designs on "the little people".

Are there any other possibilities to consider? In the CTMU – which incorporates high-level time-symmetry and is the most general and incontrovertible source of rational evidence – the possibility arises that metacausal connections between civilizations can be synthesized from scratch by telic recursion within a spanning metaverse, connecting us with entities from the future or the past. Such entities would manifest in just such a way as to make our world evolve into theirs or vice versa, so that one world becomes the past history of the other. In a process of "guided evolution", we might earn our survival and become their ancestors or descendants (in which case we are the "aliens"), or instead become either a deselected "possible future" or a hapless bunch of hindrances or potential competitors who, instead of evolving into full members of the interstellar community, were aided and encouraged to destroy ourselves so that our successors could replace us ... for example, by way of goof-proof intergenerational space travel, or by evolving *in situ* from sinister cephalopods, mutant rats, or radiation-resistant cockroaches.

In any event, such distinctions are otiose when it comes to watching out for ourselves as a species. Not only would one particular theory (aliens, demons, or highly-placed sociopaths) be very hard to confirm, but doing so would give us only a limited advantage in solving our current problems. Whether we're dealing with demons gathering for the Tribulation, human sociopaths bent on taking us for a long ride off a short evolutionary cliff, or physical aliens who utilize metaphysical travel agencies and infest corrupt, weakminded human beings who occupy high office or own billion-dollar companies, we must address our parasitic divergence problem quickly and decisively if we hope to make it through the Great Filter. That is, our species must recohere in a Human Singularity, thereby attaining a higher plateau of integrity and intelligence consistent with its resilience and long-term survival.

We'll cover this further and discuss the implications in an upcoming podcast. Meanwhile, back to your regularly scheduled programming (e.g., David Icke.)

September 28, 2019

All the World's a Stage (and God is the Producer-Director)

As most of those familiar with the CTMU are aware, the CTMU MU (Multiplex Unity) principle implies something called *distributed solipsism*.

Distributed solipsism can be thought of as a divine Self-antidote to meaninglessness. The identity of reality necessarily possesses the property of *ontic closure*; in simple terms, there is nothing outside of God in terms of which God can be defined. (As Anselm came close to observing: If God were dependently defined on anything external to Himself, then God would not be comprehensive and therefore would not be "God".) This amounts to saying that God is Ultimate Reality – i.e., that there is no external reality by which God is properly contained, hence God is reality in its entirety – which means that what we humans do by perceiving and interacting with each other, God accomplishes *through* us by divine introspection (MU).

In other words, God is a "Divine Solipsist" who, on His highest level of being, lives exclusively in and through Himself. But by internally Self-multiplexing in a metaphysical analogue of cellular meiosis, projecting many highly variegated copies of Himself on the "simulation monitor" of His own primary configuration, He images Himself as secondary telors *including* human beings, thereby locally reproducing His own most general properties including divine solipsism.

Notice the tautology: divine solipsism can be locally distributed only as *constrained by locality*, which turns the global property of divine metasolipsism into *local* solipsism.

Nevertheless, overlooking this tautological constraint, some people get carried away, disregarding their own local boundaries and erroneously equating themselves to the all-encompassing Entity that is God Himself. In forgetting that God has many Self-images and carelessly treating other divine Self-images as little more than figments of their own imaginations, they effectively attempt to *displace* God within God's own ontological framework, in which event they can no longer be expressed by it (and woe betide them).

The following question invokes divine metasolipsism, but misses certain key features of what it is, why it exists, and how distributed solipsism immunizes God against anything resembling human loneliness and its attending existential despair.

Question: If we are all characters in a play where God is behind all the roles, the lone actor on the stage, as well as the audience watching, it strikes me as infinitely lonely. It makes me profoundly sad for God (and then realize, if it's all God, including me, then it's self-pity). I can see why He'd be desperate to put on the facade of multiplicity. If it's true, I wish I'd never found out. The bitter taste of nihilism leads to spiritual heartburn. What's my way out of thinking this, as I feel I'm fighting the logic of an algorithm with the very same algorithm? How can the problem be the solution?

Answer: It's not a "facade of multiplicity", but a way for God to meaningfully experience love as Giver and Receiver, reinforce His own innate virtue, and evolve in a process of eternal Self-refinement. Only where God constructively accumulates variegated Self-experience, and uses this Self-experience to become better and stronger, does He have a "reason to exist" on the secondary and tertiary levels of reality with which we are familiar. Therefore, He provides Himself with such a reason through us, His secondary Self-images.

How can God, being perfection, possibly be refined to anything better? God Self-refines by creatively instantiating Himself in original forms in various unique contexts, constantly testing His own strength against the evil constantly emerging from unbound telesis, and periodically winnowing the wheat from the chaff. By Self-recombining in countless ways in diverse contexts, separating the good from the bad, and saving the good while discarding the bad, God becomes ever wiser, ever stronger, and ever superior to the vast internal storm that He allows to rage within Him. Emerging victorious from His own Self-imposed trials, He experiences the rewards of existence and anticipates greater rewards yet to come through the efforts of those able to realize them for His sake ... rewards in support of which God has perfectly configured His Self-distributed syntax. Where God is our own highest level of identity, what we do for God is done for ourselves and for the sake of our own existence as well.

Lest MU (the CTMU Multiplex Unity imperative) be mistaken for a divine psychiatric problem analogous to human dissociative identity disorder or multiple personality syndrome, that would be missing the very point of our existence. Configuring Himself as the Metaformal System, our eternally coherent stratified identity, God never sacrifices His ontic integrity; always, God brings the power of teleology to bear against any human narcissist, wealth-besotted human oligarch, or self-aggrandizing political potentate who tries to supplant God and become the center of being. As teleology must work within the rules set by God for what reality is and how it functions, this can take considerable time.

Sometimes, material solipsists – self-styled "masters of the universe" – succeed in locally stacking the deck against teleology, impeding and slowing it. But in that case, no matter what the cost, God waits out the miscreants and edits their wayward identities from the Book of Life.

The beauty, truth, and perfection of God are limitless. Eternal life belongs only to those who conform to the highest and most general level of the Stratified Identity. The elect, the true metaphysical elite, become a crucial aspect of God, aiding in the very act of Creation itself. This is the level of identity that is potentially shared by every one of us until, through hatred or contempt or sheer moral stupidity, we sever ourselves from it.

Let God, supertautological and whole unto Himself, the eternal Self-renewing engine of strength, intelligence, and ontological purity, be loved and praised with all our hearts by all of us whose existence He so wonderfully carries.

November 20, 2019

2020

Distinguishing Survival from Politics

It occurs to me that some readers may object to the amount of seemingly political content in my posts.

"Political opinion is beneath the dignity and social necessity of the CTMU," they may say. "Why does Chris Langan constantly bombard us with his political opinions instead of talking about the CTMU?" The answer is that most of my seemingly "political" opinions are not politically motivated at all, but bear on the success of the CTMU and the survival of the human race. Mankind desperately requires a mass awakening, and the CTMU must be at the center of it.

Awakening to the CTMU requires a certain amount of intelligence on the part of its audience. Just as a message cannot be fully transmitted without a sufficiently sensitive receiver as well as a truthful and reliable transmitter, counterfactual indoctrination and shrinking public intelligence work against high-level communication and understanding in general, especially with respect to highly abstract concepts like the CTMU.

Political indoctrination is now inescapable, and it tends to be atheistic, materialistic, and radical-egalitarian in nature. It treats human beings as interchangeable "consumers" – physical-economic stimulus-response automata – while denying the existence of innate differences among genetic, geographic, and cultural branches of the human species.

Meanwhile, ongoing demographic changes in Western societies are rapidly diminishing the overall intelligence of the populace as Third-World populations, burdened with (largely heritable) low mean IQ and other traits not necessarily adaptive to Western Civilization, are encouraged to flood into Western nations as refugees or economic migrants from their poor and overcrowded countries.

Once here, the newcomers are granted benefits and subsidies which reduce any adaptive pressure they might otherwise have felt, letting them refuse assimilation while aggressively outbreeding and gradually interbreeding with higher-IQ host populations already subject to intense academic and mass-media indoctrination favoring interracial and intercultural marriage and miscegenation. The inevitable effect is to alter patterns of reproduction and national population genetics while disrupting social cohesion. Host populations are systematically disparaged and scapegoated as "oppressors" and sporadically victimized by interracial violent crime, demoralizing them, degrading their self-images, discouraging in-group procreation, and eroding their reproductive aspirations and family values, thus reducing the frequency of their genes in the resulting hybridized population.

Logic and biology are very clear on the effect of such immigration and breeding policies on the distribution of heritable traits. Absent something analogous to natural selection, they bestow a non-adaptive genetic advantage on those strains which reproduce more rapidly and prolifically while disadvantaging those which reproduce deliberately and responsibly to maintain the long-term carrying capacity of the environment. The inevitable result that the traits of the former, including a proclivity for impulsiveness and excessive procreation, come to dominate those of the latter, including the capacity for rational deliberation and reproductive restraint and responsibility.

In the case of human populations, this is greatly exacerbated by self-reinforcing feedback involving counterfactual propaganda, institutionalized indoctrination, educational deterioration, and official pandering to more primitive, less refined populations through lopsided identity politics on grounds of past and ongoing "oppression". This amounts to groundlessly penalizing more adaptive stock, who may then be conveniently scapegoated by the actual oppressors.

To have the desired beneficial impact, absolute truth must be comprehensible. But to dumb it down and reduce it to sound bites is potentially to denature it and lessen or preclude its intended effect. An idea can only be made so simple, and once the mean intelligence level of the public falls too far below the level required to comprehend it, its impact may be so attenuated that it fails its purpose. Where its purpose is to catalyze the evolutionary transformation of mankind and ward of the looming specter of human extinction, policies which make the public less aware of it and less receptive to it threaten the survival of humanity.

In my view, the prevalence of such policies is in every way intentional. Those who presently reside atop the global heap don't want people to know the truth. Instead, they want people to regard themselves as helpless pawns in a dualistic theater, mere passive spectators with no ability to participate in the onstage drama. Leave that to the experts, the technocrats, the faux-elected and unelected puppets of those behind the Curtain of Oz! Above all know your place, and leave the thinking to the big shots. (Don't worry – you're in the best of hands, even if they're uncommonly grasping and often bloody.) In other words, the reason I sometimes seem preoccupied with "politics" is that unless we maintain and uphold reasonable standards of intelligence and sanity by conscientiously resisting the dysgenic policies of our would-be global masters, they will continue to plot our undoing and downgrade our moral and intellectual faculties until we can no longer resist them. And then, having reduced us to mindless, docile, amoral drones, they will establish their insectile world-hive and center it exclusively on themselves until it implodes under the force of their unremitting assault on human destiny.

Am I displaying national favoritism? Perhaps, but not baselessly. The United States still has an important role to play in fulfilling the destiny of mankind. The Third World will never pay attention to the CTMU without having it impressed on them from without, and European oligarchs no longer care whether their own people live or die, let alone understand ultimate reality and how they relate to it. EU oligarchs are fouling their nest as quickly and irreversibly as they can, inundating the little people with foreign invaders while forbidding them to defend themselves from attack or even complain without fear of incarceration. Here across the Atlantic, one can almost hear their well-armed security personnel yelling "Kneel, Swine! A EUrocrat is coming!" as they ride around in their gilded coaches with the black velvet curtains drawn over the windows.

Nor do I intentionally display political bias. Neither mainstream liberals nor mainstream conservatives have favorably distinguished themselves by recognizing the CTMU or contributing to its dissemination, so I owe allegiance to neither. If it were possible to productively align with the leaders of either major US political party while maintaining the integrity of absolute truth, I might do so if only for the sake of expedience. But mainstream political hacks and hucksters seem to want nothing to do with absolute truth or anyone claiming knowledge of it. This makes them worse than useless not only to me, but ultimately to the rest of mankind.

In short, everyone can set his/her mind at ease – I'm not playing politics. I'm merely objecting to, and warning others against, those trying to cheat mankind of its rightful destiny by playing a very aggressive and self-serving kind of politics indeed.

January 27, 2020

On to Logical Proof

I was recently sent an essay entitled "Gödel's Ontological Proof"[2] in reference to Kurt Gödel's modal-logic elaboration on Saint Anselm's original version of the ontological argument for the existence of God.

Concisely, the ontological argument goes like this:

"By definition, God is that with respect to which nothing greater is conceivable. But if God were to exist only as a concept, His greatness would be less than if He also existed in objective reality. So given the meaning of 'greatest', God must exist not just conceptually but objectively, and is therefore both conceptually and objectively real."

Gödel, working with a version of the argument due to Leibniz, was attempting to clarify it using modal logic. But in the synopsis, the author (Jack Dikian) points out that despite its use of modal logic, Gödel's ontological proof boils down to Anselm's original version after all:

> ...in essence [Gödel] was arguing that, by definition, God is that for which no greater can be conceived. And while God exists in the understanding of

> the concept, we could conceive of him as greater if he existed in reality. Therefore, he must exist.

The author continues:

> ...the mathematical model composed by Gödel proposed a proof... its theorems and axioms, assumptions which cannot be proven, can be expressed as mathematical equations, and that means they can be proven. [my italics]

Perhaps so. The trick, however, is to come up with a true logical ontology with respect to which an "ontological proof" is valid. **No logical ontology, no proof.**

Standard and modal logic by themselves do not comprise a viable ontology; extra metalogical scaffolding is required. Like Gödel's "metamathematical" proof of the incompleteness theorems, a true ontological proof must be metalogical in character. But ontology requires more structure still.

If anyone is interested in what kind of extra structure is required, see the paper "The Metaformal System"[3] in *Cosmos and History: The Journal of Natural and Social Philosophy*. It describes the fully articulated skeleton of a true ontology, and therefore of a valid ontological proof of God's existence.

In this paper, the G.O.D. (Global Operator-Descriptor) denotes a Self-distributive Self-identification Operator which turns out to bear description as God, which of course is partially why the acronym was chosen.

For those who care about the social and spiritual destination to which the above paper leads, try "Metareligion as the Human Singularity"[4] in the same journal.

Of course, I've been writing about proving God's existence for decades, and I'll be writing yet another paper reviewing how the proof is constructed, exactly what it proves, and where it points. But for now, the true ontological proof can be condensed as follows:

"God, as the ultimate language of Self-identification and thus the eternal essence of being, is ontological by nature, attributing existence to Himself and all that He contains by identifying Himself and all that He contains." (Or as John 1:1 puts it: "In the beginning was the Word, and the Word was *with* God, and the Word *was* God." In this metalogical interpretation of "the Word", it equates to the supertautological intrinsic identity-language of reality, i.e., the Metaformal System.)"

In short, as the "language of being", God is the infinite wellsprings from which existence flows; without God, there can be no existence at all. Therefore, God has dominion over all that exists. And lest these observations be thought too suggestive of Abrahamic monotheism, the same kind of reasoning carries over to much of Asian philosophy and spirituality as well. Mutual adjustments are sometimes required, but the core elements of Eastern and Western belief systems are seamlessly accommodated in one and the same metaphysical (ontological and epistemological) framework.

As anyone who reads "Gödel's Ontological Proof" quickly learns, it's not really about Gödel's ontological proof at all. Aside from the synopsis, the author has very little to say about Gödel's theological reasoning; he seems more concerned with Gödel the Man, and his interests, beliefs, and accomplishments. The proof itself is summed up like this:

> Ultimately, the formalization of Gödel's ontological proof is unlikely to win over many atheists, nor is it likely to comfort true believers, who might argue the idea of a higher power is one that defies logic by definition. For mathematicians looking for ways to break new ground, however, the news could represent an answer to their prayers.

Very insightful. While those denying the reality of God and/or rejecting logic for faith cannot be moved by mere logical argumentation, such argumentation is nonetheless worthwhile.

Or in other words, if all of this God business offends one's fideistic or atheistic sensibilities, or if one distrusts explicit theological reasoning as a matter of principle, that's neither here nor there. The matter is objectively decidable after all.

On behalf of all who prefer to stand on a real ontology, let God be praised.

February 11, 2020

CTMU for Dummies

It seems that some people may still labor under the delusion that the CTMU is too complex for ordinary minds to handle on any level of description.

That's not actually the case. Here are three ways that the CTMU can be simply described to people who lack extensive education and/or technical sophistication. First, the CTMU can be succinctly described as a language that is structured in such a way as to talk *to* itself *about* itself. That's how it exists and evolves. This reality-language is of a particular form required by its functionality, but still bears description as a language. The form required is that of the Metaformal System, the supertautological intrinsic identity-language of reality (the linguistic self-configuration through which reality identifies itself, and which allows us to identify reality with certainty by direct replicated perception; its existence is implied because *that's what we, and reality, actually do*).

When we say that reality talks *to* itself, it means that its ingredients, syntactic identification operators, are communicating and interacting with each other, "talking and listening". Reality uses internal images of itself to self-communicate and self-transform. When we say that reality talks *about* itself, it means that the formal and substantive aspects of identification operators are inseparably coupled within them; the formal intension of each operator couples to its extensional substance in the (external and internal) states of the operators. Thus, what reality says to itself is bound to the actual content of reality.

Don't see how language and reality could have anything in common even though you and everyone else uses language to describe various parts of reality every single day? Then let's focus on the identification operators themselves and call reality a kind of *operator algebra*, an "identity system" consisting of identity operators that identify the states of other identity operators. You can think of these operators as a swarm of interpenetrating spheres which assimilate and locate each other – or internally "simulate" each other – in space and time, but extend into metaphysical reality in order to do so. They interpenetrate and assimilate through an operation called *conspansion*, and intersect and refresh themselves through an operation called *telic recursion*.

Too many syllables in "operator algebra"? Alright then, the CTMU can even be explained as something like a reflexive computer simulation – sort of like "The Matrix", but without the need for an artificial host system. Its host system is distributed among the points of its manifold – the "space" which serves as its medium – and uses its past states to "simulate" its future states by "self-modeling", converting input to output and then iterating in a vast feedback loop. (The CTMU arrived a decade before the movie, so don't bother trying give Hollywood the credit. The relatively unsophisticated Hollywood idea goes back even farther than this particular movie.)

I could go on. On this level of description, there is nothing so complex about the CTMU that it couldn't be explained to a hand-signing monkey, or close to it. Anyone who denies it or its comprehensibility is, in effect, calling himself a dunce (or worse). If there's a problem, it's that the modern education (indoctrination) system has literally scrambled the brains of ordinary people, rendering them all but impervious to reason.

This problem is systemic, and it has gotten much worse over the last 30 years. That's why I sometimes remark that those who insulted and lied about the CTMU and its author (me) in order to make themselves look and feel smart, thus denying it the head start it needed to light the way forward for humanity, are bound straight for hell.

(I'm not kidding about that, by the way. If you're one of the culprits – in which case you probably signed up here at Patreon to get the latest and then duck out before your account can be debited, as most trolls don't want to anger their sacred money-idol by paying for anything opposed to atheism or materialism – your upside is that "hell" has only one syllable, and you can probably pronounce it correctly if you try really hard.)

The only way to improve on *that* would be to turn over a new leaf and try to undo the systemic damage your disinformation has already caused. At this point, I suggest that you try very diligently.

March 14, 2020

Q&A on God

Transcribed from a Podcast Channel recording session.

Johnny Wunder: In regards to God, where do we have it right and where are we so wrong?

Chris Langan: God is complex, and can be fully explained only using a very sophisticated kind of logical theology that most people are presently far from understanding. Sincere and dedicated people spend entire lifetimes in pursuit of mere scraps of spiritual wisdom; the idea that there's a royal road to God, or that God can be captured in sound bites for the lazy and preoccupied, is absurd. With questions like this, in a venue like this one, we can't afford to do more than skim the surface.

Surprisingly, much of what people think they know about God is correct. God is Absolute. God is the Ultimate Reality, more real than the physical world and anything in it. God is the Creator; God is omnipresent, omniscient, omnipotent, and omnibenevolent. God is a judge before whom all must stand; He is sentient or conscious in a higher sense than we can imagine, and knows our hearts and minds as well as we do. God has made man in His image, and has allowed man to inherit free will and other generic properties of God Himself, but expects man to instantiate them wisely and responsibly and does not condone laziness, stupidity, or irresponsibility. In making each of us a potential conduit for good, God has unavoidably also made us potential conduits of evil, and expects us to resist its tendency to be realized through us. God can be as *personal* as He can be *impersonal*, spanning all changes and variations of perspective. No matter what man does, God preserves His own integrity at all costs, and sacrifices it for no one.

What do people get wrong about God? God is not uncritical and all-loving like a cosmic nanny; God loves those who love God, and does not violate this symmetry for haters and blasphemers. As previously mentioned, God is a judge, and woe betide us if His judgment of us is negative. He is forgiving, but only to those willing to pay the price of forgiveness (rebirth and remaking as opposed to empty promises and backsliding). God is patient, but not a patsy; He is sympathetic, but not a coddler. God is not a player of favorites, as reputed in some non-Christian strains of Abrahamic monotheism – he knows the human heart and mind as images of Himself, and expects to see one's merit reflected in one's deeds. God is a pacifist when He can afford to be, but in the end – when patience and reason fail – charges man to fight against evil rather than surrender to it. The chronic failure to properly identify and fight evil on God's behalf is deemed moral and physical cowardice, and cannot be rewarded.

The fact that God gives man free will means that He lets man make his own decisions, and where necessary, suffer the consequences. God expects man to equip himself morally and intellectually for life in the world, and has no sympathy for those who never accept blame and resort to parasitism on grounds of oppression. God does not choose those in power; he lets us choose them, and expects us to hold them to account. God is not a respecter of laws or the human beings who make them – too many of them are hypocrites who use the law to control that which they hate – and is not on the side of any government that is not on the side of God. God expects nations to obey the will of their rightful citizens, not the will of parasites who use them and interlopers who despise them. He expects governments to promote the overall utility and advancement of mankind and the planet that we share with each other and with all future generations. Governments must act on behalf of the future as well as the present.

March 30, 2020

Logos, Jesus, and the CTMU

As everyone here should know, the CTMU – also referred to as Logos or Absolute Truth – is, among other things, a metareligious framework in which various religions, insofar as they are valid, can be convergently interpreted and thereby reconciled.

The CTMU qualifies as "Absolute Truth" because truth and reality are synonymous, and the CTMU explains what reality is and how it is logically or "truthfully" generated (logic can be described as "the structure of truth"). The CTMU constitutes a "metareligious framework" because it consists of the metaphysical logic without which the metaphysical content of scripture is essentially meaningless. Without the CTMU, God – or equivalently, ultimate reality – cannot be meaningfully defined. While the CTMU is seemingly opposed by philosophical relativism, the supposed mutual inconsistency of absolutism and relativism is as illogical and illusory in metaphysics as it is in physics. But as this is lost on most philosophical relativists, we now have a kind of spiritual relativism according to which every set of spiritual beliefs is as good as any other. Thus, some might expect that the "convergent interpretation of religion" amounts to a form of egalitarianism which completely levels the spiritual playing field, putting all religions and their central figures on a par. However, the aim of metareligion is not to hide the differences among religions and their founders and adherents, but to merge their strengths and resolve their disagreements.

As for the deeds, habits, and personalities of religious leaders, no attempt is made to obscure their differences. In particular, the CTMU does not attempt to hide the uniquely exemplary nature of Jesus. Jesus differs from Buddha and Mohammed, for example, in that he was neither a hebephilic warlord nor the scion of a rich and powerful family who was generally silent about God, and therefore qualifies not only as a divinely certified leader of the faithful, but as a true universal exemplar as well.

What does this mean? It means that although the CTMU does not dispute the status of Buddha and Mohammed in their respective religious frameworks, neither does it claim that their personal profiles "distribute" over the members of any healthy society in a truly exemplary way. Not everyone can be born into wealth and power as was Buddha; moreover, although the existence of God as a hyperconscious entity is an unavoidable entailment of certain aspects of Buddhist philosophy, he was all but silent on the issue of God's existence. Similarly, a society in which everyone is an aggressive and acquisitive warlord who lives by the sword and uses it mercilessly to enforce his will would quickly descend into Somalia-style chaos. This implies that despite their elevated spiritual status, neither Buddha nor Mohammed qualifies as a universal physical, moral, and spiritual exemplar for mankind.

Ethics is not about mere adherence to a legalistic code of behavior. Although deontology and consequentialism are often contrasted in philosophical reasoning, it is ideally consequential as well as deontological. It is supposed to conform to a set of absolute rules that are good in themselves – for example, the Ten Commandments, the Golden Rule, or the Kantian categorical imperative, which rely on Logos insofar as absolute ethical rules are justified only in terms of absolute truth and reality – but it is no less true that the expected consequences of good actions conforming to good rules are supposed to be beneficial rather than harmful, leading to positive rather than negative outcomes for all concerned. The rules, absolute though they may be, should conduce in the real world to a rational "utilitarian" expectation of as much good as possible for as many as possible without egregiously violating individual rights.

However, ethics is also about obligation, or what human beings owe to society and to each other. It is thus about fairness, justice, and moral symmetry, whereby (1) everyone dutifully discharges one's moral debts to others, (2) individuals discharge their debts to self, family, and society (along with all other levels of stratified identity), and (3) society discharges its debt to individuals under a rational "social contract". In other words, moral symmetry applies between individuals and various levels of society, making society a moral entity in its own right; otherwise, society becomes abusive and degenerate. The cosmic ledger must be everywhere balanced on all levels of obligation.

Ethically, an exemplar is a person who serves as a desirable model for others in thought, attitude, and behavior, representing the best that a person can be. Such a real-world model enables people to "live right" and improve themselves by emulation, conforming to the model's example. It follows that the exemplar should be someone to whose model everyone can conform on a real-world basis; everyone should be able to identify with the exemplar and emulate his/her thought, attitude, and behavior. Where ethics are universal, as they are expressly held to be by most religions, the relationship of a universal moral exemplar to the rest of humanity must mirror the symmetry of morality itself; the exemplar's circumstances and personality must be amenable to symmetric distribution along with the morality concept associated with the operative belief structure.

According to Christians, Jesus is one aspect of a triune (3-aspect) God Who conceived him in the womb of a human mother; in addition, he was the perfect blood sacrifice to God, serving as an eternal human "sacrificial lamb" who offered himself to God the Father in atonement for original sin. (By the implications of triunity, God conceived or originated Himself as the physical being Jesus, sacrificed Himself, accepted the sacrifice, found within Himself the divine empathy to forgive mankind after living among men and sharing their lot, and distributed His forgiveness over the human species as a "medium of salvation".) The post-sacrificial resurrection of Jesus is then offered as "proof" of the possibility of the miracle of salvation (extraordinary divine intervention on behalf of individual human beings) and the divinity of Jesus and the ultimacy of his sacrifice, principles crucial to the ideology and historical spread of Christianity. Although the death and resurrection of Jesus has receded into ancient history and can no longer be witnessed by human eyes, it remains a primary article of faith that continues to inspire Christians throughout the world.

Of course, Jesus is not claimed only by Christians; he is also frequently claimed by Jews as one of their own. According to some he was a reprehensible apostate; to others he was a good and observant (but widely misinterpreted) orthodox Jew who fought against the Roman occupation of Palestine; to adherents of Messianic Judaism, a modern religious movement combining certain aspects of Christianity with elements of Jewish belief and tradition, he is the Jewish messiah. Unfortunately, none of these viewpoints is entirely consistent with the status of Jesus in Christianity. To Christians, Jesus was blameless, a harsh critic of certain Jewish beliefs and practices (e.g., rigid Jewish exceptionalism, monopolistic money-changing in the temple), and the messiah for everyone as opposed to only the Jews.

As much as possible, the CTMU dispenses with such disagreements. Instead, it emphasizes the present function of Jesus as the ideal of human perfection – wise, compassionate, altruistic, and open to human empathy and emulation – and the logico-metaphysical structure of Logos, with which Jesus is identified. In stark contrast to Buddha and Mohammed, Jesus lived his life out in the real world as an ordinary working man, and took the hits as they came without becoming apathetic or sociopathic. That's what **we're** supposed to do.

Thus, Jesus qualifies as a true moral exemplar for all of mankind.

August 15, 2020

Is it Fair to Say, All Lives Matter?

We all like to think that human life is very dear. But have you ever heard the phrase "life is cheap"? It means that in a certain situation from a certain perspective in a certain time and place, life is of no special significance and can be freely spent with impunity. It rubs many people the wrong way because it seems to put a price on human life, which many of us like to think is priceless, and a low one at that. But it has to do with the nature of utility, which is synonymous with value, and where money is a repository and scale of value, it falls subject to general principles having to do with "pricing" and scales of utility.

Of course, money is actually a very poor scale of value for most purposes, as it reflects only the individual utility of those who have it. People with money can spend it however they like, being limited only by what others are willing to sell them in return for it. But while they love to claim that "this is how economic reality works", that says nothing of how *well* it works in any given instance on any given scale. As we all know, money is often used for evil, and those who don't have any are likely to be among those to whom this evil is done. Given that we are all members of society and microcosmic images of the cosmic identity, society and the cosmos are among its victims as well.

Utility exists on other levels including that of society, and that of the cosmos as a whole. (While there are those who deny that these higher forms of utility exist, or insist that they exist only insofar as they align with individual utility, there is no coherent way to make a case for it.) Given the existence of individual, social, and cosmic utility, the value of human life can be evaluated from three perspectives belonging to (1) the individual, (2) others in the environment, and (3) the cosmos.

Regarding level 1, every sane human being has value to oneself. One is valuable to oneself because everyone is his or her own potential for self-actualization and the realization of utility (pleasure, benefit, gain). But levels 2 and 3 are a little more complex. Regarding level 2, everyone is of potential value to other people, but can actualize that value only insofar as one is loyal to them and supports their (socially and teleologically consistent) self-actualization. Regarding level 3, while everyone is of potential value to God, that value is actualized only insofar as one remains connected to God as a faithful image thereof, or at least somewhat amenable to God's will.

Denying the connection of value to utility functions extinguishes value completely – value and utility are basically synonymous – so it is not an option. This leaves us with no choice but to look at the utility of human life from the perspectives just enumerated – from one's own perspective, the perspectives of others, and the supreme perspective of God. Reality is conscious and has value to itself. This value is introspectively apprehended. This self-registered value is "intrinsic", constituting the generic utility of God. Because God generatively Self-distributes to create secondary Self-images, this intrinsic self-utility distributes to the images, i.e., to mankind and to life in general.

Individual self-utility: If one did not exist as a coupling of being and sentience, one would have no utility at all. Hence, one has unlimited self-value. Yet one can harm oneself, diminishing or reversing one's future self-utility.

Interpersonal utility: One is useful or potentially useful to other individuals, or to a coherent group. (An incoherent group is just the sum of its individuals, while a coherent group is also their product.) Obviously, although people and groups can work to mutual benefit, they can just as easily have negative impact on each other.

God's utility: One is useful to God only insofar as one's contributions to His teleological Self-actualization exceed one's interference. But in order to contribute positively, one must acknowledge that to which one is contributing. If one does not, then one has no reliable way of knowing whether one's contribution is positive or negative.

Both individuals and coherent groups, including "races", can be evaluated on all three levels – personal, interpersonal, and divine. So on each level, the question is whether an individual and/or group contributes positively or negatively to individual, group, and cosmic self-actualization. Implicit in this statement is that each level of grouping possesses some level of statistical or psycho-sociological coherence. To be statistically coherent, a group must display characteristic aggregate properties which allow it to be objectively distinguished from other groups "in the large". In contrast, psycho-sociological coherence, including economic and political coherence, comes down to a state of mind and is therefore subjective.

Unfortunately, it is quite obvious that people and groups of people very often have negative impact on themselves, on other people and groups, and even on the state of the Earth and thus the state of the cosmos as a whole. In this case, their actual utility is negative in those respects, and this is how they "matter": negatively. So the question is, does a continuing potential for positive utility outweigh the negative actual utility of one's life?

Generic utility has two aspects. One aspect is quality of life as immediately realized by the subject, including level of need and satiety; the other aspect is more like a vector pointed in the direction of prospective improvement or positive evolution, including one's own future self and future generations of human beings and ultimately on teleology. The vector (or tensor)-like aspect is at least if not more important than the "living-in-the-present" aspect, as it applies to a larger set of individuals and groups extending far into the future.

To say that "all lives matter" is usually taken to mean that all life has positive value. Since this assertion is false in physical reality – some people and groups hate themselves, hate others up to and including society as a whole, and even hate God and the reality with which He is properly identified – it must be interpreted in terms of potential utility. This is a mathematical statement subject to mathematical analysis. To have value, something must be positioned on a scale of value, and a scale of value can exist only under certain conditions.

The innate value of X is "the value of X regardless of how it is actualized", which means that innate value is a generic potential rather than a specific state. Under telic monism, value – like every other property – belongs to telesis. On the global level, it belongs to global telesis or teleology, which exists as a potential until it is bound or actualized as a state or pattern of states. Once it is actualized, the state can be analyzed deontologically or consequentially.

A life matters, or has innate worth, to the extent that it has the *potential* to be useful to God. However, not all lives are *actually* useful to God. In fact, the life of an evil person has negative actual value. So does that of a rioter who derives pleasure from burning buildings and attacking innocent people, or harming other human beings and their property out of sheer spite. In the extreme case, such a life is worth less than nothing. In other words, its value is negative.

This gives rise to a question: can an entire race be of negative value? A case can be made that the answer is yes. For example, an entire race which is coherent (displays internal solidarity) and most of whose members are evil has negative value – its overall utility is just the sum of the utility of its members, plus nonlinear terms quantifying the degree to which they aid each other in evil. The coherence of the group spreads the evil of its majority to other members who side with them out of solidarity and thus support their criminality.

The same applies to a coherent race that blames all of its problems on the builders of the culture on which it depends and thus attempts to destroy it, or a coherent race that is totally atheistic and thus hates God and reality and tries to spread its ideology by harming theists, or a race which believes in communism and communistic revolution no matter the cost (as in "Anything is justified in bringing about the Dictatorship of the Proletariat!", where "the Proletariat" turns out to be a self-serving gaggle of banksters and party elite). In fact, the larger such negative-value races grow numerically, the more dangerous they become to teleology, and the lower or more negative their value.

Identity is stratified. People are all equal with respect to the fact that they have an opportunity to connect with the highest level of identity, namely God. But this does not imply that they have equal tendencies to do so. Some people have combinations of traits – stupidity, low impulse control, and violent behavior patterns, or sociopathy, deceitfulness, and narcissism – which render them highly unlikely to make positive contributions to others or to society as a whole. It has now emerged that such people have actually been determining the course of society, systematically degrading it to weaken it for the kill (complete global domination).

The idea that everyone has equal worth and equal rights is an idealization which assumes that all people are good and responsible. For example, there was a time when even savages – or what we would call "Third World primitives" today – were considered inherently noble. (The "noble savage" concept is usually attributed to the 18th-century Enlightenment philosopher Jean Jacques Rousseau, who thought that primitive man was originally free from sin, appetite, and the right-wrong distinction.) Unfortunately, it has emerged that on the whole, this is simply not the case. The Third World is a hellhole not because its inhabitants are "noble", but because some are greedy and bloodthirsty while others are weak and fearful. Nobility has nothing to do with it.

A person whose life has negative value can always turn over a new leaf and attain positive value. It is in this ubiquitous potential that the "equal value" of human life resides. What does NOT exist is some kind of ethical free lunch where good and evil are effectively equated by attributing equal rights to those who subscribe to one or the other. **That's out**, as it leads to a highly destructive form of moral relativism that puts good and evil on an equal moral footing while effectively advantaging evil against good, and thus promotes the destruction of human lives and societies and the planetary environment.

September 17, 2020

The Limits of Disruption: Yo, I Gotcha "Disruptive Tech" Right Here

Ever been around ambitious techies who like to imagine that they are "reimagining the world" in order to make the Big Bucks with their genius-caliber keyboard-tapping? The excitement of being in the rarified presence of such people, and thus possibly witnessing the very inception of a brand new techie revolution (or getting in on the ground floor), can be electrifying!

Much of the excitement comes from the fact that sometimes, the reimaginings of ambitious techies or entrepreneurs pay off big. Such an outcome is especially likely where they tap into the "network effect" to "virally" magnify the innate significance of their "innovations", e.g., search engine page ranking algorithms, social media platforms, or "cool apps". (I put "innovations" in quotes because sometimes they are unoriginal – others have already thought of them, as many had thought of individualized two-way social media before they were commercialized – and/or intellectually vapid, especially when considered on their merits against the technology they displace.) In this case, their "reimagining" of the world can change almost everything almost overnight, not necessarily for the good of all.

Economically, the network effect is sometimes defined to be present wherever increasing the number of buyers / users for goods or services increases their value. However, this definition is too general. Under this overly broad definition, the network effect can be one-sided; more buyers equals greater demand, and this lets the seller raise his prices and his profit (more sellers works the other way around; while more buyers = more demand = higher prices, more sellers = greater supply = lower prices). But this is economically trivial, as the good or service may be bought out of necessity or for strictly personal reasons. The relationship between supply and demand is ubiquitous in capitalistic economies.

The network effect goes beyond mere supply and demand. It is a self-amplifying feedback cycle between supply and demand in which users attract other users to the product or service. Thus, their value to the seller increases along with the value of the service to its users as their numbers grow. When a service or commodity attracts more buyers / users for reasons of peer approval, fashion, socialization, or market access, a symmetry emerges: instead of the product simply having value to its buyers, the buyers add value to the product. Thus, the value of the good or service being sold is not a mere linear function of the number of buyers, but an exponential one. Increasing the number of buyers causes an explosive increase rather than a mere proportionate increase in demand. Each buyer becomes an integral part of the product, and thus literally adds to its value.

A ready example of the two-sided network effect is the bar business – a bigger crowd means more social value and thus more demand, which leads to more supply (because the patrons literally add to the social content of the bar), which leads to more demand (more people want to come to the bar because it's "hot"), and so on in an exponential growth pattern until a limit is reached (e.g., the walls of the bar and the local fire code).

Another example is the Internet. Initially, as the Internet was available only to the military and research scientists, it was of little value to the public for such things as socializing. But the more people gained access to the Internet, the more valuable it became to the public. The Internet then gave rise to other salient and increasingly problematic examples including online social media like Facebook and Twitter and search engines like Google.

The network effect can lead to explosive profitability for those who exploit it. But eventually, the effect reaches a point of diminishing returns. At some point, nearly everyone is using the good or service in question and has thus become a part of it. The market has been exhausted, and growth is no longer possible. Thus, the value of disruptive tech diminishes as prior goods and utilities are disrupted and displaced; once everyone is using it, the social and competitive advantages it initially offered its users are gone. At this point, it becomes clear that where a good or service is necessary for basic social functions essential to the sound governance of society, the statement "everyone is using it" equates to "this has become essential infrastructure".

And here we have the crux of the problem: once the proprietors of services like Facebook, Twitter, and Google achieve a virtual monopoly on various aspects of the Internet, which is now easily recognizable as a crucial public utility, they tend to become dictatorial. With FB and Twitter, this takes the form of censorship and deplatforming; with Google, it takes the form of banishment from search pages. In other words, there is a down side to exploiting the network effect: it can turn you from an old-fashioned "laissez faire" capitalist into a power-drunk monopoly capitalist while changing your private company into public infrastructure that you may not be entitled to control as you did when you were merely disrupting established technology in the name of free enterprise.

Sometimes the world changes so much that old ways and old habits disappear. New technology proves so "disruptive" that people are bedazzled, bewitched, and mass-hypnotized by it, and it expands like a runaway fungus. Eventually, society is locked into a "new normal" to which it is essential. Competing technologies vanish, people change their habits and adapt to the new normal, and we who live in the dramatically reimagined, newly renormalized world are stuck using the techies' "innovations" and thus obliged to make them rich beyond the wildest dreams of avarice, at least from the modest perspective of past generations.

In short, viral disruptive tech can have the troubling and often inconvenient effect of creating monsters. Sometimes, when the IPO is over and the fat cats are invested to their gills, an apple-cheeked starry-eyed techie who started out with the best of intentions – "I just wanted to use my special genius to change the world for the better!" – finds himself running his very own software gulag, having trapped the rest of us in what amounts to a technological prison. The older technology he replaced is now obsoleted and out of fashion, and people simply can't compete unless they do what everyone else in the box is doing. Everyone must adjust to "the new normal".

At this point, the public is just starting to realize its predicament. When we find ourselves walled in by some new technology and imprisoned in one of these boxes, a certain sequence of events is reliably triggered.

1. The head of the daring techie imagineer responsible for putting us in the box swells like an inflating weather balloon as he groks the inevitability with which Fate has placed him in the catbird seat.

2. The techie, having already reimagined the world, now reimagines himself as God. He is reinforced in this delusion first by his sycophants, and then by all of the other disruptive techie gazillionaires and assorted big shots with whom he now hangs. Why, together they can do more than just reimagine the world; through the triumph of globalism, corporate-fascist

political standardization, and Maoist PC relinguification, they can save the world!

3. The gazillionaire techie, having let it all go to his head, becomes a full-fledged megalomaniac and waxes authoritarian. He decides to elevate his own beliefs and opinions above all others, tyrannically imposes rules constraining the behavior of his "clients", and defines certain outcomes for them according to whether or not they conform to those rules, no matter how baseless and/or whimsical they may be. He censors their speech, denies them basic financial services, cancels them, even supports mobs who hate them and do terrible things to them like loot their stores, burn their property, even assault and murder them.

This phenomenon is not confined to nouveau riche techie gazillionaires. It is also observed in those who were born gazillionaires and merely wish to use new technologies to advance their own interests using their knowledge of human nature ... technologies like Bernaysian psychology, NLP, brain chips, and government-tested mind-control techniques that can be used to maintain control of society. The problem, of course, is that there is nothing about being a gazillionaire, techie or otherwise, that should entitle one to control society or mess with other peoples' lives and minds. (This fact should be of special interest to the international banking cartel, whose financial monopoly was achieved by applying the network effect to currency, lending, and investment.)

Here's what techies and their gazillionaire sugardaddies, venture capitalists, and investors need to understand. When your technology is so disruptive that you change the world and modify the way people do things to such an extent that it becomes socioeconomic and political infrastructure, and you use it to affect and/or directly interfere with their ability to express themselves, defend themselves, pursue happiness, or enjoy the rights which society owes them and is obligated to protect, your tech is no longer merely "disruptive". The disruption stage is ancient history, and you have now redefined your "innovation" as "infrastructure". Congratulations!

But your victory comes at a price. It means that if you are running your show like a jealous potentate and forcing everyone who uses your tech to conform to your opinions and sensibilities, you are on thin ice. You are, in fact, just another morally and/or intellectually mediocre gazillionaire who is attempting to play monopoly with public infrastructure and use it to control your fellow man, and you need to stop or be stopped.

Which is not to forget about something else you need, namely, a swift kick in your ass.

October 01, 2020

Declaration of Independence

I grew up poor in a classical liberal environment ... an environment in which there was no money even for necessities, a condition which has often inspired distaste and contempt by the rich along with those who aspire to their company. But I'm not what most people would now call a "liberal".

The reason is that since the mid-20th century, the meaning of "liberalism" has changed. Liberals have become "progressives", less like the poor and the working class and more like well-to-do part-time socialist or communist ideologues. Modern liberals care little for the US Constitution, especially where it cannot be applied to their political advantage by way of racial and/or other kinds of identity politics involving "oppressed minorities". Appearing to defend the "oppressed" lets them signal their virtue and reap the rewards. In fact, so great have the rewards become that not only liberal politicians, but conservative ones as well, spend as much time as possible claiming to defend the "Constitutional rights" of "oppressed minorities".

But this raises a question: from whom are the "oppressed" being "protected"? It can't be the government, as by definition, the government can do no wrong. It can only be the American majority. *They* must be the "oppressors"! Hence, they (and in many cases, *we*) deserve to be penalized. But how? Examples abound. One penalty is US immigration policy, which tends to dispossess the majority and therefore has never enjoyed majority support. Eternal war is another corrective measure – war may be good for the rich, but it holds no prospect of benefit for the vast majority of real Americans, who receive nothing in return for American blood and treasure. Rolling back the Constitutional rights of the majority is yet another favorite penalty – what better way to shut their mouths than passing unconstitutional laws against "hate speech" to stop them from complaining about their plight? Yet another is taxing them to oblivion in order to subsidize "the oppressed" by way of reparations for all of the horrible things that the majority has done to them! (Funny – I recall doing nothing to them at all, and have been poorer and more "oppressed" than most of them to boot.) Hence, the official preference for socialism as opposed to traditional American-dream free-enterprise laissez-faire capitalism as opposed to the monopoly kind.

Another thing that gave the "liberals" away was their profound Second Amendment flip-flop. Governments and their rich cronies have always oppressed the working man and the poor. One would therefore think, as did the US Founding Fathers, that "liberals" – who claim to stand up for these groups – would want working people and poor people to be able to meaningfully defend themselves from aggression, as for a long time they proudly did. Gun ownership was unquestioned, as it is the last resort for responsible citizens who need to protect themselves and their families from criminal aggression, whether the criminals be individual, organizational, or governmental.

But then came the turnaround. Suddenly, in defiance of the Second Amendment, people calling themselves "liberals" were saying that *only the police and military* should have firearms. In other words, they were talking about taking the right of self-defense away from working and poor people while leaving the rich and powerful with hundreds of thousands of heavily armed mercenaries at their disposal for use against the public. In other words, they **switched sides** and became "controlled opposition" for the overclass. (In order to look themselves in the mirror, many of them styled themselves as lovers of the Constitution who respect law and order and "support our troops/our police". But of course, modern soldiers and police are no longer "ours". At best, they are deluded mercenaries for the oligarchy who love their paychecks and jealously band together in fear of the public and contempt for "civilians".)

In short, "democratic" politicians, like their "conservative" counterparts, had finally figured out where the money and power are – not with the American majority any more, but with the oligarchy, whose fortunes automatically swell like blimps under their own rules. Savvy democrats learned how to personally acquire significant chunks of this money while conning the impotent and relatively penniless majority into thinking that their interests were being decently represented. They became the paid shills of globalist oligarchs, forcing many of us to pull away from them as though they were crawling with lice (which, in a sense, they are). We've watched too many of them get too rich too fast to credit them with fooling anyone but the useful idiots who stupidly continue to vote for them.

Does this mean that I'm a conservative? In some ways, yes. I want to *conserve* what makes American life worth living for actual Americans. But at the same time, I don't regard just anyone as an American. Many people, especially those who have come here since 1965 in unvetted mass Third-World economic migrations, have neither love nor understanding for America, American culture and ideals, or real Americans. They are hungry nomads, hostile and rapacious demographic invaders on the take, and will do whatever they can to fatten in the American orchard while doing all they can to destroy it for everyone else. As for voting, they reliably vote for whichever politician promises them the greatest windfall in special privileges and free money picked from majority pockets. (This, by the way, is what triggered the fall of Rome.)

A country knows it's in trouble when its alleged "two-party" political class is unified in selling it out by systematically flooding it with indigent, fast-breeding migrants bent on displacing its majority citizens from their jobs and/or their neighborhoods, and ultimately from their territory. At this point, both parties are pandering for minority votes while begging like dogs under the table of the corporate/financial class, and this means supporting economic migration regardless of party affiliation. The American overclass has grown dependent on plentiful cheap labor, and will do whatever it can to obtain it even if only one immigrant in ten is willing to work and assimilate. That includes feeding the lust of democratic politicians for a never-ending flood of "minority" voters, who actually belong to reproductively incontinent global majorities bent on demographic conquest.

Governments and corporations can't get along without each other these days. Needing very large amounts of money to function, they have merged under the aegis of those who control the world banking system. These oligarchs have made it clear that they hold America in bondage and contempt and intend to subjugate its majority to involuntary, unelected, technocratic "global governance". Members of the political class or "PC" – whose native language, appropriately enough, is also known as "PC" – have learned that even if they have not yet taken their cut, they must pander to these invaders and other minority malcontents in order to impress or appease other members of their class, i.e., other politicians, and above all, their oligarchic paymasters, the only people who actually count in their treasonous, parasitic, bubble-enshrouded fantasy world. In short, they have learned to "cuck out".

We have a very important election coming up. It's time to face facts: most political candidates might as well be foreign agents, and most of "our" "leaders" are self-dealing frauds who function autocratically while stuffing their pockets and lying to our faces about it as a matter of course. Obviously, that's no way to run a constitutional republic.

That's why I'm an independent – I generally vote the candidate, not the party. I advise other clear-thinking people to do the same.

Unfortunately, that doesn't leave us with many options. Look very closely before you pull that lever.

October 20, 2020

2021

PC Wolves and Hyenas in the Danger Zone

The following quote is taken from a group email discussion. I'm replying to a single comment by one of the participants, Keith, a physicist by profession.

Keith: "I don't view [weeding out troublesome hardcore political leftists] so much as culling the herd, but as getting rid of the hyenas and wolves constantly attacking the herd. There are definitely some sick members of the herd that need to be quarantined to prevent them from further infecting the herd with their pernicious, highly infectious mental diseases. Some are rabid enough that they need to be removed from the herd permanently."

My Response: Well-said, mostly. However, there are a few other things to say that some, in their indoctrinated states, may not appreciate. I'll say them anyway.

There seems to be a tendency in these threads to adhere to political correctness, and to avoid any inference about who the supreme hyenas and wolves actually are. Unfortunately, this prevents us from getting rid of the most dangerous wolves and hyenas, who must be correctly identified before they can be neutralized. Misidentifying them as mere politicians, even Red Chinese politicians, would be the height of naivety. Cull a few troublemakers from the political left, and all else being equal, it won't be long before another gaggle takes up where the first gaggle left off.

A few years ago, I was somewhat sympathetic to PC. After all, peoples' feelings are important to them, so why hurt them unnecessarily with insensitive speech? But then, much to my disappointment and indignation, I came to see that the point was not to soften the wording of interpersonal communications, but to control their cognitive and behavioral correlates through humiliation, demoralization, and threats of coercion. PC is an application of Wittgenstein's observation that the limits of one's language are the limits of one's world; its point is to limit one's world by limiting one's language, so that the only world that one is allowed to have is a world owned and controlled by "language oligarchs".

Let's use a bit of abstraction to generalize the situation. PC is not about getting people in group X to respect the feelings of those in groups Y and Z, but simultaneously restricting and weaponizing the medium of communication (language) in order to control the thought and behavior of X, Y, and Z. Upon any accused deviation from PC – i.e., any expression of "forbidden thought" – trigger words like "bigot", "racist", "slavery", and "supremacism" are used as weapons against X and as triggers to provoke and rationalize hatred and violence toward X by Y and Z, and eventually to justify "hate speech" laws criminalizing members of X for exercising their right of free expression. Linguistic rules allegedly intended to limit hatred of Y and Z become a means to demonize and inspire hatred of X, so that members of X are socially ostracized, berated, harangued, threatened, de-platformed, disinvited from events, dropped from engagements, fired from their jobs, physically attacked, legally prosecuted, and thus totally demoralized and neutralized as a political force.

When the abstraction is reversed by the most obvious substitution – X = {Whites}, Y = {Blacks}, and Z = {Browns} – several realizations emerge.

1. PC has been forced on Whites in particular because, given that language entrains cognition, forbidding them to speak of their ethnic level of identity, or to speak of it only as evil while speaking of other ethnicities only as good, amounts to dissolving it. This subjects Whites to group decoherence so that they can be individually isolated, divided from other groups against which they are verbally and cognitively defenseless, and dominated. Thus isolated, they let the world be captured and ruled by others whom they do not question, criticize, blame, or resist.

2. Doing this to Whites targets the traditional majority population of the United States, which in turn amounts to targeting and weakening the United States itself. This removes the population of the United States as a coherent political force in global politics, allowing US power and influence to be captured by internal and external enemies using Bernaysian mass media propaganda, academic indoctrination, legal coercion, social media censorship, political subversion, technological trickery including surreptitious tracking and surveillance, peer pressure, the desire for approval, and plain old fear.

3. Obviously, the enemies in question are those who already possess the money and power to dominate and control academia, the mass media and social media, the government and legal system, and the development and manufacturing of advanced technology. Money has long been monopolistically controlled by the international banks, and money controls governments, corporations, universities, and media conglomerates. The mere knowledge of its power is often enough to ensure compliance. The universal need and desire for money allows individuals to be controlled through bribery, and money pays for uniformed and civilian enforcers to be used for coercion, extortion, and blackmail.

4. The fact that PC is coherently pursued, and lopsidedly used against Whites in particular is enough to imply either an anti-White, anti-US conspiracy among those responsible – who, per 3 above, include international bankers – or a socio-political pathology implicitly driven by factors of sociobiology/evolutionary psychology, or both. It is important to understand that this is not just a speculative "conspiracy theory". This is a chain of reasoning which, while technically probabilistic, is nonetheless inevitable. There are no credible alternative explanations. The idea that PC originated accidentally and evolved without direction is absurd on its face. A large and highly successful ethnic group cannot be *accidentally* turned against itself and rendered nonviable as other groups are *accidentally* encouraged in the opposite

direction. Such phenomena exhibit sustained asymmetry, are non-entropic, and are necessarily systematic and intentional on the part of those running the system.

In this clip[5], journalist Glenn Greenwald explains the connection between the democratic party and the information oligarchs of Silicon Valley.

It remains only to observe that despite their hundreds of billions (or in some case, trillions) in personal wealth, these oligarchs – like the rest of the global corporatocracy – are themselves dependent on international bankers for operating capital, management of their investment portfolios, and the exchanges on which their stock grows and multiplies.

The entire PC agenda, including racial integration, diversity and multiculturalism, affirmative action, and various other leftist desiderata and social engineering goals, is grossly lopsided, anti-meritocratic, and clearly designed to isolate and destroy one ethnic group in particular, namely, the White American and European majorities. This is the only possible explanation for why Whites, a worldwide minority evidently slated for demographic extinction, stand accused of excessive privilege and inordinate guilt for all the world's ills including war, plague, famine, poverty, oppression, and alleged Judeo-Christian "religious denialism" of the findings of modern science, and why every other racial identity group is officially encouraged to favor their own over Whites but never vice versa.

Social systems exhibit two opposing tendencies, *centralization* and *distribution*. The distribution tendency leads toward individual autonomy and capitalism (utility can be individually generated, controlled, and exchanged), while the opposing centralization tendency leads toward a dualistic hive-like class stratification where oligarchs rule over the masses. (Believe it or not, back in 1925, this distinction explicitly took the form of a human breeding program designed to separate a so-called "Master Race" from a "Eurasian-negroid Race of the Future", and it was funded by royalists and "meta-capitalistic" global financiers usually regarded as capitalists. This was to be achieved through the dissolution of borders, mass Third World migration, and compulsory miscegenation. This program is presently in full swing. At no point did Hitler have anything to do with it.)

From the complementarity of centralization and distribution, we obtain a generalization of the Marxian economic-sociopolitical dialectic: centralized feudal oligarchy dissolves into distributed capitalism, which then evolves into neo-feudal oligarchy which takes on the pyramidal form of a corporate technocracy. The top and bottom of the pyramid have complementary perspectives: *fascism* and *communism*, corresponding to the centralization and (corrupted) distribution imperatives respectively. The oligarchs naturally recognize themselves as despotic oligarchs or fascist dictators, but the bottom level, consisting of the powerless, oppressed, impoverished masses, recognizes the oligarchs only as "party officials" representing "the people". For privacy and security, the oligarchs may conceal themselves behind an intermediate "governmental" layer consisting of figureheads and highly trained "experts", but lurk behind it in the shadows and secretly pull its strings like spiders on a web. Whether the single ruling party is called "Communist" or "Fascist / National Socialist" is quite irrelevant.

The problem that we now face is that despite their unshakeable intellectual narcissism – a Social Darwinian form of narcissism in which money and power equate to moral, intellectual, and genetic merit – the oligarchs occupying the top of the pyramid and responsible for our current predicament understand very little of reality and human identity. Thus, they do not understand the entities and phenomena they are pretending to master. Moreover, they are congenitally unable to recognize the merit of any idea that they did not generate and pay to have fed back to them by experts trained in educational institutions that they alone control. This artificial control loop has the potential to create a self-reinforcing delusion which, via PC and other cognitive control mechanisms, becomes a *mass* self-reinforcing delusion which places the entire world in grave and immediate danger.

This is in fact what has occurred. PC, along with the other control mechanisms mentioned above, is that of which this self-reinforcing mass delusion is made.

January 22, 2021

***

Last time around, we discussed the "danger zone" oligarchy and its strategy to rule the world. In this continuation, we'll discuss the mass self-reinforcing delusion it has imposed on us using PC (political correctness) and other means of cognitive control.

The natural question is "Why are they doing this to us?", and the answer is really very simple. Perhaps some readers remember the 1980's song "Everybody Wants to Rule the World" (by the talented but pathetically named English duo "Tears for Fears"). The titular assertion is true insofar as it is *rational*, i.e., insofar as a desire to rule the world and thus to maximize one's control of the environment reflects the maximization of personal utility, and this is how rationality is conventionally defined in economics and decision theory.

The oligarchic global elite, who have most of the wealth and power and thus are the only ones in a position to seize and control the entire planet, are nothing if not "rational" in this sense. They very much have the advantage in the race to rule the world and have repeatedly voiced their intention to do so. Because money = power, the cream of the global socioeconomic elite are the proprietors of the World Banking System. It follows that the international bankers are the people of whom we should most beware with respect to global tyranny. (Note that every economy in the world, including those of China and Russia, is wired into the international banking system and is to that extent controlled by it. Such "evil empires" act not in defiance of the world banking system, but as integral parts of it.)

These oligarchs – the kinds of people who jet around in CO_2-belching private airliners to Davos and Bilderberger conferences at which they secretly agree among themselves to shove the Great Reset down our throats – have a problem, which of course means that we have a problem. That is, even the best of them inhabit what we call the intellectual "danger zone". This is the range of intelligence (or IQ) in which one recognizes oneself as much smarter than average, but in a high-IQ analog of the Dunning–Kruger effect, is insufficiently intelligent to fully recognize one's own intellectual limitations or the true intellectual capacities of others. In fact, the elite seem to regard themselves and their hired experts as intellectually infallible, mistaking their own beliefs and opinions for truth itself. This allows them to become locked in mutual feedback cycles incorporating some very strange and dangerous assumptions.

Many such elite assumptions should be easily recognizable.

The elite assume their own intellectual and genetic superiority (Social Darwinism).

They assume that money, power, and membership in the socioeconomic elite equal intelligence.

They assume that the non-elite portion of humanity is intellectually inferior and therefore disposable, and that academia has infallibly identified those worth using.

They assume that their own knowledge is reliable. Many assume that beyond their own mental states, there is no such thing as truth.

Many assume that God does not exist and that human beings are inefficient and therefore disposable stimulus-response machines begging for mechanical control and/or elimination.

Most embrace unproven hypotheses like AGW and assume that chemtrails and other easily observable phenomena are mass delusions.

Many regard political conservatives and especially Trump supporters as deranged White Supremacists and "domestic terrorists".

They believe that they alone know how to change the world for the better and to determine social value, essential functionality, and essential personnel in times of crisis.

They assume that it makes sense to inject the populace with improperly tested and therefore highly dangerous experimental mRNA vaccines and other potentially harmful substances, and that the public is stupid, fearful, or trusting enough to submit.

They assume that they possess the moral standing to access, manipulate, and freeze the bank accounts and credit lines of those of whom they disapprove.

They assume that only they should own private property.

And of course, they embrace, or at least seem to embrace, a comical range of PC absurdities: race does not exist, there is an infinite number of human genders, academics are the smartest people of all and know everything there presently is to know, liberals are talented and intelligent whereas conservatives are boring dunces, the PRC is a model society to be emulated across the globe, and so on.

The United States was founded as a refutation of tyranny, and the United States alone continues to stand against a tyranny of globalist despots. The founding of the United States was the culmination of the (nearly exclusive) development of Western Civilization by indigenous Europeans. It follows that those who want to displace America and Americans in the global power structure would seek to interfere with ethnic solidarity among those of American and European ancestry. All other groups are therefore pitted against this particular group (as is a twisted, tortured, self-hating subset of Whites themselves), resulting in a twisted, PC-based system of identity politics masquerading as "morality".

To dismiss this obvious fact as a baseless "conspiracy theory" is to deny human nature, the mathematics of decision theory, and past and current events. If you doubt it, just stand in front of the mirror with a straight expression on your face and say: "Nobody in a position to seize and control the entire planet would dare try to do so!"

If you can do this without cracking a smile, then you must be totally ignorant of the long string of empires that spanned the world as it was then known, and which is generally called "world history" by the few who still bother to study it.

We can no longer afford dishonesty on this score. Even if one likes to think that "race does not exist", or that one's own group will be rewarded by the corporate-financial oligarchs for its fealty, one may rest assured that in the end, the oligarchs care only about themselves. Intense self-interest is their primary trait, accounting for how they got where they are. As confirmed hundreds of times over the millennia, people and ethnic groups can at most be their useful idiots, serving them as scapegoats and human shields. Always, those who have impoverished and oppressed everyone else, and who caused all the misery and thus saw chaos approaching, scurry away as chaos erupts, leaving their ethnic pets-of-convenience holding the bag. Then they reemerge when the coast is clear, funding memorials for the dead, hypocritically appropriating their victimhood, and demanding all the retribution and "reparations" they can skim in the most sanctimonious PC language they can muster.

This is a cycle we can no longer tolerate.

In short, PC kills. At this point, anyone aware of recent history is forced to conclude that according to official Western-style "identity politics", Whites are nothing but whale dung on the ocean floor, everyone else is morally and/or intellectually superior, and population replacement by mass nonwhite immigration is therefore richly in order. Hence, for example, the inadequate turnout of certain so-called "minorities" for President Trump despite what often seemed his unabashed pandering, and their increasing anti-White bellicosity and support for verminous rabble like BLM and AntiFa against their own interests. Under cover of PC, ethnic identity politics has become a key component of elite divide-and-conquer strategy, and it has been devastatingly effective in demoralizing the American majority and attenuating its survival imperative.

Not to belabor my personal history, but the reason I couldn't miss any of this despite my best efforts to do so comes down to many years of hard and privational experience in the monetary and intellectual economies. I have been "negatively preferred" in academia, the general workforce, and the intellectual economy alike, and I couldn't help but notice as I was being methodically "cancelled" that of those who achieved better outcomes with less ability, certain others were heavily favored. For example, I was unable to work as a civil servant because I lacked the black or brown "complexion for the connexion". Such a favored (dark) complexion gave test subjects a 30-point advantage on the civil service exam, so that a black or brown person scoring 70 percent effectively beat out any White person scoring 100 percent (unless this criterion was "manually overridden" by the nepotistic bureaucratic connections of the applicant, of which I had none).

This agenda has been painfully obvious for at least half a century, and it is destroying our moral, physical, and economic viability. Slavery came and went, but this insidious, faux-empathic, nation-killing kind of anti-meritocracy is still going strong even as our security and infrastructure disintegrate around our ears. Historically, this trend goes all the way back through the nonwhites-only Immigration Act of 1965, the LBJ Great Society initiative, and the Civil Rights Era to Marxism (Bolshevism, Leninism, Stalinism, Maoism, etc.) and even farther. But economically and politically, it goes straight up the ladder to the globalist elite, who have wanted total control from time immemorial and the death of America from its very inception. This is the hilltop from which all of the proverbial sh*t has been rolling down upon us.

Although it is politically incorrect to use the phrase "White genocide" to refer to population replacement by mass economic migration from the Third World, the former is definitely implied by the latter. I will therefore continue to use it.

January 29, 2021

Camel, Needle, Cloak

There is a tendency among Christians and other religious believers to "absolutize" scripture. Insofar as scripture qualifies as absolute truth, this is a natural thing to do. However, several things can go wrong for the following reasons.

1. There is no "pure literal interpretation" of any written work. Until interpretative constraints are established and an interpretation is given, it's just a meaningless sequence of symbols on a page... "chicken scratches", as one might call them. (This is why, for example, scripture usually comes with authoritative translations, concordances, and commentaries.) Yet beyond the rules of syntax of the natural cognitive language in which scripture is expressed, interpretation has degrees of semantic freedom, allowing interpretations to diverge. Some interpretations of scripture are mutually contradictory. In the overall logical system containing all possible interpretations of scripture, contradictions mean that nothing can be reliably inferred.

   This is a problem, and without a viable solution, scripture has no model even in thought let alone behavior.

2. Scripture is said to be "absolute and inerrant" because it is supposedly delivered by God in completed form, containing all that it requires to be perfectly understood. But in practical terms, it is inerrant only because at its low syntactic and semantic resolution, any errors it contained could not be definitively discerned. By the same token, neither can its full meaning be discerned. To make scripture syntactically and semantically discernible, it must be properly interpreted. The first level of interpretation must be the mapping of its symbols and constituent concepts into a well-defined logical structure. (Logic is defined to encapsulate the rules of consistency and thus of sane human thought.) As this structure is not explicit in scripture itself, scripture is not "self-interpreting". Yet it is in this interpretative structure – call it logic, or reality, or Logos – that Absolute Truth inheres. Therefore, scripture must be coupled with this structure

   In other words, Absolute Truth (logic itself), which is ascribed to scripture but not explicit in scripture itself, has a particular form in terms of which all scripture must be interpreted. Inasmuch as a mere book labeled "scripture" is nothing but a series of pages containing a meaningless sequence of symbols, the true meaning of scripture inheres in the combination of the book and the logical structure into which its symbols are mapped. (Note that scripture itself is not edited, amended, or extended by adjoining this structure to it.)

3. Attempts to "literally interpret" Scripture are often devoid of common sense. For example, the Christian proverb about "turning the other cheek" has been transformed by "literal interpretation" into an absurd command that one passively invite yet another blow from a vicious criminal who has

> already beaten one half to death and whose confederates are violently raping one's wife and mother in front of one's face. (The truth is that Jesus Christ would call one a coward or worse for passively spectating such an attack.)
>
> Ethics involves symmetry, as this property equates to fairness and justice. The ethical core of Christianity includes such precepts as the Golden Rule, *Love your neighbor as you love yourself*, and *Love God above all else*. Christian ethics, while a valid characterization of a perfect heavenly world over whose inhabitants it distributes, does not describe a world largely populated by "defectors" from its ethical system. Christian rules are therefore inapplicable to people who simply refuse to respect Christian ethics and moral symmetry. They can be tentatively applied in the hope that this will cause others to see the light and change their ways of thought and behavior, but beyond a certain point, this is futile.

There is a disquieting tendency among some who call themselves "Christians" to reject every insight regarding Christianity that is not contained in the Bible itself regardless of its validity. "There's only one spiritual authority I listen to, and that's Jesus himself!" they say. "If it looks spiritual but isn't in the Holy Bible, it's of the devil!" Essentially, such people – whose sole moral authority is a man who, though Godlike, is no longer physically present to explain and interpret his own words – are asserting complete moral autonomy, which amounts to no morality at all.

We have just given some of the reasons why this viewpoint is untenable. It is not something that can be legitimately condoned in the name of Christianity. Those who attempt to do so anyway are called "fake Christians". At best, they are virtue-signaling fools who secretly arrogate total moral autonomy; at worst, they are dissemblers trying to make Christianity unworkable so that it fails and is destroyed by inferior worldviews.

In this light, let us consider the following two questions:

Q1: Jesus said: "Whoever has two tunics is to share with him who has none, and whoever has food is to do likewise." Is the CTMU in accordance with this urge?

Q2: And what about this Jesus quote: "it is easier for a camel to go through the eye of a needle than for a rich man to enter the kingdom of God". Is it unethical to be wealthy? The answers are as follows.

A1: Jesus was talking to his followers. To respect and obey Jesus' commands is to become a Christian. Thus, the sharing rule applies only to those who respect one's generosity and reciprocate. It does not apply to those who have previously demonstrated that they will not reciprocate, and that they will merely use one's cloak to stay alive until they have an opportunity to slaughter Christians.

A2: Not necessarily. It depends on how one gets wealthy, and how one uses one's wealth after getting it. One must respect Christian principles and interpret them logically in both stages. Giving your wealth to the wrong people, or to people who will use it to do evil, can be worse than hedonistically squandering it on wine, women, and song.

Egalitarianism notwithstanding, not all people are functionally interchangeable. There are some who would be quite out of line to give their limited resources away to people who cannot perform their functions and do not require those resources to perform their own functions. (Many people claim, in effect, to further teleology in essential ways that cannot be duplicated by others. Sometimes they do so in good faith, sometimes out of ignorance or delusion, and sometimes for evil purposes. We're talking about the first group only.)

That's what the CTMU says about the proper interpretation of these dictums.

[date unknown]

***

Question: Is political correctness the end-times delusion discussed in 2 Thessalonians 2:10-12?

> 10 and all the ways that wickedness deceives those who are perishing. They perish because they refused to love the truth and so be saved. 11 For this reason God sends them a powerful delusion so that they will believe the lie 12 and so that all will be condemned who have not believed the truth but have delighted in wickedness.[6]

Answer: The New Testament contains eschatological prophecies that are in some ways like logical inferences. What they foretell, in CTMU language, is terminal parasitic divergence. In the Bible, this is led by an Antichrist or "man of lawlessness" who opposes all for which Jesus stands. Like the devil himself, the father of lies (John 8:44) and deceiver of nations (Revelation 20:3, 8), this apocalyptic figure – a man, or a class of men – will deceive mankind and lead it toward perdition.

The Antichrist is not necessarily a single human being, but an egregore, a higher-order entity that can acquire and coordinate the thoughts and activities of many people in order to use them as resources in accomplishing its ends. This description would apply even if its power were concentrated in a single all-powerful individual; that individual would simply become a tool through which the egregore controls others.

For the sake of efficiency, this egregore has already concentrated itself in a plutocratic elite through whom it exercises ownership and control of the channels of indoctrination and communication. Once the entity acquires and controls these channels through its "human resources", it uses them to engineer the replacement of reality as it actually exists with a false antiteleological narrative suited to its own purposes, including domination and destruction.

In effect, the egregore uses the government, the mass and social media, and the education system to propagate a false narrative which, on mental ingestion by the populace, amounts to a deceptive "simulated reality" from which escape is all but impossible and of which the Satanic egregore believes it can remain forever in control. This simulation can be likened to a mass psychosis or delusion, and political correctness can be considered a part of it ... the part that protects it from erosion by truth. PC, a calculated restriction of cognitive language which, in keeping with the linguistic dimension of cognition, has the effect of restricting cognition itself, is a regulatory mechanism on which the egregore relies to control academia, the media, the government, the populace, and thus the world, protecting its false narrative from criticism while forcing it on the masses.

Human beings are images of God. They possess, in appropriate measure, the freedom of God Himself. In principle, this gives them the power to resist and even to defeat the Antichrist, which implies that the Antichrist is not necessarily destined for victory. But sadly, free will has been surrendered by many who, in effect, have sold their souls in return for wealth and power, or merely for comfort or sheer survival. What the prophecy says is that so many people are willing to sell their souls, to be acquired and controlled by the egregore in exchange for advantage, that the process is for all practical purposes deterministic and thus predictable.

It is on this predictability that the prophecy relies. However, due to the free will of human beings, the outcome is not yet written in stone. If we can all come together around a true understanding of reality – Absolute Truth or Logos, as it is called in the Bible – then by cooperating under its aegis, we can escape terminal parasitic divergence and save the world from destruction.

March 07, 2021

***

Question: A question arises from this next passage indicating that all oaths are from the devil. Is this because the devil seeks to bind our free will and our governance should be strictly "self-governance" based on morality and trust?

Here is the passage [Matthew 5:33-37]:

> 33 "Again you have heard that it was said to those of old, 'You shall not swear falsely, but shall perform your oaths to the Lord.' 34 But I say to you, do not swear at all: neither by heaven, for it is God's throne; 35 nor by the earth, for it is His footstool; nor by Jerusalem, for it is the city of the great King. 36 Nor shall you swear by your head, because you cannot make one hair white or black. 37 But let your 'Yes' be 'Yes,' and your 'No,' 'No.' For whatever is more than these [deviates from these] is from the evil one.[7]

Answer: The oath is meant to "qualify" or attach conditions to truth, or be substituted for truth, as though presently having an intention is equivalent to keeping it; the oath becomes a provisional salve of conscience and eventually a deceit, and one's deeds can depart from it. Affirmation and negation must apply on the telic level, thereby controlling what one does in the terminal realm (where changing one's mind for the sake of convenience or reward is otherwise possible).

Remember that this kind of telic invariance requires adaptation to circumstances outside one's control, basically meaning that one's commitment to teleology must remain in effect even if it means starting over from scratch. This passage targets each of us as a "self-programmer in N" as opposed to the physical homunculus programmed in T and subject to aleatory circumstance.

The important thing is observing one's obligations to God, not talking about them. No matter what one may swear on or to, talking can always be reversed; deeds and sincere commitments cannot. The only oaths that make a difference are those which are not merely sworn, but kept. To say "yes" is to affirm and keep one's commitment; to say "no" is to reject a proposed commitment so that one may choose a better one. The yes | no dichotomy is just an encapsulation of 2VL, the generic syntax of identification, and we are enjoined to apply it consistently with teleology, never deviating from teleology but only adapting to new conditions as they become evident.

The meaning of the following chapter is that human beings have a common generic (high-level) description as a "teloric image of God", and it is to the rectification of antiteleological deviance that prayers must be directed. One should not pray to reward or comfort evildoers unless the rewards are made contingent on reformation and moral rehabilitation. The wayward telor must stop transgressing the teleological constraints on which "teleological consistency" is defined. Feeding telesis into a miscreant who will remain outside these constraints and merely use it for more evil is not permitted. No valid prayer takes the form "Gee, I hope God remembers to cherish and comfort our parasitic elite as they destroy the world, because that's what Christ would do!"

(...)

May 09, 2021

A Few CTMU Group Questions

Q1: Is it important for a theory of reality to be testable or falsifiable (in the Popper sense)?

A1: No. It is in fact impossible. This is due to the universal scope of such a theory and the inherent weakness of the falsifiability concept (see A2–A4).

Q2: If yes, is CTMU testable or falsifiable?

A2: Again, the answer is "Not in the empirical sense." But then again, neither is anything subject to the problem of empirical induction. As for theories and statements necessarily true of reality as a whole, it's clearly out of the question – the very attempt would undermine its own support and thereby falsify *itself*.

While the CTMU is testable and verifiable by metaformal reasoning, it is not falsifiable in Popper's sense. Rather, it is logically necessary and universally distributed. By the very definition of such a theory, falsification is a no-go.

Q3: What would be the best approach towards scientifically verifying its validity?

A3: "Science", every intelligent practitioner of which should know by now that its methodology has been stretched past the breaking point, must be redefined to accommodate metaphysical reasoning and satisfy the intelligibility criterion of reality by incorporating the supertautological CTMU Metaformal System.

Remember, the CTMU *must be true* because reality would not be intelligible without it, and reality *must be intelligible* in order to exist. Self-intelligibility is a requirement of coherent structure and dynamics for any self-contained reality. Equivalently, ontology and epistemology are everywhere coupled.

Q4: If no: Why is Karl Popper wrong about the importance of falsifiability?

A4: Falsifiability is just fine for statements that are not necessarily true ("not necessarily true" is what "falsifiable" means). In the absence of proof, they must be tested using an appropriate combination of empirical / observational and/or mathematical / inferential methodology.

However, here we run into such problems as (1) the model-dependence of scientific reasoning, and (2) model-theoretic ambiguity (theories do not in general fully determine their universes or vice versa).

Regarding 1, empirical induction is model-dependent. In the sciences, empirical induction usually begins with the implicit assumption of axioms that, while true for some abstract mathematical universe and/or some subset of the physical universe, need not be true for the entire physical universe of replicable observations. That is, reality may not model the theory from which the axioms are "borrowed". Such assumptions affect the course of inductive reasoning, often deceptively. Moreover, some models exist as limits in extended models that effectively suspend some of their constraints, opening new possibilities not previously considered.

Regarding 2, the Löwenheim–Skolem theorem and Duhem–Quine thesis show that there exists a form of model-theoretic ambiguity expressed as follows: theories do not completely determine universes, and universes do not completely determine theories. (In contrast, the CTMU is a metatheory or "metamodel" supertautologically inducible from any coherent universe whatsoever.)

Now let's consider what might be called the Paradox of Falsifiability: Where falsifiability is an empirical-scientific attribute which means "not logically necessary", this says that necessary truths regarding empirical reality do not exist. But then science is pointless, as it can never yield a final (comprehensive and fully ascertained) theory of reality, let alone one which accommodates metaphysics. But then reality does not have an ascertainably valid definition. So the definition of falsifiability has underdefined definientia (definitional parameters), and is therefore itself underdefined.

The bottom line is that as far as reality as a whole is concerned, the CTMU has a *monopoly* on proof, which is why it can also be said to have a monopoly on absolute truth. Its continued "cancellation" by the venal, meretricious nincompoops of mainstream science journalism is the Crime of the Millennium. Hence, *Scientific American* and John Horgan.

April 12, 2021

***

Question: Is time indivisible yet continuous? I know one of the basic assumptions in the CTMU is that there is no continuum.

Answer: As Bergson observed, the continuum is infinitely divisible. So the short answer is "no".

The usual conception of a "continuous manifold" is self-contradictory. Continuous manifolds must be generative and hological, i.e., self-dually self-similar. Otherwise, as Bergson also noted, one ends up with an infinite regress and gets hammered by Zeno's paradoxes (because any pair of distinct points implies a continuous and therefore uncountably infinite interval between them).

By the way, **the CTMU contains no assumptions**, let alone an assumption like "there is no continuum". Reality is discrete in some respects and continuous in others. Reality makes use of both concepts.

Spacetime is devoid of motion in the sense that paths must be terminated before they can be traversed. When we see a classical object "moving through space", we are actually seeing a discrete sequence of states, a series of ocular snapshots. Conversely, to say that the object is in continuous motion is to disregard the precise states thereby traversed; after all, *states* are *static* – characterized by precise values of their dimensional attributes – while motion is dynamical or responsive to forces (thanks to relativity, inertial motion is static within an inertial frame, but a "kinematic" reflection of spacetime geometry as observed from other frames). An object can be in motion or stasis, but not both at once in the same frame of reference; we must consider them in alternation, just as indicated by the uncertainty principle of quantum mechanics. So insofar as spacetime consists of states and their intervening geometry, it is static and adynamical. (It is sometimes said to be characterized by "geometro-dynamics", but this requires a higher-order characterization of the medium.)

To this extent, the idea that spacetime as usually understood – a "dynamical continuum" that simultaneously accommodates state and motion – is a convenient fiction. The dynamics of reality, how it really evolves, is generative and telic-recursive. True dynamics require a *hological* continuum incorporating the nonterminal realm and thus accommodating nonlocality, where *hology* can be simplistically understood as a self-dual counterpart of "holographic" self-similarity.

[date unknown]

Holism vs Trollism: Another CTMU Critique Answered

I'm always being asked for short, easy-to-ingest descriptions of the CTMU. Here's one of the ways it can be described.

But first, a little background. Readers are no doubt aware that I've received what I consider far too much attention from trolls. The troll routine seems to have begun for two reasons: (1) I was discovered and publicized for having a high IQ yet being a solid working-class guy like the movie character "Good Will Hunting" (I don't know anyone connected with that movie and feel no need to change that), and (2) I stated at the very outset that my theory of reality, the CTMU, has a theological aspect and can be used to prove the existence of God.

If there are two non-PC concepts in the world, they are God and IQ differentials. In fact, a cosmos-class IQ seems to be one of the main reasons that God has been going through a PR crisis of late. Surely anyone so smart that He can run an entire universe in His mind is some kind of cosmic know-it-all, and in all likelihood a big bully as well! But of course, God – in addition to having a very thick skin when it comes to blowing off His critics, so to speak – does not allow Himself to be directly accessed by trolls bent on insulting Him, lying about Him, and turning others away from Him.

Needless to say, if you're an evil blasphemous troll, this access-denied policy can be super-frustrating – after all, we're talking about nothing less than divine "cancel culture"! Thus, it came as no surprise that once I'd become accessible through the Internet and the mainstream media as a theistic high-IQ poster boy, I – though a human being just like everyone else – became a kind of emergency scapegoat for the Almighty, exactly the kind of "crank" and purveyor of "pseudoscience" from whom pseudo-skeptical "debunkers" and trolls in general love to bite big bloody chunks! Divine cancel culture had gone around, and then it had come around.

The comments answered here were posted in a CTMU forum on Facebook. I was about to reply to them there, but decided against it for two reasons: they were presented in a way that violated the forum rules, and I didn't want to reward what seemed to be a troll by giving him a personalized response as a payoff for his insolence. That would be reinforcing bad behavior.

So I'm posting it right here instead.

***

Comment: Wow Chris, I find it interesting how most if not all of your written communications are constructed in such a way that the meaning of your point is obfuscated by connected word syntax that is difficult to follow and understand!

Response: First, I'm not to be tagged here. See the forum rules if you're in doubt.

Secondly, communication is a 2-way street. Unfortunately, most of those who have criticized my CTMU presentation have neglected to criticize their own involvement in the communication breakdowns they deplore. So instead of sucking it up and admitting any fault whatsoever, they blame it all on me.

Accusing me of "obfuscation" invites a reciprocal objection, namely, that you're intellectually lazy and refuse to even use an English dictionary. I've written several papers on the CTMU. Reading a few words of one of them and making no real effort to understand what one has read is a bad way to come up with a meaningful critique. Not everything comes in bite-sized chunks like popcorn and pork rinds.

As for briefly and transparently explaining what the CTMU is, it's a perfectly self-contained language that reality uses for self-communication and self-control. This language functions as (1) an abstract theory, (2) the concrete universe thereof, and (3) the model (interpretative mapping) which couples and identifies them. This 3-in-1 property is called *triality*, and the trialic CTMU language is a *supertautology* in which the signs or symbols of the language correspond to the points of its universe. Signs and points are coupled in a *generative manifold* intrinsic to the language, making it a self-referential cosmic *identity-language*. The CTMU is unique in these features – many of its concepts are contained in no convenient "For Dummies" manual, as I had to create them myself.

If that's too "obfuscative" and thus too hard a row to hoe, try this. Take a good hard look at the computer on which you've been typing. There's a display monitor connected by all sorts of cables to a computer case full of hardware and software, which is itself connected to a wall outlet. Imagine that the display is even better than a touch-screen monitor; you can directly feed it all kinds of input. The entire monitor is thus a kind of universal input-output device. It is thus loosely analogous for present purposes to Kantian "phenomenal reality", as opposed to the inscrutable "noumenal" reality on which Kant supposed that it rests.

There are also conventional input devices like a mouse, a keyboard, a card-reader, even a microphone and video camera, along with output devices like a printer and external speakers and even the monitor itself, connected to the computer by cables and wireless transmitter-receivers. But for the sake of the analogy under construction, just imagine that you and these various devices are internally displayed on the monitor. Instead of being connected directly to the computer, everything now connects through the monitor. Moreover, because input and output events all occur within the monitor, the monitor – with help from the computer to which it is connected – generates its own input and output.

Inside the computer case, analogous to what might be called "deep reality", there's a CPU, RAM, an HDD for storage, a power supply, and other hardware. There's also programming or "software", for some of which you are personally responsible. The whole setup is made out of one basic meta-substance called "telesis", a convergent generalization of matter and energy and everything it takes to make them work together in overall reality. Telesis is the most primitive and fundamental form of reality, having all of the innate properties which let it configure itself just as we see in the universe around us, even functioning as the wall socket and ultimate power source of the system.

This is where Kant was probably trying to go with his metaphysical distinction between phenomenal (sensible, conceivable) reality and noumenal reality (unknowable but still identified with phenomenology in some deep unknowable way). He seemed to look at them in a manner suggestive of the distinction between a computer monitor and the computer to which it is connected, regarding one as "phenomenal" and knowable, and the other as an unknowable kind of "deep reality" (just as you can't see the specific operations that are being performed inside your computer or cell phone CPU to convert your input signals to output that you can sense). He was even correct in his implicit conviction that "reality" is basically a convergent generalization of both of these aspects.

But to pursue the computational analogy, Kant made a mistake: he assumed that the display monitor in which we reside can tell us nothing about the hardware and software inside the computer case. As we now know, he had no sound basis for such an assumption; we can in fact make inferences about what's going on inside the computer case using basic logic and mathematics alone, as did Alan Turing and other early computation theorists, and perhaps a little classical physics. While remaining true to the best aspects of the Kantian perspective, the CTMU thus departs from it by adjoining monitor to computer using a quantum interface between terminal and nonterminal stages of existence, in effect "putting the monitor (and all of its contents including external reality) inside the computer". (Of course, you'd have to read my papers and use the forum search function to understand the details and all of the rather advanced reasoning behind them.)

The CTMU does something else as well: it reduces all of the structure and causality derived from telesis – the most essential stuff of reality – to the language described above in the paragraph following "popcorn and pork rinds". In a straight computer analogy, hardware is fixed and immutable while only software and input data are seen as linguistic and mutable. The CTMU reduces hardware, software, data, and users themselves to the trialic supertautological identity language mentioned here and explicated in every one of my papers, where it is also referred to in various contexts as *Self-Configuring Self-Processing Language* (1999-2017), the *Metaformal System* (2018b), *Quantum Metamechanics* (2019) (a "quantum meta-machine"), the *Reality Self-Simulation* (2020), and (you guessed it) the **CTMU**. The end result is much more precise and profound that anything you'll see in a movie like *Tron* or *The Matrix*.

Remember, this is just a computer *analogy*, not a literal computer system made of silicon, copper, steel, and plastic. The CTMU reality-system or "reality self-simulation" requires no external power source, requires no external medium, and does not rely on conventional computation. Its display monitor differs profoundly from the discretely pixelated device at which you are presently staring, which enables you to be *inside* the display rather than *outside* of it, and lets the display and computer be merged into one big self-distributed freedom-supporting "metamachine". The system merely provides you with the "illusion" that you and it are somehow embedded in an "external reality", except that it's more than just an illusion. Whereas an illusion is usually understood to contradict physical reality, physical reality is exactly what *this* "illusion" *really is*. The CTMU merely explains how an "illusory" universe is created and sustained in a metaphysical self-extension or "metaverse".

I hope this helps. If not, don't be afraid to do some homework.

May 30, 2021

Four Questions With Answers

[**Note** (6/16/2021): I've reworded some of this for clarity. I have a habit of slapping up a post in the middle of the night and scheduling it for publication a week or so later, intending to come back and proofread it. Sometimes I don't get around to the proofreading and the post goes up as a rough (sometimes very rough) draft. When I notice it's there, I fix it. That's what happened here.]

Here are four questions that were asked the other day by a curious participant in the Facebook CTMU Group who is presently reading the 2002 CTMU paper "The Cognitive-Theoretic Model of the Universe: A New Kind of Reality Theory".

Questioner: I am reading CTMU ANKORT and, instead of great contributions, I have petty feedbacks and silly questions (let's hope for a contribution later). Looking forward to hearing your thoughts on the below.

My Response: Unfortunately, Facebook is a volatile platform on which I'm throttled to the point of absurdity, and therefore not the place to publish new technical information. (I infer that I'm being throttled there by comparing my slow-growing number of followers to the numbers I acquire in much less time on other platforms.) So I'll publish the answers on Patreon instead.

Meanwhile, please don't be misled into thinking the CTMU is wrong in any important way. Trying to prove such a doomed assertion is something that many have tried; just that many have died (so to speak). There is no escape.

Q1: First, in the "no continuum" paragraph there is some vagueness: lots of reference to the continuum of real numbers (as in a complete uncountable field, with irrational numbers such as √2), but explanation of continuum as in something that could be the countable rational numbers (infinite number of points between any two points) – which I believe could be what is meant and criticized for the sake of "conspansive computability" as time progress. There is sort of a clear definition that is obscured by the examples. Any thoughts on what is exactly the continuum criticized?

A1: Defined in the standard linear way, the continuum is a problematical concept for a number of reasons. Perhaps the most obvious is that it is oxymoronic; although it is assumed to consist of exact locations, the exact *location* concept is itself oxymoronic. That's because (1) to make a location on a number line "exact", an infinite decimal would in effect have to be "completed" (this is the formal counterpart of the geometric location of an *exact* point), which is impossible, and (2) an exact location, considered as a "point", would have a "size" (diameter, volume) of 0 or "nothing". But an extended space cannot consist of things which have 0 extent and thus amount to "nothing" therein. That is, given that 0 + 0 + 0 + ... + 0 = 0, we'd have no space at all. No number of 0D cuts or "limit points" add up to anything with nonzero extent.

There's another way to look at it. The property of having an "infinite number of points between any two points" (as allegedly satisfied by the rationals and the reals) is not equivalent to the more intuitive "no gaps between successive points" criterion (as allegedly satisfied by the reals, but not really). As 0D (extensionless, zero-diameter, zero-dimensional) points cannot be "in contact" without collapsing to the same 0D point, there would have to be nonzero gaps between successive points, which would imply discreteness. Points would thus be countable. This violates the definition of a "continuum", of which the points are *uncountable*. In short, 0D points lack the finite extensionality criterion for existence in space: *to exist in* a space something must have *nonzero extent therein*. This, in part, is why the real continuum was extended to form the CTMU conspansive manifold.

As for the distinction between the reals and the rationals, consider rationals of the form 1/n between 0 and 1, n an integer greater than 1. As n = 2, 3, 4, ..., ∞, there are infinitely many. Given a linear correspondence between the interval 0 ... 1 and the interval between arbitrary rational numbers a/b and c/d, linearity ensures that the image of each rational number 1/n is rational. So no matter what the rational numbers a/b and c/d may be, there is an infinite number of rational numbers between them, just as there is between 1 and 0. Thus, if we simply define the continuum on the "infinite number of points between any two points" criterion, we have what might be called the "rational continuum". Note, however, that we still have gaps between successive rational points; those gaps never go away, because no matter how many rationals we have, there is still an infinite number of rationals between any pair thereof. So we don't actually have the "smooth", infinitely differentiable, totally gapless continuum we *assumed* we had.

Extending the rationals to the reals, $Q \rightarrow R$, adds points, namely, those which are irrational by definition. (Bear in mind that locations specified by either rationals or reals cannot be fully resolved in a geometric space like a number line; again, this would require the completion of a nonterminating decimal, and nobody has the time to explicitly write one out or execute an algorithm to do so. Geometric locations are inevitably inexact and therefore merely approximate.) In effect, adding the reals to the rationals generates an infinitesimal "microcontinuum" between each pair of rationals, which however are already spanned by a "rational continuum". The space simply becomes "denser", with the real continuum properly including the rational continuum. But unfortunately, the definition of "continuum" remains oxymoronic due to the failure of the "no gaps between successive points" criterion.

Q2: Second, in section 4.6, there seems to be a typo. The sentential tautology "X ∨ ~X" is presented as a principle of non-contradiction. ("Applied to perception, this means that when something is seen or observed, it is not seen in conjunction with its absence; ..."). However, under usual interpretation of sentential logic, this refers to the law of excluded middle (according to which you could have something and its absence), while the principle of non-contradiction would be ~(X ∧ ~X)[8]. Quite a difference and slightly annoying in the context of a logic paper. The joke of the logician answering "Yes" to "is it a boy or a girl?" reminds of the difference of OR for logicians and natural language, and they might find it surprising that the layman's interpretation is used to interpret standard logic symbols in the paper. Is it a typo or something that I do not understand?

A2: Regarding "X ∨ ~X", "∨" = XOR (obviously). At the time I wrote this paper, there was no standard symbol for XOR anywhere near my keyboard, so I hoped that the meaning would be obvious. Just as obviously, this applies primarily to observable reality; by definition, replicable observations do not differ for any given event.

Note that in the CTMU, the law of excluded middle applies to the terminal domain, where it amounts to an existential exclusion principle. It does not apply in the nonterminal domain, which is where true quantum potentials exist. (True quantum potentials are generative; as the quantum wave function "propagates", no it-from-bit question has yet been asked. Hence there's nothing to which a QWF can collapse; the potential is not sufficiently well-defined.)

Q3: Third, I wonder about a consequence of the Super-Copernican Principle:

> So although the "bit-size" of the universe is too great to have been completely generated by the observer-participants who have thus far existed, future generations of observer-participants [...] have been and are now weighing in from the future.

As far as I can tell, the Super-Copernican principle [rejecting "now-centeredness."] is re-appropriated in the CTMU.

A3: No, the future cannot be algorithmically predicted with certainty. Reality is generative rather than deterministic. That's why we have free will.

Q4: I wonder if this means that there is actually a predictable future in the universe, i.e. the world would not be "about to be wiped" due to local telors since future telors would already be "confidently" weighting in. The proposed relationship between time, retro-causality and free will is a hard piece that I have not yet wrapped my head around, so an open question.

A4: As in boxing, "weighing in" does not mean winning. Future telors can be frustrated despite having weighed in, largely because different telors in different possible futures – different "blank universes" which they seek to fill in with real content using resources from an extant evolving universe – want different histories to exist (each wants a history leading to the future state in which it has achieved real, fully embodied existence). Telors need not have mutually consistent self-configurations; they can compete for existence. Whether or not we like it, we are forever in the middle of such contests.

June 13, 2021

Marxism and Critical Race Theory

This one is pretty much self-explanatory. It's in response to a post by one "Professor Vortex" trying to explain the difference between Marxism and Critical Race Theory, two related aspects of the ongoing run-up to the "Great Reset", a proposed world-transformation after which everything – and that means *everything* – will be run exclusively by the likes of Rothschild, Gates, and Schwab. After nearly 60 quarrelsome posts in which no one seemed to be catching on, I felt compelled to intervene.

Professor Vortex: I am sorry but none of you know what you're talking about: Anti-Fascist and Black Lives Matter has not one thing to do with Friedrich Engels and Karl Marx or any other doctrine-ideology that evolved and developed from them. CRITICAL RACE THEORY is not ENGELSIST-MARXIST, it is ANTI-WHITE RACE! No to European Man, no to Caucasian Man. I do not think many of you know your political philosophy. And the Conservatives are cowards because they do not want to face this fact. For your own good, wake up!

My response: There's something to be said for this, but there's some confusion as well. In critical times like these, such confusion is painful to behold. So let me explain once more what's going on here.

The relationship between Marxism and CRT (Critical Race Theory) is one of convenience. To find out *whose* convenience, one first asks the question "Cui bono?" Obviously, the nonwhite races of the world have **not** spontaneously overcome all their gut-level distrust of "the other" and internecine tribal hatred and banded together against Whites; the entire thing can only have been engineered, and when you answer the question "who benefits?", there are your social engineers. So the question is whether there exists a coherent group that can use both Marxism and CRT to its benefit – there's the motive – and already has the wealth, power, and organizational structure to provide itself with means and opportunity.

There is such a group: **the powers and principalities that be**... oligarchs, elites, monopoly capitalists unified by a web of financial and political connections and determined to lord it over the little people and rule the world. (Don't bother trying to rationalize the absurd claim that people who spend their entire lives accumulating and enjoying wealth, power, and privilege have some kind of inherent internal mechanism that stops them from conspiring on a global level; whether their various conspiracies are called "commercial" or "political", **conspiracy is what they do best**. In their positions, it's decision-theoretically rational.) Their strategy is called "Ordo ab chao" (order from chaos) and/or "Divide et impera" (divide and conquer), meaning that a prospective slave-class can be much more easily defeated and subjugated once it has been deprived of racial, cultural, and national cohesion, reduced to anarchy and chaos, and reordered from scratch to the liking of the conquerors.

Marxism is preferred for societal breakdown because it inspires class war and fractures society in the economic dimension while setting the stage for a "dictatorship of the proletariat" which is actually the dictatorship of elite puppeteers and their chosen high party officials. CRT is preferred for the societal breakdown stage because it inspires race war and fractures society in the racial and cultural dimensions. CRT is especially convenient because it has what mathematicians call "symmetry"; it makes no difference which races band together and which race is scapegoated as a common target. If your goal is first to dominate the Western World, which is predominantly White, you simply rotate the underlying object of the "CRT group" until the vertex "White" lines up with the pointer "Scapegoat". Then you align with the most advantageous nonwhite vertices – e.g., the Chinese vertex (high population, lots of wealth, preexisting antipathy toward Western culture, values, and ethnicity) – and with their help, consolidate all of the nonwhite vertices around yourself and your primary allies using the shared hatred you've engineered against Whites, marshalling their combined power against Whitey using the mass media and global education system (Academia Inc.).

See? It's not rocket science. Marxism and the CRT are unified not just by common elite interests, but history: CRT is a form of Critical Theory, which emerged from the Frankfurt school and later the Institute for Social Research, a pretentious pack of Marxist pseudointellectuals trying to figure out, way back in the 1920's, how their elite masters could weaken, undermine, and take over the West before symbiotically unifying with the East in a global socialistic tyranny. That's why they're now called "globalists". And of course, communism has the added convenience of being naturally global in scope; on the limiting global scale, fascism is pragmatically indistinguishable from it. It's a pre-packaged ideological bait-and-switch. If you're stupid enough to look forward to being a good little minion or minor party official of the communist Mega-State, you may never even awaken to the fact that you're really just brain-dead stuffing for a global fascisto-communistic corporatocracy which has transformed itself around your ears into a neofeudal aristocracy for whose miserly crumbs you and your comrades must stand in line forever after.

Voilà: global tyranny dominated by sheer money-power ... the power of usury, as old-school Christian rulers used to put it when righteously signaling their virtue by surrendering the all-important power of money-creation to this or that gang of morally uninhibited central banksters in order to fund their wars of self-enrichment in the names of their scapegoated subjects. The relationship between the elite and their scapegoats is analogous to that between certain parasitic organisms and their hosts, especially organisms which exert behavioral control over host organisms in a way suggestive of modern Bernaysian psychology. Hence, it is properly described as a **parasitic divergence** (my neologism), frozen forever in place by the elite monopoly on advanced surveillance-and-control technology.

If any of this still comes as a surprise to you, wake the @#$% up. No, you're not in the club. No, you're not going to be inducted into "the Elite" due to your being Jewish (but strangely white in color anyway, in case you haven't checked recently), or their grateful recognition of your enormous brainpower. (I've looked at the distribution, and most of you – with at most half a dozen exceptions – don't even begin to make it. Even if you're smart enough, you're not sufficiently rich, influential, and well-connected.) You're going to be just another lonely, forlorn wastrel, not even allowed to sit on the curb with an empty beggar's cup. And that's only if you're not caught up in a mass culling according to the depopulation agenda of a superfluous gaggle of money-shuffling Social-Darwinistic useless-eating vultures who, notwithstanding their ponerological "dark tetrad" traits (sociopathy, narcissism, sadism, and Machiavellianism), aren't smart enough to wipe their own butts without skimming the cream of their own parrot-centric indoctrination mills ("universities") to show them how.

Yes, yes, I can hear the objections already. For example:

1. "But what if the elite are really in the know, and are doing all of this for the good of mankind?"

While the elite admittedly have a lot to hide and therefore keep a lot of secrets ("national security", you see), no one in his right mind would ever accuse them of excessive concern for "the good of mankind". If that's what they were into, and if they had the brains to see it through, then this would have emerged in their speech and behavior. **It hasn't.** The elite unquestionably have a plan, but it's not about "the good of mankind". It's about the good of them, personally. Specifically, it's about keeping their own asses in a luxurious state of leisure guaranteed by total control. If this were not the case, they'd be rationally, coherently, scientifically justifying themselves to the rest of humanity. **They aren't.**

At one time, I thought I might be getting close enough to a couple or half a dozen of them to enjoy an honest, truthful "disclosure session". Unfortunately, it transpired that I wasn't (although I might not have been disposed to go along with what was disclosed, my spirit was one of constructive cooperation, and if necessary, confidentiality). And if I don't qualify, then neither do any of the rest of you. At most, those in this distribution are only partially informed, and should think twice if you've been led to believe otherwise. This is not a meritocracy you're in, and in all likelihood, the leading proprietors of the New World Order see you as no more than a minor functionary and "useful idiot".

2. "What if we've already been taken over by aliens, and unbeknownst to us (but not to the elites), the aliens have a gun to our heads?"

This objection might look like a joke, but it isn't. In fact, it's a perfectly respectable objection given the inhuman behavior of the elite, presently in full swing (biowarfare, mask-mania, toxic "vaccine"-mania, lockdowns and economic strangulation, the usual eternal wars, US presidential election theft, the Great Reset/ID2020/Agenda 2030, etc.). But as the aliens, such as they may be, remain behind the curtain of Oz, we have no choice but to blame our escalating misery on their elite tools of conquest and oppression, who at best would amount to high-dollar inter-species prostitutes.

3. "What if no one really knows the truth? Moral and cognitive relativism would reign supreme, it *really would be* all about money and power, and one group of elites would be no better or worse than any other!"

As I'm perhaps in the best possible position to know, the elite place no value on truth, preferring to cancel it at their convenience. Or rather, they're not into sharing it with the likes of you. The only kind of truth they're ever out in the open about is the kind that enriches them, empowers them, and makes you more dependent on and subservient to them ... and even *that* kind of truth is mostly wrapped up in trade secrets, classified research, and black budget projects.

It doesn't take a genius, kiddies. At this point, **it's the nose on your face**. The problem is that mankind deserves a chance, and as it stands, it has almost none. (The elite, insofar as they technically belong to mankind, don't have a chance either, but lack the intelligence to understand why not.) Nobody's doing anything to help because the people and agencies in positions to help are getting conned, bribed, blackmailed, and extorted to sell your asses down the creek into the sewage reservoir, and being for the most part of extremely low quality as human beings, they can't meaningfully resist.

Anyone else who goes along with any of this has the IQ of dog dirt. We're in biblical territory now. Any questions?

(No, I didn't think so.)

June 21, 2021

The Alpha and the Omega

If there's one thing that bugs me, it's people who critique or disparage the CTMU without having read the CTMU ... or in short, trolls and/or dummies.

Here's my response to a CTMU critique on the Facebook CTMU Group page along with an explanatory addendum.

Comment: "No rational mind can conceive of the possibility that language existed prior to human thought. ... It is outrageously absurd for a scientific theory to claim that concepts of time and matter are dependent on the nature of a language."

Response: This is untrue. My mind is fairly rational (some would say very rational), yet I can conceive of a metaformal language that exists "prior to" the emergence of human language, and which defines the nature of space and time.

Remember, language is an algebraic system. If one can imagine a primeval reality with any algebraic (or dually, geometric) structure whatsoever – and without it, all we have is an unbound structural potential with no means of actualization until it configures itself accordingly – then there is no way to rule out linguistic structure and dynamics. In fact, there is no way to do without them, as they are necessities of existence.

On the other hand, if one believes that reality emerged spontaneously from the impenetrable fog of unbound telesis, then emergence cannot occur before there is a channel of emergence. This channel amounts to structure, and it turns out that language is the most primitive and necessary structure of all. No language = no structure.

Indeed, every mathematical theory and/or structured description is a language. That's why a TOE, a metatheory of all valid theories, can *only* be a language; no language = no theory. Hence, reality must self-configure as a language from the beginning – it's a tautological implication of any configuration and any theory thereof.

We may therefore surmise that the ideas expressed in the above comment (part of a very long post) are not only inconsistent with ontic criteria, but antithetical to them. The CTMU is an ontic supertautology. There's no way out.

It's an interesting post, but flawed. People like Wheeler, and me for that matter, are not so easily critiqued. In the spirit of the CTMU, try not to make assumptions about reality simply because they seem obvious or inevitable, or because their negations seem unthinkable. If reality were all that obvious, we'd have known all about it millennia ago.

ADDENDUM: I see some circular argumentation below against what has been called the Principle of Linguistic Reducibility (Langan, 2002). Let's ask and answer a couple of simple questions about it.

Q1: When we subtract all the language from the argumentation below, what remains of an intelligible nature?

A1: Nothing at all. Intelligibility entails intelligence, which requires a self-consistent cognitive language.

Q2: But if language cannot emerge except as a language, then how can it emerge at all?

A2: As a perfectly self-referential language – i.e., **as itself**, or in other words, **as a comprehensive identity language**. We can continue this "language game" ad infinitum, but the answer is always the same.

The only question that means anything is this:

Q3: What kind of language can give rise to itself and instantiate itself?

A3: A self-originating language that is referentially and logically idempotent – a perfectly self-derived and self-induced language that can serve as its own "cause" and its own "effect", or in other words, its own Alpha and its own Omega. This is the one possibility ignored by Kant, Wittgenstein, and so on – the language without which nothing they ever said could possibly be intelligible.

That's right – **we'd need the CTMU**. There is no escape. I'm not going to explain this again before I start shutting down threads. When something this inarguable is expressed in this inevitable a way, those who argue with it are merely wasting time. If wasting time is what you're all about, then go waste someone else's.

Thanks for your attention.

July 10, 2021

***

Comment from Patreon member: "It is THINGS which are evil, never people."

Answer from Chris Langan: On the contrary, evil emerges through people because they fail to properly gate it, and once they succumb to it, it owns them. No need to blame yourself for that, or for the confusion of others (especially after trying to help). The person about whom you are worrying is not completely lost; he can still get hold of himself and correct his own confusion. But until he wants to, no one else can do it for him.

[date unknown]

Quick Q&A: A Question About the CTMU

Question: What is the meaning of your phrase "general reductive covariance"?

Answer: *Covariance* usually refers to the constancy or integrity of laws of nature; thus, Einstein's Theory of Relativity (Special and General) refers to the covariance of physical laws. *Invariance*, on the other hand, usually refers to instances of physical laws, e.g., physical events in causal chains over which energy is transferred and which are therefore *governed* by laws of physics.

Within the context of invariance, covariance is defined a bit differently, being juxtaposed to a complementary concept called *contravariance*.

As any high school student knows, classical physics models reality in terms of vector fields attaching to each point of a space a *vector* possessing magnitude and direction. A *tensor* is a mathematical entity that "generalizes" the vector concept in such a way that the entity, of which a vector is a special instance, displays invariance with respect to changes of coordinate system. Instead of attaching vectors to points to define a vector field, one may attach tensors to obtain a tensor field.

In tensor language, *covariance* is a property of dual vectors or "covectors" in which gradient-like quantities change with the size of an individual unit. For example, the change of temperature you experience per unit of displacement as you walk toward a heat source is covariant – the greater your unit of distance as you advance toward the source, say one foot rather than one inch, the more the temperature changes per unit distance. Thus, 1 degree per inch becomes 12 degrees per foot (i.e., 12 degrees per 12 inches), with both expressions describing the same rate of change. In contrast, *contravariance* refers to vector-like quantities that change *oppositely* to the size of a unit. For example, when the unit of distance is reduced from "1 foot" to "1 inch", the coefficient 1 expands to 12 in order to leave overall distance invariant as the unit shrinks. Thus, 1 foot becomes 12 inches, with both expressions describing the same physical distance.

The phrase "general reductive covariance" obviously has context. For example, it could have come from a Chris Langan quote. In the future, specify the source, and your question may receive a more specific answer (if I see it and decide to answer it, of course).

ADDENDUM: The questioner has added sourcing:

> In other words, (1) where informational distinctions regarding a system X are regarded as instantiations of law, they can also be regarded as expressions conforming to syntax; and (2) the expression of differences requires a unified expressive syntax (or set of "laws"), and this syntax must distribute over the entire set of differential expressions (or "instantiations of law"). E.g., where X is a "perceptual intersect" consisting of generally recognizable objects, attributes and events, the laws of perception must ultimately be constant and distributed. Where a putative nomological difference exists for some pair of loci (A, B), REDUCTIVE SYNTACTIC COVARIANCE [questioner's emphasis] applies due to the need for an expressive medium, and where no such difference exists for any pair of loci (A, B), syntactic covariance applies *a fortiori* with no need for reduction (Langan 2002, p. 18).

Answer (part 2): Yes. This is a description of the CTMU concept of syndiffeonesis. If you "reduce" laws of nature or "physical laws" to the more general (and fundamental) term (SCSPL) "syntax" in my above explanation, replacing one with the other, everything remains consistent.

Remember, the CTMU has been structured to seamlessly assimilate physical theories like Relativity and quantum mechanics, so the terminology is often interchangeable. Because the CTMU absorbs both of these theories, it can be described – whenever I feel like it – as a theory of "quantum gravity". In fact, the CTMU is the only correct and comprehensive theory of "physics" in general, and much else in the bargain... in fact, *of everything*.

That's why, in addition to calling it "mathematical metaphysics", we call it a "TOE" or "theory of everything". Academic physicists or philosophers may think they have a monopoly on this term, but they couldn't be more wrong.

July 18, 2021

When Law Enforcement Becomes Thuggery

Some things are good for their own sake – things like freedom, justice, virtue, and brotherly love. Contrary to the official narrative, **law enforcement is not one of them**.

Law enforcement is only as good or just as the laws and policies that are enforced, the means and methods of enforcement, and the people who issue the commands. Where the chain of command is corrupt, and bad or unjust laws and policies are badly enforced, law enforcement cannot function as intended.

Sadly, it has become increasingly clear that the police and military chains of command are corrupt, and that as a result, many bad and unjust laws are being badly and unjustly enforced. The border is being held open against the will of the American majority so that unvetted criminal, disease-ridden, and/or unassimilable illegal aliens can be bused into clean, crime-free neighborhoods. Any attempt by citizens to protect our territorial sovereignty is declared "vigilantism" and crushed. Masks and lockdowns are being forced on citizens without scientific support on the command of tyrannical political puppets calling themselves "mayors" and "governors". BLM and AntiFa rioters, looters, and murderers are being allowed to run amok without any sort of protection for majority citizens. Heroic protestors are being unlawfully detained and tortured for lawfully gathering in opposition to election fraud. Young children are being openly sexualized in the schools. Fake presidents and vice presidents are being installed directly under our noses. **It's all bad**.

As long as this persists, law enforcement personnel – even those who credibly claim to have good intentions – are nothing but thugs holding us at gunpoint as our nation, culture, and racial identity are looted out from under us by everyone from globalist bankster-oligarchs to violent, race-hating, drug-addled street trash. At the very least, this makes it hard to disagree with cries for "defunding the police". So perhaps others can understand why I'm getting tired of assuring "our" boys in blue that I support them.

In order to support the police in good conscience, I must believe that they support me and other citizens. But that's not usually what they do. Yes, some police are exceptions ... until, that is, they refuse a bad order and are fired by the higher-ranking personnel who issue their paychecks. That's not good enough to earn my respect – I've had too many jobs that are riskier and pay less than being a cop. (At one time, I might even have *become* a cop were it not for the fact that "oppressed minorities" were being spotted 30 points on the civil service exam, so that affirmative--action candidates with scores as low as 70 beat out White men with scores of 100, making it nearly impossible for a White male to be fairly, non-nepotistically hired.) Yes, we need good cops, but we don't have enough of them to turn the tide. A police applicant is not hired if psych exams reveal that he/she has the kind of personality that questions orders and/or resists orders that cannot be justified. In fact, one is not hired if one's measured intelligence (IQ score) is too high. That's because a high-IQ person is more likely to sport a functioning conscience and question orders on that basis. Applicants are now chosen for their suitability as mindless grunts, thugs, and hired muscle.

Yes, yes, I know. There are objections to be made in defense of the status quo.

Objection: "But we're in a war – a war against crime!"

Response: Actually, the "war" seems to be against law-abiding majority (White) citizens. Everyone else seems to be getting a pass no matter what kind of crime one may commit.

Objection: "Without the police, we'd be at the mercy of the criminals!"

Response: We're at the mercy of the criminals *anyway*, and the police often protect them and even give majority citizens up to them like sacrificial lambs.

Objection: "The thin blue line is all that separates us from societal collapse!"

Response: Then exactly what are the police doing for us as society collapses around our ears? It's not clear that they're helping us at all. Sometimes, they make the situation worse than it would be without them.

Objection: "You wouldn't talk like that if you understood what it means to be without police protection!"

Response: Nonsense. I was a bar bouncer in New York for 25 years, working in about 50 different bars and nightclubs. I saw much more violence, including gun violence, than the average cop, I wasn't allowed to carry a gun myself, but there was never, ever a cop around when one was needed. When they finally showed up after the coast was clear, they usually turned out to be more trouble than they were worth, often blaming the victims for whatever had transpired and failing to apprehend the assailants.

Yes, police are necessary to keep order. Even a just and orderly society can't get by for long without them. But where **the prevailing order is nothing but organized crime**, keeping it is not a viable option. Being corrupt, it must be reformed or replaced, and if the police aren't up to helping, then nobody needs them or wants them around. They're as useless as paps on a boar for anything but persecuting citizens for victimless infractions and filling citation quotas, and worse, they are obstacles and deterrents to improvement. Miscreants often rely on corrupt or neglectful police for protection while committing crimes.

Some parts of the country are better than others in this respect. Here in North Missouri, we're not doing too badly – the local police and sheriff departments are mostly into doing their jobs, and I suspect that very few of their members would ever participate in, say, an illegal attempt to disarm the public. But this is deep rural America, frozen in time. In the cities, chaos reigns, corruption rules, and all the cops ever seem to do is bully majority citizens and scramble to cover their asses when, for example, a media-sainted minority drug dealer like George Floyd swallows too much Chinese fentanyl in order to avoid being caught with it.

Why beat around the bush? The police are now taking political prisoners, robbing citizens of their money and property via civil asset forfeiture, providing armed security for BLM looters, and brutalizing those who decline to wear ineffective masks or take dangerous experimental vaccines. In the "bluest" and most populous parts of the country, **they are increasingly indistinguishable from criminal thugs**, and they must be pressured to change their evil ways.

Those who are unwilling to press for police reform are in default of their civic duty, and thus cannot credibly support "our boys in blue". To do so would be to advocate for continued corruption and abuse.

August 01, 2021

The "Theory" of Transhumanism

While pointedly ignoring the CTMU, which was designed for the cognitive rehabilitation of the human species, the proprietors of security-minded organizations like NATO have found a new focus for their investigations of human cognition: **cognitive warfare**.

Who needs cognitive rehab when you can have cognitive warfare instead?

Internet source: "Military Is Developing 'Cognitive Warfare' Weapons" (Technocracy News & Trends 2021)[9]

Cognitive Warfare is defined as "the art of technological alteration of human cognition in target populations". The technologies to be used for this purpose are the "same ones that are driving Transhumanism on a global scale, namely, NBIC – nanotechnology, biotechnology, information technology, and cognitive science." Cognitive warfare is further described as "waging a battle for your brain in the human domain to make everyone a weapon" through the "weaponization of brain sciences" and "hacking the individual" by exploiting the "vulnerabilities of the human brain" in order to implement more sophisticated "social engineering".

Notice the telltale phrase "driving Transhumanism". *Transhumanism* is capitalized like the name of a religion or proprietary ideology or theory of philosophy. And indeed, it turns out to have both religious and proprietary aspects. Notice also the conflation of "weaponizing the human brain" with "social engineering" – it's almost as though the brain is to be weaponized against society and humanity themselves. And last but not least, notice the classification of cognitive warfare as an "art", meaning the expression or application of human creativity, imagination, and skill in a beautiful and emotionally powerful way. What higher form of art could there be than "attacking the human brain using weaponized brain science for purposes of social engineering"? If that's "art", then obviously it's a black art.

Transhumanists believe, often with religious fervor, that we are approaching a "singularity" or technological merging of man and machine. This is supposed to launch a "posthuman era" in which humanity has taken charge of its own evolution, transcended its lowly organic roots, and executed a Great Leap

Forward to a status somewhere between cyborg and full robot, placing the human brain and body squarely in the "Internet of Things" and perhaps even uploading human consciousness to one or more computerized virtual realities as illustrated in sci-fi movies like "Tron", "Freejack", and "The Matrix".

While Transhumanists vary in their specific beliefs and expectations regarding the technological augmentation of human biology and cognition, all agree that we're on the verge of something new, marvelous, and liberating. Heady stuff indeed. It's positively Millennial!

On a related note, we're all aware of the sinister potpourri of strange ingredients that have been found in "Covid vaccines" by various reputable experts. The general consensus is that some appear designed to "hardwire" human beings for real-time monitoring and remote control. Or at any rate, their inclusion is hard to explain in any other way.

If this is true – and it seems rather probable given the dearth of alternative explanations – then our leaders and health authorities, or at least those from whom they take their orders, would appear to be waging a sort of "cognitive warfare" on the human species, or at least that part of it receiving the "vaccines". It would also seem that they subscribe to Transhumanism, which is admittedly "driven by" the same technologies.

"Wow!" we're evidently supposed to say. "These Trans-humanists must have a truly advanced theory of reality, life, and consciousness! Otherwise, how could they even dream about all of that? They're clearly much smarter than mere organics like us! Like powerful human computers, they've already made the leap to a superior machine-like state of consciousness! Aren't we lucky to be in the hands of techie geniuses with minds vastly better than ours? Let us bow down and lavish praise upon our enlightened Techie Overlords and the Financial Stakeholders in their companies!"

But actually, the part about having a theory is not quite accurate, for Transhumanism has no supporting theory at all. It's just a set of beliefs and speculations. If there were really a well-formed Theory of Transhumanism relating reality, life, and consciousness, we'd be hearing all about it. Surely the Transhumanists would want to tell us about it, if only to ease acceptance of the weird ingredients in these so-called "vaccines". Yet none of them seems to know any more about these subjects than was known half a century ago. They just seem to go around in circles, alternately speculating and complaining about the lack of progress.

Not, mind you, that such a theory is impossible. Actually, such a theory has existed for decades. It's called the CTMU. However, the CTMU was "cancelled" for pointing toward a Human Singularity instead of the better-publicized Technological Singularity preferred by Transhumanists, and also because it has an ingredient equivalent to God, a big no-no from the atheist-materialist Transhumanist perspective. Thus, even though the CTMU is the unifying theory of reality, life, and consciousness that the Transhumanists desperately need, they reject it because it deviates from the Official Transhumanist Narrative: "Technology will free us of our biological chains at last! Do what thou wilt, Humanity!". So they've cancelled it, leaving themselves with – drum roll, please – no theory at all.

Nevertheless, onward they gamely proceed. How could this be? It's simple: they're not really interested in a true Theory of Transhumanism at all, or even a theory successfully relating reality, life, and consciousness. For those who own and control the technology on which cognitive warfare and the Transhumanist Singularity rely, truth and understanding have nothing whatsoever to do with it. They're after something else entirely.

In short, Transhumanism isn't about humanity taking control of its own evolution at all. It's about billionaire techies and their trillionaire global financiers / world banking sugardaddies *taking control of humanity*. It's about providing themselves with a centralized control panel from which they can micromanage humanity like a vast swarm of radio-controlled androids.

The end goal is quite obvious: it's a globalist-run World-Hive of which they occupy the very center as an all-powerful royal caste, and at the periphery of which everyone else has been reduced to a completely mechanized worker, soldier, or drone.

People have long complained about being mere "cogs", trivial machine parts in the corporate machinery they inhabit. As it turns out, they didn't know the half of it. It's really much worse than that.

Transhumanistic cognitive warfare is not about defeating the enemies of humanity. It's about *defeating humanity* and ensuring victory for those waging it against us.

October 25, 2021

***

Comment from Patreon member: Their "theory" seems to be, the brain is just a computer so let's make a better, internet-connected computer for our brains so we can outlive our 1.0 bodies.

Answer from Chris Langan: Right. Perhaps they think that once their "neural computers" are wired to the WWW, they'll be able to upload themselves to the 'net just like the "Lawnmower Man", the central character in yet another movie that really lit their fire, especially because it ended with the Lawnmower Man installing himself as World Dictator through the telephone system. In short, their "theory" is pure Hollywood. Logic need not apply.

[date unknown]

The "Immigrant Invasion"

Many people think that the "Great Reset" oligarchs who commissioned our undoing with the C19 plandemic, along with clever people calling themselves "scientists" and easily hired for thirty pieces of silver each, are too clever for the good of themselves and everyone else.

Either that, or they are not clever enough. Although they typically claim to be doing what they do for the good of mankind, thus allegedly precipitating all of the wealth into their very own pockets due to their undeniable merit, value to society, and moral rectitude, there are many things that belie this claim.

For example, take ladybugs (*please*). As a small child, I remember admiring these pretty little red-and-black-spotted beetles for their gentleness and aesthetics. Mom would sing the ladybug song to my brothers and me, explaining on demand that ladybugs protect the plants by eating all the bad little aphids that suck them dry. How unjust, we thought, that their houses were on fire and their children alone! But one could always offer them consolation for their misfortune, letting them crawl on one's skin or clothing while appreciating their shiny gemlike beauty. For anyone but an aphid, there seemed to be no down side to them at all.

But things have changed enormously since then... so enormously that I was just driven inside by a veritable blizzard of them, and seeking my consolation was the last thing on their minds. Being insects and therefore cold-blooded, they have low affinity for cold weather. So although millions of them inhabit the fields during the warm months, at this time of year they begin looking for warm places to spend the winter. That means farmhouses, of which they sometimes cover the entire south-facing (sunny) exterior walls as they look for a way in.

If ladybugs behaved like proper little ladies, one might try to be charitable about it. After all, here in the Midwest, where bitterly cold arctic storms can park for weeks when the days are short, everyone needs a place to spend the winter. Besides, ladybugs are small, and as long as they confine themselves to the exterior walls, one doesn't see them. But if there's anything these bugs are not, it's "proper little ladies". People are warm, so they land on people for roughly the same reason they land on houses, and once they land, why not grab a quick bite? The truth of it is, they bite like hell, and that's just how badly it hurts when they do. Asian lady beetles *ain't no ladies*, and the brew of concentrated plant toxins in their saliva burns like a red-hot needle.

No doubt this seems strange to most readers. Up until the mid-20th century, they were gentle little creatures about whom mothers sang songs to their children. What could possibly have made them go so bad? Was it pollution? Mutagenic farm chemicals? An entomological zombie plague? An arthropodal strain of women's lib?

Close enough. The change was brought about when genius government scientists brought over a new and allegedly more efficient kind of "ladybug" from a faraway place to aid us in our agricultural endeavors. It seems that our indigenous ladybugs had been lying down on the job, failing to gobble up enough aphids. So the pencil-neck nerds in lab coats decided, quite on their own, that we needed "outside help". Just as one might expect, they went looking for it outside the US, and just as government authorities do when a domestic labor shortage allegedly rears its head, they opened the floodgates to "immigrants" from a foreign country.

The foreign country was China, and the immigrants were an invasive species known as Asian lady beetles. Close relatives of American ladybugs, they too are small, red or orange, have little black spots, and eat aphids. The problem is that aphids are not all they eat. They also eat living human flesh, or at least give it the old college try. As invasive species tend to do, they bred and bred and ate and ate until, in this part of the country anyway, there was nary an American ladybug in sight. And now every autumn, as they seek to enter the domiciles of human beings to hunker down for the winter, they attempt to grab a final meal from the tenants to tide them over until spring.

Today's lady beetle attack reminded me of trying to repair fences around here. Fences are important in this part of the country. While good fences make good neighbors, bad fences make for disagreements and somebody else's cows trampling your ground and eating all your hay. But unfortunately for some, fences don't last forever. The posts rot and rust, the wires corrode and break, and large, free-ranging farm animals barrel right through the holes in them without a thought, moving from one property to the next as they please.

Mending and maintaining fences takes work. It's a constant and onerous task. Seeing a problem in need of a solution, the scientific brain trust of the US government soon lit on a solution analogous to the one they found for the ladybug problem: immigration. This time, the foreign immigrant they recruited was a pretty, sweet-smelling little flower called multiflora rose. The USDA sold it to farmers and ranchers as "natural fencing". The idea was to plant it along the fencelines so that thanks to the miracle of biology (and the basic unpalatability of multiflora to livestock), a neat fencelike hedge of multiflora would spring up where the barbed wire had rotted and fallen down.

One might wonder why a farm animal weighing the better part of a ton should pay any attention to such a flower, let alone treat it like a "fence". According to the theory, it's because the pretty little flowers bloom from ropey, incredibly tough, densely thorned vines that are tougher and sharper than steel barbed wire and physically impenetrable with very much less than a bulldozer equipped with a flamethrower. So dense and tangled are the vines and so vicious are the thorns that when even the biggest strongest animals see them, they turn away. Unfortunately, the plan then goes completely awry: they simply walk around the multiflora and break through the fence in another spot. (Not only does the multiflora "fortify" the fence only sporadically, it quickly spreads and displaces native flora, compromising access to interior sections of the property.)

Thus, the fence must be repaired anyway. But in order to effectively repair or replace a stretch of fencing, one needs access to the entire fence, to which the messy clumps of multiflora jealously deny access. One can think of multiflora like a slightly smaller version of the gigantic spiked vines which choke the Enchanted Forest depicted in "Sleeping Beauty", the animated Disney feature that has enchanted (and scared) generations of young children. Get anywhere near them, and you're sure to be sorry. They literally seem to reach out and grab your clothing, and the thorns are so sharp that using your hands to detach yourself leaves them painful and bleeding even if you're wearing thick leather gloves. In fact, once you've been caught, any kind of movement can prove disastrous, as it gives the thorns another chance to catch you, wind the vines around you like snakes, and puncture and lacerate you through your clothing. It's enough to make you rethink the whole concept of fencing.

This reminded me in turn of another plant called tall fescue. Tall fescue is a kind of grass, originally imported from the UK, that is used by US state highway departments to line the roadways, carpeting the strip of land between the road and the perimeter fencing of farms and ranches. Unlike most grasses, it adapts well to clay soil, likes both sun and shade, often stays green in winter, needs little mowing and fertilization, and is tolerant of drought and heat stress. It's a miracle grass that grows fast, and every bit as tall as advertised!

But watch out if you have horses, as it can kill them. Tall fescue is prolific, and quickly finds its way from the shoulders of the highways to fields and pastures where it is consumed by grazing livestock. Unfortunately, tall fescue can be infected with a certain endophyte, *neotyphodium coenophialum*, which produces toxic chemicals that can be lethal to horses. Entire horse pastures have had to be replanted with less poisonous species of grass to protect the horses grazing therein.

Like Asian lady beetles and multiflora rose, tall fescue is an invasive immigrant. But whereas tall fescue immigrated from Europe, Asian lady beetles and multiflora immigrated from China, some of whose main imports to the US have been invasive pests. Many of these immigrant species were eagerly welcomed as decorative plants for peoples' yards and gardens, becoming invasive only when they encountered no natural enemies in their new environments. However, Chinese immigration was not limited to decorative plants. Other invasive Chinese immigrants receiving considerable press lately are gypsy moths, giant carp, and murder hornets. Yet another was bubonic plague, which immigrated to Europe in fleas immigrating on rats immigrating on merchant ships starting in the early to mid-1300s. And let's not forget another Chinese product, influenza. China was the origin of Asian Flu (1957), Hong Kong Flu (1968), highly pathogenic avian influenza A(H5N1) (1996), A H7N9 avian influenza/bird flu (2017), and many other (usually less lethal) strains.

It almost sounds like the main immigrant from China has been influenza. But that would be neglecting another important Chinese import, e.g., new strains of coronavirus including SARS (Severe Acute Respiratory Syndrome), caused by SARS-CoV-1 (2002), and what is now purported to be yet another such immigrant: Covid 19. Depending on whom you believe, C19 originated in either Chinese cave bats or a Chinese bioweapons laboratory in Wuhan. (It is only fair to observe that not all invasive immigrants are from China. Smallpox is thought to have originated in India or Egypt at least 3,000 years ago, and ancient Egyptian paintings and carvings depict otherwise healthy people with symptoms of polio, including withered limbs and young children walking with canes. Polio remains a huge problem throughout Africa to this day.)

Until recently, most of our pathogenic immigrants have arrived in the Western World by accident. Nevertheless, the associated dangers should by now be obvious to everyone, especially as we are now experiencing many dangerous outbreaks due to an illegal government-imposed open-borders policy under which illegal immigrants are not only admitted without papers, but not even tested for dangerous diseases like antibiotic-resistant tuberculosis, leprosy, and Chagas disease. Why, then, do they not seem obvious to supposedly super-clever scientists employed by government and government-partnered corporations including Big Pharma, not to mention the entire government-regulated medical establishment?

The reason, of course, is "evil". Certain corporations have long been understood as evil (consumed by self-interest at the expense of the common good), and we now live under what amounts to thinly disguised government-corporate fascism in which governments "partner" with corporations and financial institutions that do not always take our best interests to heart (which is no surprise, as they have no hearts at all). It is now obvious, as it was obvious to the US Founding Fathers, that governments – and by extension, government-corporate partnerships and government-employed scientists – are not the benign entities for which they have long been mistaken. Centuries after the proper form of government was described in the US Constitution and other founding documents, we find ourselves stuck with something else entirely.

Now, with the autonomous Titanism of Prometheus himself, bio-scientific idiots even worse than the ones who fed the United States to invasive species like Asian lady beetles and multiflora rose have engaged in criminal "gain-of-function research" on pathogenic viruses, using sophisticated technologies like PCR and CRISPR to amplify the virulence and lethality of already horrific microbes. In so doing, they have simply followed the Transhumanist agenda of their masters: mutagenic, teratogenic, and carcinogenic substances are to be forcibly injected into the human population and the global ecosystem, and the more the merrier.

In short, the faux-Promethean government-controlling globalist billionaires and trillionaires allegedly using Science to "Save the World", along with the amoral money-grubbing university-indoctrinated, corporate-acculturated scientists who work for them, have themselves become invasive pests of the worst kind. They've been costing us far more than they're worth, and the damage they've already done will cost us far more in the future.

In the right hands, science can be a boon and a credit to mankind. The problem is that it is presently in the wrong hands, and its accelerating misuse is literally killing us.

November 10, 2021

Great Resetism and the Meaning of Being Human

Michael Rectenwald of the Mises Institute, a Libertarian think-tank, has been running a series on the "Great Reset". According to his latest installment (Part 6, 11/02/2021) on the "Fourth Industrial Revolution" (4-IR) and its associated form of governance, Rectenwald asserts that Klaus Schwab and his "Great Reset" cronies aim to **redefine what it means to be human**:

> According to Klaus Schwab, the founder and executive chair of the World Economic Forum (WEF), the 4-IR follows the first, second, and third Industrial Revolutions—the mechanical, electrical, and digital, respecttively. The 4-IR builds on the digital revolution, but Schwab sees the 4-IR as an exponential takeoff and convergence of existing and emerging fields, including Big Data; artificial intelligence; machine learning; quantum computing; and genetics, nanotechnology, and robotics. The consequence is the merging of the physical, digital, and biological worlds. The blurring of these categories ultimately challenges the very ontologies by which we understand ourselves and the world, including "what it means to be human".[10]

First, ontology is a branch of metaphysics. It is neither properly pluralized as "ontologies", nor is it internally pluralistic. It centers on the most general and basic kind of existence, *existence in general*, which by definition is monic – it's the homogeneous property which characterizes "everything that exists in any way, shape, or form". Secondly, Klaus Schwab is not a metaphysician. He's an engineer-turned-economist. As usually conceived, neither engineering nor economics has anything to do with metaphysics; hence, the associated college curricula contain little or no mention of it. Mr. Schwab's notion that he is somehow qualified to pontificate on the subject is therefore rather mysterious. Accordingly, as a metaphysician, let me see if I can demystify it a little.

Putting it as simply as possible, Klaus Schwab, like his peers Soros, Gates, and Kissinger, has not produced and cannot produce a coherent philosophical theory of "What It Means to Be Human". How do we know this? We know it because such people constantly say and do ridiculous things, can't get their brains to function without denominating their thoughts in currency, lack the intellectual power and profundity to philosophize their way out of a wet paper bag, and can't even bring themselves to recognize, recruit, or honestly collaborate with people who are verifiably smarter and more knowledgeable on such matters. If anything, their metaphysical pretensions confirm the widespread idea that they are simply danger-zone Dunning–Kruger victims who can't tell bank balances from IQ scores, and who therefore behold Aristotle, Newton, or Einstein every time they see their reflection in the bank window.

The absurd theory of humanity favored by Schwab and company might as well be named after its key event, the Great Reset. "Great Resetism", as we therefore call it, is a chaotic hodgepodge of atheism, physicalism, scientism, Transhumanism, mangled Nietzscheanism, Social Darwinism, and Kalergian Zionist / neofeudalist pseudoanthropology. Putting it all together, we get something like this:

"There are two races of humanity, a parasitic Bankster-Aristocratic **Master Race** monopolizing survelliance and control technology, and its host, a servile and intellectually debilitated **Eurasian-negroid race of the future** (EANRF) formerly referred to as the 'little people' and free game for unlimited genetic, chemical, mechanical, and electronic modification by their Masters according to the principles of Transhumanism. This will bring about **The Singularity**, a global, Earth-shattering tech-human hybridization event in which *technology* supersedes *humanity* (insofar as tech machinery responds much more readily and predictably to Master-Race commands than do mere organisms)."

Basically, the plutocrats think that they alone are "human", whereas the rest of us are biological stimulus-response machines configured at their convenience as automated production-consumption units in the emergent World Hive of which they are the "royal caste". Transhumanism is just a convenient Frankensteinian pseudo-philosophy through which they are announcing their intention to upgrade themselves as "Masters" while downgrading the rest of humanity to make it easier to "master". (They didn't popularize self-aggrandizing descriptions like "Masters of the Universe" just for fun, you see.) **Great Resetism** simply identifies *them* as the Master Race, i.e., the ones who have all the money, own all the tech, and will therefore control the Tech Singularity.

The EANRF, to which the vast majority of people belong whether or not they know it yet, has its own set of directives chosen by the Master Race. These include such mainstays of ideological rectitude as PCanity, multiculturalism, socialism / communism, CRT, LGBTQism, America / White people / Christ-hatred, and counterfactual radical egalitarianism. (None of that egalitarianism nonsense for the Master Race, mind you!) In short, the Master Race requires that all members of the EANRF accept their basic worthlessness, expendability, and congenital peon-status while obsequiously groveling and supplicating to their Masters, only to be eternally pissed and shat on by their Masters for their trouble. And lest you become disgruntled due to your lack of any meaningful reward for your service, bear in mind that for a grubby, insignificant "little person" like you, unstinting service to your Masters *is* reward!

This is called a **parasitic divergence**. It's what you get when you let a despicable pack of amoral sociopathic dunces use their unlimited concentration of capital, and the $2 whores they've bought with it, to run things. They let their egos run away with their tiny pea-sized brains, style themselves as world-class philosophers, and explode in a global blizzard of sh*t as they retreat to safety in their fortified palaces, private islands, cloud-clad skyscrapers, and armored redoubts appropriately called "DUMBs" (Deep Underground Military Bases... and yes, these certainly exist, as venerable installations like Cheyenne Mountain, Mount Weather, and other COG facilities attest).

[**Continuity of government** (COG) is a set of policies and procedures under which government operations are to be maintained in the event of a disaster such as nuclear war. Largely developed during the Cold War, it was later extended by the post-911 "Continuity of Operations Plan" to cover non-nuclear disasters as well. While COG was initially sold to the masses as a way to maintain stability and public safety, perceived government abuses and "corporate-government partnerships" have resulted in its being widely regarded as the corporatocracy's way of maintaining *its own* power and safety *from* the public in the wake of its own gross malfeasance. Obviously, COG requires safe hiding places for the rich and powerful, the best of which tend to be deep underground. Verified examples include the Cheyenne Mountain Complex in Colorado Springs, Colorado, which serves as the Alternate Command Center and training site for NORAD and USNORTHCOM (some of whose regular operations take place at nearby Peterson Air Force Base), and the Mount Weather Emergency Operations Center in Mt. Weather, Virginia, which is reportedly now run by FEMA. There are no doubt many more such facilities, the locations of which are kept secret for obvious reasons.]

In a state of nature, races tend to evolve in coupling with cultures and ideologies under geographic containment. In the advanced stage of social pathology called parasitic divergence, a parasitic subspecies goes nature one better. It emerges not just by incest and inbreeding under rigorous self-sequestration in expensive gated communities, secure centers of power, secluded globalist hideaways, and inaccessible COG facilities, nor just by developing its own exclusive culture and ideology (e.g., Transhumanistic Social Darwinism). It also becomes deceitful and highly aggressive, attacking its host by:

1. Preventing geographic containment of the host by opening its borders and incentivizing mass migration, thus re-engineering the demographic.

2. Normalizing the modified demographic through intensive mass indoctrination and forced Covid 19 lockdowns on the pretext of public safety.

3. Destroying the traditional culture and values of its host-subspecies to create a psycho-sociological *tabula rasa* amenable to reprogramming.

4. Starting with the blank slate it has made of the host's mind, redefining the host's ideology as (e.g.) communism or some easy-to-manage variation thereof.

In short, **the Great Reset is intended to finalize the parasitic divergence**. Its purpose is to split one species into two, with all wealth and power securely consolidated under the exclusive control of the parasitic subspecies while members of the host subspecies line up to be numbered and chipped like dehumanized robotic zombies, blinking stupidly at their television screens / cerebral imaging devices while dutifully repeating "We will own nothing and be happy!" This, along with plain old livestock culling where deemed appropriate, is how Transhumanistic parasitic divergence impacts "the meaning of being human." It is impossible to miss the resemblance of the modern global elite to a writhing mass of tapeworms... imbecilic tapeworms which, once they have completely demoralized and disincentivized their hosts, will be forced to infest and feed on each other as the pickings grow increasingly slim. It couldn't be more obvious if one caught them wriggling out of a sewer pipe with little "Made in Hell"/"Property of Satan" tags stitched into their foreheads.

Obviously, they should leave the theorizing about "what it means to be human" to actual humans actually capable of it. **They don't qualify**, and neither it seems do their ridiculous acadummy sycophants.

December 01, 2021

On the Survival of Mankind

I'm afraid this will sound a little "Philip K. Dick". I suggest that you entertain it anyway.

The global oligarchy is now engaged in profoundly inhuman behavior, and is thus best regarded as inhuman. Certain dangerous trends in biological and medical research indicate that with all the rationality of a suicidal drug addict, its members are on a path that threatens to eradicate the human species by irreversibly contaminating the human genome. As usual, their primary motives are to maintain or increase their own wealth and power. But beyond that, their motives seem even less benign: to (1) eliminate the free resistance-capable human population while (2) "reengineering" those humans who remain, possibly hybridized or "chimerized" with other species, as remote-control golems via the "Tech Singularity".

Obviously, such a program is self-defeating in that it poses grave danger to the oligarchs themselves. Certain genies, once released from their bottles, can be impossible to put back in. "Life finds a way," as they say, and debilitating genetic modifications can be hard or impossible to contain. It follows that doing deliberate, methodical harm to the genome of one's own species is symptomatic of profound irrationality or total insanity, including cult psychosis associated with a combination of "transhumanism" and narcissistic Social Darwinism. What else could prompt the elite to tamper destructively with the genome of their own species when they and their academic hirelings lack a full understanding of the genetic code, and lack the knowledge to genetically modify the human organism in a safe and controllable way?

This suggests that the oligarchs may be controlled by more rational (coherently self-interested) entities which, insofar as they have yet to emerge, "come from the future" via the advanced semimodel of CTMU metacausal theory. By definition, events that occur in the spatiotemporal display D of the CTMU Reality Self-Simulation must be generated in the processor P before appearing in the display, which means that the future of the display is contained in P and in this sense involves "advance causation". The future of D resides in P and is "sent back in time" from P to D. This differs from the "determinism | nondeterminism" dichotomy of standard causal theory. In a Laplacian deterministic display, the present state of the display determines all past and future states and realizes future states by localistic "retarded" D-to-D causation. In a random nondeterministic display, systemic evolution – if it could occur at all – would be acausal and chaotic. But in the CTMU, where reality is generative, self-deterministic, and amenable to free will and meaningful purpose, future states are selectively advance-modeled in P before appearing in D. This changes the nature of causality, making it self-dual in the advanced | retarded sense and giving it new dimensions in the bargain.

In the CTMU, the term "entity" may be replaced by "telor", which describes not just physical entities capable of intentionality, but entities which have not yet achieved full physical actualization in this reality and are thus "pre-physical", existing in some possible future. By definition, telors seek actualization, and may thus deem a habitable planet like the Earth an evolutionary or adaptive niche which, once cleared of its existing dominant species, could serve them as a resource for self-actualization. From the vantage of some telors, the presence of a dominant species like humanity might interfere with such exploitation, but extinction or debilitation could clear the way. Once human extinction were assured or humanity incapacitated by irreparable genetic disruption or contamination, colonization could proceed by means of advanced causation, through which terrestrial biology would be made to converge inexorably on their emergence.

Humanity can avoid this only by coming together to generate its own preemptive telon (metacausal template or "coherent intention"): racial survival and passage through the Great Filter. In other words, humanity must "pre-displace" its future competitor-species by retaking control of terrestrial biocausation and evolving to a sustainable configuration. This, of course, is what we call the Human Singularity, an alternative to the nefarious Tech Singularity favored by the likes of Kurzweil, Gates, Musk, and Schwab. Unfortunately, the competitors of humanity are causing it to "de-evolve" so that it cannot manifest the required cohesion.

This analysis breaks down into several alternative scenarios.

1. The aliens are already here (e.g., in subterranean, submarine, or extra-planetary / Martian or farside lunar bases), and bent on killing off and/or enslaving the competition (humanity) by threatening, manipulating, or psychically coercing the mind-controlled elite. The current presence of aliens is perhaps the most common take on UFO's.

2. The aliens are physically or metaphysically *en route* to Earth as future occupants, but want the planet clear of meaningful resistance, or humanity ready for enslavement, by the time they arrive. The human oligarchy is being controlled by psychic means (acquisition by an alien telon, forming an anti-human egregore). Physical UFO scout vessels may already be present. This scenario, possibly in combination with scenario 1, is a bit reminiscent of the television series "The X-Files", but with a much fuller view of the true nature of causation.

3. The aliens will exploit what remains of the mutated human race to mutate themselves into emergence by advanced causation. Time is no object, as their terminal consciousness (timelike cognition) can jump directly to their evolutionary point of emergence. (UFO's and aliens may be detected in the present as metaformal ideations, i.e., "quasiperceptions" or individualized percepts or limited activations of perceptual syntax stimulated by alien telesis.) This third possibility is in some ways the most natural means of "alien conquest". It accommodates the assumption that the human oligarchy is simply criminally insane and will exterminate humanity without external assistance or encouragement. In this case, another species will end up reusing the niche thereby vacated, provided that the Earth has not been rendered permanently uninhabitable. However, this new terrestrial species would not be distinguishable from a future species which has back-engineered itself into place.

4. Aliens will metacausally back-engineer symbiotic alien-human hybrids (as opposed to a nonsymbiotic parasite-host relationship) by merging the genetic codes of the two species. Alien terraforming will be rendered unnecessary by alien-human spiritual cohabitation in the bodies of native terrestrial organisms with terrestrial biology. This scenario, often discussed in connection with the so-called UFO abduction phenomenon, is a more benign form of scenario 3 in which humanity survives in engineered symbiosis with alien spiritual identities.

5. Humanity will come together in a human singularity and successfully pass through the Great Filter. (Again, UFO's and aliens may be detected as metaformal ideations, or individualized activations of perceptual syntax akin to "predictive programming", i.e., the involuntary telonic acquisition

of human subjects. However, the alien contamination of humanity will be interdicted.) This set of possibilities covers the entire range of personal expectations regarding the anti-human behavior of the global elite, human survival, and UFO-related phenomena. In scenarios 1-3, the aliens are malevolent, as suggested by the inhumane character of what is now afoot (benign aliens are not ruled out, but thus far have failed to break the parasitic hold of their malign counterparts on humankind). Scenario 4 arguably involves a less objectionable sort of alien, and scenario 5 is preferable by any measure.

Although one's personal expectation may seem to differ in some meaningful way from all of these possibilities, any such differences would probably be illusory. Scenarios 1-4 can occur in any combination among nonhuman telors competing for human resources.

December 20, 2021

***

Comment from Chris Langan: Perhaps there's a little confusion on what constitutes a "parody of David Icke". Mr. Icke talks about shapeshifting reptilians on television. The above post, on the other hand, talks about possible advanced-causal explanations for certain observations. To deny these possibilities with a wave of the hand amounts to embracing the usual physicalistic retarded-causation-only model of reality... a model which preempts any meaningful kind of metaphysics and misses the true order of things. While I appreciate everyone's support, I wouldn't expect to see such an offhand dismissal here. Could there be a logical argument behind it? If so, feel free to share it.

[date unknown]

2022

On Motion and Certainty

Question: Does the CTMU agree with the Taoist principle that God is the movement of the universe as a whole?

Answer: Yes, but only where "motion" means something like "cosmic evolution". Obviously, the universe cannot have external motion, as it must internally contain the medium in which motion occurs. Even internally to the universe, motion cannot be directly observed – one merely sees two states in rapid succession, with continuous motion interpolated between them.

Consider the following objection: "When one looks at an object changing its position, the object never leaves one's sight. Hence, one *has to be* seeing it move!"

While there is something to be said for this objection, one must regard the motion as a series of identification events in which the successive states of the object are successively identified. This sequence of identifications is limited not only by the human visual system, which can only take so many "snapshots" in one unit of time, but by logic itself.

Objects coincide with, and are identified in terms of, their states. One's ability to identify the object is thus limited in resolution by one's ability to register and describe its states. But accurately registering even a single state by direct observation or logical inference requires exact, or at least geometrically discernible, values for its variables or coordinates.

In a standard continuum, there is an infinite number of exact values (sets of coordinates) to identify along any given linear trajectory. Hence, the process of specifying them all is interminable and therefore a logical impossibility. It follows that *seeing* them all is also a logical impossibility, and likewise for any meaningful perception or conception of "continuous motion".

Let's look at it a little more closely. To be observed, an object X must occupy a definite state s (X) in which one can assign values to its variables. To see a "quantum of motion" of X is to see a binary sequence $s_1(X)$, $s_2(X)$ of adjacent states with no states in between; the "motion" that occurs between $s_1(X)$ and $s_2(X)$ is witnessed only as a blur for which $s_1(X)$ and $s_2(X)$ constitute "boundary conditions".

Motion means change of state in a metric space, and change requires at least two successive states of the same entity separated by distance > 0. Without two bounding states, one on each end of the state-transition, a linear (1D) change of state is indiscernible and in fact undefined. On the other hand, no matter how close together or far apart we put these bounding states – no matter how large our "quantum of motion" – we have only a blur between them.

Suppose that we subdivide the linear continuum between $s_1(X)$ and $s_2(X)$ to get one or more intermediate states, e.g., $s_{1.5}(X)$. Then we're simply adding and/or seeing another discrete state halfway between $s_1(X)$ and $s_2(X)$, and instead of one blur between $s_1(X)$ and $s_2(X)$, we now have two blurs: one blur between $s_1(X)$ and $s_{1.5}(X)$, and another blur between $s_{1.5}(X)$ and $s_2(X)$. Instead of eliminating the original blur, we've simply made two successive blurs out of it.

As subdivision doesn't work to eliminate the blur, could we instead make a "composite state" out of it by directly combining $s_1(X)$ and $s_2(X)$ in some kind of superposition? Unfortunately, perceiving both $s_1(X)$ and s2(X) as a single combined state precludes any change at all. The single composite state "$s_1(X)$ and $s_2(X)$" coincides with both $s_1(X)$ and $s_2(X)$ and thus cannot be intermediate. With no intermediate state through which X can "move" *en route* from one state to the other, continuous linear motion is out of the question.

Because continuous linear motion entails continuous change in both time and space, subdivision and superposition also disrupt the concept of spatiotemporal continuity and thus time and space themselves. Even the notion of causal ordering is disrupted. Any linear subdivision of the completed interval $[s_1(X), s_2(X)]$ comes "after the fact" of state-transition, which means that motion is determined ex post facto and is no longer a true dynamical property in the usual sense.

The CTMU restores spatiotemporal continuity and causal dynamism with features like triality, hology, conspansive duality, and metacausation.

January 02, 2022

***

Comment: Scientists and physicists who may often attempt to discredit the CTMU do not like to acknowledge or admit that their opposing arguments cannot really carry as much weight as they would have us believe.

Response: There have been no cogent, let alone valid, arguments presented against the CTMU. The CTMU is a supertautology (comprehensive metatautology), and there can be no valid argument against a supertautology. All attempts to produce one have amounted to troll droppings.

Comment: What's important is to realize that various conceptions of proof don't all mean the same thing and cannot be used interchangeably.

Response: The key property is *certainty*. Certainty is available only through logical inference and direct replicated perception. These are based on the CTMU concept of *metatautology*, which describes CTMU quantization in terms of identity operators and their internal and external state-transition events and provides a basis for certainty.

Let's all be grateful that there's a reliable way out of the many paradoxes plaguing science, mathematics, and every other discipline. The CTMU resolves such paradoxes with a whole new level of metaphysical certainty.

No CTMU = no certainty, and so much for any alternative conceptions of truth, proof, and science.

[date unknown]

The Intentional Aspect of Semantic Entropy

Comment: The cat's out of the bag. Covid-19 is an extermination project.

Response: Hasn't that been obvious all along?

Perhaps not to everyone. Much of the problem has been sheeple who foolishly assumed that the oligarchs and plutocrats and their political whores and media pimps ever intended to vaccinate the *underclass* against *Covid 19*.

Make no mistake, the vaccines were definitely intended to kill germs and cure or prevent disease. The plan was always:

"**X vaccinates Y against Z with V on behalf of Y** (including X)"

or more specifically:

"**X** (overclass medical authorities) **vaccinates Y** (the general population including overclass and underclass) **against Z** (a disease) **with V** (a vaccine) **on behalf of Y** (including X)."

But there was a hidden semantic switcheroo from:

1. "X = the overclass *vaccinates* Y = the underclass *against* Z = Covid 19 *with* V = vaccine *on behalf of* Y (including X)"

to:

2. "X = the overclass *vaccinates* (Y-X) = the underclass *against* Z = the underclass Y-X *with* V = vaccine (and Covid 19) *on behalf of* X (*excluding* the underclass Y-X)."

In other words, the overclass intended to genocidally "meta-vaccinate" *themselves* against *us* (the little people), classifying the underclass not as "potential disease victims" but as the *disease itself*, generalizing vaccine to mean "something that kills undesirable organisms and/or protects their victims", and limiting the intended benefits to **only themselves** (and of course "the Earth", of which they think as their own private property).

What slick semantic weaselry! And the beauty part is that "X vaccinates Y against Z with V on behalf of X" remains just as true for interpretation 2 as it originally was for interpretation 1, which conveniently reduces the likelihood of extreme cognitive dissonance.

Vaccination used to be about injecting a vaccine into potential disease victims, including both overclass and underclass. Now it's about injecting the "vaccine" directly into upscaled "germs", namely, members of the underclass themselves. The overclass views common people as "germs" and themselves as the only important potential "victims", and is merely exploiting a semantic transposition across the difference of scale whereby human "germs" are large enough to be directly injected with the "vaccine". (Obviously, instead of getting the "vaccine", the overclass stages photo ops in which a few of them get injected with saline solution instead of toxic mRNA and graphene hydrogel.)

Trust overclass scoundrels and their hired academic, media, and legal / political word-weasels to crablouse the language with this kind of sneaky semantic symmetry transformation. The basic idea is that by modifying language within given semantic symmetry constraints (the symmetry eases cognitive dissonance), they can modify human cognition and perception themselves. This, of course, comes from the *Sapir–Whorf hypothesis*, which simply formalizes Wittgenstein's statement (1933, p. 149):

> *The limits of my language* mean the limits of my world.

This deliberate kind of semantic word-wizardry is a mainstay of law and politics, which is why national constitutions suffer from what I've previously called "semantic entropy". It's usually as deliberate as coldblooded murder, but mere statistical entropy is what it looks like when everything is attributed to randomness and the intentional aspect is ignored.

January 20, 2022

CTMU Group Q&A

Here are several questions that were presented earlier today, mostly on the CTMU Facebook Group.

Q1: What does the CTMU say about trinitarianism?

A1: God is an Identity Operator, the GOD (Global Operator-Descriptor). This Operator Self-configures as a mapping from Alpha to Omega (the beginning and the end, or the "boundary conditions" of the causal differential equation sequentially integrated to produce spacetime and associated with the "Standard Model"). This mapping has a domain, a codomain, and a transformation from one to the other.

This is CTMU triality, an advanced form of "trinitarianism". Triality is a symmetry of the GOD, and each aspect plays each role from an appropriate vantage.

From perhaps the most useful vantage, the domain is "the Father", the codomain is "the Son" (represented by the terminally embodied man Jesus, the ideal of human perfection), and the transformation or metaformal-grammatical syntax-semantics coupling is "the Holy Ghost". Alternatively, we can call the domain "the Holy Ghost", the codomain "the Son", and the transformation "the Father" (God is the Primary Telor, and telesis is transformational by nature).

Note also that as Jesus is both the Son of God and the Son of Man, there is a telic symmetry between God the Father and Man the terminal phenomenon, each being the image of the other by topological and descriptive inclusion.

Q2: How does someone determine when life actually begins? 9 months? 6 months? 6 days? It is impossible to determine. Therefore, it is morally safest to assume life begins at conception [and form abortion policy and other aspects of reproductive policy on that basis].

A2: The universe is already everywhere alive. It is alive because it is embedded in a living God, and secondarily in our minds. Its conception and birth are distributed over it as its origin, providing any given moment of organic conception with a medium of origination. Reproductive morality resides in the embedment of conception (an organic origination metaevent) in the cosmic/divine origin or reality-creation metaevent.

This comes down to teleological consistency, which is defined not on the premise that every life is equally valuable – e.g., "a bacterium has every bit the right to live that any person does" – but on the teleological consistency and global utility of any given instance of conception and birth. The conception (metaphysical Self-configuration) and birth (physical Self-actualization) of God has priority over that of any given organism.

Unlike a zygote, God accommodates all future consequences of the existence and development of the zygote. Teleology thus seeks to guide reproduction in such a way that positive consequences outweigh negative consequences. If you fail to allow for (and preferentially distinguish between) positive and negative consequences, then you have no coherent basis for your "morality" whatever it may be.

Q3: According to the CTMU, can a human mind merge with the mind of God? Can a mind whose syntax is limited "expand" its syntax until it erases itself and merges into the absolute syntax of generalized cognition? If so, how should this be done? Some spiritual traditions affirm the possibility of destroying the distance between the human mind and the mind of God. These (esoteric) traditions include Taoism, Christianity, Sufism, Advaita Vedanta. Each tradition has its own methods.

A3: Expansion of the human mind to coincide with that of God comes at a price unpayable by most secondary telors. That price is the virtual annihilation of one's specific human identity. Because the identity of God exceeds human identity, identifying with God shatters the boundaries in terms of which one's human identity is defined.

Moreover, because the identity of God is the highest level of identity and thus shared by every telor with every other, one's individual identity becomes mixed with that of every other (intact, non-excluded) telor. As a result, one's ego and linear cognition are replaced by a universal form of cognition in which every lesser cognitive process is embedded, so that one can no longer focus exclusively on one's own individual self and personal priorities.

In switching from the individual to the divine perspective, everything changes. Most of one's self-image, the core of individualized terminal consciousness, is trivialized to the point of virtual invisibility (unless one is very "special" with respect to teleological necessity, in which case one's configuration and capacities are more consistent than usual with those of God, and less like those of ordinary human beings).

This "level change" of identity requires a mutable but nonetheless persistent identity to fill the role of "that which is transformed (changed)", and a mere human identity is too weak to support so profound a transformation. Thus, a true God's-eye view is reserved to God alone. Men can be, and are, localized images of God with more or less teleological consistency than others, but to be God in all His glory is a prerogative belonging only to God.

Even the Eastern mystics who practice meditation and (in effect) claim to achieve something approaching this state are not actually achieving it. Their minds remain physically instantiated and uniquely localized by Pauli exclusion, incomparable to the comprehensive, hyper-parallelized Mind of God.

Q4: Is man engulfed in chaos and misery due to interference by aliens or demons?

A4: There is only one way to preclude aliens or demons as causes of human misery, and that is to "generalize beyond them", replacing their associated causal hypotheses with a perfectly generic tautological hypothesis.

One simply completes the "alien | demon" distinction by incorporating it in an exhaustive disjunction of possible specific causes, and then transcends it by thinking strictly in terms of the generic structure of reality.

March 04, 2022

CTMU Group Q&A: The Real Universe

These very good questions were posted to the Facebook CTMU Group by member Michał Szczęsny. I was responding to them there before I realized that the answers would be too long and involved for the forum. I then transferred them to Patreon, and as I often do, scheduled them with the intention of coming back to edit and improve them. As usual, they were automatically posted before I got back to them.

I've made a few little tweaks and improvements since they posted here a couple of hours ago.

Q1: Regarding "the real universe timelessly emerges from a background of logically unquantified potential," could it be said that "the real universe" signifies the SCSPL syntactic METAVERSE – physical UNIVERSE coupling that emerges from the ontic ground state, aka the UBT (Unbound Telesis)?

A1: Yes. As reality is stratified, we make a distinction between the physical (terminal) universe and its source. The source is "real universe" as a whole, or in CTMU terminology, the metaverse. Because it consists of pure syntax, it is called the "syntactic metaverse", which means that while its syntax distributes over every possible universe, it is semantically uninstantiated on its own level of generality. Its instances are universes which are specified by instantiation and thus occupy lower levels of generality.

The metaverse is that level of syntax which distributes over any possible terminal universe like the one we see around us. These levels – metaverse and universe, syntax and its semantic instantiations – are coupled by dual T|D (topological | descriptive) inclusion. The origin of the terminal universe in the source implies that the source *topologically* includes the terminal world we see, which then regards the source as an ambient medium conforming to or *described by* its syntax. (This dual-inclusive interplay of syntax and semantics is called "hology", and a perfectly hological reality is called "supertautological".) The syntactic UDF is inwardly, T-inclusively distributed by the primary id-operator to secondary id-operators as a fundamental aspect of systemic structure, and by hological symmetry, the secondaries reciprocate by outwardly, D-inclusively distributing it to the surrounding medium and all of its contents.

D-inclusion and T-inclusion are combined when a language is "objectified" in terminal (sensible, e.g., written or spoken) form, and represented to the reader or observer as the states and processes of tertiary syntactors in the course of terminal communication (which is conspansive rather than merely ectomorphic as usually assumed). For example, when a sentence is printed on a page or fixed array, it is an aggregation of ink particles on paper particles. The expression consists of printed symbols consisting of ink-stained paper fibers contained in a piece of paper. This is topological inclusion (T-inclusion), and decoherent or physically differentiated on the quantum scale.

After being intentionally realized as thought and then written down, a printed sentence "becomes physical" and loses coherence. It then regains coherence when read by a secondary telor (but not by a machine, which is incoherent and thus cannot restore its coherence by coherent assimilation). The mind of the secondary telor then descriptively includes (coherently D-includes) the sentence as a pure abstract attribution, restoring its cognitive value as a "message" for the purpose of conscious communication.

Q2: Remembering that Creation is distributed and ongoing, would it be correct to say it is the SCSPL syntactic metaverse that is TIMELESS, and it is the physical universe that is bound in time, and therefore has a "date of birth" and beginning of sorts? How well-grounded is the claim about the age of the (physical) universe (≈13B)? Or is the concept of age simply illusory in this context?

A2: Yes. *Timeless* and *atemporal* are synonymous, and the CTMU has always been described as globally atemporal. There is no external measure of time; time is generated internally to reality. This requires syntax, and the syntax in which time is generated obviously defines and distributes over all time. Thus, it is not itself temporal, but timeless, atemporal, or metatemporal (with usage depending on context). Time, the sequencing of events, is thus generated by a "metaevent" in which all lesser events are metasimultaneous.

Bear in mind that while the universe has a "point of origin", this is not a "date of birth" in the usual extrinsic sense. Being background-free on the syntactic level (i.e., having no background but itself and the metaverse), the universe is largely independent and incommensurate with respect to any other universe. Thus, there is no external metric in which the origin of the universe can be located or in which the universe has absolute extent in any external time and space. Size and duration – and related quantities like shape and motion – are strictly intrinsic. (The "age of the universe", or the temporal distance of the current state of the universe from the cosmic origin, is intrinsically estimated in the Standard Model.)

Q3: Would it be useful to distinguish two types of unboundedness: physical and metaphysical? The first one referring to the yet undetermined, potential states of the physical universe, i.e. physically unconstrained and inaccessible to the senses, yet existing in the metaverse, and the second one signifying an absolute unboundedness on the metaphysical level, a realm in which identification is impossible, i.e. an unidentifiable realm of UBT, "inaccessible" (irrelevant?) even to the timeless level of G.O.D's Consciousness, or in other words unreal, absent from the metaverse.

A3: This distinction has been there from the start. UBT is a "cognitive limit" insofar as reaching it would unbind cognitive syntax and disable cognition, removing all cognitive support. It can be understood only as pure conspansive (self-configurative, self-identificative) potential.

Insofar as it can be defined, UBT couples with syntax. In fact, syntax can only exist by the self-actualization (self-identification) of telesis, and by means of a logico-inductive bootstrap, syntax can thus recognize telesis as self-actualizative potential from which extensional reality emerges (i.e., as "pure intentionality", the primitive drive to self-originate or "self-actualize from scratch" by telic alpha-omega feedback). Conversely, UBT can be described as the "self-potentialization of extensional (physical) reality", thus reflecting conspansive duality.

However, insofar as UBT "exists in pre-real form", it has a very important "metaproperty": the capacity to bridge the gap from "the structureless and undefined" (UBT, infinity, ontic potential, total freedom devoid of informational constraint) to the G.O.D. or syntactic metaverse (syntax-level structure) and finally to terminal instantiation (terminal spacetime structure, with finitude and linearity). Syntax then supports transfinite induction in the opposite direction, letting us conceptualize the leap from finitude to infinity.

March 30, 2022

Consciousness as Illusion

This is in response to a paper posted on academia.edu on which I was asked to comment. Its title is interrogative: "Mental Consciousness – Truth or Illusion?"[11] by Yoram Har-Lev:

> **Abstract** A common belief is that the senses reflect the real world to the brain, which in turn, presents it to our mental consciousness. People consider mental consciousness to be their own living self (first person) where we have mental experiences. An independent non-physical mental entity where we think, feel, and control our actions. It was found that there is specific brain activity for each conscious perception we have. The problem is that the nature of the interaction between brain activity and consciousness is still a mystery, coined the "hard problem" – meaning that science can never explain how physical activity in the brain presents mental consciousness experiences.
>
> But perhaps we bark at the wrong tree. Suppose we have no mental consciousness and what we think of as mental experiences are illusions our brain produces. The belief that there is a mental entity is based solely on intuition and long-standing tradition. But many intuitive beliefs were proven wrong in the past.

Is it not possible that the belief in mental consciousness is also wrong?

In his article I try to lay reasons supporting the idea that there is no mental consciousness.

My Comments: Pardon my interjection, but this can be a very confusing topic. Here are a couple of salient points for consideration:

1. Consciousness is subjective; one "feels" it and must consciously, introspectively look inward to apprehend it. On the other hand, if it were objective in the empirical sense, it could be directly perceived in other supposedly conscious beings, when in fact it can only be indirectly inferred through their appearance and behavior. You seem to be positing an alternative: discounting its subjective aspect completely. This would make it "objective", or as some would say, "physical". Unfortunately, that's a rather ill-defined term as well these days, as physics has been going in a direction increasingly divergent from traditional empirical objectivism.

2. Consciousness obviously has a subjective or "mental" aspect. However, the idea that it is EXCLUSIVELY mental comes down to idealism, which comes with no proper definition of consciousness, or Cartesian (mind-body) dual-

ism, which is logically insupportable on its face. Defining consciousness in such a way that its mental (subjective) aspect is negated therefore comes down to physicalism, and specifically the supervenience of mind on matter. You thus seem to be saying that consciousness is "physical" regardless of the fact that "physical", like "consciousness", is semantically ambiguous.

As a side note, I wonder if you've heard of the cognitive scientist Donald Hoffman. He's been getting quite a bit of air time regarding the illusory nature of consciousness as exemplified by optical illusions and other sensory quirks. His idea is that the senses and even the mind deceive the conscious observer about what it is seeing, ostensibly enhancing the survival prospects of the observer or organism and thus comprising an evolutionary adaptation. But he stops short of saying that consciousness itself is an illusion, as this would moot his entire perspective.

1. The idea that consciousness is an illusion is implicitly based on the premise that there exists something on which the illusion is imposed. Insofar as we can easily define that "something" as consciousness, the idea that consciousness is essentially an illusory concept is self-contradictory. Like illusion, physical reality is also registered by something like conscious observers. While one might respond that it need be registered by/impress itself on nothing at all, that doesn't work. Physical objects are observed to interact, and this means that they are "impressing themselves on each other". Basically, reality would therefore be "an illusion imposed by reality on reality itself". That's a rather deep assertion that requires considerable unpacking.
2. Therefore, let me unpack it just a little. The words "illusion" and "simulation" are closely related. There is already a well-developed theory of reality self-simulation, or with a bit of semantic transposition, "reality self-illusion". It's called the CTMU (Cognitive/Consciousness-Theoretic Model of the Universe), and I'm the author. (I was the first to render a detailed application of the so-called "simulation hypothesis" back in 1989, in a paper called "The Resolution of Newcomb's Paradox". I believe it's referenced in the Wikipedia article on Newcomb's paradox.) There are several academic papers on it along with numerous essays dating back over 30 years. The CTMU goes into the illusory aspects of consciousness, but shows that consciousness itself is not illusory. Rather, as one might surmise from its name, it shows that consciousness is the very basis of reality. This is the future of consciousness theory.

In addition to the words "mental" and "consciousness", the title of your paper also includes the word "truth", which is sometimes considered to have several definitions or interpretations. Actually, all valid (truthful) definitions of "truth" come together in its full definition, which relates itself to reality using the concept of dual inclusion: "Truth is the property of an attribution that denotes its inclusion in the set of true attributions, or a valid and comprehensive (hence retroscopic) theory of reality including all of its physical and metaphysical states." Hence, truth and reality are synonymous. This leads to a global reflexive structure called a "supertautology" which couples ontology and epistemology. This structure is a system of cognitive-perceptual id-operators which is closed with respect to intelligibility; self-intelligibility is its intension. It is comprehensive because it accommodates everything intelligible, excluding only the unintelligible. It is reflexively consistent or self-validating in the sense that consistency is everywhere enforced by INT | EXT self-duality, i.e., by hology, under which what is perceived is just what is consistent with overall structural and dynamical patterning or syntax, a normative kind of self-similarity. This, it functions as a cognitive identity language whose syntax forbids the perception, conceptualization, or intelligible referencing of any would-be counterexample. This makes it "absolute truth", or as it is sometimes called, "logos". It can be regarded as a metaphysical extension of logic which extends the tautological axioms of propositional logic to the CTMU supertautology.

[Minds become coherent metatautological quantum identity operators – id-operators are conscious by definition, as they internally convert external input to external state and thus form INT | EXT identities which "model themselves in external reality by descriptive inclusion" – which contain propositional tautologies as ingredients of their accepting / generative syntax, but can also err and embrace pseudotautologies that end up violating the tautologies, sometimes leading to catastrophe.]

I hope this helps.

April 16, 2022

Syndiffeonesis and Replication

Here are a couple of questions recently asked in the Facebook CTMU Group. I was going to post answers there, but they got a little too involved (this happens rather often). So put on your thinking cap and prepare for another exciting journey into the CTMU!

Q1: In the CTMU the concept of syndiffeonesis explains that when the difference between object A and object B can be proven, that difference also simultaneously proves that both objects are rooted in (and therefore extensions from) an overarching reality. Does this mean that the human experience is an indicator of what everything else in reality is experiencing, yet on a level proportionate to its particular substance?

A1: Yes, but just to the extent that the experience is true and replicable by other inhabitants of reality.

Let's rephrase a little: "Syndiffeonesis means that when the difference (diffeonesis) between object A and object B can be discerned by observer O, this proves that both objects are instantiations of the same (synetic) supporting syntax. This O-syntax distributes over the objects and thus comprises a joint medium associated with O."

Say that you distribute your awareness (accepting syntax) over A and B in a syndiffeonic relationship. Can you assume that the person next to you does the same, in just the same way that you do?

Actually, we have two syndiffeonic relationships, one associated with you (acceptor X) and the other with your neighbor (acceptor Y). The question is, does each relationship *replicate* the other? Are you both seeing the same objects A and B in basically the same way?

X and Y (you and your neighbor) can establish such agreement by matching your descriptions of what you see. That's called *replication*, or as a capacity rather than a process, *replicability*, and it amounts to a syndiffeonic metarelation between two syndiffeonic relationships, yours and your neighbor's. Given this syndiffeonic metarelation, we can be sure that neither you nor your neighbor has previously gobbled up a handful of 'shrooms and is seeing two little gray aliens instead of A and B as perceived by the other.

To some extent, you and your neighbor can replicate each other's observations. You can both see each other looking directly at A and B, implying that you both see the same two objects. Moreover, you can carefully describe the objects to each other, enumerating their properties and differences. One is blue and one is red, one is in the corner and one is in the middle of the room, and so on. This suggests that the same syntactic "categories of perception", or qualia, or Parmenidean or Platonic forms, are being activated in both of you by the percepts A and B.

But what if one or both of you have just killed a pint of Pfizer or Moderna, are seizing and gurgling and turning uranium-green, and have no idea what you're saying, thinking, or seeing? Then your agreement means nothing, and that kills the replication. In this case, you need a third observer. So here I am, ready to volunteer! This makes things slightly (but not too much) more complicated.

We now have two syndiffeonic relationships $X(A, B)$ and $Y(A', B')$. I'll be Z, and let A'' and B'' be the two objects as they appear from my perspective. Replication is then an equivalence relation $Z : X(A, B) = Y(A', B')$. Z (that is, I) must confirm that X and Y are seeing the same pair of objects, $(A, B) = (A', B')$, thus confirming that X and Y are using the same perceptual syntax to perceive $A=A'$ and $B=B'$.

So what we have is this: the observers X, Y, and Z (you, your neighbor, and I) are distributing our perceptual syntax(es) over each other, and to the extent that they match, we share the same medium (syntactic distribution). The common syntax of X, Y, and Z, i.e. that of the medium, distributes identically over both of the objects, yielding basically identical descriptions of the same two objects $A = A' = A''$ and $B = B' = B''$ (and also of X, Y, and Z).

When X and Y (you and your neighbor) replicated each other's observations, we had two replications. Now we have three, one for each of us. Every time we add an observer, we get another replication. With n observers, we get n replications of (n-1) other observations. In fact, given a set of observers $R = \{O_1, O_2, \ldots, O_n\}$, we can form 2^n (2 to the nth power) groups of observers who can compare notes from their individual and/or consensual perspectives, that is, as many groups as there are subsets in the power set of R (minus one for the empty set, which is in the power set, and which in a sense represents the "objective medium" or syntactic distribution considered independently of subjective observers). In this sense, replication is exponentially strengthened, or perhaps weakened, as n increases.

How could replication possibly be weakened? Once we must take into account the interactions among observers in various groups, it becomes hard to keep everything confined to raw perception. When we form groups of observers who compare notes, the observations and/or the notes may be contaminated with deductive or inductive inferences and opinions. This can toss a monkey wrench into the entire process, causing one group C to reject the observations or conclusions of another group D because D made observational errors or allowed confounding factors to creep into the experiment, and/or made questionable deductions from what C considers bad axioms or induced dubious explanations from what C regards as questionable data, and/or has what C looks down on as inadequate pedigrees, shady affiliations, or bad attitudes. Hence, the chaos of modern "peer reviewed" academic journals.

"Peer review" is synonymous with "replication" as discussed above, except that the replication is not just perceptual, observational, or experimental, but cognitive and inferential as well. While we can confine syndiffeonesis to raw perception by being strict about making no inferences, this becomes impossible once we start "doing science" and engaging in theorization along with observation and experimentation. Theorization is inference from observation, and when it comes to replication, it's definitely the fly in the ointment (and is often tacitly associated with the "problem of induction").

How does the CTMU get around the problem of induction? By logically inducing a supertautological identity-language with a self-distributed UDF master-syntax, without which we couldn't connect with each other in one unified reality...the reality we replicate for each other every time we are influenced by or acknowledge each other's existence or our common percepts. In fact, without this UDF, reality could not "perceive itself through us", or for that matter evolve according to laws of nature that can acquire their physical arguments. This supertautology is the minimal and therefore necessary model of mutual perception, and by virtue of it, all sane observers can agree on the existence of reality and our mutual perception thereof.

Q2: For instance, are we justified in assuming that, because a human being experiences consciousness and thought, a rock also experiences consciousness and thought, yet in an almost inconceivable primitive degree because of its lack of intelligence in comparison to more developed forms of life?

A2: A rock *instantiates* primary and secondary consciousness as an object thereof, but only "on a level proportionate to its particular substance". Its consciousness is not "experienced" in the same way.

The brain and body together comprise the physical vehicle of experience, and a rock lacks that kind of complexity and self-modeling capacity. Its level of consciousness, is more elementary. It consists of molecules locked into place by molecular forces and "experiencing" each other as neighbors, but totally determined and lacking telic freedom. That's a relatively weak kind of "experience". But this is just as well from a practical viewpoint, as a rock is very limited in the actions it might "freely" perform even if it had brain-like complexity.

In other words, a rock doesn't make much of a telor. The rock's functional complexity and self-modeling capacity are insufficient to experience primary or secondary consciousness. Yet it cannot be denied that consciousness is in some sense present, as the system identity – the Primary Telor – distributes restricted versions of its own essential (global) properties over everything in the system.

April 25, 2022

References

Articles and papers

Langan, C. M. (1989) The Resolution of Newcomb's Paradox. *Noesis*, № 44.

Langan, C. M. (1999) Introduction to the CTMU. *Ubiquity*, Vol. 1, № 1.

Langan, C. M. (2002) *The Cognitive-Theoretic Model of the Universe: A New Kind of Reality Theory*. Princeton, MO: Mega Foundation Press. Originally published in *Progress in Complexity, Information, and Design*, Double Issue, Vols. 1.2-3.

Langan, C. M. (2003) Cheating the Millennium: The Mounting Explanatory Debts of Scientific Naturalism. In W. A. Dembski (Ed.) *Uncommon Dissent: Intellectuals Who Find Darwinism Unconvincing*. Wilmington, DE: ISI Books.

Langan, C. M. (2017) An Introduction to Mathematical Metaphysics. *Cosmos and History: The Journal of Natural and Social Philosophy*, Vol. 13, № 2, pp. 313-330.

Langan, C. M. (2018a) Metareligion as the Human Singularity. *Cosmos and History: The Journal of Natural and Social Philosophy*, Vol. 14, № 1, pp. 321-332.

Langan, C. M. (2018b) The Metaformal System: Completing the Theory of Language. *Cosmos and History: The Journal of Natural and Social Philosophy*, Vol. 14, № 2, pp. 207-227.

Langan, C. M. (2019) Introduction to Quantum Metamechanics (QMM). *Cosmos and History: The Journal of Natural and Social Philosophy*, Vol. 15, № 1, pp. 265-300.

Langan, C. M. (2020) The Reality Self-Simulation Principle: Reality is a Self-Simulation. *Cosmos and History: The Journal of Natural and Social Philosophy*, Vol. 16, № 1, pp. 466-486.

Other authors

Einstein A. (1920) *Relativity: The Special and the General Theory: A Popular Exposition*. London: Methuen & Co., Ltd.

Feynman, R. (1967) *The Character of Physical Law*. Cambridge, MA: MIT Press.

Wittgenstein, L. (1933) *Tractatus Logico-Philosophicus*. London: Kegan Paul, Trench, Trubner, & Co., Ltd.

Internet sources

At the time the book went to press, the following links worked.

[1] Association for Psychological Science, "Why Delaying Gratification is Smart: A Neural Link Between Intelligence and Self-control"

Related: Sage Journals, "Individual Differences in Delay Discounting: Relation to Intelligence, Working Memory, and Anterior Prefrontal Cortex"

[2] Academia.edu, "Gödel's Ontological Proof"

[3] Cosmos and History: The Journal of Natural and Social Philosophy, "The Metaformal System: Completing the Theory of Language"

[4] Cosmos and History: The Journal of Natural and Social Philosophy, "Meta-religion as the Human Singularity"

[5] "Glenn Greenwald to RT|"Silicon Valley oligarchs have no accountability'"

[6] Bible Gateway, 2 Thessalonians 2:10-12, New International Version

[7] Bible Gateway, Matthew 5:33-37, New International Version

[8] Stanford University, "An Introduction to Philosophy"

[9] Technocracy News & Trends, "Military Is Developing 'Cognitive Warfare' Weapons"

[10] The Mises Institute, "The Great Reset, Part VI: Plans of a Technocratic Elite"

[11] Academia.edu, "Mental Consciousness – Truth or Illusion?"

The Smartest Man in the World: The Interview They Didn't Want You to See

PREFACE

The Daily Wire is a conservative media company with HQ in Nashville, TN. It was founded in 2015 by Ben Shapiro and film director Jeremy Boreing. It's a leading online publisher with a notable presence on Facebook, and it produces video podcasts for people like Ben Shapiro, Jordan Peterson, and Candace Owens. It also repackages journalism from various news websites, usually with a conservative slant, and has released several feature-length films and two television series.

As the company has conservative leanings, there have been complaints that some of its content is unverified and that facts are twisted to fit its partisan perspective. But this is to be expected, as progressivists and their partisans always claim, usually on no evidence whatsoever, to have an exclusive lock on truth and reason. In fact, DW appears to be a respectable company.

I was interviewed by the DW on April 30 [2022] in Nashville. It must have cost the company several thousand dollars (including airfare, car fees, and overnight accommodations for my wife and me). They pressed me repeatedly to be interviewed on short notice by Michael Knowles, who conducted the interview in what seemed a very professional way. The idea was that Michael wanted to use the interview as the inaugural presentation in a special series he planned to introduce.

In the course of the interview, Gina and I met Michael, Jeremy (co-owner of the DW), and other members of the DW staff. Everyone claimed to love the interview, and not just a little. We expected to see it online within a couple of weeks.

But then it was pulled without explanation of any kind.

(...)

Excerpted from an article "High Strangeness at the Daily Wire" published on Chris Langan's Ultimate Reality *Substack.*

THE INTERVIEW

MK: Welcome to my extremely ethereal and trippy new set. I feel as though I am floating in a cloud here, in this all-white venue, and perhaps that is fitting for this first interview in a series of much longer, much more in-depth interviews. I'm so excited that my first guest is Christopher Langan, the smartest man in the world. I do not say that as a subjective statement or to flatter Chris. I mean that in as technical a way as possible. Chris has one of, if not the highest IQ ever recorded, somewhere between 190-195 and 210. And Chris is not here by way of some fancy, distinguished professorship at such and such brand name university, nor did Chris just get off of his private yacht, out of the south of France, and come here to leave his billion-dollar company. Chris came here from a farm in Missouri after a career as a bouncer at bars around New York.

Chris, thank you very much for coming on.

CML: Thank you for inviting me, Michael.

MK: I first stumbled onto you when I was eighteen years old, a freshman in college, and—listen, I barely got out of high school math. I barely got out of calculus.

CML: I barely got out of high school myself.

MK: That's a good point. But I said, "This guy, he's saying things that are really, really interesting, and so I want to learn more." I've wanted to talk to you ever since then. I know that before we get into metaphysics, the existence of God, free will, politics, culture, and everything in between, I know people are going to be asking, "Why is the smartest man in the world not just buying and selling all of us all the time? Why is he living on a farm in the middle of Missouri?"

CML: Well, that's a good question. And it's that I was never actually interested in money. When I was a kid, my brothers and I, my family, we were not exactly the richest folks in town. We seldom had enough money to buy food or clothes, and so I immersed myself in books and reading and decided that what I wanted to do was pursue knowledge. It costs you nothing to pursue knowledge, really, provided you can sustain yourself while you're in pursuit of it. So that's what I did. I simply focused myself on, I *want to know the truth about reality. I want to know what kind of world it is that I'm living in*. And that's what I went for. Now, as far as the making of money is concerned, one thing that I found out is that there are certain ingredients, certain advantages, that you need in order to become rich. Of course, it helps to be born with money—

MK: (*laughs*) That's the easiest way.

CML: —and it helps to have a lot of connections, the right kind of connections, and not to alienate the people who have all the money, because then they'll exclude you and cancel you. That's what they do. That's what cancel culture is. Basically, people are being frozen out of the economy. And I found myself getting frozen out of the economy that way from an early age. I tried to go to college but ran into a couple of problems, personnel problems, on the faculties of the colleges in question, and that stopped me. Basically, when you can't get a college education, you are canceled economically.

MK: Presumably though you show up to college. Even though you've got a tough upbringing and not any real advantages in terms of family or society you've got a higher IQ than anybody in the room. You're obviously extremely smart, extremely self-educated, so you get into college. It should be a total breeze for you.

CML: Well, it was a total breeze for me, too much of a breeze for me. You ask the wrong kinds of questions of people who are full of themselves and think they have all the answers. Like in a calculus class, "Why don't you explain what an infinitesimal interval is and how you can traverse from one end to the other?" and they'll look at you as though you've got two heads.

MK: I ask that all the time. I ask my waiters.

CML: (*laughs*) Well, I got a very poor reaction out of that. There was a math instructor, Albert Leisenring, I think, who decided that I must be absolutely brain-dead. This guy was a very strange character. He was about six and a half feet tall. He'd come into class. Everybody would be waiting there for him. He'd make a late entrance. We'd be waiting for 10–15 minutes. He'd walk in with this great big stack and mimeograph sheets, and then he'd hand them out—walk around the room, methodically handing out these sheets to all of the students, and everybody would've a sheet in front of him. Then he'd walk up to the room, and symbol for symbol, everything that was written on the sheets would appear on the blackboard. Then he'd turn around and walk out. I was having a hard time with this. There were certain things that I didn't quite understand—why he was doing them the way he was. So I kept on trying to track him down to his office. He was never in his office. I would wait in the hall for hours and hours but this guy never showed up. Finally, I caught him in his office and I said:

"Hi, Professor Leisenring, can I come in?"

"Well, I'm really kind of busy right now."

"I just wanted to ask you one question. Why do you do this? Why are you taking a set-theoretical approach to calculus like this? They don't seem to be compatible. On the one hand, calculus deals with change, whereas sets are static things. Why are you taking this particular approach to it?"

And he looks at me—Asperger's victim, right?—and he looks down at his thing, "Well, some people just don't have the mental firepower to be mathematicians." What was I supposed to do? Hit the guy? I wanted to hit him.

MK: Right, right, but he says something that's sort of pathetic almost to say.

CML: It is pathetic, but it basically told me a lot about how he sees the world, how he sees other people. I don't want to take a course from a guy like that, and it offended me because I'd been a poor kid. I just came off a ranch. I'd been working all summer punching cows and there I was. I meet all these—there was a bunch of hippies in there—

MK: Right.

CML: —basically New York hippies. Now, I wouldn't call them hippies. These are basically affluent kids from New York. The place where I went was basically—that was the clientele, that was a student body. And they all were constantly talking, asking questions—bright for the most part, but I felt like a fish out of water. It was a culture shock for me. What is this? I'm used to being around a bunch of hay seeds. Cowboys—punching cows, going to the bar at night or whatever, drinking beer, and that was nothing like that. These kids were sitting around smoking pot, doing drugs, psychedelics—

MK: Now, just to clarify, when he says that not everyone has the power to be a mathematician—was this an admission of his own failure to explain his process, or is he calling you stupid?

CML: He was claiming that I didn't have the intellectual firepower to be a mathematician, or at least that's the way I interpreted it. And because I was used to being slighted in that way—I kind of grew up in a rough and tumble—if people feel intellectually threatened by you, you get a lot of this kind of thing up. So I assumed that that's what was happening and I've never seen or heard anything after that that would lead me to believe anything different.

MK: So, you leave college, presumably you're much more intelligent than anyone that you're going to meet on the faculty or in the students. So you leave college—what is it about being a bar bouncer? What is it about that physical activity? Because presumably even without a college degree, you could have done some middling paper-pushing job and it probably wouldn't have been very lucrative or fulfilling for you but, presumably you could have done something like that instead of a tough physical, potentially dangerous job, like your being a bouncer.

CML: I could have, but there are certain problems—for example, when I was in New York, I got a job in a grommet factory. It was Stimpson grommets. You had Grumman Aircraft there that got military defense contracts, and I think they were working on F-16's at one point. And then there was Stimpson grommets, which produced aircraft rivets for airplanes. And I had this machine—the sound, the noise from this machine—wham! wham! wham!—no hearing protection, and no nothing was issued to anybody. I started losing my hearing and so forth, and I figured, *Well, I can't stand this anymore. I got to get out of here.* I was about to leave, but I had a girlfriend, and she said, "No, I want you to stay." So I figured, *Okay, let me see if I can get another kind of job.* So I went and took the civil service exam and was offered a job by the IRS. And that of course was a moral dilemma.

MK: That's a fate worse than hell. (*laughs*)

CML: Exactly! How awful do I want to be as a person, you know? (*laughs*) And I decided that I needed to go home anyway, so I went back to Montana at that point. But basically then what I found out—I went back to New York when I was in my late twenties. That's when I started doing the bar bouncing thing. I wasn't making that much money. I was working for $40 a night, coming out of there bloody, with shirts ripped off my back. I couldn't even pay for the shirts. So I figured, *Okay, what I'll do is I'll take the civil service exam again*. Now—I don't want to sound insensitive, but at that point in New York, there was a protocol whereby you take the civil service exam, and if you are a minority, if you are non-white, you get 30 extra points.

MK: Well, there's nothing insensitive in it and this is just a fact. It's a fact of our law—

CML: They've been doing it for a long time. It's called affirmative action of course. And when that happens—I'm applying to be, let's say, a police officer, and all of these other guys, these non-white guys are looking also to be police officers. You learn that there's a line of three thousand guys in front of you, so then you give up your idea of being a police officer, and you give up your idea of ever succeeding, getting a job on the basis of a civil service exam. Now, they do have white police officers in New York, but almost all those guys are connected. They've got some kind of uncle or acquaintance or somebody who's on the police force that will put in a good word for them. I didn't have anybody like that, and merit made no difference whatsoever. This is not a meritocracy we live in. You can take any number of these tests and outscore everybody else and get nothing and nowhere.

MK: And it's especially true of civil service. It's true in a lot of fields in the economy, but especially in civil service, that has been a pronounced issue for a long time. So, presumably during all this time though, you're not just saying, as many people do when they leave college, whether they graduate or not, "Okay, well, that's it. I'm never reading a book again. I'm done with all that book learning." Something tells me that's not your mindset, even as you're doing these physical jobs.

CML: It certainly wasn't. And I would go to a library, sales—and found a little bookstore that had some academic books in it, and just get whatever I could. I always had to, basically, read whatever I found. I couldn't afford to order a book. Even back then, books were expensive. I couldn't go to a bookstore and pay full price for a book, so I was constantly buying used books, which were—when they were textbooks—used and therefore outmoded—the field advances, the book stays the same. But nevertheless, it's got some of the stuff that I need in it and then I can absorb that. So that's what I did. I just, basically, worked on my own ideas, trying to apply what I read in these books.

MK: Was there any field that attracted you in particular? I mentioned that I don't have anything past high-school math, and even that I was pretty sketchy on. Math, physics, philosophy, theology?

CML: Well, when I was fourteen, I was working on a ranch in Wilsall, Montana, which is just across Bridgers from Bozeman. I don't know if you're familiar with Bozeman, but anyway—I was punching cows, stacking hay, and irrigating on this ranch. I took two books with me. One of them was a book by Albert Einstein [Einstein, 1954] on the theory of relativity, and the other one was a book by Bertrand Russell [Russell, 1959] I would read these books—I was living in a covered wagon. They call it a sheep wagon these days, but literally—I'm not kidding—that's what it was: a kerosene lamp in a sheep wagon, out in the middle of a field.

MK: Reading Bertrand Russell and Albert Einstein—

CML: Reading Bertrand Russell and Albert Einstein, and it occurred to me, *These two things really need to be put together*. Then, once I decided that I started putting them together, and then I found out about Kurt Gödel and the undecidability theorem—and yes, absolutely. See, because reality isn't just geometric, which is what Einstein thought it was, nor is it just linguistic. It's a blend of the two. Russell saw it as being linguistic. Einstein saw it as being geometric. So I decided that reality must be logico-geometric, putting the two of them together, and of course, then I realized I've got to put together a theory—construct a theory in which reality is actually logico-geometric. That's where the CTMU came from.

MK: The CTMU is the Cognitive-Theoretic Model of the Universe. [Langan, 2002]

CML: The Cognitive-Theoretic Model of the Universe.

MK: This is your theory of everything.

CML: My theory of everything and it's all in the name. If you take a good, close look at that, you've got a cognitive theory, and of course, you know a theory is a kind of language, theoretic language. Then you've got a cognitive-theoretic model. You've got a model, and then you've got a universe—you've got a language, you've got a universe, and then the model is the mapping between them. The CTMU says that those are all the same thing. All of those terms, all of those properties, are distributed everywhere over reality. Reality can have only one structure once you realize that and you implement it in theoretical form.

MK: So then, the question I asked you is kind of a stupid question, because I said which field was it that attracted you. And your answer is, "Yes."

CML: (*laughs*) Well, I would have to say it would have to be logic and language, and then physics and mathematics. So those are the fields, and that's what I thought I was conveying, but apparently I need to spell it out.

MK: You do! This is so important that you're not just talking about this siloed aspect of thought, or this philosophy over here, language over here, math over here, physics over here, but you are presenting something that is universal.

CML: Correct. Absolute and universal. To get the absolute invariance, the absolute truths of reality, things that are true everywhere you go, no matter at what time you exist, or in what place you exist—it's the same.

MK: So for those of us who have an IQ that's a little bit lower than yours—I'll admit it with no false modesty and no undue confidence.

CML: Well, IQ is not the last word on intelligence by any means. IQ is where you focus. You can focus, marshal all your intellectual energy, and focus it very tightly on one item that you've been presented with. Okay? Those tests contain items and you're focusing on each one of those items. You're not seeing anything else—and that's what IQ is. But in addition to that depth and that focus, there is also aperture. Think of the mind as a kind of camera. What a lot of high IQ people have a lot of difficulty doing is widening their mental aperture. You've got to be flexible. You've got to be able to widen and narrow that aperture at will as you're doing the depth perception too. So you've got the focus, depth of focus, the magnification as it were, plus the aperture. Most high IQ people, they have the magnification, but they don't have the aperture.

MK: So then from the perspective—simultaneously, I suppose, both of depth and breadth here—if we're talking about a theory of everything, the first question we have to establish: Does God exist?

CML: Yes.

MK: Simple as that.

CML: Reality has an identity. The identity is that as which something exists. As a matter of fact, when you say the word *reality*, you are naming an identity. You're identifying something. (*looks around and opens his arms*) This. [referring to reality]

MK: I'm smiling because your answer on this is so beautiful. It just reminds me of Moses at the Burning Bush, and Moses at the Burning Bush says, "Who shall I tell the people that you are?" talking to God, and God says, "Tell them, 'I am, that I am.'"

CML: That's right.

MK: I am Identity Itself. I'm Being Himself.

CML: That's exactly right and that's what the CTMU says. It just comes up with the mathematical structure that you need to build a reality out of. So you come up with that identity and then you search it for its properties. Once you've built the preliminary framework, then you start deducing the properties of this identity and you find out that those properties match those of God as described in most of the world's major religions.

MK: Just the theistic religions? I'm thinking of Judaism, Christianity, and Islam, or are you talking also of say—

CML: Buddhism, Daoism, Hinduism, Vedism. Of course Hinduism and Buddhism have a God. Daoism—their central principle is the Way or the *Dao*, and they don't see *Dao* as God. And then in Buddhism, of course, they are trying basically to achieve *Śūnyatā* or "emptiness." A lot of Buddhists don't even understand what that's supposed to mean, but once again, there's no God there—you can kind of read God implicitly. Some Buddhists—I've talked to Buddhists who actually think that there is a God in Buddhism of a sort. That concept of pure consciousness is what it is, and if you ask them, "Well, whose consciousness are you talking about?" they will point at themselves and say, "My consciousness." In a way, they kind of attribute the existence of everything to themselves.

MK: I know a lot of people in Hollywood and Washington, DC, who do the same thing, actually.

CML: Well, that's right and that's why Buddhism is very fashionable among some of those people.

MK: (*laughs*) That's a good point.

CML: When you look at what they've all got, you come down to the same thing. Everybody has the Dao or the Way—that's the way reality works. Everybody has Śūnyatā, which is pure syntax—pure cognition with no instantiation, no content. Then you've got what the Abrahamic religions called God. It's all the same thing. But—what are its properties? Are its properties such that you can deny the existence of God, or are its properties such that God definitely has to exist? The answer is God exists. Properties of the central substance and central principle of reality, those properties are attributed to God, including of course things like—you have the three O's: omniscience, omnipotence, and omnipresence, but then you've also got consciousness—God has to be sentient.

MK: So we're not just defining God out of existence. Sometimes you'll hear people say, "God exists," but they'll give God such a weak and shallow definition that the God that they're describing has no relation to the God that we conceive of. You're saying God Himself is conscious and therefore personal?

CML: Yes. You can establish a personal relationship with God. We're images of God. You know what an image is—it's basically the product of a mapping. God maps himself into each human being. That's a very personal thing that God is doing for us. And I don't understand how anybody can say that it's any different. We reflect the structure of the universe, each one of us. We're carried by it, everything we do. We exist in a medium. What is that medium? Where did it come from? What holds it together? What is the unifying coherence, the source of coherence of that medium?

MK: Is your claim a pantheistic claim that God is the universe, or the universe is God, and that's that? Or no—or is God outside of the universe and created—

CML: God is greater than—well, what is the universe?

MK: Damned if I know.

CML: Have you ever heard of the simulation hypothesis?

MK: Yes.

CML: Okay. Well, the simulation hypothesis is basically the idea that the reality we see around us—physical reality—is simulated on some sort of an automaton or a computer.

MK: Yeah. Some aliens somewhere have just fooled us.

CML: Right, right, exactly—I'll eat a piece of liquorice.

MK: (*laughs*) Before one launches into the simulation, one needs a little sustenance, you know.

CML: The idea is that you've got some kind of an automaton running. You've got a simulation running on it. And God—it's more panentheistic. You know what panentheism is?

MK: I do. I think I do. That means that we are in God—let me know how I've gone wrong. We are in God and God is in us, and it is not merely that God in the creation are one and the same, but they are quite related. Is that something approaching panentheism?

CML: A little bit, yes. The idea is that you've got the physical universe that you see around you, but God is not confined to the physical universe. An ordinary pantheist thinks—assumes—that God is somehow confined to the universe, that there is just what we see around us and God is in every piece of it—God is distributed over it. But it's a little bit more complex than that because this part of the universe that we see around us, cannot exist just by itself. Okay? There are certain things that it entails. And when you go into those entailments, that's how you get to God. That's how you get to the identity of reality. Now, to get back to the reality of Self-simulation, or at least that's what I call it: Self-simulation. [Langan, 2020] But to get back to the simulation hypothesis, we are living in the display of that simulation. In addition to the display, there is also a processing aspect, and God captures both of those things. He captures both the display and the processor.

MK: What do you mean? I hate to put it in—

CML: Well, I mean, okay—here's the display. (*looks around*) You realize the display contains states. You see things, objects. States are static. That's why they're called states. Static. How do they change? Well, they have to be processed. Something has to be processing them, and in calculus, for example, those are tiny little infinitesimal intervals. But they're not actually contained in the states themselves. They have a neighborhood, a little tangent space or what have you, where you can sort of draw little vectors that suggest that some kind of processing is going on, but the idea of being a state and being a process, those are two different things in the ordinary way of looking at it. It turns out that you can't properly describe reality and causation at all, unless you put those things together somehow, and that's what it takes God to do. Okay? God provides the processing functionality for your state. You have an internal state and an external state. You're a material human being. To explain how that is changing through time and maintaining its coherence through time, even as it changes, that's what you need God for.

MK: I certainly agree with that entirely. I might not be sophisticated enough to parse all of the quibbles that there might be, but broadly speaking as a Christian, so much of what you're saying resonates as obviously true for me. The idea that, well, I'm a member of the body of Christ. The idea that God creates the world in this great act of love, this great act of charity.

CML: Self-love. Don't forget you're an image of God, so when loving Himself, God loves you, and you're supposed to love God back.

MK: Right, and this ties into something like the Trinity, right? The idea that God is three persons in one divine unity. So all of this is making a lot of sense to me. So now, how do I make sense of consciousness?

CML: Well, ordinarily—you know what quantization is?

MK: You know I know the word. (*smiles widely*)

CML: You decide what the ultimate irreducible objects are, those are the quanta in terms of which your reason. It turns out that in order to quantize that theory that I was talking about—that theory of identity where you've got the display, and you've got the processor, and it's handling both—it turns out that in order to handle both of those things, you need a certain kind of quantum. That quantum is called an identity operator. God is the identity, so obviously these little quanta, they're doing things, they're processing, so we can call them operators, right? They are identity operators. The identity operator, basically, takes input from the outside world, recognizes it, or accepts it using syntax, processes it, and then returns it to the world as an external state. So things come in, then they're processed, there's throughput—you could call that the subjective or internal state of the identity operator—and then it's returned to the external universe.

MK: But are you attributing now—

CML: What I'm saying is that's consciousness, and I'm saying that consciousness exists in every part of the universe, because those are the quanta.

MK: That's what I'm asking. Are you telling me that this table is conscious?

CML: In that sense? Yes. Generically conscious, but it's relying on our consciousness to do it. There's levels of quanta. These are tertiary quanta. They're all put together using physical localistic forces, but those are underdeterminative. They don't fully determine what happens. Why? The Heisenberg uncertainty principle, for example, tells you that the quantum rules are probabilistic, they don't actually determine events. So what determines events? We do. We don't know how we do it, but we do it. When everybody's will is put together, we're all creating the best possible universe we can for ourselves. And God is what harmonizes all of our different perspectives and makes things happen for all of us at the same time. And if we were doing things correctly, this would be the best of all possible worlds. Sadly however, we oftentimes make mistakes. And that's what we have to get out of doing, but we can't get out of doing it until we understand what reality is, what we are, and what the relationship between those two things is.

MK: Now you've mentioned two things that raise a new question for me. You mentioned this idea of simulation—just the simulation or Self-simulation—and you've mentioned us doing things. So then it would seem to me, we have to tackle the question, do we really do much of anything at all in the sense: do we have free will?

CML: Yes, we do.

MK: I'm glad to hear it. I always thought we did.

CML: Well, yes, we have to have free will. I was discussing with you earlier the idea of a fixed array—now, modern physics. Basically, you've got a bunch of quantum fields and superposition, and then those fields consist of little fluctuations, little quantum fluctuations, right? Where is the fixed array? We were talking about a manifold with a bunch of zero dimensional points, okay? Those two things are not compatible. Quantum field theory and that fixed array manifold, where you can parameterize all of the causal functions using the manifold: that doesn't work. Those two things don't fit together.

MK: When we were speaking about this earlier you put this into even more layman's terms and I've somehow—it has already flown out of my head. Can you put that into more basic terms? What you've just said.

CML: But what is it you need to understand about this?

MK: Why are these two concepts you're describing not reconcilable? What is the problem with these—?

CML: I explained that to you already. Alright. It consists of zero dimensional points, limit points, or cuts. I told you what a Dedekind cut was, right? These limit points have zero extent. They're exact locations, and that's a cut. You've got something on one side of the point, and then you've got something on the other side of the point, and any line that you draw through the point itself is going to be cut by the point. So we call that a cut. Okay? And the cut is zero dimensional, which means that it has no extent at all. It's an exact location. It's precise. No extent. It's not smeared out, okay? If you take all of those points and you add them together, you get zero. Because no matter how many times you add zero to itself, you just get zero. And because the manifold consists of those points, that's what you have to do to find out the extent of the manifold. So, what this is telling you is that the manifold itself has zero extent. There's nothing there. So the real manifold of classical physics is a paradoxical construct. Well, that's why that doesn't work with quantum field theory and the idea that things are quantum fluctuations and fields—as a matter of fact, that particular concept of the real manifold, that doesn't work for anything at all. It's a conceptual convenience. We can reason about reality in terms of it, and we can actually get to some very interesting conclusions using it, but it doesn't work in the long run. Most causal functions of course are parameterized in terms of this manifold. In other words, you've got the x, y, and z axis, and the forces that exist, that cause things to happen, those forces are all directed along one of those axes. That's what a force is. When you take away the points, those zero dimensional points, now, all of that disappears. The basis of causation has just fled the coop on you, right? Because you no longer have little points, the tails of the little vectors that point there, you can no longer parameterize your causal functions using those points and vectors. So what do you do to get causal functions? Well, it turns out you have to use something called advanced causation, and you combine that with ordinary retarded causation, and you get something else entirely. It's called meta-causation in the CTMU. It's referred to using two operations, one of which is called conspansion, and the other—sub-operation—which is called telic recursion.

MK: Because when we're talking about free will—often the conversation, especially these days, becomes this sort of shallow discussion of, "Well, this caused this, and I'm going to describe a totally deterministic system, and so as a result of this causing this, causing this, causing this, you don't have free will," and you're saying cause is actually more complicated than just—cause.

CML: That's correct. In other words, talking about free will on those terms was otiose—means nothing. You can't get anywhere with it. Reality is actually generative. It's not just a fixed manifold. Everything is being created all the time. Not just our states—our states are being recreated, right? I can cross my legs. I can uncross my legs. That's the changing state, but the medium around us is changing. When I look at you, I'm seeing Michael Knowles. I'm seeing you sitting there, but that means that I'm seeing your boundary. I'm seeing what distinguishes you from the external environment. There's a medium around you. So I have to be regenerating that at the same time as I'm regenerating your state in my head. Okay? And when I say regenerating—there's a reason I'm doing that—I could also say I'm recognizing Michael Knowles. I'm recognizing your state right now, but I'm also recognizing the state of the medium around you, because otherwise, I wouldn't be able to distinguish you from the medium, and you wouldn't exist at all.

MK: Right. Well, it's sort of like with a little baby that has trouble recognizing the limits of things and recognizing what some individual object might be—the glass on the table—they have trouble distinguishing those things.

CML: Precisely. The baby has to learn to distinguish those boundaries, and it has to receive the right visual cues at the right age, so that it can actually learn how to do that.

MK: So now we're at meta-causation. So what does meta-causation—how—?

CML: Causation works from past to future. Meta-causation works from past to future and from future to past, in a closed loop, it's called a conspansive cycle. Okay? So that's basically what we have to do to quantization in order to make causality wor—and of course, when I say causality, I mean meta-causation. Because ordinary past-to-future causality parameterized by a fixed array—that doesn't work.

MK: So I get past-to-future. This seems pretty simple. I pick up the glass of water. I put it over here, and now I remember two seconds ago that water was over here, and then I caused it to go over there. So how do you cause something to go from the future to the past?

CML: It can't go anywhere unless there's some place for it to go. That's all I'm saying. In the universe, everything changes with time. In the theory of relativity, for example, all the points are events and that means that there's a time parameter involved. So when you take that glass and you move it from there to there, you think that there's a point right there where you're going to move that glass. That's false. The point to which you are going to move that glass is actually in the future. When you pick up that glass, it's still in—you understand what I'm saying?

MK: Yes. Okay. That makes sense. And then getting back to God. Yes, we're always trying to get back to God. God is outside of time and space.

CML: God distributes over time and space, and there's some left over. Time and space is static. It's a display. Imagine that you're a little homunculus inside a computer display, like *The Matrix*, for example. Okay? God not only distributes over that, but there's a whole other domain where God exists, and that's the processing domain. That's the non-terminal domain in the CTMU. We're in the terminal domain. Right? And most physicists when they try to reason about the terminal domain, they reason under physical confinement: "Well, I can only look at physics, and I have to use what I know about physics, what I can observe about physics, to explain everything else that I explain." So they have that kind of explanatory closure going right there.

MK: So speaking of this non-terminal domain. In a really basic question—I'm not going to ask you if I'm going to go to Heaven or Hell, but will I go to either Heaven or Hell?

CML: You will persist after you die. Where you go, depends on who Michael Knowles really is (*both laugh*) and you would know that better than anyone.

MK: Yeah, I hope I know that better than—but you're telling me, I'm going somewhere.

CML: Yes.

MK: You're confident of that? I don't just evaporate. I don't just turn into oblivion.

CML: Well, you can. If you displease God, that's exactly what's going to happen to you. God is going to cut you off and He is going to say, "I can't see him anymore." He's going to turn away from you and then you won't be able to reunite. Salvation will be impossible for you because salvation means that God has got to pull you back into Himself. But God doesn't want to see you anymore. He doesn't even know you exist. He knows your physical body is there, but He's not interested anymore because you hate Him. You deny His existence, you offend Him, so He's not going to look at you, right? So now what happens? Well, you're dead. You still want to live. There's something in you that still desperately wants to live, so it's still going to be there. What happens now? Well, you try to create your own world for yourself, but if you're a bad person, an evil person, what kind of world is that going to be? It's going to be an evil world, and that's what we call—Hell.

MK: This is what John Milton says in the mouth of Satan. He says, "The mind is its own place, and it can make a Hell of Heaven, or Heaven a Hell."[1] [paraphrase]

CML: Now you got it.

MK: It all reminds me of Dante, also. This idea of God turning away. At the very deepest part of hell is Satan frozen in a lake of ice of his own making because of the flapping of his own wings, because he's apart from the warmth of God.

CML: Precisely.

MK: Okay.

CML: That's the way it has to work.

MK: So we've gotten through death, judgment, Heaven and Hell, free will, and God.

CML: Not completely. There's a lot more to be said.

---

[1] "The mind is its own place, and in it self; Can make a Heav'n of Hell, a Hell of Heav'n."

MK: And we still have some time. I mean, fortunately on my usual show, there is about thirty seconds to come to any conclusion about anything, and thankfully I am not bringing Chris Langan into Nashville to talk for ten minutes. This is going to be a much longer discussion. So—for this discussion, would you like a cigar?

CML: (*laughs*) Sure, I haven't smoked a cigar in a long time.

MK: Oh, excellent. You know, the body is a temple. The temple needs incense.

CML: Back in the day, I used to enjoy a cigar every now and then.

(CLASSICAL MUSIC PLAYS)

CML: Good cigar.

MK: I'm glad you like it. This is one of my favorite cigars that's come out. You know, if you were on the Joe Rogan show, they would offer you something a little stronger, but (*Chris laughs*) we're going to keep it to tobacco on this show, I think.

CML: I think I've seen a couple of—yeah, I've seen the Rogan show a couple of times. He was talking about some kind of drugs. At that point in his career, I think that's all he was talking about.

MK: (*laughs*) Did you ever get into drugs? I'm not saying that just because you were around bars in New York, but even the drugs that everyone says expand your mind, and anything like that?

CML: Well, you couldn't really grow up in the era when I grew up, around the kind of people I was surrounded with, without trying some drugs—I don't have a substance abusive tendency in my body. I just don't have any bad habits except for this. (*reaches for liquorice*)

MK: (*laughs*) This is again liquorice and candy.

CML: Oh man, I love that stuff. It's addictive.

MK: (*laughs*) It is.

CML: But I have to hide it from myself. I don't have any in the house right now.

MK: So you never—because I would always have people tell me, especially in college, they'd say, "Michael, you have to drop acid. You have to smoke this. You have to take mushrooms. It's going to expand your mind," and I thought, *My mind—my brain, such as it is, is pretty much all I got. I'm not exactly the captain of the football team, and if some drug messes up my brain, I'm sunk.*

CML: That's correct. But of course, what they say is that some of these medications can actually be good for you in terms of—you may have certain things, certain mental routines that are destructive and that aren't good for you that can be interdicted by psychedelics, for example.

MK: What's your take on it?

CML: My take is that could very well be true. You know, the psychedelics could open up a gap between your temporal consciousness and your non-terminal consciousness, right? Your mind is an extended stratified thing. We're just using our terminal consciousness right now, but there are other aspects to your consciousness. You can actually get certain kinds of insights and communications if you just open up a gap. You understand? Let them come in and fill the gap and then you can see them as they would ordinarily—

MK: So the things that people see when they're on psychedelics—I usually write them off and I tell people, "You are just hallucinating," but you're saying maybe that's not all it is. They might be seeing something real.

CML: Well, once again, what is reality? Is reality just stuff out there. No, reality has a mental aspect, right? And once you admit that basically everything has a metal aspect, then of course what's going on in your mind is real. It takes on a kind of reality. It's not the same as physical reality. It is nevertheless real.

MK: Are angels and demons real?

CML: Yes—yes.

MK: I think so too. Is there a fear that if you take some of these drugs, you might be letting in the wrong guys.

CML: That's a problem, isn't it? And that's a problem that I think a lot of people have encountered. You have to be a certain kind of person to be able to handle these drugs and not be sucked under by them, okay? Because once your mind is messed with in that way—it's weakened, you're not exactly in control anymore—something else can come in and grab it. And if you open up that gap that I was talking about, what can come into that gap might not be good for you.

MK: God is real. Angels are real. Demons are real. Is the Devil real?

CML: Oh yes. Yes. Well, it has to be. We were talking about Michael Knowles being surrounded by the medium—you've got a boundary. Well, God has a boundary too. He's got a very tight boundary. He's a perfect—He's perfect. He can't take anything resembling imperfection. He can't take it into Himself because that would be a contradiction, okay? So, God needs an antithesis in order to be properly defined. What is that antithesis? Anti-God or Satan. So it definitely exists. Now, Satan isn't coherent because he basically hates existence. Nevertheless, he gains coherence through human beings, through secondary telors as they're called in the CTMU. In other words, Satan can nucleate power structures, for example, things like corporations and governments, where you've got people in there that can be acquired as resources. And there's a kind of skeleton, a corporate organization, a governmental organization, that's holding them together, holding them in place, that can be exploited by Satan.

MK: So you're not describing Manichaeism. You're not saying there's God and the opposite of God, and there's some maybe equivalence between the two. You're saying that God—obviously there is an antithesis—Christ has an anti-Christ—but it's incoherent. Are you saying that the Devil sort of lacks substance, that's why he needs the humans?

CML: I'm saying the Devil lacks coherence. Coherence is what brings everything into superposition with itself. In other words, it allows—this is going to sound a little bit paradoxical—something to communicate non-locally with itself. All of its possible states are in superposition. They exist all at once. And this is pretty much inescapable.

MK: I'm reminded of a writer, René Girard, who had this idea that the Devil, being who he is, is a kind of contradiction of Being. And it seems to me what Girard says is something similar to what you're saying, which is that he requires us to kind of do his dirty work.

CML: That is correct, yeah. We give existence to the Devil, to Satan. Now, you have to make a distinction, however, between Satan and Lucifer, for example. Now, Lucifer is an angel. That's what he is supposed to—a fallen angel, but nevertheless, an angel, right? The angel of light, the morning star, whatever you want to call him. That is not Satan. Those are two different things.

MK: What's the difference?

CML: This would no doubt be classified as a heresy, for example, in the Church.

MK: I did not bring the inquisition with me yet. (*Chris laughs*) They're not behind the curtains here. So what is your idea then? Because traditionally it's understood that Satan and Lucifer are the same person. You're saying they're different.

CML: Yes. Yes I am.

MK: How are they different?

CML: Well, Lucifer walks the fence. Basically, he's the angel of light. What does light do? Basically light is what brings reality to us. He's walking the fence between good and evil. We're all in the same boat, we all walk the fence between Good and Evil. How can we do that? Well, through Lucifer—

MK: And the grace of God.

CML: —through Lucifer—basically Lucifer has been exiled here because God can't take—won't tolerate the imperfections of this place. He has Lucifer handle that for Him, and that's why Lucifer is what he is. That's how he fell down here to Earth. He's performing a function here. That function is light. I mean—do you follow what I'm—?

MK: I do. I'm still trying to parse the difference a little. If Christ says, "I saw Satan fall like lightning from heaven," is he describing Satan here, or he's describing Lucifer, or both?

CML: If Christ were here and I could hear his exact words, I could perfectly interpret them.

MK: Right, okay.

CML: But it's very hard to do that two thousand years after the fact, with all the different translations and interpretations that the Bible has undergone. What I like to do is I like to approach it from first principles and look at it logically, rationally. And what does logic tell us, if anything? Well, logic tells us that basically God needs a negation, but that negation is too incoherent to function in this world except through human beings. What's the rest of it? What's the rest of it all about? Lucifer is a concept that is better suited than Satan to that purpose.

MK: I love this point that he's incoherent, that the old Devil is incoherent. I read some poem—I think it's a modern poem—that says sin is a big problem because it's evil, but it's also a big problem because it's such a waste of time. It's so incoherent. When you take the suffering out of it and you just think of people committing any particular sin, it just doesn't make any sense, does it?

CML: The sense that it makes to them is they're enjoying it. They're hedonistic, deriving pleasure from the sin. And of course we all, you know, have the same pleasure mechanisms, so we're all tempted to commit sin, but we've got to keep a lid on it. There have to be limits. And what I'm worried about is that now Satan controls the world, and there are certain people involved in running things, who don't have those limits. They're so rich, they're so powerful that they don't need those limits anymore. So they've more or less thrown them off.

MK: Now there's nothing new to the idea that Satan or Lucifer—or let's call him just the old Devil himself—is the prince of this world, and that the rich and the powerful do his bidding, and they're bad people, and they follow their lowest appetites and their pleasures, and they can be sadistic—so that's been going on for a long time. Is there something particular about the moment we are living in that says it's actually gotten much worse?

CML: Yes. We've had so many technological advancements that now the technology of surveillance and coercion are such—and these people are so rich—they're like black holes gravitating all the money to themselves—that they're unstoppable, and because they're unstoppable, because they actually run everything, we are endangered by them now.

MK: It sounds to me like you're saying we don't live in *Schoolhouse Rock!*—I'm a Bill up on Capitol Hill. We're not living in the Republic that a lot of us say that we're living in.

CML: That's correct. Basically the world is globalistic now. It's run by globalists. That's their goal. That's what *global* means, "We're going GLOBAL." (*laughs*)

MK: That's oligarchic—and it would therefore be oligarchic.

CML: That's absolutely correct. Oligarchic, because very few of them actually exist.

MK: And this is beyond a US Senator. This is beyond the structures.

CML: Those are puppets. It's fairly common knowledge now, I suppose, that I regard most politicians as being one step removed from prostitutes.

MK: (*laughs*) That's been true for—they're two oldest professions.

CML: There are a few exceptions, but let me tell you, they're on the run. So it's a very serious situation. And if you take a look at these globalists, the ones that have thrown off all the limits, and they're fantastically rich and powerful now, and you take a look at what they actually have upstairs, I'm afraid it's not very impressive.

MK: Right.

CML: I mean if you really look at their intellectual produc—Soros might be the best of them. He's at least got a theory of economics. It's flawed, but he at least has that. The rest of them are just a bunch of sybarites.

MK: What about, I guess even outside of banking or finance per se, what about someone like Bill Gates? That guy's everywhere. He seems to be—I don't know why he is an expert on healthcare but my television tells me that he is, why is he— (*Chris makes a money gesture and laughs*)—so someone like that. But is he a—he's a smart guy. He's got to be, right?

CML: Oh, he's not stupid. Bill Gates has got a respectable intellect, but there's something misfiring there. Bill is very concerned with overpopulation and we do have an overpopulation problem.

MK: So then this brings me to—I asked friends of mine. I said, "speaking to Chris Langan, send in your questions," and most of the questions were about metaphysics, and God, and the CTMU, and all of your thoughts on all those things. But one friend of mine said, "You know, I want you to ask him about Bitcoin. I want you to ask him about money. Basically, if there is any way to break the government or super-government control of money." If I want to start my own currency I guess I could. I'd call it the Michael coin. Is this one way to push back against an increasingly technocratic global form of government?

CML: Yes there is but it has to start at the local level. You're not going to print your own world currency to compete head to head with these. You've got to go and basically attend a few city council meetings. You've got to go in there, say your piece, say, "We've got a problem: the world is run by globalists. This is America. We should have our right of self-determination here, but that's been taken away from us." See who agrees with you about it. Somebody will probably come up to you after your speech and say, "You know, I agree with you a hundred percent." Then you start—you find out who those people are, you get together, and you start putting pressure on your local politicians.

MK: Are you saying the situation is so bad, politically, that actually—people will call you a radical and an extremist and a terrorist for something like this—but are you basically saying, "Jefferson, we need to go fight another revolution or something, or is there—"

CML: I would just as soon see this whole thing resolved non-violently. Of course, I want peace on earth. I want everybody to feel as though we're in fellowship with each other—the Brotherhood of Man—you know what I'm talking about. That is the right way to do it.

MK: When you, or I, or Donald Trump for that matter, use the word *globalism*, what the liberal establishment will say is that globalism is good and we should have more of it, or they'll say that's a crazy conspiracy theory and there's no such thing as globalism.

CML: (*laughs*) What??

MK: I could point to a lot of international organizations that increasingly try to take power away from national governments.

CML: What are intelligence agencies? What are trade secrets? What's intellectual property? They're all property. They're all conspiracies, okay? People trying to get ahead by lying by omission, or lying directly to other people, the competition, right? That's what it is. You can't get away from conspiracy. It's how the world works, and—as a matter of fact—is game-theoretically rational.

MK: So you're saying not conspiracy THEORY, you're saying just FLAT OUT conspiracy.

CML: That's what makes the world go around, I'm afraid. Of course, the elite themselves realize this. They know that that's how the world works. They just want to distract you.

MK: Well, that's such a great way to put it, where you say a trade secret or just a group of people come together. They say, "I've got this good idea for a company, and we're going to come together and do that, and we're not going to let the competition know about it. We're going to start this and hopefully it'll be successful, and we'll make some money," and that is a form of—

CML: WEF and their young global leader program, young global leaders. I mean everybody's a young global leader, right? Trudeau, Macron, Merkel, Putin. For the Putin deal, that very much surprised me, but Klaus Schwab is on rec—he's on video, in fact, claiming that Vlad Putin is one of his young global leaders, trained in globalism.

MK: It's funny you mention the young wing of the World Economic Forum. When I was a freshman in college—you get all of these opportunities for internships and fellowships and you try to get a grant, and so you apply to these things. I don't know anything about any of these groups, and so I said, "I'm a politically-minded young man," very conservative, so it made me very different than my classmates, and one of the opportunities that came across was the World Economic Forum Global Changemakers fellowship.

CML: You were invited to be a part of this?

MK: No, I applied. What did I kno—

CML: You actually applied?

MK: I applied.

CML: Shame on you.

MK: Well, I didn't know what it was. They said:

"What's the experience?"

"Well, I'm very involved in this conservative group and this right-wing group, and this— "

CML: Good way to get ahead. I got to give you that.

MK: Well, but I'll tell you, I thought I had a good résumé. It wasn't like my classmates. They had all their liberal groups. So I said, "I'm a conservative, but you probably want some conservatives here, right?" (*Chris laughs*) So I apply and, you know, you're going to be shocked to hear this, Chris. I didn't hear back.

CML: (*laughs*) Yeah, I'm absolutely flabbergasted.

MK: And my liberal classmates, who I felt had a weaker application, but they were all part of the liberal groups, they did get it. I guess it makes sense, now that I know what the World Economic Forum is.

CML: What is socialism? I mean, what is communism, right? These things were actually funded. They're more or less invented by the central ban—you know who originally was paying Marks and Engels, and before that Moses Hess, and other people to come up with—it's a strategy for world domination, and it has been for the last two hundred years. What has changed?

MK: So when someone like Trump comes along and he says, "I'm not going to fight this political battle along exactly left-right lines, or Republican–Democrat, but I'm saying I'm going to stand for the American nation, and nationalism generally, against globalism," but he managed to get through. He did get elected.

CML: Yes, he did.

MK: Were you a Trump supporter? You don't need to say if you don't want to.

CML: As a matter of fact, back when people were saying, "Trump can't read. Look at the way he hesitates. He's got trouble with the telepro—the guy can't read." I actually stood up and I said, "No, I've met Donald Trump and I think his IQ is probably equal to that of the average Harvard professor."

MK: Really?

CML: Oh yeah. I actually stood up for Do—of course, I never got anything out of it, but I thought I was doing—you know, telling people that, basically, he is competent, he's a good businessman and therefore could be good for the country, and I wanted people to understand that.

MK: So how did this guy—I've asked this to my friends and I said, "How did this guy who I think was basically on the level, and really did at least try to pose a stopping block to the globalism program, how did he ever get elected? How'd they let him through?"

CML: Well, he was a little bit of a wild card, right? Trump is not as easily controlled as some of them are. So he stood up there and he spoke his mind. He's got good instincts in that respect because what he said resonates with what is in the minds of a lot of American citizens. We want to have our own country. We want America to have for us. We don't want our borders to be open. We don't want millions of third-world migrants coming here every year and displacing us from our own territory. And Trump actually did a good job of enunciating that. That's how he got elected. The first time around, you know, they weren't prepared, so they couldn't do anything about it. The second time around— *(laughs)*

MK: Are you suggesting, Chris, that there were some questions about the 2020 presidential election?

CML: *(laughs)* Anyone who doesn't understand that election was, shall we say, not quite up to snuff is some kind of moron.

MK: *(laughs)* Because I was told by all the fact-checkers, and all those social media censors that if I raised any question—I said, "You know, it seems like they kind of violated the law in Pennsylvania, and it seems like actually they kind of extended the voting periods." They said, "That's crazy. You're a conspiracy theorist. How dare you?"

CML: Well, that's exactly what they do. You see? You've violated group-think, so you're out. Your opinion means nothing.

MK: How would you describe your politics? Populist, conservative—?

CML: I was always pretty much of an independent. When I was a kid, I was basically a democrat, working—my family was democratic. I think my stepfather was a—you ever hear of the Wobblies?

MK: Yeah, yeah.

CML: He was an organizer for the Wobblies out of Chicago. Always I was raised around this liberal attitude, so, yeah, we were democrats, and I thought, *Yeah, the working man, the people that actually hold this country together, that keep everything moving, those are the ones that we should really care about, and if we're not careful, the rich will walk all over them, so they need somebody to defend them and to take care of them*. I'm one of those people. I wasn't rich. My family had no money at all, so naturally I gravitated to that. Then the Democratic Party changed. No longer was it about the working man. Suddenly it was full of these billionaire techies that have never done an honest day's labor in their entire miserable lives, and I realized—wait a minute—things have changed here. There is no more Democratic Party of the kind that I used to know about, that I used to belong to, and that's when I became a conservative, and a conservative in a stronger sense than most people who call themselves conservatives today, because they're really just cuckolded by the liberals. (*Michael laughs*) They're afraid to say boo to them. I mean, look!

MK: You know what's funny? I was told, some years ago—this was probably 2016 or 2017—someone said, "Michael, you can't use the word *cuckold* as an insult because that's racist or sex—" and I said, "Look, I'm of Italian extraction and cuckold is the oldest insult in the Italian culture that—"

CML: The Sicilians have a gesture for that. (*laughs*)

MK: Yes. (*he extends his index and little fingers while bending the middle ones and the thumb*) And it's the rock-star gesture. I forget the name of the rock-star—also Sicilian who popularized it—very famous and I can't believe his name escapes me—but that sign is the symbol of the cuckold. I said, "There's nothing wrong about using it. It's a very funny insult and it's descriptive."

CML: Apparently, there are married couples that actually—that's their hobby. I can't see it myself, but—

MK: Not my thing. It's obviously—it's disordered. Nobody wants, you know—I don't think rationally anyone would say that's a good way to live your life. And what you're saying is that there are people who call themselves conservatives and really they're just squish, cuckolded, or they just give in.

CML: It's a figure of speech. We're not accusing them of literal cuckoldry.

MK: No, they might be, I don't know. I'm not in their bedroom.

CML: But basically they are letting the other party wear the pants, and they're putting on their little panties, and doing whatever they're told. That's what I'm saying. That's the modern rhino, the modern Republican, you know. I mean, that's what most of them are, and believe you me, most of them wouldn't do a damn thing for anybody else. They're basically into their own thing, their own self-interest. And they just never take any risks on behalf of the American people anymore. And that offends me—because we're paying them and they've promised to represent our interests and they're not doing it.

MK: Well, you look at really basic issues, something like immigration. Well, and Trump really won on the immigration issue. Something like immigration—this is no insult to anyone around the world who wants to come to the United States. Frankly, it's not even an insult to the illegal aliens who are pouring across the border. But I think anyone can agree, taking in one million legal immigrants, and two million illegal immigrants annually—it's going to be more than that this year probably—on top of that, three million people per year—

CML: Taking them in? Whose houses are they living in? You're see them living in anybody from Washington DC's house?

MK: No, it's not Martha's Vineyard. It's not Southampton.

CML: You don't see too many of them there.

MK: No, but it's a political strategy that is opposed. If you look at public opinion polls, the vast majority of Americans, including lots and lots of Democrats want to drastically reduce this immigration issue. And yet, doesn't matter which party gets elected, the border remains open. The West—especially after World War II, you've got this huge boom in the West. You've got the second half of the 20th century, now we're into the 21st century, that's where all the money is, that's where all the opportunity is, and so just naturally people want to flood in there, and if you've got relatively lax immigration policies—

CML: Well, in order to get into this country, you used to have to prove that you were of value, that you could offer it something. It wasn't just about us giving to you. If you were an immigrant and you wanted to come here, you had to show that you had something to offer yourself. Okay? You had to go through Ellis Island, or wherever, you had to learn English, you had to learn about the Constitution, you had to buy in to our political sys—

MK: But now it seems you're incentivized to hate America (*laughs*) or to hate Western civilization.

CML: That's right. And now who would be doing that? Any real American? Not that I know of. It's a contradiction in terms.

MK: So what's the end game here? I mean, if America—

CML: World domination. They're going global. That's what it's all about.

MK: So, if we've got this oligarchic group, this self-appointed oligarchic group that I in my naiveté in freshman year of college, I say, "Oh, the World Economic Forum," or forget even WEF. There are other international organizations, the United Nations, obviously, the European Union, which the Brits have pulled theirselves out of. But there are these groups throughout the West that want to erase national borders, essentially.

CML: That's correct. And they want to deracinate everybody that lives inside.

MK: Yes, erase culture. I say this as the relatively mocha-skinned Southern Italian here. But there is a sense that—even just look at Europe. We don't want there to be French culture, and German culture, and British culture.

CML: I know. What a tragedy that is. Look at those cultures. They're beautiful.

MK: All the most beautiful cultures in the world end at the French, and the French on top of all those beautiful cultures too.

CML: (*laughs*) You know what the Brits always say, "France is much too good for the French."

MK: (*laughs*) Yeah. That's a good way to put it. So then what is your thought? Moving forward, we're in this very bad political spot right now. It would seem to me we've been in bad spots before—tower of Babel didn't seem particularly pleasant. The solution to that didn't seem particularly pleasant to live either.

CML: Divide and conquer.

MK: So what happens now? Do we just need to field a good candidate, or no, since it's too far gone for that?

CML: Well, the political system is kind of shot. Okay? So I'm afraid it's going to take some activism. We're going to have to—look, I don't want anything violent to happen. I don't want a new revolution, but I will tell you this: that if people don't act as though they are willing to fight for their own rights, they're going to lose those rights. Founding fathers knew it, you know, it's written right into the constitution. If we don't fight for our rights, we're going to lose them—simple as that.

MK: You mentioned everyone getting stuck with the Fauci ouchie for the last two years, and, you know, it's so extremely effective that we all need to take fifty-five boosters too, (*Chris laughs*) just to marvel at the perfection and the efficacy of it. Are you anti-vaccine, generally? I take it you're anti-COVID.

CML: Well, I bought into the COVID thing, at first. And Gina and I bought gas masks for us—those N100 masks, and I advised people what to do, so as not to get infected with the deadly COVID. And then I noticed that it wasn't really kill—I live in northern Missouri. Nobody up there was wearing masks and nobody was dying of COVID-19, so I realized, *Well, there's got to be something a little bit off about this, right?* and so I kind of like started getting away from it, and then I noticed that it was being used as a pretext for something called the Great Reset.

MK: And again, I mean, to bring back the World Economic Forum, this is a page on the WEF website: the Great Reset of the Global Economies and the Global Political Order.

CML: Right. They're very open about it. Okay? And there have been white papers written in the past about using just this methodology to get everybody behind the globalist agenda, right? Operation Warp Speed was part of that, you know?

MK: So then, does Trump bear some responsibility for giving Fauci the microphone?

CML: He definitely bears responsibility. My question is: does he really know what's going on, or has he surrounded himself with people that are misleading him about what's going on? I would like to believe that Donald Trump has been misled, but if he knows how much damage those so-called vaccines have actually caused, and he's still pushing them, then we have a problem.

MK: You know, if you raise any questions about even the COVID vaccines, which at the very least, they don't do what Biden, and Fauci, and Walensky told us they would do, namely to stop you from contracting the virus or spreading it—at the very least, it doesn't do that—you're called a crazy anti-vaxxer.

CML: That's called cognitive dissonance, right? I mean, who's crazy? You know, if you can look at the reality of what is happening and you can deny it, and call anyone who believes in it a conspiracy theorist, what kind of person are YOU? Well, you're obviously a nutjob, a kook, a moron. You're something like that, right? And that's what all these people are. As I say, "Conspiracy makes the world go around."

MK: Now, beyond finance or beyond these NGOs, you mentioned the intelligence agencies. Something that occurs to me is: if I were working at an intelligence agency—they never invited me—never got that invite on ZipRecruiter.

CML: Perhaps you should consider yourself lucky. (*laughs*)

MK: (*laughs*) I think I do. I certainly do. But if I were at an intelligence agency and I've heard there was a guy with the highest IQ ever recorded who was contradicting the liberal establishment agenda, I'd probably have a file on that guy. Have you had any run-ins with the—?

CML: Some interest has been shown. It was oblique. They're very—let's just put it this way: I managed to meet three people that are involved in that line of work when there was no actual reason that that should have happened. In other words—but they're very cagey about it—if they want to recruit you, they'll probably do it obliquely. If you don't apply to actually join an intelligence service, but they're interested in you, they will try to get you involved with someone who they control, and then he will vector you around and put you in touch with the right peop—but you've got to accept the agenda sight unseen before that happens. Right? You've more or less got to let them know that, Okay. *No matter what they're actually planning to do, I'm going to go along with it, because I really, really want to be in the CIA*, or the defense intelligence agency, in the NSA or whatever. Right?

MK: No such agency.

CML: (*laughs*) There's no such agency, I like that. (*laughs*) Yeah, but you have to somehow display some sort of willingness before you will be directly approached by them. And I guess I never displayed that kind of willingness, so I'm not there. I'm not in the intelligence community. I am, you know, periodically in touch with people who are in the intelligence community in various email distributions and things like that, but that's it.

MK: Since we're already in it, there are going to be people out there who'll say, "This guy is talking about every conspiracy theory I ever heard in the whole book." Since we're already in it, to take a slightly different tack, but still within the realm of what Wikipedia is going to call conspiracy theories—do you believe in aliens?

CML: Well, I will say this: the intelligence community, large sectors of it, believe in aliens.

MK: Really?

CML: Yes. Oh, absolutely. It's a big—I mean, they're constantly discussing these things. Yes. And if you look at the global elite, you know, and you want to—"Well, I don't want to blame the world banksters for this, that's a conspiracy theory." Then who is pulling all those strings? Could there be another kind of entity: aliens, demons, whatever. Could there be something that's pulling their strings—that the global banksters know about and they're taking orders from—but they're being totally concealed and hidden from the public? This is a viable hypothesis, and it's one that the intelligence communities don't reject—

MK: Hmm, it's funny you—

CML: —which is kind of neither here nor there, because the intelligence communities themselves could be run by aliens, for all we know.

MK: People are going to be laughing when you say that there are alien—

CML: Oh, let them laugh, but they won't be able to come up with any supporting argument against what I'm saying. They don't know enough. They're ignorant.

MK: It's funny that you mention angels and demons—pardon— aliens and demons in the same breath, because I don't particularly believe in aliens. I mean, what do I know? There are a lot of things I don't know about. But I certainly believe that demons exist. I'm pretty confident that demons exist. I'm less confident that aliens exist.

CML: Well, but the thing is a lot of people are reporting that they're seeing these UFOs, and other strange alien-suggestive phenomena.

MK: So what is—something like—they call it the Tic Tac: these weird UFOs that seem to violate the laws of physics.

CML: Correct. Basically they look like little blobs of light or spheres that can commit maneuvers that are totally against the laws of physics as we understand them. They look like they have mass, but somehow they're not subject to inertia. They can turn on a dime, on a right angle, and continue at the same speed. They can accelerate at tremendous g-forces. How can that happen?

MK: Do you have any theory?

CML: Well, yes I do, but it's an involved—it's fairly involved. Basically, when you reduce the world to telesis and you understand that reality consists of a merging of mind and matter—it's an interplay of the two—

MK: It's not just stuff and it's not just pure abstraction.

CML: It can be real without necessarily being a completely physical phenomenon. Okay? In other words, there can be a mental component to it. Right? And that is what I think is happening here. I think that actually there's something to it. The number of people that report UFO incidents is simply too large. People aren't all liars. They don't run around risking their reputations by saying, "Okay, I'm going to make a nut out of myself and say these UFOs exist."

MK: But couldn't they just be nuts?

CML: Well, I suppose they could—

MK: But you're saying there is too many of them.

CML: I'm saying it's very improbable that they are, because not that many people are that nutty.

MK: You know, my friend Andrew Klavan who you met earlier, he has this party trick—he's a mystery writer and he has written ghost stories and things—and at dinner parties, he'll ask a random person, "Hey, have you ever seen a ghost?" and he says a lot of the time people will say yes. And throughout history people report having seen a ghost.

CML: That is correct.

MK: And he says, "You know, too many people are saying they've seen a ghost for their not to be ghosts, or something like a ghost."

CML: That's correct. That is absolutely correct.

MK: Have you ever seen a ghost?

CML: I've seen things that look like ghosts—

MK: Such as?

CML: I've seen things that look like UFOs too. They're there. I mean, if you're not a very observant person, you'll never see anything like that. If your mind is completely closed, and you just screen all that stuff out, you'll never see anything like that. But if you're an open-minded person and you're observant, you will see things like that, eventually. Okay?

MK: Yeah. I don't want to put you on the spot to give an example, because if I describe an experience that I would call religious or numinous, sometimes people say, "Well, describe it." I can go around the edges, but it's somewhat ineffable. Could you describe such an experience?

CML: Well, yes, I could describe such an experience. On the UFO thing, I was working for the forest service near Lewistown, Montana, in the Lewis and Clark National Forest, not too far from Malmstrom Air Force Base. And I was in charge of a mountain range called the Little Snowies. I was the regional fire guard for that area, and I was alone up there. There were campgrounds and I'd go up and tend to the campgrounds and make sure that they were all cleaned up and everything, and that was one of my duties. And I had to be a fire guard, which means spotting and putting out forest fires and things like that. One day I was up there in a forest service pickup truck at a certain campground. I was there and suddenly—I look up—and I say suddenly—it was just in the sky. I became aware that it was up there and I looked up. It was this huge spheroidal, but elliptical, not a perfect sphere, like a saucer that was turned partially on its side. And I thought, *Well, that must be one of those lenticular clouds that I've heard of*, so I looked at it and I figured, *Okay, now I wonder how this evolves in time*. Totally changeless. I actually took sticks and I tried to use parallax to figure out, you know—*what is this thing? Why isn't it changing? Why isn't it moving? Why is it completely featureless and metallic looking?* Well, it wasn't any lenticular cloud. I literally stared at it for a long, long time before I finally just got, *Okay. It's there*. I got back in my pickup and drove away.

MK: Did you call it in or anything?

CML: I've told a number of people about it. Yeah. I didn't call it in. No, it was irrelevant to the state of my campgrounds, you know?

MK: It's funny you mentioned because—sometimes people say—this is a little different than seeing a UFO, but somewhat related—people will say, "I don't believe in God. I don't believe in Christianity because I don't believe in miracles. If miracles really happened, we'd all know about it, and—" One, if you look at the gospel accounts of miracles happening, the wildest part of it is not the miracle happening. It's the fact that when the miracle happens, everyone treats it as this miraculous event, and then five seconds later, they move on and people treat it as totally ordinary. And this happens again and again, and again, it's even happened—I would say that I've experienced things that are miraculous.

CML: Yes.

MK: And even in my own life, not one week later, I'm just going about my business as though it never happened. Or I'll forget that it happened or anything—it was a miracle! Shouldn't that—?

CML: But you don't deny that they happened?

MK: No.

CML: Right.

MK: I don't deny.

CML: That's the important thing, because then you know, you're really nuts.

MK: Yeah. (*laughs*) When you deny that it happened, right?

CML: That's right.

MK: Because you're so beholden to some preconceived notion of the way things are.

CML: That's correct. So, who are the real nuts? Obviously, they're the people that are doing all the talk about the alien conspiracy theories. Okay? There's obviously something there, or the number of people that report them wouldn't be reporting them. There are people in the CIA who actually claim to have been abducted. I'm not going to tell you who they are, but they're there. Okay? And we're not talking about small individuals. We're not talking about low-ranking individuals. We're talking about high-ranking individuals.

MK: These people—not just that you've read a book about—you're saying these people you've talked to.

CML: People that—there was one of them in particular that I was actually—somebody that I knew was trying to arrange an introduction to this person, and I said:

"Well, so he's in the CIA. Is he going to tell me the truth? I'm not in the CIA. Is he going to tell me the truth about anything?"

"Well, no, actually what he wants to talk about is abductions, because, you know, he and his wife were actually abducted by aliens and he wants to talk to you about—"

"Is he going to tell me the truth about anything?"

"I don't know."

"Then I don't want to talk to him. I don't want to talk to him at all. There has to be some kind of understanding that I'm not going to get lied to. Otherwise, I'm wasting my time, because I'll never know whether he is lying or telling the truth."

So if somebody isn't going to commit upfront to telling me the truth, I don't have time for him.

MK: If he were a really smart liar, he would've just said, "I will tell you the truth," and lied even if he wanted to.

CML: I suppose so. I mean, that's what they do in the CIA.

MK: Yeah. It's sort of the job, right? If you're a spy—

CML: That's correct. Than you lie your ass off at every available opportunity about everything. They don't all do that. I think there are a couple of people that I sort of trust when they talk about certain subjects. But by no means are they entirely in the know even. I mean, if you're in the CIA, you're basically a bureaucrat. You look up, you see—"Well, I've got my superior and then I'm looking up farther. Now, I see the director of the CIA—" Who's giving that guy his orders? It ain't the president. I mean, suddenly you're in a void. Who's running things? As far as most of these guys are concerned, it could be the Devil himself that's running things. They just don't know.

MK: You know, it's so amazing that you say that because it takes it full circle to the point that when I hear, "Okay. This so and so is actually really beholden to so and so, and this person is actually beholden to this person—" and I know that that's a fact because you can just see it in politics. You can actually kind of go to an org chart and kind of point to it, but then eventually say, "And so and so, if in a really powerful position, they're actually responsible to so and so, and accountable—" and you think, okay, at a certain point, you think, *I don't know. I have no way of verifying this. This could be it*, and then when you go all the way back to "and so and so is really just answering to the Devil." Now you've got me again, because I read that in my Bible. I know that that's true. I know that he is the prince of this world.

CML: Well, yeah, yes. By the way, I mean, there are accounts from biblical times and pre-biblical times that involve UFOs, things in the sky that show up during critical battles, for example, and turn the tide, right? This goes all the way back to the Vedic religion, the pre-Hindu religion. There are all kinds of people who have reported these things and they end up in various religious scriptures, including Christian scriptures as well, in some cases.

MK: I can't think of an example of an extraterrestrial appearance.

CML: Well, no, they're basically attributed to God. These are the interventions of angels or something. You even can go back and look at the Mayans, and the Incas, and people like that. There are pictures on temple walls—rather they're carvings—that seem to suggest that these things actually existed or at least were perceived.

MK: Right. There is a great meme that I'm really taken with—it actually has to do with IQ, which is a nice coincidence, but I love this meme, because it's got the really stupid guy at the one end of the bell curve and he is sort of drooling, and he says, "Duh, all the bad stuff is because of the Devil." Right? And then you get up on the bell curve into—I don't know—say IQ of a 120 or a 130, and it's the really smart guy, the egghead with the glasses. And he says—

CML: No, that's just the start of the danger zone.

MK: The danger zone: when you're just smart enough, but not smar—and those guys, they say, "No, actually the Devil doesn't exist, and actually we're all just sort of bags of chemicals and it's all rationally explainable," and blah-blah-blah-blah-blah—and then you get up to the guy at the really high end of the IQ curve in this meme, and he says, "No, actually the bad stuff's just from the Devil." He agrees with the guy at the other end of the— (*laughs*) but it strikes me that there's really something to that. Roger Scruton, the late conservative philosopher, said that the job of the conservative intellectual is to articulate things that the common man knows intuitively.

CML: That's right.

MK: And it's especially interesting when it comes to you, because you're a guy at the very highest end of the intelligence spectrum, and you've lived the life of a blue-collar guy.

CML: That's right. I wanted meaning in my life. It's like I say, I wanted meaning above all things. You got to make sure that you decide upfront when you're a young man, if you want to accomplish anything—who do you serve? Truth or Mammon? You don't have a choice—you do have a choice, but you can't do both at once. It's either going to be truth, which is the same as God—I mean, truth and God are to some extent synonymous—

MK: Yeah. The good, the true, and the beautiful.

CML: —or it's going to be money. And if you choose money, if you choose Mammon, then you're going to be serving Mammon, who happens to be the globalists at this point, because they control all the wealth and power.

MK: So, you mentioned—I think we totally agree Marxism is a very terrible ideology that's caused a lot of suffering in the world. You would agree.

CML: I'm not wasting this cigar.

MK: Absolutely not. Take your time, please. But so, we would agree with that: we're pretty anti-Marxist. Yes.

CML: I'm very anti-Marxi—it's a terrible philosophy. I mean, it's as full of holes as you can imagine.

MK: Even though all the geniuses at Princeton or whatever—there's all these Marxists.

CML: Look, that's a closed—that's a club. If you don't have the key to the clubhouse, you're not getting in. They won't even talk to you. No academic will take me on—will actually start arguing, even in his own field at this point, because they know they're going to lose, and they will. Any academic anytime.

MK: I'm as anti-higher-education cartel as there can be—

CML: Right. They're indoctrination mills. You have to look at the entire educational system as being one great big indoctrination factory, and the people that work in it—those faculty members—they're chosen, they're selected as indoctrinators of the youth.

MK: Yes. And they pedal Marxism. But I like what you just said. You boiled it down even more simply. A lot of politics comes down to, "Well, you are a socialist and I'm very anti-socialist," and "Well, you're a capitalist," or this or that. But you seem to be raising some problems with capitalism. Man cannot serve two masters.

CML: Well, there are two kinds of capitalism. One of them is, laissez-faire capitalism: you're supporting your family, you're trying to get ahead, you're trying to enjoy a good and comfortable life, but you're not forgetting about God, while you do it. Then you've got monopoly capitalism, and this is what libertarians and other people sometimes forget: that laissez-faire capitalism. Because money gravitates—it's like a black hole—it takes money to make money. You got to have a bunch and once you have a bunch of it, it comes more and more easily to you—I think you're out of butane.

MK: I know. (*trying to light a cigar*) Let me see if I can will—there we go! Hey, how about that? Not quite miraculous, but good. So, monopoly capitalism is the sort of thing we call crony capitalism, crooked capitalism, but it's—

CML: Basically, you're not just controlling the supply side, you're controlling the markets. Okay? You're told what you can buy: "We've decided that everybody should be using electric cars. We're not going to sell the other kind anymore." That's monopoly capitalism, and because if you control the manufacturing, the industry, right? If you're a trade organization or a trade association and you control that industry, you can literally control the market. You can only buy what you have the opportunity to buy, what they're giving you the opportunity to buy. These days, they're only giving you what they want you to have. Those are the only choices you have as a consumer.

MK: They're doing this with light bulbs to it. It drives me nuts. They tried to do it ten years ago. They are trying to ban incandescent light bulbs, which I find to be much more beautiful than those ugly LED, or any of the other light bulbs. I just—in my home, I just find it's warmer and more beautiful to have, and they are trying to outlaw.

CML: Well, my wife agrees with you.

MK: Really?

CML: Well, we have those old-style incandescent bulbs in the bedroom, for example.

MK: I think of my Christmas tree. I don't want those little LEDs. They're harsh. They're ugly. They're blue. I want the warmer light and yet—

CML: But they do make LEDs, by the way, and almost all frequencies these days. You can get daylight LEDs. That's costly though. I mean, they're way more expensive than an incandescent bulb.

MK: So, this actually raises even a different question here. We've talked about truth. We've talked about goodness. There's one more transcendental: beauty.

CML: Correct.

MK: Where does beauty factor into your thoughts about everything?

CML: Well, here we get into the CTMU—we all know about the monetary economy, right? You get money, you buy things, you increase your wealth. The economy is actually telic. It's a vast favor-bank. I do for you, you do for me. This is actually the trading system that exists in nature.

MK: Can you define the word *telic*?

CML: Yes. *Telesis* is the monic substance of which everything else is composed.

MK: Okay, I think I've got it.

CML: In other words, some people—a physicist says, "Well, that's energy. Energy is the basic substance of which everything is—" Mass-energy, right? I'm saying, "No, you've got to go up a step." You need something more general with other properties that, for instance, can accommodate information. You need something more general than energy to explain that. Or, if you want to, you can play a little semantic game and then expand, gradually, the definition of energy, so that it takes everything in. But I call the ultimate extension of that telesis. Basically, it's a fundamental form of intentionality. And if you want to talk about the intentionality of God, then you're talking about a dirty word in academia and what is that? Teleology. The Will of God.

MK: Right. The things have a purpose.

CML: That's correct.

MK: Yes. And so you're saying that there is this telic aspect to the economy, too.

CML: That is fundamentally the ONLY aspect. Money is an abstraction of that. It's a very specified, controllable, physically-commoditized version of telesis, and you have a certain group, a certain crowd that has established a monopoly on that. Now, when you establish a monopoly on money, you're actually monopolizing the telic economy as well. How? People invest their telesis in what can make them money. Unless you're like me and you prefer meaning in your life. Truth. All right? That's what telesis is and it encompasses not just money, but truth and meaning.

MK: That the way that money and the monetary economy is manipulated will take into it many more aspects of life that—and maybe steer them the wrong way—let me try and bring it to Earth, at least in the way I was thinking about it, which is: I HATE modern buildings. I hate them. They're so ugly.

CML: Drab and Soviet style.

MK: They are. And they bring me down, and I walk into Grand Central, a nice old building in New York, and I feel like a king, and I feel dignified, and I feel like a full human being, and then I walk into tha—they've actually recently tried to make it better, but—

CML: Imagine how you'd have felt in the Notre Dame cathedral before they burned it down.

MK: Which I never got to see. I never got to go in.

CML: Me neither.

MK: So, what is that experience like? I wonder for someone who's just got this perhaps quirk of birth, where you just have this extraordinarily high IQ, what is the experience of life? Is it kind of funny and amusing, or is it extremely annoying?

CML: I hope somebody got a little charge out of it, got a little amusement out of it, because I sure as hell didn't, most of the time, but I don't complain. This is God's a world. God gave me my lot, knew that I was going to go after meaning instead of money, and everything that came after was more or less like clockwork.

MK: If you had any say in the matter, which you don't, and in which you certainly did not, would you choose to be born very intelligent or of middling intelligence, or of low intelligence?

CML: What do you think the answer is? I wouldn't trade my mind for anyone else's. Nobody. Simply because I know, I understand what the difference is now. If you're not intelligent, you don't cultivate your intelligence and make the most of it, eventually you lose sight of truth entirely, and that's what has happened to most of these people. They're adapting, you know? "Okay. We've got a narrative. I'm not going to get ahead in life, unless I accept the narrative," so they accept the narrative. It's called adaptation. It's artificial. It's not like natural selection where nature selects you. You're being selected now by a bunch of ideologues, bunch of people who want you to adopt a certain way of looking at things, because they have all the money and power, and they want to keep it forever. That's why the Christian Church has fallen so radically out of favor with most of these people, even among certain Christians, high-ranking Christians, all right? It shows up in everything they say, and I don't want to mention any particular Pope, on the other hand—

MK: You're referring to a certain Mr. Francis, present Vicar of Christ? Well, he says these things that are so puzzling and potentially heretical, and certain cardinals have raised dubia—doubts and said, "Could you please answer this?

It sounds like you've said something heretical." And the answer is always (*imitates an Italian accent*), "Well, you see, the Papa was'a misunderstood. He was'a misinterpreted (*Chris laughs*) in *la Repubblica* newspaper." And you say, "Well, you know, if he keeps being misreported and misinterpreted, why does he keep giving the interviews to these atheists and far leftist journalists?" I don't know. It's not mine to answer, I guess.

CML: Exactly. And I mean, this is something that I think needs to be addressed by the Catholic Church itself, if it wants to stop shedding Christians, right and left, you see. Because people are not stupid. I mean, some of them can convince themselves of things that aren't true, but basically human beings are intelligent. And you're cheating your intelligence when you subscribe to these narratives, and apparently the Pope who communicates directly with God, he buys into these narratives and that's a contradiction in terms.

MK: You know, it's funny you mention Papa Francesco because he's come after the Latin Mass and a lot of young—

CML: Do you know he claims to have been a bouncer at one point in his life.

MK: I read that somewhere. He doesn—you know, you look like you could be a bouncer. Papa Francesco does not look like he was a boun—I don't know. Do you buy it?

CML: (*laughs*) Maybe in a really light-duty nightclub, he might have made it, you know what I'm saying? Certainly not in most of the ones that I worked in.

MK: He's gone after the Latin Mass, and I've noticed that young Catholics—you know, I fell away. I was basically an atheist for about ten years. Something that drew me—there were a lot of things that drew me back—but something that drew me back was the Latin Mass and specifically the beauty, the smells and the bells, and the idea that the priest is facing the altar, and we're all facing together, and he's leading us in our worship of God. And it's not just acoustic guitars and feel good and be a nice person, and making it all about us.

CML: It's the sense of the sacred.

MK: The sense of the sacred and a sense of the beautiful. It's why I ask that question about the role of beauty in your ideas.

CML: That's once again a function of the telic economy. Basically, it's synonymous with teleology. Something cannot be anti-teleological or deviate from the Will of God and still be objectively beautiful, all right? I mean, there are people who can take a religious icon or a picture of Jesus Christ, for example, and urinate on it, like Larry David, I mean, I don't know.

MK: Or even there was the famous picture *Piss Christ*, right? It was a photograph of a crucifix.

CML: Isn't that disgusting?

MK: Disgusting. It was funded by the federal government. I think it received funding—

CML: They paid for that?

MK: I believe it received some amount of funding from—I could be getting this wrong, so don't quote me—but I thought it was the National Endowment for the Arts, or one of those organisations.

CML: Really??

MK: I could be wrong, but—

CML: That's bad, real bad. So—I don't know what to say in defense of these people. I don't think there is a defense for them. But if that's the kind of thing that they regard as art, then they are not in touch with beauty, and the reason they're not in touch with beauty is that they're not in touch with truth.

MK: But I thought—I was told by all my modern teachers that beauty is in the eye of the beholder. It's a subjective thing, and it's not objective.

CML: It is. But unfortunately some beholders leave something to be desired (*Michael laughs*) in their aesthetic sense.

MK: You're saying there is an objective standard.

CML: Yes. Absolutely. Is it consistent with teleology or is it not? Obviously—basically you've got this identity—God—who is perfection. Anything that deviates from that cannot, by definition, be good or beautiful. Now, nature experiments. I mean, it goes from one thing to another and tries certain things out, but once it becomes evident that something isn't working then it's not consistent with teleology. Once it proves to be degenerate and causes people to become demoralized and sick of life, then we have to do away with it.

MK: I've been told that we need to tolerate all sorts of—

CML: God does not tolerate evil and neither should we.

MK: It's a beautiful way to put it. So what do you make of the modern right? You know, the modern left is always pushing this stuff, but a lot of the modern right says—there was a prominent columnist who called himself conservative. I guess he sort of was for a while, but now he's pretty firmly on the left, and he said, "I don't like Drag Queen Story Hour, but we need to tolerate it, it's one of the blessings of liberty. If we don't tolerate drag Queen's story hour—"

CML: What?? A blessing of liberty?

MK: Look, it's in the preamble to the constitution. I think James Madison's rolling over in his grave at the very thought of that—

CML: Constitution mentions trannies, does it?

MK: Not in my last read.

CML: I was raised in an environment where we were basically taught to be tolerant of people, of their foibles, and if they wanted to be a little bit deviant—well, a little bit of deviation is good. You know, you can't just be locked into one rigid way of looking at things, so I developed tolerance, but now they've become militant. They've become rabid. They're shoving it down children's throats, and it's just not good for society. It violates the biological imperative of survival. To become totally sexual—there's a reason that men and women exist, and that they mate, and create children. It's for the survival of the human race. That has to remain paramount. That's got to be the preferred mode of existence. You've got to buy into that. Now, if you want to be—if you want to deviate a little bit in your private life, and you've got somebody who agrees to be deviant along those lines, and you want to get together in the privacy of your own abode that's—

MK: You're not going to have Chris Langan knocking on their door.

CML: I'm not going to knock on their door. It's none of my business. It's for them to work out with God Almighty, and if they can do that, more power to them. I don't know how they're going to do it, but give them a chance, okay? But when they go out and they have gay day parades where they're cavorting nude, and they're going into children's restrooms, and so forth—that's where we have to draw the line, you see?

MK: I mentioned this modern writer who is very misguided on this particular point. Are there thinkers alive today who you think are worth reading? People always say, "What's the reading list. What should I read? What should I listen to? Are there philosophers or popular writers—?"

CML: There are no academic philosophers that are worthy, as far as I'm concerned, of attention.

MK: What about non—?

CML: It's not because all of them are stupid or anything like that. It's because they bought into the party. They're academics. They cannot violate the academic, communistic, atheistic narrative, and get away with it. And if you're in that system then obviously you're playing along with it, and you can't do that and be an honest, truthful philosopher.

MK: Are there non-academic writers, thinkers, philosophers, who you would recommend people read?

CML: I'm sure they're out there. I haven't encountered any, not recently.

MK: So that gets rid of the living people. What about those dead guys, all those old dead guys, going back to ancient Greece, or anywhere in the middle?

CML: Oh, a lot of those guys had something very useful to say. (*smiles*)

MK: If a listener right now said, "I want to know who, beyond his own writing, the smartest man in the world wants me to read." Who would you recommend?

CML: Well, I would say that you should go back and definitely read Aristotle, and Plato, and Socrates. Start with that and then go up, and then work your way through there. A lot of smart guys read Plotinus, read about neoplatonism, and once you find out about neoplatonism, then you can go back to the ancient Vedic religion, because it's basically the same thing in different language. And as you know, then neoplatonism kind of became a mainstay of medieval Christianity, and there's a lot to that. So that takes us up through some of the medieval Christian philosophers: Anselm, Aquinas, Augustin, and beyond that it starts getting dicey. I mean, even when you get up to like René Descartes, for example, everything becomes dualistic. You've got these Cartesian spaces that more or less exclude the observer. There are no images of God in a Cartesian space.

MK: Chris, have you ever—because you worked at bars—you ever hear about the horse who shows up to the bar and asks the bartender for a drink? And the bartender says, "Horse, man, you've been here every night this week, I think you're an alcoholic," and the horse says, "I don't think I am!" Poof! The horse disappears. Now you see, you will understand this. This is a joke about Descartes, who says, "I think, therefore I am," but if I had told you that part first, I would've put Descartes before the horse.

CML: (*bursts out laughing*) I like it!

MK: This is the only successful circumstance in which that joke has ever landed. I'm really pleased. I finally got a laugh out of that joke.

CML: That's a very intelligent joke. Much more intelligent than the ones I usually hear.

MK: Well, thank you.

CML: Especially in a bar.

MK: (*laughs*) That at least I can believe. So, that digression aside, I'm glad you bring this up because people always ask me—they say, "What philosopher should I read?" I say, one, "Why are you asking me?" but two, they always say, "Should I read John Locke? Should I read Rousseau? Should I read John Rawls? Should I read this?" I say, "Maybe focus a little earlier first." And it's funny you mention: once you get up to Aquinas, after him things start to go a little like kooky.

CML: A little south. Until now, of course, all of that is completely out of fashion, and most of these modern philosophers are really Frankfurt-school-type people and that's cultural Marxism, which of course is just the social program of the Marxists.

MK: But, Chris, even though I just wrote a book last year on cultural Marxism, I have been told, reliably informed by Wikipedia, that's just a crazy conspiracy theory.

CML: Yes. Well, Wikipedia seems to think I'm a crazy conspiracy theorist too. You read my biography there?

MK: (*laughs*) Not recently.

CML: Everything bad that you can say, you know? Basically, I was exiled from Wikipedia years ago. I went in and said, "Now, wait a minute, here. No, I'm not an intelligent design creationist *per se*. I have a different approach to teleology," and all that, and I was immediately kicked off Wikipedia. So I can't—they can say whatever they want to in my biography in Wikipedia. There's not a thing I can do about it.

MK: Because you even suggested the idea of some sort of intelligent design.

CML: That is correct. Well, I've been getting trolled ever since around the year two thou—that goes back further than that really, but when I went on national television and said, "Well, the existence of God is actually amenable to logical analysis and you can show that there is a God provided you have the right—" That was enough: "Intelligent design creationist, get out of here." You know, and these people are—by the way, these new atheists, I mean, they're adamant.

MK: You're talking about Richard Dawkins, Daniel Dennett.

CML: I am talking about Richie Dawkins, Daniel Dennett, Sam Harris—well, I don't know, Chris Hitchens. I mean, guys of this nature, you know, pyz Myers—PZ [Paul Zachary] Myers, and people like that. I had a run-in with Myers. I mean, he hates my guts. He hated my guts twenty years ago. None of them will mention my name now though, because they know that they can be crushed. Deep down inside they know—they won't come anywhere near me. At some point, these people decided that they had to exclude anybody who could beat them. That's where the cancellation thing began, really. It's a new-atheist thing, and now since the new atheists are a lot more fashionable than Christians are, that's the order of the day.

MK: I blame my falling away from the Church on just being a punk thirteen-year old boy, but the second thing I blame is the new atheists. They were very popular at the time and I was very taken with them. Chris Hitchens. He's a funny guy. I thought, *Okay, he must be a smart guy.*

CML: Excellent speaker, incisive, witty, but obviously he just hates God. I mean, he makes that very clear: "All religions are evil because they mislead people and drag people away from—"

MK: But you didn't just say that he doesn't believe in God. You said he hates God.

CML: Hates God. Dawkins obviously hates God. "Why is there evil in the world?" usually. That's what it comes down to. "Would a just God tolerate all of the evil that is occurring in the world?" The fact of the matter is there can be no good—there can be no pleasure without pain, there can be no good without evil. God knows that there is always the possibility to commit evil—

MK: Free will, right?

CML: Right. But he has given us free will. He bequeathed that—we're his images. He is free. Therefore, we are free, and he expects us to fight against evil, to stop it from triumphing.

MK: Then, given will, and the fact that you use these words hate and love—very often when people talk about God, it's as though it were some academic exercise, "Well, do I believe in God, do I—" but you're using verbs that are more active than that—

CML: Yes.

MK: —hate and love. Do you think that our discussion or our understanding of God therefore must be active and participatory rather than merely academic?

CML: Absolutely. And furthermore, it must be rational as well as faith-based. Lot of religionists are fideistic, which means they rely completely on faith, and say that we can have no knowledge of God. Wasn't that nonsense? You have to know what you got faith in. How can you invest faith in something that you don't know what it is? You need at least an elementary description of it before you can predicate—before you can attribute faith to it, before you can invest belief in it. If you invest belief in X, but you don't know what X is, is that meaningful? Obviously not. So faith and rationality are indelibly coupled. There's just no other way to talk about it. So as soon as you hear somebody say, "Well, it's entirely a matter of faith and we can believe whatever we want to—have faith in whatever we want to," that makes no sense at all to me.

MK: I totally agree with that, and it does bother me when religious people speak in that way in modern times. It's funny though you mentioned Thomas Aquinas. Was there ever a more rational guy in the world than Tom? The guy who wrote—to every question in religion, he had three objections and then answers to the objections. Quite a rational—

CML: We've gotten to another Aristotle. But all of these early-Christian philosophers are now out of fashion. Nobody pays attention anymore—well, except Christians, real ones, not fake ones.

MK: Practicing ones.

MK: So then—I know, I've kept all of your time, but I don't care, darn it! You're here in Nashville and I've got—

CML: Yeah, I'm just fine.

MK: Do you have religious practice?

CML: There is—yes. Well, let's just put it this way: I don't currently attend church but up until several—I started doing, like, my own kind of thing with preaching, right? So I decided that I didn't—but I was for about 10–15 years—I went to a little country church up the road from us. I live on a road where there used to be three historic churches. Two of them were burned down. Arson.

MK: Not accidental.

CML: Not accidental, no. And they were historical Civil War era churches, but one was left. And it was north of us, about two and a half miles north of us, on route P there. So I attended that church pretty religiously, very regularly for a long time. Now, as it happened, the last thing that I heard—it's run by a wonderful lady that actually takes care of most of what goes on there, and it's got a good pastor, right?—but I finally decided that what I have to do is go out and start talking about these things myself, so since then, that's what I've been doing.

MK: I wish you were here on a Sunday. I could take you to my good Latin Mass parish here. I think you would like it.

CML: I was twice baptized to Catholic way back when, but it's like I say, I don't—I was a kid. One of my aunts was a nun, aunt Sybil, right? I have some fond memories of what it's like to be a real Catholic, but I don't think that that kind of Catholicism is in vogue.

MK: It's certainly not in vogue. When young people are searching for meaning—as you say, you looked for meaning in your life—reminds me of Victor Frankl, right? *Man's Search For Meaning* says, "Search for meaning is man's primary motivator, more than sex, more than anything else." [paraphrase]

CML: And that's the way it should be, but that's not the way it is anymore. They just want the money. They won't even believe that you're intelligent if you don't have a lot of money.

MK: Right. Well, that's what people always ask you, "Why aren't you a billionaire?"

CML: Why aren't I a billionaire?

MK: That's what people—I've seen people—

CML: Because billionaires hate my guts. See— (*laughs*) I don't go along with the—look, they have to give you the opportunity to make money. You need to have connections. In order to make money, you need to make friends out of people who have a lot of it.

MK: If you had the connections, would you have striven to make money?

CML: Not at the expense of truth. No, I'd have gone after meaning first. Now, I used to think that, *Well, it should be possible in theory to go after both*, then

I discovered that wasn't necessarily true. If you go after truth too hard, then you're going to make a bad impression on Mammon, on the people with all the money, and eventually you're going to be cancelled out of any sort of comfortable life.

MK: And isn't—there's also just a time issue, right? I mean, if you dedicate yourself to—I don't know—doing spreadsheets or something, you're probably not going to have a ton of time to pursue truth, or am I wrong about—am I just underestimating—?

CML: You're absolutely right. You know, eventually that's what you become. We are what we do, more or less, so you've got to make up your mind what you want to spend your time doing. I consider it to be better to seek truth, and meaning, and God, than it is to pursue money. Now, who agrees with me these days? Not many people.

MK: I'd like to have it all, but I certainly agree.

CML: Well, I would like to bring mankind back to the point where we care about those things again.

MK: Because, of course, then the question is—if you just focus your life on money, just for money's sake, just to get stuff—you can't take it with you. (*laughs*) Eventually, all that stuff is going to decay.

CML: Well, yes, of course it is. But a lot of these rich people are into something called transhumanism. They don't think that they're going to have to leave, so there's no problem (*smiles*) with not being able to take it with you, because you're not going anywhere. You're going to upload your consciousness to a computer, you see.

MK: I've heard this. It's a very popular idea. A lot of my young friends will tell me, "Michael, we might not even die," as though every generation in history hasn't had the same delusion, but it's all—I always laugh at it. But very wealthy, very powerful, ostensibly intelligent people take this seriously, "We're just going to upload our brain—"

CML: Ostensibly intelligent.

MK: Osten— (*laughs*) That's an important caveat. You don't think we're just going to upload our brains to the cloud or something?

CML: You know, that's what reality already is.

MK: We've already done it.

CML: Self-simulation. We've already done it. We are living in the display of a simulation—a Self-simulating Universe—and we are going to be uploaded when we're done with this simulat—we're going to be uploaded to the source of all truth, meaning, and power. Then depending on the decision that is made by that entity, we can be redistributed to another universe, a heaven, a hell—

MK: A new creation.

CML: Pardon me?

MK: A new creation. The idea of the resurrection of the body, right?

CML: That is correct.

MK: So before these genius gazillionaires who fancy that they're going to upload their brains to the IBM computer or something, before they meet their particular judgment, before they shuffle off this mortal coil, do you think there's any chance that they're going to figure out the secret and not have to face death?

CML: No.

MK: No way?

CML: It's like I say, Soros is probably the best among them, and he's way—he's far—you know how it is—I mean, I guess the Hindus had a better way of looking at this. In Christianity, there's one goal: salvation. In Hinduism, there are four goals. Okay. You've got *Kama*, *Artha*, *Dharma*, and *Moksha*. And *Kama* is sensual pleasure, and *Artha* is being able to acquire wealth and power. Then you've got *Dharma*, which is righteousness, and then above *Dharma*, you've got *Moksha*, which is salvation. That's actual union with the absolute or with God. The people running the world right now—all of these people that you just mentioned—are stuck in numbers one and two. They haven't made it to *Dharma* and they're never going to make it to *Moksha*. That's why they want to upload their consciousness to a machine. They know that God isn't going to take them back, if they continue doing what they're doing. So they want to displace reality itself by recreating reality in a mechanical form that then they can control.

MK: Do you think there's any relation between the trans-humanist movement and the transgender movement, the idea that we're just going to escape our bodies?

CML: Yes, I do. The idea, I think, is that you can throw off all constraint. You can throw off the constraints of reality, including constraints of gender. That's what I think the commonality is, and I think these are—they're anti-teleological, both of them.

MK: I agree.

CML: We do have certain constraints. Reality is full of constraints and we're better off recognizing what they are.

MK: Before I let you go as we sit here in the clouds—in some simulation of the clouds here—before I let you go, people are going to be asking, "So what do I do? What should I do right now?" assuming we're not going to launch the second American Revolution, or anything like that, "What should I do in my own life?" even maybe beyond any political questions.

CML: Search for God. Ask God to establish a personal relationship with you. It's available. You're attached to God by your soul. There's this soul that attaches you to God. You can receive the Will of God into yourself, and to some extent, the power of God into yourself, but you've got to be receptive. If you're not receptive, then you're cutting yourself off from God, and once you do that, then you do not share the identity of reality itself, and you're done. Once your physical body expires, there is nothing that will carry you because you have denied it and rejected it. You follow me?

MK: I do.

CML: That's what people need to do.

MK: I do. I think that's good advice.

CML: I'm glad you agree. That's good for you. (*both laugh*)

MK: It's good. It's sort of like reaching out and grabbing, (*shakes an imaginary listener*) "Do THIS. Search for God," as though to shake the audience.

Chris, this has been a true pleasure. Thank you for coming out. Thank you for making the trek.

CML: Well, it's been my pleasure. I don't travel very often. I've told you the reasons for that, but I'm glad I made this trip, Michael.

MK: Thank you, Chris. (*shake hands*)

References

Selected articles and papers

Langan, C. M. (1989) The Resolution of Newcomb's Paradox. *Noesis*, № 44.

Langan, C. M. (1999) Introduction to the CTMU. *Ubiquity*, Vol. 1, № 1.

Langan, C. M. (2002) *The Cognitive-Theoretic Model of the Universe: A New Kind of Reality Theory*. Princeton, MO: Mega Foundation Press. Originally published in *Progress in Complexity, Information, and Design*, Double Issue, Vols. 1.2-3.

Langan, C. M. (2003) Cheating the Millennium: The Mounting Explanatory Debts of Scientific Naturalism. In W. A. Dembski (Ed.) *Uncommon Dissent: Intellectuals Who Find Darwinism Unconvincing*. Wilmington, DE: ISI Books.

Langan, C. M. (2017) An Introduction to Mathematical Metaphysics. *Cosmos and History: The Journal of Natural and Social Philosophy*, Vol. 13, № 2, pp. 313-330.

Langan, C. M. (2018a) Metareligion as the Human Singularity. *Cosmos and History: The Journal of Natural and Social Philosophy*, Vol. 14, № 1, pp. 321-332.

Langan, C. M. (2018b) The Metaformal System: Completing the Theory of Language. *Cosmos and History: The Journal of Natural and Social Philosophy*, Vol. 14, № 2, pp. 207-227.

Langan, C. M. (2019) Introduction to Quantum Metamechanics (QMM). *Cosmos and History: The Journal of Natural and Social Philosophy*, Vol. 15, № 1, pp. 265-300.

Langan, C. M. (2020) The Reality Self-Simulation Principle: Reality is a Self-Simulation. *Cosmos and History: The Journal of Natural and Social Philosophy*, Vol. 16, № 1, pp. 466-486.

Other authors

Descartes, R. (1996) *Meditations on First Philosophy with Selections from the Objections and Replies*. Edited by J. Cottingham. Cambridge: Cambridge University Press.

Einstein, A. (1920) Relativity: *The Special and the General Theory: A Popular Exposition*. London: Methuen & Co., Ltd.

Einstein, A. (1954) *Ideas and Opinions*. New York: Crown Publishers.

Frankl, V. A. (1959) *Man's Search for Meaning: An Introduction to Logotherapy*. Translated by Ilse Lasch. Boston, MA: Beacon Press.

Gödel, K. (1962) *On Formally Undecidable Propositions of Principia Mathematica and Related Systems*. Translated by B. Meltzer. New York: Basic Books.

Russell, B. (1959) *The Problems of Philosophy*. New York: Oxford University Press.

Russell, B. (1979) *History of Western Philosophy and its Connection with Political and Social Circumstances from the Earliest Times to the Present Day*. London: Allen & Unwin Ltd.

Internet sources

At the time the book went to press, the following links worked.

Aristotle (350 B.C.E) The Internet Classics Archive, https://classics.mit.edu/Browse/browse-Aristotle.html

Aquinas, T. (2020) *Summa Theologiae*. [written between 1265 and 1273] Aquinas Institute, Inc., https://aquinas.cc/la/en/~ST.I

Milton, J. (1674) *Paradise Lost*. Book One. Poetry Foundation, https://www.poetryfoundation.org/poems/45718/paradise-lost-book-1-1674-version

Langan, C. M. (2022) High Strangeness at the Daily Wire. Chris Langan's Ultimate Reality, https://chrislangan.substack.com/p/high-strangeness-at-the-daily-wire

Plato (1946) *The Republic*. [written 360 B.C.E] The Internet Classics Archive. Translated by Benjamin Jowett, https://classics.mit.edu/Plato/republic.html

Plotinus (1952) *The Six Enneads*. [written 250 A.C.E.] The Internet Classics Archive. Translated by Stephen Mackenna and B. S. Page, https://classics.mit.edu/Plotinus/enneads.html

Made in United States
Cleveland, OH
29 April 2026

36361365R00453